ISCHEMIC INJURY

A Molecular Insight

ISCHEMIC INJURY

A Molecular Insight

Edited by

Thakur Gurjeet Singh, PhD

Heena Khan, PhD

AAP | APPLE ACADEMIC PRESS

First edition published 2025

Apple Academic Press Inc.
1265 Goldenrod Circle, NE,
Palm Bay, FL 32905 USA

760 Laurentian Drive, Unit 19,
Burlington, ON L7N 0A4, CANADA

CRC Press
2385 NW Executive Center Drive,
Suite 320, Boca Raton FL 33431

4 Park Square, Milton Park,
Abingdon, Oxon, OX14 4RN UK

Library and Archives Canada Cataloguing in Publication

CIP data on file with Canada Library and Archives

Library of Congress Cataloging-in-Publication Data

CIP data on file with US Library of Congress

ISBN: 978-1-77491-540-0 (hbk)
ISBN: 978-1-77491-541-7 (pbk)
ISBN: 978-1-03268-002-6 (ebk)

About the Editors

Thakur Gurjeet Singh, PhD
MPharm (Pharmaceutical Sciences – Pharmacology),
PDCR (Clinical Research and PCPV (Pharmacovigilance)

Thakur Gurjeet Singh, PhD, MPharm, has 16 years of experience in industry, research, and teaching. He is working as Dean, Chitkara College of Pharmacy, Chitkara University, Punjab, India. He is engaged with projects on interception of nicotine, morphine, alcohol dependence, ischemia, diabetic neuropathy and nephropathy toxicology, epilepsy, Alzheimer's disease, and stem cell research studies and handles affiliation formalities with the director's office. He has also attended UGC networking courses and various preclinical and clinical research training programs. He has more than 182 research publications (international and national) and 25 book chapters, eight books and 58 patents filed under his name. To date, he has 170 Scopus-indexed publications in peer-reviewed journals (with 1778 citations, an h-index 22 and i10-index 58, impact factor: 482.5 and RG score: 37). He has attended more than 60 national and international conferences. He has guided theses for 65 MPharm students and is supervising 10 PhD candidates. He is a reviewer for various highly reputed international journals. He is a member secretary of the Committee for the Purpose of Control and Supervision of Experiments on Animals and is the In-Charge of the animal house facility at Chitkara College of Pharmacy. He is also a member secretary for institutional human ethical committee registered with the Department of Health Research and CDSCO New Delhi, India.

Heena Khan, PhD
MPharm (Pharmaceutical Sciences – Pharmacology), PhD

Heena Khan, PhD, MPharm, has over five years of experience in research. She is working as a Junior Research Fellow. She is engaged with several projects on stroke, diabetes, Alzheimer's disease, and Parkinson's disease. She has more than 30 research and review publications (international), which have been cited in highly reputed journals. She is a reviewer for various highly reputed international journals. She has participated and attended several national and international conferences. She is also a member of the International Society for Pharmacoeconomics and Outcomes Research.

Contents

Contributors

Preeti Arya
Guru Gobind Singh College of Pharmacy, Yamunanagar, Haryana, India

Renu Arya
Pharmacology Division, Chandigarh College of Pharmacy, Landran, Mohali, Punjab, India

Paras Attri
Department of Pharmaceutical Sciences and Technology, Maharaja Ranjit Singh Punjab Technical University, Bathinda, Punjab, India

Vertika Awasthi
Chitkara College of Pharmacy, Chitkara University, Punjab, India

Arockia Babu
Pharmacology Division, Chandigarh College of Pharmacy, Landran, Mohali, Punjab, India

Shivani Bagga
Chitkara College of Pharmacy, Chitkara University, Rajpura, Punjab, India

Kajal Bagri
Department of Pharmaceutical Sciences and Technology, Maharaja Ranjit Singh Punjab Technical University, Bathinda, Punjab, India

Shiveena Bhatia
Chitkara College of Pharmacy, Chitkara University, Punjab, India

Benu Chaudhary
Guru Gobind Singh College of Pharmacy, Yamunanagar, Haryana, India

Deeksha
Department of Pharmaceutical Sciences, Indira Gandhi University, Meerpur, Rewari, Haryana, India

Rahul Deshmukh
Associate Professor and Head, Department of Pharmaceutical Sciences and Technology,
Maharaja Ranjit Singh Punjab Technical University, Bathinda, Punjab, India

Geeta Deswal
Guru Gobind Singh College of Pharmacy, Yamunanagar, Haryana, India

Ajmer Singh Grewal
Guru Gobind Singh College of Pharmacy, Yamunanagar, Haryana, India

Amarjot Kaur Grewal
Chitkara College of Pharmacy, Chitkara University, Rajpura, Punjab, India

Uditi Handa
Guru Gobind Singh College of Pharmacy, Yamunanagar, Haryana, India

Amteshwar S. Jaggi
Department of Pharmaceutical Sciences and Drug Research, Punjabi University, Patiala, Punjab, India

Satyan Kapoor
Department of Pharmaceutical Sciences and Technology, Maharaja Ranjit Singh Punjab Technical University, Bathinda, Punjab, India

Tajpreet Kaur
Department of Pharmacology, Khalsa College of Pharmacy, Amritsar, Punjab, India

Ayansh Kaushik
Scholar, Department of Pharmacology, ISF College of Pharmacy, Moga, Punjab, India

Peeyush Kaushik
Guru Gobind Singh College of Pharmacy, Yamunanagar, Haryana, India

Heena Khan
Chitkara College of Pharmacy, Chitkara University, Rajpura, Punjab, India

Kuldeep Kumar
Department of Pharmaceutical Sciences and Drug Research, Punjabi University, Patiala, Punjab, India

Manish Kumar
Associate Professor, Chitkara College of Pharmacy, Chitkara University, Rajpura, Punjab, India

Narinder Kumar
Department of Pharmaceutical Sciences and Technology, Maharaja Ranjit Singh Punjab Technical University, Bathinda, Punjab, India

Sunil Kumar
Department of Pharmaceutical Sciences, Indira Gandhi University, Meerpur, Rewari, Haryana, India

Vinay Kumar
Pharmacology Division, Chandigarh College of Pharmacy, Landran, Mohali, Punjab, India

Pratima Kumari
Chitkara College of Pharmacy, Chitkara University, Punjab, India

Shilpa Kumari
Department of Pharmaceutical Sciences and Technology, Maharaja Ranjit Singh Punjab Technical University, Bathinda, Punjab, India

Maanvi
Department of Pharmaceutical Sciences and Technology, Maharaja Ranjit Singh Punjab Technical University, Bathinda, Punjab, India

Jitender Madan
National Institute of Pharmaceutical Education and Research, Hyderabad, Telangana, India

Shahid Mushtaq Malik
Department of Pharmaceutical Sciences and Technology, Maharaja Ranjit Singh Punjab Technical University, Bathinda, Punjab, India

Sudhanshu Mallan
Scholar, Department of Pharmacology, ISF College of Pharmacy, Moga, Punjab, India

Manish
Department of Pharmaceutical Sciences, Indira Gandhi University, Meerpur, Rewari, Haryana, India

Sidharth Mehan
Department of Pharmacology, ISF College of Pharmacy, Ghal Kalan, Moga, Punjab, India

Sanjana Mehta
Chitkara College of Pharmacy, Chitkara University, Punjab, India

Vipasha Mehta
Maharaja Ranjit Singh Punjab Technical University, Bathinda, Punjab, India

Kajal Nagpal
Guru Gobind Singh College of Pharmacy, Yamunanagar, Haryana, India

Deepika Purohit
Department of Pharmaceutical Sciences, Indira Gandhi University, Meerpur, Rewari, Haryana, India

Khadga Raj
Scholar, Department of Pharmacology, ISF College of Pharmacy, Moga, Punjab, India

Sakshi Rana
Maharaja Ranjit Singh Punjab Technical University, Bathinda, Punjab, India

Suray Ranjan
Department of Pharmaceutical Sciences and Technology, Maharaja Ranjit Singh Punjab Technical University, Bathinda, Punjab, India

Jyoti Rathi
Department of Pharmaceutical Sciences, Indira Gandhi University, Meerpur, Rewari, Haryana, India

Rajender Sandhu
Department of Pharmaceutical Sciences and Technology, Maharaja Ranjit Singh Punjab Technical University, Bathinda, Punjab, India

Shamly
Department of Pharmaceutical Sciences and Technology, Maharaja Ranjit Singh Punjab Technical University, Bathinda, Punjab, India

Chaman Sharma
Department of Pharmaceutical Sciences and Technology, Maharaja Ranjit Singh Punjab Technical University, Bathinda, Punjab, India

Manu Sharma
Department of Pharmacy, Banasthali Vidyapith, Rajasthan, India

Priyanka Sharma
Department of Pharmaceutical Sciences and Technology, Maharaja Ranjit Singh Punjab Technical University, Bathinda, Punjab, India

Rupali Sharma
Department of Pharmaceutical Sciences and Technology, Maharaja Ranjit Singh Punjab Technical University, Bathinda, Punjab, India

Veerta Sharma
Chitkara College of Pharmacy, Chitkara University, Punjab, India

Richa Shri
Department of Pharmaceutical Sciences and Drug Research, Punjabi University, Patiala, Punjab, India

Amrit Pal Singh
Assistant Professor, Department of Pharmaceutical Sciences, Guru Nanak Dev University, Amritsar, Punjab, India

Gurpreet Singh
Department of Pharmaceutical Chemistry, ISF College of Pharmacy, Ghal Kalan, Moga, Punjab, India

Manjinder Singh
Chitkara College of Pharmacy, Chitkara University, Punjab, India

Nirmal Singh
Professor and Former Head, Department of Pharmaceutical Sciences and Drug Research, Punjabi University, Patiala, Punjab, India

Ravinder Singh
Chitkara College of Pharmacy, Chitkara University, Punjab, India

Shamsher Singh
Professor, Neuroscience Division, Department of Pharmacology, ISF College of Pharmacy, Ghal Kalan, Moga, Punjab, India

Thakur Gurjeet Singh
Assistant Dean and Head, Chitkara College of Pharmacy, Chitkara University, Rajpura, Punjab, India

Varinder Singh
Assistant Professor, Chitkara College of Pharmacy, Chitkara University, Punjab, India

Aarchi Singla
Maharaja Ranjit Singh Punjab Technical University, Bathinda, Punjab, India

Deepak Singla
Guru Gobind Singh College of Pharmacy, Yamunanagar, Haryana, India

Rupinder Kaur Sodhi
Pharmacology Division, Chandigarh College of Pharmacy, Landran, Mohali, Punjab, India

Kumari Tanuja
Maharaja Ranjit Singh Punjab Technical University, Bathinda, Punjab, India

Rajni Thakur
Maharaja Ranjit Singh Punjab Technical University, Bathinda, Punjab, India

Smriti Thakur
Department of Pharmaceutical Sciences and Technology, Maharaja Ranjit Singh Punjab Technical University, Bathinda, Punjab, India

Komal Thapa
Chitkara School of Pharmacy, Chitkara University, Baddi, Himachal Pradesh, India

Piyush Vatsha
Maharaja Ranjit Singh Punjab Technical University, Bathinda, Punjab, India

Nirmala Kumari Yadav
Department of Pharmaceutical Sciences, Indira Gandhi University, Meerpur, Rewari, Haryana, India

Snehlata Yadav
Department of Pharmaceutical Sciences, Indira Gandhi University, Meerpur, Rewari, Haryana, India

Abbreviations

5-HT	serotonin
AA	arachidonic acid
AAA	abdominal aortic aneurysm
ABI	acute brain ischemia
ACCESS	acute candesartan cilexetil evaluation in stroke survivors
ACE	angiotensin-converting enzyme
Ach	acetylcholine
AChE	acetylcholine esterase
ACS	acute coronary syndrome
ADP	adenosine diphosphate
AF	atrial fibrillation
AGE	advanced glycation end
AgNP	silver nanoparticle
AGT	angiotensinogen
AIF	apoptosis-inducing factor
AIS	acute ischemic stroke
AJ	adherens junction
AKI	acute kidney injury
ALLHAT	antihypertensive and lipid-lowering treatment to prevent heart attack trial
ALR	augmenter of liver regeneration
AMBRA1	an activating molecule in Beclin1 regulated autophagy
AMI	acute myocardial infarction
AMP	adenosine monophosphate
AMPA	alpha-amino-3-hydroxy-5-methylisoxazole-4-propionate
AMPK	5′-adenosine monophosphate-activated protein kinase
Ang II	angiotensin II
ANN	artificial neural network
ANT	adenine nucleotide translocase
AP-1	activator protein 1
APAF	apoptotic protease activating factor
Apaf1	apoptotic protease factor 1
APC	activated protein C
APC	adenomatous polyposis coli
APL	antiphospholipid
ARDS	acute respiratory distress syndrome
ARE	antioxidant response element
Arg-1	arginase-1
ASIC	acid-sensing ion channels

ASK1	signal-regulating kinase 1
AT1	angiotensin type 1
AT2	angiotensin II type 2
ATG 13	autophagy-related gene
Atg	autophagy-related proteins
ATI	angiotensin I
ATII	angiotensin II
ATP	adenosine triphosphate
AuNP	gold nanoparticle
BAK	BCL-2 homologous antagonist/killer
BAX	BLC 2 associated X protein
BBB	blood-brain barrier
BBs	beta-blockers
BChE	butyrylcholinesterase
BCL2	B cell lymphoma 2
BDNF	brain-derived neurotrophic factor
BI	brain injury
BID	BH3 interacting domain death agonist
BNIP3	BCL2 interacting protein 3
BP	blood pressure
bZIP	basic leucine zipper proteins
CA	cardiac arrest
CA	channel blocker
CABG	coronary artery bypass surgery
CAD	coronary artery disease
CAESAR	pre-clinical assessment of cardioprotective therapies
CAM	cell adhesion molecules
cAMP	cyclic adenosine monophosphate
CAPP	Captopril Prevention project
CARASIL	cerebral autosomal recessive arteriopathy with subcortical infarcts and leukoencephalopathy
CARD	caspase recruitment domain
CAS	carotid artery stents
CAT	catalase
CBF	cerebral blood flow
CCA	common carotid artery
CEA	carotid endarterectomy
CEC	circulating endothelial cells
CGRP	calcitonin gene-related peptide
CH3	cysteine/histidine-rich domain 3
ChAT	choline acetyltransferase
CHD	congenital heart disease
CHF	congestive heart failure

ChIP	chromatin immunoprecipitation
CINC	cytokine-induced neutrophil chemoattractant
CK-MB	creatinine kinase-MB
CMD	coronary microvascular dysfunction
CNC	cap-n-collar
CNS	central nervous system
CNT	carbon nanotube
CO	carbon monoxide
COMT	catechol-O-methyltransferase
COPD	chronic obstructive pulmonary disease
$CoQH_2$	ubiquinone
COVID-19	coronavirus disease 2019
COX	cyclooxygenase
CRADD	caspase requirement domain and death domain-containing protein
CREB	cAMP responsive element binding
CRP	C reactive protein
CsA	cyclosporin A
CSF	cerebrospinal fluid
CT	computed tomography
CTA	CT angiography
CVA	cerebrovascular accident
CVD	cardiovascular disease
CVD	cardiovascular disease
CVT	cerebral venous sinus thrombosis
CyP	cyclophilin
Cyt C	cytochrome c
DA	dopamine
DALY	disability-adjusted life year
DAMP	damage-associated molecular pattern
dATP	deoxyadenosine triphosphate
DED	death effector domain
DES	drug-eluting stents
DGF	delayed graft function
DIC	disseminated intravascular coagulopathy
DISC	death-induced signaling complex
DMSO	dimethyl sulfoxide
DNA	deoxyribonucleic acid
DPI	diphenyleneiodonium
DPSC	dental pulp stem cell
DRP	dynamin-related protein
DUOX	dual oxygenase
ECA	external carotid artery
ECG	electrocardiogram

ELAM	endothelial leukocyte adhesion molecule
eNOS	endothelial nitric oxide synthase
EPO	erythropoietin
EPOR	receptors increase in hypoxic tissue
ER	endoplasmic reticulum
ER	estrogen receptors
ERK	extracellular regulated kinases
EST	embryonic stem cell
ETC	electron transport chain
EVT	extravascular treatment
FAD	flavin adenine dinucleotide
FADD	Fas-associated death domain
$FADH_2$	flavin adenine dinucleotide
FCI	focal cerebral ischemia
FET	forward electron transport
FFA	free fatty acid
FHS	family history of stroke
FHs	follicle-stimulating hormone
FMN	flavin mononucleotide
FWOC	first window of cardioprotection
GABA	gamma-aminobutyric acid
GABAAR	gamma-aminobutyric acid A receptor
GAD	glutamate decarboxylase
GBD	global burden of disease
GCI	global cerebral ischemia
GCL	glutamate-cysteine ligase
GFAP	glial fibrillary acidic protein
Glu	glutamate
GMCSF	granulocyte-macrophage colony-stimulating factor
GNPs	gold nanoparticles
GPCR	G-protein coupled receptor
GPX-2	glutathione peroxidase
GR	glutathione reductase
GSH	glutathione
GSK-3β	glycogen synthase kinase-3β
GSR-1	glutathione reductase
H_2O_2	hydrogen peroxide
HDL	high-density lipoproteins
HEpEF	heart failure with a preserved ejection fraction
HFrEF	heart failure with a reduced ejection fraction
HIF	hypoxia-inducible factor
HIV	human immunodeficiency virus
HMGB1	high mobility group box 1

HO-1	heme oxygenase 1
HOPE	heart outcomes prevention evaluation
hrsACE2	human recombinant soluble ACE2
HRT	hormone replacement therapy
HSP	heat shock proteins
HSS	hepatic simulator substance
HTRA2	high-temperature requirement factor A2
I/R	ischemia reperfusion
IA	intra-arterial
ICA	internal carotid artery
ICAM-1	intercellular adhesion molecule 1
ICD	international classification of diseases
ICH	intracerebral hemorrhage
ICP	intracranial pressure
ICU	intensive care unit
IFM	interfibrillar mitochondria
IFN	interferons
IHD	ischemic heart disease
IL	interleukins
IL-1β	interleukin 1β
IL-6	interleukin-6
iNOS	inducible nitric oxide synthase
IPC	ischemia preconditioning
IPOC	ischemic postconditioning
IPSC	induced pluripotent stem cell
IR	insulin receptor
IR	ischemia-reperfusion
IRI	ischemia/reperfusion injury
IS	ischemic
IS	ischemic stroke
IV	intravenous
IVH	intraventricular hemorrhage
IκB	inhibitor of kappa B
JAK	Janus kinase
JNK	c-Jun N-terminal kinase
KCs	Kupffer cells
Keap-1	Kelch-like ECH associating protein 1
LC3II	microtubule-associated *protein* 1A/1B-light chain 3
LDH	lactate dehydrogenase
LDL	low-density lipoprotein
LFA-1	lymphocyte function-associated antigen 1
LIFE	losartan intervention for endpoint study
LMWH	low-molecular-weight heparin

LOX	lipoxygenase
LPS	lipopolysaccharide
LRP	lipoprotein receptor-related protein
LRR	leucine-rich repeat
LTs	leukotrienes
LVH	left ventricular hypertrophy
LVO	large vascular occlusion
LVO	large vessel occlusion
mAChR	nAChR and muscarinic
MAO	monoamine oxidase
MAPK	mitogen-activated protein kinase
MAS	macrophage activation syndrome
MBL	mannose-binding lectin
MCA	middle cerebral artery
MCAO	middle cerebral artery occlusion
MCP-1	monocyte chemo-attractant protein-1
MCUC	mitochondrial calcium uniporter complex
MEK1/2	mitogen-activated protein kinase
MERCI	mechanical embolus removal in cerebral ischemia
MFF	mitochondrial fission factor
MFN2	mitofusin-2
MHC	major histocompatibility complex
MI	myocardial infarction
MIRI	myocardial ischemia/reperfusion injury
miRNAs	microRNAs
ML	magnetoliposome
MLKL	mixed lineage kinase domain-like
MMP	matrix metalloproteinases
Mo-co	molybdenum-containing molybdopterin cofactor
MOF	multi-organ failure
MPO	myeloperoxidase
MPTP	mitochondrial permeability transition pore
mPTP	mitochondrial permeable transition pore
MRF-1	microglial response factor-1
MRI	magnetic resonance imaging
MSC	mesenchymal stem cell
mtDNA	mitochondrial DNA
mTOR	mammalian target of rapamycin
MVA	motor vehicle accidents
nAChRs	nicotinic acetylcholine receptors
NAD	nicotinamide adenine dinucleotide
NADH	reduced nicotinamide adenine dinucleotide
NADPH	nicotinamide adenine dinucleotide phosphate

NCX	sodium-calcium exchanger
NE	norepinephrine
NEMO	NF-κB essential modulator
NET	neutrophil extracellular traps
NF-κB	nuclear factor kappa beta
NHCP	Na^{2+}/HCO_3^- co-transporter
NHE	Na^{2+}/H^+ exchanger
NINDS	National Institute of Neurological Disorders and Stroke
NLK	nemo-like kinase
NLRP3	NLR family pyrin domain containing 3
NLS	nuclear localization sequence
NMDA	n-methyl-d-aspartate
NMDAR	n-methyl-d-aspartate receptor
nNOS	neuronal NOS
NO	nitric oxide
NOD	nucleotide-binding oligomerization domain
NOS	nitric oxide synthase
NOX	nicotinamide adenine dinucleotide phosphate oxidase
NP	nanoparticle
Nrf2	nuclear factor erythroid 2 related factor 2
NSAIDs	non-steroidal anti-inflammatory medicines
NSC	neural stem cells
NVU	neurovascular units
OCP	oral contraceptive pill
OGD	oxygen-glucose deprivation
OPA 1	optic atrophy 1
OXOPHOS	oxidative phosphorylation
PAF	platelet-activating factor
PAMP	pathogen-associated molecular pattern
PAPA-CAD	perindopril (an angiotensin-converting enzyme inhibitor) and amlodipine to atorvastatin in patients with coronary artery disease
PAR	protease-activated receptor
PCI	percutaneous coronary intervention
PDGF	platelet-derived growth factor
PGA	polyglycolic acid
PGC-1α	peroxisome proliferator-activated receptor-gamma coactivator (*PGC*)-*1 alpha*
PI3K	phosphatidyl inositol 3-kinase
PICA	posterior inferior cerebellar artery
PKB	protein kinase B
PKC	protein kinase C
PLA	polylactic acid
PLC	phospholipase C

PLGA	polylactic-co-glycolic acid
PPAR- δ	peroxisome proliferator-activated receptor delta
PPAR	peroxisome proliferator-activated receptor
PPCI	primary percutaneous coronary intervention
PROGRESS	perindopril protection against recurrent stroke study
PRR	pattern recognition receptor
PTEN	phosphatase and tensin analog
PTH	parathyroid hormone
PTT	photothermal therapy
PUMA	p53 up-regulated modulator of apoptosis
PYD	pyrin domain
QOL	quality of life
RAAA	renin/ACE1/angiotensin II/angiotensin II
RAAS	renin-angiotensin aldosterone system
RAS	renin-angiotensin system
RB1CC1	retinoblastoma, coiled-coil protein 1
RCT	randomized controlled trial
RCVS	reversible cerebral vasoconstriction syndrome
REF	reverse electron flow
REGARDS	reason for geographic and racial differences in stroke
RET	reverse electron transport
RHD	rheumatic heart disease
RIPC	remote ischemic preconditioning
RIPK	receptor-interacting protein
RIPostC	remote ischemic postconditioning
RIPre	remote limb preconditioning
RIRR	ROS-induced ROS release
RISK	reperfusion injury salvage kinase
RNS	reactive nitrogen species
ROS	reactive oxygen species
RR	respiratory rate
RXR	retinoid X receptor
S1P	sphingosine-1-phosphate
SAFE	survival activating factor enhancement
SAH	subarachnoid hemorrhage
SAPK	stress-activated protein kinase
SARS	severe acute respiratory distress syndrome
SARS-COV-2	severe acute respiratory syndrome coronavirus-2
SCAST	Scandinavian candesartan acute stroke trial
SCOPE	study on cognition and prognosis in the elderly
SDF	stromal cell-derived factor
Ser9	serine 9
SERCA2a	sarco/endoplasmic reticulum Ca^{2+} ATPase

SIC	sepsis-induced coagulopathy
SIRT1	signaling Information Receiver/Transmitter
SIRT1	sirtuin 1
SNAC	Society for Neuroscience in Anesthesiology and Critical Care
SNP	single nucleotide polymorphism
SNS	sympathetic nervous system
SOC	store-operated channels
SOD	superoxide dismutase
SPECT	single-photon emission computerized tomography
SPG7	spastic paraplegia 7
SSM	subsarcolemmal mitochondria
STEMI	ST-segment elevation myocardial infarction
SVD	small vessel disease
SWOC	second window of cardioprotection
TBI	traumatic brain injury
tBID	truncated form
TCA	tricarboxylic acid cycle
TF	tissue factor
THCY	plasma homocysteine
TIA	transient ischemic attack
TIMPs	tissue inhibitors of metalloproteinases
TJ	tight junctions
TLR	toll-like receptors
tMCAO	temporary middle cerebral artery occlusion
TMPRSS2	transmembrane protease serine 2
TNF	tumor necrosis factor
TNFR	tumor necrosis factor receptors
TNF-α	tumor necrosis factor-α
TOPK	T-LAK-cell-originated protein kinase
tPA	tissue plasminogen activator
TRADD	TNF receptor-associated death domain
TRAIL	TNF-related apoptosis-inducing ligand
TRAIL	tumor necrosis factor-related apoptosis-inducing ligand receptor
TRP	transient receptor potential
TWEAK	tumor necrosis factor-like weak inducer of Apoptosis
TXN-1	thioredoxin
Tyr216	tyrosine 216
ULK1	uncoordinated-51-like kinase 1
UPR pathway	unfolded protein response pathway
VALUE	valsartan antihypertensive long-term use evaluation
VCAM-1	vascular cell adhesion molecule 1
VD	vitamin D
VEGF	vascular endothelial growth factor

VGAT	vesicles by GABA-specific vesicular transporter
VGCC	voltage-gated Ca^{2+} channels
vWF	von Willebrand factor
WBC	white blood cell
WHO	World Health Organization
WKY	Wistar–Kyoto
WPW	Wolff-Parkinson-White
WSO	World Stroke Organization
XDH	xanthine dehydrogenase
XO	xanthine oxidase
XOR	xanthine oxidoreductase
δ-PKC	δ-protein kinase-C

Preface

In the past few decades, stroke has continued to be a global risk to health and a leading cause of death and functional decline in individuals affected. New drug discovery is essential for improving the work of researchers and pharmacologists whether in scientific work, in the scholarly world or in government and industry. This book focuses on molecular pharmacology and covers the most recent improvements in cell and receptor pharmacology, with a noteworthy emphasis on the system of activity of novel medications being worked on and imaginative pharmacological advancements.

Strokes occur at different stages of life, and there are sex-based differences in the treatment of ischemic injury. Consequently, we have included chapters that cover in-depth paradigms that encompass the prevention, evaluation, therapy selection, result projection, and long-term course of disease in affected patients. The authors have covered the effect of stroke in daily life and the therapeutics and complications for the prevention of stroke. The book also details the involvement of platelets, neurotransmitters, renin-angiotensin-aldosterone system, and vitamin D in ischemic injury.

The COVID-19 pandemic created an overwhelming burden on the healthcare system, including on the treatment of ischemic stroke; this book looks at advancements and challenges that were faced during the pandemic.

The various aspects of recognizable proof of preclinical and clinical improvement of medications that are therapeutically beneficial in human illness have also been discussed. We have also included chapters that discuss a logical approach to thoroughly examining the patient with acute stroke, providing an informative review of acute stroke disease in a clear and succinct manner.

The significant etiologies of ischemic and hemorrhagic stroke are reviewed in detail. The book also includes information on the improvement of novel reagents or advancements that are of significance and offers new ways of thinking regarding medication advancement. We have also included chapters that discuss nanotechnology and biosensors applications for human health

This volume provides a holistic look at pharmacological points that incorporate aspects of the physiology of disease, receptor-connected intracellular weak structures, different transduction pathways, and multiple I/R injuries.

This volume is the culmination of our attempt to provide basic and advanced knowledge on the recent approaches in molecular pharmacology to students medical professionals, scientists, medical students, medical and neurology residents, and allied healthcare. The book is written in simple and clear text, and we also used many figures and tables to make the book easier to understand.

We express our appreciation to all of the contributing authors who helped us tremendously with their contributions, time, critical thoughts, and suggestions to put together this peer-reviewed edited volume.

The editors are also thankful to Apple Academic Press and their team members for the opportunity to publish this volume. Lastly, we thank our family members for their love, support, encouragement, and patience during the entire period of this work.

CHAPTER 1

Stroke Insight

HEENA KHAN,[1] THAKUR GURJEET SINGH,[1] and RAHUL DESHMUKH[2]

[1]*Chitkara College of Pharmacy, Chitkara University, Punjab, India*

[2]*Maharaja Ranjit Singh Punjab Technical University, Bathinda, Punjab, India*

ABSTRACT

Stroke is a significant cause of death in the developed world. Ischemic or hemorrhagic acute strokes cause neuronal injury and neurological impairments by reducing blood flow, nutrients, and oxygen to a brain region. Recovery rates can be maximized with early, oriented treatment, rehabilitation, and lifestyle changes. While controlling arterial risk factors is critical for secondary prevention of all types of ischemic stroke, there should be ways to prevent specific ischemic stroke subtypes. A multidisciplinary strategy for improving nutrition, exercise, and medication adherence can also benefit ischemic injury. This chapter gives detailed information on the evaluation and management of strokes. The critical objective of stroke prevention is to determine the prevalence of lesions for therapeutic purposes and to prevent and cure stroke.

1.1 INTRODUCTION

A stroke is a sudden neurological impairment that lasts at least 24 hours; additionally, it is referred to as a "cerebrovascular accident (CVA) or apoplexy" [1]. Stroke is the top cause of disability and the fifth-highest cause of death in the United States. Strokes are divided into two classes. An ischemic stroke is a more common type, which occurs when blood flow to a specific brain area is cut off; an ischemic stroke causes 85% of all acute strokes. About 15% of acute strokes are hemorrhagic, exacerbated by rupturing of a blood vessel [2, 3]. Hemorrhagic strokes are classified into two types, i.e., "intracerebral and subarachnoid hemorrhage," which account for around 5% of every stroke. Stroke can be caused by a range of factors in the heart, including "chronic hypertension, arteriosclerosis, and emboli" as an outcome of atrial fibrillation (AF). In younger patients, the range of available causes may include clotting abnormalities, cervical arterial dissection, and different aspects of vasculitis [4]. Before administering any form of stroke treatment, a thorough history and

Ischemic Injury: A Molecular Insight. Thakur Gurjeet Singh & Heena Khan (Eds.)

physical examination must be conducted with emergent neurological imaging. Individuals can try to increase their chances of substantive recovery with early therapeutic strategies based on the etiology of the stroke, rehab programs, and lifestyle choices [5, 6].

1.1.1 CLASSIFICATION

1. **Ischemic Stroke:** An atherosclerotic plaque ulcerates in a cerebral artery, causing platelets to aggregate and fibrin to coagulate, forming the thrombus that occludes the artery. Fewer than 20% of cases progress ulceration but gradually obstruct blood flow, manifesting as TIAs. Neurological symptoms of an ischemic stroke typically manifest abruptly, but they can also manifest (Figure 1.1). The elderly suffer from "atherosclerosis ischemic stroke" and occurs without warning in more than 80% of cases. A TIA is considered a critical warning sign within the next few months of the stroke [7].

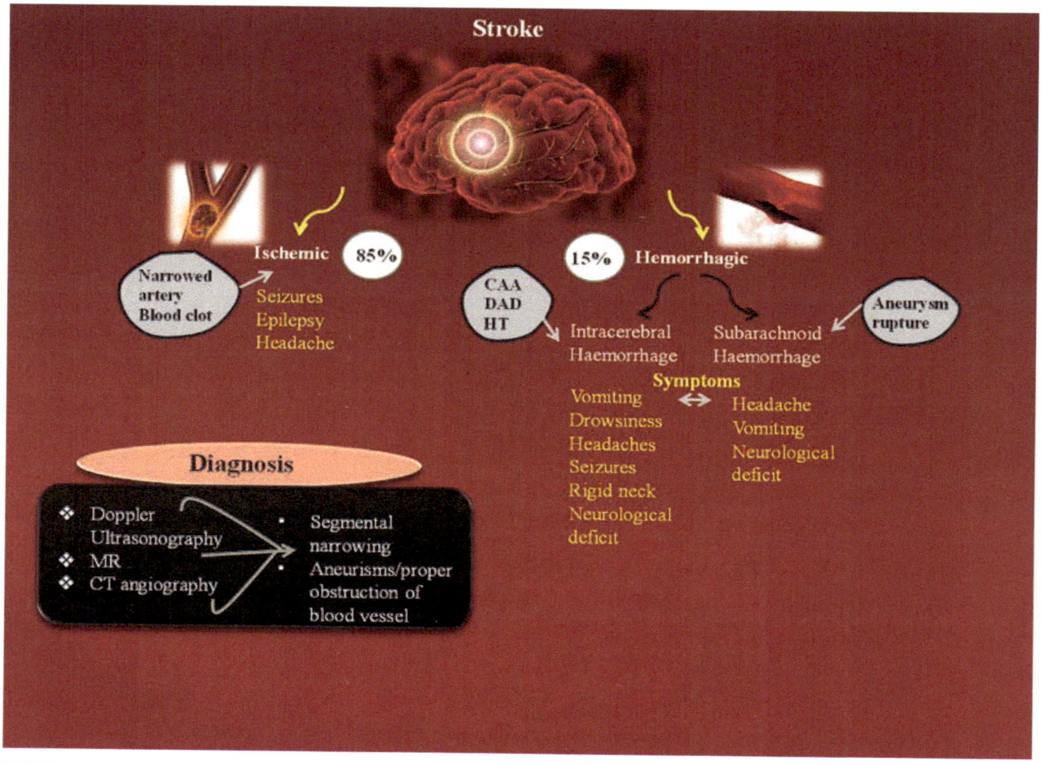

FIGURE 1.1 The types of strokes and diagnosis in detail [CAA: Cerebral amyloid angiopathy; DAD: Degenerative arterial disease; and HT: hypertension].

2. **Embolic Ischemic Stroke:** It is even more common in patients with rheumatic heart disease, AF, prosthetic valves, myocardial infarction, and atheroma of the

larger arteries [8]. Embolisms are frequently multifocal and may result in minor hemorrhages around the obstruction—a cerebral artery occlusion results in reduced blood flow and ischemia. If the episode only takes a few moments, recovery is complete and immediate. Infarction (cell death) occurs within minutes, depending on the severity of the ischemia, which leads to irreversible damage, including the restoration of blood flow known as the infarct's 'core' [9]. The core is the tissue impaired functionally by the absence of blood flow but can recover when the blood supply is restored, referred to as a stroke's "ischemic penumbra." Most patients will have a treatable ischemic penumbra for 3 hours, but it can be referred to as the therapeutic thrombolytic window for up to 12 hours. Therefore, it is critical for effective intervention to identify the treatable patients correctly. Infarction seizures occur at 3–5%, more often than thrombosis after embolism. Between 6 and 18 months after a stroke, the same proportion of patients develop epilepsy. Thus, idiopathic epilepsy may be caused by a silent cortical infarction in the elderly [2, 7].

3. **Hemorrhagic Stroke:** There are two kinds: those caused by intracerebral hemorrhage due to hypertension, degenerative arterial disease, or cerebral amyloid angiopathy, and those caused by subarachnoid hemorrhages resulting from aneurysm rupture [10]. In certain instances, the source of the bleeding cannot be confirmed due to the rarity of other potential causes. Advanced age, excessive alcohol consumption, and hypertension are the primary risk factors; however, cocaine abuse is a significant risk factor for cerebral hemorrhage in adolescents [11]. The majority of intracerebral hemorrhagic strokes occur over 30–90 minutes. Symptoms will vary according to the site and extent of the hemorrhage. It is common to experience focal neurological symptoms, vomiting, and drowsiness. While headaches are possible, rigid necks and seizures are relatively rare. Significant hemorrhages may result in stupor or coma. Most subarachnoid hemorrhages present suddenly with severe headache, vomiting, and neurological deficits, with approximately 50% of patients experiencing altered consciousness [10].

1.1.2 DIAGNOSIS

The cerebral vasculature is visualized using MR or CT angiography, revealing additional information such as segmental narrowing, aneurisms, or proper obstruction of blood vessels. Doppler ultrasonography of the "carotid and vertebral arteries" provides additional details, notably beneficial when recommending intravascular thrombolysis patient care. One analysis discovered a correlation between immediate or long thrombolysis success and the occlusion site as determined by "Doppler ultrasonography" [12].

1.2 PATHOPHYSIOLOGY

Atherosclerosis has been the most prevalent and significant underlying pathology that results in forming an atherothrombotic plaque due to "low-density lipoprotein" cholesterol

(LDL) buildup in the cerebral arteries. Those certain plagues could very well constrict or restrict the neck or intracranial arteries, likely to result in distal cerebral ischemia, which might also rupture more frequently [13]. When the plague breaks out, the innate cholesterol crystals are exposed, attracting blood cells and fibrin. When an artery becomes blocked, the neurons immediately adjacent to it lose their oxygen and nutrient supply. Inability to undergo aerobic metabolism and produce ATP results in the failure of the Na+/K+ ATPase pumps and accumulation of Na+ within the cells and K+ outside the cells. Cells become depolarized as a result of the accumulation of Na+ ions, which results in glutamate release. Glutamate acts as a receptor activator for NMDA and AMPA, enabling calcium ions to enter cells. The ongoing flow of calcium leads to constant neuronal firing, and finally, cell mortality through excitotoxicity (Figure 1.2) [14, 15].

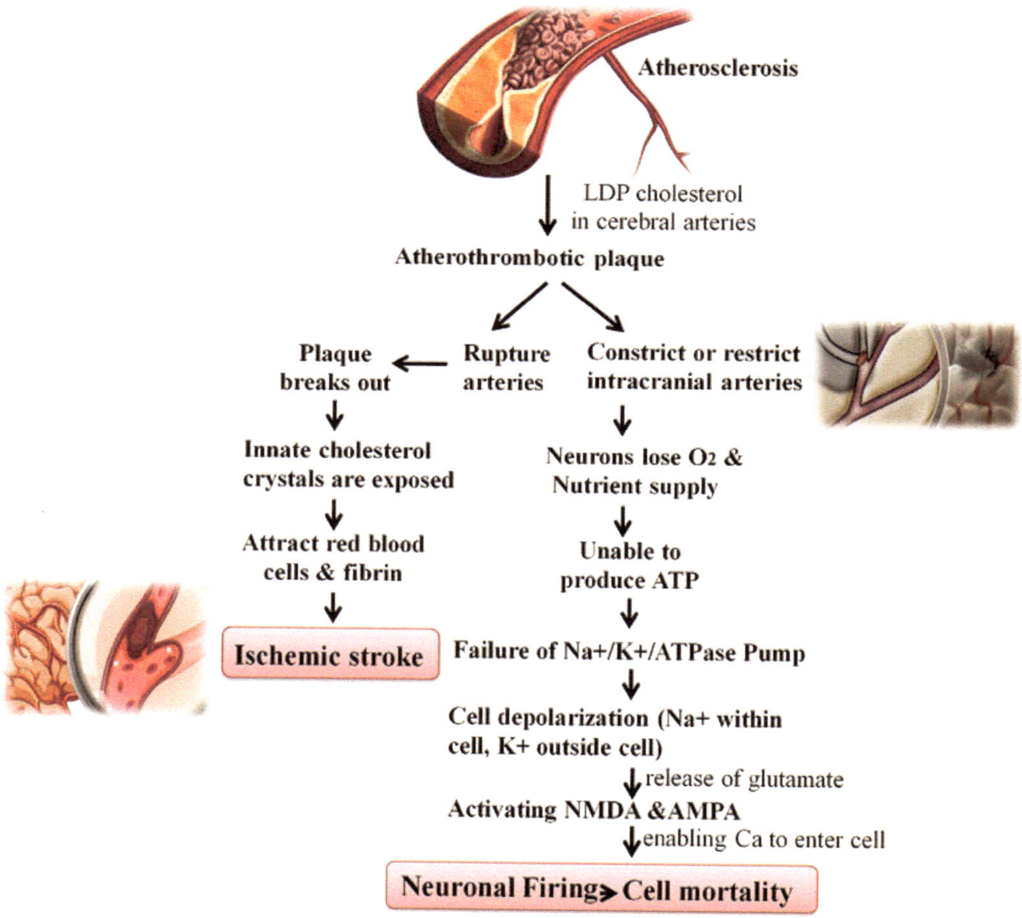

FIGURE 1.2 Pictorial representation of pathophysiology of stroke.

There are no discernible macroscopic alterations during the first 10–12 hours. Cytotoxic edema occurs as a result of failure in energy production and neuronal cellular swelling.

Diffusion-weighted MRI can visualize this early infarction state by demonstrating controlled diffusion due to a cellular bulge of neurons. Cytotoxic cerebral edema develops within 6–12 hrs. of the stroke onset. FLAIR sequence MRI may be the best method for visualizing this phase. Cytotoxic and vasogenic edema results in swelling/inflammation of the infarcted area and elevations of intracranial pressure (ICP), followed by an invasion of phagocytic cells that strive to detox the body cells [16]. Several months following a stroke, astrocytes form a dense network of capillaries, glial fibers, and connective tissue [14]. Hemorrhagic strokes result in a similar type of cell death and systematic repair event by having access to blood fluid overload and resorption [10].

1.3 PREVENTION

Stroke prevention remains critical, as a significant primary and secondary risk factor for modifiable stroke is high blood pressure. While there are effective and widely available medications for treating hypertension, they are frequently not used [17]. According to one study, just 60% of patients took their antihypertensives as described, owing to various factors, including misleading disease information and negative medication views [18]. A decrease in Lipid levels with statins can lessen the stroke risk in patients with high cholesterol—one meta-analysis discovered a 21% relative risk reduction [19]. Endarterectomy can treat atherosclerosis surgically, but the procedure is frequently inconvenient due to the associated burdens. AF, a significant risk factor for ischemic stroke (over the age of 60), can be treated with warfarin, aspirin, or a pacemaker. Stroke risk can be dropped significantly in populations by addressing certain risk factors such as smoking, diabetes, and excessive alcohol abuse [20]. Even when all other variables are controlled besides, the incidence of stroke fatalities rises with declining socioeconomic status. Prevention strategies targeted at specific geographic areas have been suggested to be the most economically efficient intervention [21]. Risk reduction is critical in low and middle-income countries if effective interventions such as stroke intensive care are mostly impractical, unacceptable, or otherwise unobtainable [22].

1.3.1 SECONDARY PREVENTION

There are now various strategies available based on secondary prevention that can help lower the prevalence of stroke recurrence. Aspirin is also one of the simplest and safest strategies to minimize secondary stroke. Compared to placebo, long-term aspirin monotherapy decreases the risk of recurrent stroke by 18.1% ($p = 0.013$) [23, 24]. Dipyridamole treatment also results in a significant risk reduction. Additionally, the combination of extended-release dipyridamole and aspirin led to a tremendous reduction in stroke recurrence. This is advised as the recommended standard of therapy for recurrent stroke prevention (37.0% risk reduction, p 0.001) [24]. Clopidogrel was demonstrated to be as efficient at preventing ischemic stroke recurrence in aspirin-allergic patients [25]. Aspirin and clopidogrel together aren't effective than either medication alone and increase bleeding risk [26, 27]. A study determined that "aspirin and aspirin plus dipyridamole" were equally cost-effective, even though combination therapy improved results at the expense of cost. Clopidogrel treatment was found to be the least damaging of the three [28, 29].

Warfarin can significantly decrease the relative risk of subsequent stroke by approximately 70% in patients with AF; however, a meta-analysis of trials showed no statistically significant change between warfarin and aspirin's effects [30]. Additionally, any potential benefit of warfarin is counterbalanced by an increased risk of bleeding (0.3%–0.6% per year). Carotid endarterectomy is a secondary preventative treatment for patients with a 70% stenotic carotid artery. For optimal benefit, carotid endarterectomies should be performed within 12 weeks of acute stroke [2]. According to the evidence, treating blood pressure medications may also serve as a secondary preventive health measure. The PROGRESS Study of "Perindopril Protection against Recurrent Stroke" demonstrated an approximately 30% decline in recurring stroke over five years. This decrease in blood pressure was witnessed regardless of the patient's baseline value [1, 2]. Statins can also be used to decrease the incidence of stroke recurrence. This benefit is not related to the patient's baseline cholesterol level and may be due to statins' anti-inflammatory or neuroprotective properties [3, 31].

1.4 THERAPY

Before administering any treatment, the patient's airway, breathing, and circulation must be assessed for stability. The following step is to determine whether they are an alteplase candidate (rt-PA). The criteria for exclusion are based on AHA/ASA guidelines. The goal of fibrinolytic therapy is to dissipate the clot and reintroduce blood flow to the affected areas. To be effective, the fibrinolytic must be given within three to four hours of symptom onset, depending on the eligibility requirements. "Time is brain" is crucial for stroke care, as it is for MI and sepsis and necessarily involves a rapid and collaborative treatment approach [32–34]. The time between treatment windows has evolved in "telemedicine and mobile stroke" units. The timeline for therapy was recently extended due to the advent of endovascular treatment [35–37]. There is a criterion for thrombolytic therapy which includes that the symptoms must be consistent with an ischemic stroke; there have been no recent head injuries; no MI in the preceding three months; 21 days without GI bleeding; no arterial rupture in the previous seven days; there has been no major surgical procedure in the preceding two weeks; there is no record of intracranial bleeding; systolic Blood pressure less than 185 millimeters mercury, diastolic blood pressure less than 111 millimeters mercury; there has been no recent trauma and perhaps bleeding; not currently taking any oral anticoagulant; glucose levels greater than 50 mg/dl and platelet count greater than 1,00,000 [38].

Hemorrhage, a potential severe fibrinolytic therapy complication, and its transformation is classified into hemorrhagic infarction and parenchymal hematoma. There are more hemorrhagic infarctions than parenchymal hematomas. An increase in infarction area, gray matter, cerebral embolism, and AF, as well as acute hyperglycemia, poor collateral circulation, and low platelet count, all predict the occurrence of this complication [38]. There have been considerable advancements in acute stroke care in recent years. Multiple trials published in recent years have demonstrated that in patients with anterior circulation major artery occlusion, endovascular thrombectomy performed during the first six hours is vastly superior to routine medical care. These benefits were consistent regardless of geographic location or patient individualities [39, 40]. In Stroke Care 2018, "the DAWN and DEFUSE 3" trial used perfusion imaging to extend the stroke window in selected patients from 16 to 24 hours. As a result, more patients were treated for up to 16–24 hours [41]. Patients should begin antiplatelet therapy, such as aspirin, between 24 and 48 hours of the onset of symptoms. For the first couple of days following a stroke, blood pressure should be slightly increased to improve oxygenation to the peri-infarction zones. Comorbidities must also be addressed and managed during the hospital stay of the patient [42, 43]. If the saturation of the room's air is less than 95%, oxygen should be provided. In addition, during a patient hospital stay, comorbidities must be addressed and managed (Figure 1.3) [44].

Regarding managing cerebral edema, cerebral edema peaks 72–96 hours after a stroke attack and is detected using a non-contrast CT scan. A change in mental state, loss of consciousness, or a change in pupillary size may all be indicators. Mannitol is used to treat it, but the use of corticosteroids is debatable. In most patients, positioning, hyperventilation, hypothermia, and hyperosmolar treatment will be adequate.

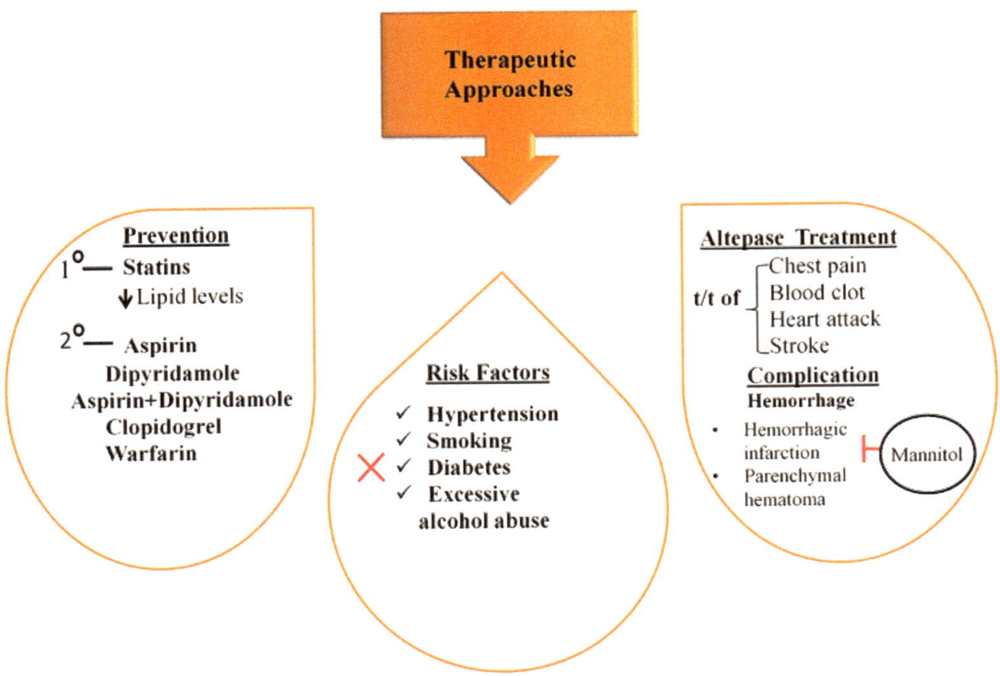

FIGURE 1.3 Depiction of therapeutic approaches for stroke, i.e., prevention, risk factors, treatment.

1.5 SUPPORTIVE CARE

The immediate goal of cerebral infarction is to improve cerebral perfusion in the ischemic area, monitor potential stroke complications (cerebral edema, hemorrhagic transformation, seizures, cardiovascular, and pulmonary difficulties, fever, and malignant hypertension), and prevent the frequent complications of bedridden patients such as malnutrition, infection, pulmonary damage embolism. In preventing these complications, initial treatment is beneficial. Early physical therapist involvement may reduce the risk of long-term stroke health problems and accelerate the recovery process [19]. A good diet, paralyzed extremities support, and the patient's overall positioning can also minimize complications immediately after the stroke. In the period following a stroke, there are also ongoing trials assessing the effectiveness of active cooling [45]. Airway support and ventilatory aid may be needed for acute stroke patients with depressed levels of conscious awareness or airway obstruction. A further recommendation is that only hypoxic patients should receive additional oxygen. Control of hypoglycemia or hyperglycemia after the stroke is generally recommended. A reasonable target would be to decrease significantly higher glucose levels to <300 mg/dL [45, 46]. If necessary, a dietetic modification or a nasogastronic feeding tube should also monitor and treat the swallowing function. Such patients showed a reduction in death and dependency 90 days after a stroke [19, 47]. Cardiovascular diseases (primarily myocardial infarction and arrhythmias) are potential causes and problems of acute stroke and should be evaluated and treated carefully in stroke patients according to established protocols.

Use of anticoagulants within the first 14 days should be avoided. Doubts remain within the scientific community on whether or not hypertension can be systematically treated during the initial diagnosis of the stroke. Studies have proved both points; thus, further research is needed regarding the same. Hypotension can also be a complication and should be closely controlled [48]. After stabilizing the patient, family education, depression screening, treatment, and physical and functional rehabilitation should begin immediately. After that, the patient should be evaluated further to ascertain the stroke cause, and surgical and medical therapies to prevent recurrent ischemic events should be administered. Patients who undergo hemispheric decompression within 48 hours of an infarction benefit from the procedure "malignant middle-cerebral-artery territory infarction and space-occupying" brain edema [49]. Around 75% of patients who underwent surgery had better results, compared to 24% of controls, according to a combined analysis of three clinical trials [50, 51].

1.6 CONCLUSION

Stroke prevention and management are most effectively accomplished through the collaboration of integrated care. Healthcare facilities with stroke teams that have field expertise and shorter "door-to-needle" times have a lower mortality rate and better outcomes. Stroke patients require close collaboration between neurologists, radiologists, emergency nurses, emergency department physicians, and pharmacists. Each of these professionals contributes to the diagnosis and treatment of patients who have suffered an acute stroke. Frequently, the emergency nurse is the first to contact a stroke patient, typically during triage. It is critical to recognize the condition and enlist the assistance of a stroke team. The emergency pharmacist's role is to assist in the rapid review of acute treatments provided for potential drug interactions and accurate dosing. After a stroke's acute diagnosis and treatment, the patient may need intense physical rehabilitation, speech management, and dietary counseling. The prognosis is favorable for those who regain function within three months but is uncertain for those who retain neurological deficits. At this point, physical therapists and nursing rehabilitation are critical in delivering family and patient education training, as daily living activities are frequently difficult. A multimodal approach that includes an antiplatelet agent, improved blood pressure control, better blood sugar level, a high statin, tobacco cessation, and a healthy lifestyle can protect at least 80% of successive strokes. Stroke management has been one of the utmost perplexing and challenging issues confronting health professionals. The best outcomes will be obtained only through a multidisciplinary approach.

KEYWORDS

- **CNS injury**
- **ischemic**
- **neurological deficit**
- **pathophysiology**
- **secondary prevention**
- **stroke**
- **thrombolysis**

REFERENCES

1. Donnan, G. A., Fisher, M., Macleod, M., & Davis, S. M., (2008). Secondary prevention of stroke–authors' reply. *The Lancet, 372*(9643), 1036.
2. Khan, H., Gupta, A., Singh, T. G., & Kaur, A., (2021). Mechanistic insight on the role of leukotriene receptors in ischemic–reperfusion injury. *Pharmacological Reports, 73*(5), 1240–1254.
3. Warlow, C. P., Van, G. J., Dennis, M. S., Wardlaw, J. M., Bamford, J. M., Hankey, G. J., Sandercock, P. A., et al., (2011). *Stroke: Practical Management*. John Wiley & Sons.
4. Malika, K., & Shavkat, I., (2020). Morphological features of ischemic and hemorrhagic brain strokes. *JCR, 7*(19), 7906–7910.
5. Grewal, A. K., Singh, T. G., & Singh, N., (2020). Potential herbal drugs for ischemic stroke: A review. *Plant Arch, 20*(1), 3772–3783.
6. Khan, H., Kashyap, A., Kaur, A., & Singh, T. G., (2020). Pharmacological postconditioning: A molecular aspect in ischemic injury. *Journal of Pharmacy and Pharmacology, 72*(11), 1513–1527.
7. Tadi, P., Lui, F., & Budd, L. A., (2021). *Acute Stroke (Nursing)*.
8. Khan, H., Grewal, A. K., & Singh, T. G., (2022). Pharmacological postconditioning by protocatechuic acid attenuates brain injury in ischemia–reperfusion (I/R) mice model: Implications of nuclear factor erythroid-2-related factor pathway. *Neuroscience, 491*, 23–31.
9. Adams, Jr. H. P., Adams, R. J., Brott, T., Del, Z. G. J., Furlan, A., Goldstein, L. B., Grubb, R. L., et al., (2003). Guidelines for the early management of patients with ischemic stroke: A scientific statement from the stroke council of the American stroke association. *Stroke, 34*(4), 1056–1083.

10. Hanley, D. F., Awad, I. A., Vespa, P. M., Martin, N. A., & Zuccarello, M., (2013). Hemorrhagic stroke: Introduction. *Stroke*, *44*(6_suppl_1), S65–S66.
11. Sharma, V., Singh, T. G., & Mannan, A., (2022). Therapeutic implications of glucose transporters (GLUT) in cerebral ischemia. *Neurochemical Research,* doi: 10.1007/s11064-022-03620-1. Epub ahead of print. PMID: 35596882.
12. Saqqur, M., Uchino, K., Demchuk, A. M., Molina, C. A., Garami, Z., Calleja, S., Akhtar, N., et al., (2007). Site of arterial occlusion identified by transcranial doppler predicts the response to intravenous thrombolysis for stroke. *Stroke*, *38*(3), 948–954.
13. Khan, H., Sharma, K., Kumar, A., Kaur, A., & Singh, T. G., (2022). Therapeutic implications of cyclooxygenase (COX) inhibitors in ischemic injury. *Inflammation Research*, 1–16.
14. Ding, S., Zhang, M., Zhao, Y., Chen, W., Yao, G., Zhang, C., Zhang, P., & Zhang, Y., (2008). The role of carotid plaque vulnerability and inflammation in the pathogenesis of acute ischemic stroke. *The American Journal of the Medical Sciences*, *336*(1), 27–31.
15. Adams, Jr. H. P., & Gordon, D. L., (1991). Epidemiology of and stroke-preventive strategies for atherothromboembolic brain infarction in the elderly. *Clinics in Geriatric Medicine*, *7*(3), 401–416.
16. Mărgăritescu, O. T. I. L. I. A., Mogoantă, L., Pirici, I., Pirici, D., Cernea, D., & Mărgăritescu, C. L., (2009). Histopathological changes in acute ischemic stroke. *Rom. J. Morphol. Embryol.*, *50*(3), 327–339.
17. Bornstein, N., Silvestrelli, G., Caso, V., & Parnetti, L., (2006). Arterial hypertension and stroke prevention: An update. *Clinical and Experimental Hypertension*, *28*(3, 4), 317–326.
18. Kalra, P., Khan, H., Kaur, A., & Singh, T. G., (2022). Mechanistic insight on autophagy modulated molecular pathways in cerebral ischemic injury: From preclinical to clinical perspective. *Neurochemical Research*, 1–19.
19. Amarenco, P., Labreuche, J., Lavallée, P., & Touboul, P. J., (2004). Statins in stroke prevention and carotid atherosclerosis: Systematic review and up-to-date meta-analysis. *Stroke*, *35*(12), 2902–2909.
20. Benjamin, E. J., Virani, S. S., Callaway, C. W., Chamberlain, A. M., Chang, A. R., Cheng, S., Chiuve, S. E., et al., (2018). Heart disease and stroke statistics—2018 update: A report from the American Heart Association. *Circulation*, *137*(12), e67–e492.
21. Tiwari, P., Khan, H., Singh, T. G., & Grewal, A. K., (2022). Poly (ADP-ribose) polymerase: An overview of mechanistic approaches and therapeutic opportunities in the management of stroke. *Neurochemical Research*, 1–23.
22. Krishnamurthi, R. V., & Feigin, V. L., (2021). 14 global burden of stroke. *Stroke E-Book: Pathophysiology, Diagnosis, and Management*, 163.
23. Tornyos, D., Bálint, A., Kupó, P., El Abdallaoui, O. E. A., & Komócsi, A., (2021). Antithrombotic therapy for secondary prevention in patients with non-cardioembolic stroke or transient ischemic attack: A systematic review. *Life*, *11*(5), 447.
24. Diener, H. C., Cunha, L., Forbes, C. E., Sivenius, J., Smets, P., & Lowenthal, A., (1996). European stroke prevention study 2. Dipyridamole and acetylsalicylic acid in the secondary prevention of stroke. *Journal of the Neurological Sciences*, *143*(1, 2), 1–13.
25. Wang, Y., Johnston, C., Bath, P. M., Meng, X., Jing, J., Xie, X., Wang, A., et al., (2021). Clopidogrel with aspirin in high-risk patients with acute non-disabling cerebrovascular events II (CHANCE-2): Rationale and design of a multicenter randomized trial. *Stroke and Vascular Neurology*, svn-2020.
26. Chien, L. N., Liu, H. Y., Chiou, H. Y., & Chi, N. F., (2020). Efficacy and safety of clopidogrel and aspirin do not differ in patients with stable ischemic stroke. *Journal of the Chinese Medical Association*, *83*(7), 651–656.
27. Patel, E. C. B. R., & Beth, J., (2020). A randomized, blinded, trial of clopidogrel versus aspirin in patients at risk of ischemic events (CAPRIE). *50 Landmark Papers Every Vascular and Endovascular Surgeon Should Know*, 40.
28. Bayer, S., Eom, K., Sivapragasam, N., Silva, D. A. D., Choon, G., Koh, H., Tan, K. B., et al., (2020). Estimating costs and benefits of stroke management: A population-based simulation model. *Journal of the Operational Research Society*, 1–13.
29. Matchar, D. B., Samsa, G. P., & Liu, S., (2005). Cost-effectiveness of antiplatelet agents in secondary stroke prevention: The limits of certainty. *Value in Health*, *8*(5), 572–580.

30. Katsanos, A. H., Kamel, H., Healey, J. S., & Hart, R. G., (2020). Stroke prevention in atrial fibrillation: Looking forward. *Circulation, 142*(24), 2371–2388.
31. Rihal, V., Khan, H., Kaur, A., & Singh, T. G., (2022). Vitamin D as therapeutic modulator in cerebrovascular diseases: A mechanistic perspectives. *Critical Reviews in Food Science and Nutrition*, 1–23.
32. Khan, H., Grewal, A. K., Kumar, M., & Singh, T. G., (2022). Pharmacological postconditioning by protocatechuic acid attenuates brain injury in ischemia-reperfusion (I/R) mice model: Implications of nuclear factor erythroid-2-related factor pathway. *Neuroscience, 491*, 23–31. doi:10.1016/j.neuroscience.2022.03.016.PMID: 35314251.
33. Kim, J., Thayabaranathan, T., Donnan, G. A., Howard, G., Howard, V. J., Rothwell, P. M., Feigin, V., et al., (2020). Global stroke statistics 2019. *International Journal of Stroke, 15*(8), 819–838.
34. Li, L., Scott, C. A., Rothwell, P. M., & Oxford Vascular Study, (2020). Trends in stroke incidence in high-income countries in the 21st century: Population-based study and systematic review. *Stroke, 51*(5), 1372–1380.
35. Prabhakar, N. K., Khan, H., Grewal, A. K., & Singh, T. G., (2022). Intervention of neuroinflammation in the traumatic brain injury trajectory: *In vivo* and clinical approaches. *International Immunopharmacology, 108*, 108902.
36. Schuler, F., Rotkopf, L. T., Apel, D., Fabritius, M. P., Tiedt, S., Wollenweber, F. A., Kellert, L., et al., (2020). Differential benefit of collaterals for stroke patients treated with thrombolysis or supportive care. *Clinical Neuroradiology, 30*(3), 525–533.
37. Saklani, P., Khan, H., Gupta, S., Kaur, A., & Singh, T. G., (2022). Neuropeptides: Potential neuroprotective agents in ischemic injury. *Life Sciences, 288*, 120186.
38. Wang, W., Li, M., Chen, Q., & Wang, J., (2015). Hemorrhagic transformation after tissue plasminogen activator reperfusion therapy for ischemic stroke: Mechanisms, models, and biomarkers. *Molecular Neurobiology, 52*(3), 1572–1579.
39. Golnari, P., Nazari, P., Ansari, S. A., Hurley, M. C., Shaibani, A., Potts, M. B., & Jahromi, B. S., (2021). Endovascular thrombectomy after large-vessel ischemic stroke: Utilization, outcomes, and readmissions across the United States. *Radiology, 299*(1), 179–189.
40. Goyal, M., Menon, B. K., Van, Z. W. H., Dippel, D. W., Mitchell, P. J., Demchuk, A. M., Dávalos, A., et al., (2016). Endovascular thrombectomy after large-vessel ischemic stroke: A meta-analysis of individual patient data from five randomized trials. *The Lancet, 387*(10029), 1723–1731.
41. Thomalla, G., & Gerloff, C., (2019). Acute imaging for evidence-based treatment of ischemic stroke. *Current Opinion in Neurology, 32*(4), 521–529.
42. Hacke, W., Kaste, M., Bluhmki, E., Brozman, M., Dávalos, A., Guidetti, D., Larrue, V., et al., (2008). Thrombolysis with alteplase 3 to 4.5 hours after acute ischemic stroke. *New England Journal of Medicine, 359*(13), 1317–1329.
43. Katzan, I. L., Furlan, A. J., Lloyd, L. E., Frank, J. I., Harper, D. L., Hinchey, J. A., Hammel, J. P., et al., (2000). Use of tissue-type plasminogen activator for acute ischemic stroke: The Cleveland area experience. *JAMA, 283*(9), 1151–1158.
44. Hasan, T. F., Rabinstein, A. A., Middlebrooks, E. H., Haranhalli, N., Silliman, S. L., Meschia, J. F., & Tawk, R. G., (2018). Diagnosis and management of acute ischemic stroke. In: *Mayo Clinic Proceedings* (Vol. 93, No. 4, pp. 523–538). Elsevier.
45. Marler, J. R., Tilley, B. C., Lu, M., Brott, T. G., Lyden, P. C., Grotta, J. C., Broderick, J. P., et al., (2000). Early stroke treatment associated with better outcome: The NINDS rt-PA stroke study. *Neurology, 55*(11), 1649–1655.
46. Powers, W. J., Alejandro, A. R., Teri, A., Opeolu, M. A., Nicholas, C. B., Kyra, B., José, B., et al., (2019). Guidelines for the early management of patients with acute ischemic stroke: 2019 update to the 2018 guidelines for the early management of acute ischemic stroke: A guideline for healthcare professionals from the American Heart Association/American Stroke Association. *Stroke, 50*(12), e344–e418.
47. Khan, H., Singh, A., Thapa, K., Garg, N., Grewal, A. K., & Singh, T. G., (2021). Therapeutic modulation of the phosphatidylinositol 3-kinases (PI3K) pathway in cerebral ischemic injury. *Brain Research, 1761*, 147399.

48. Taylor, F. C., Cohen, H., & Ebrahim, S., (2001). Systematic review of long-term anticoagulation or antiplatelet treatment in patients with non-rheumatic atrial fibrillation. *BMJ, 322*(7282), 321–326.
49. Berge, E., & Sandercock, P. A., (2002). Anticoagulants versus antiplatelet agents for acute ischemic stroke. *Cochrane Database of Systematic Reviews*, (4).
50. Nepal, P. R., & Rijal, S., (2020). Outcome following decompressive surgery for malignant middle cerebral artery infarction. *Eastern Green Neurosurgery*, *2*(1), 23–26.
51. Mohan, R. K., Crocker, M., & Moynihan, B., (2017). Decompressive craniectomy for the treatment of malignant middle cerebral artery infarction. *British Journal of Neurosurgery*, *31*(4), 401–409.

CHAPTER 2

Myocardial Ischemic-Reperfusion Injury

HEENA KHAN,[1] THAKUR GURJEET SINGH,[1] and RAHUL DESHMUKH[2]

[1]*Chitkara College of Pharmacy, Chitkara University, Punjab, India*

[2]*Maharaja Ranjit Singh Punjab Technical University, Bathinda, Punjab, India*

ABSTRACT

Myocardial ischemia/reperfusion injury (MIRI) is a severe medical problem referred to as lethal reperfusion injury. It is caused by the restoration of coronary blood flow following an ischemic episode. It is the most damaging outcome of I/R injury and is the primary reason for the ischemic myocardium's failure to retrieve following reperfusion therapy. Reperfusion has been and still is the only effective strategy for the salvage of ischemic myocardium in the disease acute phase infarction, but it also results in substantial damage. Survival of cardiomyocytes has also been facilitated via a variety of signaling pathways. This chapter discusses certain drugs acting via signaling pathways in detail, providing insight into potential future therapies. By gaining insights into the mechanisms, mediators, and risk factors associated with I/R injury, researchers can increase their chances of developing and implementing novel strategies for I/R injury in clinical patient care.

2.1 INTRODUCTION

The myocardium is thought to have a negligible capacity for regeneration, and scarring occurs when large amounts of cardiac muscle are lost [1]. To avoid the death of harmed cardiomyocytes, timely recovery of blood flow to the myocardial tissue is the best treatment. Reperfusion is the only strategy for reducing the infarct size and lowering mortality and morbidity in current clinical practice. Myocardial ischemia develops when coronary blood flow is impaired by coronary vessel obstruction due to coronary atherosclerosis. Myocardial infarction is irreversible damage caused by severe and prolonged myocardial ischemia. It is caused by the rupture of an epicardial coronary artery blockage, which causes recurrent thrombosis and blockage of the coronary artery. Despite massive use, enhanced logistics, and an improved approach to coronary reperfusion, the disease burden related to the development of heart failure due to an acute myocardial infarction (AMI)

Ischemic Injury: A Molecular Insight. Thakur Gurjeet Singh & Heena Khan (Eds.)

stays high [2]. According to a study, the fatality rate for heart disease is about 9 million globally and decreased by 10%, in line with the increase in successful reperfusion. However, a recent study demonstrated that reperfusion on its own injures the ischemic myocardium [3]. These insults significantly reduce the therapeutic benefits of reperfusion. Jennings et al. first described myocardial I/R injury in 1960 [4]. It is categorized by cell swelling, myofibril contracture, and sarcolemma disruption in the myocardium that has been subjected to ischemia accompanied by reperfusion. Following the initial report of I/R injury, evidence has accumulated in the published research over the last few decades. The chapter summarized the clinical manifestations and various basic mechanisms and mediators of myocardial I/R injury. There could be more effective ways to reduce myocardial damage if a better insight into the mediators and mechanisms of I/R injury has been known [5].

2.2 MECHANISMS AND MEDIATORS IN MYOCARDIAL ISCHEMIA/REPERFUSION INJURY

Following identifying the significance of myocardial I/R injury, piling up research has focused on the mechanisms and mediators causing cardiac injury during I/R. Numerous pathological methods and mediators have been proposed to be vital in cell injury caused by ischemia and reperfusion. These mediators play a role in the development of distinct pathological changes during each stage of I/R. These pathological changes are interconnected and contribute to cell damage. Below mentioned processes include the critical pathological changes that occur during cardiac I/R [4, 6]:

1. **pH Paradox:** Ischemia increases tissue acidosis due to the accumulation of intracellular hydrogen, sodium, and calcium ions. When blood flow is reinstated, acidosis is quickly adjusted by activating the Na+/H+ exchanger and the Na^+-HCO-symporter [7]. Rapid pH adjustment during reperfusion has been associated with increased cytotoxicity [8], possibly because this rapid pH shift facilitates the opening of "mitochondrial permeability transition pores" (MPTP) [9], which results in cardiomyocyte death.
2. **Calcium Overload:** Ischemia causes intracellular and mitochondrial calcium overload, which is aggravated after reperfusion [10]. The Na+/H+ exchanger and the Na^+–HCO_3^-transporter are activated during reperfusion, resulting in the accumulation of sodium intracellularly. On the other hand, a high sodium level stimulates cytosolic calcium accumulation via reverse Na^+–Ca^{2+} exchange [11]. Due to ATP deficiency during the I/R process, cytosolic calcium accumulates further by increasing calcium entry via sarcolemmal L-type Ca^{2+} channels. An increase in calcium may cause myofibrillar hypercontractility [12], mitochondrial dysfunction [13], and myocardial stunning [14].
3. **Reactive Oxygen Species (ROS):** These are essential mediators of myocardial I/R injury, and as evidenced by interventions aimed at improving ROS scavenging to prevent reperfusion injury, artificial ROS generation perpetuates the I/R injury

response. It increases ROS production and its results can be identified in post-ischemic tissues [14]. There is a splash of ROS [15] during the first few minutes of myocardial reperfusion that could be from multiple sources. The most likely sources of reperfusion-induced oxidative stress are "xanthine oxidase, nicotinamide adenine dinucleotide phosphate oxidase, mitochondria, and uncoupled nitric oxide synthase" [16]. All such oxygen species are incredibly reactive and quickly engulf the cell's endogenous free radical scavenging system. Through a variety of means, this detrimental oxidative stress leads to myocardial injury and cardiomyocyte mortality. They might not only cause immediate damage to membrane integrity, but they can also cause indirect harm by opening the MPTP and activating pro-apoptotic pathways [14].

4. **Mitochondria Dysfunction:** Cardiomyocytes use a tremendous amount of energy, so the density of mitochondria within the cells is significantly higher. Several factors, including Ca^{2+}, regulate mitochondrial function [16]. Mitochondrial Ca^{2+} is a positive activator in a normal physiological state that can activate mitochondrial metabolic machinery to produce more ATP [17]. The cytosolic calcium overload, however, results in consequent mitochondrial calcium overload during the I/R process. Transfer of Ca^{2+} across the mitochondrial membrane can amplify mitochondrial Ca^{2+} overload [18], causing mitochondrial dysfunction. Mitochondrial Ca^{2+} overload causes the electrical membrane potential to diminish [19], favoring MPTP opening and impairing ATP production. MPTP, a non-selective channel located within the inner mitochondrial membrane, has been implicated in I/R injury [20]. The mitochondrial inner membrane, which is usually impervious to proteins and ions, is essential for keeping the transmembrane capability of the mitochondria [21]. The formation of pores results in the formation of a nonselective channel and the diffusion of electrical potential throughout the membrane. MPTP tends to remain close during ischemia and reopen during reperfusion in reaction to a state when there is an overload of oxidative stress, mitochondrial calcium and phosphate, and rapid pH correction in the management of acute I/R [22, 23]. This causes loss of the electrochemical gradient, ROS generation, cytochrome c release, and the activation of the pro-apoptotic pathway, which eventually leads to cell death [24].
5. **Inflammation:** It has also been hypothesized to play a significant role in I/R injury [25], although the I/R approach is generally performed in a sterile environment. Such sterile inflammation, marked by the buildup of inflammatory cells, has a role in stimulating innate and adaptive immunological responses and cardiac injury [26]. Many phenotypic aspects of I/R stimulate the host immune response to invading pathogens [27]. The binding of ligands triggers the innate immune system to pattern recognition molecules like Toll-like receptors, referred to as "damage-associated molecular patterns" [28]. Mechanisms named "Antigen-specific and antigen-independent" activate lymphocytes throughout the I/R process [29]. During the I/R process, the innate and adaptive immune systems recruit and activate immune cells, which harm the damaged myocardium (Figure 2.1).

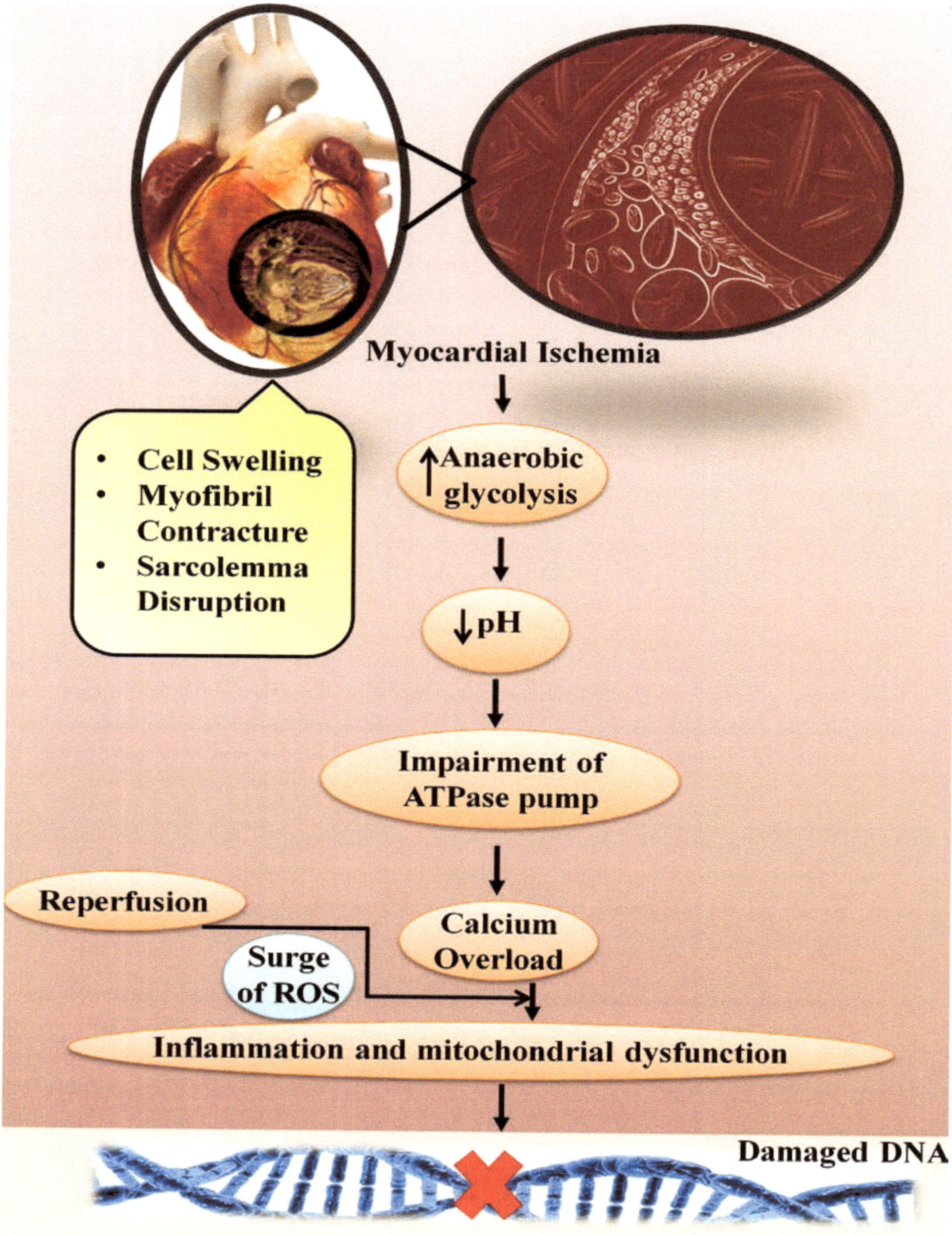

FIGURE 2.1 The pathophysiology of myocardial ischemic-reperfusion injury.

2.3 FACTORS LEADING TO WORSENING OF MYOCARDIAL ISCHEMIA

There are various risk factors for AMI. Some are modifiable (treatable), and others are non-modifiable (can not be changed). The significant risk factors of AMI are described hereunder:

1. **Physical Activity:** Individuals who are inactive and have several cardiac risk factors are more prone to developing AMI [30]. Physical activity has been shown to reduce the risk of heart attack by up to 20%–30% [31]. For instance, certain leisure activities such as walking and cycling can help protect against cardiovascular disease [32]. In contrast, others, such as vigorous household physical activity, may not prevent CVD [33]. The association between physical activity and AMI was examined and concluded that inactivity is a risk factor for consequent AMI [34]. Similarly, it has been proposed that a low level of indoor activity is associated with a decreased risk of AMI [35].
2. **Smoking:** Tobacco use is a significant risk factor for premature atherosclerosis, myocardial infarction, and sudden cardiac arrest. Smoking is linked to an increased risk of AMI and STEMI through various intricate mechanisms [36–38]. Regarding atherogenesis, smoking raises serum LDL-cholesterol and triglyceride levels while decreasing HDL-cholesterol levels, which is further likely to result in a buildup of oxidized LDL-cholesterol inside the arterial wall. Smoking activates the sympathetic nervous system (SNS) primarily via its nicotine content, increasing heart rhythm and systolic blood pressure. This increased heart rate surges myocardial oxygen demands. Elevated SNS activity not only results in coronary arterial vasoconstriction but [39] reduces myocardial blood flow during times of increased oxygen demand. In response to surged myocardial oxygen demand and decreasing coronary blood flow, smoking also increases carboxyhemoglobin levels in the blood, potentially reducing myocardial oxygen delivery from oxyhemoglobin even further [40].
3. **Alcohol Consumption:** People who do not drink alcohol regularly have a significantly increased risk of AMI in the later hour. There is reliable evidence that moderate routine binge drinking is linked with a decreased risk of CVS [41, 42], whereas frequent binge drinking is implicated in the development of cardiovascular events [43]. The majority of studies demonstrated that alcohol consumption could increase HDL-C and adiponectin levels while decreasing fibrinogen concentrations. Further raising the chances of atherosclerosis, anoxia, and atrial fibrillation (AF) are causing damage to cardiac muscle cells and the cardiovascular system. Additionally, some researchers discovered that binge drinking could raise prostacyclin levels in the blood vessel wall, improve the condition of vascular endothelial cells, improve insulin sensitivity, and increase resistance to thrombosis [44, 45]. Likewise, long-term frequent binge drinking may focus on improving heart rate variability [46, 47], thereby lowering the risk of MI.
4. **Dyslipidemia:** Another significant risk factor for CVD, is usually defined as total cholesterol, LDL, triglycerides, apo B, levels exceeding 90% of the general population or HDL and apo A levels below the 10% [48] contributing

to myocardial infarction risk. Decreased HDL-C and elevated triglyceride levels disrupt the metabolic process, resulting in severe consequences. Dyslipidemia is prevalent in young patients presenting with AMI, with hypertriglyceridemia being the most prevalent, followed by hypercholesterolemia, elevated LDL, and low HDL [49]. The Framingham study established the link between dyslipidemia and coronary artery disease through randomized trials. High total cholesterol, LDL cholesterol, and a low HDL cholesterol level are all solid predictors of coronary atherosclerosis. Adjusting dyslipidemia could perhaps help reduce the risk of cardiovascular attack [50].

5. **Diabetes Mellitus:** Diabetes is a chronic condition caused by the body's inability to efficiently and successfully use insulin. A genetic predisposition causes it in combination with environmental factors [51–53]. Since type 2 diabetes recognizes several risk factors with coronary artery diseases (CAD), such as hypertension, age, dyslipidemia, physical inactivity, obesity, and stress, an increase in diabetes prevalence implies an increase in the risk of CAD as well [54, 55]. Diabetes reduces people's life expectancy by nearly eight years due to higher death rates [51]. Diabetes increases the risk of myocardial infarction by speeding up the progression of atherosclerosis, altering the lipid profile, and facilitating the formation of atherosclerotic plaque [56]. Diabetes mellitus is the leading individual risk factor for heart disease. Diabetes is also a possible cause of fatal myocardial infarction cases: myocardial infarction is more severe in people with diabetes than in those without diabetes [57].
6. **Hypertension:** Blood pressure and myocardial infarction are inextricably linked. The risk of a myocardial infarction gets increased by hypertension, and the higher the blood pressure, the greater the risk [58]. It is a significant risk factor for coronary artery atherosclerosis, which can result in a heart attack. Numerous mechanisms may contribute to hypertensive patients' increased coronary risk. Hypertension speeds up the progression of atheroma, raises shear stress on plaques, has a detrimental functional effect on the coronary arteries, and disrupts endothelial function and sympathetic tone control '72.' Controlling hypertension with strict adherence to prescribed medications and lifestyle modifications significantly reduces the risk of myocardial infarction.
7. **Obesity/BMI:** The risk of myocardial infarction is directly proportional to someone's BMI. Since substantial obesity is a known risk factor for myocardial infarction, it considerably increases the chance of infarction. Dwivedi et al. [59] conducted a meta-analysis of prior studies and concluded that obesity and overweight are linked to an increased risk of AMI. Weight gain and obesity can harm one's health; thus, it's essential to keep one's BMI under control to avoid AMI, as being overweight is an early marker for MI [60]. Study analysis found that abdominal obesity increases the probability of AMI in both men and women of all ages and sexes [61].
8. **Other:** Risk factors for myocardial infarction include stress, gout, age, gender, family history, and periodontal disease. Heart attacks and stroke are increased by chronic life stress, social isolation, and anxiety [38]. It's believed to be linked to

coronary artery hemodynamic stress and atherosclerotic plaque rupture, which leads to thrombosis [62]. Also, Gout patients are also more likely to suffer from a heart attack. The inflammatory response associated with gout plays a significant role in the onset and progression of atherosclerosis and the development of a prothrombotic milieu that leads to acute coronary events like angina or myocardial infarction [63, 64]. In AMI, advanced age is linked to a higher rate of mortality [65, 66]. About 80% of heart disease deaths occur in people aged 65 or older [38]. Men are more likely than women to experience a heart attack at a younger age. After menopause, women's heart attack rates rise, although they do not reach the same level as men's. Despite this, heart disease is the top cause of death in men and women alike [38]. Periodontal disorders are an inflammatory group of disorders caused primarily by bacteria and associated by-products [64]. There is mounting evidence that poor tooth health, particularly periodontal disease, raises the risk of CHD [67]. A link between periodontal disease and AMI has been found in multiple research. An independent risk factor for AMI is a family history of myocardial infarction. AMI risk is doubled when a first-degree relative has had an AMI [31] (Figure 2.2).

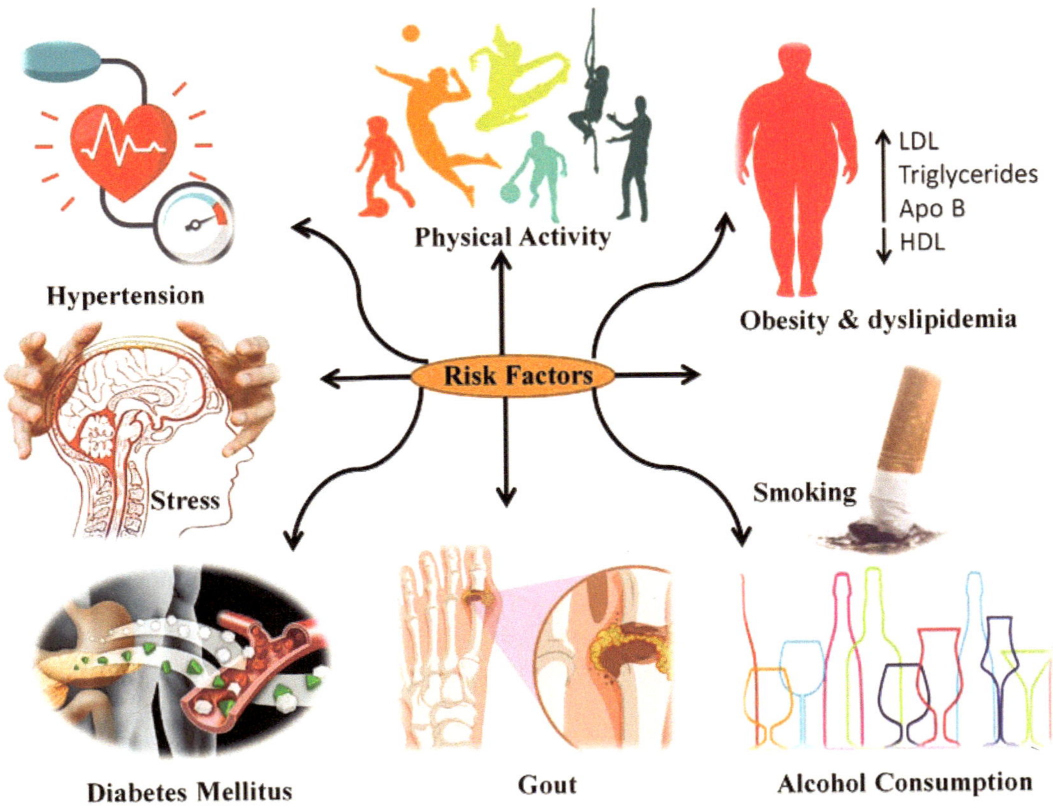

FIGURE 2.2 Pictorial representation of the risk factors associated with myocardial ischemic-reperfusion injury.

2.4 CARDIOPROTECTIVE STRATEGIES

2.4.1 ISCHEMIC CONDITIONING

This ischemic preconditioning phenomenon became the dominant paradigm for cardioprotection. Notably and frequently overlooked, this pioneering study on ischemic conditioning established that protection occurs only with subsequent reperfusion; when sustained coronary occlusion was not followed by reperfusion within 3 hours, the conditioning protection was diminished [68]. Ischemic preconditioning gained widespread acceptance and was overextended. The unique method of IpreCo is protected for only a limited time. When the time-lapse between the brief coronary occlusion that results in protection and the prolonged coronary occlusion that results in infarction was lengthened from 5 minutes to 2 hours, the protection was dramatically diminished. A delayed form of ischemic preconditioning was later identified, which begins 24 hours after preconditioning and lasts 1–3 days. In patients with heart failure, ischemic preconditioning is not possible due to the unpredictable nature of the ischemic event. However, the protective effects of ischemic preconditioning may play a role in patients with pre-infarction angina ensuring more substantial results [69]. IPreCo can be generated only in patients undergoing "elective percutaneous coronary intervention (PCI) or surgical coronary revascularization" [70–72].

2.4.2 ISCHEMIC POSTCONDITIONING

Ischemic postconditioning, which involved periods of "brief coronary reocclusion–reperfusion," tried to apply early for the duration of myocardial reperfusion following a continued coronary occlusion, decreased infarct size to a certain level as IpreCo. Additionally, this finding established unequivocally that reperfusion injury relates to infarct size. Unlike IPreCo, IPoCo can be used in patients undergoing primary PCI for ST-segment elevation myocardial infarction (STEMI) [73, 74]. In patients with STEMI [75], clinical trials have shown that cycles of short coronary reocclusion–reperfusion (1 minute each) minimize infarct size [76, 77], edema, coronary microvascular obstruction [78], and strengthen left ventricular contractile function [79].

2.4.3 REMOTE ISCHEMIC CONDITIONING

RIC is a brief "cycle of occlusion–reperfusion" in one coronary vascular region, reducing the infarct size due to prolonged occlusion-reperfusion of a nearby coronary artery [80]. Later work demonstrated that RIC is a systemic phenomenon that can be triggered from greater distances. Local injury caused by I/R in the "extremities or distinct parenchymal organs" can induce myocardial protection and protection in other parenchymal organs [81, 82]. RIC can be induced before, during, or following coronary occlusion. RIC is a non-invasive and precise procedure that can be initiated during coronary occlusion prior to primary PCI. Three cycles of RIC applied to the leg in addition to standard treatment improved the outcome of patients with STEMI compared to standard treatment alone, according to a clinical trial published in 2018 [83]. In a study involving patients with STEMI who underwent PCI with or without RIC before the change, only those diagnosed with cardiac arrest before the PCI benefited clinically from RIC [84].

2.4.4 PHARMACOLOGICAL CARDIOPROTECTION

The drugs being evaluated for cardioprotection are derived directly from the identified cardioprotective signaling pathways [85]. They are focused on either inhibiting harmful mechanisms (e.g., caspase activation inhibition and ROS scavenging) or activating protective mechanisms (such as the excessive release of adenosine or NO). Since this infarction is unexpected and treatment after the ischemic event no longer provides protection [86], only the intake of drugs before or at way earlier reperfusion is appropriate in heart failure patients. Many drugs have been demonstrated to lessen infarct size in preclinical models after exogenous delivery, although other researchers have not consistently replicated these findings. In preclinical studies, administration of adenosine reduced infarct size by [87] or did not reduce infarct size by [88]; thus, not surprisingly, intravenous or intracoronary adenosine therapy as an adjuvant to reperfusion did not lower infarct size in STEMI patients [89].

Other medicines that target the mitochondria, when given at reperfusion, did not have a consistent cardioprotective effect in individuals with AMI [90]. The administration of the β-blocker metoprolol or the GLP-1 (glucagon-like peptide-1) analog exenatide is unrelated to cardioprotective signal transduction. Metoprolol injected before reperfusion lowered infarct size, as did patients with reperfused acute STEMI who took intravenous metoprolol before reperfusion in a short proof-of-concept study [91, 92]. The therapeutic effect of metoprolol was linked to neutrophils and the reduction of neutrophil-mediated microvascular plugging [93], rather than a direct action on cardiomyocytes. Nevertheless, the results of a follow-up phase III trial on metoprolol therapy in STEMI patients were neutral, possibly due to the lower dose and delayed administration [68, 94].

In preclinical trials, exenatide, metoprolol, melatonin, and ARP-4 lowered coronary microvascular damage and infarction size in mice but didn't affect clinical outcomes [95–98]. Caspase inhibitors [95] and exosomes, which can contain cardioprotective signals [99, 100], are two further potential treatments demonstrated to decrease infarct size in preclinical trials and are worth further exploration. The use of medicines that suppress necroptosis and apoptosis together lowers infarct size in guinea pig hearts more often than each agent alone [101].

2.4.5 HYPOTHERMIA

Body temperature is a significant predictor of myocardial infarct size, and even small changes within the range of normothermia affect infarct size [102]. Notably, according to preclinical studies [103], hypothermia must be achieved during the ischemic episode to benefit myocardial infarct size. However, delayed hypothermia after established reperfusion diminished the no-reflow phenomenon. While relatively mild hypothermia significantly delayed ATP depletion during ischemia in a rabbit model, it was associated with activation of survival pathways, particularly the extracellular signal-regulated kinase pathway [104]. Compared to a sham protocol, vagal induction via low-level electrical transcutaneous activation at the right auricular tragus decreased infarct size and arrhythmias and strengthened ventricular function in STEMI patients [105].

2.5 COMPLICATIONS ASSOCIATED WITH MYOCARDIAL ISCHEMIA-REPERFUSION INJURY

1. **Mechanical:** Cardiogenic shock, VSR, acute MR, FWR, and true ventricular aneurysm.
2. **Electrical:** Bradyarrhythmias, bundle branch blocks, tachyarrhythmia, and fascicular blocks.
3. **Inflammatory:** Dressler syndrome and Peri-infraction pericarditis.
4. **Ischemic:** Post-infarction angina (infarct extension and reinfarction).
5. **Embolic:** Systemic embolism and Mural thrombus [106].

2.6 MECHANISTIC OVERVIEW

Myocardial ischemic reperfusion injury (MIRI) is associated with a high morbidity and mortality rate, and it continues to be the leading cause of death worldwide [1]. It is critical to restore blood circulation to the myocardium as soon as feasible to minimize myocardial infarction. However, reperfusion generates many problems, such as ROS overproduction, mitochondrial dysfunction, apoptosis, inflammation, and calcium overload [2, 3]. To date, considerable attention has been dedicated to excessive ROS generation, and it has been shown that significantly reducing ROS may significantly reduce myocardial damage [4]. However, there are presently few effective medications to relieve ischemia-reperfusion harm. To avoid cellular apoptosis and necrosis, reperfusion of the ischemic myocardium must be undertaken (Table 2.1).

2.7 CONCLUSION AND FUTURE DIRECTIONS

One of the most suitable approaches for preserving the ischemic myocardium and lowering mortality and morbidity from AMI is to restore blood flow to the infarct region as soon as possible. On the other hand, reperfusion can conversely result in tissue injury, negating the perks of therapy. In the last few decades, studies have been conducted to understand the mechanisms underlying cardiac I/R injury. Numerous cellular signaling transduction is needed for cell growth and is closely regulated, according to available evidence. Ischemia accompanied by reperfusion disrupts this meticulously maintained equilibrium. Rapid pH shifts, calcium overload, mitochondrial malfunction, inflammation, and protein

TABLE 2.1 Preclinical Studies Summarize Drugs Explored in Myocardial Ischemia-Reperfusion Injury to Date

SL. No.	Drug	Pathway Involved	Result	References
1.	Asiatic acid	Modulation Akt/GSK-3β/HIF-1α signaling pathway.	Pretreatment with Asiatic acid provided protective effects against myocardial ischemia by improving mitochondrial function and reducing IC Ca^{2+} release.	[107]
2.	GYY4137	Activation of PHLPP-1/Akt/Nrf2 signaling pathway.	GYY4137 exerts protection against MIRI by attenuating cellular injury.	[108]
3.	Salvianolate	Modulation of the ERK1/2 pathway.	Salvianolate pretreatment mitigates MIRI by modulating the ERK1/2 pathway.	[109]
4.	Syringic acid	Activating PI3K/Akt/GSK-3β signaling pathway.	Syringic pretreatment mitigated MIRI via preventing mitochondrial-induced apoptosis regulated by PI3K/Akt/GSK-3β.	[110]
5.	Rapamycin	JAK2-STAT3 signaling pathway.	Rapamycin protected against MIRI by mitochondrial K_{ATP} channels opening and limiting the apoptosis and necrosis of cardiomyocytes.	[111]
6.	Melatonin	Activation of SIRT-1 and 3.	Melatonin treatment attenuated the MIRI by upregulation of Bcl-2 and SIRT-1 expression and downregulation of Bax, Caspase-3.	[112, 113]
7.	Vitexin	MAPK pathway and Epac1-Rap1 signaling pathway.	Vitexin exhibited protection against MIRI via regulating various inflammatory cytokines and mitochondrial dysfunction.	[114, 115]
8.	Sappanone	Modulation of Nrf2	Sappanone prevented MIRI by maintaining redox balance and cellular antioxidant defense.	[116]
9.	Lipopolysaccharide (LPS) low dose	Modulation of IRE-1 signaling pathway.	Pretreatment with LPS attenuated the myocardial apoptosis by downregulating caspase-3 expression in MIRI.	[117]
10.	Vaspin	Modulation of TLR/NF-κB signaling pathway.	Vaspin exhibit a protective effect against MIRI by reducing pro-inflammatory mediators, e.g., IL-1β, IL-18, and TNF-α.	[118]
11.	Hesperidin	Activation of PI3K/Akt/mTOR.	Hesperidin attenuated the MIRI by inhibiting autophagy.	[119]
12.	Gastrin	Activation of RISK and SAFE pathways.	Gastrin improved myocardial function by reducing myocardial injury markers and cardiomyocyte apoptosis.	[120]
13.	JLX001	Activation of the JAK2-STAT3 pathway.	JLX001 improved left ventricular function and reduced apoptosis.	[121]
14.	Dexmedetomidine (DEX)	Modulating PI3K/Akt signaling pathways.	DEX protected against myocardial infarction by reducing the caspase-3 expression and modulating PI3K/Akt pathway.	[122]
15.	Propofol	Inhibition of JAK/STAT signaling pathways.	Propofol protected against MIRI by inhibiting myocardial cell apoptosis.	[123]

TABLE 2.1 *(Continued)*

SL. No.	Drug	Pathway Involved	Result	References
16.	Erythropoietin (EPO)	Activation of PI3K/Akt signaling pathway.	EPO pretreatment inhibited the release of pro-inflammatory cytokines.	[124]
17.	RP105	Suppression of TLR4/TRIF signaling pathway.	RP105 attenuated myocardial injury by inhibiting TLR4, NF-κB, TNF-α, and IL-6 expression.	[125, 126]
18.	Ulinastatin (UTI)	JNK and P38 MAPK signaling pathways.	UTI rendered the MIRI by downregulating the expression of TNF-α and IL-6.	[127]
19.	A-Lipoic acid (α-LA)	Activation of PI3K/Akt/Nrf2 pathway.	Lipoic acid attenuated MIRI by inhibiting the cardiomyocyte's necrosis, inflammation, and apoptosis.	[128]
20.	Morphine	Inhibition of TLR4/NF-κB pathway.	Morphine protected MIRI by reducing myocardial infarction and necrosis.	[129]
21.	Clemastine fumarate (CF)	Activation of TLR4/PI3K/Akt pathway.	CF attenuated MIRI by suppressing the inflammatory response, cardiomyocyte damage, and oxidative stress.	[130]
22.	Hydrogen sulfide (H_2S).	Modulation of JNK signaling pathway.	H_2S provided myocardial protection by improving the damaged mitochondrial cristae and membrane.	[131]
23.	Trimetazidine	Activation of Akt/mTOR pathway.	TMZ showed to be cardioprotective by inhibiting autophagy and the apoptosis process.	[132]
24.	Isoflurane	Inhibition of the p38 MAPK pathway.	Isoflurane protected cardiac function by reducing myocardial infarction and oxidative stress.	[133]
25.	Lumbrokinase (LBK)	Inhibition of TLR4-JNK and NF-κB signaling pathway.	LBK protected against MIRI by inhibiting COX-2, iNOS, and MMP-9 levels.	[134]
26.	Alpha-linolenic acid (ALA).	Modulation of P2X7R/NF-κB signaling pathway.	ALA protected the myocardium against the ischemic injury via inhibiting TNF-α and IL-1β production.	[135]
27.	Pravastatin	Modulation miR-93/Nrf2/ARE Signaling pathway.	Pravastatin acted as a protective agent in MIRI by promoting HO-1 and NQO1 genes.	[136]
28.	Irisin	Modulation of AMPK signaling pathway.	Irisin attenuated MIRI by improving mitochondrial function, and cellular ATP biogenetics.	[137]
29.	Metformin	Modulation of AMPK/NLRP3 signaling pathway and upregulation of STEAP4 signaling pathway.	Metformin showed protective effects against MIRI by decreasing apoptosis and pro-inflammatory cytokines.	[138, 139]

TABLE 2.1 *(Continued)*

SL. No.	Drug	Pathway Involved	Result	References
30.	Remifentanil	Suppression of Fas apoptosis signaling pathway.	Remifentanil showed protection against MIRI by decreasing oxidative stress and inflammation.	[140]
31.	Neuregulin-1 (NRG-1)	Modulating ErbB4 receptor.	NRG-1 attenuated MIRI by inhibiting ER stress and apoptosis.	[141]
32.	PIAS1	Modulation of PIAS1-mediated PPARγ SUMOylation.	PIAS1 inhibited inflammation and apoptosis of cardiomyocytes.	[142]
33.	Fisetin	Activation of RISK signaling pathway.	Fisetin decreased MIRI by decreasing myocardial infarct size, LDH, and apoptosis.	[143]
34.	Artemisinin	Suppression of NLRP3 inflammasome mechanism.	Artemisinin protected myocardial cells against MIRI by inhibiting of cardiac autophagy.	[144]
35.	Cilostazol	Activation of the cAMP-dependent pathway.	Cilostazol proved to have a protective role in MIRI by promoting angiogenesis.	[145]
36.	Resveratrol	Activation of PI3K/AKT.	Resveratrol mitigated MIRI by reducing myocardial cell death and mitochondrial oxidative damage.	[146]
37.	Glutamine	Activation of PI3K/AKT pathway.	Glutamine protects against myocardial injury by reducing inflammatory cytokines.	[147]
38.	Adrenomedullin	PI3K/AKT-dependent pathway.	Adrenomedullin prevented the MIRI by inhibiting the myocardial apoptosis and decreasing left ventricular end-diastolic pressure (VEDP).	[148]
39.	Bortezomib	Activation of Mrf2/HO-1 signaling pathway.	Bortezomib alleviates MIRI by inhibiting the oxidative damage.	[149]
40.	Visfatin	Activation of PI3K/AKT/HSP70 signaling pathway.	Visfatin decreased inflammation and apoptosis after MIRI.	[150]
41.	Butorphanol	Inhibition of MAPK pathway.	Butorphanol reduced caspase-3, IL-6, and IL-1α levels after MIRI.	[151]
42.	$PSMB_4$	Activation of NF-κB signaling pathway.	$PSMB_4$ decreased cardiomyocyte apoptosis during MIRI.	[152]
43.	Ligustrazine	Activation of PI3K/AKT pathway.	LAZ diminished MIRI by reducing oxidative stress and inflammation.	[153]
44.	Adiponectin	AMPK and COX-2 dependent pathways.	Adiponectin protected MIRI by diminishing myocardial apoptosis and TNF-α expression.	[154]
45.	Nicorandil	PI3K/AKT signaling pathway.	Nicorandil protected the myocardium by reducing ER-inducing apoptosis.	[155]

TABLE 2.1 *(Continued)*

SL. No.	Drug	Pathway Involved	Result	References
46.	Apelin	PI3K/AKT and P70S6 pathway.	Apelin reduced MIRI by reducing the infarct size.	[156]
47.	Cilostazol	Activation of PPARγ/JAK2/STAT3 signaling pathway.	CSZ protected cardiac function, inflammation, and apoptosis following MIRI.	[157]
48.	22-oxacalcitriol	Suppression of NF-κB/TNF-α pathway.	22-OC decreased inflammation, thereby protecting against MIRI.	[158]
49.	ZFP36L2	LncRNA PVT1	ZFP36L2 attenuated mitochondrial fusion-fission during MIRI.	[159]
50.	Diallyl trisulfide	AMPK-mediated AKT/GSK-3β/HIF-1α activation.	DATS exerted cardioprotection by attenuating cellular apoptosis.	[160]
51.	FK506	Suppression of CaN and ASK1 signaling pathway.	FK506 reduced the infarct size and apoptosis following MIRI.	[161]
52.	Ligustrazine	Inhibition of P38MAPK, NFκB, activation of Nrf2/HIF-1α and PI3K/AKT signaling pathway.	Inhibiting apoptosis, TNF-α expression, and enhancing NO release.	[162]

phosphorylation imbalance are all events that emerge from the I/R processes, and they all work together to create adverse effects in damaged cardiomyocytes. Cardioprotective methods have often failed to translate from good studies to clinical practice. Additional knowledge of the molecular events activated by the I/R pathway is required. There is a great need to develop remedies to guard against I/R injury if we investigate the mechanisms of I/R and better understood the underlying signaling system. We are excited to research and implement innovative therapeutics to reduce myocardial I/R injury in clinical practice. The use of additive cardioprotective therapies in the future and focusing on patients who genuinely require cardioprotection (those with cardiogenic shock or in Killip class III-IV) are recommended.

KEYWORDS

- **cardioprotective strategies**
- **coronary artery damage**
- **ischemia/reperfusion**
- **ischemic postconditioning**
- **myocardial injury**
- **myocardial ischemia**
- **reperfusion injury**

REFERENCES

1. Heusch, G., & Gersh, B. J., (2017). The pathophysiology of acute myocardial infarction and strategies of protection beyond reperfusion: A continual challenge. *European Heart Journal, 38*(11), 774–784.
2. Khan, H., Sharma, K., Kumar, A., Kaur, A., & Singh, T. G., (2022). Therapeutic implications of cyclooxygenase (COX) inhibitors in ischemic injury. *Inflammation Research*, 1–16.
3. Lesnefsky, E. J., Chen, Q., Tandler, B., & Hoppel, C. L., (2017). Mitochondrial dysfunction and myocardial ischemia-reperfusion: Implications for novel therapies. *Annual Review of Pharmacology and Toxicology, 57*, 535–565.
4. Heusch, G., (2020). Myocardial ischemia–reperfusion injury and cardioprotection in perspective. *Nature Reviews Cardiology, 17*(12), 773–789.
5. Yang, C. F., (2018). Clinical manifestations and basic mechanisms of myocardial ischemia/reperfusion injury. *Tzu-Chi Medical Journal, 30*(4), 209.
6. Khan, H., Kashyap, A., Kaur, A., & Singh, T. G., (2020). Pharmacological postconditioning: A molecular aspect in ischemic injury. *Journal of Pharmacy and Pharmacology, 72*(11), 1513–1527.
7. Saklani, P., Khan, H., Gupta, S., Kaur, A., & Singh, T. G., (2022). Neuropeptides: Potential neuroprotective agents in ischemic injury. *Life Sciences, 288*, 120186.
8. Lillo-Moya, J., Rojas-Solé, C., Muñoz-Salamanca, D., Panieri, E., Saso, L., & Rodrigo, R., (2021). Targeting ferroptosis against ischemia/reperfusion cardiac injury. *Antioxidants, 10*(5), 667.
9. Kim, J. S., Jin, Y., & Lemasters, J. J., (2006). Reactive oxygen species, but not Ca^{2+} overloading, trigger pH-and mitochondrial permeability transition-dependent death of adult rat myocytes after

ischemia-reperfusion. *American Journal of Physiology-Heart and Circulatory Physiology, 290*(5), H2024–H2034.
10. Garcia-Dorado, D., Ruiz-Meana, M., Inserte, J., Rodriguez-Sinovas, A., & Piper, H. M., (2012). Calcium-mediated cell death during myocardial reperfusion. *Cardiovascular Research, 94*(2), 168–180.
11. Murphy, E., & Steenbergen, C., (2008). Mechanisms underlying acute protection from cardiac ischemia-reperfusion injury. *Physiological Reviews, 88*(2), 581–609.
12. Kohlhauer, M., Pell, V. R., Burger, N., Spiroski, A. M., Gruszczyk, A., Mulvey, J. F., Mottahedin, A., et al., (2019). Protection against cardiac ischemia-reperfusion injury by hypothermia and by inhibition of succinate accumulation and oxidation is additive. *Basic Research in Cardiology, 114*(3), 1–9.
13. Yang, M., Linn, B. S., Zhang, Y., & Ren, J., (2019). Mitophagy and mitochondrial integrity in cardiac ischemia-reperfusion injury. *Biochimica et Biophysica Acta (BBA)-Molecular Basis of Disease, 1865*(9), 2293–2302.
14. Granger, D. N., & Kvietys, P. R., (2015). Reperfusion injury and reactive oxygen species: The evolution of a concept. *Redox Biology, 6*, 524–551.
15. Khan, H., Singh, A., Thapa, K., Garg, N., Grewal, A. K., & Singh, T. G., (2021). Therapeutic modulation of the phosphatidylinositol 3-kinases (PI3K) pathway in cerebral ischemic injury. *Brain Research, 1761*, 147399.
16. Xu, H. X., Cui, S. M., Zhang, Y. M., & Ren, J., (2020). Mitochondrial Ca^{2+} regulation in the etiology of heart failure: Physiological and pathophysiological implications. *Acta Pharmacologica Sinica, 41*(10), 1301–1309.
17. Boyman, L., Karbowski, M., & Lederer, W. J., (2020). Regulation of mitochondrial ATP production: Ca^{2+} signaling and quality control. *Trends in Molecular Medicine, 26*(1), 21–39.
18. Jung, H., Kim, S. Y., Cecen, F. S. C., Cho, Y., & Kwon, S. K., (2020). Dysfunction of mitochondrial Ca^{2+} regulatory machineries in brain aging and neurodegenerative diseases. *Frontiers in Cell and Developmental Biology, 8*.
19. Delierneux, C., Kouba, S., Shanmughapriya, S., Potier-Cartereau, M., Trebak, M., & Hempel, N., (2020). Mitochondrial calcium regulation of redox signaling in cancer. *Cells, 9*(2), 432.
20. Huang, G., & Docampo, R., (2020). The mitochondrial calcium uniporter interacts with subunit c of the ATP synthase of trypanosomes and humans. *MBio, 11*(2), e00268–20.
21. Naumova, N., & Šachl, R., (2020). Regulation of cell death by mitochondrial transport systems of calcium and Bcl-2 proteins. *Membranes, 10*(10), 299.
22. Akande, O., Chen, Q., Toldo, S., Lesnefsky, E. J., & Quader, M., (2020). Ischemia and reperfusion injury to mitochondria and cardiac function in donation after circulatory death hearts-an experimental study. *PLoS One, 15*(12), e0243504.
23. Morciano, G., Bonora, M., Campo, G., Aquila, G., Rizzo, P., Giorgi, C., Wieckowski, M. R., & Pinton, P., (2017). Mechanistic role of mPTP in ischemia-reperfusion injury. *Mitochondrial Dynamics in Cardiovascular Medicine*, 169–189.
24. Khan, H., Tiwari, P., Kaur, A., & Singh, T. G., (2021). Sirtuin acetylation and deacetylation: A complex paradigm in neurodegenerative disease. *Molecular Neurobiology, 58*(8), 3903–3917.
25. Khan, H., Gupta, A., Singh, T. G., & Kaur, A., (2021). Mechanistic insight on the role of leukotriene receptors in ischemic–reperfusion injury. *Pharmacological Reports*, 1–15.
26. Frangogiannis, N. G., (2014). The inflammatory response in myocardial injury, repair, and remodeling. *Nature Reviews Cardiology, 11*(5), 255–265.
27. Otani, K., & Shichita, T., (2020). Cerebral sterile inflammation in neurodegenerative diseases. *Inflammation and Regeneration, 40*(1), 1–8.
28. Xiong, Y. Y., Gong, Z. T., Tang, R. J., & Yang, Y. J., (2021). The pivotal roles of exosomes derived from endogenous immune cells and exogenous stem cells in myocardial repair after acute myocardial infarction. *Theranostics, 11*(3), 1046.
29. Linfert, D., Chowdhry, T., & Rabb, H., (2009). Lymphocytes and ischemia-reperfusion injury. *Transplantation Reviews, 23*(1), 1–10.
30. Rathore, V., Singh, N., Mahat, R. K., Kocak, M. Z., Fidan, K., Ayazoglu, T. A., Aydin, K. Y. G., et al., (2018). Risk factors for acute myocardial infarction: A review. *EJMI, 2*(1), 1–7.

31. Mritunjay, M., & Ramavataram, D. V. S. S., (2021). Predisposing risk factors associated with acute myocardial infarction (AMI): A review. *Indian Journal of Forensic Medicine & Toxicology, 15*(2), 407.
32. Yamaji, T., Harada, T., Hashimoto, Y., Nakano, Y., Kajikawa, M., Yoshimura, K., Chayama, K., et al., (2021). Stair climbing activity and vascular function in patients with hypertension. *Hypertension Research*, 1–9.
33. Michael, E., White, M. J., & Eves, F. F., (2021). Home-based stair climbing as an intervention for disease risk in adult females; a controlled study. *International Journal of Environmental Research and Public Health, 18*(2), 603.
34. Shajrawi, A., Granat, M., Jones, I., & Astin, F., (2021). Physical activity and cardiac self-efficacy levels during early recovery after acute myocardial infarction: A Jordanian study. *The Journal of Nursing Research, 29*(1), e131.
35. Yerramalla, M. S., McGregor, D. E., Van, H. V. T., Fayosse, A., Dugravot, A., Tabak, A. G., Chen, M., et al., (2021). Association of daily composition of physical activity and sedentary behavior with incidence of cardiovascular disease in older adults. *International Journal of Behavioral Nutrition and Physical Activity, 18*(1), 1–13.
36. Khan, H., Sharma, R., Kaur, A., & Singh, T. G. (2018). *The Endocannabioids System and Their Implications in Various Disorders. 49*(2), 97–106. Article No. 20.
37. Xu, H. B., Wang, J., Chen, J. L., Guo, C., Yuan, J. S., Hu, F. H., Yang, W. X., et al., (2020). Impacts of smoking status on the clinical outcomes of coronary non-target lesions in patients with coronary heart disease: A single-center angiographic study. *Chinese Medical Journal, 133*(19), 2295.
38. Huma, S., Tariq, R., Amin, F., & Mahmood, K. T., (2012). Modifiable and non-modifiable predisposing risk factors of myocardial infarction: A review. *Journal of Pharmaceutical Sciences and Research, 4*(1), 1649.
39. Rastogi, A., Talwar, T., Shah, P., Gupta, N., Singh, V. V., & Manzoor, S., (2020). Smoking and involvement of right coronary artery in ischemic heart disease. *RAMUS, 2*(3), 49–52.
40. Kim, S. T., & Park, T., (2020). Acute and chronic effects of cocaine on cardiovascular health. *Nutrition and Cardiovascular Health, 20*, 169.
41. Gupta, A., Khan, H., Kaur, A., & Singh, T. G., (2020). Novel targets explored in the treatment of alcohol withdrawal syndrome. *CNS & Neurological Disorders-Drug Targets (Formerly Current Drug Targets-CNS & Neurological Disorders).*
42. Kurth, T., Rist, P. M., Ridker, P. M., Kotler, G., Bubes, V., & Buring, J. E., (2020). Association of migraine with aura and other risk factors with incident cardiovascular disease in women. *JAMA, 323*(22), 2281–2289.
43. Deleuran, T., Schmidt, M., Vilstrup, H., & Jepsen, P., (2020). Time-dependent incidence and risk for myocardial infarction in patients with alcoholic cirrhosis. *European Journal of Clinical Investigation, 50*(4), e13205.
44. Khadse, N. A., Wankhade, A. M., & Gaiki, A. G. (2020). *Myocardial Infraction: Etiology, Risk Factors, Pathophysiology, Diagnosis and Management.*
45. Larsson, S. C., Burgess, S., Mason, A. M., & Michaëlsson, K., (2020). Alcohol consumption and cardiovascular disease: A mendelian randomization study. *Circulation: Genomic and Precision Medicine, 13*(3), e002814.
46. Golan, R., Gepner, Y., & Shai, I., (2019). Wine and health–new evidence. *European Journal of Clinical Nutrition, 72*(1), 55–59.
47. Peterson, H., Mayhugh, R. E., Bahrami, M., Rejeski, W. J., Simpson, S. L., Heilman, K., Porges, S. W., & Laurienti, P. J., (2021). Influence of heart rate variability on abstinence-related changes in brain state in everyday drinkers. *Brain Sciences, 11*(6), 817.
48. Castaldo, L., Narváez, A., Izzo, L., Graziani, G., Gaspari, A., Di Minno, G., & Ritieni, A., (2019). Red wine consumption and cardiovascular health. *Molecules, 24*(19), 3626.
49. Ali, S. N., Bashir, M., & Sherwani, M., (2016). Pattern of dyslipidemia in young patients with acute ST elevation myocardial infarction. *J. Sheikh Zayed Med. Coll., 7*, 998–1001.
50. Patti, A. M., Giglio, R. V., Papanas, N., Rizzo, M., & Rizvi, A. A., (2019). Future perspectives of the pharmacological management of diabetic dyslipidemia. *Expert Review of Clinical Pharmacology, 12*(2), 129–143.

51. Firdous, S. M., Ghosh, S., Nath, P., & Khan, H. (2021). *Molecular Basis of Diabetic Nephropathy,* pp. 63–91.
52. Firdous, S. M., Dhang, P., Koley, A., & Khan, H. (2021). *Diabetic Retinopathy: Pathogenesis and Therapeutic Management.* pp. 124–139.
53. Magliano, D. J., Zimmet, P., & Shaw, J. E., (2015). Classification of diabetes mellitus and other categories of glucose intolerance. *International Textbook of Diabetes Mellitus*, 1–16.
54. Zadeh, F. J., Mohammadtaghizadeh, M., Bahadori, H., Saki, N., & Rezaeeyan, H., (2020). The role of exogenous fibrinogen in cardiac surgery: Stop bleeding or induce cardiovascular disease. *Molecular Biology Reports*, 1–10.
55. Pintaudi, B., Scatena, A., Piscitelli, G., Frison, V., Corrao, S., Manicardi, V., Graziano, G., et al., (2021). Clinical profiles and quality of care of subjects with type 2 diabetes according to their cardiovascular risk: An observational, retrospective study. *Cardiovascular Diabetology, 20*(1), 1–7.
56. Khan, M. Z., Pervaiz, M. K., & Javed, I., (2016). Biostatistical study of clinical risk factors of myocardial infarction: A case-control study from Pakistan. *PAFMJ, 66*(3), 354–360.
57. Kodama, S., Tanaka, S., Heianza, Y., Fujihara, K., Horikawa, C., Shimano, H., Saito, K., et al., (2013). Association between physical activity and risk of all-cause mortality and cardiovascular disease in patients with diabetes: A meta-analysis. *Diabetes Care, 36*(2), 471–479.
58. Hajar, R., (2017). Risk factors for coronary artery disease: Historical perspectives. *Heart Views: The Official Journal of the Gulf Heart Association, 18*(3), 109.
59. Garg, N., Singh, T. G., Khan, H., Arora, S., Kaur, A., & Mannan, A., (2021). Mechanistic interventions of selected *Ocimum* species in management of diabetes, obesity, and liver disorders: Transformative developments from preclinical to clinical approaches. *Biointerface Res. Appl. Chem., 12*(1), 1304–1323.
60. Kalra, P., Khan, H., Kaur, A., & Singh, T. G., (2022). Mechanistic insight on autophagy modulated molecular pathways in cerebral ischemic injury: From preclinical to clinical perspective. *Neurochemical Research*, 1–19.
61. Hankey, G. J., (2020). Population impact of potentially modifiable risk factors for stroke. *Stroke, 51*(3), 719–728.
62. Rihal, V., Khan, H., Kaur, A., & Singh, T. G., (2022). Vitamin D as therapeutic modulator in cerebrovascular diseases: A mechanistic perspectives. *Critical Reviews in Food Science and Nutrition,* 1–23.
63. Prabhakar, N. K., Khan, H., Grewal, A. K., & Singh, T. G., (2022). Intervention of neuroinflammation in the traumatic brain injury trajectory: *In vivo* and clinical approaches. *International Immunopharmacology, 108*, 108902.
64. Liu, S. C., Xia, L., Zhang, J., Lu, X. H., Hu, D. K., Zhang, H. T., & Li, H. J., (2015). Gout and risk of myocardial infarction: A systematic review and meta-analysis of cohort studies. *PLoS One, 10*(7), e0134088.
65. Holder, S. M., Bruno, R. M., Shkredova, D. A., Dawson, E. A., Jones, H., Hopkins, N. D., Hopman, M. T., et al., (2021). Reference intervals for brachial artery flow-mediated dilation and the relation with cardiovascular risk factors. *Hypertension, 77*(5), 1469–1480.
66. Migliacci, R., Guglielmini, G., Busti, C., Falcinelli, E., Minuz, P., & Gresele, P., (2021). Walking-induced endothelial dysfunction predicts ischemic cardiovascular events in patients with intermittent claudication. *Vascular Medicine*, 1358863X211001927.
67. Larvin, H., Kang, J., Aggarwal, V. R., Pavitt, S., & Wu, J., (2021). Risk of incident cardiovascular disease in people with periodontal disease: A systematic review and meta-analysis. *Clinical and Experimental Dental Research, 7*(1), 109–122.
68. Hausenloy, D. J., Barrabes, J. A., Bøtker, H. E., Davidson, S. M., Di Lisa, F., Downey, J., Engstrom, T., et al., (2016). Ischemic conditioning and targeting reperfusion injury: A 30 year voyage of discovery. *Basic Research in Cardiology, 111*(6), 1–24.
69. Rout, A., Tantry, U. S., Novakovic, M., Sukhi, A., & Gurbel, P. A., (2020). Targeted pharmacotherapy for ischemia reperfusion injury in acute myocardial infarction. *Expert Opinion on Pharmacotherapy, 21*(15), 1851–1865.
70. Garcia-Dorado, D., Andres-Villarreal, M., Ruiz-Meana, M., Inserte, J., & Barba, I., (2012). Myocardial edema: A translational view. *Journal of Molecular and Cellular Cardiology, 52*(5), 931–939.

71. De Waha, S., Manesh, R. P., Christopher, B. G., Magnus, O. E., Akiko, M., Ingo, E., Ben-Yehuda, O., et al., (2017). Relationship between microvascular obstruction and adverse events following primary percutaneous coronary intervention for ST-segment elevation myocardial infarction: An individual patient data pooled analysis from seven randomized trials. *European Heart Journal, 38*(47), 3502–3510.
72. Eriksson, K. E., Eidhagen, F., Liska, J., Franco-Cereceda, A., Lundberg, J. O., & Weitzberg, E., (2021). Effects of inorganic nitrate on ischemia-reperfusion injury after coronary artery bypass surgery. *British Journal of Anaesthesia.*
73. Zhang, X. Y., Huang, Z., Li, Q. J., Zhong, G. Q., Meng, J. J., Wang, D. X., & Tu, R. H., (2020). Ischemic postconditioning attenuates the inflammatory response in ischemia/reperfusion myocardium by upregulating miR-499 and inhibiting TLR2 activation. *Molecular Medicine Reports*, *22*(1), 209–218.
74. Wu, Y., Liu, H., & Wang, X., (2020). Cardioprotection of pharmacological postconditioning on myocardial ischemia/reperfusion injury. *Life Sciences*, 118628.
75. Araszkiewicz, A., Grygier, M., Pyda, M., Rajewska, J., Michalak, M., Sławek-Szmyt, S., & Lesiak, M., (2019). Ischemic postconditioning reduces infarct size and microvascular obstruction zone in acute ST-elevation myocardial infarction–a randomized study. *Postępy W Kardiologii Interwencyjnej= Advances in Interventional Cardiology*, *15*(3), 292.
76. Kaur, A., Singh, T. G., Khan, H., Kumar, M., Singh, N., & Abdel-Daim, M. M., (2022). Neuroprotective effect of piclamilast-induced post-ischemia pharmacological treatment in mice. *Neurochemical Research*, 1–14.
77. Vidal-Calés, P., Cepas-Guillén, P. L., Brugaletta, S., & Sabaté, M., (2021). New interventional therapies beyond stenting to treat ST-segment elevation acute myocardial infarction. *Journal of Cardiovascular Development and Disease*, *8*(9), 100.
78. Rezkalla, S. H., Stankowski, R. V., Hanna, J., & Kloner, R. A., (2017). Management of no-reflow phenomenon in the catheterization laboratory. *JACC: Cardiovascular Interventions*, *10*(3), 215–223.
79. Thibault, H., Piot, C., Staat, P., Bontemps, L., Sportouch, C., Rioufol, G., Cung, T. T., et al., (2008). Long-term benefit of postconditioning. *Circulation*, *117*(8), 1037–1044.
80. Heusch, G., Bøtker, H. E., Przyklenk, K., Redington, A., & Yellon, D., (2015). Remote ischemic conditioning. *Journal of the American College of Cardiology*, *65*(2), 177–195.
81. Kleinbongard, P., Skyschally, A., & Heusch, G., (2017). Cardioprotection by remote ischemic conditioning and its signal transduction. *Pflügers Archiv.-European Journal of Physiology*, *469*(2), 159–181.
82. Ho, A. F. W., & Ong, M. E. H., (2020). Remote ischemic conditioning in acute myocardial infarction–implications of the CONDI-2/ERIC-PPCI trial for prehospital and emergency medicine. *Prehospital Emergency Care*, *24*(6), 862–864.
83. Gaspar, A., Lourenço, A. P., Pereira, M. Á., Azevedo, P., Roncon-Albuquerque, R., Marques, J., & Leite-Moreira, A. F., (2018). Randomized controlled trial of remote ischemic conditioning in ST-elevation myocardial infarction as adjuvant to primary angioplasty (RIC-STEMI). *Basic Research in Cardiology*, *113*(3), 1–10.
84. Cheskes, S., Koh, M., Turner, L., Heslegrave, R., Verbeek, R., Dorian, P., Scales, D. C., et al., (2020). Field implementation of remote ischemic conditioning in ST-segment–elevation myocardial infarction: The FIRST study. *Canadian Journal of Cardiology*, *36*(8), 1278–1288.
85. Rehni, A. K., Singh, T. G., Kalra, R., & Singh, N., (2009). Pharmacological inhibition of inducible nitric oxide synthase attenuates the development of seizures in mice. *Nitric Oxide, 21*(2), 120–125.
86. Nepper-Christensen, L., Lønborg, J., Høfsten, D. E., Sadjadieh, G., Schoos, M. M., Pedersen, F., Jørgensen, E., et al., (2021). *Clinical Outcome Following Late Reperfusion with Percutaneous Coronary Intervention in Patients with ST-Segment Elevation Myocardial Infarction. 10*(5), 523–531.
87. Caccioppo, A., Franchin, L., Grosso, A., Angelini, F., D'Ascenzo, F., & Brizzi, M. F., (2019). Ischemia reperfusion injury: Mechanisms of damage/protection and novel strategies for cardiac recovery/regeneration. *International Journal of Molecular Sciences*, *20*(20), 5024.
88. Lasley, R. D., (2018). Adenosine receptor-mediated cardioprotection—Current limitations and future directions. *Frontiers in Pharmacology*, *9*, 310.
89. Kingma, J. G., (2018). Myocardial infarction: An overview of STEMI and NSTEMI physiopathology and treatment. *World Journal of Cardiovascular Diseases*, *8*(11), 498.

90. Bøtker, H. E., Cabrera-Fuentes, H. A., Ruiz-Meana, M., Heusch, G., & Ovize, M., (2020). Translational issues for mitoprotective agents as adjunct to reperfusion therapy in patients with ST-segment elevation myocardial infarction. *Journal of Cellular and Molecular Medicine, 24*(5), 2717–2729.
91. Lorca, R., Jiménez-Blanco, M., García-Ruiz, J. M., Pizarro, G., Fernández-Jiménez, R., García-Álvarez, A., Fernández-Friera, L., et al., (2020). Coexistence of transmural and lateral wavefront progression of myocardial infarction in the human heart. *Revista Española de Cardiología (English edn.).*
92. Hoedemaker, N. P., Vincent, R., Robbert, J. De. W., Niels, V. R., Valentin, F., García-Ruiz, J. M., Fikret, E., et al., (2020). Early intravenous beta-blockers in patients undergoing primary percutaneous coronary intervention for ST-segment elevation myocardial infarction: A patient-pooled meta-analysis of randomized clinical trials. *European Heart Journal: Acute Cardiovascular Care, 9*(5), 469–477.
93. García-Prieto, J., Villena-Gutiérrez, R., Gómez, M., Bernardo, E., Pun-García, A., García-Lunar, I., Crainiciuc, G., et al., (2017). Neutrophil stunning by metoprolol reduces infarct size. *Nature Communications, 8*(1), 1–15.
94. Roolvink, V., Ibáñez, B., Ottervanger, J. P., Pizarro, G., Van, R. N., Mateos, A., Dambrink, J. H. E., et al., (2016). Early intravenous beta-blockers in patients with ST-segment elevation myocardial infarction before primary percutaneous coronary intervention. *Journal of the American College of Cardiology, 67*(23), 2705–2715.
95. Khan, H., Bangar, A., Grewal, A. K., Bansal, P., & Singh, T. G., (2022). Caspase-mediated regulation of the distinct signaling pathways and mechanisms in neuronal survival. *International Immunopharmacology, 110*, 108951.
96. Bouleti, C., Mathivet, T., Serfaty, J. M., Vignolles, N., Berland, E., Monnot, C., Cluzel, P., et al., (2015). Angiopoietin-like 4 serum levels on admission for acute myocardial infarction are associated with no-reflow. *International Journal of Cardiology, 187*, 511–516.
97. Hausenloy, D. J., Garcia-Dorado, D., Erik, B. H., Davidson, S. M., Downey, J., Engel, F. B., Jennings, R., et al., (2017). Melatonin as a cardioprotective therapy following ST-segment elevation myocardial infarction: Is it really promising? Reply. *Cardiovascular Research, 113*(11), 1418, 1419.
98. Kyhl, K., Lønborg, J., Vejlstrup, N., Kelbaek, H., Helqvist, S., Holmvang, L., Jørgensen, E., et al., (2016). A post hoc analysis of long-term prognosis after exenatide treatment in patients with ST-segment elevation myocardial infarction. *Euro Intervention: Journal of EuroPCR in Collaboration with the Working Group on Interventional Cardiology of the European Society of Cardiology, 12*(4), 449–455.
99. Davidson, S. M., Andreadou, I., Barile, L., Birnbaum, Y., Cabrera-Fuentes, H. A., Cohen, M. V., Downey, J. M., et al., (2019). Circulating blood cells and extracellular vesicles in acute cardioprotection. *Cardiovascular Research, 115*(7), 1156–1166.
100. Yadid, M., Lind, J. U., Ardoña, H. A. M., Sheehy, S. P., Dickinson, L. E., Eweje, F., Bastings, M. M., et al., (2020). Endothelial extracellular vesicles contain protective proteins and rescue ischemia-reperfusion injury in a human heart-on-chip. *Science Translational Medicine, 12*(565).
101. Koshinuma, S., Miyamae, M., Kaneda, K., Kotani, J., & Figueredo, V. M., (2014). Combination of necroptosis and apoptosis inhibition enhances cardioprotection against myocardial ischemia–reperfusion injury. *Journal of Anesthesia, 28*(2), 235–241.
102. El Farissi, M., Keulards, D. C., Zelis, J. M., Van't, V. M., Zimmermann, F. M., Pijls, N. H., & Otterspoor, L. C., (2021). Hypothermia for reduction of myocardial reperfusion injury in acute myocardial infarction: Closing the translational gap. *Circulation: Cardiovascular Interventions, 14*(8), e010326.
103. Yamada, K. P., Kariya, T., Aikawa, T., & Ishikawa, K., (2021). Effects of therapeutic hypothermia on normal and ischemic heart. *Frontiers in Cardiovascular Medicine, 8*, 72.
104. Hjortbak, M. V., Jespersen, N. R., Jensen, R. V., Lassen, T. R., Hjort, J., Povlsen, J. A., Støttrup, N. B., et al., (2021). Cardioprotective effect of combination therapy by mild hypothermia and local or remote ischemic preconditioning in isolated rat hearts. *Scientific Reports, 11*(1), 1–14.
105. Yu, L., Huang, B., Po, S. S., Tan, T., Wang, M., Zhou, L., Meng, G., et al., (2017). Low-level tragus stimulation for the treatment of ischemia and reperfusion injury in patients with ST-segment elevation myocardial infarction: A proof-of-concept study. *JACC: Cardiovascular Interventions, 10*(15), 1511–1520.
106. Bajaj, A., Sethi, A., Rathor, P., Suppogu, N., & Sethi, A., (2015). Acute complications of myocardial infarction in the current era: Diagnosis and management. *Journal of Investigative Medicine, 63*(7), 844–855.

107. Huang, X., Zuo, L., Lv, Y., Chen, C., Yang, Y., Xin, H., Li, Y., & Qian, Y., (2016). Asiatic acid attenuates myocardial ischemia/reperfusion injury via Akt/GSK-3β/HIF-1α signaling in rat H9c2 cardiomyocytes. *Molecules, 21*(9), 1248.
108. Qiu, Y., Wu, Y., Meng, M., Luo, M., Zhao, H., Sun, H., & Gao, S., (2018). GYY4137 protects against myocardial ischemia/reperfusion injury via activation of the PHLPP-1/Akt/Nrf2 signaling pathway in diabetic mice. *Journal of Surgical Research, 225*, 29–39.
109. Qi, J. Y., Yu, J., Huang, D. H., Guo, L. H., Wang, L., Huang, X., Huang, H. D., et al., (2017). Salvianolate reduces murine myocardial ischemia and reperfusion injury via ERK1/2 signaling pathways *in vivo*. *Chinese Journal of Integrative Medicine, 23*(1), 40–47.
110. Liu, G., Zhang, B. F., Hu, Q., Liu, X. P., & Chen, J., (2020). Syringic acid mitigates myocardial ischemia reperfusion injury by activating the PI3K/Akt/GSK-3β signaling pathway. *Biochemical and Biophysical Research Communications, 531*(2), 242–249.
111. Das, A., Salloum, F. N., Durrant, D., Ockaili, R., & Kukreja, R. C., (2012). Rapamycin protects against myocardial ischemia–reperfusion injury through JAK2–STAT3 signaling pathway. *Journal of Molecular and Cellular Cardiology, 53*(6), 858–869.
112. Yu, L., Sun, Y., Cheng, L., Jin, Z., Yang, Y., Zhai, M., Pei, H., et al., (2014). Melatonin receptor-mediated protection against myocardial ischemia/reperfusion injury: Role of SIRT 1. *Journal of Pineal Research, 57*(2), 228–238.
113. Zhai, M., Li, B., Duan, W., Jing, L., Zhang, B., Zhang, M., Yu, L., et al., (2017). Melatonin ameliorates myocardial ischemia reperfusion injury through SIRT 3-dependent regulation of oxidative stress and apoptosis. *Journal of Pineal Research, 63*(2), e12419.
114. Dong, L. Y., Li, S., Zhen, Y. L., Wang, Y. N., Shao, X., & Luo, Z. G., (2013). Cardioprotection of vitexin on myocardial ischemia/reperfusion injury in rat via regulating inflammatory cytokines and MAPK pathway. *The American Journal of Chinese Medicine, 41*(6), 1251–1266.
115. Yang, H., Xue, W., Ding, C., Wang, C., Xu, B., Chen, S., Zha, B., et al., (2021). Vitexin mitigates myocardial ischemia/reperfusion injury in rats by regulating mitochondrial dysfunction via Epac1-Rap1 signaling. *Oxidative Medicine and Cellular Longevity, 2021*.
116. Shi, X., Tao, G., Ji, L., & Tian, G., (2020). Sappanone A protects against myocardial ischemia reperfusion injury by modulation of Nrf2. *Drug Design, Development and Therapy, 14*, 61.
117. Wu, T., Jiang, N., Ji, Z., & Shi, G., (2019). The IRE1 signaling pathway is involved in the protective effect of low-dose LPS on myocardial ischemia-reperfusion injury. *Life Sciences, 231*, 116569.
118. Yuan, L., Dai, X., Fu, H., Sui, D., Lin, L., Yang, L., Zha, P., et al., (2018). Vaspin protects rats against myocardial ischemia/reperfusion injury (MIRI) through the TLR4/NF-κB signaling pathway. *European Journal of Pharmacology, 835*, 132–139.
119. Li, X., Hu, X., Wang, J., Xu, W., Yi, C., Ma, R., & Jiang, H., (2018). *Inhibition of Autophagy Via Activation of PI3K/Akt/mTOR Pathway Contributes to the Protection of Hesperidin Against Myocardial. 42*(4), 1917–1924.
120. Yang, X., Yue, R., Zhang, J., Zhang, X., Liu, Y., Chen, C., Wang, X., et al., (2018). Gastrin protects against myocardial ischemia/reperfusion injury via activation of RISK (reperfusion injury salvage kinase) and SAFE (survivor activating factor enhancement) pathways. *Journal of the American Heart Association, 7*(14), e005171.
121. Yin, Q., Zhao, B., Zhu, J., Fei, Y., Shen, W., Liang, B., Zhu, X., & Li, Y., (2020). JLX001 improves myocardial ischemia-reperfusion injury by activating Jak2-Stat3 pathway. *Life Sciences, 25*.
122. Cheng, X. Y., Gu, X. Y., Gao, Q., Zong, Q. F., Li, X. H., & Zhang, Y., (2016). Effects of dexmedetomidine postconditioning on myocardial ischemia and the role of the PI3K/Akt-dependent signaling pathway in reperfusion injury. *Molecular Medicine Reports, 14*(1), 797–803.
123. Chen, X., Wang, Y., Xiao, Z. Y., Hou, D. N., Li, D. B., & Zhang, X. P., (2019). Effect of propofol on myocardial ischemia/reperfusion injury in rats through JAK/STAT signaling pathway. *Eur. Rev. Med. Pharmacol. Sci., 23*(14), 6330–6338.
124. Rong, R., & Xijun, X., (2015). Erythropoietin pretreatment suppresses inflammation by activating the PI3K/Akt signaling pathway in myocardial ischemia-reperfusion injury. *Experimental and Therapeutic Medicine, 10*(2), 413–418.

125. Li, X., Yang, J., Yang, J., Dong, W., Li, S., Wu, H., & Li, L., (2016). RP105 protects against myocardial ischemia–reperfusion injury via suppressing TLR4 signaling pathways in rat model. *Experimental and Molecular Pathology*, *100*(2), 281–286.
126. Yang, J., Yang, C., Yang, J., Ding, J., Li, X., Yu, Q., Guo, X., et al., (2018). RP105 alleviates myocardial ischemia reperfusion injury via inhibiting TLR4/TRIF signaling pathways. *International Journal of Molecular Medicine*, *41*(6), 3287–3295.
127. Yang, Z. H., Lu, Y. J., Gu, K. P., Xiang, Z. Y., & Huang, H. M., (2019). Effect of ulinastatin on myocardial ischemia-reperfusion injury through JNK and P38 MAPK signaling pathways. *Eur. Rev. Med. Pharmacol. Sci.*, *23*(19), 8658–8664.
128. Deng, C., Sun, Z., Tong, G., Yi, W., Ma, L., Zhao, B., Cheng, L., et al., (2013). α-Lipoic acid reduces infarct size and preserves cardiac function in rat myocardial ischemia/reperfusion injury through activation of PI3K/Akt/Nrf2 pathway. *PloS One*, *8*(3), p.e58371.
129. Wang, Y., Wang, L., Li, J. H., Zhao, H. W., & Zhang, F. Z., (2019). Morphine alleviates myocardial ischemia/reperfusion injury in rats by inhibiting TLR4/NF-kappaB signaling pathway. *Eur. Rev. Med. Pharmacol. Sci.*, *23*(19), 8616–8624.
130. Yuan, X., Juan, Z., Zhang, R., Sun, X., Yan, R., Yue, F., Huang, Y., et al., (2020). Clemastine fumarate protects against myocardial ischemia reperfusion injury by activating the TLR4/PI3K/Akt signaling pathway. *Frontiers in Pharmacology*, *11*, 28.
131. Li, H. W., & Xiao, F. Y., (2020). Effect of hydrogen sulfide on cardiomyocyte apoptosis in rats with myocardial ischemia-reperfusion injury via the JNK signaling pathway. *Eur. Rev. Med. Pharmacol. Sci.*, *24*(4), 2054–2061.
132. Wu, S., Chang, G., Gao, L., Jiang, D., Wang, L., Li, G., Luo, X., et al., (2018). Trimetazidine protects against myocardial ischemia/reperfusion injury by inhibiting excessive autophagy. *Journal of Molecular Medicine*, *96*(8), 791–806.
133. Zhou, Y., Peng, D. D., Chong, H., Zheng, S. Q., Zhu, F., & Wang, G., (2019). Effect of isoflurane on myocardial ischemia-reperfusion injury through the p38 MAPK signaling pathway. *Eur. Rev. Med. Pharmacol. Sci.*, *23*(3), 1342–1349.
134. Wang, Y. H., Chen, K. M., Chiu, P. S., Lai, S. C., Su, H. H., Jan, M. S., Lin, C. W., et al., (2016). Lumbrokinase attenuates myocardial ischemia-reperfusion injury by inhibiting TLR4 signaling. *Journal of Molecular and Cellular Cardiology*, *99*, 113–122.
135. Wang, S., Liu, C., Gong, C., Li, T., Zhao, J., Xiao, W., Liu, Y., et al., (2018). Alpha linolenic acid intake alleviates myocardial ischemia/reperfusion injury via the P2X7R/NF-κB signaling pathway. *Journal of Functional Foods*, *49*, 1–11.
136. Liu, Z., Zhang, F., Zhao, L., Zhang, X., Li, Y., & Liu, L., (2020). Protective effect of pravastatin on myocardial ischemia reperfusion injury by regulation of the miR-93/Nrf2/ARE signal pathway. *Drug Design, Development and Therapy*, *14*, 3853.
137. Xin, C., Zhang, Z., Gao, G., Ding, L., Yang, C., Wang, C., Liu, Y., et al., (2020). Irisin attenuates myocardial ischemia/reperfusion injury and improves mitochondrial function through AMPK pathway in diabetic mice. *Frontiers in Pharmacology*, *11*.
138. Zhang, J., Huang, L., Shi, X., Yang, L., Hua, F., Ma, J., Zhu, W., et al., (2020). Metformin protects against myocardial ischemia-reperfusion injury and cell pyroptosis via AMPK/NLRP3 inflammasome pathway. *Aging (Albany NY)*, *12*(23), 24270.
139. Luo, T., Zeng, X., Yang, W., & Zhang, Y., (2019). Treatment with metformin prevents myocardial ischemia–reperfusion injury via STEAP4 signaling pathway. *Anatolian Journal of Cardiology*, *21*(5), 261.
140. Qu, Z. J., Qu, Z. J., Zhou, H. B., Xu, C. S., Zhang, D. Z., & Wang, G., (2019). Protective effect of remifentanil on myocardial ischemia-reperfusion injury through Fas apoptosis signaling pathway. *Eur. Rev. Med. Pharmacol. Sci.*, *23*(13), 5980–5986.
141. Fang, S. J., Li, P. Y., Wang, C. M., Xin, Y., Lu, W. W., Zhang, X. X., Zuo, S., et al., (2017). Inhibition of endoplasmic reticulum stress by neuregulin-1 protects against myocardial ischemia/reperfusion injury. *Peptides*, *88*, 196–207.
142. Xie, B., Liu, X., Yang, J., Cheng, J., Gu, J., & Xue, S., (2018). PIAS1 protects against myocardial ischemia-reperfusion injury by stimulating PPARγ SUMOylation. *BMC Cell Biology*, *19*(1), 1–12.

143. Shanmugam, K., Boovarahan, S. R., Prem, P., Sivakumar, B., & Kurian, G. A., (2021). Fisetin attenuates myocardial ischemia-reperfusion injury by activating the reperfusion injury salvage kinase (RISK) signaling pathway. *Frontiers in Pharmacology, 12.*
144. Wang, F., Gao, Q., Yang, J., Wang, C., Cao, J., Sun, J., Fan, Z., & Fu, L., (2020). Artemisinin suppresses myocardial ischemia–reperfusion injury via NLRP3 inflammasome mechanism. *Molecular and Cellular Biochemistry, 474*(1), 171–180.
145. Li, J., Xiang, X., Xu, H., & Shi, Y., (2019). Cilostazol promotes angiogenesis and increases cell proliferation after myocardial ischemia–reperfusion injury through a cAMP-dependent mechanism. *Cardiovascular Engineering and Technology, 10*(4), 638–647.
146. Yu, D., Xiong, J., Gao, Y., Li, J., Zhu, D., Shen, X., Sun, L., & Wang, X., (2021). Resveratrol activates PI3K/AKT to reduce myocardial cell apoptosis and mitochondrial oxidative damage caused by myocardial ischemia/reperfusion injury. *Acta Histochemica, 123*(5), 151739.
147. Cui, Z. H., Zhang, X. J., Shang, H. Q., Wang, X., & Rong, D., (2020). Glutamine protects myocardial ischemia-reperfusion injury in rats through the PI3K/AKT signaling pathway. *Eur. Rev. Med. Pharmacol. Sci., 24*(1), 444–451.
148. Okumura, H., Nagaya, N., Itoh, T., Okano, I., Hino, J., Mori, K., Tsukamoto, Y., et al., (2004). Adrenomedullin infusion attenuates myocardial ischemia/reperfusion injury through the phosphatidylinositol 3-kinase/ Akt-dependent pathway. *Circulation, 109*(2), 242–248.
149. Liu, C., Zhou, J., Wang, B., Zheng, Y., Liu, S., Yang, W., Li, D., et al., (2021). Bortezomib alleviates myocardial ischemia reperfusion injury via enhancing of Nrf2/HO-1 signaling pathway. *Biochemical and Biophysical Research Communications, 556*, 207–214.
150. Xin, B. R., Li, P., Liu, X. L., & Zhang, X. F., (2020). Visfatin relieves myocardial ischemia-reperfusion injury through activation of PI3K/Akt/HSP70 signaling axis. *European Review for Medical and Pharmacological Sciences, 24*(20), 10779–10789.
151. Wang, H., Wang, J. L., Ren, H. W., He, W. F., & Sun, M., (2019). Butorphanol protects on myocardial ischemia/reperfusion injury in rats through MAPK signaling pathway. *Eur. Rev. Med. Pharmacol. Sci., 23*(23), 10541–10548.
152. Yang, C., Yu, P., Yang, F., He, Q., Jiang, B., Zheng, L., Wang, Q., et al., (2021). PSMB4 inhibits cardiomyocyte apoptosis via activating NF-κB signaling pathway during myocardial ischemia/reperfusion injury. *Journal of Molecular Histology*, 1–11.
153. Lv, L., Meng, Q., Xu, J., Gong, J., Cheng, Y., & Jiang, S., (2012). Ligustrazine attenuates myocardial ischemia reperfusion injury in rats by activating the phosphatidylinositol 3-kinase/Akt pathway. *Annals of Clinical & Laboratory Science, 42*(2), 198–202.
154. Shibata, R., Sato, K., Pimentel, D. R., Takemura, Y., Kihara, S., Ohashi, K., Funahashi, T., et al., (2005). Adiponectin protects against myocardial ischemia-reperfusion injury through AMPK-and COX-2–dependent mechanisms. *Nature Medicine, 11*(10), 1096–1103.
155. Wu, H., Ye, M., Yang, J., Ding, J., Yang, J., Dong, W., & Wang, X., (2015). Nicorandil protects the heart from ischemia/reperfusion injury by attenuating endoplasmic reticulum response-induced apoptosis through PI3K/Akt signaling pathway. *Cellular Physiology and Biochemistry, 35*(6), 2320–2332.
156. Kleinz, M. J., & Baxter, G. F., (2008). Apelin reduces myocardial reperfusion injury independently of PI3K/Akt and P70S6 kinase. *Regulatory Peptides, 146*(1–3), 271–277.
157. Li, J., Xiang, X., Gong, X., Shi, Y., Yang, J., & Xu, Z., (2017). Cilostazol protects mice against myocardium ischemic/reperfusion injury by activating a PPARγ/JAK2/STAT3 pathway. *Biomedicine & Pharmacotherapy, 94*, 995–1001.
158. Zhou, C. N., Yao, W., Gong, Y. N., Li, Y., Wang, C. H., & Huo, Y. F., (2019). 22-Oxacalcitriol protects myocardial ischemia-reperfusion injury by suppressing NF-κB/TNF-α pathway. *Eur. Rev. Med. Pharmacol. Sci., 23*(12), 5495–5502.
159. Wu, F., Huang, W., Tan, Q., Guo, Y., Cao, Y., Shang, J., Ping, F., et al., (2021). ZFP36L2 regulates myocardial ischemia/reperfusion injury and attenuates mitochondrial fusion and fission by LncRNA PVT1. *Cell Death & Disease, 12*(6), 1–13.
160. Yu, L., Di, W., Dong, X., Li, Z., Xue, X., Zhang, J., Wang, Q., et al., (2017). Diallyl trisulfide exerts cardioprotection against myocardial ischemia-reperfusion injury in diabetic state, role of AMPK-mediated AKT/GSK-3β/HIF-1α activation. *Oncotarget, 8*(43), 74791.

161. Feng, X., Li, J., Liu, J., Jin, M., Liu, X., Du, H., Zhang, L., et al., (2011). Protective effect of FK506 on myocardial ischemia/reperfusion injury by suppression of CaN and ASK1 signaling circuitry. *Cardiovascular Toxicology*, *11*(1), 18–27.
162. Zheng, Q., Huang, Y. Y., Zhu, P. C., Tong, Q., Bao, X. Y., Zhang, Q. H., Zheng, G. Q., & Wang, Y., (2018). Ligustrazine exerts cardioprotection in animal models of myocardial ischemia/reperfusion injury: Preclinical evidence and possible mechanisms. *Frontiers in Pharmacology*, *9*, 729.

CHAPTER 3

Preclinical and Clinical Aspects of I/R Injuries

SHILPA KUMARI,[1] RAHUL DESHMUKH,[1] and THAKUR GURJEET SINGH[2]

[1]*Department of Pharmaceutical Sciences and Technology, Maharaja Ranjit Singh Punjab Technical University, Bathinda, Punjab, India*

[2]*Chitkara College of Pharmacy, Chitkara University, Punjab, India*

ABSTRACT

Ischemic/Reperfusion injury (I/R injury) is the leading cause of death and illnesses throughout the world. Reduced blood flow or ischemic/hypoxic circumstances characterize IR injury. IR injury is linked to various clinical symptoms, including stroke, myocardial infarction, gastrointestinal dysfunction, systemic inflammatory syndromes, and other neurological problems. Ischemia-reperfusion damage is a severe medical issue that physicians have to address immediately. For I/R damage and related illnesses, there are numerous etiological factors, pathological mechanistic methods, diagnostic approaches, and accessible therapies. Recent research has concentrated on the mechanical systems to ischemia-reperfusion injury pathology at the molecular level, particularly the role of oxidative stress and apoptotic pathways. The role of mitochondrial NADPH oxidase, nitrative stress, and xanthine oxidase is also explored. Local inflammation and ROS generation rise when blood flow is restored after extended ischemia, leading to secondary damage. Thus this chapter is designed to explore recent mechanistic and pharmacological approaches that have been validated in preclinical and clinical reports, which will be helpful to readers and researchers all around the world.

3.1 INTRODUCTION

Ischemic/Reperfusion injury is a multifactorial process comprised of ischemia and reperfusion characterized by the lack of blood supply that does not meet the demands for normal cellular and metabolic functions [1]. Reperfusion, i.e., the establishment of proper blood supply or reoxygenation, further exacerbates the hypoxic insult. Ischemic reperfusion injury is a paradoxical manifestation that contributes to cellular death, metabolic derangements, ischemic tissue death, and further organ damage including heart, kidney, lung, gut,

Ischemic Injury: A Molecular Insight. Thakur Gurjeet Singh & Heena Khan (Eds.)

skeletal muscle, and brain [1–3]. I/R injuries can cause severe damage to various organs, resulting in illnesses. Reactive oxygen species (ROS), tissue destruction by proteases and metalloproteinases, changes in calcium handling, endothelial dysfunction, microvascular injury, drug-mediated toxic insults, and other pathogenic variables all contribute to the pathophysiology of many illnesses that anaesthesiologists deal with, such as sudden cardiac death, shock, and stroke. Although fulfilling the need for oxygen and blood flow to an oxygen-deprived organ is necessary to avoid cellular damage, reperfusion may cause tissue damage that is worse than that caused by ischemia alone. The histopathological evidence suggests that ischemia-reperfusion is worse than ischemia alone.

Ischemic reperfusion injuries lead to various comorbidities such as CVS, CNS dysfunctions, renal system failure, and other organ system failures [4, 5]. The inflammatory response following I-R, if severe enough, can cause vascular damage and multiple organ failure [6–9]. CVS disorders initiated by ischemic-reperfusion injuries include myocardial infarction, arrhythmias, stroke, and other peripheral vascular diseases. Cerebral ischemia is linked with many diseases or irregularities such as difficulty speaking, vertigo, blurred vision, loss of coordination, and other neurodegenerative disorders. Many hypotheses suggest the mechanisms and etiologies of ischemic/reperfusion injuries and their associated disorders. Various treatments approach different pathologies of CNS, CVS, and other I/R injury-related disorders [10]. This chapter will provide a detailed outlook of both preclinical and clinical therapies available and could be targeted in future research for treating I/R injury-associated disorders. A complete overview of preclinical and clinical drugs for I/R injury and its corresponding disorders will help readers, healthcare professionals, and future researchers learn, set up a hypothesis, and develop newer therapeutics for such conditions.

3.2 ISCHEMIC/REPERFUSION INJURY (I/R INJURY) PATHOLOGIES

Cerebral ischemia is the most common and critical form of brain insult that occurs due to improper blood supply to the brain and other associated regions. Cerebral ischemic injury results in different disorders such as stroke cerebral infarctions and hypoxic-ischemic encephalopathies that can ultimately lead to death or any lifetime ailment [11]. Cerebral ischemia can affect either the specified region or the whole brain part. The spread and induction of cerebral ischemia is mainly due to structural and physiological modifications in the various CNS parts, mostly arrhythmias, and other autonomic alterations (Figure 3.1). Stroke is one of the most common causes of mortality worldwide and almost about 80% of all strokes are ischemic [12, 13].

The vascular and molecular processes of ischemic brain injury, as well as prospective therapeutic techniques such as reperfusion, symptomatic neuroprotection, and other therapeutic strategies for functional recovery such as neural plasticity and stem cell transplantation, are discussed here [14]. Vascular dysfunctions increase the risk of cellular, tissue, and organ damage during stroke because of the lack of ATP and oxygen stores. These dysfunctioning led to the increased oxidative stress and mitochondrial dysfunctioning. Blood reperfusion to the oxygen deprived organs comparatively increases the organ damage and system failure. The vital organs such as heart and brain are prone to the ischemic damage.

After reperfusion there is increased production of ROS, mitochondrial ETC, upregulation of NADPH oxidase, and iNOS makes the cells more susceptible to injuries [15, 16].

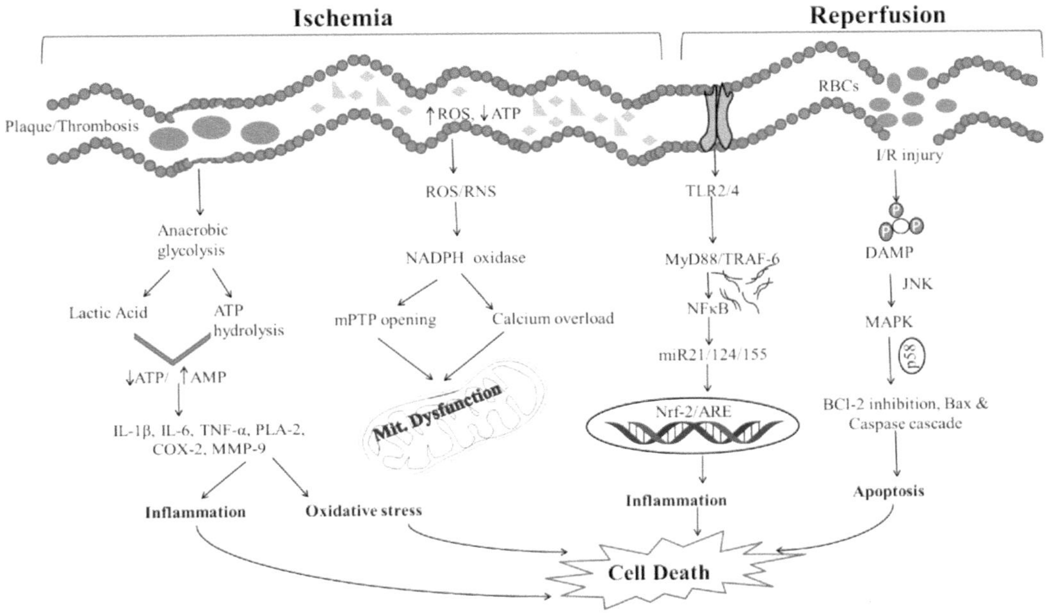

FIGURE 3.1 Pathology of ischemia-reperfusion injury.

3.2.1 OXIDATIVE STRESS AND INFLAMMATION: MAJOR PATHOLOGIES OF I/R INJURIES

Ischemia/reperfusion injury tends to result in irreversible neuronal damage and death, including various pathological pathways such as oxidative stress, glutamate toxicity, apoptosis, endogenous molecule secretion, and inflammation. I/R damage in the brain is complicated by the interaction of oxidative stress and inflammation. Increased ROS is one of the major causes responsible for the development of ischemia damage and also other pathological processes of ischemic stroke [17, 18]. During cerebral ischemia, reactive oxygen species (ROS) generate oxidative stress, which increases the likelihood of tissue damage, eventually, neuronal death followed by brain after reperfusion [19].

3.2.1.1 ROS/RNS IN I/R INJURY

The reactive oxygen species (ROS) and free radicals are created during reperfusion, that causes peroxidation of various molecules such as lipids, nucleic acids, and initiates many cell death pathways. In CNS I/R injury, inflammation is caused by increased proinflammatory factors and decrease in anti-inflammatory factors. As a result, inflammatory substances can promote or increase inflammation, aggravating ischemia injury [20]. Excessive generation of

ROS and decreased clearance of ROS causes tissue damage, and irreversible injuries [21, 22]. Mitochondrial malfunction and increased oxidative stress from ETC are two major variables that exacerbate the permanent damage induced by cerebral ischemia/reperfusion. Neuronal mitochondria create enormous levels of ROS and hydroxyl radicals and thus increased oxidative stress disrupts the body's oxidative and antioxidant equilibrium, resulting in neuronal loss and, ultimately, tissue damage [23, 24].

Antioxidative therapies and strategies targeting mitochondrial dysfunctioning all have proven to be effective techniques for protecting brain cells from the degenerative pathways of CNS I/R [25]. When lipids peroxide or ROS levels rise, they react to macromolecules such as lipids, DNA, proteins, sugar molecules, etc., exacerbating the situation. Antioxidant levels, on the other hand, diminish those outcomes. Mitogen and other stress-activated protein kinase protect rats with localized ischemia-reperfusion injury by lowering inducible nitric oxide synthase (iNOS) levels and boosting the interleukin-10 levels have an anti-apoptotic and anti-inflammatory effect on astrocytes [26].

3.2.1.2 I/R-INDUCED OXIDATIVE STRESS AND INFLAMMATION: THE ROLE OF NRF2/ARE AND SIRTUINS

Increased oxidative stress and mitochondrial damage affect various activities, including altered energy metabolism, apoptosis, increased ER stress, and altered Ca^{2+} balance via activation of Nrf2/ARE and other regulatory pathways, including MAPK, PKC, and PI3K/AKT [27, 28]. SIRT1 has a broad role in cellular physiological functions, such as energy metabolism, ETC, and apoptosis. In brain I/R injury, their ability to tolerate oxidative stress has been extensively explored [29]. SIRT1 is responsible for activating the FOXO family, and FOXO3a can withstand increased oxidative stress-induced injury and boost ROS clearance activity of antioxidant enzymes [30, 31]. Inflammatory cytokines can increase SIRT2 activity, increases NF-κB deacetylation via decreasing NF-κB transcriptional activity, and suppress the expression of inflammatory mediators like IL-6, TNF-α, etc. [32], and have antioxidant and anti-inflammatory properties [33].

3.2.1.3 TLRS AND CYTOKINES IN I/R DAMAGE

After increased oxidative stress amplifies the impact of damage, the equilibrium between anti-inflammatory and proinflammatory cytokines is disrupted [34]. The transcription factors NF-κB are primarily responsible for producing proinflammatory cytokines such as TNF-α, IL-1β, IL-6, and other chemokines. Ischemia increased the release of substances such as hyaluronic acid, fibronectin, heat shock proteins (HSP), and nucleic acids from injured tissue, activating Toll-like receptors (TLR) 2, 4, and 5, a kind of pattern recognition receptor (PRR) [35]. PRR activation can trigger the death signaling pathway as well as the generation of proinflammatory cytokines. TLRs also triggers the release of a variety of proinflammatory cytokines and chemokines, as well as neutrophil and macrophage invasion. Adaptive proteins mediate signal transduction in response to TLR activation, the most important of which is MyD88, which activates the transcription factor NF-B, resulting in the production of proinflammatory cytokines and chemokines. The expression of TLR2 and TLR4 is increased after the I/R damage, specifically in macrophages and dendritic cells [36].

3.2.2 I/R-INDUCED APOPTOSIS

It is commonly accepted that mitochondria and nitric oxide (NO) play a critical role in programed cell death. Various signaling molecules have been shown to influence cell death mechanisms through methods akin to preconditioning. The role of apoptosis in cell death during ischemia and reperfusion has been shown *in vivo* mice models [37]. Increased ischemia exposure causes an increase in necrosis, whereas reperfusion causes an increase in apoptosis. The necessities for the survival of viable cells are given as oxygen and glucose supply is restored. Reperfusion speeds up the apoptosis process and aids in the completion of apoptosis. Current preventative medicines that minimize intrinsic apoptosis following I/R can target other signaling pathways and molecules that regulate BCL2, including as PI3K/AKT and MEK1/ERK1/2, JAK2/STAT3, eNOS, and PTEN [38].

3.2.3 ROLE OF MITOCHONDRIAL NADPH OXIDASE SYSTEM IN IR INJURIES

Other signaling pathways and molecules that modulate BCL2, such as PI3K/AKT and MEK1/ERK1/2, JAK2/STAT3, endothelial nitric oxide synthase, and PTEN, can be employed as targets for existing preventive medicines.

Hypoxia also activates NOX enzymes by inducing hypoxia inhibitory factor-1 (HIF-1). To create a positive feedback loop, oxidative stress stimulates HIF-1 synthesis. Furthermore, after restoring blood flow to ischemic tissue, cells release a variety of chemical mediators that activate NADPH oxidase, including phospholipase A2, TNF-α, IL-1β, IFN-Ɣ, and angiotensin II [39]. Platelet-activating factor is created when phospholipase A2 is released, which causes tissue levels of thromboxane and leukotrienes to rise, causing local inflammation [40, 41]. NOX derived ROS can cause reperfusion injury in different organs, including the heart, brain, and other peripheral organs, by causing inflammatory cell build-up in post-ischemic tissue. Overexpression of NADPH oxidase is promoted by cytokines generated by macrophages and mast cells. Through the angiotensin converting enzyme, angiotensin II stimulates local angiotensin II receptors and promotes NADPH oxidase expression and ischemia-reperfusion damage. NOX inhibition reduces NADPH oxidase expression and superoxide generation in endothelial cells [42]. Figure 3.2 provides thorough information on the many pathogenic pathways that could be targeted with pharmacological therapies.

3.3 PATHOLOGIES OF CNS I/R INJURIES

Cerebrovascular illnesses include ischemic and hemorrhagic strokes, which are caused by anomalies in the cerebral circulation, primarily changes in the cerebral blood supply.

Brain ischemia followed by reperfusion has been shown to trigger a cascade of pathogenic events that contribute to irreversible tissue destruction [43]. Increased oxidative stress, neuroinflammation, reduced vascular responsiveness, and a progressive loss in microvascular density pose negative effects on the brain and its associated parts. These pathological adaptations wreak havoc on cerebral vascular autoregulation and blood flow reserve, resulting in cognitive decline and dementia [44].

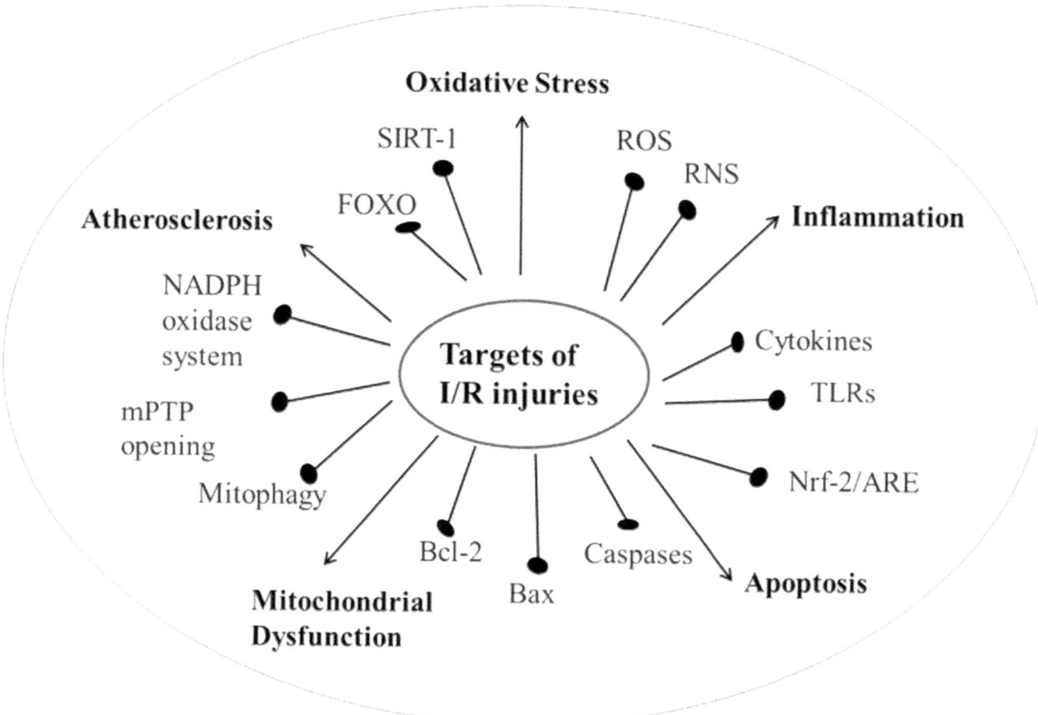

FIGURE 3.2 The figure depicts thorough information on the many pathogenic pathways that could be targeted with pharmacological therapies.

Histologic alterations in global ischemia injury appear between 12 and 24 hours of injury [45]. The first alterations noticed are neurons shrinking or swelling. "Red neurons" can arise as a result of cytoplasmic eosinophilia, nuclear pyknosis, and other necrotic changes [46, 47]. Over time, this results in the loss of neurons and gliosis. Reperfusion damage is the strange and complicated phenomenon of increased cellular malfunction and cell death once blood flow is restored to previously ischemic tissues. It is characterized by biochemical and physiological changes that result in oxidant generation and complement activation culminating in an inflammatory response driven by neutrophil and platelet cell contacts with the endothelium and among the cells themselves [39, 48].

3.3.1 PATHOLOGICAL EVENTS DURING CNS/IR INJURIES

Atherosclerosis and acute myocardial infarction (AMI) are two physiological mechanisms that cause ischemia and contribute to hypoxia and hypoperfusion [49]. Hypoxia is caused by a blockage in arterial blood flow, which leads to mitochondrial electron transport chain malfunction. Anaerobic metabolism, sodium-potassium pump failure, and ribosome separation are all caused by reduced ATP generation in mitochondria (Figure 3.3). Anaerobic metabolism in cells produces less adenosine triphosphate (ATP) and antioxidants. In addition, lactic acid retention could lead to metabolic acidosis. On the cell surface, sodium-potassium pumps (Na+-K+-ATPase pumps) and calcium pumps (Ca^{2+}-ATPase pumps) may fail. When the Na+-K+-ATPase pumps fail, sodium is retained in the cells while potassium is liberated [50]. The activity of sodium–hydrogen exchanger pumps (Na+-H+ pumps) is reduced when sodium levels in cells are high. Calcium reuptake is reduced when calcium pumps in the endoplasmic reticulum (Ca^{2+}-ATPase pumps) become faulty. Water flows into the cytoplasm and cell expansion occurs when hydrogen, sodium, and calcium ions pile up in cells [51].

Because there are less antioxidative agents in ischemia cells, the formation of reactive oxygen species (ROS) increases. Endothelial dysfunction, DNA damage, and a local inflammatory response are all promoted by ROS. A cytokine storm can be triggered by inflammatory cascades and oxidative stress, resulting in cell death due to structural damage. Reperfusion is a dynamic procedure that can take many days. Understanding the precise process of ischemia-reperfusion injury could lead to new therapeutic approaches and injury prevention. When brain oxygen flow is reduced, neuronal ischemia ensues, and neuronal death occurs within minutes. The brain's nutritional, glycogen, and energy resources are depleted. Adenosine triphosphate depletion causes anaerobic metabolism, cerebral lactate accumulation, and intracellular acidosis [52].

Furthermore, cellular ischemia produces intracellular Ca^{2+} influx via N-methyl-D-aspartate channels, activating lytic enzymes and promoting mitochondrial dysfunction, further depleting adenosine triphosphate [53]. Finally, excitatory neurotransmitters activate lipases and proteases, which trigger apoptosis. After global cerebral ischemia, clinically, loss of neurologic function is characterized by a diminished state of awareness. A range of events occur when the blood flow to the brain is diminished, which are discussed in subsections.

3.3.1.1 MICROVASCULAR DYSFUNCTION

Vascular endothelial cells line the inside of blood arteries and perform a number of functions, including maintaining vascular tonicity, altering microvascular barrier function in capillaries and veins, and controlling proinflammatory and prothrombotic processes [54]. These endothelial cells become dysfunctional when tissues are subjected to ischemia/reperfusion (I/R), resulting in impaired vasoconstriction and vasodilation. In addition to these effects, I/R-induced reactive species activate NLRP3 signaling, modify connexin/pannexin signaling, and trigger mitochondrial fission processes, all of which cause arterioles, capillaries, and venules to behave abnormally. In tissues exposed to ischemia and reperfusion

(I/R), endothelial cell function in all segments (arterioles, capillaries, and venules) of the microvasculature is reduced [49]. Endothelial cell dysfunction is manifested by decreased arteriole vasodilation, increased fluid filtration and leukocyte blockage in capillaries, and increased leukocyte-endothelial cell adhesion and protein extravasation in venules [49, 55].

The bulk of these reactions are triggered by an imbalance in reactive oxygen species and nitric oxide production. Cardiovascular disease risk factors such as hypercholesterolemia, diabetes, and hypertension tend to increase the microvascular responses to I/R [56]. Endothelial cells encounter elevated oxidative stress, which causes leukocyte-endothelial cell adhesion via transcription-independent (early reaction) and regulatory (delayed response) processes. Enhanced leukocyte recruitment is connected to high endothelial barrier dysfunction in hypercholesterolemic mice, whereas increased protein extravasation in diabetic animals' postischemic venules occurs independently of leukocyte recruitment.

However, there is considerable evidence that hepatic microcirculation irregularities and the formation of oxygen-free radicals play a role in this clinical state. Endothelial cell damage and an imbalance of vasoconstrictor and vasodilator molecules such as endothelin, reactive oxygen species (ROS), and nitric oxide (NO) appear to be contributors to vascular microcirculation dysfunction [57]. This may result in the generation of proinflammatory cytokines, sinusoidal obstruction of neutrophils, oxidative stress, hypoxic cell damage, and parenchymal failure. Ischemic preconditioning is a powerful inducible endogenous process that protects cells from ischemia/reperfusion harm. Endothelin, tumor necrosis factor, and oxygen free radicals are all involved in microcirculatory dysfunction, and preconditioning has been hypothesized to limit their effects. Kupffer cells (KCs) have been linked to leukocyte recruitment and microvascular dysfunction associated with liver inflammation [34].

3.3.1.2 CEREBRAL EDEMA

A well-known complication of cerebral ischemia-reperfusion is cerebral/brain edema. A severe and often life-threatening consequence of several brain traumas is cerebral edema [58]. The most prevalent kind of brain edema is late-stage edema, which is characterized by a breach of the blood-brain barrier. For cerebral edema to arise, several physiological factors must be present, including altered cerebrovascular permeability, hydrostatic pressure, and adequate osmotic and colloidal gradients. VEGF is a potent vascular permeability mediator. Reduced oxygen supply, on the other hand, has been linked to increased levels of VEGF and its receptors after focal cerebral ischemia. Aquaporin-4 regulates brain water transport, and endothelin-1 causes cerebral edema, according to previous research. However, the role of the NLRP3 inflammasome in regulating brain edema and BBB breakdown during cerebral ischemia-reperfusion remains unknown [34].

3.3.1.3 ANEMIA

Anemia is common in people with chronic heart failure and is linked to poor clinical outcomes. Multiple factors contribute to anemia in chronic heart failure, with one contributing component being a subnormal compensatory surge in endogenous erythropoietin levels in response to anemia. I/R damage is a key vascular pathological paradigm that leads to morbidity and death in a variety of disorders, including myocardial infarction, stroke, acute renal failure, and transplantation [48]. I/R damage is characterized as a restriction of blood supply to an organ, which can cause ischemia, followed by perfusion restoration and reoxygenation. Many of the repercussions of this condition can be explained by mechanisms related with I/R injury, thanks to technological developments in the pathophysiology of I/R injuries.

3.3.1.4 IMPAIRED AUTOREGULATION

Finally, there is evidence of a malfunctioning cerebrovascular autoregulatory mechanism, implying that the preterm brain has limited ability to regulate blood flow when oxygen saturation or cardiac output is insufficient. Repeated ischemia events caused by any of these pathways will result in an inflammatory response, which will increase metabolic demand, use a higher fraction of given oxygen, and potentiate further harm during the next ischemic period. I/R damage targets the brain vasculature as well as neurons. Localized ischemia causes cerebrovascular damage, which increases with longer I/R durations [59]. Reperfusion after a brief period of ischemia can help salvage damaged brain tissue, although it is not always effective. Post-ischemia reperfusion disrupts autoregulatory mechanisms, resulting in a lack of control over cerebral blood flow, which exacerbates ischemic injury. As a result, I/R has an effect on the cerebral arteries and arterioles, which determine the degree of ischemia and reperfusion.

3.3.1.5 CARBON DIOXIDE (CO_2)

In all aerobic living species, carbon dioxide is a waste product of aerobic cellular respiration. When the balance between carbon dioxide created and carbon dioxide removed is disrupted, ischemia symptoms might get worse. Hypocapnia is a common and underestimated complication of a variety of conditions, such as early asthma, high-altitude pulmonary edema, and acute lung damage. Hypocapnia is still a common, albeit contested, therapy choice for adults and children with traumatic brain injury. Hypercapnia can occur for a variety of reasons [58]. CO_2 has previously been used to treat carbon monoxide poisoning, enhance cerebral perfusion during carotid endarterectomy, speed up general anesthesia recovery, and treat retinal artery blockage. Inducing hypercapnia rapidly in a critically ill patient may result in greater complications. Myocardial function is harmed by acute hypercapnia. Increased CO_2 inhibits the ability of vessels to contract and relax, resulting in damage. Hypercapnia has been shown to hurt skeletal muscle function, particularly diaphragmatic function [60]. Increases in the brain and cerebral blood volume caused by hypercapnia may harm intracranial pressure, resulting in BBB breakdown, vasoconstriction, hypercapnia, enhanced ROS production, and apoptotic induction. Furthermore, decreased cerebral perfusion pressure, increased excitatory neurotransmitter release, neuronal apoptosis, increased cerebrovascular resistance, energy/ATP crisis, and cell death occur.

3.3.1.6 HYPEROXIA

Atherosclerosis causes coronary arteries to thin and occlude, resulting in insufficient oxygen delivery for appropriate oxidative metabolism. Blood flow must be restored via thrombolysis, percutaneous coronary intervention, or surgical revascularization to avoid severe ischemia and consequent cardiomyocyte destruction. In addition to restoring oxygen supply to the cells, introducing molecular oxygen to the ischemic tissue causes a series of adverse reactions known as reperfusion damage. Before ischemia, a brief exposure to hyperoxia causes low-grade oxidative stress and an (ischemic) preconditioning-like effect in the myocardium, protecting the heart against further injury [34].

For individuals with AMI, revascularization and early reperfusion are the optimal therapy options. Reperfusion, on the other hand, could be a "double-edged sword." In post-ischemic tissue, restoring blood flow produces irreversible cellular damage. The generation of reactive oxygen species (ROS) is a prominent pathogenic element in reperfusion injury, despite the fact that the causes are varied and complex. In small amounts, ROS are waste products from normal metabolism. Superoxide (O_2), hydrogen peroxide (H_2O_2), the hydroxyl radical (HO•), and singlet oxygen (O_2•) are ROS that play a role in cellular signaling and microbicidal defense. On the other hand, excessive production of ROS increases lipid peroxidation and oxidative damage to membranes, proteins, and DNA [50].

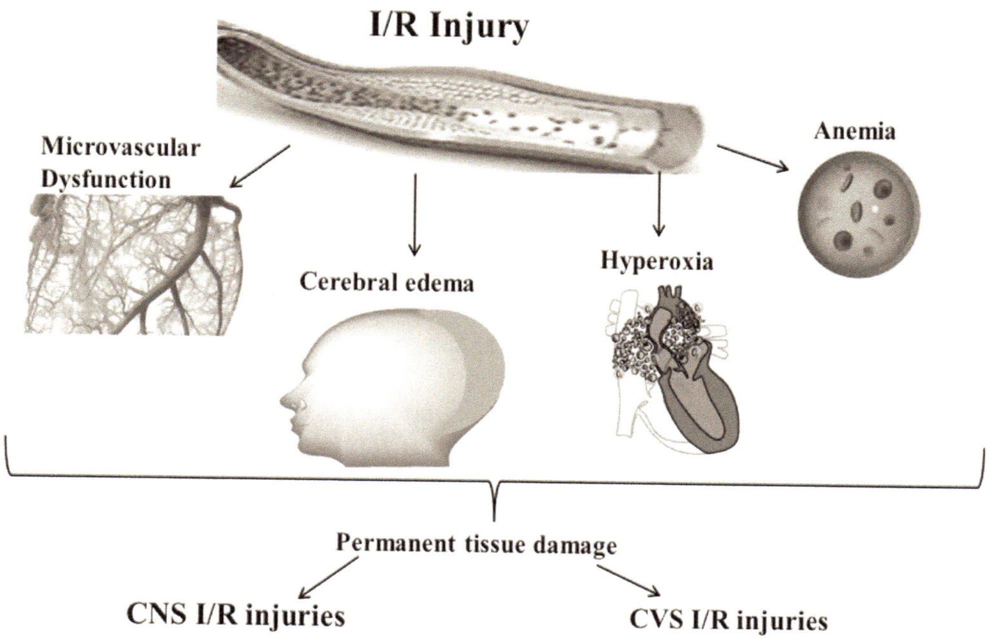

FIGURE 3.3 Complications associated with I/R injuries.

3.4 CVS I/R INJURY PATHOLOGIES

Early coronary artery reperfusion remains the most effective method of lowering infarct size after an AMI. Both sepsis and ischemia/reperfusion injury cause endothelial dysfunction, cellular injury, and multiple organ failure due to high levels of reactive oxygen species (ROS). This massive generation of reactive oxygen species (ROS) has been demonstrated *in vitro* [61]. The plasma of cardiac arrest subjects exhibited an immediate pro-oxidant status in endothelial cells, resulting in substantial endothelial damage. In the case of brain parenchymal damage during ischemia and reperfusion injury (like in other organ systems), a complicated chain of events occurs [62, 63]. At first, blood flow to the affected tissue is disrupted, followed by a depletion of cellular energy resources and anaerobic substrate glycolysis, lactic acidosis, sodium potassium pump failure, glutamate release, cytotoxic edema, free radical formation, and activation of both innate and adaptive immune responses [51].

Autophagy is one of the processes involved in myocardial I/R injury pathogenesis. Under normal conditions, autophagy, which helps maintain cellular homeostasis by eliminating old organelles and excessive or long-lived proteins, occurs at a low level in the heart [64]. Autophagy inhibition, on the other hand, might be harmful to the myocardium. In cardiovascular diseases such IHD, cardiac hypertrophy, and heart failure, autophagy flux has been reported to fluctuate. The brain spends lot of energy to phosphorylate ATP and regulate and maintain ionic gradient equilibrium oxidatively. When blood flow is disrupted, mitochondrial ATP synthesis is inhibited, allowing for the entry of Na^+ and efflux of K^+ ions from the cell. As a result, the entire homeostasis is disrupted, resulting in cell membrane depolarization. Increased intracellular Ca^{2+}, Na^+, and ADP induce mitochondria to generate harmful amounts of reactive oxygen species, resulting in oxidative damage. Various cytokines induce the production of matrix metalloproteinases in astrocytes and microglia [20]. It disrupts the BBB, generating vascular edema and hemorrhage and destroying capillaries, collagen, and laminins in the basal lamina, further degrading the composition of the vascular wall and increasing BBB permeability, resulting in neuronal death via necrosis and apoptosis.

3.5 AVAILABLE TREATMENTS FOR I/R INJURY

There are various strategies available for the treatment of I/R injury. The prime focus of all the therapies is to promote blood flow to the ischemic tissues and prevent disease progression. Both types of pharmacological and non-pharmacological treatments are available for the I/R injuries as discussed in Figure 3.4 [42, 65]. The non-pharmacological treatments include the methods such as ischemic preconditioning, plaque stabilization, plaque regression, collateralization, hemodynamic remodeling, surgical procedures, etc. On the other hand, pharmacological treatments comprised of drugs belonging to different classes with varying mechanisms of action, such as antioxidants, anti-hyperlipidemic, anti-platelets, TNF-α inhibitors, analgesics, statins, etc. [42, 58, 66]. Various other natural compounds and herbs are also available to minimize the extent of tissue injury during ischemia and reperfusion.

Treatment Approaches for I/R injuries

Pharmacological Treatments

- Anti-oxidants
- Anti-hyperlipidemics
- Fibrinolytics
- Analgesics
- TNF-α inhibitors
- Others (Monoclonal antibodies, anaesthetics, antibiotics, nitrates etc.)

Non-Pharmacological Treatments

- Vascular remodelling
- Ischemic pre & post conditioning
- Plaque stabilisation & regression
- Collateralization
- Hemodynamic changes
- Surgeries & transplants

FIGURE 3.4 Pharmacological and non-pharmacological treatments are available for the I/R injuries.

3.5.1 NON-PHARMACOLOGICAL STRATEGIES FOR I/R INJURIES

Based on clinical manifestations, strategies for treating I/R injury are designed in three stages, i.e., before ischemia, during ischemia, and following ischemia.

3.5.1.1 ISCHEMIC PRECONDITIONING

Ischemic preconditioning (IPC) is widely used to prevent the progression of I/R injury by making tissues more resistant to the lost cause by reperfusion followed by ischemia [67]. IPC should be ischemia followed by reperfusion for the short period generally performed before prolonged ischemia. IPC takes place in two stages, i.e., early preconditioning and late preconditioning. Early pre-conditioning lasts up to 1–2 hours, whereas late conditioning performs within 24 hours and lasts up to 48–72 hours. IPC triggers the release of chemokines, neurotransmitters, and adhesion molecules, releases calcium ions, and potentiates ROS release [68]. IPC works by the activation of various receptors like G-protein coupled receptors, adenosine receptors, bradykinin receptors, and adrenoceptors.

Intracellular signal transduction pathways are activated when these receptors are activated. This is due to the involvement of the nitric oxide synthase pathway, reperfusion injury salvage kinase (RISK), and survival activating factor enhancement (SAFE) [69, 70]. The activation of this signaling pathway enhances the cell survival signaling pathways. The sympathetic nervous system is activated when adrenoceptors are activated. GPCR activation activates PI3K/AKT signaling, which activates cGMP downstream and results in the production of nitric oxide (NO) [71, 72]. PKG and PKC are further activated by NO, resulting in the generation of ROS and the activation of mitochondrial ATP dependent K channels (ATPK), which normalize mitochondrial membrane depolarization. MEK1/2/ERK1/2 and RISK pathways are simultaneously activated by PI3K/AKT, with downstream control of p70S6K and GSK3 activation.

The SAFE pathway is activated under inflammatory conditions, mainly interacts with type 2 TNF-α receptor, and upregulates the JAK/STAT3 pathway. Upregulation of the JAK/STAT3 pathway completely blocks the mitochondrial permeable transition pore (mPTP) opening and prevents further deterioration [73]. Moreover, it also enhances the release of anti-apoptotic agents like Bcl2, Bcl-Xl, and cyclins. It also promotes the release of antioxidants like glutathione, heme oxygenase-1, etc. Thus IPC serves as I/R defensive strategy [57]. The clinical implication of IPC is known as Remote ischemic preconditioning (IRC). This is known for its promising role in the prevention of I/R injury. RIC is a reversible technique in which ischemia followed by reperfusion is an offer to a single organ to prevent the subsequent insult to the other organs [51, 74]. Thus pharmacological activation of receptors involved in the IPC may work as a protective strategy toward the I/R injury.

3.5.1.2 PLAQUE STABILIZATION AND PLAQUE REGRESSION

As discussed above, Ischemia is mainly characterized by decreased blood flow. This primarily happens due to the plugging of arteries due to embolism, thrombosis, and atherosclerosis [75]. This tends to lead to plaque formation inside the arteries, narrowing blood vessels, and reducing blood flow. Thus the prime focus is to remove the plaque deposition and widen the blood vessels to increase blood flow. Plaque stabilization refers to stabilizing the plaque content inside the blood vessels and strengthening the endothelium wall. It means promoting the thickening of the fibrous cap, reduction in inflammatory events, and preventing the enlargement of plaque size.

This depends on the balance between vascular smooth muscle cells and macrophages and T cell activation [76]. Vascular smooth muscle activation promotes plaque stabilization, but an increase in the activity of macrophages and t-cells leads to plaque rupture. Plaque regression means a reduction in the volume of plaque deposited inside the wall and reverses the morphology of blood vessels to a normal state from a diseased condition. Plaque stabilization can be achieved by using anti-platelets, anti-coagulants, and anti-inflammatory drugs. ACE inhibitors are also used as they inhibit plasminogen activation. Plaque regression mainly depends upon the reduction of plaque lipid core, macrophage activation, and thickening of the fibrous cap. This requires long-term therapy, including lipid-lowering drugs like statins, HMG-CoA reductase, etc. [77, 78].

3.5.1.3 COLLATERALIZATION

Collateralization is a non-invasive strategy to slow down the progression of I/R injury. This is also termed collateral artery remodeling, which includes transforming pre-existing non-functional collateral arteries into functional ones. Native collateral arteries do not protect against ischemic injury, but their remodeling offers much protection to tissues. Remodeling increases the vascular diameter of blood vessels and the blood flow capacity of ischemic areas and prevents disease progression [49].

3.5.1.4 HEMODYNAMIC CHANGES

1. **Ischemic Preconditioning:** Ischemic per conditioning is generally performed after initiating ischemia and lasts till reperfusion. This condition includes the placement of blood pressure cuff inflation on the arm to increase the salvage index.
2. **Ischemic Postconditioning:** Ischemic Post Conditioning (IPosC) is performed after the prolonged ischemic insults and just before the complete reperfusion. IPosC provides repetitive ischemic insult to the ischemic tissues followed by interrupted reperfusion after the complete and final reperfusion. IPosC prevents the activation of cell death mechanisms after reperfusion. IPosC reduces apoptotic cell death by activating the JAK/STAT pathway. IPosC activates various receptors like GPCR, adrenoreceptors, adenosine receptors, etc., leading to upregulation of the PI3K/Akt pathway. Moreover, IPosC reduces infarct size and volume by promoting the phosphorylation of Akt. Phosphorylated Akt inhibits the phosphorylation of GSK-3β [69, 74, 79]. Moreover, lessening the release of calcium ions reduces mitochondrial membrane potential and prevents depolarization.

This also reduces the release of cytochrome c by inhibiting the opening of mPTP. This minimizes the caspase activation and release of pro-apoptotic factors. This reduces the activity of intrinsic and extrinsic apoptotic events and prevents the activation of cell death mechanisms. IPosC also potentiates the release of anti-apoptotic factors like Bcl_2, Bcl-xl, etc. IPosC also reduces the activity of various oxidative systems like NADPH oxidase, Xanthine oxidase, etc., to prevent ROS formation. The activity of these systems is more pronounced during the reperfusion state and the rise in the production of ROS. Nonetheless, IPosC potentiates the PKC/cGMP activity, promotes nitric oxide synthesis, and reduces the production of reactive nitrite species [80]. Moreover, NO leads to the dilation of the endothelial wall and lowers blood pressure leading to lessen infarct volume.

3.5.2 PHARMACOLOGICAL APPROACHES FOR I/R INJURIES

1. **Antioxidants:** Our body has its antioxidant defensive system that protects us from oxidative damage from ROS and RNS. When this balance between oxidative and antioxidative mechanisms got disrupted, various pathologies associated with increased oxidative stress discussed earlier begin [49, 61, 81]. Multiple compounds are studied for their antioxidant potential in I/R injuries.
2. **Anti-Hyperlipidemics:** It is helpful in case of I/R injuries. They work through a variety of methods, including activation of reperfusion injury salvage kinase, enhanced phosphorylation of Akt, ERK1/2, and GSK3β downregulation. There are no. Compounds including statins, cholesterol absorption inhibitors, and fibrates are studied in the I/R injuries [47, 82].
3. **Others:** Other than both the classes, there are various other drugs, such as anesthetics, monoclonal antibodies, nitrates, etc.), which are already studied for the management of I/R injuries.

3.6 RECENT DATA AVAILABLE ON PRECLINICAL AND CLINICAL REPORTS ON I/R INJURIES

Plenty of pharmacological and non-pharmacological treatments are available for reversing ischemic conditions in case of I/R injuries [57, 69, 75, 79, 83]. But still, there is no complete cure for I/R injury or any adverse condition produced by ischemia. The non-pharmacological treatments only help reverse ischemic conditions but not the damage caused to the body systems. Here we will summarize the available preclinical and clinical reports for the I/R injuries.

3.7 CONCLUSION

I/R injuries are the major leading cause of CVS and CNS disorders leading to higher morbidity and mortality rates. The I/R injuries affect both the significant parts of the body, i.e., heart and brain. The damage caused by these may lead to permanent disabilities or death. There are multiple reports available on preclinical and clinical studies of I/R injuries. The above data suggests that currently, the damage caused by I/R is not completely curative; only supportive treatments and therapies are available. The need of the hour is to understand the pathological mechanisms of I/R injuries and explore newer therapeutics for managing IR injuries. This chapter will help readers and researchers in this field understand the existing pathologies and various targets for I/R injuries.

KEYWORDS

- **carbon dioxide**
- **clinical manifestations**
- **ischemia/reperfusion injury**
- **mitochondria**
- **NADPH oxidase system**
- **oxidative stress**
- **preclinical studies**
- **reactive oxygen species**

REFERENCES

1. Testa, A., Frati, G., Versaci, F., & Biondi-Zoccai, G., (2021). Ischemia-reperfusion injury: Can we stop the curing-hurting paradox? *J. Cardiovasc. Pharmacol., 77*, 427–429.
2. Yang, C. F., (2018). Clinical manifestations and basic mechanisms of myocardial ischemia/reperfusion injury. *Tzu-Chi Med. J., 30*, 209.

3. Tiwari, P., Khan, H., Singh, T. G., & Grewal, A. K., (2022). Poly (ADP-ribose) polymerase: An overview of mechanistic approaches and therapeutic opportunities in the management of stroke. *Neurochem. Res.*, 1–23.
4. Vigil, T. M., Frieler, R. A., Kilpatrick, K. L., Wang, M. M., & Mortensen, R. M., (2022). Aconitate decarboxylase 1 suppresses cerebral ischemia-reperfusion injury in mice. *Exp. Neurol., 347*, 113902.
5. Wu, L., Xiong, X., Wu, X., Ye, Y., Jian, Z., Zhi, Z., & Gu, L., (2020). Targeting oxidative stress and inflammation to prevent ischemia-reperfusion injury. *Front. Mol. Neurosci., 13*, 28.
6. Buelna-Chontal, M., García-Niño, W. R., Silva-Palacios, A., Enríquez-Cortina, C., & Zazueta, C., (2021). Implications of oxidative and nitrosative post-translational modifications in therapeutic strategies against reperfusion damage. *Antioxidants, 10*, 749.
7. Hochrainer, K., (2018). Protein modifications with ubiquitin as response to cerebral ischemia-reperfusion injury. *Transl. Stroke Res., 9*, 157–173.
8. Kamel, M., Ahmed, S. M., & Abdelzaher, W., (2020). The potential protective effect of modafinil in intestinal ischemic reperfusion-induced in rats. *Int. Immunopharmacol., 88*, 106983.
9. Khan, H., Sharma, K., Kumar, A., Kaur, A., & Singh, T. G., (2022). Therapeutic implications of cyclooxygenase (COX) inhibitors in ischemic injury. *Inflamm. Res.*, 1–16.
10. Ma, L. L., Li, Y., Yin, P. P., Kong, F. J., Guo, J. J., Shi, H. T., Zhu, J. B., Zou, Y. Z., & Ge, J. B., (2018). Hypertrophied myocardium is vulnerable to ischemia/reperfusion injury and refractory to rapamycin-induced protection due to increased oxidative/nitrative stress. *Clin. Sci., 132*, 93–110.
11. Gauberti, M., Lapergue, B., Martinez De, L. S., Vivien, D., Richard, S., Bracard, S., Piotin, M., & Gory, B., (2018). Ischemia-reperfusion injury after endovascular thrombectomy for ischemic stroke. *Stroke, 49*, 3071–3074.
12. Mandalaneni, K., Rayi, A., & Jillella, D. V., (2020). *Stroke Reperfusion Injury*. PMID: 33232020.
13. Saklani, P., Khan, H., Gupta, S., Kaur, A., & Singh, T. G., (2022). Neuropeptides: Potential neuroprotective agents in ischemic injury. *Life Sci., 288*, 120186.
14. Perez-Vizcaino, F., Duarte, J., & Andriantsitohaina, R., (2006). Endothelial function and cardiovascular disease: Effects of quercetin and wine polyphenols. In: *Free Radical Research* (pp. 1054–1065). https://doi.org/10.1080/10715760600823128.
15. Bardaweel, S. K., Gul, M., Alzweiri, M., Ishaqat, A., ALSalamat, H. A., & Bashatwah, R. M., (2018). Reactive oxygen species: The dual role in physiological and pathological conditions of the human body. *Eurasian J. Med., 50*, 193.
16. Kapoor, M., Sharma, N., Sandhir, R., & Nehru, B., (2018). Effect of the NADPH oxidase inhibitor apocynin on ischemia-reperfusion hippocampus injury in rat brain. *Biomed. Pharmacother., 97*, 458–472.
17. Khan, H., Garg, N., Singh, T. G., Kaur, A., & Thapa, K., (2022a). Calpain inhibitors as potential therapeutic modulators in neurodegenerative diseases. *Neurochem. Res.*, 1–25.
18. Khan, H., Bangar, A., Grewal, A. K., Bansal, P., & Singh, T. G., (2022b). Caspase-mediated regulation of the distinct signaling pathways and mechanisms in neuronal survival. *Int. Immunopharmacol., 110*, 108951.
19. Zhou, H., & Toan, S., (2020). Pathological roles of mitochondrial oxidative stress and mitochondrial dynamics in cardiac microvascular ischemia/reperfusion injury. *Biomolecules, 10*, 85.
20. Zhou, Q., Gong, X., Kuang, G. E., Jiang, R., Xie, T., Tie, H., Chen, X., et al., (2018). Ferulic acid protected from kidney ischemia reperfusion injury in mice: Possible mechanism through increasing adenosine generation via HIF-1α. *Inflammation, 41*, 2068–2078.
21. González-Montero, J., Brito, R., Gajardo, A. I., & Rodrigo, R., (2018). Myocardial reperfusion injury and oxidative stress: Therapeutic opportunities. *World J. Cardiol., 10*, 74–86. https://doi.org/10.4330/wjc.v10.i9.74.
22. Saklani, P., Khan, H., Singh, T. G., Gupta, S., & Grewal, A. K., (2022a). Demethyleneberberine, a potential therapeutic agent in neurodegenerative disorders: A proposed mechanistic insight. *Mol. Biol. Rep.*, 1–13.
23. Prabhakar, N. K., Khan, H., Grewal, A. K., & Singh, T. G., (2022). Intervention of neuroinflammation in the traumatic brain injury trajectory: *In vivo* and clinical approaches. *Int. Immunopharmacol., 108*, 108902.
24. Sharma, A., Khan, H., Singh, T. G., Grewal, A. K., Najda, A., Kawecka-Radomska, M., Kamel, M., et al., (2021). Pharmacological modulation of ubiquitin-proteasome pathways in oncogenic signaling. *Int. J. Mol. Sci., 22*(21), 11971.

25. Kaur, A., Singh, T. G., Khan, H., Kumar, M., Singh, N., & Abdel-Daim, M. M., (2022). Neuroprotective effect of piclamilast-induced post-ischemia pharmacological treatment in mice. *Neurochem. Res.*, 1–14.
26. Zhang, L., Sui, R., & Zhang, L., (2022). Fingolimod protects against cerebral ischemia reperfusion injury in rats by reducing inflammatory cytokines and inhibiting the activation of p38 MAPK and NF-κB signaling pathways. *Neurosci. Lett., 771*, 136413.
27. Garg, C., Kaur, A., Singh, T. G., Sharma, V. K., & Singh, S. K., (2022). Therapeutic implications of sonic hedgehog pathway in metabolic disorders: Novel target for effective treatment. *Pharmacol. Res.*, 106194.
28. Khan, H., Grewal, A. K., & Singh, T. G., (2022d). Pharmacological postconditioning by protocatechuic acid attenuates brain injury in ischemia–reperfusion (I/R) mice model: Implications of nuclear factor erythroid-2-related factor pathway. *Neuroscience, 491*, 23–31.
29. Song, J., Zhang, W., Wang, J., Yang, H., Zhou, Q., Wang, H., Li, L., & Du, G., (2019). Inhibition of FOXO3a/BIM signaling pathway contributes to the protective effect of salvianolic acid A against cerebral ischemia/reperfusion injury. *Acta Pharm. Sin. B., 9*, 505–515.
30. Grewal, A. K., Singh, N., & Singh, T. G., (2019). Effects of resveratrol postconditioning on cerebral ischemia in mice: Role of the sirtuin-1 pathway. *Can. J. Physiol. Pharmacol., 97*(11), 1094–1101.
31. Khan, H., Tiwari, P., Kaur, A., & Singh, T. G., (2021a). Sirtuin acetylation and deacetylation: A complex paradigm in neurodegenerative disease. *Mol. Neurobiol., 58*(8), 3903–3917.
32. Garg, N., Singh, T. G., Khan, H., Arora, S., Kaur, A., & Mannan, A., (2021). Mechanistic interventions of selected *Ocimum* species in management of diabetes, obesity, and liver disorders: Transformative developments from preclinical to clinical approaches. *Biointerface Res. Appl. Chem, 12*(1), 1304–1323.
33. Jacob, B., Stock, D., Chan, V., Colantonio, A., & Cullen, N., (2020). Predictors of in-hospital mortality following hypoxic-ischemic brain injury: A population-based study. *Brain Inj., 34*, 178–186. https://doi.org/10.1080/02699052.2019.1683897.
34. Sekhon, M. S., Ainslie, P. N., & Griesdale, D. E., (2017). Clinical pathophysiology of hypoxic ischemic brain injury after cardiac arrest: A "two-hit" model. *Crit. Care, 21*, 1–10. https://doi.org/10.1186/s13054-017-1670-9.
35. Khan, H., Singh, T. G., Dahiya, R. S., & Abdel-Daim, M. M., (2022c). α-lipoic acid, an organosulfur biomolecule a novel therapeutic agent for neurodegenerative disorders: An mechanistic perspective. *Neurochem. Res.*, 1–12.
36. Yi, Q., Tan, F., Tan, J., Chen, X., Xiao, Q., Liu, Y., Zhang, G., & Luo, J., (2019). Minocycline protects against myocardial ischemia/reperfusion injury in rats by upregulating MCPIP1 to inhibit NF-κB activation. *Acta Pharmacol. Sin., 40*, 1019–1028.
37. Andreadou, I., Schulz, R., Papapetropoulos, A., Turan, B., Ytrehus, K., Ferdinandy, P., Daiber, A., & Di Lisa, F., (2020b). The role of mitochondrial reactive oxygen species, NO and H_2S in ischemia/reperfusion injury and cardioprotection. *J. Cell. Mol. Med., 24*, 6510–6522.
38. Milano, G., Morel, S., Bonny, C., Samaja, M., Von, S. L. K., Nicod, P., & Vassalli, G., (2007). A peptide inhibitor of c-Jun NH_2-terminal kinase reduces myocardial ischemia-reperfusion injury and infarct size *in vivo. Am. J. Physiol. – Hear. Circ. Physiol., 292*, 1828–1835. https://doi.org/10.1152/ajpheart.01117.2006.
39. Davidson, S. M., Ferdinandy, P., Andreadou, I., Bøtker, H. E., Heusch, G., Ibáñez, B., Ovize, M., et al., (2019). Multitarget strategies to reduce myocardial ischemia/reperfusion injury: JACC review topic of the week. *J. Am. Coll. Cardiol., 73*, 89–99. https://doi.org/10.1016/j.jacc.2018.09.086.
40. Moreira, P. I., Zhu, X., Liu, Q., Honda, K., Siedlak, S. L., Harris, P. L., Smith, M. A., & Perry, G., (2006). Compensatory responses induced by oxidative stress in Alzheimer disease. *Biol. Res., 39*, 7–13.
41. Sharma, V. K., Singh, T. G., & Mehta, V., (2021). Stressed mitochondria: A target to intrude Alzheimer's disease. *Mitochondrion, 59*, 48–57.
42. Hayes, C. A., Valcarcel-Ares, M. N., & Ashpole, N. M., (2021). Preclinical and clinical evidence of IGF-1 as a prognostic marker and acute intervention with ischemic stroke. *J. Cereb. Blood Flow Metab., 41*, 2475–2491. https://doi.org/10.1177/0271678X211000894.
43. Gornall, A. G., Bardawill, C. J., & David, M. M., (1949). Determination of serum proteins by means of the biuret reaction. *J. Biol. Chem., 177*, 751–766.
44. Francischetti, I., Moreno, J. B., Scholz, M., & Yoshida, W. B., (2010). Leukocytes and the inflammatory response in ischemia-reperfusion injury. *Rev. Bras. Cir. Cardiovasc., 25*, 575–584.

45. Wang, M., Zhang, J., Zhang, J., Sun, K., Li, Q., Kuang, B., Wang, M. M. Z., et al., (2021). Methyl eugenol attenuates liver ischemia reperfusion injury via activating PI3K/Akt signaling. *Int. Immunopharmacol., 99*, 108023.
46. Fisher, W. S., (1992). Cerebral ischemia. *Neurosurgery, 30*, 810. https://doi.org/10.1227/00006123-199205000-00060.
47. Tong, Q., Zhu, P. C., Zhuang, Z., Deng, L. H., Wang, Z. H., Zeng, H., Zheng, G. Q., & Wang, Y., (2019). Notoginsenoside R1 for organs ischemia/reperfusion injury: A preclinical systematic review. *Front. Pharmacol.*, 10. https://doi.org/10.3389/fphar.2019.01204.
48. Sharma, D., Maslov, L. N., Singh, N., & Jaggi, A. S., (2020). Remote ischemic preconditioning-induced neuroprotection in cerebral ischemia-reperfusion injury: Preclinical evidence and mechanisms. *Eur. J. Pharmacol., 883*, 173380. https://doi.org/10.1016/j.ejphar.2020.173380.
49. Paul, S., & Candelario-Jalil, E., (2021). Emerging neuroprotective strategies for the treatment of ischemic stroke: An overview of clinical and preclinical studies. *Experimental Neurology*. Elsevier Inc. https://doi.org/10.1016/j.expneurol.2020.113518.
50. Mokhtari-Zaer, A., Marefati, N., Atkin, S. L., Butler, A. E., & Sahebkar, A., (2019). The protective role of curcumin in myocardial ischemia–reperfusion injury. *J. Cell. Physiol., 234*, 214–222.
51. Leurcharusmee, P., Sawaddiruk, P., Punjasawadwong, Y., Chattipakorn, N., & Chattipakorn, S. C., (2018). The possible pathophysiological outcomes and mechanisms of tourniquet-induced ischemia-reperfusion injury during total knee arthroplasty. *Oxid. Med. Cell. Longev., 2018*. https://doi.org/10.1155/2018/8087598.
52. Liu, Z., Fan, Y., Won, S. J., Neumann, M., Hu, D., Zhou, L., Weinstein, P. R., & Liu, J., (2007). Chronic treatment with minocycline preserves adult new neurons and reduces functional impairment after focal cerebral ischemia. *Stroke, 38*, 146–152. https://doi.org/10.1161/01.STR.0000251791.64910.cd.
53. Hernandez-Resendiz, S., Chinda, K., Ong, S. B., Cabrera-Fuentes, H., & Zazueta, C., (2018). The role of redox dysregulation in the inflammatory response to acute myocardial ischemia-reperfusion injury-adding fuel to the fire. *Curr. Med. Chem., 25*, 1275–1293.
54. Zhong, J., Ouyang, H., Sun, M., Lu, J., Zhong, Y., Tan, Y., & Hu, Y., (2019). Tanshinone IIA attenuates cardiac microvascular ischemia-reperfusion injury via regulating the SIRT1-PGC1α-mitochondrial apoptosis pathway. *Cell Stress Chaperones, 24*, 991–1003.
55. Luo, J., Chen, J., Yang, C., Tan, J., Zhao, J., Jiang, N., & Zhao, Y., (2021). 6-Gingerol protects against cerebral ischemia/reperfusion injury by inhibiting NLRP3 inflammasome and apoptosis via TRPV1/FAF1 complex dissociation-mediated autophagy. *Int. Immunopharmacol., 100*, 108146.
56. Vega, J., (2020). *Brain Ischemia Types and Causes*, 1–7.
57. Chen, T., & Vunjak-Novakovic, G., (2019). Human tissue-engineered model of myocardial ischemia-reperfusion injury. *Tissue Eng. – Part A., 25*, 711–724. https://doi.org/10.1089/ten.tea.2018.0212.
58. Sorby-Adams, A. J., Vink, R., & Turner, R. J., (2018). Large animal models of stroke and traumatic brain injury as translational tools. *Am. J. Physiol. – Regul. Integr. Comp. Physiol., 315*, R165–R190. https://doi.org/10.1152/ajpregu.00163.2017.
59. Crofts, A., Kelly, M. E., & Gibson, C. L., (2020). Imaging functional recovery following ischemic stroke: Clinical and preclinical fMRI studies. *J. Neuroimaging, 30*, 5–14. https://doi.org/10.1111/jon.12668.
60. Busl, K. M., & Greer, D. M., (2010). Hypoxic-ischemic brain injury: Pathophysiology, neuropathology, and mechanisms. *Neuro Rehabilitation, 26*, 5–13. https://doi.org/10.3233/NRE-2010-0531.
61. Olugbodi, J. O., Samaila, K., Lawal, B., Anunobi, O. O., Baty, R. S., Ilesanmi, O. B., & Batiha, G. E. S., (2021). Computational and preclinical evidence of anti-ischemic properties of L-carnitine-rich supplement via stimulation of anti-inflammatory and antioxidant events in testicular torsed rats. *Oxid. Med. Cell. Longev*. https://doi.org/10.1155/2021/5543340.
62. Bir, S. C., Khan, M. W., Javalkar, V., Toledo, E. G., & Kelley, R. E., (2021). Emerging concepts in vascular dementia: A review. *J. Stroke Cerebrovasc. Dis., 30*, 105864.
63. Rihal, V., Khan, H., Kaur, A., & Singh, T. G., (2022). Vitamin D as therapeutic modulator in cerebrovascular diseases: A mechanistic perspectives. *Crit. Rev. Food Sci. Nutr.*, 1–23.
64. Kalra, P., Khan, H., Kaur, A., & Singh, T. G., (2022). Mechanistic insight on autophagy modulated molecular pathways in cerebral ischemic injury: From preclinical to clinical perspective. *Neurochem. Res.*, 1–19.
65. Khan, H., Kashyap, A., Kaur, A., & Singh, T. G., (2020). Pharmacological postconditioning: A molecular aspect in ischemic injury. *J. Pharm. Pharmacol., 72*(11), 1513–1527.

66. Black, S. C., (2000). *In vivo* models of myocardial ischemia and reperfusion injury: Application to drug discovery and evaluation. *J. Pharmacol. Toxicol. Methods, 43*, 153–167. https://doi.org/10.1016/S1056-8719(00)00102-7.
67. Chen, Z. X., Xu, Q. Q., Shan, C. S., Shi, Y. H., Wang, Y., Chang, R. C. C., & Zheng, G. Q., (2019). Borneol for regulating the permeability of the blood-brain barrier in experimental ischemic stroke: Preclinical evidence and possible mechanism. *Oxid. Med. Cell. Longev., 2019*. https://doi.org/10.1155/2019/2936737.
68. Rehni, A. K., & Singh, T. G., (2012). Involvement of CCR-2 chemokine receptor activation in ischemic preconditioning and postconditioning of brain in mice. *Cytokine, 60*(1), 83–89.
69. Deng, J., Feng, J., Liu, T., Lu, X., Wang, W., Liu, N., Lv, Y., et al., (2018). Beraprost sodium preconditioning prevents inflammation, apoptosis, and autophagy during hepatic ischemia-reperfusion injury in mice via the P38 and JNK pathways. *Drug Des. Devel. Ther., 12*, 4067.
70. Sun, X., Wang, D., Zhang, T., Lu, X., Duan, F., Ju, L., Zhuang, X., & Jiang, X., (2020). Eugenol attenuates cerebral ischemia-reperfusion injury by enhancing autophagy via AMPK-mTOR-P70S6K pathway. *Front. Pharmacol., 11*, 84.
71. Rehni, A. K., Singh, T. G., Kalra, R., & Singh, N., (2009). Pharmacological inhibition of inducible nitric oxide synthase attenuates the development of seizures in mice. *Nitric Oxide, 21*(2), 120–125.
72. Khan, H., Singh, A., Thapa, K., Garg, N., Grewal, A. K., & Singh, T. G., (2021). Therapeutic modulation of the phosphatidylinositol 3-kinases (PI3K) pathway in cerebral ischemic injury. *Brain Res., 1761*, 147399.
73. Khan, H., Gupta, A., Singh, T. G., & Kaur, A., (2021b). Mechanistic insight on the role of leukotriene receptors in ischemic–reperfusion injury. *Pharmacol. Rep., 73*(5), 1240–1254.
74. Hausenloy, D. J., & Yellon, D. M., (2007). Reperfusion injury salvage kinase signaling: Taking a RISK for cardioprotection. *Heart Fail. Rev., 12*, 217–234. https://doi.org/10.1007/s10741-007-9026-1.
75. Candelario-Jalil, E., & Paul, S., (2021). Impact of aging and comorbidities on ischemic stroke outcomes in preclinical animal models: A translational perspective. *Exp. Neurol., 335*, 113494. https://doi.org/10.1016/j.expneurol.2020.113494.
76. Andreadou, I., Bell, R. M., Bøtker, H. E., & Zuurbier, C. J., (2020a). SGLT2 inhibitors reduce infarct size in reperfused ischemic heart and improve cardiac function during ischemic episodes in preclinical models. *Biochim. Biophys. Acta – Mol. Basis Dis., 1866*, 165770. https://doi.org/10.1016/j.bbadis.2020.165770.
77. Cruz, S. A., Qin, Z., Stewart, A. F. R., & Chen, H. H., (2018). Dabrafenib, an inhibitor of RIP3 kinase-dependent necroptosis, reduces ischemic brain injury. *Neural Regen. Res., 13*, 252–256. https://doi.org/10.4103/1673-5374.226394.
78. Watson, S., (2018). *Microvascular Ischemic Disease* (pp. 1–14). Heal. Media a Red Ventur. Co.
79. Jiang, T., Zhan, F., Rao, Z., Pan, X., Zhong, W., Sun, Y. U., Wang, P., et al., (2019). Combined ischemic and rapamycin preconditioning alleviated liver ischemia and reperfusion injury by restoring autophagy in aged mice. *Int. Immunopharmacol., 74*, 105711.
80. Pandya, R. S., Mao, L., Zhou, H., Zhou, S., Zeng, J., Popp, A. J., & Wang, X., (2011). Central nervous system agents for ischemic stroke: Neuroprotection mechanisms NIH public access. *Cent. Nerv. Syst. Agents Med. Chem., 11*, 81–97.
81. Pan, H., Xu, Y., Cai, Q., Wu, M., & Ding, M., (2021). Effects of β-asarone on ischemic stroke in middle cerebral artery occlusion rats by an Nrf2-antioxidant response elements (ARE) pathway-dependent mechanism. *Med. Sci. Monit. Int. Med. J. Exp. Clin. Res., 27*, e931884–1.
82. Liu, D. Q., Chen, S. P., Sun, J., Wang, X. M., Chen, N., Zhou, Y. Q., Tian, Y. K., & Ye, D. W., (2019). Berberine protects against ischemia-reperfusion injury: A review of evidence from animal models and clinical studies. *Pharmacol. Res., 148*, 104385. https://doi.org/10.1016/j.phrs.2019.104385.
83. Zheng, Q., Zhu, J. Z., Bao, X. Y., Zhu, P. C., Tong, Q., Huang, Y. Y., Zhang, Q. H., et al., (2018). A preclinical systematic review and meta-analysis of astragaloside IV for myocardial ischemia/reperfusion injury. *Front. Physiol., 9*, 795.
84. Ballestín, A., Casado, J. G., Abellán, E., Vela, F. J., Álvarez, V., Usón, A., López, E., et al., (2018). Ischemia-reperfusion injury in a rat microvascular skin free flap model: A histological, genetic, and blood flow study. *PLoS One, 13*, e0209624.

CHAPTER 4

Molecular Aspects of Ischemic Reperfusion Injury

MAANVI,[1] SHILPA KUMARI,[1] CHAMAN SHARMA,[1] RAHUL DESHMUKH,[1] and THAKUR GURJEET SINGH[2]

[1]*Department of Pharmaceutical Sciences and Technology, Maharaja Ranjit Singh Punjab Technical University, Bathinda, Punjab, India*

[2]*Chitkara College of Pharmacy, Chitkara University, Punjab, India*

ABSTRACT

Ischemic Reperfusion Injury (IRI) is a multi-factorial condition that arises due to the restoration of blood flow to ischemic tissues. Ischemic injury is the 2nd most widely accepted reason for human death. Ischemia happens either due to an increase in oxygen demand by the tissues or plugging of the arteries leads to a reduction in the blood supply. Ischemia can be corrected by immediate restoration of blood flow to the ischemic area. But, delayed reperfusion instead gives positive results, but may worsen the condition of the patient. The prime focus of reperfusion is to restore blood flow and promote the production of molecular oxygen. Ischemia promotes electrolyte imbalance. Reperfusion results in the upregulation of various cell survival signalings. This enhances the production of physiological reactive oxygen species in response to enhanced metabolic activities. This tends to activate various oxidative systems like the NADPH oxidase system, and Xanthine oxidase system and disrupts the mitochondrial transport chain. This condition potentiates the ischemic insult and reduces the life span of the organ. Ischemic reperfusion injury promotes the expression of apoptosis, mitoptosis, necrosis, and necroptosis pathways. Thus reperfusion tends to increase the severity of disease after ischemia. In this chapter, we will discuss all the possible mechanistic pathways to understand the pathophysiology of Ischemic reperfusion injury mediated cell death which will enhance the knowledge of the reader about IRI and helps to develop newer approaches for the treatment of IRI.

4.1 INTRODUCTION

Ischemic reperfusion injury (IRI) is a multi-factorial condition that arises during the restoration of blood flow to the ischemic tissues [1]. Epidemiological study on Ischemia suggests that 87% of total death cases are due to ischemia and become the 2nd most

Ischemic Injury: A Molecular Insight. Thakur Gurjeet Singh & Heena Khan (Eds.)

common reason for death worldwide [2]. Ischemic injury is either due to rupture of blood vessels or plugging of arteries due to embolism, occlusion, thrombosis, etc. [3]. Moreover, Hypoperfusion is the main cause of Ischemia [4]. There are various conditions linked to hypoperfusion are sepsis, organ transplantation, acute coronary syndrome (ACS), etc. [5]. Several guidelines describe that timely resuscitation of the blood flow to the ischemic tissues reconciles the patient's condition (6–8). However, resuscitation or reperfusions of blood to the ischemic areas do not give satisfactory results. Resuscitation of blood flow to the ischemic tissue after prolonged ischemia further potentiates the degree of ischemic insult [9]. This gives rise to a secondary illness termed Ischemic reperfusion injury (IRI). IRI progresses in two stages, i.e., ischemic stage and the reperfusion stage. The ischemic stage comprises electrolyte imbalance and ATP depletion, whereas the reperfusion stage comprises activation of various oxidative stress systems like NADPH oxidase and xanthine oxidase leads to mitochondrial disruption [3]. This tends to increase ROS production and contributes to the activation of various cell death mechanisms. This includes autophagy, apoptosis, necrosis, and necroptosis [10]. Thus ischemia followed by reperfusion is the major challenge for the physician to restore the ischemia in patients. However, there are various pathologies linked to IRI that needs to be explored. Thus in this chapter, we will try to cross-link all the possible mechanisms involves in the progression of Ischemic reperfusion injury which helps in the development of a more precise strategy to manage ischemic conditions in patients.

4.2 MECHANISTIC CONCEPTS OF ISCHEMIC REPERFUSION INJURY

There are various mechanisms involved in the progression of ischemic insult of tissues followed by reperfusion or reoxygenation. Hypoxia followed by reperfusion potentiates the ischemic insult by activating various deleterious pathways like mitochondrial dysfunction, calcium overload, excitotoxicity, and production of reactive oxygen species, etc. [11]. Hypoxia enhances anaerobic respiration and lowers ATP production. In anaerobic condition, lower ATP and pH promotes the stabilization of HIF-α. The HIF-α stabilization promotes oxygen-independent respiration and metabolism, thus regulating the progression of ischemia. The HIF-α stabilization promotes the upregulation of various cell survival signaling and promotes the autophagy-mediated removal of damaged cells [12]. Moreover, the ischemic stage has been well-known for electrolyte imbalance. Ischemic conditions reverse the activities of the Na^+-K^+ ATPase pump and Ca^{2+}-ATPase pump. The decreased pH reverses the Na^+-Ca^{2+} exchanger pump (NCX) activity. This ultimately leads to the excessive intracellular release of Ca^{2+} ions from the L-type voltage-gated ion channel. The excessive release of Ca^{2+} ions increases the osmolarity of cells leads and promotes cell swelling [13]. Excessive calcium release leads to the activation of calpains. Calpain activation offers the rupture of the sarcolemmal wall and further promotes the release of calcium ions, thus promoting calcium overload [14]. On the other hand, excessive calcium ion raises the mitochondrial membrane potential resulting in mitochondrial depolarization. This promotes the opening of mitochondrial permeable transition pore (mPTP) [15]. The mPTP leaks out cytochrome c which binds to apoptotic protease factor 1 (Apaf1) and promotes the activation of various caspases. This leads to the promotion of transcription of pro-apoptotic factors and reduces the production of anti-apoptotic factors. This offers the upregulation of apoptosis-mediated cell death [16].

The prime focus of reperfusion is to restore blood flow to the ischemic tissues and promotes the production of ATP and restores the cellular pH levels. Reperfusion promotes aerobic respiration and promotes metabolic activities [17]. Early-stage reperfusion somewhere contributes to the upregulation of cell survival signalings but later on, reperfusion enhances the production of reactive oxygen species and contributes to the oxidative burden. Enhanced metabolic activities during reperfusion promote the production of reactive oxygen species as an end product. Thus reperfusion stage is more known to increase the production of ROS and oxidative burden. This further increases lipid peroxidation and reduces the release of various pro-oxidants [18]. Apart from this, reperfusion promotes the release of pro-inflammatory mediators in response to tissue repair. This leads to the upregulation of various deleterious pathways and promotes the gene transcription of pro-inflammatory cytokines. Elevation in the secretion of pro-inflammatory markers upregulates the expression of necroptosis-mediated cell death [19]. Moreover, the elevated level of ROS and pro-inflammatory cytokines act as a substrate for the upregulation of various oxidative systems like NADPH oxidase, and Xanthine oxidase leading to the disruption in the mitochondrial transport chain. The activation of the oxidative stress system consequently offers the activation of caspases, an essential enzyme required for the activation of various cell death mechanisms like apoptosis [20]. Thus reperfusion promotes the severity of ischemic insult and deteriorates the condition of the patient. This mainly happens due to the activation of various cell death mechanisms like autophagy, apoptosis, necrosis, necroptosis, etc., as well depicted in Figure 4.1.

4.3 OXIDATIVE STRESS SYSTEM INVOLVED IN ISCHEMIC REPERFUSION INJURY

Oxidative stress is described as an imbalance between the activity of pro-oxidants and anti-oxidants leading to the generation of reactive oxygen species (ROS) [21]. Prolonged hypoxia enhances oxidative burden and induces reversible cell injury. This condition can be restored by correcting blood flow to the ischemic areas. During reperfusion, ROS produced as metabolic end products of enhanced aerobic activity. This further activates various oxidative stress systems like the NADPH oxidase system, and Xanthine oxidase system leading to the disruption in the mitochondrial transport chain [20]. This consequently promotes the release of free radicals and promotes the production of reactive oxygen species. This is known as ROS-induced ROS release (RIRR) [22] is well explained in the next section.

4.3.1 NADPH OXIDASE SYSTEM

Nicotinamide adenine dinucleotide phosphate (NADPH) oxidase (NOX) is a membrane-bound enzyme of the NOX family. This family comprises about 7 members of two categories, i.e., NOX1-NOX5 and dual oxygenase (DUOX) 1&2. All the members of the NOX family possess the same structural and functional features [23]. Structurally NOX consists of six transmembrane domains, a flavin adenine dinucleotide (FAD), and NADPH binding cytosolic domain. NOX is highly found in the colon, phagocytes, inner ear, lungs, kidney, lymphoid tissues, and thyroid tissues. NOX is physiologically present in the inactive form of naive monomer. All members of the NOX family are dependent on interacting proteins for their stabilization, maturation, heme incorporation, and their translocation [24]. Different NOX interacts with different interacting proteins like NOX1-3 interacts with p22phox transmembrane protein along with p47phox/NOXO1, activator subunits like NOXA1/p40phox, p67phox/NOXA2, and the G-protein mainly Rac. Whereas, NOX4 only requires p22phox for its activity, while NOX5, DUOX1, and DUOX2 activation depends on direct Ca^{2+} binding [25]. All NOXs possess catalytic activity whose main function is to transfer electrons from NADPH to molecular oxygen from their FAD and heme prosthetic domain. Thus their prime action is to produce superoxide ions and release hydrogen peroxide, thus producing a little number of reactive oxygen species (ROS) [26]. About 90% of ROS is generated under aerobic respiration to produce ATP can be internalized into mitochondria [27].

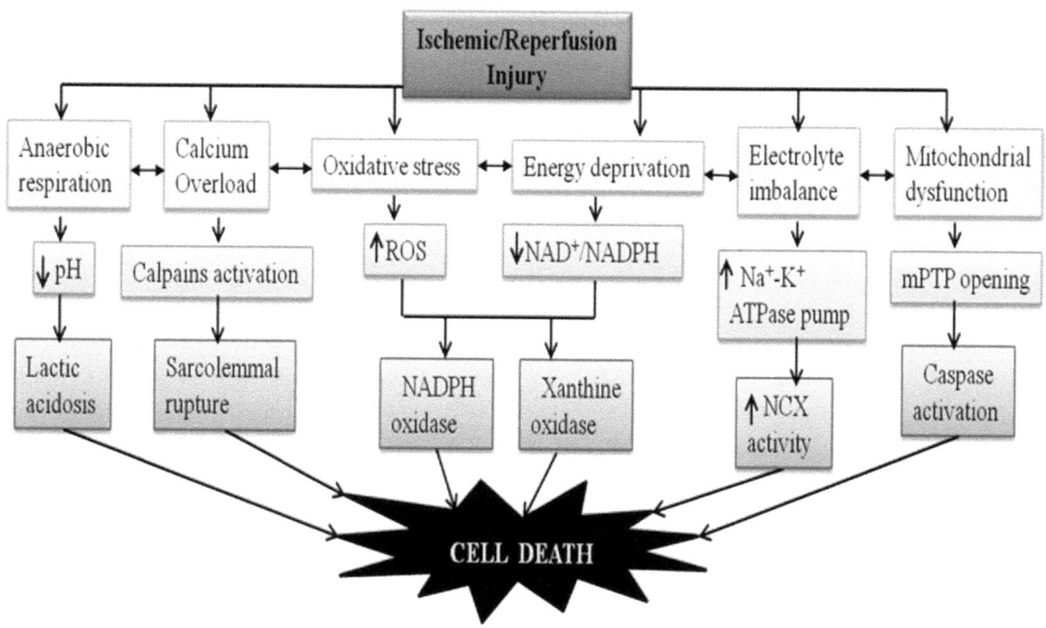

FIGURE 4.1 Mechanism of ischemic reperfusion injury.

It has been recently observed that NOX elevates oxidative burden under increased tissue oxygen demand. Thus NOX acts as a major intermediate to produce ROS under ischemic/reperfusion conditions [28]. Ischemia reverses the activity of NCX and potentiates the extracellular release of calcium ions. NOX system activation requires excessive calcium inside the cells. Ischemic condition promotes the stabilization of hypoxic inhibitory factor-1α (HIF-1α) under anaerobic conditions to promote oxygen-independent respiration [29]. Whereas, HIF-1α stabilization upregulates the activity of NOX [30]. This leads to the production of ROS and initiates ROS-mediated signalings. ROS produced during ischemia activates p38/MAPK/JNK pathways whose activation leads to translocation of NF-κB from cytosol to nucleus and promotes the gene transcription. This further raises the level of pro-inflammatory mediators. On the other hand, upregulation of the p38/MAPKJNK pathway promotes lipid peroxidation and raises pro-oxidants level [19]. This signaling further downregulates the production of anti-oxidants thus reducing the anti-oxidant signaling and raising the oxidative burden [31].

Apart from this, NOX activation is mainly linked to the reperfusion stage. Reperfusion enhances aerobic respiration and promotes ATP production. This enhances metabolic activity and produces reactive oxygen species as metabolic end products. This ROS promotes the ROS-mediated signalings and further promotes the production of ROS. This is known as ROS-induced ROS release (RIRR). This RIRR can upregulate the activity of NADPH oxidase and increase oxidative burden [22]. Further, reperfusion promotes the cell survival signalings in its early stage and activates PI3K/Akt mediated signaling. But activation of this signaling works as protective for a short period and gives detrimental effects later on. Reperfusion promotes cell proliferation and angiogenesis by the release of angiotensin II. Angiotensin II promotes the release of vesicular endothelial growth factor (VEGF) [32]. Elevation in VEGF expression enhances the NOX-mediated oxidative stress. Moreover, reperfusion enhances the production of arachidonic acid (AA) in response to tissue repair. Metabolism of AA is mainly taken up by cyclooxygenase (COX) and lipoxygenase (LOX). Upregulation of COX and LOX enzymes was best observed in the myocardial I/R injury model of rats. These enzymes are highly known for the production of eicosanoids, prostaglandin, leukotrienes, thromboxane, etc. [33]. This can act as pro-inflammatory markers and leads to the production of reactive oxygen species. Pro-inflammatory markers and ROS can act as a substrate for the activation of NADPH oxidase. NADPH oxidase is the major source of ROS production during reperfusion [34]. On the other side, NADPH oxidase activation leads to phenotypic changes in microglial activity. This results in the activation of microglia which promotes phagocytosis [35]. This ultimately leads to cell death as depicted in Figure 4.2. Apart from the NADPH oxidase, Xanthine oxidase and disruption in mitochondrial electron transport chain also contributes the oxidative stress and promote cell death is clearly will be explained in the latter sections. Thus it indicates that reperfusion rather promotes ischemia and further deteriorates the condition of the patient.

Ischemia and reperfusion promote the activation of the NADPH oxidase system (NOX). NOX can upregulate the release of ROS/RNS and give rise to oxidative burden. ROS further activates the p38/MAPK/JNK pathway which leads to the translocation of NF-κB. This promotes the release of pro-inflammatory mediators. On the other side, ROS promotes the metabolism of AA with the help of cyclooxygenase and lipoxygenase. This further promotes the release of pro-oxidants. Upregulation of pro-inflammatory mediators and pro-oxidants promotes the activation of the cell death mechanism.

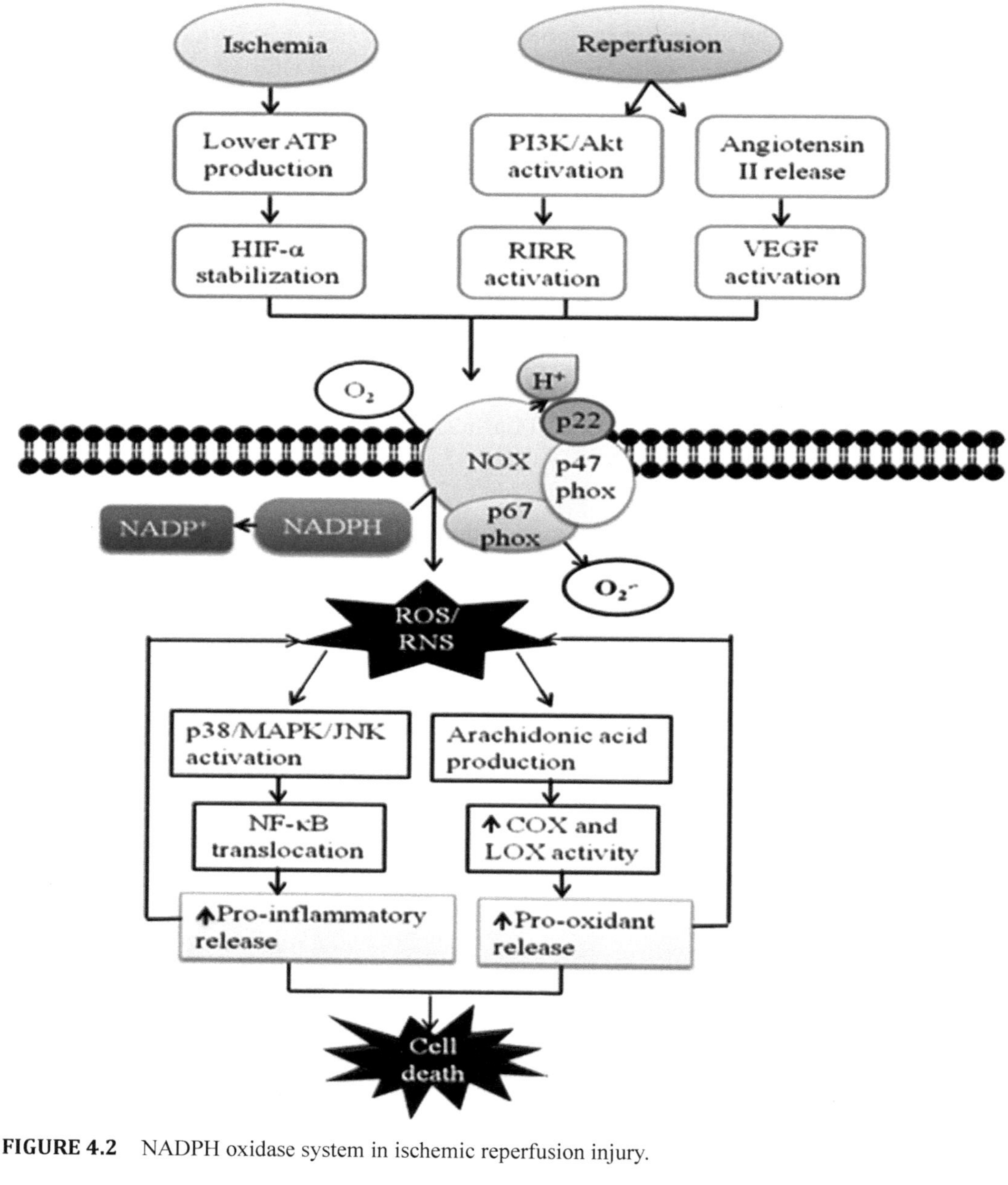

FIGURE 4.2 NADPH oxidase system in ischemic reperfusion injury.

4.3.2 XANTHINE OXIDASE SYSTEM

Xanthine oxidase system (XOS) is composed of an oxidoreductase enzyme belonging to the metaloflavioid family. The Xanthine oxidase system includes two isoforms of enzymes, i.e., Xanthine oxidase (XO) and Xanthine dehydrogenase (XDH). Xanthine oxidase is of 290 kD molecular mass and contains one molybdenum-containing molybdopterin cofactor

(Mo-co), flavin adenine dinucleotide (FAD), and four iron-sulfur attachment site domains [36]. XOS is mainly involved in the metabolism of purines XDH performs the oxidation of hypoxanthine and converts hypoxanthine to xanthine whereas xanthine oxidase converts xanthine to uric acid. The purine oxidation begins at the Mo-co site whereas FAD is oxidized to nicotinamide adenine dinucleotide (NAD^+) and O_2 at its reduction sites. XOS is mainly responsible for the reduction of NAD^+ to NADPH and molecular oxygen [37]. Their conversion releases superoxide ions and hydrogen peroxide, the major substrate for ROS signaling. Thus XOS is a major source of physiological ROS generation after NADPH oxidase [38]. Under physiological conditions, xanthine dehydrogenase uses NAD^+ as an electron acceptor, whereas xanthine oxidase uses O_2 as the final electron acceptor and produces superoxide ions and hydrogen peroxide ions [39]. But these superoxide ions and hydrogen peroxide ions give rise to oxidative stress under hypoxic conditions.

Ischemic condition results in lower pH and ATP production. This leads to the conversion of XDH to XO [40]. XO is the major site for the binding of xanthine but under a lower pH level, reversible competition takes place between nitric oxide and xanthine for the binding to xanthine oxidoreductase at the Mo-co site. Thus hypoxic conditions lead to the production of reactive nitrite species (RNS) mainly peroxynitrite [41]. The RNS further aids in the production of superoxide ions and hydrogen peroxide. This begins ROS-mediated signaling and deteriorates the condition of the patient. Ischemia gives rise to nutrient-deficient conditions and prevents the inhibition of the autophagy complex by inhibiting the mTOR complex. The mTOR complex activation promotes the phosphorylation of uncoordinated-51-like kinase 1 (ULK1) complex and autophagy-related proteins (Atg) mainly Atg13. Thus ischemic condition promotes the activation of autophagy by upregulation of XOS in response to the promotion the cell survival [42].

Reperfusion leads to the recovery of the blood flow to the ischemic tissues and promotes energy production. Reperfusion increases the pH level and ATP production which reverses the conversion of xanthine dehydrogenase to xanthine oxidase [43]. Their conversion further leads to the reduction of hypoxanthine to xanthine and further xanthine to uric acid. This release superoxide ions and hydrogen peroxide as an intermediate product. Both ions act as a substrate for the activation of the xanthine oxidase system [39]. XO acts as an upstream regulator of PI3K/Akt/mTOR signaling. XO is involved in the production of ROS which promotes the oxidation of the repressor molecule of PI3K, i.e., PTEN (Phosphatase and tensin homolog). PTEN is responsible for the dephosphorylation of PI3K and converts PIP3 to PIP2, thus downregulating the PI3K-mediated cell survival signaling. But XO potentiates the oxidation of PTEN and upregulates PI3K/Akt signaling [44]. PI3K/Akt works by the inhibition of the mTOR complex and promotes autophagy.

Moreover, reperfusion leads to the release of local inflammatory mediators which acts as an upstream regulator for the xanthine oxidase system. Xanthine further binds to the toll-like receptors (TLR) mainly TLR4. The activation of TLR4 promotes the formation of the inflammasome by the activation of NLRP3 [45]. Apart from this, ROS produced during the reperfusion stage promotes upregulation of the JNK pathway. JNK activation promotes the phosphorylation of NLRP3, essential for the deubiquitination of NLRP3. NLRP3 activation enhances the activity of caspase1. Caspase1 can act as intermediate for the activation of interleukin 1β (IL-1β). This caspase activation and IL-1β activation act as pro-apoptotic agents and promotes apoptotic mediated cell death as indicated in Figure 4.3. Moreover, XOS inhibition mainly by allopurinol prevents the activation of NLRP3 and caspase1, thus reducing the chances of apoptotic cell death [46]. Moreover, pretreatment with febuxostat, a xanthine oxidase inhibitor provides cardioprotection to ischemic reperfusion injury by inhibiting ROS production in the I/R model of mice [47]. XOS produced further activates its various substrates like adhesion proteins, angiotensin II, cytokine cascade, and apoptotic events [48]. All this ultimately leads to promotes inflammation and apoptosis ultimately leads to cell death.

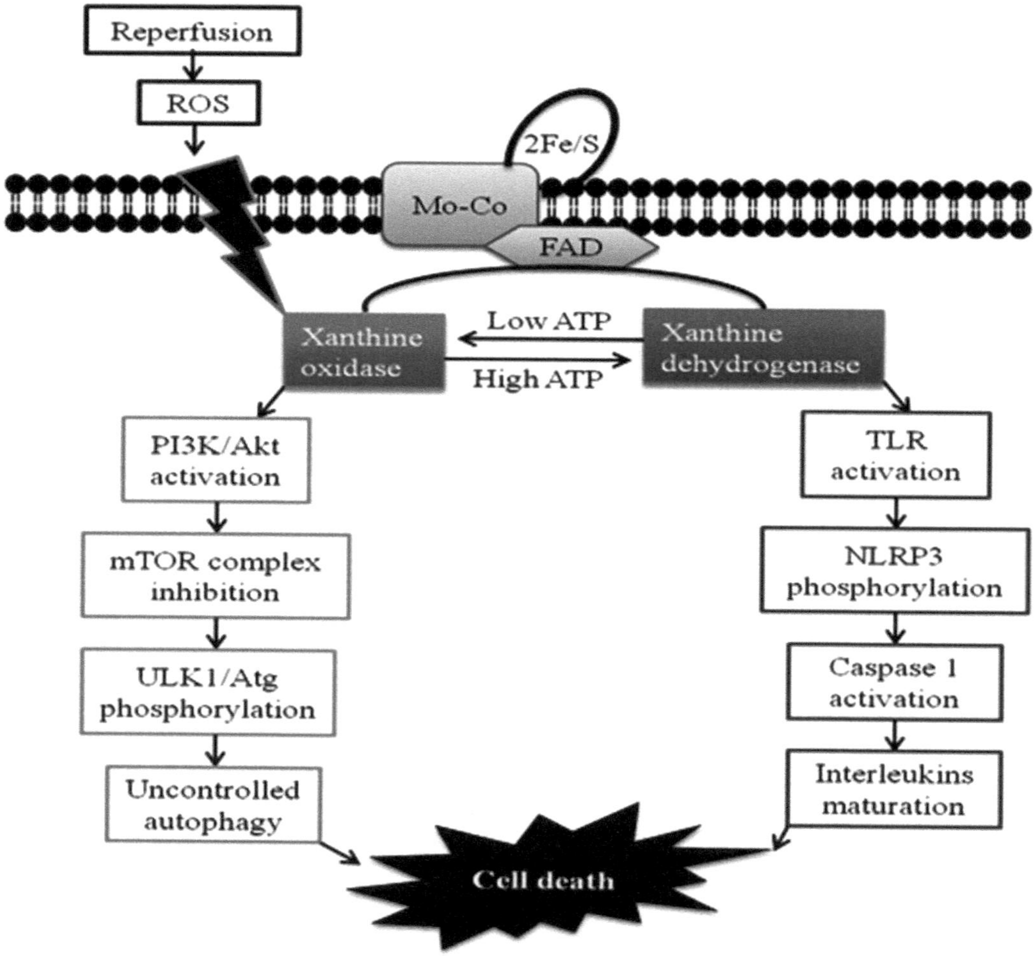

FIGURE 4.3 Role of xanthine oxidase system in IRI.

The Xanthine oxidase system (XOS) is mainly activated under reperfusion conditions. XOS upregulates the PI3K/Akt pathway under the reperfusion stage and inhibits the mTOR complex. Inhibition of the mTOR complex promotes the phosphorylation of ULK1/Atg and results in uncontrolled autophagy. On the other side, reperfusion promotes ROS generation and leads to the activation of TLRs. Their activation promotes the phosphorylation of NLRP3. NLRP3 contributes to the activation of caspase 1 and promotes interleukin maturation. They all together acts as pro-apoptotic agents and promote cell death.

4.3.3 MITOCHONDRIAL ELECTRON TRANSPORT CHAIN

The mitochondrial electron transport chain (ETC) is the main site for oxidative phosphorylation in mammals which produces more than 90% of total ATP [49]. ETC runs in the inner mitochondrial membrane which uses intermediate products formed during the tricarboxylic acid cycle (TCA) and glycolysis via reduction/oxidation for the formation of water and energy [50]. ETC produces an electrochemical gradient by a series of a redox reactions to produce ATP. ETC is mainly composed of 4 complexes, i.e., Complex I (NADPH dehydrogenase), Complex II (Succinate dehydrogenase), Complex III (Cytochrome c reductase), and Complex IV (Cytochrome c oxidase) with two additional electrons transportes, i.e., cytochrome c and ubiquinone [51]. Electrons in the ETC can be entered through matrix and intermembrane via NADPH and $FADH_2$ [52].

- NADH; $NAD^+ + H^+ + 2e^- \leftrightarrow NADH$;
- $FADH_2$; $FAD^+ + 2H^+ + 2e^- \leftrightarrow FADH_2$.

Both the metabolites act as a final acceptor for oxygen and reduce it in the water and produce ATP [53]. Ischemia can shift aerobic respiration to anaerobic respiration. This enhances the proton motive forces and impairs the activity of complex I which leads to the accumulation of ubiquinone ($CoQH_2$) [54]. Increased proton motive forces enhance the reverse flow of electrons resulting in reverse electron transport from complex II to complex I. Moreover, the anaerobic condition prevents the reversible oxido-reduction of succinate to fumarate. Ischemia promotes the catalysis of succinate dehydrogenase which leads to the inhibition of oxido-reduction of succinate to fumarate, resulting in the accumulation of succinic acid. Succinic acid accumulation promotes the generation of ROS via reverse electron transport (RET) flow from complex I [55]. This promotes the accumulation of succinic acid and impairs the activity of the mitochondrial transport chain. Ischemia also reduces the NAD^+/NADPH ratios major substrate for the upregulation of PGC-1α/SIRT1 signaling [56]. Ischemia reduces the mitochondrial DNA transcription via reducing PGC-1α/SIRT1 and reduces the mitochondrial biogenesis [57]. Apart from this, reduction in mitochondrial DNA transcription affects Nrf2/Keap1 signaling and reduces the production of anti-oxidants. Nrf2/Keap1 signaling is mainly responsible for gene transcription. But, ischemia results in the downregulation of PI3K/Akt signaling which produces the signal for dissociation of the Nrf2/Keap1 complex [58]. Dissociation of Nrf2/Keap1 promotes the translocation of Nrf2 from the cytosol to the nucleus. Nrf2 then binds to the antioxidant response element (ARE) and promotes the gene transcription of various anti-oxidants like heme oxygenase 1 (HO-1), glutathione, catalase, etc. But ischemic conditions promote the generation of ROS which inhibits the dissociation of the Nrf2/Keap1 complex and promotes the proteasomal degradation of Nrf2 [59]. Moreover, reduction in Nrf2/Keap1 signaling is linked with the higher activity of caspases. Thus this contributes to the oxidative burden and promotes the various cell death mechanisms.

Reperfusion corrects the electrolyte imbalance and promotes the production of ATP. This rapidly oxidized accumulated succinate by succinate dehydrogenase, which burst a large number of superoxide ions. Reperfusion promotes the forward electron transport (FET) flow from complex I to complex IV within seconds of reoxygenation and normalize the complex I activity [60]. This reduces the $CoQH_2$ pool and lowers Δp to generate ATPs. Reperfusion promotes ATP production and promotes the generation of superoxide ions [61]. The superoxide ions further react with nitric oxide and produce reactive nitrite species (RNS) mainly peroxynitrite ($ONOO^-$) [62]. Thus reperfusion somewhere corrects the electrolyte imbalance but also promotes the excessive production of reactive oxygen species in response to enhanced metabolic activities as depicted in Figure 4.4. The excessive ROS production further activates various deleterious pathways and contributes to the activation of cell death mechanisms have been explained is next sections.

4.4 CELL DEATH MECHANISM INVOLVED IN ISCHEMIC REPERFUSION INJURY

Ischemic condition results in lower ATP production and higher nutrient demand. Ischemia promotes the phosphorylation of AMPK which acts as a sensor for nutrient deprivation. This leads to activation of autophagy and autophagic mediated clearance of damaged organelles. Thus autophagic activation during the initial phase of ischemia works as a protective pathway [5]. But, reperfusion enhances the production of reactive oxygen species which tends to activate various cell death mechanisms like apoptosis, mitoptosis, necrosis, and necroptosis [63]. Reperfusion also enhances the expression of autophagy which leads to disruption of healthy cells too [64]. Thus IRI deteriorates the condition of the patient by the activation of various cell death mechanisms which have been explained in the next section.

Figure 4.4 shows the role of the mitochondrial transport chain in ischemic reperfusion injury. Ischemia leads to the accumulation of succinic acid which reverses the flow of electrons from complex II to complex I. This increases the production of free radicals and generates superoxide ions. This reduces the activity of PGC-1α/SIRt1 signaling and impairs mitochondrial biogenesis. Reperfusion decreases the proton motive forces (Δp) and promotes the forward flow of electrons by oxidizing succinic acid. This results in a burst of superoxide ions which contributes to the upregulation of pro-oxidants. This contributes to the oxidative burden and promotes cell death.

4.4.1 AUTOPHAGY

Autophagy is defined as a conserved biological process responsible for the degradation of damaged organelles and macromolecules from the cytoplasm to the lysosome [65]. Autophagy is activated under stressful conditions that upregulate the expression of cell signaling pathways. Ischemic condition enhances the AMP/ATP ratio to activate AMPK. AMPK act as a nutrient sensor and activates under nutrient-deprived condition [66]. Moreover, Ischemic insults lead to the inactivation of the mTOR (mammalian target of rapamycin) pathway and activate AMPK [67]. Hypoxic conditions enhance the phosphorylation of AMPK at its Thr172 site. The mTOR negatively regulates autophagy by inhibiting phosphorylation of the ULK1 complex, whereas AMPK promotes the autophagy counteracting the mTOR activity (68, 69). ULK1 complex consists of Atg101, Atg13, Fip200, and Ulk1, phosphorylates Beclin1 and AMBRA1 (the activating molecule in Beclin1 regulated autophagy) [70].

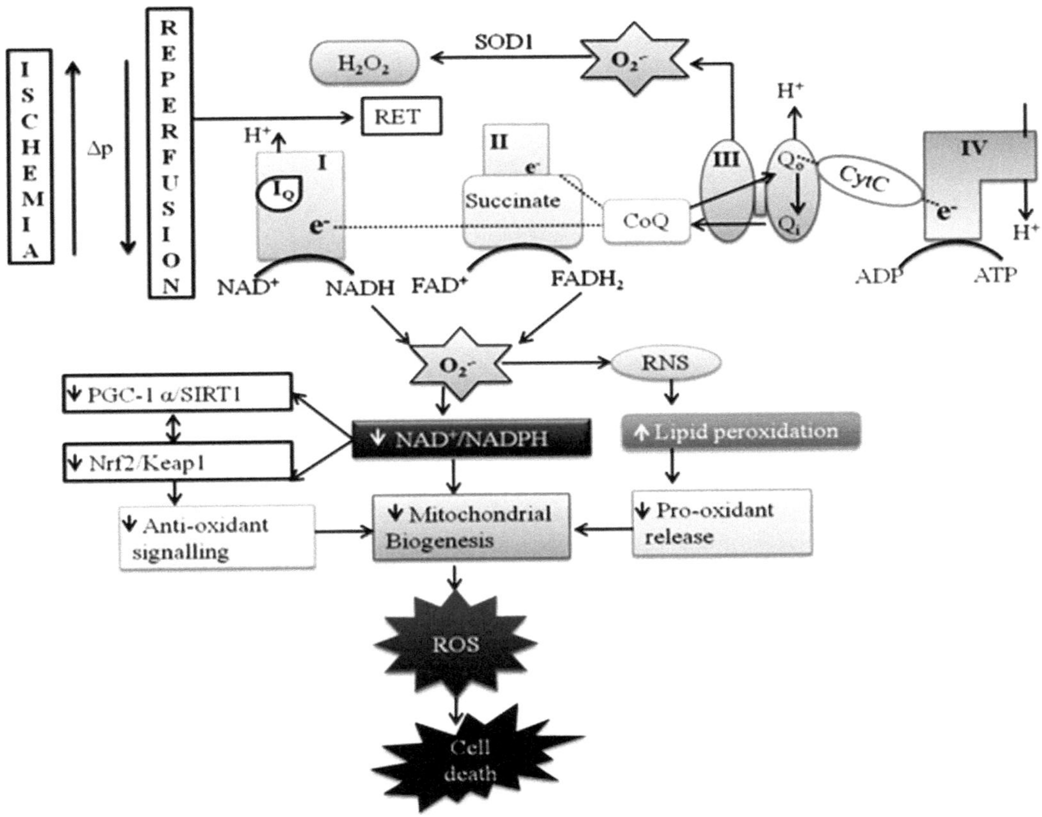

FIGURE 4.4 Mitochondrial electron transport chain in IRI.

Ischemia also leads to electrolyte imbalance which increases the extracellular release of calcium ions. Free cytosolic calcium ions upregulate the expression of AMPK and promote autophagic mediated clearance of damaged organelles [71]. Autophagic mediated degradation releases several amino acids and fatty acids which are further recycled by the tricarboxylic acid cycle and generates ATP. Ischemia leads to the stabilization of HIF-1α and promotes oxygen-independent autophagy. Ischemia enhances the expression of BNIP3 and promotes autophagy [72]. HIF-1α activation contributes to the activation of BNIP3 and NIX whereas hypoxia promotes the dephosphorylation of FUNDC1 and promotes the LC3-mediated formation of autophagosomes. Moreover, HIF-1α mediated activation of autophagy requires BNIP3, Atg's, and Beclin1 was well observed in SH-SY5Y neuroblastoma cells [73]. Add on to this, HIF-1α/BNIP3 upregulation also promotes autophagy in I/R and OGD/R was well established in H9C2 cells and potentiates the cell survival signaling [74]. Thus it can be suggested that autophagy activation during ischemia provides a protective role whereas its detrimental effects have been seen during the reperfusion stage.

The reperfusion stage leads to the inactivation of AMPK due to enhanced nutrient supply to the ischemic tissues. Thus it is clear that activation of autophagy during reperfusion is independent of AMPK activation. Reperfusion promotes the PI3K/Akt mediated cell signaling pathways. PI3K/Akt activation promotes the activation of beclin1 and reduces the expression of pro-apoptotic factors. Furthermore, autophagic mediated cell death during the reperfusion stage is highly linked with the over-activity of Beclin1. Beclin1 promotes the phosphorylation of the LC3II complex which initiates the formation of autophagosomes [75]. Moreover, enhanced pH during the reperfusion stage also contributes to the maturation of autophagosomes, thus leading to uncontrolled autophagy. Reperfusion enhances ROS generation and oxidative stress which acts as a signal for autophagy [76]. Reperfusion enhances cell proliferation and angiogenesis by the release of angiotensin II. Angiotensin II also promotes the formation of autophagosomes by promoting the activity of Atg12-Atg5 [77]. Apart from this, ROS signaling increases the release of superoxide ions and hydrogen peroxide ions which potentiates the activation of Beclin1 and leads to uncontrolled autophagy as well depicted in Figure 4.5. Simultaneously, ROS reduces the expression of anti-apoptotic factors and promotes the release of pro-apoptotic factors [78]. Their upregulation leads to the activation of the apoptotic cell death mechanism which contributes to irreversible cell death has been explained in the next section.

Autophagy plays a dual role in ischemic reperfusion injury. During ischemia, autophagy works as protective signaling. Ischemia promotes the stabilization of HIF-1α which promotes the phosphorylation of AMPK and activates autophagy. Reperfusion leads to the activation of Beclin1 which leads to uncontrolled autophagy and results in cell death.

4.4.2 APOPTOSIS/MITOPTOSIS

Apoptosis is defined as a programed cell death mechanism. Apoptosis is generally activated under stressful conditions to promote the cell death of damaged tissues or organelles. Prolonged ischemic conditions reduce the expression of anti-apoptotic factors and promote the release of pro-apoptotic factors, thus promoting apoptosis-mediated cell death [79]. Based on initiating factors, apoptosis is classified into two subclasses, i.e., intrinsic apoptosis and extrinsic apoptosis [80]. Both the pathways work differently but ultimately lead to the activation of caspase3, which leads to apoptosis. The intrinsic apoptotic pathway is known as the mitochondrial cell death mechanism because this pathway works by the changes in the integrity of the mitochondrial membrane and its permeability. The excessive release of calcium ions promotes the activation of Calpains. Calpains further promote the oligomerization of Bax, a pro-apoptotic factor [81]. The elevated levels of pro-apoptotic factors enhance the release of cytochrome c and promote apoptosis. Moreover, enhanced calcium release increases the mitochondrial membrane permeability leading to activation of mitochondrial permeable transition pores (mPTP) [82]. Prolonged opening of mPTP leaks out cytochrome c. This cytochrome c binds to the apoptotic protease activating factor 1 (Apaf1) [83]. The former formed complex further activates the inactive caspases. Firstly, the formed complex recruits the active form of caspase9 from its naïve forms, i.e., procaspase9. Caspase9 is crucial for the activation of caspase3 and converts procaspase3 to caspase3. Activation of the caspase3 is the final step in the apoptosis activation [84]. Moreover, excessive release of cytochrome c leads to impaired mitochondrial DNA activity. Thus it reduces the mitochondrial transcription and decreases mitochondrial biogenesis.

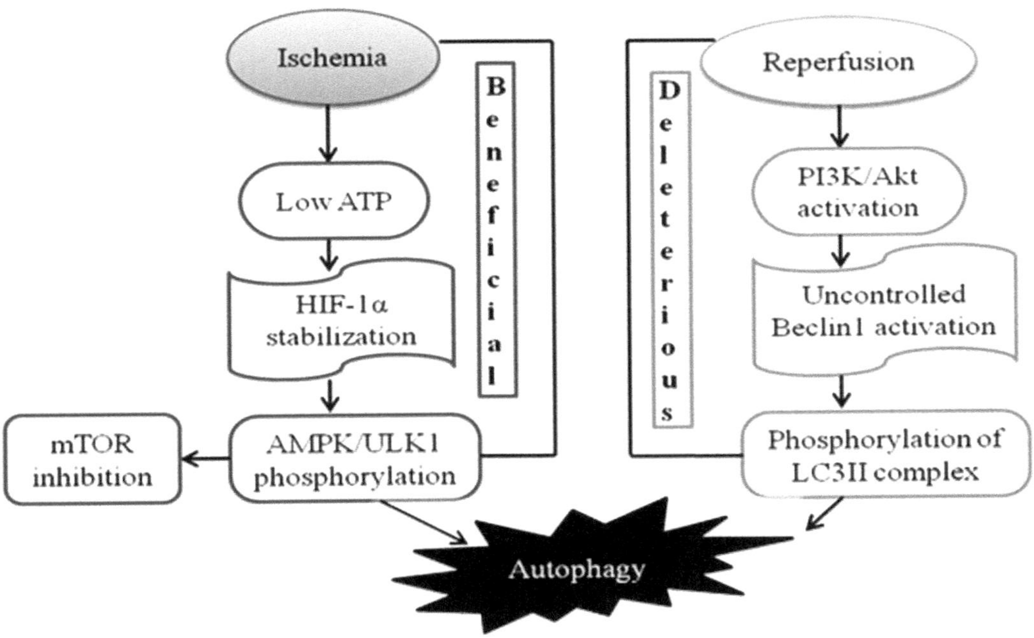

FIGURE 4.5 Autophagy contribution in IRI.

The extrinsic pathway is known as the death receptor pathway. This pathway is activated by various ligands like Fas ligands, and TNF-α binds to the death ligand domain of various receptors [85]. These receptors are tumor necrosis factor-related apoptosis-inducing ligand receptor 1 (TRAIL R-1/2), tumor necrosis factor receptors 1&2 (TNFR-1/2), cytokine receptors, etc. [86]. As discussed earlier, reperfusion enhances the activity of various oxidative systems which tends to increase oxidative stress and leads to ROS production. This ultimately leads to the initiation of ROS signaling. ROS signaling further upregulates the expression of the p38/MAPK/JNK pathway. This pathway promotes the translocation of NF-κB and promotes gene transcription. This leads to the upregulation of TNF-α, IL-1β, and other cytokines. This TNF-α can bind to the death domain of various receptors like TRAIL R-1/2 and TNFR-1/2. Their binding promotes the conversion of procaspase8 to caspase8. Caspase8 further acts as an intermediate for the activation of caspase3 and promotes the extrinsic apoptotic cell death mechanism [87]. Reperfusion enhances the PI3K/Akt mediated cell survival signaling. But this signaling only gives positive effects during the initial phase of reperfusion. Further, PI3K/Akt activation acts as an early mediator for the release of pro-inflammatory mediators and pro-apoptotic factors [33]. This elevates the release of various pro-apoptotic factors like Bax, Bak, PUMA, etc. [16]. It has been observed in the myocardial I/R mice model that, ischemia can be prevented by inhibiting the intrinsic apoptotic pathway but this effect cannot be achieved by the inhibition of the extrinsic apoptotic mechanism. Apart from this, mitochondria also contribute to the cell death mechanism termed mitoptosis [88]. Bax/Bak regulates mitochondrial fission by interacting with Drp1, an essential protein for mitochondrial fission located in the outer mitochondrial membrane [89]. Drp1 is required for the oligomerization of Bax/Bak to the mitochondrial membrane which increases mitochondrial outer membrane permeability and enhances the release of cytochrome c as depicted in Figure 4.6 [90]. Therefore, inhibition of Drp1 prevents mitochondrial fission at the very beginning of the reperfusion [91]. Thus apoptosis and mitoptosis processes are a topic of cross-talk and influence the activation of each other.

4.4.3 NECROSIS/NECROPTOSIS

Necrosis is defined as an unregulated cell death mechanism initiates in response to excessive external stress. Necrosis may be initiated by changes in physical, chemical, and biological processes including mitochondrial dysfunction, organ swelling, and cell disintegration [92]. Necrosis is thought to be uncontrolled and unregulated whereas necroptosis is a controlled programmed cell death mechanism that works without the caspase activation [93]. Moreover, Caspase activation can inhibit necroptosis and activate apoptosis. Thus necroptosis and apoptosis negatively regulate each other [94]. Necroptosis activation requires kinase receptor-interacting protein 1 (RIPK1), and RIPK3 [95]. Necroptosis is mainly observed in response to the excessive release of inflammatory mediators like TNF-α and inhibition of caspase activity. TNF-α binds to its receptor TNFR and promotes the dimerization of RIPK1 and RIPK3 [96]. The formed dimer can phosphorylate mixed lineage kinase domain-like protein (MLKL) and activate peptidyl arginine-induced necroptosis [97]. The necroptosis-mediated cell death can be observed during the hypoxic condition. During reperfusion, caspase activity increases in post-ischemic tissue due to reperfusion which favors the condition to apoptotic cell death mechanism. Caspase activation can hydrolyze RIP1, thus inhibiting the dimerization of RIP1 and RIP3 required for the formation of necrosomes [98]. Thus apoptosis is described as the preferred cell death mechanism during reperfusion to ischemic tissues whereas inhibition of necroptosis may work as a therapeutic target for preventing inflammatory activities during ischemic tissue injury [99].

Ischemic reperfusion injury promotes apoptosis-mediated cell death. IRI promotes the release of cytochrome c by prolonging the opening of mPTP. IRI can also upregulate the p38/MAPK/JNK pathway which leads to the translocation of NF-κB. NF-κB translocation promotes the activation of the TNF-α receptor. On the other side, IRI results in upregulation of Drp1 expression which promotes mitochondrial fission. All this together enhances the Bax/Bak ratio and promotes the activation of caspases. Caspase activation is the final step in the activation of apoptosis.

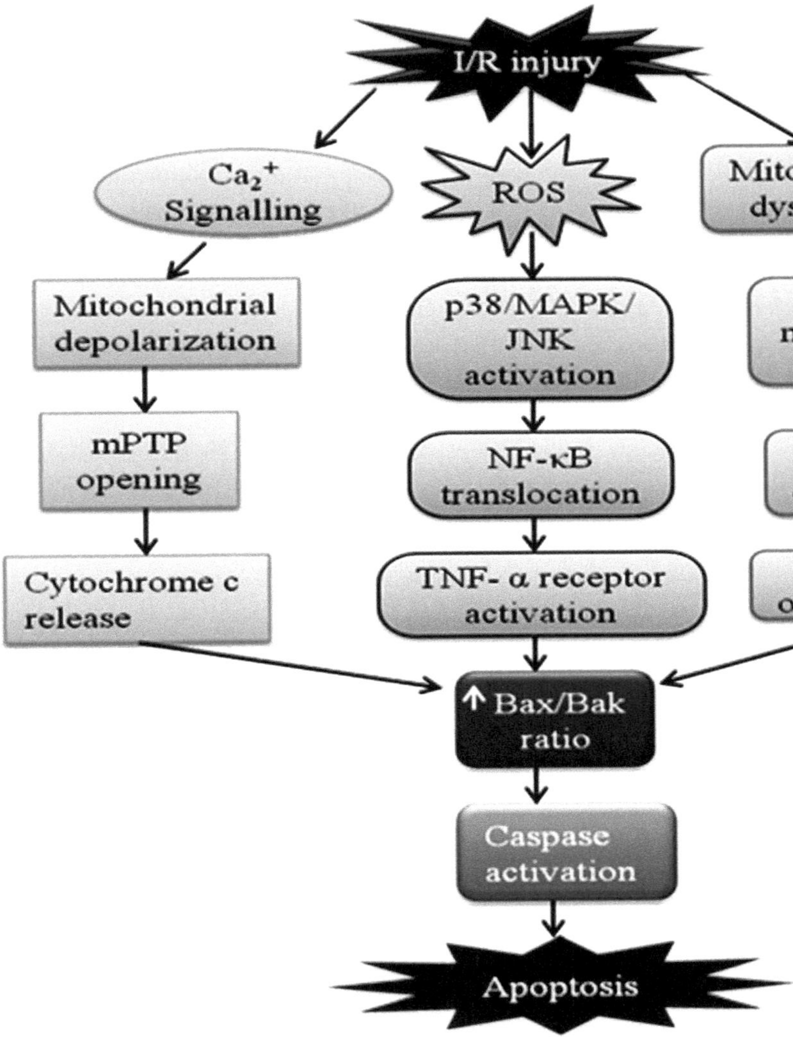

FIGURE 4.6 Apoptosis-mediated cell death in IRI.

4.5 CONCLUSION

Ischemic/Reperfusion Injury is a multifactorial condition that arises during the reinforcement of molecular oxygen to ischemic tissues. Reperfusion after prolonged ischemia leads to far worst conditions than untreated ischemia. This chapter provides basic mechanistic concepts of IRI which run around the activation of oxidative systems due to ROS generation which further leads to cell death. Ischemia shifts aerobic respiration to an anaerobic condition leading to electrolyte imbalance, decrease pH, impaired calcium signaling, mitochondrial dysfunction, and reduced ATP production. Further, reperfusion activates various oxidative systems like NADPH oxidase and Xanthine oxidase and produces more reactive oxygen species, and reduces anti-oxidant signaling. This triggers various cell death mechanisms like autophagy, apoptosis/mitoptosis, and necrosis/necroptosis. Apoptosis is the most studied cell death mechanism of IRI whereas and their activation is associated in a time-dependent manner. Nonetheless, the pathophysiology of IRI is very vast and requires more attention from researchers for its elucidation. A complete understanding of the pathophysiology of IRI has the potential to provide newer therapeutic targets for the development of newer strategies for the treatment of IRI.

KEYWORDS

- **apoptosis**
- **autophagy**
- **ischemia**
- **ischemic reperfusion injury**
- **necroptosis**
- **necrosis**
- **oxidative stress**

REFERENCES

1. Li, J., Rogers, N. M., & Hawthorne, W. J., (2021). Chapter 1 – ischemia-reperfusion injury. In: Orlando, G., & Keshavjee, S. (eds.), *Organ Repair and Regen* (pp. 1–42). Academic Press. Available from: https://www.sciencedirect.com/science/article/pii/B9780128194515000123 (accessed on 9 November 2023).
2. Saini, V., Guada, L., & Yavagal, D. R., (2021). Global epidemiology of stroke and access to acute ischemic stroke interventions. *Neurology [Internet], 97*(20 Supplement 2), S6 LP–S16. Available from: http://n.neurology.org/content/97/20_Supplement_2/S6.abstract (accessed on 9 November 2023).
3. Barthels, D., & Das, H., (2020). Current advances in ischemic stroke research and therapies. *Biochim. Biophys. Acta – Mol. Basis. Dis. [Internet], 1866*(4), 165260. Available from: https://www.sciencedirect.com/science/article/pii/S0925443918303478 (accessed on 9 November 2023).
4. Saklani, P., Khan, H., Gupta, S., Kaur, A., & Singh, T. G., (2022). Neuropeptides: Potential neuroprotective agents in ischemic injury. *Life Sciences, 288*, 120186.
5. Wu, M. Y., Yiang, G. T., Liao, W. T., Tsai, A. P. Y., Cheng, Y. L., Cheng, P. W., et al., (2018). Current mechanistic concepts in ischemia and reperfusion injury. *Cell Physiol. Biochem. [Internet]., 46*(4), 1650–1667. Available from: https://www.karger.com/DOI/10.1159/000489241 (accessed on 9 November 2023).
6. Levine, G. N., Bates, E. R., Blankenship, J. C., Bailey, S. R., Bittl, J. A., Cercek, B., et al., (2011). ACCF/AHA/SCAI guideline for percutaneous coronary intervention. A report of the American college of cardiology foundation/American heart association task force on practice guidelines and the society for cardiovascular angiography and interventions. *J. Am. Coll. Cardiol., 58*(24), e44–122.

7. Jones, A. E., & Puskarich, M. A., (2014). The surviving sepsis campaign guidelines 2012: Update for emergency physicians. *Ann. Emerg. Med., 63*(1), 35–47.
8. Rhodes, A., Evans, L. E., Alhazzani, W., Levy, M. M., Antonelli, M., Ferrer, R., et al., (2017). Surviving sepsis campaign: International guidelines for management of sepsis and septic shock: 2016. *Intensive Care Med., 43*(3), 304–377.
9. Naito, H., Nojima, T., Fujisaki, N., Tsukahara, K., Yamamoto, H., Yamada, T., et al., (2020). Therapeutic strategies for ischemia reperfusion injury in emergency medicine. *Acute Med. Surg., 7*(1), e501.
10. Villalpando-Rodriguez, G. E., & Gibson, S. B., (2021). Reactive oxygen species (ROS) regulates different types of cell death by acting as a rheostat. In: Tejero, J., (ed.), *Oxid. Med. Cell Longev. [Internet], 2021,* 9912436. Available from: Https://doi.org/10.1155/2021/9912436.
11. Sharma, V. K., Singh, T. G., & Mehta, V., (2021). Stressed mitochondria: A target to intrude Alzheimer's disease. *Mitochondrion, 59,* 48–57.
12. McNamara, P. J., & El-Khuffash, A., (2017). 71–oxygen transport and delivery. In: Polin, R. A., Abman, S. H., Rowitch, D. H., Benitz, W. E., & Fox, W. W. (eds.), *Fetal and Neonatal Physiology* (5th edn., pp. 724–737. e2). Elsevier. Available from: https://www.sciencedirect.com/science/article/pii/B9780323352147000718 (accessed on 9 November 2023).
13. Lumb, A. B., (2017). Chapter 22 – hypoxia. In: Lumb, A. B. (ed.), *Nunn's Applied Respiratory Physiology* (8th edn., pp. 327–334.e1). Elsevier. Available from: https://www.sciencedirect.com/science/article/pii/B9780702062940000228 (accessed on 9 November 2023).
14. Sutanto, H., Lyon, A., Lumens, J., Schotten, U., Dobrev, D., & Heijman, J., (2020). Cardiomyocyte calcium handling in health and disease: Insights from *in vitro* and in silico studies. *Prog. Biophys. Mol. Biol. [Internet], 157*, 54–75. Available from: https://www.sciencedirect.com/science/article/pii/S007961072030016X (accessed on 9 November 2023).
15. O-Uchi, J., Jhun, B. S., Mishra, J., & Sheu, S. S., (2018). 7 – organellar ion channels and transporters. In: Zipes, D. P., Jalife, J., & Stevenson, W. G. (eds.), *Cardiac Electrophysiology: From Cell to Bedside* (7th edn., pp. 66–79). Elsevier. Available from: https://www.sciencedirect.com/science/article/pii/B9780323447331000079 (accessed on 9 November 2023).
16. Redza-Dutordoir, M., & Averill-Bates, D. A., (2016). Activation of apoptosis signaling pathways by reactive oxygen species. *Biochim. Biophys. Acta–Mol. Cell Res. [Internet], 1863*(12), 2977–2992. Available from: https://www.sciencedirect.com/science/article/pii/S0167488916302324 (accessed on 9 November 2023).
17. Shen, L., Gan, Q., Yang, Y., Reis, C., Zhang, Z., Xu, S., et al., (2021). Mitophagy in cerebral ischemia and ischemia/reperfusion injury [Internet]. *Frontiers in Aging Neuroscience* (Vol. 13). Available from: https://www.frontiersin.org/article/10.3389/fnagi.2021.687246 (accessed on 9 November 2023).
18. Sun, M. S., Jin, H., Sun, X., Huang, S., Zhang, F. L., Guo, Z. N., et al., (2018). Free radical damage in ischemia-reperfusion injury: An obstacle in acute ischemic stroke after revascularization therapy. Giustarini, D., (ed.), *Oxid. Med. Cell Longev. [Internet]* (Vol. 2018, p. 3804979). Available from: https://doi.org/10.1155/2018/3804979.
19. Prabhakar, N. K., Khan, H., Grewal, A. K., & Singh, T. G., (2022). Intervention of neuroinflammation in the traumatic brain injury trajectory: *In vivo* and clinical approaches. *International Immunopharmacology, 108*, 108902.
20. Forrester, S. J., Kikuchi, D. S., Hernandes, M. S., Xu, Q., & Griendling, K. K., (2018). Reactive oxygen species in metabolic and inflammatory signaling. *Circ. Res. [Internet], 122*(6), 877–902. Available from: https://doi.org/10.1161/CIRCRESAHA.117.311401.
21. Pizzino, G., Irrera, N., Cucinotta, M., Pallio, G., Mannino, F., Arcoraci, V., et al., (2017). Oxidative stress: Harms and benefits for human health. *Oxid. Med. Cell Longev., 2017*, 8416763.
22. Di Meo, S., Reed, T. T., Venditti, P., & Victor, V. M., (2016). Role of ROS and RNS sources in physiological and pathological conditions. *Oxid. Med. Cell Longev., 2016*, 1245049.
23. Shen, J., Rastogi, R., Geng, X., & Ding, Y., (2019). Nicotinamide adenine dinucleotide phosphate oxidase activation and neuronal death after ischemic stroke. *Neural Regen. Res., 14*(6), 948–953.
24. Vermot, A., Petit-Härtlein, I., Smith, S. M. E., & Fieschi, F., (2021). NADPH oxidases (NOX): An overview from discovery, molecular mechanisms to physiology and pathology. *Antioxidants, 10.*

25. Touyz, R. M., Anagnostopoulou, A., Rios, F., Montezano, A. C., & Camargo, L. L., (2019). NOX5: Molecular biology and pathophysiology. *Exp. Physiol. [Internet], 104*(5), 605–616. Available from: https://doi.org/10.1113/EP086204.
26. Ibi, M., & Yabe-Nishimura, C., (2020). Chapter 1 – The role of reactive oxygen species in the pathogenic pathways of depression. In: Martin, C. R., & Preedy, V. R. (eds.), *Oxidative Stress and Dietary Antioxidants in Neurological Diseases* (pp. 3–16). Academic Press. Available from: https://www.sciencedirect.com/science/article/pii/B9780128177808000013 (accessed on 9 November 2023).
27. Tarafdar, A., & Pula, G., (2018). The role of NADPH oxidases and oxidative stress in neurodegenerative disorders. *Int. J. Mol. Sci., 19*(12).
28. Burtenshaw, D., Hakimjavadi, R., Redmond, E. M., & Cahill, P. A., (2017). Nox, reactive oxygen species and regulation of vascular cell fate. *Antioxidants (Basel, Switzerland), 6*(4).
29. Iommarini, L., Porcelli, A. M., Gasparre, G., & Kurelac, I., (2017). Non-canonical mechanisms regulating hypoxia-inducible factor 1 alpha in cancer [Internet]. *Frontiers in Oncology, 7*. Available from: https://www.frontiersin.org/article/10.3389/fonc.2017.00286 (accessed on 9 November 2023).
30. Chen, R., Lai, U. H., Zhu, L., Singh, A., Ahmed, M., & Forsyth, N. R., (2018). Reactive oxygen species formation in the brain at different oxygen levels: The role of hypoxia inducible factors [Internet]. *Frontiers in Cell and Developmental Biology, 6*. Available from: https://www.frontiersin.org/article/10.3389/fcell.2018.00132 (accessed on 9 November 2023).
31. Dumanović, J., Nepovimova, E., Natić, M., Kuča, K., & Jaćević, V., (2021). The significance of reactive oxygen species and antioxidant defense system in plants: A concise overview [Internet]. *Frontiers in Plant Science, 11*. Available from: https://www.frontiersin.org/article/10.3389/fpls.2020.552969 (accessed on 9 November 2023).
32. Khan, H., Singh, A., Thapa, K., Garg, N., Grewal, A. K., & Singh, T. G., (2021). Therapeutic modulation of the phosphatidylinositol 3-kinases (PI3K) pathway in cerebral ischemic injury. *Brain Research, 1761*, 147399.
33. Khan, H., Sharma, K., Kumar, A., Kaur, A., & Singh, T. G., (2022). Therapeutic implications of cyclooxygenase (COX) inhibitors in ischemic injury. *Inflammation Research*, 1–6.
34. Hernández-Espinosa, D. R., Massieu, L., Montiel, T., & Morán, J., (2019). Role of NADPH oxidase-2 in the progression of the inflammatory response secondary to striatum excitotoxic damage. *J. Neuroinflammation [Internet]., 16*(1), 91. Available from: https://doi.org/10.1186/s12974-019-1478-4.
35. Dong, R., Huang, R., Wang, J., Liu, H., & Xu, Z., (2021). Effects of microglial activation and polarization on brain injury after stroke [Internet]. *Frontiers in Neurology, 12*. Available from: https://www.frontiersin.org/article/10.3389/fneur.2021.620948 (accessed on 9 November 2023).
36. Khan, H., Gupta, A., Singh, T. G., & Kaur, A., (2021). Mechanistic insight on the role of leukotriene receptors in ischemic–reperfusion injury. *Pharmacological Reports, 73*(5), 1240–1254.
37. Wang, C. H., Zhang, C., & Xing, X. H., (2016). Xanthine dehydrogenase: An old enzyme with new knowledge and prospects. *Bioengineered, 7*(6), 395–405.
38. Xie, N., Zhang, L., Gao, W., Huang, C., Huber, P. E., Zhou, X., et al., (2020). NAD+ metabolism: Pathophysiologic mechanisms and therapeutic potential. *Signal Transduct. Target Ther [Internet], 5*(1), 227. Available from: https://doi.org/10.1038/s41392-020-00311-7.
39. Bortolotti, M., Polito, L., Battelli, M. G., & Bolognesi, A., (2021). Xanthine oxidoreductase: One enzyme for multiple physiological tasks. *Redox Biol. [Internet], 41*, 101882. Available from: https://www.sciencedirect.com/science/article/pii/S2213231721000306 (accessed on 9 November 2023).
40. Tejero, J., Shiva, S., & Gladwin, M. T., (2018). Sources of vascular nitric oxide and reactive oxygen species and their regulation. *Physiol. Rev. [Internet], 99*(1), 311–379. Available from: https://doi.org/10.1152/physrev.00036.2017.
41. Rehni, A. K., Singh, T. G., Kalra, R., & Singh, N., (2009). Pharmacological inhibition of inducible nitric oxide synthase attenuates the development of seizures in mice. *Nitric Oxide, 21*(2), 120–125.
42. Villa-González, M., Martín-López, G., & Pérez-Álvarez, M. J., (2022). Dysregulation of mTOR signaling after brain ischemia. *International Journal of Molecular Sciences, 23*.
43. Johnson, T. A., Jinnah, H. A., & Kamatani, N., (2019). Shortage of cellular ATP as a cause of diseases and strategies to enhance ATP [Internet]. *Frontiers in Pharmacology, 10*. Available from: https://www.frontiersin.org/article/10.3389/fphar.2019.00098 (accessed on 9 November 2023).

44. Mohamed, M. Z., Baky, M. A. El., Hassan, O. A., Mohammed, H. H., & Abdel-Aziz, A. M., (2020). PTEN/PI3K/VEGF signaling pathway involved in the protective effect of xanthine oxidase inhibitor febuxostat against endometrial hyperplasia in rats. *Hum. Exp. Toxicol., 39*(9), 1224–1234.
45. Pratomo, I. P., Noor, D. R., Kusmardi, K., Rukmana, A., Paramita, R. I., Erlina, L., et al., (2021). Xanthine oxidase-induced inflammatory responses in respiratory epithelial cells: A review in immunopathology of COVID-19. In: Moshage, H. J., (ed.), *Int. J. Inflam [Internet]* (Vol. 2021, p. 1653392). Available from: https://doi.org/10.1155/2021/1653392.
46. Ives, A., Nomura, J., Martinon, F., Roger, T., LeRoy, D., Miner, J. N., et al., (2015). Xanthine oxidoreductase regulates macrophage IL1β secretion upon NLRP3 inflammasome activation. *Nat. Commun. [Internet], 6*(1), 6555. Available from: https://doi.org/10.1038/ncomms7555.
47. Wang, S., Li, Y., Song, X., Wang, X., Zhao, C., Chen, A., et al., (2015). Febuxostat pretreatment attenuates myocardial ischemia/reperfusion injury via mitochondrial apoptosis. *J. Transl. Med. [Internet], 13*(1), 209. Available from: https://doi.org/10.1186/s12967-015-0578-x.
48. Moris, D., Spartalis, M., Spartalis, E., Karachaliou, G. S., Karaolanis, G. I., Tsourouflis, G., et al., (2017). The role of reactive oxygen species in the pathophysiology of cardiovascular diseases and the clinical significance of myocardial redox. *Ann. Transl. Med., 5*(16). [Internet]. Available from: https://atm.amegroups.com/article/view/16040 (accessed on 9 November 2023).
49. Bhatti, J. S., Bhatti, G. K., & Reddy, P. H., (2017). Mitochondrial dysfunction and oxidative stress in metabolic disorders—A step towards mitochondria based therapeutic strategies. *Biochim. Biophys. Acta–Mol. Basis Dis. [Internet]., 1863*(5), 1066–1077. Available from: https://www.sciencedirect.com/science/article/pii/S0925443916302927 (accessed on 9 November 2023).
50. Ahmad, M., Wolberg, A., & Kahwaji, C. I., (2022). *Biochemistry, Electron Transport Chain.* In Treasure Island (FL).
51. Sharma, A., Khan, H., Singh, T. G., Grewal, A. K., Najda, A., Kawecka-Radomska, M., Kamel, M., et al., (2021). Pharmacological modulation of ubiquitin-proteasome pathways in oncogenic signaling. *International Journal of Molecular Sciences, 22*(21), 11971.
52. Xiao, W., Wang, R. S., Handy, D. E., & Loscalzo, J., (2018). NAD(H) and NADP(H) redox couples and cellular energy metabolism. *Antioxid Redox Signal, 28*(3), 251–272.
53. Nolfi-Donegan, D., Braganza, A., & Shiva, S., (2020). Mitochondrial electron transport chain: Oxidative phosphorylation, oxidant production, and methods of measurement. *Redox Biol. [Internet], 37*, 101674. Available from: https://www.sciencedirect.com/science/article/pii/S221323172030879X (accessed on 9 November 2023).
54. Dambrova, M., Zuurbier, C. J., Borutaite, V., Liepinsh, E., & Makrecka-Kuka, M., (2021). Energy substrate metabolism and mitochondrial oxidative stress in cardiac ischemia/reperfusion injury. *Free Radic. Biol. Med. [Internet], 165*, 24–37. Available from: https://www.sciencedirect.com/science/article/pii/S0891584921000538 (accessed on 9 November 2023).
55. Tretter, L., Patocs, A., & Chinopoulos, C., (2016). Succinate, an intermediate in metabolism, signal transduction, ROS, hypoxia, and tumorigenesis. *Biochim Biophys Acta–Bioenerg. [Internet], 1857*(8), 1086–1101. Available from: https://www.sciencedirect.com/science/article/pii/S0005272816300597 (accessed on 9 November 2023).
56. Koh, J. H., & Kim, J. Y., (2021). Role of PGC-1α in the mitochondrial NAD(+) pool in metabolic diseases. *Int. J. Mol. Sci., 22*(9).
57. Zhou, Y., Wang, S., Li, Y., Yu, S., & Zhao, Y., (2017). SIRT1/PGC-1α signaling promotes mitochondrial functional recovery and reduces apoptosis after intracerebral hemorrhage in rats. *Front. Mol. Neurosci., 10*, 443.
58. Chen, Q. M., & Maltagliati, A. J., (2018). Nrf2 at the heart of oxidative stress and cardiac protection. *Physiol. Genomics, 50*(2), 77–97.
59. Khan, H., (2021). PND55 protocatechuic acid induced pharmacological postconditioning ameliorates cerebral ischemia reperfusion injury in MICE by modulation of nuclear factor erythroid 2-related factor pathway. *Value in Health, 24*, S169.
60. Scialò, F., Fernández-Ayala, D. J., & Sanz, A., (2017). Role of mitochondrial reverse electron transport in ROS signaling: Potential roles in health and disease [Internet]. *Frontiers in Physiology, 8.* Available from: https://www.frontiersin.org/article/10.3389/fphys.2017.00428 (accessed on 9 November 2023).

61. Onukwufor, J. O., Berry, B. J., & Wojtovich, A. P., (2019). Physiologic implications of reactive oxygen species production by mitochondrial complex I reverse electron transport. *Antioxidants, 8.*
62. Rafael, R., (2018). Oxygen radicals, nitric oxide, and peroxynitrite: Redox pathways in molecular medicine. *Proc. Natl. Acad. Sci. [Internet], 115*(23), 5839–5848. Available from: https://doi.org/10.1073/pnas.1804932115.
63. Khan, H., Kashyap, A., Kaur, A., & Singh, T. G., (2020). Pharmacological postconditioning: A molecular aspect in ischemic injury. *Journal of Pharmacy and Pharmacology, 72*(11), 1513–1527.
64. Xiang, M., Lu, Y., Xin, L., Gao, J., Shang, C., Jiang, Z., et al., (2021). Role of oxidative STRESS in reperfusion following myocardial ischemia and its treatments. In: Tocchetti, C. G., (ed.), *Oxid. Med. Cell Longev.* [Internet] (Vol. 2021, p. 6614009). Available from: https://doi.org/10.1155/2021/6614009.
65. Kalra, P., Khan, H., Kaur, A., & Singh, T. G., (2022). Mechanistic insight on autophagy modulated molecular pathways in cerebral ischemic injury: From preclinical to clinical perspective. *Neurochemical Research*, 1–9.
66. Roca-Agujetas, V., De Dios, C., Lestón, L., Marí, M., Morales, A., & Colell, A., (2019). Recent insights into the mitochondrial role in autophagy and its regulation by oxidative stress. In: Peluso, I., (ed.), *Oxid. Med. Cell Longev. [Internet]* (Vol. 2019, p. 3809308). Available from: https://doi.org/10.1155/2019/3809308.
67. Cork, G. K., Thompson, J., & Slawson, C., (2018). Real talk: The inter-play between the mTOR, AMPK, and hexosamine biosynthetic pathways in cell signaling [Internet]. *Frontiers in Endocrinology,* 9. Available from: https://www.frontiersin.org/article/10.3389/fendo.2018.00522 (accessed on 9 November 2023).
68. Ma, J., Yang, L., Ren, J., & Yang, J., (2018). Chapter 20–autophagy, oxidative stress, and redox regulation. In: Ren, J., Sowers, J. R., & Zhang, Y. (eds.), *Autophagy and Cardiometabolic Diseases* (pp. 237–251). Academic Press. Available from: https://www.sciencedirect.com/science/article/pii/B9780128052532000201(accessed on 9 November 2023).
69. Chun, Y., & Kim, J., (2021). AMPK-mTOR signaling and cellular adaptations in hypoxia. *Int. J. Mol. Sci., 22*(18).
70. Zachari, M., & Ganley, I. G., (2017). The mammalian ULK1 complex and autophagy initiation. *Essays Biochem., 61*(6), 585–596.
71. Jiang, L. B., Cao, L., Yin, X. F., Yasen, M., Yishake, M., Dong, J., et al., (2015). Activation of autophagy via Ca(2+)-dependent AMPK/mTOR pathway in rat notochordal cells is a cellular adaptation under hyperosmotic stress. *Cell Cycle, 14*(6), 867–879.
72. Tan, S., Kennedy, B., & Wong, E., (2021). Chapter 2 – mitochondrial biogenesis and mitophagy. In: De Oliveira, M. R. (ed.), *Mitochondrial Physiology and Vegetal Molecules* (pp. 35–90). Academic Press. Available from: https://www.sciencedirect.com/science/article/pii/B978012821562300037X (accessed on 9 November 2023).
73. Lu, N., Li, X., Tan, R., An, J., Cai, Z., Hu, X., et al., (2018). HIF-1α/Beclin1-mediated autophagy is involved in neuroprotection induced by hypoxic preconditioning. *J. Mol. Neurosci., 66*(2), 238–250.
74. Zhang, Y., Liu, D., Hu, H., Zhang, P., Xie, R., & Cui, W., (2019). HIF-1α/BNIP3 signaling pathway-induced-autophagy plays protective role during myocardial ischemia-reperfusion injury. *Biomed. Pharmacother. [Internet], 120*, 109464. Available from: https://www.sciencedirect.com/science/article/pii/S0753332219339976 (accessed on 9 November 2023).
75. Ma, S., Wang, Y., Chen, Y., & Cao, F., (2015). The role of the autophagy in myocardial ischemia/reperfusion injury. *Biochim Biophys Acta–Mol. Basis Dis. [Internet], 1852*(2), 271–276. Available from: https://www.sciencedirect.com/science/article/pii/S0925443914001380 (accessed on 9 November 2023).
76. Bugger, H., & Pfeil, K., (2020). Mitochondrial ROS in myocardial ischemia reperfusion and remodeling. *Biochim. Biophys. Acta–Mol. Basis Dis [Internet], 1866*(7), 165768. Available from: https://www.sciencedirect.com/science/article/pii/S0925443920301137 (accessed on 9 November 2023).
77. Hassanpour, M., Rezabakhsh, A., Pezeshkian, M., Rahbarghazi, R., & Nouri, M., (2018). Distinct role of autophagy on angiogenesis: Highlights on the effect of autophagy in endothelial lineage and progenitor cells. *Stem. Cell Res. Ther. [Internet], 9*(1), 305. Available from: https://doi.org/10.1186/s13287-018-1060-5.
78. Khan, H., Bangar, A., Grewal, A. K., Bansal, P., & Singh, T. G., (2022). Caspase-mediated regulation of the distinct signaling pathways and mechanisms in neuronal survival. *International Immunopharmacology*, 110.

79. Kaczanowski, S., (2016). Apoptosis: Its origin, history, maintenance and the medical implications for cancer and aging. *Phys. Biol. [Internet], 13*(3), 31001. Available from: http://dx.doi.org/10.1088/1478-3975/13/3/031001.
80. Obeng, E., (2020). Apoptosis (programmed cell death) and its signals: A review. *Brazilian J. Biol., 81.*
81. Domínguez-Fernández, T., Rodríguez, M. A., Sánchez, M. V., Gómez, G. C., Medel, O., & Pérez, I. D. G., (2018). A calpain-like protein is involved in the execution phase of programmed cell death of entamoeba histolytica [Internet]. *Frontiers in Cellular and Infection Microbiology, 8.* Available from: https://www.frontiersin.org/article/10.3389/fcimb.2018.00339 (accessed on 9 November 2023).
82. Wei, Z. Z., & Wei, L., (2017). In: Caplan, L. R., Biller, J., Leary, M. C., Lo, E. H., Thomas, A. J., Yenari, M., et al., (eds.), *Chapter 47–Mitochondrial Mechanisms During Ischemia and Reperfusion* (pp. 230–234). San Diego: Academic Press. Available from: https://www.sciencedirect.com/science/article/pii/B9780128030585000473 (accessed on 9 November 2023).
83. Kim, S. H., & Kim, H., (2021). Chapter 27–The therapeutic effects of lycopene on mitochondrial function in human diseases. In: De Oliveira, M. R. (ed.), *Mitochondrial Physiology and Vegetal Molecules* (pp. 593–612). Academic Press. Available from: https://www.sciencedirect.com/science/article/pii/B9780128215623000411 (accessed on 9 November 2023).
84. Yang, T., Guo, R., Ofengeim, D., Hwang, J. Y., Zukin, R. S., Chen, J., et al., (2022). In: Grotta, J. C., Albers, G. W., Broderick, J. P., Day, A. L., Kasner, S. E., Lo, E. H., et al., (eds.) *5–Molecular and Cellular Mechanisms of Ischemia-Induced Neuronal Death* (pp. 57–73.e6). Philadelphia: Elsevier. Available from: https://www.sciencedirect.com/science/article/pii/B9780323694247000053 (accessed on 9 November 2023).
85. Lossi, L., (2022). The concept of intrinsic versus extrinsic apoptosis. *Biochem. J. [Internet], 479*(3), 357–384. Available from: https://doi.org/10.1042/BCJ20210854.
86. Kucka, K., & Wajant, H., (2021). Receptor oligomerization and its relevance for signaling by receptors of the tumor necrosis factor receptor superfamily [Internet]. *Frontiers in Cell and Developmental Biology, 8.* Available from: https://www.frontiersin.org/article/10.3389/fcell.2020.615141 (accessed on 9 November 2023).
87. Gough, P., & Myles, I. A., (2020). Tumor necrosis factor receptors: Pleiotropic signaling complexes and their differential effects [Internet]. *Frontiers in Immunology, 11.* Available from: https://www.frontiersin.org/article/10.3389/fimmu.2020.585880 (accessed on 9 November 2023).
88. Lyamzaev, K. G., Knorre, D. A., & Chernyak, B. V., (2020). Mitoptosis, twenty years after. *Biochemistry (Mosc)., 85*(12), 1484–1498.
89. Jenner, A., Peña-Blanco, A., Salvador-Gallego, R., Ugarte-Uribe, B., Zollo, C., Ganief, T., et al., (2022). DRP1 interacts directly with BAX to induce its activation and apoptosis. *EMBO J. [Internet]*, n/a(n/a), e108587. Available from: https://doi.org/10.15252/embj.2021108587.
90. Peña-Blanco, A., & García-Sáez, A. J., (2018). Bax, Bak and beyond—Mitochondrial performance in apoptosis. *FEBS J. [Internet], 285*(3), 416–431. Available from: https://doi.org/10.1111/febs.14186.
91. Sharp, W. W., Beiser, D. G., Fang, Y. H., Han, M., Piao, L., Varughese, J., et al., (2015). Inhibition of the mitochondrial fission protein dynamin-related protein 1 improves survival in a murine cardiac arrest model. *Crit. Care Med., 43*(2), e38–47.
92. Kim-Campbell, N., Gomez, H., & Bayir, H., (2019). Chapter 20 – cell death pathways: Apoptosis and regulated necrosis. In: Ronco, C., Bellomo, R., Kellum, J. A., & Ricci, Z. B. T., (ed.), *Critical Care Nephrology* (3rd edn., pp. 113–121.e2) Philadelphia: Elsevier. Available from: https://www.sciencedirect.com/science/article/pii/B9780323449427000200 (accessed on 9 November 2023).
93. Dhuriya, Y. K., & Sharma, D., (2018). Necroptosis: A regulated inflammatory mode of cell death. *J. Neuroinflammation [Internet], 15*(1), 199. Available from: https://doi.org/10.1186/s12974-018-1235-0.
94. Tummers, B., & Green, D. R., (2017). Caspase-8: Regulating life and death. *Immunol. Rev., 277*(1), 76–89.
95. Liu, Y., Liu, T., Lei, T., Zhang, D., Du, S., Girani, L., et al., (2019). RIP1/RIP3-regulated necroptosis as a target for multifaceted disease therapy (Review). *Int. J. Mol. Med. [Internet], 44*(3), 771–786. Available from: https://doi.org/10.3892/ijmm.2019.4244.
96. Galluzzi, L., Kepp, O., Chan, F. K. M., & Kroemer, G., (2017). Necroptosis: Mechanisms and relevance to disease. *Annu. Rev. Pathol., 12*, 103–130.

97. Zhan, C., Huang, M., Yang, X., & Hou, J., (2021). MLKL: Functions beyond serving as the executioner of necroptosis. *Theranostics., 11*(10), 4759–4769.
98. Kaur, A., Singh, T. G., Khan, H., Kumar, M., Singh, N., & Abdel-Daim, M. M., (2022). Neuroprotective effect of piclamilast-induced post-ischemia pharmacological treatment in mice. *Neurochemical Research*, 1–4.
99. Del Re, D. P., Amgalan, D., Linkermann, A., Liu, Q., & Kitsis, R. N., (2019). Fundamental mechanisms of regulated cell death and implications for heart disease. *Physiol Rev. [Internet], 99*(4), 1765–1817. Available from: https://doi.org/10.1152/physrev.00022.2018.

CHAPTER 5

Role of Mitochondria in Ischemic-Reperfusion Injury

NARINDER KUMAR,[1] KAJAL BAGRI,[1] SHAHID MUSHTAQ MALIK,[1] RAHUL DESHMUKH,[1] and THAKUR GURJEET SINGH[2]

[1]*Department of Pharmaceutical Sciences and Technology, Maharaja Ranjit Singh Punjab Technical University, Bathinda, Punjab, India*

[2]*Chitkara College of Pharmacy, Chitkara University, Punjab, India*

ABSTRACT

Ischemia/reperfusion (IRI) injury occurs when cellular injury which is initiated by hypoxia or anoxia in an organ is aggravated when tissue pH and oxygen supply are restored. Initially, it starts as a process that is confined to a region that results in the death of parenchymal cells and then escalates to a severe inflammatory response involving both Cytotoxic processes, both direct and indirect. Ischemia-reperfusion injury in organ transplantation is a complex process that results in damage to organs and primary graft failure. The injury to the organ is caused by a complex series of events, involves the energy substrate depletion, as well as the modification of the two kinds of cell death, are apoptosis and necrosis., as well as ionic balance and there is the development of reactive oxygen species. A rising body of data suggests mitochondria are involved in disease. This procedure is required due to the significant changes that occur during ischemia and reperfusion. Understanding the pathways that cause mitochondrial damage could be useful in creating measures to prevent it.

5.1 INTRODUCTION

5.1.1 *MECHANISM OF ISCHEMIA-REPERFUSION INJURY*

Atherosclerosis and acute myocardial infarction (AMI) are two physiological processes that enhance ischemia and contribute to hypoxia and hypoperfusion [1, 2]. Hypoxia is initiated by a blockage in the arterial blood flow, which leads to mitochondrial electron transport chain malfunction. Reduced Anaerobic metabolism, sodium-potassium pump failure, and ribosome

Ischemic Injury: A Molecular Insight. Thakur Gurjeet Singh & Heena Khan (Eds.)

separation are all caused by ATP production in mitochondria. Adenosine triphosphate (ATP), antioxidative agents are produced at lower levels in cells through anaerobic metabolism. Furthermore, retention of lactic acid might result in metabolic acidosis. NA^+/K^+ pump and Ca^+ pump may also malfunction on cell's exterior. When the sodium-potassium ATPase pumps stop working, Na^+ is held in the cells and K^+ is released Na^+/H^+ exchanger pumps activity is decreased when Na^+ ranks in cells are increased. Calcium ATPase pumps in the ER also become defective, reducing the reuptake of calcium.

The cause of hyperosmolarity is the build-up of calcium, sodium, and hydrogen ions in the cells, Water flows into the cytoplasm, causing cell expansion. The cellular pH is lowered as hydrogen is retained, resulting in decreased action of enzymes and nuclear chromatin clumping. Restoring blood flow to tissue having ischemia after the reperfusion phase gives oxygen through RBCs, which lowers protein formation. The production of ROS will increase in ischemic cells due to a decrease in the concentration of antioxidative agents. Oxidative stress occurs by ROS. Which can cause endothelial dysfunction, DNA injury, and native inflammation. A cytokine storm might be triggered by inflammatory pathways and oxidative stress which may result in cell death due to cellular structural damage [3]. Reperfusion is a dynamic stage that might last many days. Knowing the exact process of ischemia-reperfusion damage might offer the groundwork for new treatment options as well as injury prevention.

5.1.2 MITOCHONDRIA

Early studies showed mitochondrial activity has a role in IR damage. A rupture of ROS generation by mitochondria is the first harmful event after reperfusion. Mitochondrial Reactive oxygen species do not only cause acute injury, but also illness in the moments, days, and weeks after reperfusion. The final rupture of Reactive oxygen species manufacturing reperfusion damages mitochondria and disrupts ATP synthesis. Elevated ROS in combination with calcium dysregulation can cause mitochondrial permeability transition [4]. After reperfusion, the mitochondrial disruption induces apoptotic and necrotic cell death. Significant harm to molecular sequence molecules (DAMPs), such as mtDNA, are released by ROS-mediated mitochondrial damage and can trigger the sterile inflammatory response. Swelling as a result of the essential immune system's activation leads to IR damage and can last for days after the initial injury. As a result, fibrotic scar tissue will grow in the weeks and months following the original IR injury, replacing dead cells and undergoing extensive tissue remodeling [5]. The final result is usually a fundamentally altered and underperforming organ, which commonly leads to long-term illness, e.g., chronic heart disease.

5.2 ELECTRON TRANSPORT CHAIN DAMAGE AND ITS MOLECULAR PATHOLOGY

5.2.1 COMPLEX I

The activity of complex I diminish comparably in both Subsarcolemmal Mitochondria and Inter fibrillar mitochondria following the condition of reduced blood supply. The activity of complex I is affected by two post-translational changes: phosphorylation and S-nitrosylation. As a result of ischemia damage, there is the formation of ROS from complex I [6]. The action of NADH ferricyanide oxidoreductase is similar to that of NADH dehydrogenase, the starting formed constituent of complex I. Low blood supply has little influence on SSM or IFM activity, providing us the information about the damage that occurs elsewhere in the complex Others, on the other hand, have shown that there is loss of the flavin mononucleotide coenzyme results in a decrease in the complexes of component of NADH dehydrogenase. The differences in findings are not immediately evident, although they might be attributed to various ischemia models in which the Complex, I experience the conformational shift in reply to hypoxia and ischemia when Fe^{+}/S clusters and Flavin mononucleotide are decreased, and the terminal acceptor ubiquinol pool is lowered, transforming from usual form Active A to dormant form (dormant D). Cysteine 39 in mitochondrial-encoded NDUF3 and cysteines on NDUF9 and NDUF1 are among the sulfhydryl groups that are oxidatively sensitive when converted to the D form. As part of an antioxidant defense mechanism, reversible glutathionylation of these residues prevents the Complex I. Oxidants that trap Complex I, i.e., D form, resulting in a long-term decrease in movement. In the lack of sulfhydryl group oxidation, the development of complexes (complex I) with the substrate restores the A procedure and its activity. Connections of Complex I with thioredoxin-2 or glutaredoxin-2 in the mitochondria may increase complex I prevention. The D form suppresses ROS generation by blunting reverse electron transport, which causes superoxide to develop [7].

The impact of rotenone or amobarbital with complex I blockade in encouraging the transition, from A to D is currently being studied. Biguanide phenformin promotes the conversion of complex I to complex D [8]. When given for 5 minutes at the start of reperfusion, metformin, like amobarbital, partially inhibits complex I in ischemic hearts and minimizes myocardial damage [9]. As a result, modifying the A to D transition in complex I might become a key component in cardiac protection. Complex I generate reactive oxygen species. NADH dehydrogenase having FMN, iron-Sulfur cluster N_2 and two are securely bonded ubiquinones positioned away from the direction of electron flow are all potential sources of ROS. Although other studies have previously attributed ROS generation to this location, inhibiting NADH dehydrogenase lowers ROS production in the baseline state, showing that the place is distal to the FMN. Rotenone suppression augmented ROS formation of complex I after ischemia, showing the area is near the ubiquinone required spot. ROS generation following ischemia is probable to come either from the N_2 iron-sulfur site or from the firmly destined quinones inside complex I, according to the utilization of forward and reverse electron flow [6].

5.2.2 COMPLEX II

Ischemic injury is less likely to affect complex II. Its function reduces in mitochondria of rat heart under global ischemia for a long time, and it decreases even more after reperfusion [10]. Decreased S-glutathionylation in the 70-kDa flavin protein and enhanced nitration and oxidation of complex II, which impairs its association with complex III, is linked to lower functionality. Variation of complex II action helps to open the mitochondrial K_{ATP} channel and inhibit ROS generation during I/R, reducing cardiac damage. $FADH_2$ or FADH•-semiquinone auto-oxidation is predicted to increase ROS production in mammalian complex II defects.

With succinate as a substrate, ROS can be produced in intact cardiac mitochondria from complex II to complex I through REF [11]. Potential (m) of the inner membrane and regulators of complex I and II impact this development. Oligomycin reduces REF-mediated ROS production by raising m when succinate is used as a source. The inner membrane's depolarization, on the other hand, completely ceases REF-mediated ROS generation. After mild ischemia without depolarization of the intermembrane space, REF-mediated ROS production still can proceed in the heart.

Inhibiting REF-mediated ROS creation during I/R lowers cardiac damage. The degree of ischemia-induced mitochondrial damage may change REF-mediated Reactive oxygen species formation in cardiac damage.

5.2.3 COMPLEX III

During ischemia, the activity of Complex III reduces [12]. In myocardial injury, the degree of ischemia-induced mitochondrial damage may affect REF-mediated ROS production. Two cysteines and two histidine coordinate a 2Fe-2S cluster in the ISP. The contents of cytochromes b and c1 in Subsarcolemmal Mitochondria and IFM from the rat heart were unaffected by ischemia. Complex III is a crucial area for oxidant generation during and after ischemia. The Qo (quinol site exterior) oxidation region produces the majority of ROS in complex III, while the Qi (quinol site internal) oxidation region also contributes [13]. The oxidized cytochrome b acts as an electron entry site into complex III. Two electrons are produced when one electron passes through ISP and the other passes through the center of Qi on the cytochrome b of the matrix side. Ischemic injury to complex III permits sufficient remaining electron transport into the complex for the making of ROS since complex III is a dimer. When one of a dimer's Qo spots is inactivated, ROS formation arises from complex III Qi centers increases.

A treatment method based on transitory, reversible blocking of complex III's Qo site is appealing. Irreversible inhibitors, mitochondrial poisons, myxothiazol, and stigmatellin, suppress reactive oxygen species production in complex III mitochondria [14]. A reversible Qo site inhibitor, on the other hand, can minimize ROS production while also reducing reperfusion damage.

5.2.4 COMPLEX IV–CARDIOLIPIN

The movement of Complex IV, which is made up of subunits of 13 peptides, depends on the reliability of catalytic, regulatory, and structural subunits. Complex IV coexists with complexes I and III in respirasomes, enhancing respiratory substrate channeling subunit peptides that are neither depleted nor inactivated by ischemia. The single phospholipid whose content is reduced is cardiolipin; SSM decreases at the very same time that OXPHOS decreases due to cytochrome oxidase inhibition [17], implying a dysfunction mechanism for complex IV. After ischemia, the mitochondrial cardiolipin concentration falls in isolated rat hearts [15] and *in vivo* canine hearts. Fusing quarantined rat cardiac cardiolipin with mitochondria consisting of liposomes, restores cytochrome oxidase activity.

During ischemia, cardiolipin decline from Subsarcolemmal Mitochondria leads to mitochondrial injury. Cardiolipin is prone to oxidative damage because of the large amount of oxidatively delicate linoleic acid (C18:2) acyl groups. Nonetheless, cardiolipin's unchanged formation after ischemia rules out the existence of oxidatively changed acyl groups. Cardiolipin's oxidatively formed per-oxy groups, on the other hand, are unstable and can swiftly degrade, leading to cardiolipin degradation. However, oxidized acyl remains might respond straight with proteins, forming covalent phospholipid-protein compounds that make cardiolipin insoluble in organic solvents. As a result, the lack of oxidized cardiolipin does not rule out oxidative reactions.

5.2.5 CYTOCHROME C

Cytochrome C is a kind of red blood cell. Ischemia causes cytochrome C to leak out of mitochondria and into the cytosol. At 30 minutes of ischemia, cytochrome C declines in SSM in tandem with cardiolipin loss [6]. *In vitro*, SSM had a higher tendency than IFM to secrete cytochrome (c) and to tolerate the fast development of ischemia injury, with cytochrome c loss. Subsarcolemmal Mitochondria are much more susceptible to cytochrome c release induced by increased outer calcium [14] and are hence increased prone to trigger cardiomyocyte demise processes.

Cardiolipin supplies cytochrome C to the inner membrane has binding sites., therefore a reduction in its amount encourages cytochrome c loss, which happens during I/R [16]. Thus, cytochrome cis depletion is connected to cardiolipin depletion, The outer membrane is then permeabilized, either by proapoptotic peptides or MPT.

5.3 ROS

It is a combined term representing all reactive forms of oxygen involving both non-radical like (H_2O_2) as well as radical like superoxide (.OH) forms which participate in the initiation and/or propagation of chain reaction.

In mitochondria, regular aerobic metabolism produces ROS, and even decreasing levels of ROS can function as a mediator in a variety of signaling pathways of the cell, such as thiol group oxidation in cysteine protein residues. Mitochondrial ROS influence both intramitochondrial and extramitochondrial signaling cascades due to membrane permeability for H_2O_2. ROS can activate G-protein coupled receptors, growth factors, cytokines, and MAPKs, as well as regulate factors of transcription (Nrf2, HIF-1α, NF-κB, p53), calcium handling proteins (calmodulin, SERCA2a, L-type Ca^{2+C} channel, and ryanodine receptor) and metalloproteinases of matrix among additional targets [17]. Notably, ROS of mitochondria are likewise involved in the signaling events (e.g., protein kinase C, MAPK, and NF-κB signaling) that underpin the beneficial preconditioning ischemic effects, thereby elevating resistance to IRI. Furthermore, ROS may be involved in signaling processes that mediate the positive effects of postconditioning or distant ischemia conditioning [18]. ROS are well recognized for their ability to produce non-specific oxidative damage to lipids, proteins, and nucleic acids. Also, mitochondrial DNA appears to be particularly vulnerable to damage induced by ROS [19].

One of the principal causes of eukaryotic ROS is mitochondrial electron transport., Previous studies have shown that O_2^- generation occurs primarily in complex I at the flavin mononucleotide subunit and the ubisemiquinone bound to the Qo site at complex III and the Mitochondrial localized NADPH oxidase 4 and monoamine oxidases can also create ROS (NOX4). Several studies published in the last decade have found that RET exists *in vivo* and may have a function in a variety of physiological processes. Under physiological circumstances (typically saturating or prolonged labor). When coenzyme Q is over-reduced, it can lead to the generation O_2^- by RET resulting transfer electrons to complex I in a reverse manner and consequent ROS generation [5]. Early electron leakage from ETC enzymes to oxygen might lead to O_2^- generation [20].

Apart from generation, ROS detoxification is important for maintaining ROS homeostasis inside mitochondria. The activity of MnSOD is primarily responsible for converting O_2^- to H_2O_2. Glutathione peroxidase (Gpx) detoxifies H_2O_2 to H_2O, with Gpx activity requiring the o glutathione oxidation (GSH) to GSSG, and glutathione reductase (GsR) action restoring GSH to sustain the activity of Gpx. Persistent mitochondrial oxidative stress also affects the non-infarcted myocardium's response to MI-induced cardiac tissue loss, perhaps contributing to overall myocardial remodeling and the development of HF [21].

5.4 REACTIVE NITROGEN SPECIES

As nitric oxide is uncharged it can rapidly diffuse across the cell during ischemia, the elevated O_2 percent formation inside mitochondria leads to the creation of per-oxy nitrite (ONOO). If created within mitochondria, it oxidizes nearby mitochondrial lipids and proteins due to its relatively short half-life. As a result, mitochondria can be the target of ONOO-induced protein tyrosine nitration while also being a source of ONOO. For instance, it causes the inactivation of MnSOD by nitrating and oxidizing key tyrosine residues. Although not yet thoroughly explored, the finding that increasing ONOO worsens cardiac IR damage and inhibiting ONOO synthesis improves recovery after IR strongly suggests that nitration of MnSOD hinders detoxification of ROS.

In other studies, IR enhanced myocardial tyrosine nitration, resulting in 23 proteins being nitrated. Around 10 of these were mitochondrial proteins, including four oxidative phosphorylation system components, five matrix enzymes, and a voltage-dependent anion channel, indicating that ONOO can disrupt a wide spectrum of mitochondrial proteins [18].

5.5 DISTURBANCES OF IONS

In liver IRI, there are changes in the homeostatic levels of Ca^{2+}, H^+, and Na^+, which can lead to cellular death if they are too high. The cytoplasm, ER, and mitochondria are the main sources of Ca^{2+} in the cell. The geographical and temporal distribution of Ca^{2+} in these compartments is determined by the interaction of Ca^{2+} migration across the membranes of these structures.

IRI disrupts the Ca^{2+} regulating systems in the ER, plasma, and membranes of mitochondria. Consequently Ca^{2+} excess occurs in the mitochondrial matrix and cytoplasm. Elevated Ca^{2+} release from ER reserves as well as its entry across the cell membrane, activate SOC channels in ryanodine receptors and plasma membrane, increasing cytoplasmic Ca^{2+} [22]. Cold ischemia injury also inhibits ER and plasma membrane Ca^{2+} ATPases, which typically export cytoplasmic Ca^{2+} to the ER lumen and extracellular space, respectively, exacerbating the rise in cytoplasmic Ca^{2+}. During hepatic IR injury, mitochondrial Ca^{2+} uniporter is activated due to elevation in cytoplasmic Ca^{2+} which results in subsequent mitochondrial Calcium (mCa^{2+}) [23]. The Ca^{2+} flow regulating PLC-dependent mP2Y-like receptors is likely involved in activating these Ca^{2+} selective channels. ADP and AMP activate mP2Y1, causing the uniporter to stimulate Ca^{2+} uptake. The ATP, ADP, or AMP via activation of mP2Y2 results in inhibition of uniporter. During the ischemia phase of IR damage, there is a decrease in $[ATP]_{cytosol}$ and an increase in $[AMP\ and\ ADP]_{cytosol}$, resulting in an overall increase in the activity of the Ca^{2+} uniporter and therefore in mitochondrial calcium.

The drop in mitochondrial transmembrane potential is caused by an increase in mitochondrial $[Ca^{2+}]$ ($m\Delta\psi$) [24]. In response, ATP-synthase reverses its activity and ATP is hydrolyzed to generate energy for the various ionic pumps that keep the mitochondrial membrane potential constant. It leads to a fractional restoration of ($m\Delta\psi$) that causes an increase in Ca^{2+} input through the mCa^{2+} uniporter. As a result of the ATP synthase activity being reversed, ATP depletion and increasing mCa^{2+} overload is generated (due to net ATP consumption by mitochondria during ischemia). The mCa^{2+} excess causes translocation of Bax from the cytoplasm to the mitochondria, resulting in the formation of MPT pores, the release of cytochrome C, and cellular death.

Hydrogen (H^+) and sodium (Na^+) also have a role in influencing the liver's response to IR assaults. Through anaerobic respiration, hepatic ischemia causes intracellular acidosis [25]. The Na^+/HCO_3^- cotransporter (NHCT) and Na^+/H^+ exchanger (NHE) are activated by liver cells to return intracellular pH to normal. In return for Na input into the cell, the NHE extrudes H into the extracellular environment. Na^+ and HCO_3^- are transported intracellularly through the NHCT. Both of these methods result in intracellular Na^+ build-up as a result of the ischemically produced intracellular acidic pH. Cell death is caused by an increase in intracellular $[Na^+]$ as a result of intracellular acidosis. Intracellular acidosis is cytoprotective in and of itself.

5.6 ISCHEMIC REPERFUSION INJURY AND MITOCHONDRIAL FISSION

Mitochondria are bilayer membrane organelles with varying shapes and sizes in various cells, and mitochondria in liver cells are consistently spherical or ovoid. Furthermore, mitochondria are extremely dynamic organelles that adjust to stress via a variety of mechanisms, including proteolytic systems that destroy misfolded proteins. Proteasomes or mitochondrial autophagy might break down or remove damaged mitochondrial outer membrane proteins. Mitochondria maintain structural and functional integrity by continual fission and fusion [26].

Mitochondrial quality control is crucial for maintaining the equilibrium of mitochondria, which is maintained by diverse mechanisms including fusion and fission of mitochondria, and selective mitochondrial autophagy [27, 28]. Mitochondrial autophagy and homeostasis are required for IRI in mammals, and the damaged mitochondria can be isolated by fusion and fission of mitochondria, while maintaining the comparative equilibrium of mitochondrial components like protein, DNA, and metabolites, while the destruction and regeneration of damaged mitochondria are carried out via mitochondrial autophagy [29].

Dynamin-related proteins control mitochondrial fusion and fission (DRPs). On mitochondria, dynamin family members have four receptors: mitochondrial fission factor, mitochondrial fission protein 1 (Fis1), and dynamic proteins of mitochondria of 49 and 51 kDa and [30]. During IRI, there is decreased intracellular ATP concentration which induces mitochondrial fission to sustain the mitochondrial number by increasing the expression of Drp1 and Fis1 protein, whereas inadequate fission of mitochondria promotes mitochondrial disintegration and apoptosis of the cell. As a result, a range of compounds has been identified to limit the excessive damage of mitochondria and IRI in the Drp1-mediated cell signaling pathway. Exogenous administration of irisin, for example, can greatly suppress the expression of Fis1 and Drp1, thereby acting as a defending factor [31].

By phosphorylating Ser-616 at Drp1, the CDK1/cyclin B complex can cause fission of mitochondria and apoptosis of hepatocytes, whereas HSS can significantly inhibit the expression of Bax and CDK1, decrease phosphorylation of Drp1 on Ser-616, and reduce cyt c release to limit IRI [32]. Augmenter of liver regeneration (ALR) inhibits Drp1 phosphorylation, whereas small ubiquitin-like modifications (SUMOylation) encourage Drp1 activation and accelerate the division of mitochondria. According to a new study, ALR can suppress Drp1 SUMOylation and hence resist the action of IRI [33].

5.7 THE MITOCHONDRIAL PERMEABILITY TRANSITION

It is a pathogenic condition caused by the emergence of highly conductive permissible transition pores in the IMM, except for those for which particular carriers exist. Under normal circumstances, the membranes of mitochondria are impermeable largely to most of the solutes. However, all solutes with a molecular mass of up to 1,500 Da, diffuse non-selectively into and out of mitochondria, on opening of PT pores (Figure 5.1). Consequently, when the MPT begins, the depolarization and uncoupling of mitochondria start and initiate massive amplitude in swelling, resulting in ATP depletion and cell death [34]. Opening just one or a few of them may be sufficient enough to cause the depolarization and swelling of mitochondria due to the highly conductive nature of the PT pores.

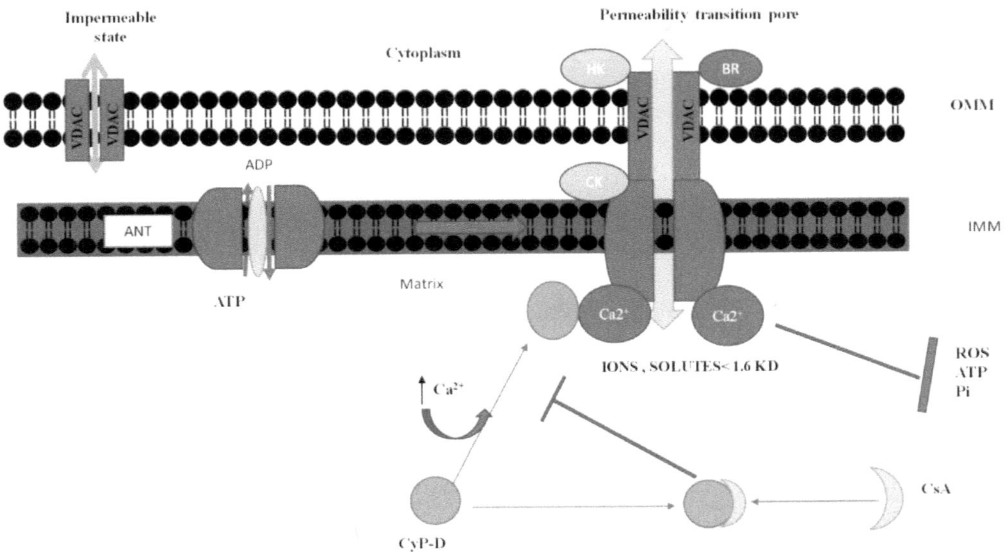

FIGURE 5.1 The common mechanism of formation of mitochondrial permeability of transition pore. The core component of mPTP is thought to be cyclophilin D, the voltage-dependent anion channel, and the adenine nucleotide transferase.

MPT is triggered by calcium, reactive oxygen species, inorganic phosphate, and alkaline pH, and, whereas cyclosporin A (CsA), Mg^{2+}, acidic pH, and trifluoperazine block permeability transition pore from opening. Once MPT starts, the permeability barrier of IMM falls resulting in rapid permeation of solutes with molecular masses of up to 1.5 kDa across the mitochondrial inner membranes. As a result, mitochondria depolarize, uncouple, and expand, depleting ATP and causing cell death by necrosis. Despite the fact that necrosis is the most prevalent kind of cell death after I/R, the MPT can also trigger apoptosis in the ischemic liver and other organs [35]. Following the onset of the MPT and subsequent mitochondrial swelling, the outer mitochondrial membranes rupture which releases cyt c into the extracellular medium triggering apoptosis dependent on caspase and ATP, because this 12 kDa protein is an integral component of the apoptosome. However, the cells do not undergo apoptosis when there is reduced availability of ATP, such as in the condition of ischemia. Instead, these ATP-deficient cells produce necrotic cell death even in the presence of upstream proapoptotic signals. As a result, ATP availability is a critical factor in the transition from necrosis to apoptosis [36]. CsA blocks reperfusion-induced apoptosis. After I/R injury, in both types of cell death, the MPT is the most common pathway. After I/R, reactive oxygen species and mitochondrial calcium are thought to trigger the MPT. After reperfusion, the onset of MPT is preceded by a rise in ROS and mitochondrial calcium, isolated rodent hepatocytes imaging [37].

The development of ROS and cell death after reperfusion can be prevented by calcium [38]. Cytosolic ROS, on the other hand, has a little role in IRI, as seen by the lack of cell protection provided by diphenyleneiodonium chloride or apocynin, both of which are inhibitors of cytosolic NADPH oxidase.

5.8 mPTP AND MITOPHAGY

In normal conditions, mitochondria, which have a complete structure and function, produce ATP as the primary energy source. As a result of the severe injury to mitochondria and fast depletion of ATP, cellular respiration switches from aerobic to anaerobic, activating the Na^+/H^+ antiporter to boost sodium concentrations and raising intracellular calcium to activate apoptotic pathways [39].

Autophagy is a critical process in cells and has been connected to some clinical disorders, including oxidative stress and inflammation. Sirtuin 1 is a nicotinamide-dependent type III deacetylase protein that regulates inflammation in the liver and metabolism of lipids, increases autophagy, and restores membrane potential and mitochondrial number [40]. Overexpression of SIRT1 inhibits abnormal Mitochondrial autophagy, permeability transitions are inhibited by SIRT1 overexpression, and its function is connected to mitofusin-2, a protein that maintains mitochondrial integrity by participating in mitochondrial fusion and fission SIRT1 overexpression results in deacetylated MFN2 that reverses a wide variety of pathogenic processes [42]. MFN2 also inhibits apoptosis by reducing calcium input and downregulating mitochondrial Ca^{2+} uptake 1/2. As a result, SIRT1 loss causes autophagy failure and mitochondrial malfunction [43]. The mPTP and MCUC are responsible for calcium absorption in the liver's mitochondria. During ischemia, MCUC accumulates enormous amounts of Ca^{2+} inside the mitochondria and after reperfusion enhances the mPTP development resulting in apoptosis [44]. Recent research suggested that EF-hand-containing transmembrane protein 1 (Letm1), leucine zipper protein, MCUC, mitochondrial ryanodine receptor type 1, and uncoupling proteins are involved in calcium absorption in mitochondria. Small molecular compounds migrate into mitochondrial inner membranes on generation mPTP, producing depolarization, swelling, and uncoupling resulting in the consumption of ATP, apoptosis, and necrosis [39]. Even though numerous processes involved in the creation of mPTP are unclear, various proteins appear to play a part, including ATP synthase cyclophilin D and adenine nucleotide translocase (CypD). CypD is one among them, and it may migrate to the inner mitochondrial membrane to induce cell death by activating mPTP, and the biggest risk factor for mPTP, CypD, may be controlled. Cyclosporin A (CsA) is a pore inhibitor.

In the Cyclophilin D signaling, many drugs are seen to decrease Cyclophilin D-mediated mitochondrial death. While reducing SPG7 synthesis raises calcium levels, p53 (which is a tumor suppressor protein) goes to the mitochondria and joins CypD to enhance the growth of mPTP, comparable to CypD deletion effects. Inhibiting SPG7-CypD binding protects cells against ROS and calcium-mediated death of cells as a consequence. Ginsenoside, the major component of ginseng, has been shown to reduce the synthesis, and cyt c, Caspase-3, and Caspase-9 and prevent cell apoptosis.

In mouse and human liver tissues C31, which is a small molecular inhibitor of cyclophilin D shown to decrease mitochondrial swelling (induced by calcium ion) and damage by hindering the mPTP opening to protect the liver from IRI. Ginsenoside has been shown to slow IRI and mitochondrial injury by obstructing the Cyclophilin D pathway and C31, which has a high affinity for CypD in mice and humans Although Rhein can cause necrosis and apoptosis in hepatocytes by inducing mPTP to open, CsA can work with CypD to lower Rhein's damaging effects by blocking mPTP from opening [45]. Furthermore, DS44170716 a small-molecule compound shows a comparable protective effect to CsA against calcium-mediated mitochondrial enlargement (Figure 5.2). It can prevent overload of calcium by inhibiting the complexes III, IV, and V, of mitochondria revealing a unique strategy for preventing apoptosis induced by calcium ions.

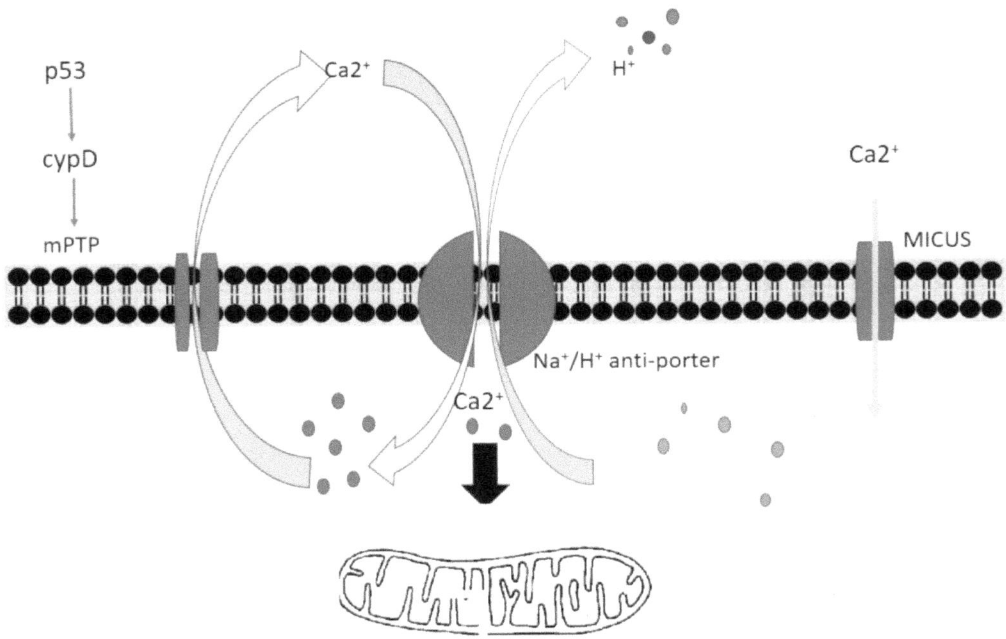

FIGURE 5.2 The calcium overload is induced by p53 CypD-mediated signaling which causes activation of mPTP thus permitting influx of calcium into the cell which further causes mitochondrial dysfunction.

5.8.1 PATHOLOGICAL IMPLICATIONS OF MPTP OPENING

The mPTP development and subsequent opening is a major role in mitochondrial malfunction and cell death caused by mitochondria. It is well-accepted that uncontrolled mPTP opening results in irreversible necrosis or apoptosis [46].

The mitochondria can swell massively and permanently when exposed to high calcium concentrations in physiological conditions, and shifts the chemiosmotic gradient across the IMM by enhancing the permeability of the membrane to small solutes resulting in mitochondrial permeability transition. The ROS level increases due to the uncoupling of oxidative phosphorylation reduction in ATP synthesis [34].

Interestingly, ischemic heart lesions and many neurodegenerative disorders have cytosolic calcium dysregulation as a pathologic element, which leads to mPTP opening via overload or uptake anomalies, despite their various etiologies [47]. Several studies have shown that activating the mPTP during initial reperfusion after ischemia is a harmful event in the heart, producing further myocardial damage. Under pathological conditions, mPTP is constantly activated in brain neurons, reducing the cells' ability to maintain calcium homeostasis and contributing to neurodegeneration. Importantly, in both circumstances, genetic and pharmacological inhibition of CypD avoids the dysfunction of mitochondria.

5.9 CELL DAMAGE IN ISCHEMIA-REPERFUSION INJURY

I/R injury causes two types of tissue damage: ischemia injury and reperfusion injury. Hypoxia and malnutrition can occur as a result of ischemic damage. The metabolic products of cells are preserved after prolonged ischemia, resulting in metabolic acidosis. Local inflammation and ROS generation increase when blood flow is restored, resulting in secondary damage. The magnitude of overall tissue damage influences the cell response. Apoptosis, autophagy, necrosis, and necroptosis are all possible outcomes of prolonged I/R injury (48, 49). Moderate ischemia-reperfusion damage may activate recovery mechanisms for subsistence and produce cell malfunction through autophagy. Cell death can be caused by apoptotic or necrotic pathways if the damage is substantial. Ischemia-reperfusion injury with a shorter period may activate cell survival mechanisms that limit ROS production and cell destruction. Depending on the pathogenic pathways of I/R damage, multiple kinds of cell death may be generated.

5.9.1 MITOPTOSIS AND APOPTOSIS

Pathophysiology of numerous disorders involves cell death. Apoptosis is triggered by hypoxia in ischemia damage and the generation of ROS in I/R injury [50]. The two primary types of apoptotic mechanisms are extrinsic and intrinsic. Both paths can influence and converse with one another. TRAIL, TL1A, TWEAK, Fas ligand, and TNF-α are death ligands and receptors that activate the extrinsic route, commonly recognized as the death receptor pathway. This signaling complex triggers caspase-8, a protease that cleaves caspase-3, resulting in cell death in damaged cells. When the integrity of the membrane of mitochondrial is compromised by cellular toxins, radiations, or hypoxia the intrinsic pathway is triggered, resulting in Bcl-2 (pro-apoptotic) protein activation. Bad increases during IRI and binds to Bcl-XL and Bcl-2. Bak and Bax are processed and integrated into the membrane of mitochondria at the same moment, resulting in the release of HTRA2/Omi, endonuclease G, apoptosis-inducing factor (AIF), Smac/Diablo, and cyt c. APAF-1 binds to cyt c and forms apoptosome by activating pro-caspase-9. Caspase-9 activation by the apoptosome initiates a cascade that promotes apoptosis [51]. Endo G also causes DNA fragmentation by interacting with AIF. By expressing PIDD and interacting with pro-caspase-2 and CRADD/RAIDD after DNA damage induced by ROS, p53 causes the creation of PIDDosome. To induce apoptosis, caspase-2 is activated by a complex that cleaves downstream caspases. I/R damage-produced apoptosis is less prevalent than necrosis [52]. Apoptosis, on the other hand, is not always followed by a local inflammatory response.

Mitochondria get fractured during mitoptosis, a process that results in programmed cell death. BAX and BAK, rather than other cleavage proteins govern the cleavage of mitochondria during mitoptosis and can be connected to the outer membrane of mitochondria via Drp1. Drp1 activity is reduced via the K38A-Drp1 pathway, resulting in mitosis, although fusion is unaffected. In cardiac I/R damage, mitochondrial morphology showed a long strip that was much longer than normal cell mitochondrial length [53]. Drp1 overexpression resulted in mitochondrial overexpression, elevated mitochondrial membrane potential, a considerable reduction in mitochondrial number, release of cyt c, and death of the cell. The Drp1/BAX not only co-splits but also harms heavier cells by creating an oligomer to increase split activity. BAX is a mitochondrial division-assist protein that promotes apoptosis. By inducing the creation of the mitochondrial fusion protein Mfn2, BAX can alter the distribution of mitochondria and promote mitochondrial fusion (mitofusion 2). In the regulation of cyt c during apoptosis, Cristae which is located in the inner membrane structure of mitochondria plays a key role.

OPA1 is a crucial protein component in cristae (optic atrophy 1). Endometrial damage may result from abnormal OPA1 function, which promotes apoptosis and mitochondrial fragmentation. The resulting mechanism uses Bid to divert cristae wrinkles and break the OPA1 oligomer creation, allowing the connection and maintaining the cristae shape. Furthermore, the released cyt c encourages the release of more cyt c. Overexpression of OPA1 slows structural changes in the mitochondrial IMM and reduces the damage caused by Bid in myocardial ischemia injury [54].

5.9.2 NECROPTOSIS AND NECROSIS

Like apoptosis Necrosis, is a type of planned cell death executionary phase, though the effects of apoptotic and necrosis cell death for an organism are significantly different. Early cell membrane permeability and organelle enlargement define necrosis, a kind of cell death. Necrosis develops when the cells are subjected to extreme extrinsic stress [55]. Necrosis is thought to be uncontrolled and inactive. Even though necrosis is common in human pathophysiology, few investigations on the mechanism of necrosis control in I/R damage have been conducted [56]. A sudden shift in the external environment produced by biological, physical, or chemical harm causes necrosis. It is characterized by organ enlargement, cell disintegration, and deprivation in the function of mitochondria, and it causes a slew of confined inflaming responses in ischemic tissue. Traditional necrosis is considered uncontrolled, even though it is seen in a range of pathologic conditions. Recent investigations in inflammatory and I/R damage animal models have verified the hallmarks of necrosis, also known as necroptosis. Similar to necrosis, Necroptosis (a planned cell death type) is regulated by death signals and characterized by a death pattern [57]. Obstruction of blood flow to tissue can cause local ischemia which damages the energy metabolism and can lead to cell death as it restricts the supply of nutrients and oxygen to the tissue [58]. Reperfusion results in the restoration of Oxygen and the generation of ROS, which leads to the inactivation and death of the cell [59]. Because the active domains of the two proteins are so similar, phosphorylation, not structural alterations, is responsible for the creation of a key necrosome protein in RIP3 and RIP1. Increasing the stability of TNFR complex II while also activating the RIP3 downstream regulatory signal route. In the absence of I/R damage, the interaction of RIP1 with RIP3 results in their dimerization that cause necroptosis induced by necrosome, whereas cell death is mediated by apoptosis or necrosis. However, in the case of RIP3 overexpression, not only was a substantial amount of ROS detected, but also it has enhanced the regulation of protein NF-κB. These findings imply that RIP3 may be involved in disease development and that RIP3 can influence the *in vivo* transition of cells into necroptosis. The phosphorylation of RIP3 at Ser161 phosphate sites can directly or indirectly engage necroptosis, including RIP1 activity. RIP3 regulates viral infection in mouse cells more than RIP1 regulates necroptosis, according to studies. However, growing data suggests that some kinds of tightly regulated intrinsic cellular mechanisms mediate necrosis which is now referred to as necroptosis [55]. The understanding of necroptosis's pathophysiological consequences and molecular mechanisms remains a critical goal of I/R damage research due to the link between inflammation and necroptosis in the etiology of human illness.

5.9.3 AUTOPHAGY

It is a biological process by which the damaged or redundant cellular components and biological macromolecules are destroyed in the membrane vesicles in the cytoplasm. The majority of cytosolic constituents and certainly injured organelles, like granules, are found in autophagic common inclusions. The presence is carried to the lysosome in dissimilar manners depending on the source of the autophagosomes, and this process is essentially split into the following types:

1. **Macroautophagy:** The ER and Golgi complex are the major sources of the membrane. The net's monolayer film generates a dual-layered parting film with a shape like a cup that fully enwraps the object to be degraded, which is subsequently bonded with the lysosomal membrane to custom an autophagosome.
2. **Microautophagy:** Lysosomal enzymes break down the degradation products.
3. **Chaperone Facilitated Autophagy:** Chaperone binds the protein to be degraded and guides it to the lysosomes, where it is broken down and deteriorates by enzymatic action.

As a result, the CMA route has a distinct perceptiveness; there is a significant difference between the first two pathways at this point. Cell autophagy is induced during ischemia-reperfusion injury [60]. Autophagy-related proteins control autophagy and they are made up of genes that are related to autophagy (ATG). Hormones have a major role in autophagy. Atg13 and RB1CC1/Atg17 form the ULK1-Atg13-RB1CC1-Atg10 complex, which acts on mTORC1 and engages in the preliminary stages of autophagy. There is decreased mTOR activity in cells during insufficient calories and energy, while ULK1 is activated, and autophagy membranes begin to develop. The Atg8/LC3 and Atg12-Atg5-Atg16 complexes, two ubiquitination combining mechanisms, are required for autophagic vesicle elongation. The cytoplasmic free LC3-I fixes to PE and is lipidated to create LC3-II, which is founded on a phagocytic membrane. The existence of certain LC3-II indicators, as well as the creation of autophagosomes, can be engaged as autophagy-induced subject matter, with autophagosomes being destroyed by the lysosomal hydrolase [61]. PI3K-Akt, mTOR, p53, CA-AMPK, and ER pressure are regulatory autophagy pathways (62, 63). mTOR promotes cell development and suppresses the earliest stages of autophagy (64, 65). It is found downstream of PI3K-Akt signaling. Rapamycin, a mTOR inhibitor, can cause autophagy. Mitochondria produce ATP at the start of reperfusion, which triggers autophagy. Reperfusion, on the other hand, can cause calcium to build up in the mitochondria. When the reperfusion pressure was not entirely neutralized by enhanced autophagy, liver cell injury occurred [66]. The usage of GFP-LC3 and chloroquine to view the picture (autophagic picture) in earlier research revealed that autophagy is promoted during the early stage of reperfusion. During the phase of intracellular flow, however, autophagy was unable to remove all defective mitochondria due to mitochondrial Ca^{+} overload and high reactive oxygen species, which surpassed the clearance autophagic capabilities, resulting in MPT. Mitochondrial permeability transition causes permanent oxidative phosphorylation decoupling, inadequate Adenosine triphosphate production, and failure of energy, resulting in hepatocellular mortality [67]. Understanding the processes that govern autophagy to decrease ischemia-reperfusion damage is crucial.

5.10 CONCLUSION

Ischemia-reperfusion injury has a wide range of clinical manifestations. Much development has been made in considering the processes underlying the development of ischemia-reperfusion damage in recent years. Mitochondrial malfunction is a key cellular event that leads to I/R damage, and eliminating aberrant and dysfunctional mitochondria as soon as possible protects not only mitochondrial quality but also cell and tissue survival. In this review, we discussed new research findings on the basic pathophysiology of ischemia-reperfusion injury., with a focus on the role of reactive oxygen species and the cell death pathway. Although the exact mechanisms driving ischemia-reperfusion damage are unclear, we hope that a better knowledge of pathophysiology will enable the development of new treatment approaches.

KEYWORDS

- **autophagy**
- **cell damage**
- **ischemic/reperfusion injury**
- **mitochondrial permeability transition pore**
- **mitoptosis**
- **necroptosis**
- **necrosis**
- **reactive oxygen species**

REFERENCES

1. Khan, H., Kashyap, A., Kaur, A., & Singh, T. G., (2020). Pharmacological postconditioning: A molecular aspect in ischemic injury. *Journal of Pharmacy and Pharmacology, 72*(11), 1513–1527.
2. Khan, H., Bangar, A., Grewal, A. K., Bansal, P., & Singh, T. G., (2022). Caspase-mediated regulation of the distinct signaling pathways and mechanisms in neuronal survival. *International Immunopharmacology, 110*, 108951.
3. Tiwari, P., Khan, H., Singh, T. G., & Grewal, A. K., (2022). Poly (ADP-ribose) polymerase: An overview of mechanistic approaches and therapeutic opportunities in the management of stroke. *Neurochemical Research, 18*, 1–23.
4. Bernardi, P., Rasola, A., Forte, M., & Lippe, G., (2015). The mitochondrial permeability transition pore: Channel formation by F-ATP synthase, integration in signal transduction, and role in pathophysiology. *Physiol. Rev., 95*(4), 1111–1155.
5. Chouchani, E. T., Pell, V. R., James, A. M., Work, L. M., Saeb-Parsy, K., Frezza, C., et al., (2016). A unifying mechanism for mitochondrial superoxide production during ischemia-reperfusion injury. *Cell Metab., 23*(2), 254–263.
6. Lesnefsky, E. J., Chen, Q., Tandler, B., & Hoppel, C. L., (2017). Mitochondrial dysfunction and myocardial ischemia-reperfusion: Implications for novel therapies. *Annu. Rev. Pharmacol. Toxicol., 57*, 535–565.
7. Pell, V. R., Chouchani, E. T., Murphy, M. P., Brookes, P. S., & Krieg, T., (2016). Moving forwards by blocking back-flow: The yin and yang of MI therapy. *Circ. Res., 118*(5), 898–906.
8. Matsuzaki, S., & Humphries, K. M., (2015). Selective inhibition of deactivated mitochondrial complex I by biguanides. *Biochemistry, 54*(11), 2011–2021.
9. Lesnefsky, E., Thompson, J., Hu, Y., & Chen, Q., (2016). Administration of metformin during early reperfusion decreases cardiac injury through inhibition of MPTP opening. *FASEB J., 30*, 724–725.
10. Khan, H., Sharma, K., Kumar, A., Kaur, A., & Singh, T. G., (2022). Therapeutic implications of cyclooxygenase (COX) inhibitors in ischemic injury. *Inflamm. Res.*, 1–16.
11. Bezawork-Geleta, A., Rohlena, J., Dong, L., Pacak, K., & Neuzil, J., (2017). Mitochondrial complex II: At the crossroads. *Trends Biochem. Sci., 42*(4), 312–325.
12. Khan, H., Gupta, A., Singh, T. G., & Kaur, A., (2021). Mechanistic insight on the role of leukotriene receptors in ischemic–reperfusion injury. *Pharmacol. Rep., 73*(5), 1240–1254.
13. Mazat, J. P., Devin, A., & Ransac, S., (2020). Modeling mitochondrial ROS production by the respiratory chain. *Cell Mol. Life Sci., 77*(3), 455–465.
14. Kausar, S., Wang, F., & Cui, H., (2018). The role of mitochondria in reactive oxygen species generation and its implications for neurodegenerative diseases. *Cells, 7*(12), 274.
15. Sheeran, F. L., & Pepe, S., (2017). Mitochondrial bioenergetics and dysfunction in failing heart. *Mitochondrial Dyn Cardiovasc Med.*, 65–80.
16. Paradies, G., Paradies, V., Ruggiero, F. M., & Petrosillo, G., (2019). Role of cardiolipin in mitochondrial function and dynamics in health and disease: Molecular and pharmacological aspects. *Cells, 8*(7), 728.
17. Kaur, A., Singh, T. G., Khan, H., Kumar, M., Singh, N., & Abdel-Daim, M. M., (2022). Neuroprotective effect of piclamilast-induced post-ischemia pharmacological treatment in mice. *Neurochemical Research*, 1–4.
18. Khan, H., Grewal, A. K., & Singh, T. G., (2022). Pharmacological postconditioning by protocatechuic acid attenuates brain injury in ischemia–reperfusion (I/R) mice model: Implications of nuclear factor erythroid-2-related factor pathway. *Neuroscience, 491*, 23–31.
19. Sharma, V. K., Singh, T. G., & Mehta, V., (2021). Stressed mitochondria: A target to intrude Alzheimer's disease. *Mitochondrion, 59*, 48–57.
20. Brand, M. D., (2016). Mitochondrial generation of superoxide and hydrogen peroxide as the source of mitochondrial redox signaling. *Free Radic. Biol. Med., 100*, 14–31.
21. Bertero, E., & Maack, C., (2018). Calcium signaling and reactive oxygen species in mitochondria. *Circ. Res., 122*(10), 1460–1478.
22. Park, S. J., Li, C., & Chen, Y. M., (2021). Endoplasmic reticulum calcium homeostasis in kidney disease: Pathogenesis and therapeutic targets. *Am. J. Pathol., 191*(2), 256–265.

23. Khan, H., Garg, N., Singh, T. G., Kaur, A., & Thapa, K., (2022). Calpain inhibitors as potential therapeutic modulators in neurodegenerative diseases. *Neurochemical Research*, 1–25.
24. Zorova, L. D., Popkov, V. A., Plotnikov, E. Y., Silachev, D. N., Pevzner, I. B., Jankauskas, S. S., et al., (2018). Mitochondrial membrane potential. *Anal. Biochem., 552*, 50–59.
25. Ogasawara, E., Nakada, K., & Ishihara, N., (2020). Distal control of mitochondrial biogenesis and respiratory activity by extracellular lactate caused by large-scale deletion of mitochondrial DNA. *Pharmacol. Res., 160*, 105204.
26. Shi, S., Wang, L., Van, D. L. L. J. W., Pan, Q., & Verstegen, M. M. A., (2021). Mitochondrial dysfunction and oxidative stress in liver transplantation and underlying diseases: New insights and therapeutics. *Transplantation, 105*(11), 2362–2373.
27. Zhou, H., Toan, S., Zhu, P., Wang, J., Ren, J., & Zhang, Y., (2020). DNA-PKcs promotes cardiac ischemia reperfusion injury through mitigating BI-1-governed mitochondrial homeostasis. *Basic Res. Cardiol., 115*(2), 1–17.
28. Wang, J., Zhu, P., Li, R., Ren, J., & Zhou, H., (2020). Fundc1-dependent mitophagy is obligatory to ischemic preconditioning-conferred renoprotection in ischemic AKI via suppression of Drp1-mediated mitochondrial fission. *Redox Biol., 30*, 101415.
29. Shin, J. K., & Lee, S. M., (2017). Genipin protects the liver from ischemia/reperfusion injury by modulating mitochondrial quality control. *Toxicol. Appl. Pharmacol., 328*, 25–33.
30. Anzell, A. R., Maizy, R., Przyklenk, K., & Sanderson, T. H., (2018). Mitochondrial quality control and disease: Insights into ischemia-reperfusion injury. *Mol. Neurobiol., 55*(3), 2547–2564.
31. Bi, J., Zhang, J., Ren, Y., Du, Z., Li, Q., Wang, Y., et al., (2019). Irisin alleviates liver ischemia-reperfusion injury by inhibiting excessive mitochondrial fission, promoting mitochondrial biogenesis and decreasing oxidative stress. *Redox Biol., 20*, 296–306.
32. Zhang, C., Huang, J., & An, W., (2017). Hepatic stimulator substance resists hepatic ischemia/reperfusion injury by regulating Drp1 translocation and activation. *Hepatology, 66*(6), 1989–2001.
33. Huang, J., Xie, P., Dong, Y., & An, W., (2021). Inhibition of Drp1 SUMOylation by ALR protects the liver from ischemia-reperfusion injury. *Cell Death Differ., 28*(4), 1174–1192.
34. Izzo, V., Bravo-San, P. J. M., Sica, V., Kroemer, G., & Galluzzi, L., (2016). Mitochondrial permeability transition: New findings and persisting uncertainties. *Trends Cell Biol., 26*(9), 655–667.
35. Mnatsakanyan, N., Beutner, G., Porter, G. A., Alavian, K. N., & Jonas, E. A., (2017). Physiological roles of the mitochondrial permeability transition pore. *J. Bioenerg. Biomembr., 49*(1), 13–25.
36. Rihal, V., Khan, H., Kaur, A., & Singh, T. G., (2022). Vitamin D as therapeutic modulator in cerebrovascular diseases: A mechanistic perspectives. *Critical Reviews in Food Science and Nutrition*, 1–23.
37. Hu, J., & Lemasters, J. J., (2020). Suppression of iron mobilization from lysosomes to mitochondria attenuates liver injury after acetaminophen overdose *in vivo* in mice: Protection by minocycline. *Toxicol. Appl. Pharmacol., 392*, 114930.
38. Sharma, V. K., & Singh, T. G., (2020). CREB: A multifaceted target for Alzheimer's disease. *Current Alzheimer Research, 17*(14), 1280–1293.
39. Zhang, H., Yan, Q., Wang, X., Chen, X., Chen, Y., Du, J., et al., (2021). The role of mitochondria in liver ischemia-reperfusion injury: From aspects of mitochondrial oxidative stress, mitochondrial fission, mitochondrial membrane permeable transport pore formation, mitophagy, and mitochondria-related protective measures. *Oxid. Med. Cell Longev., 2021*.
40. Cho, H. I., Seo, M. J., & Lee, S. M., (2017). 2-Methoxyestradiol protects against ischemia/reperfusion injury in alcoholic fatty liver by enhancing sirtuin 1-mediated autophagy. *Biochem. Pharmacol., 131*, 40–51.
41. Zhang, T., Gu, J., Guo, J., Chen, K., Li, H., & Wang, J., (2019). Renalase attenuates mouse fatty liver ischemia/reperfusion injury through mitigating oxidative stress and mitochondrial damage via activating SIRT1. *Oxid. Med. Cell Longev.*, 2019.
42. Chun, S. K., Lee, S., Flores-Toro, J., U, R. Y., Yang, M., Go, K. L., et al., (2018). Loss of sirtuin 1 and mitofusin 2 contributes to enhanced ischemia/reperfusion injury in aged livers. *Aging Cell, 17*(4), e12761.
43. Liang, R. P., Jia, J. J., Li, J. H., He, N., Zhou, Y. F., Jiang, L., et al., (2017). Mitofusin-2 mediated mitochondrial Ca^{2+} uptake 1/2 induced liver injury in rat remote ischemic perconditioning liver transplantation and alpha mouse liver-12 hypoxia cell line models. *World J. Gastroenterol., 23*(38), 6995.

44. Belosludtsev, K. N., Talanov, E. Y., Starinets, V. S., Agafonov, A. V., Dubinin, M. V., & Belosludtseva, N. V., (2019). Transport of Ca^{2+} and Ca^{2+}-dependent permeability transition in rat liver mitochondria under the streptozotocin-induced type I diabetes. *Cells, 8*(9), 1014.
45. Liu, J., Ding, D., Liu, F., & Chen, Y., (2022). Rhein inhibits the progression of chemoresistant lung cancer cell lines via the Stat3/Snail/MMP2/MMP9 pathway. *Biomed Res. Int., 2022.*
46. Bernardi, P., Di Lisa, F., Fogolari, F., & Lippe, G., (2015). From ATP to PTP and back: A dual function for the mitochondrial ATP synthase. *Circ. Res., 116*(11), 1850–1862.
47. Abeti, R., & Abramov, A. Y., (2015). Mitochondrial Ca^{2+} in neurodegenerative disorders. *Pharmacol. Res., 99*, 377–381.
48. Ling, Q., Yu, X., Wang, T., Wang, S. G., Ye, Z. Q., & Liu, J. H., (2017). Roles of the exogenous H2S-mediated SR-A signaling pathway in renal ischemia/reperfusion injury in regulating endoplasmic reticulum stress-induced autophagy in a rat model. *Cell Physiol. Biochem., 41*(6), 2461–2474.
49. Zhu, J., Yao, K., Wang, Q., Guo, J., Shi, H., Ma, L., et al., (2016). Ischemic postconditioning-regulated miR-499 protects the rat heart against ischemia/reperfusion injury by inhibiting apoptosis through PDCD4. *Cell Physiol Biochem., 39*(6), 2364–2380.
50. Liu, H., Jing, X., Dong, A., Bai, B., & Wang, H., (2017). Overexpression of TIMP3 protects against cardiac ischemia/reperfusion injury by inhibiting myocardial apoptosis through ROS/Mapks pathway. *Cell Physiol. Biochem., 44*(3), 1011–1023.
51. Chen, X., Zhang, X., Xue, L., Hao, C., Liao, W., & Wan, Q., (2017). Treatment with enriched environment reduces neuronal apoptosis in the periinfarct cortex after cerebral ischemia/reperfusion injury. *Cell Physiol. Biochem., 41*(4), 1445–1456.
52. Yu, H., Zhang, H., Zhao, W., Guo, L., Li, X., Li, Y., et al., (2016). Gypenoside protects against myocardial ischemia-reperfusion injury by inhibiting cardiomyocytes apoptosis via inhibition of CHOP pathway and activation of PI3K/Akt pathway *in vivo* and *in vitro*. *Cell Physiol. Biochem., 39*(1), 123–136.
53. Maneechote, C., Palee, S., Chattipakorn, S. C., & Chattipakorn, N., (2017). Roles of mitochondrial dynamics modulators in cardiac ischemia/reperfusion injury. *J. Cell Mol. Med., 21*(11), 2643–2653.
54. Wu, M. Y., Yiang, G. T., Liao, W. T., Tsai, A. P. Y., Cheng, Y. L., Cheng, P. W., et al., (2018). Current mechanistic concepts in ischemia and reperfusion injury. *Cell Physiol. Biochem., 46*(4), 1650–1667.
55. Saeed, W. K., Jun, D. W., Jang, K., Chae, Y. J., Lee, J. S., & Kang, H. T., (2017). Does necroptosis have a crucial role in hepatic ischemia-reperfusion injury? *PLoS One, 12*(9), e0184752.
56. Kim, C. R., Kim, J. H., Park, H. Y. L., & Park, C. K., (2017). Ischemia reperfusion injury triggers TNFα induced-necroptosis in rat retina. *Curr. Eye Res., 42*(5), 771–779.
57. Newton, K., Dugger, D. L., Maltzman, A., Greve, J. M., Hedehus, M., Martin-McNulty, B., et al., (2016). RIPK3 deficiency or catalytically inactive RIPK1 provides greater benefit than MLKL deficiency in mouse models of inflammation and tissue injury. *Cell Death Differ., 23*(9), 1565–1576.
58. Hoehn, R. S., Seitz, A. P., Jernigan, P. L., Gulbins, E., & Edwards, M. J., (2016). Ischemia/reperfusion injury alters sphingolipid metabolism in the gut. *Cell Physiol. Biochem., 39*(4), 1262–1270.
59. Jing, N., Fang, B., Wang, Z. L., & Ma, H., (2017). Remote ischemia preconditioning attenuates blood-spinal cord barrier breakdown in rats undergoing spinal cord ischemia reperfusion injury: Associated with activation and upregulation of CB1 and CB2 receptors. *Cell Physiol. Biochem., 43*(6), 2516–2524.
60. Ma, S., Wang, Y., Chen, Y., & Cao, F., (2015). The role of the autophagy in myocardial ischemia/reperfusion injury. *Biochim Biophys Acta (BBA)-Molecular Basis Dis., 1852*(2), 271–276.
61. Decuypere, J. P., Ceulemans, L. J., Agostinis, P., Monbaliu, D., Naesens, M., Pirenne, J., et al., (2015). Autophagy and the kidney: Implications for ischemia-reperfusion injury and therapy. *Am. J. Kidney Dis., 66*(4), 699–709.
62. Gao, Y., Yang, H., Chi, J., Xu, Q., Zhao, L., Yang, W., et al., (2017). Hydrogen gas attenuates myocardial ischemia reperfusion injury independent of postconditioning in rats by attenuating endoplasmic reticulum stress-induced autophagy. *Cell Physiol. Biochem., 43*(4), 1503–1514.
63. Zhu, S., Xu, T., Luo, Y., Zhang, Y., Xuan, H., Ma, Y., et al., (2017). Luteolin enhances sarcoplasmic reticulum Ca^{2+}-ATPase activity through p38 MAPK signaling thus improving rat cardiac function after ischemia/reperfusion. *Cell Physiol. Biochem., 41*(3), 999–1010.

64. Shu, L., Zhang, W., Huang, C., Huang, G., & Su, G., (2017). Troxerutin protects against myocardial ischemia/reperfusion injury via Pi3k/Akt pathway in rats. *Cell Physiol. Biochem., 44*(5), 1939–1948.
65. Zeng, Q., Fu, Q., Wang, X., Zhao, Y., Liu, H., Li, Z., et al., (2017). Protective effects of sonic hedgehog against ischemia/reperfusion injury in mouse skeletal muscle via AKT/mTOR/p70S6K signaling. *Cell Physiol. Biochem., 43*(5), 1813–1828.
66. Qin, J., Zhou, J., Dai, X., Zhou, H., Pan, X., Wang, X., et al., (2016). Short-term starvation attenuates liver ischemia-reperfusion injury (IRI) by Sirt1-autophagy signaling in mice. *Am. J. Transl. Res., 8*(8), 3364.
67. Liu, A., Huang, L., Guo, E., Li, R., Yang, J., Li, A., et al., (2016). Baicalein pretreatment reduces liver ischemia/reperfusion injury via induction of autophagy in rats. *Sci. Rep., 6*(1), 1–12.

CHAPTER 6

Stroke and Its Effects in Daily Life

KUMARI TANUJA,[1] KAJAL BAGRI,[1] RAJNI THAKUR,[1] RAHUL DESHMUKH,[1] and THAKUR GURJEET SINGH[2]

[1]*Maharaja Ranjit Singh Punjab Technical University, Bathinda, Punjab, India*

[2]*Chitkara College of Pharmacy, Chitkara University, Punjab, India*

ABSTRACT

Stroke is a syndrome characterized by developing rapidly clinical evidence of focal (or global) abnormalities of brain functioning, with signs lasting 24 hours or more or culminating in fatalities, and no known cause other than brain dysfunction. Around the world, stroke has a substantial socioeconomic impact. Strokes kill an estimated 4.5 million people worldwide each year, with more than 9 million stroke survivors. The consequences of stroke can be examined from various angles, many of which overlap: patients, their families and caregivers, and primary care. This abstract provides a thorough update on the effects of stroke, focusing on regions that still need to be addressed. Stroke rehabilitation is started early with the ultimate goal of attaining better recovery in the early phases following stroke and medical therapy to prevent future cerebral damage. Stroke patients have a variety of psychosocial problems that limit their ability to do daily chores. The monetary cost is substantial, and it is likely to increase as the population ages. As a result, information on the availability of medical expenses for elder stroke patients is required. This chapter discusses the impact of stroke on daily living, as well as emerging therapies and current treatment options for ischemic stroke.

6.1 INTRODUCTION

Ischemic stroke is a failure of functional ability caused by a sudden reduction in blood to a portion of the brain. It is due to embolism or thrombosis that inhibits a cerebral vessel that serves a specific brain region. Throughout damage to blood vessel injury, there is a core area where permanent brain damage occurs and a penumbra area where the brain loses function but is not irreparably injured due to a lack of blood supply. Arterial ischemic stroke syndromes are defined by a sudden loss of functional capacity due to brain or retinal ischemia (about 85%) or cerebral hemorrhage (15%) [1]. Venous stroke disorders are less prevalent

Ischemic Injury: A Molecular Insight. Thakur Gurjeet Singh & Heena Khan (Eds.)

than arterial strokes (about 1% of all strokes), occur sub-acutely, and are caused by thrombosis of the cerebral vascular sinus or cortical vein. Ischemic stroke is caused by a stenosis or obstruction of a single or several arteries in the brain such as (intracranial, extracranial, or extracranial cervical artery occlusion). Targeted occlusions can be triggered by cardioembolism, artery-to-artery thromboembolism, occlusive vascular disease, or small vessel disease, all of which must be distinguished for efficient secondary prevention of stroke.

6.2 NATIONAL AND INTERNATIONAL STATUS OF STROKE

Stroke is the leading major cause of morbidity and disability. The revised overall incidence of stroke in rural areas is 84262/1,00,000, while it is 334424/1,00,000 in urban areas. The overall incidence, according to recent population-based studies, is 119–145/1,00,000. There's also a big difference in rates of morbidity and mortality, with Kolkata having the largest at 42%. Stroke units are exclusively found in private clinics and are mostly located in urban areas. Intravenous and intra-arterial thrombolytics are widely used in India. The rate of IV thrombolytic is 11% in the ongoing Indo-USA National stroke registry [2].

Around the turn of the century, the age-standardized annual incidence of stroke in Europe ranged from 95 to 290 per 1,00,000, with one-month case fatality rates ranging from 13% to 35%. Each year, roughly 1.1 million Europeans suffer a stroke, with ischemic stroke accounting for nearly 80% of cases. The worldwide burden of stroke is increasing substantially as the world's population grows in size and age. Thus according to current epidemiological data, 16.9 million people have a stroke each year, resulting in a worldwide prevalence of 258 per 1,00,000 people per year, with significant disparities between low- and high counties, and an age-adjusted prevalence 1.5 times greater in men than in women. Stroke is still a deadly disease that is on the rise in aging populations globally and the demographic change in low-to-middle-income nations, despite significant developments in its treatment in the past few decades. Nonetheless, because most strokes are curable, preventive initiatives should be started right now to minimize the disease's worldwide burden.

6.3 HOW DOES STROKE AFFECT IN EVERYDAY LIFE?

6.3.1 FATIGUE EFFECT

After a stroke is a common occurrence that is linked to high mortality risk and a poor functional outcome. It has been identified as a widespread condition that can persist for up to three years following a stroke, even in those who have fully recovered their physical and cognitive abilities [3].

6.3.2 EFFECT OF ACUPUNCTURE

It was discovered that acupuncture treatment aided in the recovery of postural function, with results that were still visible two years later. In the focally injured brain, acupuncture may affect intrinsic cortical networks. Thus, modifications at the cellular level, such as neuronal and glial cell expansions and synapses, could allow brain tissue to tentatively modify itself. A higher level of neocortical plasticity should be examined than motor performance [4].

6.3.3 VISUAL LOSS

Visual loss is caused by destruction to the visual cortex that unites the visual sensory receptor cells to the optical processing centers, as a result of a stroke. Visual loss often impacts 50% of both eyes' visual fields, thereby blinding the sufferer to one side of space [5]. Stroke patients are uninformed of their visual impairment in ordinary life, which causes dread and a loss of confidence.

6.3.4 EFFECTS OF GARDENING

Therapeutic gardening has now been praised as a natural and enjoyable intervention site for enhancing the elderly's standard of living. Persons who have had a CVA would see a greater improvement in their QOL as a result of gardening activities than females. Farming activities will enhance the quality of life of individuals in CVA stage 1 recovery more than gardening activities would improve the QOL of people in CVA stage 2. Taking care of plants with short life cycles has a more enormous impact on QOL than taking care of plants with long life cycles [6].

6.3.5 INCONTINENCE

Incontinence can be induced by a variety of factors, including medicine, muscular atrophy, and changes in emotions, cognition, and memory.

6.3.6 SHOULDER PAIN

After a stroke, shoulder discomfort can manifest itself in various ways, with pain radiating to the hand and arm. Pain in the shoulder after a stroke has been linked to various factors, including diminished motor function, decreased range of motion, and sensory deficits [7] (Figure 6.1).

6.4 RISK FACTORS

6.4.1 SMOKING

Persons who chewed tobacco or smoked have a higher chance of having a stroke later in life. This is because nicotine raises blood pressure levels in the body. Smoking a cigarette for an extended period produces fatty buildup in the neck's major artery. It also thickens your blood over time, making it more prone for clots to develop, increasing your stroke risk.

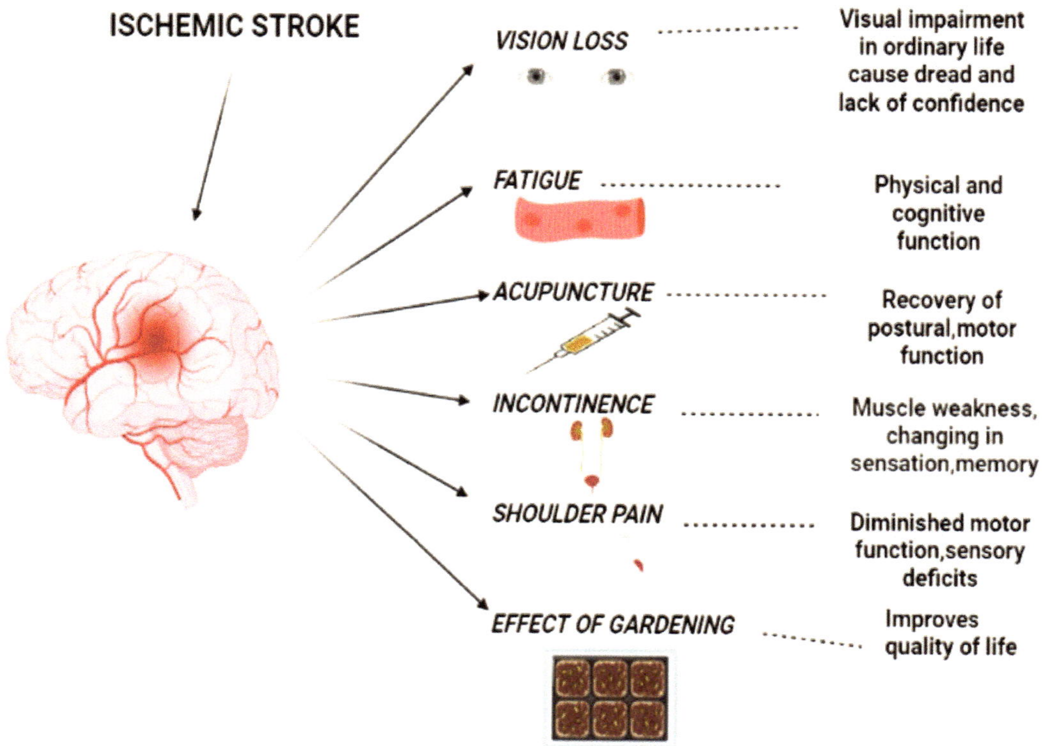

FIGURE 6.1 Effect of stroke in daily life.

6.4.2 DIABETES MELLITUS

Diabetics have a twice as high risk of stroke and have worse results than those who do not have diabetes. Excess hyperglycemia in the circulatory system results in changes in the walls of small blood vessels, resulting in inflammation and constriction, and narrowing of the arteries, cutting off blood flow and causing lasting damage to the area served by those blood vessels [8]. Uncontrolled hyperglycemia can potentially cause coronary artery damage and deadly ventricular arrhythmias.

6.4.3 ALCOHOLISM

Heavy drinking can have a bad impact on every element of one's life. This is especially true when it comes to a person's health. Alcohol abuse has been linked to a variety of health problems, ranging from minor to life-threatening [9]. Alcohol may increase the risk of Blood pressure, diabetes, liver damage, atrial fibrillation (AF), obesity, and hemostatics.

6.4.4 CHOLESTEROL

Cholesterol is a fatty substance that causes the formation of plaque deposits within capillaries. Plaques can form in the coronary arteries, which provide oxygen to the carotid arteries, and the heart, which supplies oxygen to the brain. There are two types of cholesterol present in the body that are contraindicated in stroke [10].

6.4.4.1 LOW-DENSITY LIPOPROTEIN

In terms of its ability to affect the heart and brain, low-density lipoprotein is the "bad cholesterol." It is a crucial factor in the development of arterial plaque. An increased risk of ischemic stroke has been linked to LDL cholesterol levels of more than 130 mg/dL.

6.4.4.2 HIGH-DENSITY LIPOPROTEIN

The "good cholesterol" is HDL. HDL levels of more than 35 mg/dL aid to prevent injury may occur by aiding LDL transport to the liver and out of circulation, as well as assisting in the stabilization of preexisting plaques.

6.4.5 HYPERTENSION

High BP, also known as hypertension, occurs when blood flows through vessels with too much force, putting too much strain on them and stretching them beyond their healthy limits, creating minor tears. Uncontrolled high blood pressure can cause injury or death [11]. Because it has no symptoms, it is known as "the silent killer." Because many people are unaware that they have it, it's critical to have your blood pressure checked regularly. Blood flow towards essential organs is restricted by blockages and blood clots, and without blood, the tissue dies. Because of this, high blood pressure can cause stroke, heart attack, renal failure, and heart failure.

6.4.6 DRUG ABUSE

Abuse-related drugs have been linked to stroke, particularly in young people. Addictive drugs can induce stroke by directly destroying brain capillaries or indirectly impacting other organs such as the liver (by disrupting blood-clotting pathways) or the cardiac (by negatively impacting cerebral circulation). Several mechanisms are distinct to each drug. Stimulants like amphetamines, cocaine, and their derivatives, for example, have been linked to stroke. Stimulants can cause an increase in blood pressure, which can lead to a ruptured brain artery or aneurysms and a hemorrhagic stroke. Acute ischemic stroke (AIS) is produced by stimulant-induced cerebral vasoconstriction, which reduces blood flow, promotes platelet activation, and accelerates atherosclerotic and cardiac irregularities [12] (Figure 6.2).

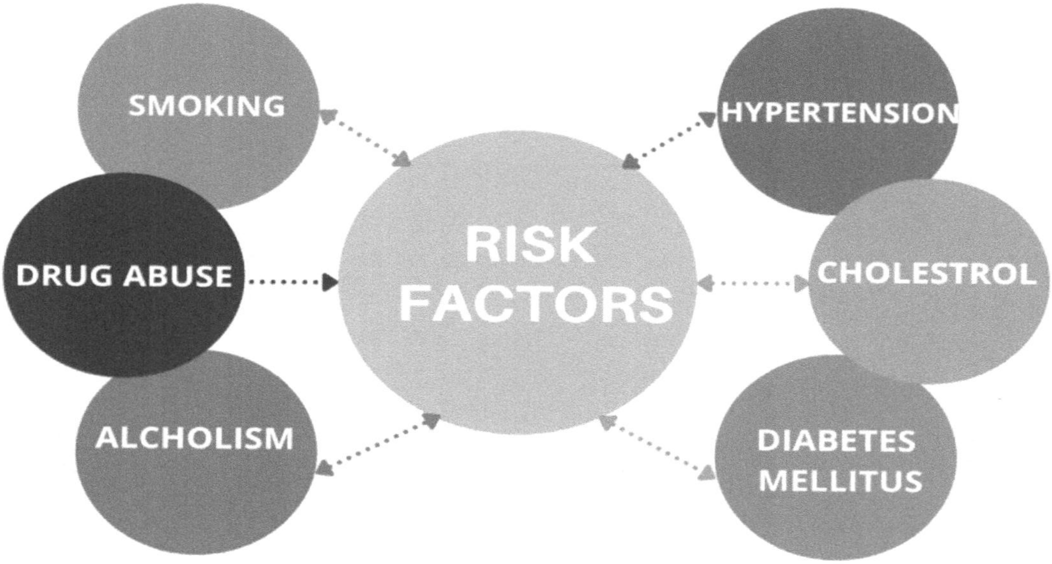

FIGURE 6.2 Risk factors associated with stroke.

6.5 PREVENTIVE MEASURES AND BIOMARKERS

6.5.1 *TELESTROKE CAPACITY*

Hospitals without being on stroke physicians may benefit from telestroke. A telestroke consultation involves a live teleconference between the patient, a stroke physician from another location, and patient healthcare professional in urgent care [13]. The stroke expert evaluates the patient, examines brain imaging, assesses whether the patient is suitable for thrombolytic or thrombectomy, appraises the need for transfer, and provides alternate treatments such as secondary stroke prevention. The benefits of having telestroke capabilities at a local hospital may extend beyond individual consultations, such as improved education and linkages to certified stroke clinics. In the United States, telestroke technology is now employed in roughly 30% of all hospital emergency rooms [14, 15].

6.5.2 *ANTIPLATELET THERAPY* (Table 6.1)

TABLE 6.1 Antiplatelet Drugs

Drug	Use	Moa	References
Aspirin	Acute phase, prevention of ischemic stroke, either alone or in combination therapy with drug.	Aspirin works by blocking (COX) irreversibly, which inhibits the formation of the procoagulant thromboxane A2 and hence lowers platelet aggregation (TXA2).	[16, 17]
Clopidogrel	If allergic to aspirin, avoid ischemic stroke in both the acute and secondary prevention periods.	An irreversible antagonist of the P2Y12 class of adenosine diphosphate (ADP) receptors on platelet surfaces prevent ADP-mediated stimulation of the downstream glycoprotein IIb/IIIa complex., resulting in decreased platelet aggregation.	[18]
Dipyridamole	Dipyridamole itself can reduce the risk of vascular events in patients who might have previous ischemia, but the effect is much greater when combined with low-dose aspirin.	Inhibiting cAMP-phosphodiesterase, limiting platelet adenosine absorption and degradation, and increasing PGI2 biosynthesis.	[18]
Prasugrel	Ischemic stroke, coronary heart disease, vascular mortality, or significant bleeding are all examples of ischemic stroke.	Inhibits the P2Y12 receptor on platelets. This blocks ADP from binding and the glycoprotein IIb/IIIa complex from activating.	[18]
Ticagrelor	Its antithrombotic properties have been widely proven in the treatment of acute coronary disease patients.	Its antithrombotic properties have been widely proven in the treatment of ACS patients.	[19]
Ticlopidine	Ischemic stroke, severe hematological adverse effects.	Inhibits ADP-induced platelet aggregation by blocking the ADP-binding site on the P2Y12 receptor in an irreversible manner.	[19]

6.5.3 *MECHANICAL THROMBECTOMY*

These are the new-generation devices used to prevent ischemic stroke caused by large vessel occlusion [20].

6.5.4 *ANTICOAGULANTS*

Rivaroxaban, Edoxaban, Apixaban, and Dabigatran, a thrombin inhibitor, have all been shown to be roughly comparable to warfarin in the prevention of stroke in persons with AF. All of them might be used as stand-ins for the originals. Apixaban is frequently prescribed due to its low risk of adverse effects and favorable pharmacokinetics for twice-daily use. Many studies have suggested that rivaroxaban and apixaban could be used to treat cancer-related hypercoagulability. Research comparing warfarin to newer medications has not yielded positive results in patients with antiphospholipid antibodies [21].

6.5.5 SURGICAL THERAPIES

6.5.5.1 CAROTID ENDARTERECTOMY

If surgeons maintain a minimal surgical risk, patients with symptomatic cervical internal carotid artery stenosis of percentage respond from carotid endarterectomy (CEA). Although CEA is effective in patients with asymptomatic carotid stenosis, endovascular insertion of carotid artery stents (CAS) may be a superior option in some circumstances. However, the risk of stroke has declined since these investigations, raising doubts regarding the usefulness of CEA or CAS in asymptomatic people. This is currently being investigated in a big clinical experiment. Until this issue is rectified, it recommends the consulting neurologist assist with patient selection for CEA or CAS.

6.5.5.2 PFO CLOSURE (PATENT FORAMEN OVALE)

PFO is reasonably common and used in benign conditions that affect roughly 25% of the population. The question of whether a PFO should be sealed with a device when individuals with low vascular risk factors have embolic strokes and are identified to have one arises. The RoPE score is a tool that can be used to evaluate a patient's attributable risk [22]. Preliminary tests of device closure for suspected symptomatic PFO showed that the procedure could be done safely, but there was no evident benefit. Consultation with a vascular neurologist and interventional cardiologist with experience in this management is now suggested in patients with embolic stroke or TIA without a clear alternative cause and PFO [22].

6.6 HERBAL DRUGS TO TREAT STROKE

6.6.1 AQUEOUS EXTRACT OF SAFFRON

Saffron and its active constituents have antitumor, antioxidant, anti-genotoxic, learning and memory booster, inflammation, convulsions, antidepressants, blood pressure, oxygen-boosting tissues, and bronchodilators properties [23].

6.6.2 TETRAMETHYLPYRAZINE

Tetramethylpyrazine is an alkaloid compound found in the rhizome of the plant herb that is Rhizoma Chuanxiong. Its medicinal effects are well-known for having neuroprotective benefits against cerebral ischemia, according to the research. Suppression of Ca^{2+} excess and activation of scavenging free radicals [24], the monitoring of mitochondrial function, the inhibition of apoptosis [25], inflammation, and the stimulation of neuronal development have all been proposed as mechanisms supporting TMP's actions.

6.6.3 GINSENOSIDE

Notoginseng (Panax notoginseng) P. Ginseng, one of the commonly used Chinese herbal remedies, have a long history of use throughout Asia for treating heart disease. *P. notoginseng* and *P. ginseng* contain saponins, which are the principal bioactive component. They have plenty of pharmacological actions, involving anticoagulant, anti-inflammatory, antioxidants, and anti-apoptotic characteristics [26].

6.6.4 TANSHINONE

Salvia miltiorrhiza's dried roots and rhizomes, known as Danshen, have long been utilized to treat cardiovascular and cerebrovascular illnesses. The primary chemical components of Danshen are salvianolic acids, tanshinones volatile oils, and inorganic substances [27]. It contains antioxidants, anti-inflammatory, and antineoplastic properties, which manifest as a reduction in mitochondrial dysfunction, microglial activation, and neuronal death. The features of TSA help to prevent the onset and progression of ischemic injury [26].

6.6.5 GASTRODIN

Gastrodin is a Chinese herb preferably known as tianma, is a well-found herbal plant that has been used for convulsions, migraines, disorientation, immobility, seizures, and other ailments [28]. By reducing apoptosis, Zn^{2+} toxicity, inflammation, and oxidative damage, gas may help to prevent ischemia injury.

6.6.6 BAICALIN

Scutellaria baicalensis is a typical herbal drug for pathogenic bacteria of the gastrointestinal and respiratory systems, as well as a variety of inflammatory conditions [28]. Baicalin is an antioxidant, anti-apoptotic, and anti-inflammatory flavonoid produced from the dried roots of Scutellaria baicalensis.

6.6.7 GINKGO BILOBA

Organic acids, flavonoids, and terpene lactones are the key active constituents of Ginkgo Biloba, a blend of several therapeutic compounds isolated from Ginkgo biloba leaves.

Extract of Ginkgo Biloba has anti-inflammatory, antioxidation, and antiapoptotic properties; as well as supports mitochondrial activity; and it has considerable therapeutic outcomes in the treatment of angina, stroke, and other cardiovascular and cerebrovascular illnesses [28].

6.6.8 SANHUA DECOCTION

Sanhua decoction is widely used to treat ischemic stroke, and it has proven to be quite successful in lowering blood viscosity and managing circulation problems. SHD inhibits p-tau AQP4 expression, improves neurological function, and lowers cerebral ischemia [29]. The chemical basis of Sanhua decoction, as well as the possible mechanism of its anti-ischemic stroke efficacy, is unknown.

6.7 BIOMARKERS FOR STROKE

Biomarkers are the external cerebral monitoring systems, which are non-invasive and can be used as stroke sensors in an emergency or at various stages of development. Differentiating stroke from non-stroke, ischemic from hemorrhagic stroke, and large vessel occlusion (LVO) from non-LVO ischemic stroke are some of the potential applications [30]. A successful stroke diagnosis before hospital admission could change the present stroke diagnostic and therapeutic strategy. As non-invasive exterior brain detection systems, a variety of technologies are continuously being developed such as EEG, near IR, accelerometers, radiofrequency, doppler ultrasound, etc. (Table 6.2).

TABLE 6.2 Biomarkers Used in Ischemia

Biomarkers	Use	References
Accelerometers	Aneurysm	[30]
	Stroke	
	Vasospasm	
	Concussion	
EEG	Head injury	[30]
	Stroke	
Microwaves	Stroke	[30]
Near-infrared	Intracranial hematomas	[30]
Doppler ultrasound	Stroke	[30]
	TBI	
	Migraine	
	Cognitive functions	
Radiofrequency	Convulsions	[31]
	Hemorrhages	

6.8 CURRENT THERAPEUTIC APPROACH

6.8.1 RECANALIZATION TREATMENT

The pillar of AIS therapy is recanalization and reperfusion, which can reduce the extent of the infarct and reverse neurologic deficits [31].

6.8.2 THROMBOLYTICS

This category covers antiplatelet treatment, thrombolytic medicines, and Plasminogen activators to prevent further cell death due to stroke by breaking blood clots and restoring regular blood flow. Plasminogen activators have also shown promise in the treatment of ischemic stroke. They work by converting plasminogen to plasmin, which decomposes clots and helps eliminate blood clots from the brain in ischemic circumstances. Mammalian activators include tissue-type Plasminogen and urokinase activators, while recombinant-type plasminogen inhibitors include tenecteplase and reteplase [32]. Whereas a single plasminogen activator is often insufficient to dissolve the clot, it is often used with a thrombin inhibitor such as argatroban.

Nonetheless, scientific investigations have shown that this treatment is ineffective due to uneven results and adverse effects. To reduce this, an embolectomy device is utilized to produce mechanical thrombolysis. It removes clots, and data suggests that when paired with Plasminogen activators, mechanical thrombolysis is more effective. The USFDA has authorized these devices as safe and feasible (e.g., MERCI). The sonothrombolysis technology, which combines plasminogen activators with an ultrasound emitting catheter, has made further progress in mechanical thrombolysis (Figure 6.3).

6.8.3 NEUROPROTECTIVE AGENTS

This series of medicines has been shown to prevent neuronal damage during ischemia. It encompasses a variety of medications that act as neuroprotectants and have complex interactions. Ischemia disrupts cell metabolism, causing toxic quantities of excitatory neurotransmitters to be released and unleashing a cascade of death signals. Cell death, membrane, and cytoskeleton (calpain) damage, as well as DNA deterioration, are all part of the death signaling cascade (endonucleases). As a result, these medications are used to treat ischemic stroke. Hypothermia, Ca^{2+} channel antagonists, glutamate receptor antagonists, scavengers of free radicals, GABA agonists, NO synthase inhibitors, and antiapoptotic medicines are all pharmaceuticals that fall into this category.

1. **Polybdetin:** It is a natural hypotensive medication that crosses the blood-brain barrier and acts as a neuroprotectant. When given orally or parenterally, it has been shown to have antioxidant effects in animal experiments. However, more research is needed to determine the specific mechanism [33].

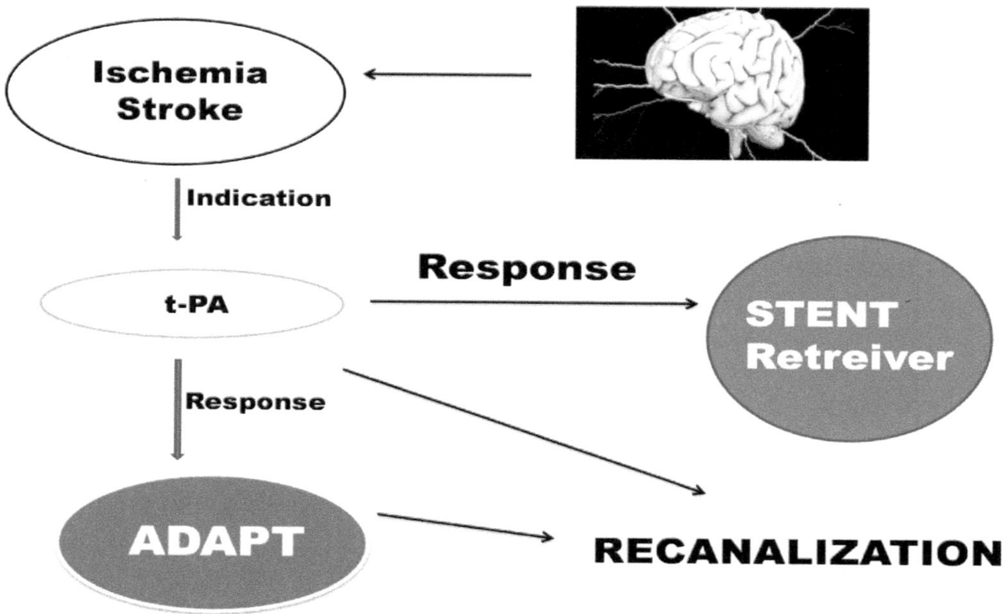

FIGURE 6.3 Recanalization process.

2. **Berberine:** It is a naturally produced drug that is useful in a variety of conditions, including fatty liver disease, glucose intolerance, inflammatory bowel disease, and more; being a robust antioxidant, it also exhibits superior outcomes in ischemic brain injury. Berberine not only lowers ROS levels but also lowers apoptosis. It has been proven in clinical tests that it lowers IL-6 levels and lessens the inflammatory response after a stroke. As a result, it suggests that this chemical could be employed to minimize ischemia injury following a stroke [34].
3. **Curcumin:** It has also been shown to help against ischemic brain injury. Curcumin exosomes significantly reduce ROS accumulation after stroke because it is an antioxidant; it also protects the integrity of the BBB; and it inhibits apoptosis mediated by mitochondrial alterations by regulating mitochondrial membrane potential [35].

6.8.4 PHOTOTHERMAL THERAPY (PTT)

Photothermal therapy has been used to treat cancer for a long time. Photothermal compounds are used in PTT to produce tumor ablation by causing localized rising temperatures. In phototherapy Graphene and black phosphorus are used for neurological illnesses, with benefits such as high therapeutic efficiency, minimum pervasiveness, and photothermal therapy's direct lethal effect. Due to its direct cytotoxic effect, PTT developed a biodegradable hydrogel as an efficient drug release platform [36]. After electrostatic adsorption had loaded the drug, phosphorus nanosheets were mixed with a low melting-point agarose solution to form a hydrogel. With an increase in temperature generated by the laser, the hydrogel can undergo reversible hydrolysis and soften, allowing the medicine to be released in a photo-controlled manner. When comparing the dark and irradiation environments, the black phosphorus hydrogel has a substantially higher drug release rate. The use of light-controlled medication release in the treatment of strokes could be useful [37].

6.8.5 NANOMEDICINES FOR ISCHEMIC STROKE

Ischemic stroke is a severe threat to human life and health, however, due to inadequate blood-brain permeability, there are very few therapies for stroke treatment. Nanotechnology might well be capable of solving some of the challenges associated with traditional drug discovery. Although nanomedicines do not exist naturally in the body, research is required to determine their biodistribution and toxicological effects following targeted delivery [38] (Figure 6.4).

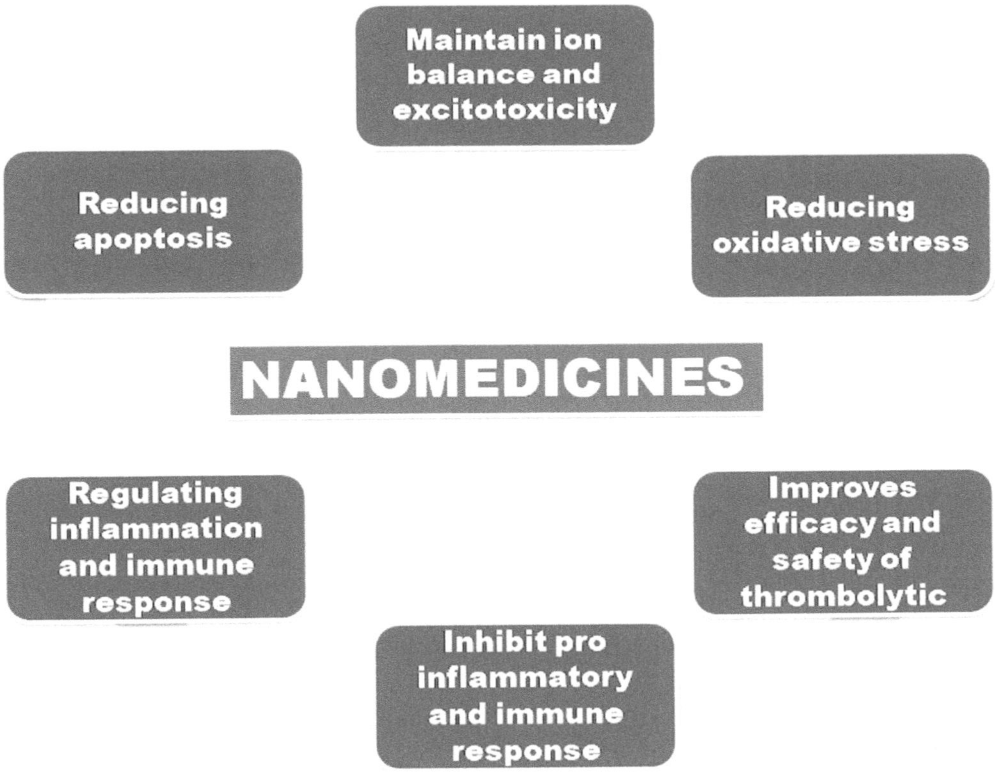

FIGURE 6.4 Nanomedicines benefits in ischemic injury.

6.8.6 PEPTIDES

Insulin would be the first pharmacological peptide to be successfully employed in diabetic treatment. Peptides function in a similar way to a natural process in the body. Because of their selectivity, efficacy, and safety, the pharmaceutical industry is increasingly focused on peptides [39]. Connexin 43 is found in the Central nervous system and acts as a gap junction protein. After an ischemia event, the amount of Connexin 43 rises dramatically, causing tissue damage. Different mimetic peptides are utilized to avoid this, and connexin mimetic peptides have been found to minimize inflammation, edema, and neuronal cell death [40, 41]. Currently, a therapeutic approach is using the size-dependent absorption properties of ruthenium carbonyl clusters, and CO on graphene oxide to create a simple CO-release platform for stroke therapy, Sonochemistry, mechanical thrombolysis, and novel neuroprotectant compounds all require greater effort to settle these treatments successfully.

6.9 FUTURE DIRECTIONS

6.9.1 NANOMEDICINES

Nanomedicines are still in the initial phases of development, and more research is needed before they can be used to cure cerebral ischemia (Table 6.3).

TABLE 6.3 Nanomedicines Used in Stroke Therapy

Category	Properties	Prime Discoveries	Advantages	References
Liposome	Single phospholipids membrane proteins make up synthetic spherical cells., with a short half-life, not cross BBB.	• Ischemic area, the rate of absorption is higher is high. • Prolonged circulation time. • Reaches the brain and decreases the infarct volume.	High efficiency. Low toxicity. Hydrophilic and lipophilic.	[42]
Polymeric nanoparticles	Encapsulated in balance expandable stents, cross BBB.	• Significant decrease the nerve injury, caspase activity, and infarct volume.	If inserted into different locations, it is possible to change drug release and boost medication stability.	[43, 44]
Metal nanoparticles	Shape, size, uniform. cross BBB	• Detection of brain disease. • Oxides of cerium and yttrium. • Nanoparticles may be used treatment of diabetic neuropathy.	MRI, free radical scavenging.	[45]
Carbon nanotubes	Composed of graphite tubes, electrical optical thermal properties, not cross BBB.	• Improve neuronal growth and differentiation, and peripheral nerve restoration. • Ischemic injury.	Due to their high penetrating strength and surface area, they can conjugate more than one molecule to their surface.	[46]
Graphene	Recently discovered nanomaterial, good performance, sensitive, cross the BBB, give good performance.	• Vasodilators treatment.	Polyaromatic, high surface area.	[47]

TABLE 6.3 *(Continued)*

Category	Properties	Prime Discoveries	Advantages	References
Black phosphorus	A wide range of extreme light absorption.	• Black phosphorus can capture Cu^{2+} and enhance BBB. • Ischemic injury.	Biodegradable selectively capture Cu^{2+}	[48]
Hydrogels	Cross BBB, noninvasive	• Hydrogels aid in the healing process by providing structural support to the wounded tissue, reducing secondary cell apoptosis. • Neuroprotective action. • Hydrogels are used in contact lenses, sanitary products, and scaffolds for tissue engineering, drug delivery systems, wound dressings, and antimicrobial agents.	Similar to real tissue in terms of flexibility; temperature or PH sensitive. Biodegradable and biocompatible.	[49]
Dendrimers	The heme oxygenase-1 gene was efficiently transported into the ischemic brain.	• Reduce blood clotting. • Stroke.	Surface modification is simple, and charged functional groups can interact with it.	[50]

6.9.2 ADVANCEMENT OF NEURONAL IMAGING TECHNIQUES

Neuroimaging has aided the anatomical and functional anatomy of the brain, and the diagnosis of stroke. Numerous advances in technology, such as the introduction of new pulse sequences and intravenous injectable agents, have improved the use of neuroimaging in clinical diagnosis over the last few decades (Table 6.4).

6.9.3 NEUROSTEROID TREATMENT

Neurosteroids are steroid hormones that have a negative impact on the CNS. They're made in the mitochondria of neurons. Cholesterol or other steroid precursors like estrogen, progesterone, dehydroepiandrosterone, allopregnanolone, and vitamin D are used to make glial cells. Among other things, neurosteroids increase, brain development, differentiation of neurons and maturation, antianxiety, myelination, antidepression, anti-epilepsy, hypnosis, and memory improvement [81]. Neurosteroids link with glial cells, endothelial cells, and the neurovascular units in the brain directly target neurons to perform their neuroprotective role. Although neurosteroids are produced by astrocytes and microglia, they are also controlled by them [81] (Table 6.5).

6.9.4 STEM CELL THERAPY

The majority of stem cell treatment research for ischemic stroke has concentrated on two processes: (i) damaged neuronal cells and tissues are replaced; and (ii) paracrine functional benefits such as immunosuppression, pro-angiogenesis, protective and neurotrophic activities [60].

TABLE 6.4 Neuroimaging Techniques

Techniques	Description	References
Ultra-high-field MR	The anatomical and functional anatomy of the brain, as well as the diagnosis of stroke, have all benefitted from neuroimaging. Throughout the previous few decades, numerous technological breakthroughs have broadened the use of neuroimaging in medical diagnoses, such as the use of intravenous contrast agents and the development of new pulse sequences.	[51]
Neural network mapping	Machine learning often referred to as artificial neural networks, is a collection of techniques that enable a computer to solve complicated problems using a large quantity of data. CNN's, a type of artificial neural network (ANN) used for image analysis, have revolutionized imaging technology, diagnostic tools, and data processing. In medical image analysis, such as lesion segmentation, anatomical segmentation, and classification, CNNs have demonstrated exceptional performance.	[52]
MR fingerprinting	MR imaging has revolutionized medical imaging and diagnosis, however, it is non-quantitative, which is a significant disadvantage. In today's scanners, physical parameters such as longitudinal relaxation time (T1) and transverse relaxation time (T2) are used to weight the contrast, yielding qualitatively weighted images (T2). MR images that provide relative information rather than absolute values of physical features are referred to as "hyper-intense" or "low intense." MR fingerprinting provides a huge step toward addressing this constraint by simultaneously providing quantitative MR imaging of many tissue characteristics such as T1 and T2, perfusion, and diffusion.	[53]

6.9.4.1 NEURAL STEM CELLS

Neural stem cells (NSCs) which are naturally found in the brain and differentiate into multiple cell types inside the CNS. NSC transplantation has been established to help brain recovery after ischemic stroke in recent investigations; however, the mechanisms are unknown, and basic/clinical studies of NSC transplantation for ischemic stroke still have to be performed. Enhance our ability of such healing properties of NSCs in ischemic stroke and appraise their potential for application in clinical stroke patients in the future [61].

6.9.4.2 EMBRYONIC STEM CELLS

Embryonic stem cells (ESCs) are produced from the mass of the blastocyst and can self-replicate and divide into cellular progenitors from all three germ layers. Human ESCs have been effectively employed to generate therapeutically essential neural lineage cells like cortical neurons, motor neurons, oligodendrocytes, astrocytes, and other neural cell types [62].

TABLE 6.5 Neurosteroids in Stroke

Category	Drug	Use	Mechanism	References
Hypothermia	–	Brain protection Stroke	Reduced metabolic demand, suppress glutamate release, decrease production of free radicals, NO, and unwanted substances.	[54]
Calcium channel antagonist	Nimodipine	Improve blood flow in tissues. Hemorrhages	Decrease calcium influx.	[54]
Glutamate receptor antagonist	Magnesium, Lamotrigine, Flunarizine, Eliprodil, Fosphenytoin	Improve cognition, neuroprotection	Induced excitotoxicity, blocks NMDA receptor	[54]
NO synthase inhibitors	Nitro-L-arginine, methyl ester and targinine	Acute stroke Low BP Cytoprotection Neuroprotection	Activation of eNOS. Inhibit the proliferation of T-lymphocytes.	[55]
Antiapoptotic agents	N-benzyloxycarbonyl, Fluoromethylketone	Acute ischemia Neurodegenerative disorders	Inhibit caspases enzyme. Interfere with an apoptotic and inflammatory mechanism.	[56–58]
Free radical scavengers	Tirilazad, Ebselen	Neuroprotection, antioxidant. Improve motor function.	Prevent ROS Damage the DNA.	[56]
GABA agonist	Clomethiazole	Neuroprotection	Increase chloride ion frequency and cause hyperpolarization.	[59]

6.9.4.3 INDUCED PLURIPOTENT STEM CELLS

Induced pluripotent stem cells (iPSTs) were produced from somatic cells such as fibroblasts and reprogrammed into a stem cell similar state to ESCs, including self-renewal and up-regulating. iPSCs, unlike ESCs, can sidestep some ethical issues and be used to generate patient-specific cells, lowering the risk of immunological rejection. Human iPSCs have been examined extensively for their ability to generate neuronal and glial pathways [63].

6.9.4.4 MESENCHYMAL STEM CELLS (MSCS)

Mesenchymal stem cells (MSCs) are stem cells derived from umbilical cords, fetal membranes, placental, bone marrow, and adipose tissue, among other organs. They're multicellular populations with various origin tissue characteristics, but they all exhibit the same phenotypic and plastic adherent growth pattern [64] (Figure 6.5).

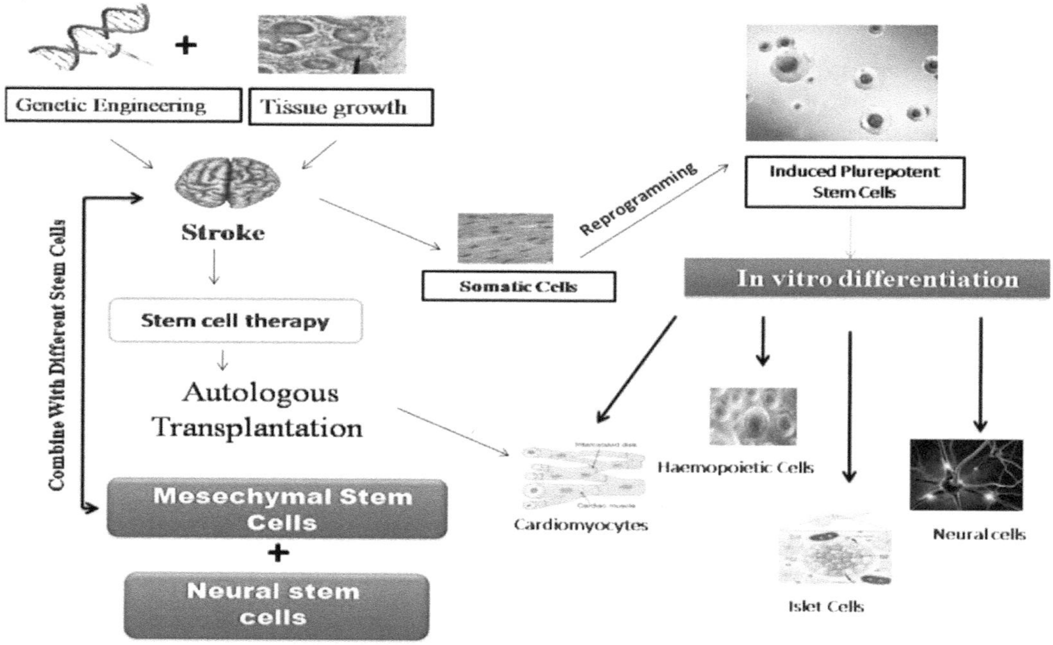

FIGURE 6.5 Mechanism of stem cell therapy in stroke.

6.10 COMPLICATIONS (TABLE 6.6)

TABLE 6.6 Neurological, CVS, Metabolic Complications After Stroke Faced by Persons

Treatment	Complication	Category of Disorder	Symptoms	Prevalence Rate	References
Tricyclic antidepressants	Depression	Neurological disorder (mood disorder)	Feeling down, anhedonia, loss of interest. Energy depletion, concentration decreases, psychotic retardation.	30%	[65]
Cyclodextrin	Dementia	Neurological disorder (brain tissue injury)	Memory impairment, confusion Disorientation	38%	[66]
Haloperidol Risperidone	Psychosis	Neurological disorder	Delusion Hallucination False beliefs	4.86%	[67]
Tricyclic antidepressants	Obsessive compulsive disorder.	Neurological disorder	Persistent thought, image unable to express.	2%	[68]
Lithium Valporic acid	Mania	Neurological disorder	Irritable mood Increased activity, energy	1–2%	[69]
Antidepressants	Anger	Neurological disorder	Aggressive mood, frustration, irritation	15–35%	[65]
Dextromethorphan	Emotionalism	Neurological disorder	Excessive laughing, crying	8%	[65]

TABLE 6.6 *(Continued)*

Treatment	Complication	Category of Disorder	Symptoms	Prevalence Rate	References
Botulinum Toxin	Oropharyngeal dysphagia	CV disorder	Difficult in swallowing	45%	[70]
Fluoroquinolones	Pneumonia	Respiratory disorder	Impairment in respiration, fever, coughing	10%	[71]
Valproic acid	Seizures	Neurological disorder	Overfiring neurons, heart rate abnormal, shivering.	50%	[72]
Levodopa	Parkinson	Movement disorder	Tremor Shaking Rigidity in posture.	10%	[73]
Metformin	Diabetes	Metabolic disorder	Hyperglycemic, insulin resistance.	70%	[74–76]

6.11 CONCLUSION

Stroke is the world's foremost cause of disability and death. Various preclinical and clinical studies, therapies, and surgical treatments are ongoing to treat ischemic stroke injury and prolonged life of the person. Despite advancements, post-stroke therapy is still uncommon in surgical and medical intervention, which has a negative influence on patients' quality of life. Stem cell transplants have just recently been identified as a possible treatment for neurological illnesses. Because NSCs are naturally found in the brain and can differentiate into a range of cell types inside the CNS, NSC transplantation is the best option for ischemia. NSC transplantation has been shown to aid brain recovery after ischemic stroke in recent research; nevertheless, the processes are unknown, and primary/clinical trials of NSC transplantation for acute stroke are currently underway.

KEYWORDS

- **apoptosis**
- **ischemia**
- **neurosteroids**
- **oxidative stress**
- **plasminogen activators**
- **stem cells**
- **thrombosis**

REFERENCES

1. Zerna, C., Thomalla, G., Campbell, B. C. V., Rha, J. H., & Hill, M. D., (2018). Current practice and future directions in the diagnosis and acute treatment of ischemic stroke. *The Lancet, 392*(10154), 1247–1256.
2. Rennert, R. C., Wali, A. R., Steinberg, J. A., Santiago-Dieppa, D. R., Olson, S. E., Pannell, J. S., & Khalessi, A. A., (2019). Epidemiology, natural history, and clinical presentation of large vessel ischemic stroke. *Neurosurgery, 85*(suppl_1), S4–S8.
3. Ezekiel, L., Field, L., Collett, J., Dawes, H., & Boulton, M., (2021). Experiences of fatigue in the daily life of people with acquired brain injury: A qualitative study. *Disability and Rehabilitation, 43*(20), 2866–2874.
4. Xu, L., Dong, Y., Wang, M., Chen, L., Zhang, Z., Su, D., & Du, Y., (2018). Acupuncture for balance dysfunction in patients with stroke: A systematic review protocol. *Medicine, 97*(31).
5. Hazelton, C., Pollock, A., Taylor, A., Davis, B., Walsh, G., & Brady, M. C., (2019). A qualitative exploration of the effect of visual field loss on daily life in home-dwelling stroke survivors. *Clinical Rehabilitation, 33*(7), 1264–1273.
6. Ho, S. H., Lin, C. J., & Kuo, F. L., (2016). The effects of gardening on quality of life in people with stroke. *Work, 54*(3), 557–567.
7. Lindgren, I., Gard, G., & Brogårdh, C., (2018). Shoulder pain after stroke–experiences, consequences in daily life and effects of interventions: A qualitative study. *Disability and Rehabilitation, 40*(10), 1176–1182.
8. Garg, N., Singh, T. G., Khan, H., Arora, S., Kaur, A., & Mannan, A., (2021). Mechanistic interventions of selected *Ocimum* species in management of diabetes, obesity, and liver disorders: Transformative developments from preclinical to clinical approaches. *Biointerface Res. Appl. Chem, 12*(1), 1304–1323.
9. Gupta, A., Khan, H., Kaur, A., & Singh, T. G., (2021). Novel targets explored in the treatment of alcohol withdrawal syndrome. *CNS & Neurological Disorders-Drug Targets (Formerly Current Drug Targets-CNS & Neurological Disorders), 20*(2), 158–173.
10. Rihal, V., Khan, H., Kaur, A., & Singh, T. G., (2022). Vitamin D as therapeutic modulator in cerebrovascular diseases: A mechanistic perspectives. *Critical Reviews in Food Science and Nutrition*, 1–23.
11. Kalra, P., Khan, H., Kaur, A., & Singh, T. G., (2022). Mechanistic insight on autophagy modulated molecular pathways in cerebral ischemic injury: From preclinical to clinical perspective. *Neurochemical Research*, 1–19.
12. Tsatsakis, A., Docea, A. O., Calina, D., Tsarouhas, K., Zamfira, L. M., Mitrut, R., & Neagu, M., (2019). A mechanistic and pathophysiological approach for stroke associated with drugs of abuse. *Journal of Clinical Medicine, 8*(9), 1295.
13. Wilcock, A. D., Schwamm, L. H., Zubizarreta, J. R., Zachrison, K. S., Uscher-Pines, L., Richard, J. V., & Mehrotra, A., (2021). Reperfusion treatment and stroke outcomes in hospitals with telestroke capacity. *JAMA Neurology, 78*(5), 527–535.
14. Zachrison, K. S., Boggs, K. M., Hayden, E. M., Espinola, J. A., & Camargo, C. A., (2020). A national survey of telemedicine use by US emergency departments. *Journal of Telemedicine and Telecare, 26*(5), 278–284.
15. Zhang, D., Wang, G., Zhu, W., Thapa, J. R., Switzer, J. A., Hess, D. C., & Ritchey, M. D., (2018). Expanding telestroke services improves the quality of care provided in super rural areas. *Health Affairs, 37*(12), 2005–2013.
16. Naqvi, I. A., Kamal, A. K., & Rehman, H., (2020). Multiple versus fewer antiplatelet agents for preventing early recurrence after ischemic stroke or transient ischemic attack. *Cochrane Database of Systematic Reviews.*
17. Khan, H., Sharma, K., Kumar, A., Kaur, A., & Singh, T. G., (2022b). Therapeutic implications of cyclooxygenase (COX) inhibitors in ischemic injury. *Inflammation Research*, 1–16.
18. Wang, D., (2021). Precision antiplatelet therapy for the prevention of ischemic stroke. *Stroke and Vascular Neurology, svn-2021.*
19. Komarova, M., Baig, S., Patel, H., Monks, K., Wasay, M., Ali, A., & Bell, S. M., (2022). Antiplatelet use in ischemic stroke. *Annals of Pharmacotherapy*, 10600280211073009.
20. Cho, Y. H., & Choi, J. H., (2021). Mechanical thrombectomy for acute ischemic stroke with occlusion of the M2 segment of the middle cerebral artery: A literature review. *Journal of Cerebrovascular and Endovascular Neurosurgery, 23*(3), 193–200.

21. Li, A., Garcia, D. A., Lyman, G. H., & Carrier, M., (2019). Direct oral anticoagulant (DOAC) versus low molecular-weight heparin (LMWH) for treatment of cancer-associated thrombosis (CAT): A systematic review and meta-analysis. *Thrombosis Res., 173*, 158–163.
22. Mas, J. L., Derumeaux, G., Guillon, B., Massardier, E., Hosseini, H., Mechtouff, L., Arquizan, C., et al., (2017). Patent foramen ovale closure or anticoagulation vs. antiplatelets after stroke. *New England Journal of Medicine, 377*(11), 1011–1021.
23. Asadollahi, M., Nikdokht, P., Hatef, B., Sadr, S. S., Sahraei, H., Assarzadegan, F., & Jahromi, G. P., (2019). Protective properties of the aqueous extract of saffron (*Crocus sativus* L.) in ischemic stroke, randomized clinical trial. *Journal of Ethnopharmacology, 238*, 111833.
24. Zhang, G., Zhang, T., Wu, L., Zhou, X., Gu, J., Li, C., Liu, W., et al., (2018). Neuroprotective effect and mechanism of action of tetramethylpyrazine nitrone for ischemic stroke therapy. *Neuromolecular Medicine, 20*(1), 97–111.
25. Zhang, H., Tang, W., Wang, S., Zhang, J., & Fan, X., (2020). Tetramethylpyrazine inhibits platelet adhesion and inflammatory response in vascular endothelial cells by inhibiting P38 MAPK and NF-κB signaling pathways. *Inflammation, 43*(1), 286–297.
26. Zhou, L., Zhang, J., Wang, C., & Sun, Q., (2017). RETRACTED: Tanshinone inhibits neuronal cell apoptosis and inflammatory response in cerebral infarction rat model. *International Journal of Immunopathology and Pharmacology, 30*(2), 123–129.
27. Zhou, X., Chen, L. L., Xie, R. F., Lam, W., Zhang, Z. J., Jiang, Z. L., & Cheng, Y. C., (2017). Chemosynthesis pathway and bioactivities comparison of saponins in radix and flower of *Panax notoginseng* (Burk.) FH Chen. *Journal of Ethnopharmacology, 201*, 56–72.
28. Zhu, T., Wang, L., Feng, Y., Sun, G., & Sun, X., (2021). Classical active ingredients and extracts of Chinese herbal medicines: Pharmacokinetics, pharmacodynamics, and molecular mechanisms for ischemic stroke. *Oxidative Medicine and Cellular Longevity, 2021*.
29. Leung, S. W., Lai, J. H., Wu, J. C. C., Tsai, Y. R., Chen, Y. H., Kang, S. J., Chiang, Y. H., et al., (2020). Neuroprotective effects of emodin against ischemia/reperfusion injury through activating ERK-1/2 signaling pathway. *International Journal of Molecular Sciences, 21*(8), p.2899.
30. Walsh, K. B., (2019). Non-invasive sensor technology for prehospital stroke diagnosis: Current status, and future directions. *International Journal of Stroke, 14*(6), 592–602.
31. Valkonen, K., Martinez-Majander, N., Sibolt, G., Tiainen, M., Räty, S., Kokkonen, T., Lappalainen, K., & Curtze, S., (2022). Recanalization treatment for acute ischemic stroke: 3-month outcome before and after implementation of routine endovascular treatment. *eNeurologicalSci, 27*, 100394.
32. Audrey, M. T., (2018). The role of plasminogen activators in stroke treatment: Fibrinolysis and beyond. *Lancet Neurol., 17*, 1121–1132.
33. Kim, S., & Jey, S., (2019). The protective mechanisms of molybdate in cerebral ischemia. *Eur. J. Pharmacol., 842*, 133–138.
34. Saklani, P., Khan, H., Singh, T. G., Gupta, S., & Grewal, A. K., (2022). Demethyleneberberine, a potential therapeutic agent in neurodegenerative disorders: A proposed mechanistic insight. *Molecular Biology Reports*, 1–13.
35. Liang, Z., & Ruyi, H., (2020). Curcumin-laden exosomes target ischemic brain tissue and alleviate cerebral ischemia-reperfusion injury by inhibiting ROS-mediated mitochondrial apoptosis. *Mat. Sci. Eng., 117*, 1–12.
36. Liang, X., Ye, X., Wang, C., Xing, C., Miao, Q., Xie, Z., Chen, X., et al., (2019). Photothermal cancer immunotherapy by erythrocyte membrane-coated black phosphorus formulation. *Journal of Controlled Release, 296*, 150–161.
37. Xie, Z., Wang, D., Fan, T., Xing, C., Li, Z., Tao, W., Liu, L., Bao, S., Fan, D., & Zhang, H., (2018). Black phosphorus analogue tin sulfide nanosheets: Synthesis and application as near-infrared photothermal agents and drug delivery platforms for cancer therapy. *Journal of Materials Chemistry B., 6*(29), 4747–4755.
38. Tian, X., Fan, T., Zhao, W., Abbas, G., Han, B., Zhang, K., & Xie, Z., (2021). Recent advances in the development of nanomedicines for the treatment of ischemic stroke. *Bioactive Materials, 6*(9), 2854–2869.
39. Saklani, P., Khan, H., Gupta, S., Kaur, A., & Singh, T. G., (2022a). Neuropeptides: Potential neuroprotective agents in ischemic injury. *Life Sciences, 288*, 120186.

40. Ugalde-Triviño, L., & Díaz-Guerra, M., (2021). PSD-95: An effective target for stroke therapy using neuroprotective peptides. *International Journal of Molecular Sciences, 22*(22), 12585.
41. Khan, H., Garg, N., Singh, T. G., Kaur, A., & Thapa, K., (2022a). Calpain inhibitors as potential therapeutic modulators in neurodegenerative diseases. *Neurochemical Research*, 1–25.
42. Zhao, Y., Jiang, Y., Lv, W., Wang, Z., Lv, L., Wang, B., Liu, X., et al., (2016). Dual targeted nanocarrier for brain ischemic stroke treatment. *Journal of Controlled Release, 233*, 64–71.
43. Lu, L., Wang, Y., Cao, M., Chen, M., Lin, B., Duan, X., Zhang, F., et al., (2017). A novel polymeric micelle used for *in vivo* MR imaging tracking of neural stem cells in acute ischemic stroke. *RSC Advances, 7*(25), 15041–15052.
44. Khan, H., Bangar, A., Grewal, A. K., Bansal, P., & Singh, T. G., (2022). Caspase-mediated regulation of the distinct signaling pathways and mechanisms in neuronal survival. *International Immunopharmacology, 110*, 108951.
45. Penninckx, S., Heuskin, A. C., Michiels, C., & Lucas, S., (2020). Gold nanoparticles as a potent radiosensitizer: A transdisciplinary approach from physics to patient. *Cancers, 12*(8), 2021.
46. Alshehri, R., Ilyas, A. M., Hasan, A., Arnaout, A., Ahmed, F., & Memic, A., (2016). Carbon nanotubes in biomedical applications: Factors, mechanisms, and remedies of toxicity: Miniperspective. *Journal of Medicinal Chemistry, 59*(18), 8149–8167.
47. Fernandes, L. F., Bruch, G. E., Massensini, A. R., & Frézard, F., (2018). Recent advances in the therapeutic and diagnostic use of liposomes and carbon nanomaterials in ischemic stroke. *Frontiers in Neuroscience, 12*, 453.
48. Li, J., Luo, H., Zhai, B., Lu, R., Guo, Z., Zhang, H., & Liu, Y., (2016). Black phosphorus: A two-dimension saturable absorption material for mid-infrared Q-switched and mode-locked fiber lasers. *Scientific Reports, 6*(1), 1–11.
49. Hong, A., Aguilar, M. I., Del Borgo, M. P., Sobey, C. G., Broughton, B. R., & Forsythe, J. S., (2019). Self-assembling injectable peptide hydrogels for emerging treatment of ischemic stroke. *Journal of Materials Chemistry B., 7*(25), 3927–3943.
50. Gumustas, M., Sengel-Turk, C. T., Gumustas, A., Ozkan, S. A., & Uslu, B., (2017). Effect of polymer-based nanoparticles on the assay of antimicrobial drug delivery systems. In: *Multifunctional Systems for Combined Delivery, Biosensing and Diagnostics* (pp. 67–108). Elsevier.
51. Balchandani, P., & Naidich, T. P., (2015). Ultra-high-field MR neuroimaging. *AJNR Am. J. Neuroradiol., 36*, 1204–1215.
52. Lundervold, A. S., & Lundervold, A., (2019). An overview of deep learning in medical imaging focusing on MRI. *Z. Med. Phys., 29*, 102–127.
53. Retico, A., (2018). Technological challenges in magnetic resonance imaging: Enhancing sensitivity, moving to quantitative imaging, and searching for disease biomarkers. *J. Instrum., 13*, C02007.
54. Buchan, A. M., & Pelz, D. M., (2021). Neuroprotection in acute ischemic stroke: A brief review. *Canadian Journal of Neurological Sciences*, 1–5.
55. Alshehri, B., Vijayakumar, R., Senthilkumar, S., Ismail, A., Abdel-Hadi, A., Choudhary, R. K., & Manikandan, P., (2022). Therapeutic potential of nitric oxide synthase inhibitor from natural sources for the treatment of ischemic stroke. *Saudi Journal of Biological Sciences, 29*(2), 984–991.
56. Luo, Y., Tang, H., Li, H., Zhao, R., Huang, Q., & Liu, J., (2019). Recent advances in the development of neuroprotective agents and therapeutic targets in treating cerebral ischemia. *European Journal of Medicinal Chemistry, 162*, 132–146.
57. Thapa, K., Khan, H., Sharma, U., Grewal, A. K., & Singh, T. G., (2021). Poly (ADP-ribose) polymerase-1 as a promising drug target for neurodegenerative diseases. *Life Sciences, 267*, 118975.
58. Thapa, K., Khan, H., Singh, T. G., & Kaur, A., (2021a). Traumatic brain injury: Mechanistic insight on pathophysiology and potential therapeutic targets. *Journal of Molecular Neuroscience, 71*(9), 1725–1742.
59. Grewal, A. K., Singh, T. G., & Singh, N., (2020). Potential herbal drugs for ischemic stroke: A review. *Plant Arch, 20*(1), 3772–3783.
60. Berlet, R., Anthony, S., Brooks, B., Wang, Z. J., Sadanandan, N., Shear, A., & Borlongan, C. V., (2021). Combination of stem cells and rehabilitation therapies for ischemic stroke. *Biomolecules, 11*(9), 1316.
61. Zhao, T., Zhu, T., Xie, L., Li, Y., Xie, R., Xu, F., Tang, H., & Zhu, J., (2022). Neural stem cells therapy for ischemic stroke: Progress and challenges. *Translational Stroke Research*, 1–11.

62. Palma-Tortosa, S., Martin, C. S., Kokaia, Z., & Tornero, D., (2021). Neuronal replacement in stem cell therapy for stroke: Filling the gap. *Frontiers in Cell and Developmental Biology, 9*, 709.
63. Duan, R., Gao, Y., He, R., Jing, L., Li, Y., Gong, Z., & Jia, Y., (2021). Induced pluripotent stem cells for ischemic stroke treatment. *Frontiers in Neuroscience, 15*, 639.
64. Guo, Y., Peng, Y., Zeng, H., & Chen, G., (2021). Progress in mesenchymal stem cell therapy for ischemic stroke. *Stem Cells International.*
65. Kim, J. S., (2016). Post-stroke mood and emotional disturbances: Pharmacological therapy based on mechanisms. *J. Stroke, 18*, 244–255.
66. Filipska, K., Wiśniewski, A., Biercewicz, M., & Ślusarz, R., (2020). Are depression and dementia a common problem for stroke older adults? A review of chosen epidemiological studies. *Psychiatric Quarterly, 91*(3), 807–817.
67. Lo Buono, V., Corallo, F., Allone, C., Bonanno, L., Sessa, E., Todaro, A., & Marino, S., (2019). Thought disorder in atypical schizophrenia-like psychosis after stroke: Analysis of clinical cases. *Neurocase, 25*(3, 4), 130–132.
68. Nemani, K., & Gurin, L., (2021). Neuropsychiatric complications after stroke. In: *Seminars in Neurology* (Vol. 41, No. 1, pp. 85–100). Thieme Medical Publishers, Inc.
69. Ferro, J. M., Caeiro, L., & Figueira, M. L., (2016). Neuropsychiatric sequelae of stroke. *Nature Reviews Neurology, 12*(5), 269–280.
70. Rofes, L., Muriana, D., Palomeras, E., Vilardell, N., Palomera, E., Alvarez-Berdugo, D., &Clavé, P., (2018). Prevalence, risk factors and complications of oropharyngeal dysphagia in stroke patients: A cohort study. *Neurogastroenterology & Motility, 30*(8), e13338.
71. Yuan, M., Li, Q., Zhang, R., Zhang, W., Zou, N., Qin, X., & Cai, Z., (2021). Risk factors for and impact poststroke pneumonia in patients with acute ischemic stroke. *Medicine, 100*(12), e25213.
72. Feyissa, A. M., Hasan, T. F., & Meschia, J. F., (2019). Stroke-related epilepsy. *European Journal of Neurology, 26*, 18–e13.
73. Defebvre, L., & Krystkowiak, P., (2016). Movement disorders and stroke. *Revue Neurologique, 172*(8, 9), 483–487.
74. Xu, J., Zhou, Y., Yan, C., Wang, X., Lou, J., Luo, Y., Gao, S., et al., (2022). Neurosteroids: A novel promise for the treatment of stroke and post-stroke complications. *Journal of Neurochemistry, 160*(1), 113–127.
75. Firdous, S. M., Dhang, P., Koley, A., & Khan, H., (2021). Diabetic retinopathy: Pathogenesis and therapeutic management. *Diabetes and Diabetic Complications.* Nova Science Publishers, Hauppauge.
76. Firdous, S. M., Ghosh, S., Nath, P., & Khan, H., (2021a). Molecular basis of diabetic nephropathy. *Diabetes and Diabetic Complications.* Nova Science Publishers, Hauppauge. pp. 63–97.
77. Dai, S., Zhao, Y., Wang, Y., Zhang, J., Fang, L., Jin, S., Shao, Y., & Huang, J., (2019). Recent advances in transistor-based artificial synapses. *Advanced Functional Materials, 29*(42), 1903700.
78. Fischer, U., Aguiar De, S. D., Norrving, B., & Caso, V., (2018). Status and perspectives of acute stroke care in Europe. *Stroke, 49*(10), 2281–2282.
79. Liu, D. Q., Chen, S. P., Sun, J., Wang, X. M., Chen, N., Zhou, Y. Q., Tian, Y. K., & Ye, D. W., (2019). Berberine protects against ischemia-reperfusion injury: A review of evidence from animal models and clinical studies. *Pharmacological Research, 148*, 104385.
80. Tan, M. J., Pan, H. C., Tan, H. R., Chai, J. W., Lim, Q. F., Wong, T. I., Zhou, X., et al., (2018). Flexible modulation of CO-release using various nuclearity of metal carbonyl clusters on graphene oxide for stroke remediation. *Advanced Healthcare Materials, 7*(5), 1701113.
81. Han, J., Yang, J.,1 & Du, Y., (2022). Adverse Reaction of Acupuncture and Antihypertensive Drugs for Treatment of Essential hypertension: A Protocol for Bayesian Network Meta-Analysis. *Computational Intelligence and Neuroscience*, 2022.

CHAPTER 7

Etiopathogenesis and Molecular Targets in Cerebral Ischemia: Current Understanding of Stroke and Therapeutic Approaches

MANISH KUMAR, SHIVANI BAGGA, and THAKUR GURJEET SINGH

Chitkara College of Pharmacy, Chitkara University, Rajpura, Punjab, India

ABSTRACT

Stroke and cerebral ischemia-related disorders are vital public health priorities and impose a huge challenge in clinical settings. Although at present stroke is the world's second-biggest cause of mortality, however, stroke triggered debility makes the life of a patient miserable. Currently, the interpretation of etiopathogenesis and therapeutic tactics in stroke are primordial despite recent advances in medical research. This review focuses on the pathogenesis, molecular targets, and new treatment approaches against impaired cerebral blood flow. The literature was investigated using the keywords "stroke," "cerebral ischemia," "molecular mechanisms," "pathogenesis," and "treatment" from Google Scholar, Scopus, Science Direct, Springer Nature, Web of Science, Bentham, and PubMed databases. Articles of the last 20 years (including book chapters) were prioritized and analyzed for their strengths and limitations, and the most suitable ones were carefully chosen. Existing cardiovascular aberrations perform a huge character in the initial stages of ischemic attack (transient ischemia) and facilitate the subsequent severe ischemic episodes that may cause mortality or permanent disability. The thromboembolic event, free radicals, inflammatory cytokines, vascular complications, and excitotoxicity are a few chief factors working at the core of this cerebrovascular disorder. The current treatment or preventive approaches are limited and are unable to revive the affected brain region wholly. The limited-time window in the treatment of stroke is a major constraint and challenge to physicians. However, investigation of diverse mechanisms of the pathogenic course of ischemia provides an insight into the molecular targets, which may be focused to develop new and effective pharmacotherapeutics in ischemic stroke.

7.1 INTRODUCTION

Cerebral ischemia (or cerebrovascular ischemia) is a region of damaging brain tissue initiated by an obstruction or contraction in the arteries that provide blood and oxygen to

Ischemic Injury: A Molecular Insight. Thakur Gurjeet Singh & Heena Khan (Eds.)

the cerebral tissues. Depreciation of the amount of blood flowing in the cerebral regions adversely affects oxygen supply, aerobic respiration, and homeostatic and survival mechanisms. A site-specific decrease in cerebral blood flow (CBF) in a brain region by occlusion of an artery or subarachnoid and intracerebral hemorrhage is often called ischemic stroke (IS). Ischemia is classified into two types: (i) ischemia that is limited to a specific area of the brain is known as focal ischemia; and (ii) global ischemia, which affects large areas of brain tissue [1]. Transient ischemic attack (TIA) frequently occurs before a full-scale IS. It is governed by autonomic and neurohormonal mechanisms (e.g., vasospasm, nitric oxide, and endothelial dysfunctions) that regulate blood pressure and heart rate. In a cerebrovascular accident, a thrombus blocks a cerebral vessel, resulting in severe regional brain ischemia. Cardiovascular emboli formation accounts for 60–70% of TIA and stroke cases. Cerebrovascular accidents reduce blood flow to a specific head region, raising the possibility of apoptosis in that area. However, if a significant amount of time elapses before restoring circulation, brain damage may be permanent [2]. Even though the ischemic core represents dead cells, cells can be salvaged in the penumbra region by restoring CBF. Even though reperfusion might be required to shield the brain tissue, albeit it can instigate reperfusion injury or cell survival impairment. Cardiovascular insufficiency leads to global brain ischemia, and it may cause permanent brain damage when there is a prolonged interruption of CBF [3].

The human brain is an exceedingly metabolically dynamic organ that constitutes 2–2.5% of entire body mass but consumes 20–25% of whole-body basal oxygen consumption. The brain is entirely dependent on the aerobic respiratory pathway of glucose metabolism for meeting the high energy demands that are mostly instigated by the generation of the action potential, transport of ions, Na^+/K^+-ATPase pump, and restoration of ion balance. Approximately 90% of energy is expended on action potential propagation and ion-fluxes, while only 10% is expended on resting membrane potential in the brain [4]. Hence, an uninterrupted blood supply is quintessential for brain existence. Many brain regions (e.g., white matter) have a higher threshold for ischemic damage. A small population of neurons is more susceptible to ischemia, such as CA1/CA3 (*Cornu Ammonis*) neurons and pyramidal neurons than dentate gyrus neurons with a higher threshold. Clinical evidence suggests a narrow therapeutic window after an ischemic attack. This therapeutic window may range from a few hours to days in patients with IS [5]. The main objectives of this therapeutic window should include (i) abolition of ischemia; (ii) restoration of CBF; and (iii) inhibition of reperfusion-induced damage.

The lack of precise pathobiological understanding of this neurological disease led to its incorrect classification in the International Classification of Diseases before 2018. Before the International Classification of Diseases-11 (ICD-11), cerebral ischemia and stroke were historically categorized under diseases of the circulatory system or cerebrovascular diseases that hampered several aspects of this neurological disorder such as research, patient care, therapy optimization, etc. In ICD-11, it is recategorized under the neurological section [6]. However, at present, several molecular mechanisms (e.g., c-Jun *N*-terminal kinase, activator protein 1, p38, nuclear factor-kappaB) and signal transduction pathways (e.g., Ca^{2+} ions, Na^+ ions, glutamate, free radicals, pro-inflammatory cytokines) are known that may interrupt CBF resulting in IS and initiate a cascade of events that has damaging influence over brain parenchyma after ischemic phase [7, 8]. The present review discussed the epidemiology, risk factors, types, recent pathophysiological aspects of cerebral ischemia, and molecular mechanisms in IS. We also discussed the possible prospective therapeutic targets that may be explored for better therapeutic outcomes in clinical settings.

7.2 EPIDEMIOLOGY

The cerebrovascular accident may represent a heterogeneous multifactorial condition characterized by the unexpected commencement of symptoms that are directly correlated to the damaged cerebral areas where the morbid development transpires. A few studies suggest that stroke and related cerebrovascular deficiencies together constitute 2nd principal reason for mortality globally. Annually stroke affects ~13.7 million population culminating in approximately 5.5 million deaths [9]. TIA (e.g., primary hemorrhage) constitutes a large portion (~75–90%) of IS. Only 10–25% of IS cases comprise full-blown ischemic attacks [10]. However, focal ischemic injury accounts for ~87% of all stroke cases. Intracerebral bleeding (hemorrhagic stroke) and global brain ischemia are responsible for remaining ischemic brain injury. The reports of the Global Burden of Disease Study (GBD) indicate an increase in the range of age, sex, and socio-economic status of the affected population, albeit there is a drop in the number of cases annually [11]. Age is the major determinant and after 55 years of age, the chances of cerebrovascular accidents amplify manifold. However, surprisingly between 1990 and 2016, in the population of 20–54 years of age, the global stroke prevalence enlarged from 12.9% to 18.6%. Concomitantly there was a decrease in the death rate by 36.2%. The highest prevalence of stroke is in China followed by Europe and the least in Latin America [12]. Young age women and older age men are more prone to stroke-related complications. Atrial fibrillation (AF) intensifies the risk of stroke in females over the age of 75 years by 20%. It has been observed that the use of contraceptive hormonal therapy, preeclampsia, and factors related to pregnancy aggravate the hazards of IS in females, which happens to be more fatal in comparison to male counterparts. The scale developed by the National Institutes of Health Stroke (0 = nil, 1–4 = trivial IS, 5–15 = modest IS, 15–20 = modest/aggravated IS, 21–42 = aggravated IS) is often used to assess the severity of ischemic attack. The mean stroke severity for men is 8.2 and for women is 10. Although infarction in cerebral tissues and intracranial hemorrhage are more prevalent in males, however, emboli formation due to cardiovascular aberrations is an aggravated IS type and is more dominant in female counterparts [13]. Higher life expectancy is the foremost reason for the higher incidence of stroke in females concerning male counterparts. However, tobacco abuse, smoking, alcohol, and substance abuse are the chief causes of stroke in men. A worldwide population-based study explored the link between behavior, demography, medical history, and participation of air pollution and ecological contaminants such as heavy metals (e.g., Pb and Cd) in precipitation of IS [14]. An investigation of risk factors in the United States (US) population revealed that hypertension is the major geographical determinant among the risk factors of stroke. Sedentary life, unhealthy diet plan, higher intake of junk food, processed food, tobacco, and substance abuse (including alcohol) accelerate stroke pathogenesis. Inadequate medical care facilities and post-stroke care deficiencies enhance the chances of a severe ischemic attack manifold. Studies in US and China revealed a contrasting relationship between financial status (i.e., medical care, health insurance) and prevention of secondary ischemic attacks. In the US, a higher financial status resulted in a higher prevention rate of secondary ischemic attacks and vice-versa in China [15, 16]. In Austria, a study investigated the association between education level and different treatments like speech therapy and echocardiography. The findings indicated no significant disparity in the prevalence of secondary ischemic attack and socioeconomic status of the patient regarding stroke care, thrombolysis treatment, occupational therapy, and physiotherapy approaches [17].

7.3 CEREBRAL BLOOD SUPPLY AND SITES OF ARTERY OCCLUSION IN ISCHEMIA

The brain receives blood supply through the internal carotid artery and vertebral artery. Left and right common carotid artery ascend where the brachiocephalic artery bifurcates. The common carotid artery (CCA) then diverges into the internal carotid artery (ICA) and external carotid artery (ECA) in the neck region. The ICA then branches into an anterior and middle cerebral artery (MCA). The Left and right vertebral arteries join to give rise to the middle basilar artery at the plane of the pons and the brainstem. ICA and basilar artery merge to form a ring called the "circle of Willis" (Figure 7.1). The posterior cerebral arteries ascend at this convergence together with two minor connecting arteries known as anterior and posterior communicating arteries [18, 19]. This circle of Willis improves the chances of uninterrupted plasma supply to the cerebral tissues in the case where any of the main arteries are occluded or damaged. Normally, the brain receives blood supply at a rate of 50–60 ml/100 g/min. A decline in this rate up to 10–20 ml/100 g/min stops brain functioning and a further decline of < 10 ml/100 g/min instigates a neurodegenerative cascade in the brain [20, 21].

7.4 TYPES OF CEREBRAL ISCHEMIA AND CLINICAL MANIFESTATIONS

Historically, cerebral ischemic conditions are classified according to the zone of the brain endangered and the reason for ischemic insult (Table 7.1). Thromboembolic occlusion of MCA is the primary cause of IS. MCA supplies blood to diverse regions of the brain such as frontal, parietal, and temporal areas. Insufficient CBF in these areas affects motor functions, sensory abilities, and speech which results in key symptoms such as ataxia, visual discrepancies, hemisensory loss, hemiparesis or monoparesis, and dysarthria [2]. In IS the core area is adversely affected resulting in the immediate death of cells within a few minutes. However, a peripheral region in the vicinity of the ischemic core (dead cells) known as penumbra still has viable cells and is supplied by collateral blood vessels. These non-functional or dormant cells in the penumbra might be saved if the primary blood supply is reinstated within a therapeutic window. In the penumbra region, cells stop functioning but are still viable and can be revived within 3 h after ischemic insult. The ischemic core is defined as the area in the brain where blood supply is < 10 ml/100 g/min. Animal models indicate that occlusion of MCA results in an ischemic core in the forebrain region. Clinical data suggest a high infraction probability (> 95%) when the CBF drops below 25% of the control value. Infarction probability is < 5% if blood flow is greater than 50% of the control value. In between 25 and 50% of CBF, there is a zone of incremental risk that indicate an increase in chances of infraction correlated with a decrease in percent blood supply [22]. Several animal models are established that mimic the pathophysiology of clinical focal cerebral ischemia (Table 7.2) [10]. Experimental animals used are Sprague-Dawley rat, C57BL/6 and SV129 mice, Wistar albino rats, SCID (CB-17/lcr-scid/scidJcl), and CB-17 (CB-17/lcr-+/+Jcl) mice. The ipsilateral cortex, neocortex, caudate, putamen, white matter, gray matter, and hippocampus are damaged in these animal models. Furthermore, based on the nature of arterial occlusion, acute IS is classified into diverse categories as described in Table 7.3 [23].

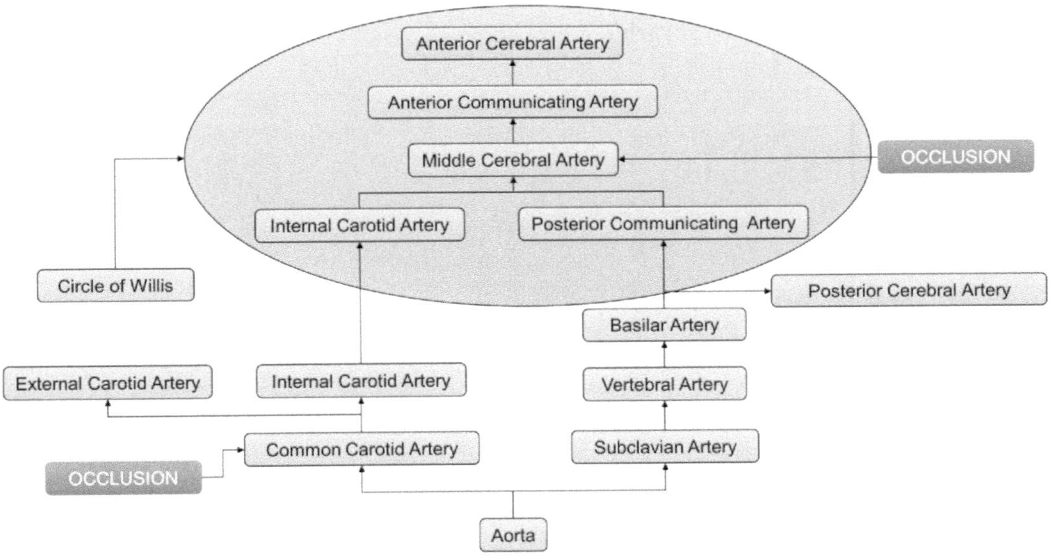

FIGURE 7.1 Cerebral blood supply and a common site of ischemia.

TIA leaves no permanent damage to the brain; however, it increases the chances of a full-blown IS manifold. A temporary blockage of the cerebral blood vessel is the cause of TIA, and this block is dislodged before any permanent damage. Normally TIA is self-resolving and within a few hours, it subsides. Approximately 30% of patients with IS received a TIA within the past few months. Hence, TIA is an indicator to seek medical care that may prove highly advantageous in the preclusion or timely alleviation of an acute ischemic attack.

Global cerebral ischemia encompasses the whole brain and correlates with cardio-vascular insufficiency. In this type of cerebral ischemia termination of blood supply in the entire brain leads to loss of neurons, impairment of the coherence of the blood-brain barrier (BBB), and activation of glial cells. Experimental models of cardiac arrest (CA) often include two- or four-vessel occlusion models (2-VO/4-VO) affecting the forebrain

TABLE 7.1 Classification of Cerebral Ischemia

Cerebral Ischemia Type	Clinical Symptoms	Diagnostic Tests	Current Therapy
Focal cerebral ischemia/IS (confined to a specific brain region). • Thrombotic (in MCA). • Embolic (in MCA). • Hemorrhagic (intracerebral or subarachnoid) stroke (feeble deteriorating vasculature, arteriovenous deformity, aneurysm).	Motor incoordination, imbalanced gait, visual and speech insufficiency, impaired judgment and understanding, limb weakness, numbness, the patient is disoriented and confused, headache, cognitive deficits, nausea, vomiting, and dizziness.	• MRI (magnetic resonance imaging), computed tomography scan (CT), single-photon emission computerized tomography (SPECT). • Blood tests (CBC). • Electroencephalography and electrocardiography. • Prothrombin time test. • Serum glucose. • Cardiac enzymes. • Neurological assessment.	• Thrombolysis (alteplase/ Tenecteplase, *i.v.*). • Thrombectomy • Antiplatelet drugs (Aspirin, clopidogrel, dipyridamole) • Anticoagulants • Antihypertensives • Statins • Non-steroidal anti-inflammatory drugs (NSAIDs)
Global cerebral ischemia (Whole brain ischemia). • Complete-brain hypoperfusion due to CA or cardiac failure, myocardial infarctions (*via* artery blockage), arrhythmias such as prolonged QT interval (LQTS) or WPW (Wolff-Parkinson-White) syndrome, shock, small vessel disease (SVD), asphyxia, coronary anomalies, arrhythmogenic right ventricular cardiomyopathy, ventricular fibrillation, hypertrophic cardiomyopathy, aortic deformity. • Incomplete-hypotension.	Loss of consciousness, syncope, electrocardiograph changes (ventricular fibrillation, asystole), asphyxia.		
TIA • Temporary blockage of CBF *via* plaque or blood coagulation.	Difficulty in speech, vision, gait, balance, strides, motor coordination impaired, lethargy, confusion, feeble limbs, headache, postural imbalance.		

TABLE 7.2 Animal Models of Cerebral Ischemia and IS

Type of Cerebral Ischemia	Animal Model	Advantages and Disadvantages
Focal cerebral ischemia/IS	**Endovascular Suture Model:** MCA occlusion *via* intraluminal suture (~60 or 90 min), MCA occlusion + ipsilateral/bilateral CCA occlusion, spontaneous brain infarction.	Minimally invasive, 90–100% successful in rodents, akin to human IS with high penumbra similarity, reproducible.
	Craniotomy Model: Electrocoagulation initiates perpetual damage/microaneurysm.	Infarct size and neurofunctional loss akin to human IS, low mortality, allows visual confirmation, CB-17 archetype has a greater survival percentage and is useful in studying reperfusion complications.
	Levine-Rice Model: Neonatal hypoxic-IS induced by one-sided ligation trailed by the reinstatement of CBF.	Hypoxic chamber required, regenerative and rehabilitative therapeutic possibilities can be investigated.
	Photo-Thrombosis Model: Illumination of the intact skull after application of a photosensitive dye (Rose Bengal) disrupts endothelial cells, platelet adherence, clotting, and cortico-striatal lesions.	Highly reproducible, less mortality percentage, no surgery, pathophysiology slightly vary from human IS, easily deployed in spontaneously moving rodents, and thus, neurological or motor functions are computable.
	Endothelin-1 (ET-1) Model: ET-1 injected in MCA using stereotaxic surgery.	70–90% reduction in CBF, minimally invasive, low mortality, appropriate for chronic lesion studies.
	Embolic Stroke Model: Microspheres (diameter 20–50 μm) or macrospheres (100–400 μm) infused in MCA/ICA *via* microcatheter that prompt multifocal infarcts.	High resemblance to human stroke, the volume of the infarct depends upon the size of the clot formed.
Global cerebral ischemia	**Two-Vessel Occlusion Model:** Occluding both CCA (2-VO) with precise hypotension *via* (i) arterial hemorrhage ~50 mmHg; or (ii) jugular vein blood withdrawal (blood flow < 15%).	Neurological and behavioral abilities are measurable but reinstating CBF initially is tough.
	Four-Vessel Occlusion Model: • Phase I: Electrocauterization in both vertebral arteries. • Phase II: Ligation of both CCA post 24 h of phase I (blood flow < 7%).	Neurological and behavioral abilities are measurable.

TABLE 7.2 *(Continued)*

Type of Cerebral Ischemia	Animal Model	Advantages and Disadvantages
	Asystole Method of CA:	
	0.5 M Potassium chloride (30 μl *i.a.* blous in jugular vein) + ventilation ceased + cardiopulmonary resuscitation (CPR) by epinephrine (3.2 μg, *i.v.* bolus).	Physical or traumatic damage in cardiomyocytes is nil, but high concentrations of KCl (potassium chloride) affect other organs, Cerebral temperature in mice preserved at 40°C, CA of 12 min is required to induce histopathological changes in the hippocampus.
	Ventricular Fibrillation Method:	
	Alternating current is delivered to the ventricular endocardium, through transesophageal cardiac pacing, or transthoracic electrical fibrillation. CPR by – (i) direct current; (ii) chest compressions; and (ii) 3.2 μg, *i.v.* bolus epinephrine.	Beneficial for studying short-term recovery, entails substantial skill and technical competency.
	Asphyxia Method:	
	The animal is paralyzed (by pharmacological interventions) and kept on a ventilator *via* an endotracheal tube, which is detached to instigate apnea. CPR by chest compressions and epinephrine.	Do not show "square wave" type brain ischemia, the considerable time lag in cessation of CBF.

region (Table 7.2) [24]. These models induce a "square wave" kind of brain mutilation that involves swift forfeiture of blood flow and reestablishment of blood flow. Commonly used experimental animals are Sprague-Dawley rats, Wistar albino rats, Swiss albino mice, and monkeys. The brain region affected are the cerebral cortex, caudoputamen, hippocampus, striatum, and cingulate cortex.

TABLE 7.3 Classification of Acute IS (TOAST Classification)

Type of Acute IS	Clinical Features	Imaging	Tests
Large-artery atherosclerosis (embolus/thrombosis).	Cortex/cerebellum impaired.	Cortex, cerebellum, brain stem, or subcortical lesions > 1.5 cm.	Stenosis of extracranial ICA
Cardioembolic (high-risk/ medium-risk).	Cortex/cerebellum impaired.	Cortex, cerebellum, brain stem, or subcortical lesions > 1.5 cm.	Cardiac basis of emboli.
Small-vessel occlusion (lacune).	Lacunar condition.	Subcortical or brain stem infarct < 1.5 cm.	–
IS due to additionally assessed etiology or some unidentified etiology.	–	–	Other abnormalities on tests.
• ≥ 2 recognized reasons.			
• Negative assessment.			
• Partial assessment.			

7.5 RISK FACTORS OF CBF DEREGULATION AND IS

Along with age, the hazards of IS increase, and these perils are amplified dangerously after 55 years of age in both males and females. Comorbid conditions (e.g., elevated blood pressure, coronary artery ailment, and dyslipidemia) further enhance the risk of stroke manifold. Approximately 60% of patients experience TIA before the onset of a full-blown stroke [25]. Some risk factors can be altered, while others cannot. The risk features that cannot be altered are:

1. **Age:** The prevalence of TIA rises with age, from 1–3 cases per 1,00,000 in those under the age of 35 to as many as 1,500 cases per 1,00,000 in those over the age of 85. Children account for less than 3% of all major cerebral infarctions. Earlier reports specified that the population between 20 and 54 years of age is more prone to ischemic insult in presence of other disorders [26]. Hispanic and black populations are more vulnerable to stroke relative to white people. The risk of hemorrhagic stroke is substantial in black individuals relative to age-matched white counterparts [27].
2. **Ancestry of the Cerebrovascular Accident:** Genetics can be classified under both amenable and unamenable hazard factors of IS. IS or some other cerebrovascular abnormality in family history magnifies the chances of cerebrovascular accidents in a person [28]. A single gene mutation for example noted in cerebral autosomal arteriopathy also contributes to the development of this neurological

disorder. Furthermore, stroke may be a consequence of other genetic mutation-induced disorders such as sickle cell anemia [29]. Genetic polymorphism in 9p21 is also associated with a rise in chances of stroke in a lifetime [30]. In a genome-wide investigation of IS, an increase in inheritance (~40%) for large blood vessel disorder, and a decline in inheritance (~16.7%) for SVD was pragmatic.

The modifiable risk factors are amenable to medical care, lifestyle dependent, and can be avoided to decrease the likelihood of IS in a person.

i. Smoking increases the likelihood of blood clots forming in arteries of the brain and heart. It increases the chances of stroke by up to 15% and a smoker is twice vulnerable to stroke in comparison to a non-smoker [31, 32].
ii. One of the greatest predisposing factors in IS is hypertension. In a research investigation of IS, it was observed that 54% of people were having a mean blood pressure of at least 160/90 mmHg [33]. A 5–6 mmHg decrease in blood pressure can reduce the chances of stroke up to 36–42% in adults as well as the aged population [10].
iii. Elevated cholesterol level is the foremost benefactor to the progress of coronary artery disease (CAD) and also happens to be a major risk factor in stroke. An increase in total cholesterol amplifies the chances of stroke, whereas an increase in high-density lipoprotein cholesterol reduces the same. In a study, an increase in total plasma triglycerides (> 2.3 mM) and a decrease in blood high-density lipoprotein cholesterol (< 0.9 mM) along with elevated blood pressure can culminate in stroke-related mortality twice [34].
iv. Diabetes increases the risk of having a stroke twice relative to non-diabetes conditions. This is because having too much sugar in blood damages the blood vessels. It can cause stiffening of the blood vessels as well as the accumulation of fatty deposits. Diabetes doubles the risk of stroke and slower down post-stroke rate of recovery [35].
v. A low to moderate alcohol consumption (30–50 ml in men and 30 ml in women) is associated with lowering of incidences of acute IS whereas it can increase the chances of hemorrhagic stroke. Regular high intake of alcohol is correlated with a higher prevalence of IS [36]. Similarly, substance abuse (e.g., amphetamines, cannabis/marijuana, cocaine, lysergic acid diethylamide, heroin, phencyclidine) is also correlated with higher chances of stroke [37].
vi. Atrial fibrillation (AF) increases the chances of stroke up to 2–5 folds and is present in 15% of all stroke cases [38]. AF potentiates debilitating factors and augments probabilities of mortality in stroke patients. A decrease in the passage of blood in the left auricle leads to thrombosis and emboli formation in AF. Other studies suggest that genetic mutations related to AF can directly predispose an individual to stroke and related cerebrovascular complications even before the onset of AF [39, 40].
vii. Physical inactivity and poor lifestyle including poor hygiene and dietary factor predispose an individual to stroke. Lack of exercise or sedentary life may aggravate or precipitate several disorders like hyperlipidemia, hypertension, diabetes, and cardiac insufficiency. Excessive salt intake is a major contributor to hypertension and processed food along with certain beverages can augment glucose intolerance and obesity [41].

7.6 PATHOPHYSIOLOGIC MECHANISMS

Ischemic lesion depends on the extent of blood flow reduction and duration of ischemia. Mild ischemia (blood flow > 30 ml/100 g/min) is nondestructive in 1st 6 h after onset ischemia. In moderate ischemia (blood flow > 12–30 ml/100 g/min) cell death has a therapeutic window of 3 h. In severe ischemia (blood flow < 10 ml/100 g/min), cell death is immediate within 1 h later to commencement of ischemia. In ischemia, a graded response towards an interruption in normal CBF (50–75 ml/100 g/min) is observed such as an increase in $[H^+]_e$ (extracellular fluid H^+ ion concentration) (35% drop in CBF) serves as a trigger [10]. Subsequently, cell swelling, cytotoxic edema, damage to the blood vessels, and vascular resistance (65% drop) ensues. Later, ionic imbalance, electrical inactivity, and acute energy depletion (85% decrease) amalgamate to cause irreversible injury to the brain structures (e.g., BBB) leading to widespread neurological deficits (Figure 7.2). Ischemic progression is divided into 3 phases [1]. (i) In the acute phase, irreversible cell injury mechanisms (e.g., mitochondrial injury, Ca^{2+}, and glutamate-dependent pathways) are activated. There is a severe decrease in blood flow, ATP (adenosine triphosphate) depletion, and necrotic mechanisms are activated. (ii) In the sub-acute phase (24–72 h after onset of ischemia), Ca^{2+}-influx further potentiates the mitochondrial damage (*via* the opening of mitochondrial permeability transition pore, leakage of cytochrome *c*, stimulation of inducible nitric oxide synthase), free radical cascade, and catabolic enzymes (e.g., phospholipase, endonuclease, caspase) that participate in chronic phase post-ischemic injury. And (iii) in the chronic phase, delayed neuronal death and specific expression of genes takes place.

IS is characterized by an acute decrease in regional CBF accompanied by hypoxia that impedes energy production by aerobic respiration. The decline in blood flow is not homogenous rather it is maximum in the ischemic core (< 7–10 ml/100 g/min) and less severe in the penumbra region (10–20 ml/100 g/min). However, restoration of blood supply within 1 h may salvage cells in the ischemic core. In positron emission tomography studies, it is observed that theoretically, a 24 h window period is available to salvage the cells in the penumbra region, however, therapeutic benefits are observable only within 3 h of onset of ischemia [8]. High energy demands due to higher metabolic activity in the brain cannot sustain this energy-deficient state. This decline in ATP generation results in derangement of ion channels and cerebral acidosis that leads to membrane depolarization and an increase in post-synaptic activation of glutamatergic (e.g., *N*-methyl D-aspartate receptors) by glutamate. Abnormal high Ca^{2+} influx *via* these ion channels activates multiple pathways that ultimately result in necrosis at the ischemic core region. Necrosis has key features such as swelling of cells and their organelles, plasma membrane withering, irreparable nuclear injury, and damage to the genetic material. If blood supply is not restored within a 3–24 h therapeutic window, cells in the peri-infarct region (penumbra) may undergo apoptotic cell death. Apoptosis is the organized way of programmed cell death to dispose of redundant cells that may be effectuated via two pathways: (i) intrinsic pathway; and (ii) extrinsic pathway. Morphologically apoptosis depicts cell shrinkage and pyknosis type features. Widespread blebbing of the cellular membrane, karyorrhexis, and parting of cell fragments results in "budding." The cytoplasm seems to be densely packed with or without nuclear fragmentation. The extrinsic and intrinsic paths involve stimulation of Cys

(Cysteine) proteases, named caspases. External stimuli and activation of death receptors in the extrinsic pathway and mitochondrial dysfunction (leakage of cytochrome C) in the intrinsic pathway can activate the caspases. Two thresholds of blood supply blockage have been proposed. In 1st threshold, blood passage is abridged up to a point where neurons are inactive electrically. In the 2nd threshold, blood flow declined up to a point where the failure of metabolic activity and functioning of ion channels is inhibited in neurons. Attainment of the 2nd threshold results in necrotic cell death whereas the 1st threshold may lead to apoptosis depending on the reperfusion time.

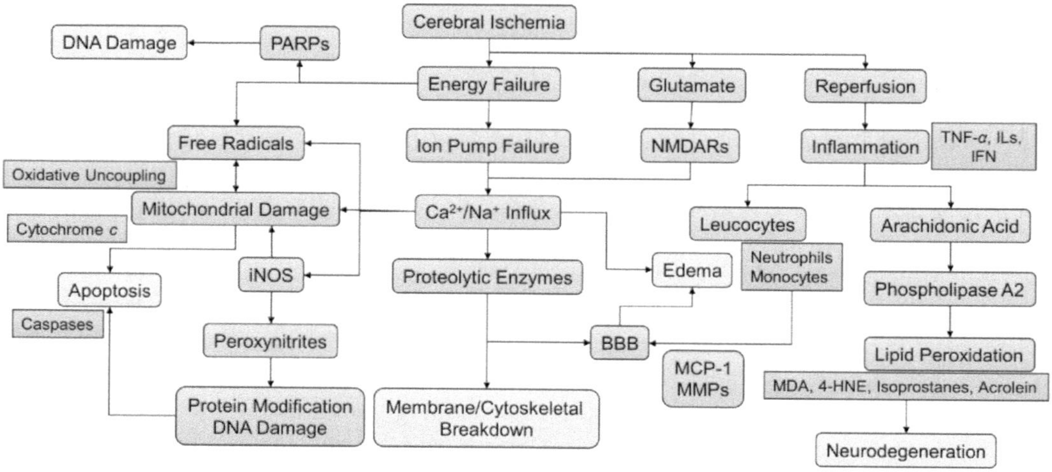

FIGURE 7.2 Pathophysiologic mechanisms in cerebral ischemia injury.

7.6.1 GLUTAMATE-DEPENDENT EXCITATORY MECHANISMS

Glutamate exerts an excitatory drive that activates glutamatergic ion channels such as *N*-methyl D-aspartate receptors (NMDARs) and *α*-amino-3-hydroxy-5-methyl-4-isoxazole propionic acid receptor (AMPAR). Exogenous glutamate in peripheral circulation cannot penetrate BBB and hence, endogenous glutamate made from *α*-ketoglutarate (from Kreb's cycle) is the only source of glutamate biosynthesis in the neurons. Albeit glia can generate glutamate from glutamine. Excitatory toxicity is the chief mechanism of the neurodegenerative cascade that takes place immediately during the ischemia phase [42]. In addition to the immediate necrosis, glutamate excitotoxicity primes the cell towards apoptotic programmed cell death in later phases of stroke. At the onset of acute ischemia, ATP depletion, membrane depolarization, and the consequent ion-channelopathies result in dysfunction of voltage-gated Ca^{2+} channels (VGCC), Na^{+} channels, and Na^{+}/K^{+} ATPase pumps. As a result, excess cytoplasmic cations such as sodium and Ca^{2+} and a decrease in K^{+} levels cause cellular swelling and excitotoxicity. Ca^{2+} triggered excess of glutamate in the synapse (accompanied by low re-uptake by presynaptic transporters) can activate post-synaptic NMDARs if the cell is depolarized and/or metabotropic receptors are activated and magnesium ion block is removed. Activation NMDARs: (i) increase intracellular Ca^{2+}

influx; and (ii) activate nitric oxide synthase (NOS) to yield a gaseous neurotransmitter, nitric oxide (NO) [43].

Ca^{2+} is extremely critical to optimal neuronal activity that prompted the evolution of multiple homeostatic pathways to regulate the subcellular localization of Ca^{2+} and $[Ca^{2+}]_i$ levels [44]. Ca^{2+} inflow *via* VGCCs and receptor-gated channels, plasma membrane and cytosolic proteins mediated Ca^{2+} buffering, subcellular organellar storage of Ca^{2+}, and Ca^{2+} outflow are all examples of these processes. These homeostatic processes keep $[Ca^{2+}]_i$ at the basal concentration in neurons, generally ~100 nM relative to a 1.0 mM in the extracellular environment. Because $[Ca^{2+}]_i$ is reserved at such a minuscule concentration in the cell, tiny or local augmentation in $[Ca^{2+}]_i$ may be employed as an activator to initiate signal transduction mechanisms that result in physiological activities like enzyme trigger or regulation of an ion channel. Ca^{2+} is regulated by (i) NMDARs; (ii) VGCCs; and (iii) inositol 1,4,5-trisphosphate regulated Ca^{2+} supplies, and ryanodine-controlled Ca^{2+} supplies in endoplasmic reticulum (ER) [45]. An equilibrium of Ca^{2+} efflux pathways through the plasma membrane, storage by Ca^{2+} affinity proteins, and absorption and storage by diverse subcellular structures (e.g., Golgi bodies, ER) are among the processes that diminish $[Ca^{2+}]_i$. ATP requires membrane outflux mechanisms, such as Ca^{2+}-ATPase localized in the plasma membrane (PMCA), and a Na^+/Ca^{2+} exchanger, which is triggered by an upsurge in $[Ca^{2+}]_i$ and is reliant on the Na^+ concentrations, which is as a consequence is reliant on the Na^+/K^+ ATPase pumps in the membrane, squeeze out Ca^{2+} from the intracellular environment to the extracellular environment. Apart from these $[Ca^{2+}]_i$ lowering efflux processes, surplus Ca^{2+} is sequestered in the cytoplasm by Ca^{2+} affinity proteins, which is thought to be the primary mechanism for lowering raised neuronal $[Ca^{2+}]_i$. A few of the proteins found in the cytosol (e.g., calbindin and parvalbumin) are hypothesized to carry Ca^{2+} ions to areas active in their outflux (e.g., PMCA and ER). Several Ca^{2+} binding proteins are found in the plasma membrane, which not only sequesters Ca^{2+} but contributes to the cellular structure by assisting in the conservation of cytoskeletal foundations. ER held Ca^{2+} reserves are built from cytosolic Ca^{2+} supplies that can be utilized to reinstate higher $[Ca^{2+}]_i$ to typical basal levels, apart from participating in an upsurge in $[Ca^{2+}]_i$ in case of signal transduction proceedings. Ca^{2+}-ATPase localized on the membrane of ER often known as sarcoplasmic ER Ca^{2+} ATPase (SERCA) is responsible for this [46].

Following ischemia, increased neuronal $[Ca^{2+}]_i$ causes injury and demise *via* a variety of pathways, with Ca^{2+}-dependent enzyme activation and mitochondrial Ca^{2+} overload proving to be primary drivers of neuronal damage [47]. High amounts of $[Ca^{2+}]_i$ activate many cysteine proteases, including calpains. Calpain's protein substrates include spectrin, tau, a microtubule-associated protein, tubulin, and neurofilaments, which are all found in the cytoskeleton. Calpain may also target the deterioration of plasma glutamate carriers, receptors, transporters, membrane proteins, cell adhesion molecules, and additional enzymes such as protein kinases and phosphatases. These modifications may threaten the cell's structural integrity, resulting in increased ion and macromolecule permeability across the membrane. Caspase's gene activation is also linked to an increase in $[Ca^{2+}]_i$. Caspases are a kind of cysteine protease that is not related to calpains. Ca^{2+}-dependent endonucleases can cause DNA damage. Phospholipases, particularly PLC (phospholipase C) and cytosolic phospholipase A2 (cPLA2) may be activated by an upsurge in $[Ca^{2+}]_i$ following injury. When cPLA2 is activated, free arachidonic acid (AA) is released and accumulates, which might change membrane permeability and increase reactive oxygen species (ROS). PLC slices phosphatidylinositol 4,5-diphosphate (PIP2) from cell membranes, releasing diacylglycerol (DAG) and inositol triphosphate (IP3), which might cause Ca^{2+} to be released from intracellular storage, contributing to an upsurge in $[Ca^{2+}]_i$. The generation of NO is another harmful enzymatic route in neurons that is controlled by Ca^{2+}. Ca^{2+} attaches with calmodulin, which subsequently controls NO synthase (NOS), generating NO from *L*-arginine, post NMDARs activation, and succeeding Ca^{2+} invasion. Peroxynitrite is formed when excess NO reacts with superoxide to form peroxynitrite, which is extremely hazardous to membrane lipids, proteins, and genetic material [48].

An abnormal increase in Ca^{2+} levels further instigates Ca^{2+} output from the ER through ryanodine receptors. This pathogenic rise in Ca^{2+} levels triggers free radicals, protease enzymes, and other Ca^{2+} dependent pathways of cell death. Although NMDARs are associated with cognitive performance through long-term potentiation and synaptic modulation, however, chronic excessive activity leads to toxic repercussions. NO is a retrograde neurotransmitter that diffuses back at the pre-synaptic membrane and augments the release of glutamate at synapse for post-synaptic activation of NMDARs. Mitochondrial dysfunction is the primary mechanism of cell death in the excitotoxic pathway. In a study, glutamate disturbed the mitochondrial membrane potential difference and caused nuclear swelling in cultured cells. Cells that recovered from the early necrotic phase had normalized membrane potential difference and ATP levels, whereas, failure to maintain this mitochondrial potential difference activated apoptotic mechanisms [49].

Despite these advances in the hypothesis and pre-clinical investigations, attenuation of glutamate-mediated excitotoxicity by NMDAR antagonists failed miserably in clinical trials [50]. Many reasons can be explored for this effect. First, NMDARs are a vital component of cognitive processing, consolidation of memory, and retrieval. This hypothesis had not taken in the fact that NMDARs are ubiquitously present in astrocytes, glia, oligodendrocytes, and neurons. Non-selective blockers can inhibit all the receptors without any differentiation. Non-selective inhibition of NMDARs resulted in several adverse effects like loss of memory, behavioral discrepancies (e.g., hallucinations), coma, etc. In a new investigation by Perez-Mato et al. [51], the findings specified that glutamate can be siphoned from the brain into blood and transported in the peripheral circulation.

Glutamate and ion-channelopathies together stitch another pathogenic sequence that causes brain edema. Edema is accomplished in three stages *viz.* cytotoxic edema, ionic, and vasogenic edema [52]. Cytotoxic edema is the earliest stage that takes place during the acute phase of ischemia. It is mostly due to ionic overloading of the cell that causes osmotic movement of solvent (water) from extracellular space to inside the cell. A decrement in extracellular ionic and water levels follows, leading to the formation of ionic edema. This results in the formation of newfangled gradients amid plasma and interstitial fluid, as well as the stimulation of transcapillary fluxes of ions and osmotically obligated water through the capillaries. Vasogenic edema results from disruption of BBB. This is followed by an invasion of water and several macromolecules from the periphery into the extracellular space of the brain through blood circulation. The primary pathways in ischemia embrace the water channel, aquaporin-4, Na^+-dependent excitatory amino acid transporters, $Na^+/K^+/$

Cl^- cotransporter 1, sulfonylurea receptor 1-regulated non-selective cation channels, and Donnan cell swelling including Cl^- and K^+ channels. Depletion of ATP levels immediately in the acute phase causes impairment of Na^+/K^+ ATPase, failure of ion gradient across the cell membrane, and membrane depolarization that results in an upsurge in $[Na^+]_i/[Ca^{2+}]_i$ (cytoplasmic levels of Na^+/Ca^{2+} ions), inward movement of Cl^- ions, and rise in $[K^+]_e$ (extracellular levels of K^+). This phenomenon is called Donnan cell swelling [53].

7.6.2 GLUTAMATE-INDEPENDENT CALCIUM TOXICITY

TRPM7 (Transient Receptor Potential Cation Channel Subfamily M Member 7) and ASIC (Acid-Sensitive Ion Channels) are two novel influx routes in ischemia [54]. TRPM is a nonselective cationic channel that is abundantly expressed in the brain and hails from the melastatin subfamily of Transient Receptor Potential (TRP). TRPM7 is a permeant ion channel for Mg^{2+}, Na^+, and Ca^{2+}, which is controlled by $[Ca^{2+}]_{i/e}$ and $[Mg^{2+}]_{i/e}$ differences. TRPM7 is a Ca^{2+}-sensing channel (nonselective) and is influenced by $[Mg^{2+}]_i$ and $[Ca^{2+}]_e$, as well as free radicals such as ROS and RNS (reactive nitrogen species) and, eventually, PIP2 [55]. Recent evidence suggests that TRPM7 maintains ROS and RNS generation by carrying an inward cationic current in retort to combined oxygen-glucose deprivation [56]. The ASICs, which, like TRPM7, are widely expressed in the CNS (central nervous system), have also garnered a lot of interest from the standpoint of non-glutamatergic causes of Ca^{2+} surplus in ischemia cell demise. The fact that these channels convey an inward Na^+ and Ca^{2+} current in retort to extracellular acidification is a unique feature [57]. Since extracellular pH declines prematurely in the progression of ischemic cerebral proceedings, ASICs have received a lot of interest as a possible Na^+ and Ca^{2+} incursion mechanism in ischemia neurons [58]. Notably, cation inflow through ASICs is important in the process of cell demise. The Na^+/Ca^{2+} exchanger (NCX) is an antiporter that, relying on $[Ca^{2+}]_{e/i}$ and $[Na^+]_{e/i}$, may serve as a Ca^{2+}-efflux/Na^+-influx channel (forward manner) or a Ca^{2+}-influx/Na^+-efflux route (reverse mode) [59]. NCX is critical in maintaining intracellular homeostasis of these two cations since the two plasma-membrane ATP-dependent pumps Na^+/K^+ ATPase and Ca^{2+} ATPase are compromised under anoxic conditions [60].

Interestingly, while TRPM7 and ASIC stimulation in cerebral ischemia are deleterious, NCX activation appears to be mostly neuro-beneficial. In fact, due to the reason that the first blockage of Na^+/K^+ ATPase – (i) $[Na^+]_i$ rises dramatically; (ii) voltage-sensitive cation channels (Na^+ and Ca^{2+}) activate in retort to the resultant depolarization; and (iii) ionotropic glutamate receptors are triggered in the primary phases of neuronal anoxic damage. NCX is compelled to act oppositely as a result of $[Na^+]_i$ surplus, outfluxing Na^+ ions whereas increasing Ca^{2+} inflow [61]. Even though this inverse approach of functioning in the initial stages of anoxia causes a rise in $[Ca^{2+}]_i$, its effect on neurons may be advantageous because it helps to reduce $[Na^+]_i$ surplus, which can else induce edema and neuronal demise *via* necrosis. During later stages of neuronal anoxia, in the presence of $[Ca^{2+}]_i$ surplus, NCX in an advancing manner, which increases Ca^{2+} efflux and Na^+ inflow, can help reduce Ca^{2+} concentrations, protecting neurons against excessive $[Ca^{2+}]_i$ triggered neurotoxicity and cell demise [62].

The presence of three distinct gene products NCX1, NCX2, and NCX3 adds to the intricacy of this antiporter's participation in the anoxic neurodegenerative chain of events. The sensitivity of these 3 isoforms to intracellular ATP levels varies. Although for NCX1 and NCX2 activity ATP is essential, NCX3 may function even if this nucleotide is not present. As a result, even while ischemia-induced ATP exhaustion in the neurons is predicted to disrupt NCX1 and NCX2 functions, it is not expected to hamper NCX3 function. When viewed in the context of various ischemic brain areas, the three NCX gene products' varied sensitivity to intraneuronal ATP alterations may have important implications [63]. Because of the Na^+/K^+ ATPase impediment that has been defined by the significant reduction of ATP content, $[Na^+]_i$ increases significantly in the ischemic core area during the early stages of brain damage. The upshot is that the sodium channel (NCX), whose means of maneuver is regulated by the electrochemical gradient of Na^+ and Ca^{2+} ions, is pushed to act in the opposite direction as a Na^+ efflux-Ca^{2+} influx pathway, so benefiting neurons in averting $[Na^+]_i$-dependent rapid necrotic cellular damage Because of the different sensitivity of the NCX gene products to ATP levels, the destiny of neurons in the ischemic core may be principally controlled by NCX3. This is because NCX3, which is typically produced in the brain areas of the ischemic core, is the only NCX isoform that is capable of maintaining its activity despite severe ATP depletion, as demonstrated in this study [64].

When mitochondrial ATP production is reduced due to a hamper of the Na^+/K^+ ATPase, a significant increase in $[Na^+]_i$ occurs, which leads to a depolarization that causes voltage-gated Ca^{2+} and Na^+ channels to open, which is trailed by (i) a fast reduction in $[Na^+]_e$ and $[Ca^{2+}]_e$; and (ii) the discharge of glutamate into the extracellular region [54]. Once released, by activating Na^+ and Ca^{2+} porous glutamate receptors, glutamate accelerates the process, reducing $[Ca^{2+}]_e$ even more. This reduction in $[Ca^{2+}]_e$ causes additional Na^+ and Ca^{2+} to flow *via* TRPM7 channels. Lactate buildup and extracellular acidification result from the shift in intracellular metabolism toward anaerobic glycolysis. ASICs detect the change in extracellular pH and activate them, allowing considerably extra Na^+ and Ca^{2+} ions to invade through the cell membrane. In a positive feedback loop, all of these events that produce intracellular Ca^{2+} burden generate an upsurge in ROS and RNS production, which magnifies $[Na^+]_i$ and $[Ca^{2+}]_i$ buildup by enhancing the TRPM7 channel function [57]. Lastly, NCX is a key component of the non-glutamatergic neurotoxicity idea. It can prevent $[Na^+]_i$ and $[Ca^{2+}]_i$ overload caused by TRPM7, ASICs, and NMDARs [63]. In fact, this antiporter may quickly enhance its function in reply to increasing $[Ca^{2+}]_i$ and oxidative and nitrosative stress. A phospholipid precursor of IP3 and DAG, plasma-membrane phospholipid PIP2 affects the activity of NCX and TRPM7 favorably. We may speculate that the phospholipid PIP2, which affects both TRPM7 and NCX in ischemic neurons, has a role in the regulation of ionic processes during ischemia by changing the concentration of PIP2.

7.6.3 MITOCHONDRIAL DYSFUNCTION AND ENERGY DEPLETION

A stroke event begins with decreased plasma in regions served by the blocked arteries. As a result of a decrease in the level of oxygen in the tissue, ATP production is hampered in cerebral ischemic cells, and thus, cytoplasmic Ca^{2+} level is affected due to the lack of energy for extrusion pumps causes "anoxic depolarization," whereas the Na^+/K^+ ATPase pumps' malfunction also disrupts resting membrane potential, leading to "anoxic depolarization." Energy failure is the prime mechanism of necrotic cell death in the ischemic core. A decline in regional CBF up to ~20% of its normal value results in loss of ion-channel (e.g., Na^+/K^+ ATPase and other energy-dependent ion pumps) functioning in the brain. Other cell-death mechanisms related to apoptosis are responsible for delayed cell death in the penumbra region if blood supply is not restored promptly [65].

Mitochondrial activity is changed in localized ischemia as a direct result of reduced glucose and oxygen supply to the tissue [66]. During the first 5 min after artery blockage, glucose and ATP levels drop dramatically. When CBF is reduced, mitochondria are damaged, and neurons switch from aerobic to anaerobic metabolism [67]. In the ischemic core, phosphocreatine serves as a relatively brief energy store, enabling a near-equilibrium mechanism associated generation of ATP from ADP (adenosine diphosphate), catalyzed by creatine kinase. Phosphocreatine follows the same trend as ATP, quickly dropping to fewer than 30% of normal levels. Due to a lack of oxygen, a part of glucose that invades the core area of ischemia is converted to lactic acid through glycolysis, resulting in a drop in pH. Lactate levels rise to more than 10 times those in non-ischemic tissue. After ischemia, approximately 2 h, phosphocreatine levels are lowered to around 70% of non-ischemic levels, while the adenylate energy component stays beyond 0.8. ATP levels drop more dramatically, nearing 50% of non-ischemic tissue after 2 h of ischemia. A part of the ADP generated through the hydrolytic breakdown of ATP is metabolized to AMP (adenosine monophosphate) or back to ATP in ischemia circumstances. Adenylate kinase catalyzes this process, which helps the brain retain the energy currency and fulfill instant energy needs. During the ischemic situation, inosine and hypoxanthine are generated by AMP at the tissue level, depleting the adenine nucleotide pool in the process. Lactate levels in the penumbra rise to levels several times greater relative to the typical brain but lower relative to the core parts. Lactate buildup shows that the supply of oxygen is much adversely constrained concerning the delivery of glucose, leading to pyruvate oxidative breakdown by the mitochondria being inhibited [68].

After 1st h of reperfusion, ATP level in core tissue improves slower relative to phosphocreatine or adenylate energy charge, reaching 50–70% of control levels. After an ischemia period of up to 3 h or more, reperfusion in the penumbral tissue restores ATP and phosphocreatine equilibrium in 1 h. Because of the delayed renewal of the reduced ATP level, the restitution of ATP level is much less effective. The capacity of mitochondria to enhance ETC (electronic transport chain) activity in retort to changes in the H^+ ion concentration through the internal membrane of mitochondria caused by the introduction of an uncoupler or ADP worsens with time [69]. Baseline respiration has been sustained to a great extent. Uncoupling of respiratory chain or ADP accrued impairment in rodents reduces respiration by 45–60% in core tissue models after 2 h of MCA blockage. In the penumbra, reductions of 15–40% are seen. Depreciation of aerobic respiration in the mitochondria of the penumbra region might amplify the impact of decreasing oxygen supply and the function of oxidative metabolism declines in this area.

The significant reductions in ATP synthesis by the mitochondria in highly ischemic core areas produce circumstances that assure cell demise *via* necrosis after 2–6 h of the commencement of the stroke [70]. The penumbra's lesser metabolic deficiencies enable cells to survive longer and give possibilities for protective treatments. Transition pore activation is mainly caused by aberrant calcium influxes and may be aided by a variety of conditions such as adenine nucleotide deficiency and oxidative stress. In the internal membrane of the mitochondria the adenine nucleotide translocase (ANT), in the exterior membrane of the mitochondria the voltage-dependent anion carrier, and a mitochondrial matrix protein, cyclophilin D, having peptidyl-prolyl cis-trans isomerase activity, have all been substantially involved [71].

The neurons scattered all over the ischemic area show cytochrome *c* release inside a few minutes in response to persistent ischemia. In the cytosol, cytochrome *c*, dATP (deoxyadenosine triphosphate), procaspase-9, and, apoptotic protease activating factor 1 (APAF-1) constitute an apoptosome. Caspase-9 is activated when an apoptosome forms, and it then cleaves other procaspases. The porousness of the mitochondrial membrane may also cause the appearance of proteins that cause apoptosis through a caspase-independent route. Under default settings, AIF (apoptosis-inducing factor) is a 62 kDa transmembrane protein that lies in the internal mitochondrial membrane. When the porousness of the outer membrane rise, proteolysis releases the part of AIF (about 57 kDa) in the intermembranous area through the activities of Ca^{2+}-stimulated calpains and Ca^{2+}-independent cathepsins. The AIF protein translocates to the nucleus and causes DNA fragmentation into bigger pieces than those generated by caspase-3 activation [65].

7.6.4 INFLAMMATION

Inflammation is an important factor that mediates the progression of neurodegenerative cascade in the later phases of ischemic brain conditions. The after-effects of these cerebrovascular diseases are so severe that ~80% of survivors suffer from some or other permanent disability. Vascular compartment and brain parenchymal cells such as neurons, glia, astrocytes, and matrix components (called neurovascular unit) together mediate the inflammatory cascade after ischemia. Microglia (resident macrophages in the brain) characterize 6–21% of the glial inhabitants and are primarily responsible for the innate immunity of the CNS. Microglia, activated by ischemic injury as early as within 30–60 min of insult, display diverse pro-inflammatory and anti-inflammatory phenotypes, i.e., M1 (neurotoxic) and M2 (neuroprotective). These are visible in the core of the infarcted area as well as the peripheral areas [72]. A significant number of cellular intermediaries such as proteases, cytokines, and ROS are produced by the M1 type, which is thought to promote classical stimulation as well as pro-inflammatory repercussions. Alternate stimulation and anti-inflammatory characteristics (production of anti-inflammatory cytokines like IL-4, and IL-10) are thought to be responsible for the phagocytosis of damaged cells or injured neurons and the secretion of numerous neurotrophic factors in the M2 role. M1 microglia phenotype is interchangeable with M2 microglia in a few instances. In this transition from M2 to M1, nuclear factor-kappaB (NFκB), Notch, mammalian target of Rapamycin (mTOR), peroxisome proliferator-activated receptor gamma (PPAR-γ), Janus kinase (JAK)-signal transducer and activator of transcription (STAT), 5′-AMP-activated protein kinase (AMPK), and cAMP-response element-binding protein (CREB) plays an important role [73]. Hence, microglia have dual functions in ischemia conditions. Activation of the M1 phenotype can increase the issue of IL-6, TNF-α, IL-1, and IFN-γ *via* surface CD16 and CD86 receptors. They also increase ROS and NO generation during ischemia. M2 phenotype expresses arginase-1 (Arg-1) and CD206 receptors and are involved in phagocytosis of dead cells and neurotoxic aggregates and also release protective influences *viz.* IL-4, TNF-β, and IL-10 may alleviate inflammation in ischemic tissues [74].

In some studies, using MCAO focal cerebral ischemia model (followed by reperfusion), inhibition of microglia activation caused a significant increase in infarct size [75]. In the MCAO model of focal cerebral ischemia, dual colony-stimulating factor-1 inhibitor, PLX3397, caused inhibition of microglia activation that resulted in enhanced infarct size, leukocyte infiltration, presence of pro-inflammatory cytokines, and programmed cell death in the brain. The findings of this study emphasized that in ischemia microglia may have an inhibitory action on the astrocyte-mediated increase in inflammatory molecules [76]. The inflammatory sequence in retort to post-ischemia events is a self-limiting technique, which is synchronized *via* a variety of intermediaries that vigorously impede detrimental inflammatory functions, comprising of eradication of dead cells, generating an anti-inflammatory environment, and expression of pro-survival mediators encouraging cerebral tissue renovation and restoration. Microglia are activated in the acute, sub-acute, and chronic stages of ischemic injury. The iNOS-triggered CD14 receptors trailed by toll-like receptor 4 (TLR4) appear in stimulated microglia in the necrotic area that might serve as a microglial activation mechanism. Interferons (IFN) are also implicated in microglia-triggered post-ischemia damage. Interferon (type 1, i.e., IFN-α/β) stimulates IFN-α/β receptor (IFNAR) complex and microglial type-1 IFN-triggered genetic function [77]. In the ischemic region, microglia move towards the ischemic lesion *via* surface G protein-coupled receptors 55 and 18 (GPR55 and GPR18) and form a close connection to trigger the "capping process." After the death of neurons, capping supports initial acknowledgment and speedy phagocytic elimination of non-functional neurons. Impairment of lymphocyte function-associated antigen 1 (LFA-1) at the gene level obliterates the capability of microglia to drift in the direction of incapacitated neurons [78].

Astrocytes are housekeeping cells involved in the homeostasis of ion transport, water, neurotrophic factors, attenuation of neurotransmitter action, and removal of waste products from the brain. Astrocytes are intricate components of the BBB capillary network and in ischemia activation of MMPs (matrix metalloproteinase) breaches the integrity of this barrier by damaging the connection of astrocytes with endothelial cells and pericytes. Normally, astrocytes take up synaptic glutamate *via* EAAT2 (excitatory amino acid transporter 2) and convert it into glutamine, hence, participating in the attenuation of glutamate excitatory drive in the brain. However, dysfunction of EAAT2 might be the reason for glutamate excitatory transmission during ischemia [79]. Astrocyte hypertrophy and proliferation in response to pro-inflammatory cytokines lead to enhanced release of monocyte chemotactic protein-1 (MCP-1), IL-1β, vimentin, nestin, and glial fibrillary acidic protein (GFAP) ensuing microgliosis. Failure of ion pumps (e.g., sodium-potassium ATPase) causes astrocytic swelling, and an increase in iNOS, NADPH (Reduced nicotinamide adenine dinucleotide phosphate) diaphorase, and GFAP in astrocytes have been observed in temporary global ischemia model [80].

Pericytes migrate from the basement membrane one hour after MCAO, resulting in BBB permeability. Pericytes, on the other hand, have been found to concentrate in the peri-infarct area of mice 7 days after an IS, possibly supporting vascular healing, and they can also segregate into diverse lineage cells like neural and vascular origins. In blood vessels, endothelium and close-fitting connections form the decisive fence, even though pericytes play an important regulatory role. Occludins and claudins are tight junction proteins that join endothelial cells, forming clefts that perform as a main barricade amid blood and the interstitial fluid. Endothelial dilation, astrocyte separation, pericyte separation, pericyte shrinkage, microglial stimulation, vasoconstriction, and blood vessel estrangement occurs as a result of dynamic changes in BBB permeability during ischemia [81]. During temporary BBB breakdown, a biphasic design develops, with a preliminary opening 2–3 h later to commencement and a second opening 24–48 h later, consequential in vasogenic edema and elevated intracranial pressure in mice. Furthermore, MMPs stimulate microglia, macrophages, and neutrophils, which release additional MMPs, additionally deteriorating cerebral arteries, 24–48 h after 2 h of MCAO with reperfusion in rodents [82].

The primary circulating immune cells to penetrate ischemic areas are neutrophils, trailed by monocytes [83]. Endogenous ligands of toll-like receptors (TLRs), also acknowledged as damage-associated molecular patterns (DAMPs), are a fast-increasing category of powerful inflammatory inducements that are extensively generated in inflammatory cores following an IS as a result of inflammation and tissue damage. HMGB1 (high-mobility group protein B1), primarily stimulates TLR2/4 on neutrophil surfaces, resulting in NADPH oxidase assemblage and ROS production, and upregulating the synthesis of pro-inflammatory cytokines and anti-apoptotic protein Bcl-xl [84]. Phagocytosis causes neutrophils to become activated, releasing antimicrobial peptides, myeloperoxidase (MPO), proteases, and superoxide radicals, all of which form ROS in a series of cascade reactions [85]. Monocytes stick to vessel walls and migrate towards ischemic regions after the neutrophil invasion (4–6 h), with maximum activity 3–7 days after the insult [86]. Neutrophils enter the injured area 5 h after reperfusion. Peak neutrophils are recorded 12 h after the injury in a permanent MCAO model [87]. First, leukocytes (white blood cells, WBCs) cling to the endothelium, which prevents erythrocytes from flowing through the microvasculature, resulting in cerebral no-reflow. Second, stimulated leukocytes yield proteases, MMPs, and ROS at the endothelium's surface, which can cause considerable damage to blood vessels and brain structures. When phospholipases activate leukocytes, physiologically active chemicals such as eicosanoids (e.g., leukotrienes and prostaglandins) and platelet-activating factors (PAFs) appear, causing narrowing of blood vessels and conglomeration of platelets. Lastly, infiltrating white blood cells aggravate neuron damage by starting proinflammatory mediators inside and around the infarct core and penumbra [88]. Following the acute stage, the infiltrating white blood cells produce additional chemokines, cytokines, and, most crucially, surplus ROS, which increases the formation of MMPs, particularly MMP-9, in the sub-acute phase. When there is surfeit protease activity and hyperstimulation of resident immune cells, the trigger of MMP function enhances the inflammatory cascade, resulting in BBB breakdown, brain edema, neurodegeneration, and hemorrhagic conversion, as well as further intensifying leukocyte infiltration [89]. Microglia cells increase excitotoxic damage and ischemia in the brain in addition to being a proinflammatory stimulator. The active division of microglia reaches its zenith after 48–72 h after ischemia and persists for some days after the preliminary damage, although circulatory white blood cells are engaged for hours to a few days next to the primary damage. Neutrophils are by far the utmost prominent white blood cells that invade the brain parenchyma later to 30 min to a few hours of ischemia, spiking during days 1–3 and then gradually declining. After cerebral ischemia, neutrophil invasion increases on day 1, peaks on day 3, and then begins to drop, but it remains detectable on days 7 and 15. Within 15 min of ischemia, neutrophils express a variety of endothelium adhesion molecules. Neutrophils have already ringed brain arteries and begun infiltration 6 to 8 h after a stroke [90]. Increased neutrophil levels have been linked to infarct volume and functional impairments. In the final phases of ischemic brain damage, T lymphocytes (T cells) are recruited. In a transient MCAO (tMCAO) mouse archetypal, T lymphocyte (CD3+) invasion peaks during days 3–4 after ischemia, but stimulated microglia (CD11b+)/macrophages and neutrophils (Ly6G+) augment very early post-ischemia. Nonetheless, few studies have shown that T lymphocytes gather in the brain before 24 h of MCAO accrued brain ischemia, which might participate in affecting tissue inflammation and damage before these cells appear in the extravascular region [91].

Cytokines participate in the upregulation of cell adhesion molecules (CAM) expression. In the core of the ischemic brain, upregulation of ICAM 1, causes BBB rupture by assisting the enrollment of WBCs, which later cause the secretion of cytokines. Diverse mediators of inflammation (e.g., Natural killer cells, macrophages, T cells, and polymorphonuclear WBCs) migrate to the ischemia region when the BBB is disrupted [92]. IL-1, TNF-α, and IL-6 are three primary proinflammatory cytokines that cause and worsen an inflammatory response after IS [93, 94]. Acute IS patients display TNF-positive (+ve) cells as early as 3 days after the stroke and remain +ve for up to 15 months, according to post-mortem investigations. TNF serum levels rise within 6 h of a stroke and remain stable for the next 10 days. The content of IL-1 is also raised in the CSF (cerebrospinal fluid), peaking on the 2nd and 3rd day after an IS. IL-1 modifies neurons, microglia, and the vasculature to mediate ischemia, traumatic, and excitotoxic brain injury [95]. In severe IS patients, IL-6 levels heighten in the CSF, just like TNF-α and IL-1. IL-6 levels escalate within 24 h in the CSF and peak on the 2nd and 4th day, according to a few pieces of research, while others find peaks on days 3 and 7 [96]. The appearance of IL-6 is potentiated by IL-1, a key mediator of neuroinflammation. As a result, anakinra, an IL-1 receptor antagonist, showed a decent clinical benefit and depressed neutrophil tally in the circulation and IL-6 content [97]. Several inflammatory mediators, their injury mechanisms, and the pharmacological interventions that can be used are mentioned in Table 7.4. These inflammatory modulators may be targeted to overcome the detrimental action of various cytokines and chemokines during ischemic injury and/or after reperfusion.

Although chemokines and their receptors, like cytokines, generally appear in little concentrations. However, later to cerebral ischemia, TNF-α and IL-1 increase the transcription and secretion of some particular chemokines like fractalkine and macrophage inflammatory protein 1 (FMIP-1 and MIP-1), cytokine-induced neutrophil chemoattractant (CINC), microglial response factor-1 (MRF-1), and MCP-1. These chemokines are elevated within 3 h of ischemia and can remain in this state for 6 h. Numerous chemokines hailing from the CXC family were elevated (> 10-fold) at 4, 24, and 72 h after reperfusion,

TABLE 7.4 Role of Immune Regulators in Ischemic Brain Cascade, Mechanism, and Possible Therapeutic Interventions

Inflammatory Modulator	Source	Mechanism	Pharmacological Interventions	References
Pro-Inflammatory Activity				
IL-1	Astrocytes, glial cells, endothelial cells.	B- and T-cell stimulation triggers inflammatory mediators (ICAM-1), controls TNF-α and IL-6, IL-1β activates NFκB *via* TLRs, inflammasome formation, prime endothelium for leukocyte adherence, leukocytes, and endothelial interaction, edema formation.	Recombinant human IL-1 receptor antagonist, Anakinra, Canakinumab, IRAK (interleukin-1 receptor-associated kinase) small molecule inhibitor, Pacritinib, R-835, PF-06650833, CA-4948 (AU-4948), BAY-1834845.	[110, 111]
IFN (Type-I: IFN-α/β/ε/κ/ω; Type-II: IFN-γ).	NK cells, T cells, monocytes, dendritic cells, macrophages.	Present antigen *via* class I/II major histocompatibility complex (MHC) molecules on the exterior of T cells and macrophages, infiltration of microglia and inflammatory cells, IFN regulatory factor-1, activate Janus kinase (JAK)/signal transducer and activator of transcription (STAT), cause atherogenesis, induces NOS mRNA.	StA-IFN-1, StA-IFN-4, glucocorticoids, Anti-IFN-γ antibodies.	[112–116]
IL-8	Microglia, astrocytes	Chemotactic function, triggers neutrophils, produces oxidative metabolites.	IL-8 receptor inhibitor repertaxin.	[117]
IL-16	Microglia, astrocytes, macrophages, dendritic cells.	Fibrin deposition, induce TNF-α and IL-1β, enhance vascular porousness, damages capillary coherence, BBB collapse, edema.	Anti-IL-16 monoclonal antibodies (mAbs).	[118]
IL-17	Microglia, astrocytes, macrophages, dendritic cells.	Epithelial, keratinocytes, endothelial, and fibroblasts secretion of ICAM-1, IL-1/6/8, TNF-α, MCP-1, PGE2, and G-CSF, NOD-like receptor protein 3 (NLRP3) inflammasome intensify IL-23/17 alliance triggered injury.	Secukinumab (human monoclonal antibody targeting IL-17A), ixekizumab (humanized IgG4 explicit against IL-17), and brodalumab (human antibody targeting IL-17 receptor A).	[119, 120]
IL-18	Microglia, astrocytes, macrophages, dendritic cells.	Rupture of atherosclerotic plaques.	Tadekinig alfa (recombinant human interleukin-18 binding protein, IL-18BP), anti-IL-18 monoclonal antibody CERC-007, GSK1070806.	[121]
ICAM-1	Endothelial cells	Adhesion of lymphocytes, eosinophils, neutrophils, and monocytes to endothelium.	Dexmedetomidine, Enlimomab (anti-ICAM-1 monoclonal antibody).	[112, 111]

TABLE 7.4 *(Continued)*

Inflammatory Modulator	Source	Mechanism	Pharmacological Interventions	References
Leukotrienes (LTs)	Lipoxygenase (LOX)	Vasoconstriction, enhance vascular porousness, and edema.	CysLTs receptor antagonists, e.g., Montelukast, HAMI 3379.	[122]
Monocyte chemotactic protein 1 (MCP-1).	Microglia, endothelial cells, astrocytes, leukocytes.	Chemotactic activity (monocytes and macrophages), increases $[Ca^{2+}]_i$, induce superoxide anions, lysozyme, and upregulates macrophage adhesion molecules (e.g., integrin family $\beta 2$ and $\alpha 4$), IL-1, IL-6.	Leukotriene receptor antagonists, Trichostatin A, Doxycycline, Propentofyllin, Propagermanium.	[123–125]
Nuclear factor-kappaB (NFκB).	Neurons and microglia *via* DAMPs, ROS, immune-related receptors, NMDARs, VGCCs, TLRs, cytokines, and growth factors.	Inflammasome protein translation, pro-IL-1β/18, excitatory drive, GABAergic inhibition, disrupt astrocytes and BBB, control caspase blockers, Bcl-2 family proteins, and TNF-receptor associate factor (TRAF).	• Pathway inhibitors: Zanubrutinib, Acalabrutinib. • Inhibitors targeting IKK complex: VGX-1027, Teglarinad chloride. • IKKε and tank binding kinase 1 (TBK1) inhibitors. • NFκB inducing kinase (NIK) inhibitors: TRC-694.	[105, 126]
P-selectin, L-selectin, E-selectin.	Endothelial Weibel-Palade bodies and platelet-associated α-granules.	Impede WBCs movement, hasten adherence to endothelium, and intravascular clogging.	Typical P- and E-selectin epitopes (HuEP5C7) antibodies, polysaccharide fucoidin (FCN), humanized anti-L-selectin antibody, sialyl-LewisX (CY-1503).	[127]
Integrins	Basal lamina, endothelial cells, leukocytes.	Modification of cell cytoskeleton, cell migration, link endothelium to extracellular matrix biomolecules (e.g., laminin and collagen), join astrocytes, endothelium, and basal lamina to breach BBB, maintain cerebral microvasculature, ICAM-1 expression.	Typical P- and E-selectin epitopes (HuEP5C7) antibodies.	[127]
Immunoglobulin supergene family [vascular cell adhesion molecule-1 (VCAM-1; CD106),	Activated endothelial cells, ROS, blood flow deregulation, and vessel injury.	Attachment and transendothelial migration of leukocytes.	Ibuprofen, LFA-1 (CD11a/CD18) and Mac-1 (CD11b/CD18) blockers, Mac-1 blocker (Leukarrest, UK-279, 276), ICAM-1 blockade (Enlimomab).	[127, 128]

TABLE 7.4 *(Continued)*

Inflammatory Modulator	Source	Mechanism	Pharmacological Interventions	References
ICAM-1 and ICAM-2, mucosal addressin cell adhesion molecule-1 (CD146), platelet-endothelial cell adhesion molecule-1 (CD31)].				
Chemokines (cytokine-stimulated neutrophil chemoattractant, MCP-1).	Cytokines	Neutrophil intrusion	Icariin (CINC-1 release inhibitor), triterpenoids.	[127, 129]
Thrombin	Vessel and tissue injury.	Monocytes and neutrophils chemotaxis, trigger adhesion molecules and endothelium communication *via* PPAR and NFκB, activates C3 and C5 complement components, and damages BBB, atherogenesis, platelet aggregation, and vascular remodeling.	Direct and indirect thrombin inhibitors (e.g., fondaparinux, desirudin).	[127]
Complement system [opsonins (iC3b, C3dg, C3d), anaphylatoxins (C3a, C5a), and terminal cytolytic membrane attack complex].	Activated intravascularly by proteases of the coagulation cascade, microglia, neurons, and neutrophils.	C1q (pro-inflammatory and anti-inflammatory), C3 (pro-inflammatory), C3a (pro-inflammatory and anti-inflammatory), C5, C5a, and C5b-9 (pro-inflammatory).	CD30 chimeric mAbs (Brentuximab vedotin), IVIg, C3aR antagonist, sCR1 (recombinant complement C3 convertase inhibitor), PMX53 (C5aR1 antagonist AcF).	[130, 131]
Danger-/damage-associated molecular patterns (DAMPs) *viz.* high mobility group binding protein 1 (HMGB1), purines (ATP, UTP, and their catabolites), heat shock peptides, PRRs on glia and astrocytes, peroxiredoxins, mitochondrial-derived *N*-formyl protein.	Injured and dying neurons.	Microglia activation.	MRP14 (myeloid-related protein) inhibitors: Paquinimod, ethyl pyruvate (REPS), anti-HMGB1 antibody.	[132–134]

TABLE 7.4 *(Continued)*

Inflammatory Modulator	Source	Mechanism	Pharmacological Interventions	References
Dual Pro-Inflammatory and Anti-Inflammatory Activity				
IL-6	Astrocytes, endothelial cells, and microglia.	Trigger appearance of adrenocorticotropic hormone and hydrocortisone, antagonize IL-1 and TNF-α, trigger soluble TNFR and IL-1 receptor blockers, endogenous neuroprotection against NMDARs, augments endothelium porousness, gliosis, and leukocyte activation induces JAK tyrosine kinase and STAT pathway.	Receptor antibodies: • Anti-IL-6 mAbs (e.g., sarilumab, tocilizumab). • Anti-IL-6 mAbs (i.e., siltuximab).	[135–137]
TNF-α	Astrocytes and microglia.	• TNF-α preconditioning impede free radical, inflammation, and apoptotic cell mutilation. • Damage to vascular endothelial cells, leukocyte infiltration, upregulation of ICAM-1 and CD11/CD18, induction of coagulation processes, increasing BBB permeability, MMPs activation.	TNF-α loaded PEG-b-(PELG-g-PLL) nanoparticles, TNF blockers (Adalimumab, Certolizumab pegol, Etanercept, and Infliximab).	[138, 139]
Anti-Inflammatory Activity				
TGF-β [inhibins, growth differentiation factors (GDFS), activins, TGF-β isoforms (1,2,3), bone morphogenetic proteins (BMPs), and glial cell-derived factors].	M2 phenotype macrophage.	Cellular division, translocation, differentiation, hemopoiesis, angiogenesis (vascular proliferation), and apoptosis, attenuate brain edema and infarct volume, repair brain tissue, reduces microglial activation, the release of IL-1Ra (IL-1 receptor antagonist).	–	[113, 140]
IL-10	–	Attenuates inflammatory release and expression of chemokines.	–	[141, 142]
IL-13	Activated Th2 cells.	Inhibits mononuclear macrophagic inflammatory cytokines and chemokines (IL-1/6/8,), inhibits macrophage inflammatory protein-1, and TNF-α, inhibits antibody-dependent cell-mediated cytotoxicity of human monocytes triggered through IFN-γ or IL-10.	–	[111, 143]
IL-4 (a relative of IL-13).	TH2 cells	T-cell differentiation, IgE release, diverges macrophages/microglia toward the M2 phenotype.	–	[113, 144]

facilitated WBCs permeation, and had substantial participation in IS etiology, according to gene analysis [98]. In an established MCAO mouse archetypal with 45-min MCA occlusion, chemokine ligand 2 (CCL-2) along with the receptor CCR2, causes severe post-ischemic inflammatory retort *via* stimulating WBCs enrollment as well as disrupting BBB and WBCs adherence to the endothelium in cerebral blood vessels [99]. CCL-2 content rise in the penumbra zone after 6 h after reperfusion, peaking at 24 and 48 h. CCL2/CCR2 levels are linked to the size of infarcts and the expansion of ischemic lesions. Furthermore, in ischemia patients, CCL2 appearance is elevated in the CSF and blood plasma [100]. CCL2 gene interruption reduced the size of the necrotic area in a focal cerebral ischemic mouse prototype (30 min blockage), while CCR2 obliteration improved motor abilities in a focal transient cerebral ischemia mouse prototype (30 min occlusion) [101]. CCL2 overexpression worsens ischemia damage in mice [102]. After cerebral ischemia, active microglia (MIP-1, MIP-2, and MRF-1) produce the first chemokines (e.g., fractalkine and MCP-1), trailed by astrocytes and injured neurons.

By eroding the basal lamina and waning the arteries, MMPs participate primarily in BBB disturbance during the acute stage of IS. The appearance of MMP-9 in rats increases significantly within 24 h after ischemia, according to studies utilizing ischemic rodent models. In the MCAO rat prototype with 2 h occlusion, the rate of MMP-9 action was extremely raised at 48 h. The disruption of the BBB by MMP-9 and tissue plasminogen activator (tPA) is documented, which corroborates hemorrhagic conversion [103].

Damage-associated molecular patterns (DAMPs) are produced early after an ischemia event. DAMPs are neuronal and glial cell components that were designed to trigger healing mechanisms when they were wounded or stressed. DAMPs are a group of chemicals that stimulate an immune response and cause cell damage. They include endogenous proteins, nucleic acids, and metabolites [104]. After cell injury, endogenous proteins such as DAMPs, HMGB1, IL-33, IL-1, Ca^{2+}-attaching S100 peptides, heat shock peptides (HSPs), and peroxiredoxin (PRXs) are produced, which cause inflammation. DAMPs also include mitochondrial DNA (mtDNA), and nucleotide derivatives including ATP, uric acid, and ROS [105]. TLRs are one of the utmost frequent transmembrane pattern-recognition receptors (PRRs), are activated by DAMPs and cytokines, and participate chiefly in the inflammatory response following brain mutilation. Through the NFκB signaling pathway, stimulated TLRs to intensify NLRP3 [nucleotide-binding oligomerization domain (NOD)-like receptor protein 3] appearance. NLRP3 constitutes a caspase-1 stimulating composite called the NLRP3 inflammasome with the adaptor ASC [apoptosis-associated speck-like protein containing CARD (caspase recruitment domain)] and procaspase-1, afterwards, the stimulated caspase-1 alters the pro-form to the established form of pro-inflammatory cytokine IL-1 and IL-18 [106]. The inflammasome is a multiprotein complex that triggers cell death in cerebral ischemia by mediating the production of inflammatory cytokines by the immune system and nerve cells in response to danger signals. NLRP3 is one of the most well-studied components of the NLR family of receptor proteins that may form inflammasomes [107]. NACHT, a nucleotide-binding and oligomerization domain, a C-terminal leucine-rich repeat (LRR), *N*-terminal adaptor protein ASC [that includes a pyrin domain (PYD)], and procaspase-1 make up the NLRP3 inflammasome. The activation of the inflammasome complex and the generation of cytokines are accomplished by a two-phase method. The priming phase includes DAMPs from injured and dead cells activating transmembrane TLRs, which leads to stimulation of the NFκB and MAPK (mitogen-activated protein kinase) signaling pathways, as well as transcription of the inflammasome protein and pro-IL-1 and pro-IL-18. The second phase involves DAMPs, such as mitochondrial ROS, activating the inflammasome complex, causing NLRP3 receptor oligomerization and ASC recruitment. The ASC attaches to procaspase-1, causing it to be cleaved and converted to active caspase-1, which then activates IL-1 and IL-18, causing inflammation [108, 109].

7.6.5 OXIDATIVE STRESS

The brain is highly susceptible to free radicals due to lesser antioxidant levels, high polyunsaturated membrane lipid concentrations, higher basal oxygen consumption, and higher levels of iron, which catalyzes the conversion of H_2O_2 (hydrogen peroxide) to highly reactive hydroxyl radicals. ROS and NOS cause severe cellular damage, ultimately leading to tissue deterioration and cell demise; these consequences comprise nucleic acid and DNA damage, lipid peroxidation, protein denaturation, cytoskeletal structure mutilation, and chemotaxis [145].

A few key mechanisms that intensify the permanent injury induced by cerebral I/R are mitochondrial malfunction and oxidative stress caused by energy metabolism failure [146]. Superoxide ($O_2^{\bullet}$), hypochlorous acid (HOCl), H_2O_2, hydroxyl radical ($HO^{\bullet}$), peroxynitrite, and NO are some of the vital ROS and RNS. The most significant ROS in the CNS is superoxide radical [147]. It is produced intracellularly as a byproduct of aerobic metabolism, primarily the mitochondrial respiratory chain. Mitochondria create superoxide as oxidative phosphorylation mediated ATP synthesis byproduct under typical cellular circumstances. Superoxide is transformed to H_2O_2 by superoxide dismutase (SOD) which acts as an intracellular messenger outside mitochondria. Oxygen levels are exhausted earlier than glucose levels in the ischemic cell, encouraging the transition to the glycolytic route of anaerobic energy synthesis. Consequently, this corroborates the accumulation of lactate acid and H^+ (hydronium ions) in the mitochondrial lumen that triggers a reversal of the mitochondrial membrane H^+ uniporter, thus, instigating surplus cytosolic H^+ accumulation and acidosis. By supplying H^+ for the transformation of superoxide into H_2O_2 or the more destructive hydroxyl radical, acidosis contributes to oxidative stress. Furthermore, the interaction of NO with superoxide produces the powerful protein and lipid oxidant peroxynitrite, quickly draining NO bioavailability in the oxygen-depleted cell. In the ATP-depleted post-synaptic cell, glutamate activation of NMDARs promotes intracellular NO and consequent $ONOO^-$ synthesis. NMDARs are physically tethered to neuronal nNOS, and when Ca^{2+} is activated and influx inside, Ca^{2+} combines with calmodulin and immediately stimulates nNOS, creating surplus NO. NO combines with superoxide anions generated by energy-depleted mitochondria's anaerobic metabolism to make peroxynitrites ($ONOO^-$). Apoptosis is mediated by these NO radicals through traditional oxidative and nitrosative stress mechanisms.

ROS is produced extracellularly as a portion of the immunological activity, principally by NADPH oxidases (NOX), MPO, and xanthine oxidase (XO). The NOX family consists of seven members (NOX1, NOX2, NOX3, NOX4, NOX5, DUOX1, and DUOX2), with the NOX2, NOX3, and NOX4 isoforms expressed mostly in the CNS. Because NADPH oxidase creates superoxide radicals, NOX inhibition has been investigated as a stroke therapeutic target. NOX2 and NOX4 are two subtypes of NOX that contribute significantly to oxidative damage after a stroke. Phosphorylation-activated p47phox and p67phox, in combination with stimulated Rac, activate and imitate NOX2. NOX3 is stimulated by NOXO1, but it is unaffected by NOXA1 or activated Rac stimulation. Superoxide degrades fast into oxygen and H_2O_2. Nevertheless, in the company of redox-active iron, the Fenton and Haber-Weiss reactions create highly reactive hydroxyl radicals from superoxide and H_2O_2 [148, 149]. XO catalyzes the change of hypoxanthine to xanthine as well as the alteration of xanthine to uric acid. XO occurs as a substitutable enzyme of xanthine hydrogenase (XDH) under normoxic circumstances. XDH is broken to the lively XO during ischemia, and cellular ATP is changed to hypoxanthine that gathers in the ischemic tissue. XO may then oxidize the change of hypoxanthine to xanthine and xanthine to uric acid during the reperfusion phase, creating molecular oxygen and H_2O_2. Endothelial NOS (eNOS), inducible NOS (iNOS), and neuronal NOS (nNOS) are the three isotypes. eNOS and iNOS are the two most significant isotopes implicated in cerebral ischemia and stroke [150]. eNOS synthesis promotes NO production in the initial periods of cerebral infarction. Despite accounting for just 10% of total NO, eNOS has crucial participation in vasodilatory effects, boosting CBF, and shielding neurons from injury. However, NO is generated by iNOS and nNOS in the late phases of cerebral ischemia, which worsens neurotoxicity and is the origin of late neuronal impairment. When superoxide is present, highly oxidative peroxynitrite may develop. It has been revealed that honokiol may lower nNOS-derived NO by inhibiting nNOS membrane translocation, hence alleviating cerebral infarction and edema following ischemia [151].

There are several antioxidant systems in the cell that work to prevent the formation of ROS or eliminate them. They might be internal or external (nutritional) in nature. Glutathione (GSH), vitamin C and E, superoxide dismutase (SOD), catalase (CAT), glutathione peroxidase (GSH-Px), heme oxygenase (HO), and glutathione reductase (GR) are some of the most significant antioxidants [152]. They add to the cell's antioxidant capacity and serve as an essential cellular defense against ROS and RNS. SOD produces H_2O_2, which CAT converts to oxygen and water. GSH is used by GSH-Px to convert H_2O_2 to water. HO-1 has antioxidant properties, helps to preserve microcirculation in ischemic damage, and has cell cycle modulatory properties [153]. In IS models, melatonin and sulforaphane improved the therapeutic outcomes (e.g., reduction in infarct size, BBB permeability, edema, and ROS) by increasing HO-1 activity. In many studies, augmenting the endogenous antioxidants, scavenging free radicals, or targeting ROS/RNS production (e.g., NOX, XO, NOS, and mitochondria) afforded much relief against cerebral ischemia and/or reperfusion injury (Table 7.5).

7.6.6 ISCHEMIC TOLERANCE

Immediate reperfusion within 1–2 min of anoxic depolarization leads to (i) an increase in neuron survival frequency; and (ii) enhance the resistance of neurons towards more

TABLE 7.5 Oxidative Stress Mediators in Cerebral Ischemia and Possible Pharmacological Interventions

Pro-Oxidant Agent	*Activation/#Pro-Oxidant Product	Injury Mechanism	Inhibitor	References
NADPH oxidases.	*#ROS	NMDAR activation, upregulates MMPs, BBB damage.	Apocynin, diphenyleneiodonium (DPI), dimethylsulfoxide (DMSO), VAS2870, gp91ds-tat, GKT136901, Vitamin C, M13, ML090, ML171, sulforaphane.	[154, 155]
Xanthine oxidase.	*#ROS, #uric acid	Oxygenation of low-density lipoproteins (LDL), oxidative stress, antioxidants depletion.	Allopurinol, TEI-6720, febuxostat, Y-700, BOF-4272, caffeic acid phenethylester (CAPE).	[154]
Lipid peroxidation	*ROS, #lipid peroxyl radical, malondialdehyde (MDA), 4-hydroxy 2-nonenal (4-HNE).	• MDA-acetaldehyde, advanced lipid peroxidation end-products, DNA cross-linking, DNA damage/mutation, DNA endonucleases activation (e.g., caspase-activated deoxynuclease, apoptosis-inducing factor, and endonuclease G). • 4-HNE regulates transcription factors (e.g., nuclear factor erythroid 2-related factor 2, activating protein-1, NFκB, PPARs), MAPK, and PI3K/AKT signaling.	Tirilazad mesylate (U-74006F), NXY-059 (Disufenton sodium), *N*-acetylcysteine (NAC), melatonin, citicoline.	[154]
Free radicals	*#ROS (superoxide, hydroxyl, hydroperoxyl radicals), RNS (peroxynitrite, NO).	Lipid, protein, and DNA modifications, break DNA double strands, protein-DNA crosslinks, DNA mutations, 8-hydroxy-2′-deoxyguanosine (8OHdG) expression, oxidative inactivation of glutamine synthetase, inhibits the metabolic conversion of glutamate to glutamine in astrocytes, excitotoxicity, increase the permeability of the mitochondrial membrane, cytochrome c release, caspase activation.	Free radical scavenger: Edaravone, Vitamin E and C, lipoic acid, ursolic acid. Free radical degradation: polyethylene glycol-conjugated SOD (PEG-SOD) and CAT (PEG-CAT).	[154]
Cyclooxygenases (COX-1/2)	*Microglia, NMDARs, neurons #ROS, prostanoids, prostaglandins, thromboxane.	Metabolism of arachidonic acid, excitotoxicity, increase infarct size, MMPs activation, BBB deterioration.	NS-398 (COX-2 inhibitor), memantine, NSAIDs (selective and non-selective COX inhibitors).	[155, 156]

TABLE 7.5 *(Continued)*

Pro-Oxidant Agent	*Activation/#Pro-Oxidant Product	Injury Mechanism	Inhibitor	References
Lipoxygenases (5-LOX, 12/15-LOX).	*Signal transducers and activators of transcription (STAT1/6), #hydroperoxy acids (HPETEs), hydroxy acids (HETEs), leukotrienes (C4, D4, E4), lipoxins (A4, B4).	Metabolize polyunsaturated fatty acids (arachidonic acid or linoleic acid), AIF activation, mitochondrial pores, MMPs activation.	LOXBlock-1 and -3 (12/15-LOX inhibitor), glutathione, α-tocotrienol, baicalein, ebselen (glutathione peroxidase mimic), curcumin, epigallocatechin gallate, AA-861 (5- and 12/15-LOX inhibitor), rosiglitazone (PPAR modulator), MK-886 (FLAP inhibitor), ML351 (12/15-LOX inhibitor), Brozopine.	[157–159]
Cytochrome P450 enzymes	#Metabolism of arachidonic acids, linoleic acids, oleic acids gluconeogenic precursors, estrogenic metabolites, hydroxyeicosatetraenoic acids (20-HETE), epoxyeicosatrienoic acids (EETs).	Generate blood flow regulators, ROS and NO through NADPH/xanthine oxidase and NOS, trigger IL-6, MCP-1, and TNF-α.	Miconazole, 17-octadecynoic acid.	[160, 161]
Mitochondria	*#ROS	ATP depletion, free radical output increase, mitochondrial pore formation, leakage of cytochrome *c*, induce apoptotic cell death.	CoQ10, MitoQ10, methylene blue, neuroglobin.	[155]
Myeloperoxidase	*Neutrophils, ROS	Oxidative stress	Curcumin, dauricine, quercetin, resveratrol, trigonelline.	[161]
Nitric oxide synthase.	*Microglia, #RNS (peroxynitrites, nitronium ions, nitrogen dioxide).	Breakdown of BBB, MMPs activation, mitochondria dysfunction, protein modifications (nitrosylation, nitration, nitrosothiols), tyrosine nitration (3-nitrotyrosine), inhibits cytochrome c oxidase in mitochondria, enhance COX-2 activity, DNA modification *via* guanine nucleotides and sugar-phosphate backbone, activation of poly(ADP-ribose) polymerase (PARP), exhausts NAD^+, cell death.	Lubeluzole, pyrroloquinoline quinone, uric acid.	[162]

severe ischemic episodes. This is called "ischemic tolerance" in neurons [163]. Ischemic tolerance is the endogenous mechanism by which neurons may protect themselves against ischemic insult during ischemia and after ischemia. Pharmacological interventions such as 3-nitropropinonic acid (succinate dehydrogenase and respiratory chain inhibitor) and haloperidol (mitochondrial inhibitor) have been able to trigger ischemia tolerance when given just 30 min before induction of ischemia. Cellular defense and stress responses against ischemia can stimulate post-translational modification of proteins or the expression of new proteins that might help in ischemic tolerance. Hypoxia-inducible factor-1 (HIF-1) is a transcription factor that senses cellular oxygen levels and plays an important role in the development of ischemic tolerance in hypoxic cells. HIF-1 can increase blood flow by augmenting vascular endothelial growth factors triggered by angiogenesis [164]. Prolyl hydroxylase 2 is attributed to the inhibition of HIF-1 action in the presence of oxygen and this enzyme can be targeted by pharmacological pre- and post-conditioning to enhance HIF-1 activity against ischemia-hypoxia conditions. Several ischemic pre-conditioning methods (e.g., hypoxia, NMDAR agonists, glutamate, inhibitor of succinic dehydrogenase, hormones, hypothermia), remote ischemia pre-conditioning, cross-preconditioning (e.g., hypothermia, hyperthermia, pharmacological, electroacupuncture, diet, exercise) are explored that confer higher cellular tolerance against ischemic insult [165]. Most notable, activation of NMDARs by agonists such as glutamate modulate several pathways of ischemia tolerance – (i) it activates Ca^{2+} dependent activation of cAMP (cyclic adenosine monophosphate) responsive element-binding protein (CREB); (ii) activates NOS-Ras-ERK pathway; and (iii) inhibits stress-activated c-Jun *N*-terminal kinase (JNK) and protein kinase B (PKB). Activation of immune mechanisms (e.g., TLRs), inflammatory pathways (e.g., cycloxygenase-2, TNF-α, ILs), thrombin (*via* extracellular signal-regulated kinase pathway), and autophagic mechanisms also enhance ischemia tolerance in cells [166].

7.7 SUMMARY AND FUTURE DIRECTIONS

Cerebral ischemia and stroke-related disorders are on the rise owing to an altered lifestyle, greater intake of processed food, beverages, alcohol, and tobacco abuse, sedentary life, and environmental factors such as pollution and toxins. Some are avoidable and some are non-avoidable. These risk factors amplify the frequency of cerebral ischemic attacks in populations already suffering from cardiovascular (e.g., hypertension, cardiac abnormalities, atherosclerosis) or metabolic (e.g., diabetes, dyslipidemia, obesity) disorders. However, a rationalized therapeutic approach (e.g., statins, antiplatelet drugs, and antithrombotics) along with required lifestyle changes (e.g., mild exercise, diet modifications) can avoid the prevalence of cerebral ischemia in these patients. Modern imaging techniques (e.g., magnetic resonance imaging, computed tomography scan, angiography, single-photon emission computed tomography, Doppler sonography) can predict the possible outcome of treatment options and might be able to assess the CBF *via* collaterals and pathogenic brain vasculature modifications in primary blood vessels. Hence, it can be inferred that under optimum medical supervision, the chances of stroke can be minimized in a patient. Nonetheless, severe stroke is preceded by mild transient ischemia that can be used as an indicator and suitable treatment or preventive strategies can prevent a full-blown ischemic attack in the majority of patients. At present, the treatment options for stroke, as well as its preventive strategies, are limited which poses a huge challenge in clinical settings. However, a deep insight into the pathogenesis of cerebral ischemia can reveal multiple targets that might be used in future anti-ischemia research. In addition to a few therapeutic approaches that might be used against IS or transient cerebral ischemia, new strategies are under investigation. Sofadil is an NMDAR antagonist having a longer neuroprotective time window relative to existing similar category drugs that have shown much promise in stroke patients. A few pharmacological interventions (e.g., AngongNiuhuang pill, BNG-1, Citicoline, Tirofiban, Tongxinluo capsule, neuroactive steroids, Dimethyl fumarate, Qizhitongluo Capsule, Ginkgolides Meglumine, and Vitamin D), biologicals (e.g., umbilical cord-derived mesenchymal stem cells, JTR-161, mutant pro-urokinase), electroacupuncture technique, mechanical thrombectomy, and remote ischemic conditioning are in various phases of clinical trials. Novel approaches such as endovascular autologous mitochondrial transplantation, normobaric oxygen, neutrophil extracellular traps (NET), biologicals (SANGUINATE™), and drugs (Divaza, Erigeron, Vorapaxar, cilostazol, clopidogrel, MCI-9042) are also being tested against cerebral ischemia. However, despite vigorous research and advances in the understanding of pathogenesis, the therapeutic outcomes are minimal for ischemic disorders. The majority of stroke patients suffer from one or more permanent disabilities which makes their life miserable. Hence, the prevention of ischemic injury must be the prime objective that should drive anti-ischemic research. Multitarget drugs can provide substantial relief against ischemic injury in comparison to drugs acting against only one pathogenic mechanism, as the pathogenesis of cerebral ischemia involves implications of several diverse pathways of cell injury.

KEYWORDS

- **calcium**
- **cerebral ischemia**
- **excitotoxicity**
- **inflammation**
- **ischemic stroke**
- **oxidative stress**
- **vascular damage**

REFERENCES

1. Ahad, M. A., Kumaran, K. R., Ning, T., Mansor, N. I., Effendy, M. A., Damodaran, T., & Hassan, Z., (2020). Insights into the neuropathology of cerebral ischemia and its mechanisms. *Rev. Neurosci., 31*, 521–538.

2. Lee, R. H. C., Lee, M. H. H., Wu, C. Y. C., Couto, E. S. A., Possoit, H. E., Hsieh, T. H., Minagar, A., & Lin, H. W., (2018). Cerebral ischemia and neuroregeneration. *Neural Regen. Res., 13*(3), 373–385.
3. Woodruff, T. M., Thundyil, J., Tang, S. C., Sobey, C. G., Taylor, S. M., & Arumugam, T. V., (2011). Pathophysiology, treatment, and animal and cellular models of human ischemic stroke. *Mol. Neurodegener., 6*(1), 1–19.
4. Heiss, W. D., (2016). The pathophysiology of ischemic stroke studied by radionuclide imaging. *J. Neurol. Neuromedicine, 1*(8), 22–28.
5. Girnar, G. A., & Mahajan, H. S., (2021). Cerebral ischemic stroke and different approaches for treatment of stroke. *Future J. Pharm. Sci., 7*(1), 1–10.
6. Rihal, V., Khan, H., Kaur, A., & Singh, T. G., (2022). Vitamin D as therapeutic modulator in cerebrovascular diseases: A mechanistic perspective. *Crit. Rev. Food Sci. Nutr,* 1–23.
7. Palencia, G., Núñez-Medrano, J. Á., Ortiz-Plata, A., Farfán, D. J., Sotelo, J., Sánchez, A., & Trejo-Solís, C., (2015). Anti-apoptotic, anti-oxidant, and anti-inflammatory effects of thalidomide on cerebral ischemia/reperfusion injury in rats. *J. Neurol. Sci., 351*(1, 2), 78–87.
8. Janardhan, V., & Qureshi, A. I., (2004). Mechanisms of ischemic brain injury. *Curr. Cardiol. Rep., 6*(2), 117–123.
9. Mukundan, G., & Seidenwurm, D. J., (2018). Economic and societal aspects of stroke management. *Neuroimaging Clin. N. Am., 28*(4), 683–689.
10. Kuriakose, D., & Xiao, Z., (2020). Pathophysiology and treatment of stroke: Present status and future perspectives. *Int. J. Mol. Sci., 21*(20), 7609.
11. GBD 2016 Neurology Collaborators, (2019). Global, regional, and national burden of neurological disorders, 1990–2016: A systematic analysis for the Global Burden of Disease Study 2016. *Lancet Neurol., 18*(5), 459–480.
12. Boehme, A. K., Esenwa, C., & Elkind, M. S., (2017). Stroke risk factors, genetics, and prevention. *Circ. Res., 120*(3), 472–495.
13. Girijala, R. L., Sohrabji, F., & Bush, R. L., (2017). Sex differences in stroke: Review of current knowledge and evidence. *Vasc. Med., 22*(2), 135–145.
14. Chen, J. C., (2010). Geographic determinants of stroke mortality: Role of ambient air pollution. *Stroke, 41*(5), 839–841.
15. Karoly, P., & Ruehlman, L. S., (2006). Psychological "resilience" and its correlates in chronic pain: Findings from a national community sample. *Pain, 123*(1, 2), 90–97.
16. Wang, Y. L., Wu, D., Nguyen-Huynh, M. N., Zhou, Y., Wang, C. X., & Zhao, X. Q., (2010). Prevention of recurrences of stroke study in China (PRESS-China) investigators. Antithrombotic management of ischemic stroke and transient ischemic attack in China: A consecutive cross-sectional survey. *Clin. Exp. Pharmacol. Physiol., 37*(8), 775–781.
17. Arrich, J., Müllner, M., Lalouschek, W., Greisenegger, S., Crevenna, R., & Herkner, H., (2008). Influence of socioeconomic status and gender on stroke treatment and diagnostics. *Stroke, 39*(7), 2066–2072.
18. Shuaib, A., Butcher, K., Mohammad, A. A., Saqqur, M., & Liebeskind, D. S., (2011). Collateral blood vessels in acute ischemic stroke: A potential therapeutic target. *Lancet Neurol., 10*(10), 909–921.
19. Rea, P., (2015). *Essential Clinical Anatomy of the Nervous System*. Academic Press.
20. Sakoh, M., Østergaard, L., Gjedde, A., Røhl, L., Vestergaard-Poulsen, P., Smith, D. F., & Gyldensted, C., (2001). Prediction of tissue survival after middle cerebral artery occlusion based on changes in the apparent diffusion of water. *J. Neurosurg., 95*(3), 450–458.
21. Sobesky, J., Weber, O. Z., Lehnhardt, F. G., Hesselmann, V., Thiel, A., Dohmen, C., & Heiss, W. D., (2004). Which time-to-peak threshold best identifies penumbral flow? A comparison of perfusion-weighted magnetic resonance imaging and positron emission tomography in acute ischemic stroke. *Stroke, 35*(12), 2843–2847.
22. Saklani, P., Khan, H., Gupta, S., Kaur, A., & Singh, T. G., (2022). Neuropeptides: Potential neuroprotective agents in ischemic injury. *Life Sci., 288*, 120186.
23. Adams, Jr. H. P., Bendixen, B. H., Kappelle, L. J., Biller, J., Love, B. B., Gordon, D. L., & Marsh, 3.rd E. E., (1993). Classification of subtype of acute ischemic stroke. Definitions for use in a multicenter clinical trial. TOAST. Trial of org 10172 in acute stroke treatment. *Stroke, 24*(1), 35–41.

24. Neumann, J. T., Cohan, C. H., Dave, K. R., Wright, C. B., & Perez-Pinzon, M. A., (2013). Global cerebral ischemia: Synaptic and cognitive dysfunction. *Curr. Drug Targets, 14*(1), 20–35.
25. Easton, J. D., Saver, J. L., Albers, G. W., Alberts, M. J., Chaturvedi, S., Feldmann, E., & Sacco, R. L., (2009). Definition and evaluation of transient ischemic attack: A scientific statement for healthcare professionals from the American Heart Association/American Stroke Association Stroke Council; Council on Cardiovascular Surgery and Anesthesia; Council on Cardiovascular Radiology and Intervention; Council on Cardiovascular Nursing; and the Interdisciplinary Council on Peripheral Vascular Disease: The American Academy of Neurology affirms the value of this statement as an educational tool for neurologists. *Stroke, 40*(6), 2276–2293.
26. George, M. G., Tong, X., Kuklina, E. V., & Labarthe, D. R., (2011). Trends in stroke hospitalizations and associated risk factors among children and young adults, 1995–2008. *Ann. Neurol., 70*, 713–721.
27. Cruz-Flores, S., Rabinstein, A., Biller, J., Elkind, M. S., Griffith, P., Gorelick, P. B., & Valderrama, A. L., (2011). Racial-ethnic disparities in stroke care: The American experience: A statement for healthcare professionals from the American Heart Association/American Stroke Association. *Stroke, 42*(7), 2091–2116.
28. Seshadri, S., Beiser, A., Pikula, A., Himali, J. J., Kelly-Hayes, M., Debette, S., & Wolf, P. A., (2010). Parental occurrence of stroke and risk of stroke in their children: The Framingham study. *Circulation, 121*(11), 1304–1312.
29. Touzé, E., & Rothwell, P. M., (2008). Sex differences in heritability of ischemic stroke: A systematic review and meta-analysis. *Stroke, 39*(1), 16–23.
30. Matarin, M., Brown, W. M., Singleton, A., Hardy, J. A., & Meschia, J. F., (2008). Whole genome analyses suggest ischemic stroke and heart disease share an association with polymorphisms on chromosome 9p21. *Stroke, 39*(5), 1586–1589.
31. Bösel, J., Schiller, P., Hook, Y., Andes, M., Neumann, J. O., Poli, S., & Steiner, T., (2013). Stroke-related early tracheostomy versus prolonged orotracheal intubation in neurocritical care trial (SETPOINT) a randomized pilot trial. *Stroke, 44*(1), 21–28.
32. Song, Y. M., & Cho, H. J., (2008). Risk of stroke and myocardial infarction after reduction or cessation of cigarette smoking: A cohort study in Korean men. *Stroke, 39*(9), 2432–2438.
33. O'donnell, M. J., Xavier, D., Liu, L., Zhang, H., Chin, S. L., Rao-Melacini, P., & Yusuf, S., (2010). Risk factors for ischemic and intracerebral haemorrhagic stroke in 22 countries (the INTERSTROKE study): A case-control study. *The Lancet, 376*(9735), 112–123.
34. Denti, L., Cecchetti, A., Annoni, V., Merli, M. F., Ablondi, F., & Valenti, G., (2003). The role of lipid profile in determining the risk of ischemic stroke in the elderly: A case–control study. *Arch. Gerontol. Geriatr., 37*(1), 51–62.
35. Garg, C., Kaur, A., Singh, T. G., Sharma, V. K., & Singh, S. K., (2022). Therapeutic implications of sonic hedgehog pathway in metabolic disorders: Novel target for effective treatment. *Pharmacol. Res.*, 106194.
36. Klatsky, A. L., Armstrong, M. A., Friedman, G. D., & Sidney, S., (2001). Alcohol drinking and risk of hospitalization for ischemic stroke. *Am. J. Cardiol., 88*(6), 703–706.
37. Gupta, A., Khan, H., Kaur, A., & Singh, T. G., (2021). Novel targets explored in the treatment of alcohol withdrawal syndrome. *CNS Neurol. Disord. – Drug Targets, 20*(2), 158–173.
38. Romero, J. R., Morris, J., & Pikula, A., (2008). Stroke prevention: Modifying risk factors. *Ther. Adv. Cardiovasc. Dis., 2*(4), 287–303.
39. Brambatti, M., Connolly, S. J., Gold, M. R., Morillo, C. A., Capucci, A., Muto, C., & Healey, J. S., (2014). Temporal relationship between subclinical atrial fibrillation and embolic events. *Circulation, 129*(21), 2094–2099.
40. Disertori, M., Quintarelli, S., Grasso, M., Pilotto, A., Narula, N., Favalli, V., Canclini, C., et al., (2013). Autosomal recessive atrial dilated cardiomyopathy with standstill evolution associated with mutation of natriuretic peptide precursor A. *Circ. Cardiovasc. Genet., 6*(1), 27–36.
41. Li, X. Y., Cai, X. L., Bian, P. D., & Hu, L. R., (2012). High salt intake and stroke: Meta-analysis of the epidemiologic evidence. *CNS Neurosci. Ther., 18*(8), 691–701.
42. Jurcau, A., & Simion, A., (2021). Neuroinflammation in cerebral ischemia and ischemia/reperfusion injuries: From pathophysiology to therapeutic strategies. *Int. J. Mol. Sci., 23*(1), 14.

43. Khan, H., Singh, A., Thapa, K., Garg, N., Grewal, A. K., & Singh, T. G., (2021). Therapeutic modulation of the phosphatidylinositol 3-kinases (PI3K) pathway in cerebral ischemic injury. *Brain Res., 1761*, 147399.
44. Kumar, V. S., Gopalakrishnan, A., Naziroglu, M., & Rajanikant, G. K., (2014). Calcium ion–the key player in cerebral ischemia. *Curr. Med. Chem., 21*(18), 2065–2075.
45. Singh, V., Mishra, V. N., Chaurasia, R. N., Joshi, D., Pandey, V. (2019). Modes of calcium regulation in ischemic neuron. *Indian J. Clin. Biochem., 34*(3), 246–253.
46. Ludhiadch, A., Sharma, R., Muriki, A., & Munshi, A., (2022). Role of calcium homeostasis in ischemic stroke: A review. *CNS Neurol. Disord. Drug Targets, 21*(1), 52–61.
47. Ouh, I. O., Kim, Y. M., Gim, S. A., & Koh, P. O., (2013). Focal cerebral ischemic injury decreases calbindin expression in brain tissue and HT22 cells. *Lab. Anim. Res., 29*(3), 156–161.
48. Kawamoto, E. M., Vivar, C., & Camandola, S., (2012). Physiology and pathology of calcium signaling in the brain. *Front. Pharmacol., 3*, 61.
49. Kalra, P., Khan, H., Kaur, A., & Singh, T. G., (2022). Mechanistic insight on autophagy modulated molecular pathways in cerebral ischemic injury: From preclinical to clinical perspective. *Neurochem. Res.*, 1–19.
50. Domercq, M., & Matute, C., (2019). Excitotoxicity therapy for stroke patients still alive. *Ebio Medicine., 39*, 3–4.
51. Sharma, A., Khan, H., Singh, T. G., Grewal, A. K., Najda, A., Kawecka-Radomska, M., Kamel, M., et al., (2021). Pharmacological modulation of ubiquitin-proteasome pathways in oncogenic signaling. *Int. J. Mol. Sci., 22*(21), 11971.
52. Ng, F. C., Churilov, L., Yassi, N., Kleinig, T. J., Thijs, V., & Wu, T. Y., (2021). EXTEND-IA TNK Part 1 and 2 Investigators. Microvascular dysfunction in blood-brain barrier disruption and hypoperfusion within the infarct posttreatment are associated with cerebral edema. *Stroke*, STROKEAHA-121036104.
53. Khan, H., Bangar, A., Grewal, A. K., Bansal, P., & Singh, T. G., (2022). Caspase-mediated regulation of the distinct signaling pathways and mechanisms in neuronal survival. *Int. Immunopharmacol., 110*, 108951.
54. Annunziato, L., Cataldi, M., Pignataro, G., Secondo, A., & Molinaro, P., (2007). Glutamate-independent calcium toxicity: Introduction. *Stroke, 38*(2), 661–664.
55. Jeong, J. H., Lee, S. H., Kho, A. R., Hong, D. K., Kang, D. H., Kang, B. S., & Suh, S. W., (2020). The transient receptor potential melastatin 7 (TRPM7) inhibitors suppress seizure-induced neuron death by inhibiting zinc neurotoxicity. *Int. J. Mol. Sci., 21*(21), 7897.
56. Lin, J., & Xiong, Z. G., (2017). TRPM7 is a unique target for therapeutic intervention of stroke. *Int. J. Physiol. Pathophysiol. Pharmacol., 9*(6), 211–216.
57. Huang, Q., Wang, X., Lin, X., Zhang, J., You, X., & Shao, A., (2020). The role of transient receptor potential channels in blood-brain barrier dysfunction after ischemic stroke. *Biomed. Pharmacother., 131*, 110647.
58. Xiong, Z. G., & Xu, T. L., (2012). The role of ASICS in cerebral ischemia. *Wiley Interdiscip. Rev Membr. Transp. Signal., 1*(5), 655–662.
59. Li, M. H., Leng, T. D., Feng, X. C., Yang, T., Simon, R. P., & Xiong, Z. G., (2016). Modulation of acid-sensing ion channel 1a by intracellular pH and its role in ischemic stroke. *J. Biol. Chem., 291*(35), 18370–18383.
60. Chassagnon, I. R., McCarthy, C. A., Chin, Y. K. Y., Pineda, S. S., Keramidas, A., Mobli, M., & King, G. F., (2017). Potent neuroprotection after stroke afforded by a double-knot spider-venom peptide that inhibits acid-sensing ion channel 1a. *Proc. Natl. Acad. Sci., 114*(14), 3750–3755.
61. Shenoda, B., (2015). The role of $Na+/Ca^{2+}$ exchanger subtypes in neuronal ischemic injury. *Transl. Stroke Res., 6*(3), 181–190.
62. Cerullo, P., Brancaccio, P., Anzilotti, S., Vinciguerra, A., Cuomo, O., Fiorino, F., & Pignataro, G., (2018). Acute and long-term NCX activation reduces brain injury and restores behavioral functions in mice subjected to neonatal brain ischemia. *Neuropharmacology, 135*, 180–191.
63. Rehni, A. K., & Singh, T. G., (2012). Involvement of CCR-2 chemokine receptor activation in ischemic preconditioning and postconditioning of brain in mice. *Cytokine, 60*(1), 83–89.
64. Pignataro, G., Brancaccio, P., Laudati, G., Valsecchi, V., Anzilotti, S., Casamassa, A., & Vinciguerra, A., (2020). Sodium/calcium exchanger as main effector of endogenous neuroprotection elicited by ischemic tolerance. *Cell Calcium, 87*, 102183.

65. Sharma, V. K., Singh, T. G., & Mehta, V., (2021). Stressed mitochondria: A target to intrude Alzheimer's disease. *Mitochondrion, 59*, 48–57.
66. Busija, D. W., Rutkai, I., Dutta, S., & Katakam, P. V., (2016). Role of mitochondria in cerebral vascular function: Energy production, cellular protection, and regulation of vascular tone. *Compr. Physiol., 6*(3), 1529–1548.
67. He, J., Gao, Y., Wu, G., Lei, X., Zhang, Y., Pan, W., & Yu, H., (2018). Bioinformatics analysis of microarray data to reveal the pathogenesis of brain ischemia. *Mol. Med. Rep., 18*(1), 333–341.
68. Du, F., Zhu, X. H., Zhang, Y., Friedman, M., Zhang, N., Ugurbil, K., & Chen, W., (2008). Tightly coupled brain activity and cerebral ATP metabolic rate. *Proc. Natl. Acad. Sci. USA, 105*(17), 6409–6414.
69. Luo, C., Bian, X., Zhang, Q., Xia, Z., Liu, B., Chen, Q., & Zhao, Y., (2019). Shengui Sansheng San ameliorates cerebral energy deficiency via citrate cycle after ischemic stroke. *Front. Pharmacol., 10*, 386.
70. Liang, J., Han, R., & Zhou, B., (2021). Metabolic reprogramming: Strategy for ischemic stroke treatment by ischemic preconditioning. *Biology (Basel), 10*(5), 424.
71. Tiwari, P., Khan, H., Singh, T. G., & Grewal, A. K., (2022). Poly (ADP-ribose) polymerase: An overview of mechanistic approaches and therapeutic opportunities in the management of stroke. *Neurochem. Res.*, 1–23.
72. Qin, C., Zhou, L. Q., Ma, X. T., Hu, Z. W., Yang, S., Chen, M., & Tian, D. S., (2019). Dual functions of microglia in ischemic stroke. *Neurosci. Bull., 35*(5), 921–933.
73. Yu, L., Su, X., Li, S., Zhao, F., Mu, D., & Qu, Y., (2020). Microglia and their promising role in ischemic brain injuries: An update. *Front. Cell. Neurosci., 14*, 211.
74. Jiang, C. T., Wu, W. F., Deng, Y. H., & Ge, J. W., (2020). Modulators of microglia activation and polarization in ischemic stroke. *Mol. Med. Rep., 21*(5), 2006–2018.
75. Lalancette-Hébert, M., Swarup, V., Beaulieu, J. M., Bohacek, I., Abdelhamid, E., Weng, Y. C., & Kriz, J., (2012). Galectin-3 is required for resident microglia activation and proliferation in response to ischemic injury. *J. Neurosci., 32*(30), 10383–10395.
76. Jin, W. N., Shi, S. X. Y., Li, Z., Li, M., Wood, K., Gonzales, R. J., & Liu, Q., (2017). Depletion of microglia exacerbates postischemic inflammation and brain injury. *J. Cereb. Blood Flow Metab., 37*(6), 2224–2236.
77. McDonough, A., Lee, R. V., Noor, S., Lee, C., Le, T., Iorga, M., & Weinstein, J. R., (2017). Ischemia/reperfusion induces interferon-stimulated gene expression in microglia. *J. Neurosci., 37*(34), 8292–8308.
78. Denes, A., Vidyasagar, R., Feng, J., Narvainen, J., McColl, B. W., Kauppinen, R. A., & Allan, S. M., (2007). Proliferating resident microglia after focal cerebral ischemia in mice. *J. Cereb. Blood Flow Metab. 27*(12), 1941–1953.
79. Bylicky, M. A., Mueller, G. P., & Day, R. M., (2018). Mechanisms of endogenous neuroprotective effects of astrocytes in brain injury. *Oxid. Med. Cell. Longev.*, 6501031.
80. Wang, H., Song, G., Chuang, H., Chiu, C., Abdelmaksoud, A., Ye, Y., & Zhao, L., (2018). Portrait of glial scar in neurological diseases. *Int. J. Immunopathol. Pharmacol., 31*, 2058738418801406.
81. Shi, Y., Zhang, L., Pu, H., Mao, L., Hu, X., Jiang, X., & Chen, J., (2016). Rapid endothelial cytoskeletal reorganization enables early blood–brain barrier disruption and long-term ischemic reperfusion brain injury. *Nat. Commun., 7*(1), 1–18.
82. Yang, Y., & Rosenberg, G. A., (2015). Matrix metalloproteinases as therapeutic targets for stroke. *Brain Res., 1623*, 30–38.
83. Martynov, M. Y., & Gusev, E. I., (2015). Current knowledge on the neuroprotective and neuroregenerative properties of citicoline in acute ischemic stroke. *J. Exp. Pharmacol., 7*, 17–28.
84. El-Benna, J., Hurtado-Nedelec, M., Marzaioli, V., Marie, J. C., Gougerot-Pocidalo, M. A., & Dang, P. M. C., (2016). Priming of the neutrophil respiratory burst: Role in host defense and inflammation. *Immunol. Rev., 273*(1), 180–193.
85. Chen, R., Zhang, X., Gu, L., Zhu, H., Zhong, Y., Ye, Y., & Jian, Z., (2021). New insight into neutrophils: A potential therapeutic target for cerebral ischemia. *Front. Immunol., 12*, 692061.
86. Jickling, G. C., Liu, D., Ander, B. P., Stamova, B., Zhan, X., & Sharp, F. R., (2015). Targeting neutrophils in ischemic stroke: Translational insights from experimental studies. *J. Cereb. Blood Flow Metab., 35*(6), 888–901.

87. Tu, X. K., Yang, W. Z., Shi, S. S., Wang, C. H., Zhang, G. L., Ni, T. R., & Song, Q. M., (2010). Spatio-temporal distribution of inflammatory reaction and expression of TLR2/4 signaling pathway in rat brain following permanent focal cerebral ischemia. *Neurochem. Res., 35*(8), 1147–1155.
88. Mannan, A., Garg, N., Singh, T. G., & Kang, H. K., (2021). Peroxisome proliferator-activated receptor-gamma (PPAR-ɣ): Molecular effects and its importance as a novel therapeutic target for cerebral ischemic injury. *Neurochem. Res., 46*(11), 2800–2831.
89. Zinnhardt, B., Viel, T., Wachsmuth, L., Vrachimis, A., Wagner, S., Breyholz, H. J., & Jacobs, A. H., (2015). Multimodal imaging reveals temporal and spatial microglia and matrix metalloproteinase activity after experimental stroke. *J. Cereb. Blood Flow Metab., 35*(11), 1711–1721.
90. Perez-de-Puig, I., Miró-Mur, F., Ferrer-Ferrer, M., Gelpi, E., Pedragosa, J., Justicia, C., & Planas, A. M., (2015). Neutrophil recruitment to the brain in mouse and human ischemic stroke. *Acta Neuropathol., 129*(2), 239–257.
91. Xie, L., & Yang, S. H., (2015). Interaction of astrocytes and T cells in physiological and pathological conditions. *Brain Res., 1623*, 63–73.
92. Reyes, R., Cardeñes, B., Machado-Pineda, Y., & Cabañas, C., (2018). Tetraspanin CD9: A key regulator of cell adhesion in the immune system. *Front. Immunol., 9*, 863.
93. Yang, C., Hawkins, K. E., Doré, S., & Candelario-Jalil, E., (2019). Neuroinflammatory mechanisms of blood-brain barrier damage in ischemic stroke. *Am. J. Physiol. Cell Physiol., 316*(2), C135–C153.
94. Ramiro, L., Simats, A., García-Berrocoso, T., & Montaner, J., (2018). Inflammatory molecules might become both biomarkers and therapeutic targets for stroke management. *Ther. Adv. Neurol. Disord., 11*, 1756286418789340.
95. Sobowale, O. A., Parry-Jones, A. R., Smith, C. J., Tyrrell, P. J., Rothwell, N. J., & Allan, S. M., (2016). Interleukin-1 in stroke: From bench to bedside. *Stroke, 47*(8), 2160–2167.
96. Gredal, H., Thomsen, B. B., Boza-Serrano, A., Garosi, L., Rusbridge, C., Anthony, D., & Berendt, M., (2017). Interleukin-6 is increased in plasma and cerebrospinal fluid of community-dwelling domestic dogs with acute ischemic stroke. *Neuroreport., 28*(3), 134.
97. Singh, N., Hopkins, S. J., Hulme, S., Galea, J. P., Hoadley, M., Vail, A., & Tyrrell, P. J., (2014). The effect of intravenous interleukin-1 receptor antagonist on inflammatory mediators in cerebrospinal fluid after subarachnoid haemorrhage: A phase II randomized controlled trial. *J. Neuroinflammation, 11*(1), 1–8.
98. Brait, V. H., Rivera, J., Broughton, B. R., Lee, S., Drummond, G. R., & Sobey, C. G., (2011). Chemokine-related gene expression in the brain following ischemic stroke: No role for CXCR2 in outcome. *Brain Res., 1372*, 169–179.
99. Fang, W., Zhai, X., Han, D., Xiong, X., Wang, T., Zeng, X., & Zhao, H., (2018). CCR2-dependent monocytes/macrophages exacerbate acute brain injury but promote functional recovery after ischemic stroke in mice. *Theranostics, 8*(13), 3530–3543.
100. García-Berrocoso, T., Giralt, D., Llombart, V., Bustamante, A., Penalba, A., Flores, A., & Montaner, J., (2014). Chemokines after human ischemic stroke: From neurovascular unit to blood using protein arrays. *Transl. Proteomics., 3*, 1–9.
101. Hammond, M. D., Taylor, R. A., Mullen, M. T., Ai, Y., Aguila, H. L., Mack, M., & Sansing, L. H., (2014). CCR2+ Ly6Chi inflammatory monocyte recruitment exacerbates acute disability following intracerebral hemorrhage. *J. Neurosci., 34*(11), 3901–3909.
102. Guo, Y. Q., Zheng, L. N., Wei, J. F., Hou, X. L., Yu, S. Z., Zhang, W. W., & Jing, J. M., (2014). Expression of CCL2 and CCR2 in the hippocampus and the interventional roles of propofol in rat cerebral ischemia/reperfusion. *Exp. Ther. Med., 8*(2), 657–661.
103. Golab, P., Kielbus, M., Bielewicz, J., & Kurzepa, J., (2015). The effect of recombinant tissue plasminogen activator on MMP-2 and MMP-9 activities *in vitro*. *Neurol. Res., 37*(1), 9–13.
104. Patel, S., (2018). Danger-associated molecular patterns (DAMPs): The derivatives and triggers of inflammation. *Curr. Allergy Asthma Rep., 18*(11), 63.
105. Jover-Mengual, T., Hwang, J. Y., Byun, H. R., Centeno, J. M., Burguete, M. C., & Zukin, R. S., (2021). The role of NF-κB triggered inflammation in cerebral ischemia. *Front. Cell. Neurosci., 15*, 633610.
106. Kumar, V., (2019). Toll-like receptors in the pathogenesis of neuroinflammation. *J. Neuroimmunol., 332*, 16–30.

107. Gao, L., Dong, Q., Song, Z., Shen, F., Shi, J., & Li, Y., (2017). NLRP3 inflammasome: A promising target in ischemic stroke. *Inflamm. Res., 66*(1), 17–24.
108. Voet, S., Srinivasan, S., Lamkanfi, M., & Van, L. G., (2019). Inflammasomes in neuroinflammatory and neurodegenerative diseases. *EMBO Mol. Med., 11*(6), e10248.
109. Feng, Y. S., Tan, Z. X., Wang, M. M., Xing, Y., Dong, F., & Zhang, F., (2020). Inhibition of NLRP3 inflammasome: A prospective target for the treatment of ischemic stroke. *Front. Cell. Neurosci., 14*, 155.
110. Diaz-Cañestro, C., Reiner, M. F., Bonetti, N. R., Liberale, L., Merlini, M., Wüst, P., & Camici, G. G., (2019). AP-1 (activated protein-1) transcription factor JunD regulates ischemia/reperfusion brain damage via IL-1β (interleukin-1β). *Stroke, 50*(2), 469–477.
111. Lakhan, S. E., Kirchgessner, A., & Hofer, M., (2009). Inflammatory mechanisms in ischemic stroke: Therapeutic approaches. *J. Transl. Med., 7*, 97.
112. Li, Y., & Liu, S., (2017). The effect of dexmedetomidine on oxidative stress response following cerebral ischemia-reperfusion in rats and the expression of intracellular adhesion molecule-1 (ICAM-1) and S100B. *Med. Sci. Monit., 23*, 867–873.
113. Liu, R., Pan, M. X., Tang, J. C., Zhang, Y., Liao, H. B., Zhuang, Y., & Wan, Q., (2017). Role of neuroinflammation in ischemic stroke. *Neuroimmunol. Neuroinflammation, 4*, 158–166.
114. Gage, Z. O., Vasou, A., Gray, D. W., Randall, R. E., & Adamson, C. S., (2016). Identification of novel inhibitors of the type I interferon induction pathway using cell-based high-throughput screening. *J. Biomol. Screen., 21*(9), 978–988.
115. Skurkovich, B., & Skurkovich, S., (2006). Inhibition of IFN-γ as a method of treatment of various autoimmune diseases, including skin diseases. *Ernst Schering Res. Found Workshop, 56*, 1–27.
116. Wanve, M., Kaur, H., Sarmah, D., Saraf, J., Pravalika, K., Vats, K., & Bhattacharya, P., (2019). Therapeutic spectrum of interferon-β in ischemic stroke. *J. Neurosci. Res., 97*(2), 116–127.
117. Connell, B. J., Gordon, J. R., & Saleh, T. M., (2015). ELR-CXC chemokine antagonism is neuroprotective in a rat model of ischemic stroke. *Neurosci. Lett., 606*, 117–122.
118. Chen, A., Oakley, A. E., Monteiro, M., Tuomela, K., Allan, L. M., Mukaetova-Ladinska, E. B., & Kalaria, R. N., (2016). Multiplex analyte assays to characterize different dementias: Brain inflammatory cytokines in poststroke and other dementias. *Neurobiol. Aging, 38*, 56–67.
119. Wang, H., Zhong, D., Chen, H., Jin, J., Liu, Q., & Li, G., (2019). NLRP3 inflammasome activates interleukin-23/interleukin-17 axis during ischemia-reperfusion injury in cerebral ischemia in mice. *Life Sci., 227*, 101–113.
120. Wasilewska, A., Winiarska, M., Olszewska, M., & Rudnicka, L., (2016). Interleukin-17 inhibitors. A new era in treatment of psoriasis and other skin diseases. *Postepy. Dermatol. Alergol., 33*(4), 247–252.
121. Zhang, N., Yu, J. T., Yu, N. N., Lu, R. C., Ma, T., Wang, N. D., & Tan, L., (2010). Interleukin-18 promoter polymorphisms and risk of ischemic stroke. *Brain Res. Bull., 81*(6), 590–594.
122. Garcia, J. M., Stillings, S. A., Leclerc, J. L., Phillips, H., Edwards, N. J., Robicsek, S. A., & Doré, S., (2017). Role of interleukin-10 in acute brain injuries. *Front. Neurol., 8*, 244.
123. Yuan, B., Shi, H., Zheng, K., Su, Z., Su, H., Zhong, M., & Yang, Z., (2017). MCP-1-mediated activation of microglia promotes white matter lesions and cognitive deficits by chronic cerebral hypoperfusion in mice. *Mol. Cell. Neurosci., 78*, 52–58.
124. Yokochi, S., Hashimoto, H., Ishiwata, Y., Shimokawa, H., Haino, M., Terashima, Y., & Matsushima, K., (2001). An anti-inflammatory drug, propagermanium, may target GPI-anchored proteins associated with an MCP-1 receptor, CCR2. *J. Interferon Cytokine Res., 21*(6), 389–398.
125. Deshmane, S. L., Kremlev, S., Amini, S., & Sawaya, B. E., (2009). Monocyte chemoattractant protein-1 (MCP-1): An overview. *J. Interferon Cytokine Res., 29*(6), 313–326.
126. Ramadass, V., Vaiyapuri, T., & Tergaonkar, V., (2020). Small molecule NF-κB pathway inhibitors in clinic. *Int. J. Mol. Sci., 21*(14), 5164.
127. Huang, J., Upadhyay, U. M., & Tamargo, R. J., (2006). Inflammation in stroke and focal cerebral ischemia. *Surg. Neurol., 66*(3), 232–245.
128. Yilmaz, G., & Granger, D. N., (2008). Cell adhesion molecules and ischemic stroke. *Neurol. Res., 30*(8), 783–793.
129. Li, K., Yu, W., Cao, R., Zhu, Z., & Zhao, G., (2017). Microglia-mediated BAFF–BAFFR ligation promotes neuronal survival in brain ischemia injury. *Neuroscience, 363*, 87–96.

130. Ma, Y., Liu, Y., Zhang, Z., & Yang, G. Y., (2019). Significance of complement system in ischemic stroke: A comprehensive review. *Aging Dis., 10*(2), 429–462.
131. Khan, H., Garg, N., Singh, T. G., Kaur, A., & Thapa, K., (2022). Calpain inhibitors as potential therapeutic modulators in neurodegenerative diseases. *Neurochem. Res.*, 1–25.
132. Khan, H., Kashyap, A., Kaur, A., & Singh, T. G., (2020). Pharmacological postconditioning: A molecular aspect in ischemic injury. *J. Pharm. Pharmacol., 72*(11), 1513–1527.
133. Masouris, I., Klein, M., Dyckhoff, S., Angele, B., Pfister, H. W., & Koedel, U., (2017). Inhibition of DAMP signaling as an effective adjunctive treatment strategy in pneumococcal meningitis. *J. Neuroinflammation, 14*(1), 1–9.
134. Land, W. G., (2020). Use of DAMPs and SAMPs as therapeutic targets or therapeutics: A note of caution. *Mol. Diagn. Ther., 24*(3), 251–262.
135. Wytrykowska, A., Prosba-Mackiewicz, M., & Nyka, W. M., (2016). IL-1β, TNF-α, and IL-6 levels in gingival fluid and serum of patients with ischemic stroke. *J. Oral Sci., 58*(4), 509–513.
136. Suzuki, S., Tanaka, K., & Suzuki, N., (2009). Ambivalent aspects of interleukin-6 in cerebral ischemia: Inflammatory versus neurotrophic aspects. *J. Cereb. Blood Flow Metab., 29*(3), 464–479.
137. Feng, Q., Wang, Y. I., & Yang, Y., (2015). Neuroprotective effect of interleukin-6 in a rat model of cerebral ischemia. *Exp. Ther. Med., 9*(5), 1695–1701.
138. Xu, G., Gu, H., Hu, B., Tong, F., Liu, D., Yu, X., Zheng, Y., & Gu, J., (2017). PEG-b-(PELG-g-PLL) nanoparticles as TNF-α nanocarriers: Potential cerebral ischemia/reperfusion injury therapeutic applications. *Int. J. Nanomedicine, 12*, 2243–2254.
139. Warrington, J. P., Drummond, H. A., Granger, J. P., & Ryan, M. J., (2015). Placental ischemia-induced increases in brain water content and cerebrovascular permeability: Role of TNF-α. *Am. J. Physiol. Regul. Integr. Comp. Physiol., 309*(11), R1425–R1431.
140. Bonaventura, A., Liberale, L., Vecchié, A., Casula, M., Carbone, F., Dallegri, F., & Montecucco, F., (2016). Update on inflammatory biomarkers and treatments in ischemic stroke. *Int. J. Mol. Sci., 17*(12), 1967.
141. Liang, Q. J., Jiang, M., Wang, X. H., Le, L. L., Xiang, M., Sun, N., & Chen, S. F., (2015). Pre-existing interleukin 10 in cerebral arteries attenuates subsequent brain injury caused by ischemia/reperfusion. *IUBMB Life, 67*(9), 710–719.
142. Khan, H., Sharma, K., Kumar, A., Kaur, A., & Singh, T. G., (2022). Therapeutic implications of cyclooxygenase (COX) inhibitors in ischemic injury. *Inflamm. Res.*, 1–16.
143. Prabhakar, N. K., Khan, H., Grewal, A. K., & Singh, T. G., (2022). Intervention of neuroinflammation in the traumatic brain injury trajectory: *In vivo* and clinical approaches. *Int. Immunopharmacol., 108*, 108902.
144. Khan, H., Gupta, A., Singh, T. G., & Kaur, A., (2021). Mechanistic insight on the role of leukotriene receptors in ischemic–reperfusion injury. *Pharmacol Rep., 73*(5), 1240–1254.
145. Manzanero, S., Santro, T., & Arumugam, T. V., (2013). Neuronal oxidative stress in acute ischemic stroke: Sources and contribution to cell injury. *Neurochem. Int., 62*(5), 712–718.
146. Şeneş, M., Kazan, N., Coşkun, Ö., Zengi, O., Inan, L., & Yücel, D., (2007). Oxidative and nitrosative stress in acute ischemic stroke. *Ann. Clin. Biochem., 44*(1), 43–47.
147. Sanderson, T. H., Reynolds, C. A., Kumar, R., Przyklenk, K., & Hüttemann, M., (2013). Molecular mechanisms of ischemia–reperfusion injury in brain: Pivotal role of the mitochondrial membrane potential in reactive oxygen species generation. *Mol. Neurobiol., 47*(1), 9–23.
148. Thapa, K., Khan, H., Singh, T. G., & Kaur, A., (2021). Traumatic brain injury: Mechanistic insight on pathophysiology and potential therapeutic targets. *J. Mol. Neurosci., 71*(9), 1725–1742.
149. Vergeade, A., Mulder, P., Vendeville, C., Ventura-Clapier, R., Thuillez, C., & Monteil, C., (2012). Xanthine oxidase contributes to mitochondrial ROS generation in an experimental model of cocaine-induced diastolic dysfunction. *J. Cardiovasc. Pharmacol., 60*(6), 538–543.
150. Picón-Pagès, P., Garcia-Buendia, J., & Muñoz, F. J., (2019). Functions and dysfunctions of nitric oxide in brain. *Biochim. Biophys. Acta (BBA) Mol. Basis Dis., 1865*(8), 1949–1967.
151. Hu, Z., Bian, X., Liu, X., Zhu, Y., Zhang, X., Chen, S., & Wang, Y., (2013). Honokiol protects brain against ischemia–reperfusion injury in rats through disrupting PSD95–nNOS interaction. *Brain Res., 1491*, 204–212.
152. Khan, H., Grewal, A. K., & Singh, T. G., (2022). Pharmacological postconditioning by protocatechuic acid attenuates brain injury in ischemia–reperfusion (I/R) mice model: Implications of nuclear factor erythroid-2-related factor Pathway. *Neuroscience, 491*, 23–31.

153. Jelinek, M., Jurajda, M., & Duris, K., (2021). Oxidative stress in the brain: Basic concepts and treatment strategies in stroke. *Antioxidants (Basel)*, *10*(12), 1886.
154. Shirley, R., Ord, E. N., & Work, L. M., (2014). Oxidative stress and the use of antioxidants in stroke. *Antioxidants (Basel)*, *3*(3), 472–501.
155. Li, W., & Yang, S., (2016). Targeting oxidative stress for the treatment of ischemic stroke: Upstream and downstream therapeutic strategies. *Brain Circ.*, *2*(4), 153–163.
156. Candelario-Jalil, E., & Fiebich, B. L., (2008). Cyclooxygenase inhibition in ischemic brain injury. *Curr. Pharm. Des.*, *14*(14), 1401–1418.
157. Zheng, Y., Liu, Y., Karatas, H., Yigitkanli, K., Holman, T. R., & Van, L. K., (2019). Contributions of 12/15-Lipoxygenase to bleeding in the brain following ischemic stroke. *Adv. Exp. Med. Biol.*, *1161*, 125–131.
158. Van, L. K., (2013). Lipoxygenase: An emerging target for stroke therapy. *CNS Neurol. Disord. Drug Targets.*, *12*(2), 191–199.
159. Gao, Y., Cao, X., Zhang, X., Wang, Y., Huang, H., Meng, Y., & Chang, J., (2020). Brozopine inhibits 15-LOX-2 metabolism pathway after transient focal cerebral ischemia in rats and OGD/R-induced hypoxia injury in PC12 cells. *Front. Pharmacol.*, *11*, 99.
160. Yu, J., Zhu, H., Kindy, M. S., & Taheri, S., (2021). Cytochrome P450 CYP2E1 suppression ameliorates cerebral ischemia reperfusion injury. *Antioxidants (Basel)*, *10*(1), 52.
161. Davis, C. M., Liu, X., & Alkayed, N. J., (2017). Cytochrome P450 eicosanoids in cerebrovascular function and disease. *Pharmacol. Ther.*, *179*, 31–46.
162. Sun, M. S., Jin, H., Sun, X., Huang, S., Zhang, F. L., Guo, Z. N., & Yang, Y., (2018). Free radical damage in ischemia-reperfusion injury: An obstacle in acute ischemic stroke after revascularization therapy. *Oxid. Med. Cell. Longevity,* 3804979.
163. Zemke, D., Smith, J. L., Reeves, M. J., & Majid, A., (2004). Ischemia and ischemic tolerance in the brain: An overview. *Neurotoxicology*, *25*(6), 895–904.
164. Lehotský, J., Burda, J., Danielisova, V., Gottlieb, M., Kaplán, P., & Saniová, B., (2009). Ischemic tolerance: The mechanisms of neuroprotective strategy. *Anat. Rec. (Hoboken)*, *292*(12), 2002–2012.
165. Albert-Weissenberger, C., & Sirén, A. L., (2010). Experimental traumatic brain injury. *Exp. & Trans. Stroke Med.*, *2*, 16.
166. Hao, Y., Xin, M., Feng, L., Wang, X., Wang, X., Ma, D., & Feng, J., (2020). Review cerebral ischemic tolerance and preconditioning: Methods, mechanisms, clinical applications, and challenges. *Front. Neurol.*, *11*, 812.

CHAPTER 8

Expatiating the Pivotal Role of Matrix Metalloproteinases and Blood-Brain Barrier Disruption in Ischemic Stroke: Molecular Insights and Therapeutic Targets

SHIVEENA BHATIA,[1] VARINDER SINGH,[1] MANJINDER SINGH,[1] RAVINDER SINGH,[1] RICHA SHRI,[2] and THAKUR GURJEET SINGH[1]

[1]*Chitkara College of Pharmacy, Chitkara University, Punjab, India*

[2]*Department of Pharmaceutical Sciences and Drug Research, Punjabi University, Patiala, Punjab, India*

ABSTRACT

Globally, ischemic stroke (IS) is among the major causes of permanent disability and represents a major challenge for translational neurology. Despite detailed insights into the pathophysiological events (vasogenic edema, blood-brain barrier (BBB) disruption, and intracranial hemorrhage) being known, to date, the only treatment paradigms available for IS are intravenous tissue plasminogen activator (tPA) and/or mechanical interventions which are aimed to restore cerebral blood flow. However, this restoration further exaggerates the ischemic injury via the plethora of biological events, of which worsening of BBB disruption remains crucial. This deterioration of events is quite intricately modulated by the interplay of various matrix metalloproteinases (MMP). Thus, understanding the mechanisms associated with MMPs mediated BBB damage is crucial for developing neuroprotective drugs. This chapter provides an insight into the role of MMPs in the pathology of IS and highlights the recent findings in this arena paving the way for future drug discovery for IS. The analysis of the literature revealed MMP's dual role in both worsening the condition of the patient as well as later recovery post-ischemic insult. The chapter also discusses combination therapies explored for enhancing the tPA treatment outcome and the potential of MMPs as a wonderful target for upcoming stroke treatments.

Ischemic Injury: A Molecular Insight. Thakur Gurjeet Singh & Heena Khan (Eds.)

8.1 INTRODUCTION

Ischemic stroke (IS) accounts for 71% of the clinical cases worldwide, making it one of the major causes of permanent disability [1]. The occurrence of this challenging and devastating disorder is attributed to a sudden and progressive decrease in the cerebral blood flow in certain regions due to the occlusion by a thrombus or embolism eliciting a sudden loss of oxygen and glucose supply to the brain tissues advancing towards cell death [2]. Despite the interplay of various mechanisms in the progression of a stroke, literature points towards cerebral ischemia being the point of initiation of a cascade of pathogenic events including intracranial hemorrhage, edema, blood-brain-barrier (BBB) deterioration, and glial activation leading to irreversible neuronal injury [3, 4]. BBB being a gateway of the brain governs the passage of compounds in the brain and is disrupted as a result of stroke injury which increased the permeability of the membrane encouraging the neuroinflammatory response along with the development of hemorrhagic edema and mortality. This disruption of the membrane is found to be closely interlinked to the subsequent alterations in the expression of the matrix metalloproteinases (MMPs) which can be correlated to the progression of the disease. MMPs also known as matrixins are endopeptidases that function in the extra-cellular environment and break down both matrix and non-matrix proteins. MMPs, take part in various metabolic processes that have a pivotal involvement in the whole ischemic stroke pathology.

The only FDA-approved treatment paradigm that exists for treating acute ischemic stroke (AIS) cases is tissue plasminogen activator (tPA), but the complete efficacy is restricted due to certain limitations attached to its usage which will also be explored in detail in this chapter later on. Emerging studies also point towards the plausible role of MMPs and tPA in modulating the BBB disruption in the ischemic episode. The MMPs are known to have therapeutic, protective as well as deteriorative roles in this story of BBB being the major event in the occurrence of ischemic stroke. The chapter explores the distinctive part played by various MMPs in the progression of stroke injury and some drugs targeting this pathway to enhance the treatment outcomes of the thrombolytic treatment and in turn improvement in the overall patient condition. The MMPs can serve as a wonderful target site for the development of a concrete supplementary or even complementary treatment regimen. The chapter gives insights into all the progress made till now in this arena.

8.2 BLOOD-BRAIN BARRIER DISRUPTION AND ISCHEMIC STROKE CROSS TALK

BBB is a gateway that connects the CNS and the peripheral circulation and regulates the entry and exit of the compounds imperative for undergoing brain metabolic processes and other associated neurological processes. The functional strength and structural integrity are therefore prerequisites for maintaining brain homeostasis [5, 6]. The disruption of this interface serves as one of the key concerns in the stroke injury cascade as it increases the permeability of peripheral immune cells causing vasogenic and hemorrhagic changes [7–13]. The BBB is made up of tight junctions (TJ) and the adherens junction (AJ), both of which play their part in restricting the endothelium permeability. The TJ is a continuous strand of membrane situated at the apical site, comprising transmembrane proteins like claudins, occluding, and cytoplasmic proteins like cingulin, zona occludens, 7h6 and AF-6 connected to the actin cytoskeleton [12, 14]. These auxiliary proteins function as the recognition sites for the placement of tight junctions and also act as a support system for holding signal transduction proteins. Any disturbance in the BBB TJ can hamper the brain's homeostasis by impairing its permeability and proper functioning. The AJ is made up of vascular endothelial cadherin protein which is structurally linked to TJs and functions in modulating their integrity [11, 12]. Since AJs contribute to regulating the structure of the TJs, their disruption can also lead to BBB permeability. The pathology of stroke follows the development of a biphasic opening. Morphologically, the disruption of BBB means the occurrence of redistribution, rearrangement, and reorganization of the proteins of the TJ, AJ, and actin cytoskeleton. These processes start within 1 hour of stroke insult [15]. The exact mechanism associated with these changes is yet to be elucidated, however, a spike in levels of MMP-9 is reported immediately post-onset of the ischemic injury [11, 16]. Research has widely explored the role of matrix metalloproteins in the BBB disruption since the variations in the permeability of the membrane was in tune with the upregulation of the MMPs expression [17–21]. The following sections comprehensively examine various MMPs' roles in the pathogenesis and management of IS.

8.3 UNFOLDING MATRIX METALLOPROTEINASES IN ISCHEMIC STROKE

Matrix metalloproteinases (calcium-dependent zinc-endopeptidases) belong to the metzincin superfamily which cleaves protein substrate by triggering a site-bound water molecule using zinc ion [22, 23]. These critically modulate the metabolic processes by extensive protein degradation and selective bond hydrolysis and play a pivotal role in tissue morphogenesis, cell migration, angiogenesis, wound healing, bone resorption, inflammation, and cancer [24, 25].

The MMPs comprise of 24 distinct proteases divided into subfamilies which are metallo-elastase (MMP-12), gelatinases (MMPs-2 and -9), endometase (MMP-26), membrane-type metalloproteinases (MMPs-14, -15, -16, -24, and -25), stromelysins (MMPs-3, -10, and -11), epilysin (MMP-28), matrilysins (MMPs-7 and -26), collagenases (MMPs-1,8 and 13) and enamelysin (MMP-20) whose activity is tightly monitored by transcription factors [15, 26, 27].

The expression of MMPs is quite minimal to undetectable levels in adults under normal conditions, but clinical studies have illustrated the upregulated expression of MMPs including MMP-2 and MMP-9 after IS [28–30]. However, MMPs are shown to exhibit both deteriorative and protective roles in patients suffering from ischemic stroke (Figure 8.1). MMPs, interfere with the permeability of BBB by degrading TJ proteins and basal lamina proteins causing the leakage and hampering of BBB, infiltration of leucocytes, brain edema, and hemorrhage [31, 32]. Whereas these also augment the recovery of patients by regulating the neurovascular remodeling [27, 31, 32].

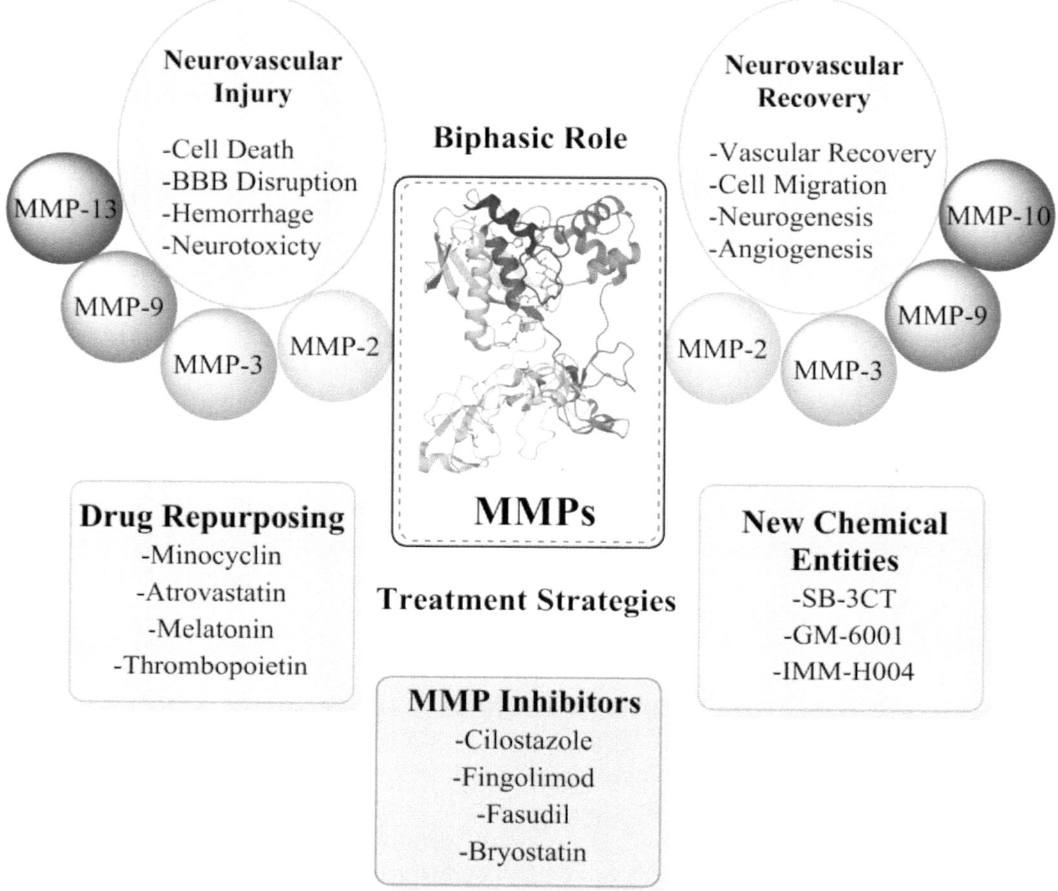

FIGURE 8.1 Biphasic role of MMPs and treatment strategies.

8.4 DETERIORATIVE ROLE OF MMP IN STROKE

Matrix metalloproteinase-2: is the widest study 72 kDa carbohydrate residue-free enzyme [33] and the most ubiquitous MMP enzyme due to its intricate involvement in various ailments like modulating the formation and remodeling of blood vessels, tissue repair, and regeneration as well as exhibition of antitumor properties [34]. The idea of the prominent role of MMP-2 in ischemic stroke came from the identification of its elevated activity in the human brain after IS. The increase in activity of MMP2 (gelatinase A) was significant at 2–5 days whereas a marked increase in its activity was reported in the subjects dying at 4 months or later [35] and this increase was related to the degradation of constituent proteins of the TJs which are occluding and claudin-5 along with collagen, laminin, and fibronectin causing BBB opening leading to infarction and hemorrhagic volume [15, 36] Thus estimating the levels of MMP-2 expression can be used as a biomarker for characterizing the severity and progression of IS.

A recent research study illustrated the aggravation of BBB damage by endothelial cell death caused by the MMP-2/9 cleaved occludin [37]. The literature reports evidence of MMP-2 involvement in the formation of glial scar within the ischemic region [33]. Expression of MMP-2 in and around amyloid-β compromised vessels is also said to be a reason for the vasculature fatal fate, leading to eventual bleeding in patients suffering from cerebral amyloid angiopathy-associated hemorrhagic stroke [38]. The correlation of MMP-2 levels with the white matter hyperintensity and elevated oxidative stress levels in IS patients was found and MMP-2 levels were identified to be associated with the pre-existing burden of the white matter hyperintensity [39].

1. **Matrix Metalloproteinases-3:** Or stromelysin-1 is a 51 kDa protein that can be differentiated from the other collagenases due to its incapacity in cleaving type-1 collagen. But it does possess the ability to degrade fibrinogens, denatured collagens, laminin, and proteoglycans. The major role of MMP-3 appears to be there in intracranial hemorrhage induced due to the tissue plasminogen treatment of ischemic stroke in mice, the dearth of which can decrease the risk of tPA treatment associated with ICH due to its possible involvement in blood vessel barrier deterioration [40]. tPA injection triggers an elevation in endothelial MMP-3 expression which is regulated via nuclear factor-kappa-B activation or lipoprotein receptor-related pathway [41, 42]. A research setup established the evidence of involvement of MMP-3 in BBB disruption during neuroinflammation and thus, an influx of neutrophils. The study found that the degradation of TJ proteins, occluding and laminin-α1 was lower in the MMP-3 knockout mice [43].
2. **Matrix Metalloproteinases-9:** Also known as gelatinase B is a 92-kDa proenzyme that not only causes the breakdown of BBB but also promotes the hemorrhagic transformation during tPA therapy [16, 33, 44, 45]. A study demonstrated the prominent role of MMP-9 in BBB opening after 3 hours of IS [15]. Investigations have revealed the elevation in serum MMP-9 levels following stroke attack with maximal elevation being observed at 48 hours post-onset. The most severe increase appears on the seventh day and is correlated to the most severe cases [15]. The studies highlighted the increased MMP-9 expression being in perfect coordination with the BBB permeability, increased infarct volume, hemorrhagic transformation, stroke severity, and poor outcomes [46, 47]. Evidence indicates the intricate role of MMP-9 in degrading both TJ and basal membrane proteins leading to BBB permeability and subsequent development of brain infarction and edema both preclinically and clinically [16, 18]. The role of MMP-9 has also been identified in the increased systemic inflammation during the BBB disruption (Figure 8.2). Systemic inflammation induced in mice caused enhanced neutrophil infiltration into the brain hampering BBB kinetics by sustained disruption of the occluding-5 and collagen-IV. The study indicated infiltrating neutrophils to be the major cause behind the remarkable elevation (up to five-fold) in MMP-9 activity in the ischemic brain. All these alterations were attributed to the improved neutrophil-derived neurovascular MMP-9 activity [48].
3. **Matrix Metalloproteinases-10:** Also known as stromelysin-2, structurally related to MMP-3 [49, 50], has more of a therapeutic role in IS. MMP-10 was

identified to be able to improve tPA-induced fibrinolysis and decrease the infarct size in experimental stroke [51]. In MMP-10 knockout mice, administration of active recombinant MMP-10 reduced the fibrinolysis against IS. Similarly, MMP-10 showed a remarkable reduction in the blood reperfusion time and infarct size in the thrombin-induced stroke model but it led to comparatively lower bleeding time along with no intracranial hemorrhage [3, 51, 52]. Also, MMP-10 thrombolytic treatment was reported to be better than tPA at halting IS-induced neurodegeneration [53].

4. **Matrix Metalloproteinases-13:** Also known as collagenase-3 are found to be related to tissue injury post-stroke [54]. The literature lists the pivotal correlation of MMP-13 to the aggressiveness of tumors implicating its role as a biomarker for tumors. The activation and nuclear translocation of MMP-13 were confirmed as the early consequence of an ischemic stimulus in the mice subjected to oxygen and glucose deprivation [55]. The role of MMP-13 was investigated in acute (24 hours and 3 days) and delayed (2 weeks) stroke phases. In the mouse with transient ischemic injury, a direct association of reduced MMP-13 expression and decreased infarct tissues and along with improved functional outcomes was reported. In the delayed phase, normal neuroblast migration was also noticed. But the dearth of MMP-13 notably downregulated the count of newborn neuroblasts (DCX+/BrdU+) in the cortical peri-infarct region. The outcomes evidenced the intricate role of MMP-13 as the central protease in infarct development and cortical remodeling during post-stroke neuro repair, which is vital for the prime angiogenic and neurogenic responses [56, 57].

8.5 MMPS AND STROKE: FOOTPRINTS OF OXIDATIVE STRESS

The oxidative stress [reactive oxygen species (ROS) and reactive nitrogen species (RNS)] pathology is intricately interlinked with the MMPs, BBB damage, and IS. ROS is elucidated to have deleterious effects on the clinical IS manifestation [58, 59]. Both ROS and RNS make the neural tissues highly susceptible to ischemic damage along with triggering the activation of MMPs. Superoxide is one of the prime reactive species demonstrated to cause BBB breakdown and development of focal destructive lesions by causing vasodilation which occurs due to the opening of potassium channels, and altered vascular reactivity [60]. Many experimental animal models of ischemia demonstrate the upregulated RNS and tyrosine nitration levels in the affected brain sections of stroke patients [3, 61]. The published literature points towards oxidative stress being the preliminary trigger for increasing the expression of MMPs after cerebral ischemia-reperfusion which further exacerbates the BBB permeability and migration of monocyte. The ROS generated triggered the upregulation of several MMPs like MMP-1, and MMP-9 and downregulated the tissue inhibitors of MMPs like TIMP-1 and TIMP-2 in a protein kinase-dependent manner [3, 32]. All these changes directly correlated to the degradation of basement membrane proteins and aggravated tyrosine phosphorylation of TJ proteins. All these occurrences were found to be controlled by administering the MMP or PTK inhibitors or antioxidant therapies suggesting oxidative stress to be pivotal in causing BBB injury by activating MMP or TIMP pathways [3, 32]. Many of the investigated scavengers to eliminate these reactive oxygen species like superoxide scavengers, and nitric oxide synthase inhibitors alleviate the BBB disruption, improve neurological outcomes and decrease the infarction volumes [62]. The role of ROS and its associated signaling pathways was also evident in tPA-mediated MMP-9 production. The tPA when administered alone caused remarkable elevation in the translocation of NF-κB to the nucleus. When administered along with hydrogen peroxide, tPA further increased the translocation. The combination also notably increased I-κB degradation. The experiment concluded that ROS enhances the tPA-modulated MMP-9 production in the brain endothelial cells [63]. Edaravone, a scavenger for free radicals was demonstrated to decrease this tPA-mediated MMP-9 production [63].

8.6 MMPS AND TPA–THE CULPRIT PARTY

Till now, thrombolysis using tPA remains the first-choice treatment for IS. tPA helps in clot solubilization by digesting fibrin and restores the blood supply [31]. But the prime limitation associated with the treatment with tPA is its narrow therapeutic window, which requires administration within the initial 3 hours to ensure therapeutic benefit. Moreover, only less than 5% of patients experience its beneficial effect. If initiated post 3 hours of the onset of symptoms, the administration may lead to neurotoxic effects [11, 31, 42, 64, 65]. The tPA is a double-edged sword possessing thrombolytic and neurotoxic effects during stroke [11]. The tPA, be it endogenous or exogenously administered can penetrate the BBB, and can potentially interact with the neurovascular unit. The tPA is secreted by the neurons and microglia in various parts of the brain like the hippocampus, hypothalamus, cerebellum, and amygdala [66, 67]. The endogenous tPA modulates the neuronal death and microglial activation post excitotoxic injury. It is also said to influence the permeability of the neurovascular unit which is linked to the occurrence of ischemia [31, 68]. The tPA action is terminated when it binds to protein PAI-1 [69, 70]. This process regulates the fibrinolytic efficacy of the tPA. The formed tPA/PAI-1 complex containing inactivated tPA is cleared from circulation by lipoprotein receptor-related protein-1 (LRP-1) [71]. This clearance mechanism ensures the systemic half-life of tPA to be only 5–10 minutes duration. Another potent endogenous inhibitor of tPA that modulates its catalytic action in the brain is neuroserpin. The literature demonstrates the neuroserpin to exert a neuroprotective role as it decreases the activation of microglia and tPA activity [68, 72]. Thus, increased expression of neuroserpin in the brain prevents the BBB leakage, and edema and decreases the size of the infarct thus widening the therapeutic window for the tPA administration [73].

The studies point out the pivotal role of MMP-9 in tPA-mediated neurotoxicity (Figure 8.2). The activated MMP-9 causes degradation of the neurovascular unit and extracellular matrix and tPA causes an elevation in the MMP-9 levels post-ischemia. The tPA perfusion disrupts BBB and degrades basal membrane protein laminin in rodents [29, 74]. The MMP-9 is found to be elevated in the plasma and brain of patients with ischemic stroke and delayed tPA perfusion causes these levels to boost further [29, 74, 75]. Since the MMP-9 is a prime mediator of the hemorrhagic transformation, therapeutic inhibition of

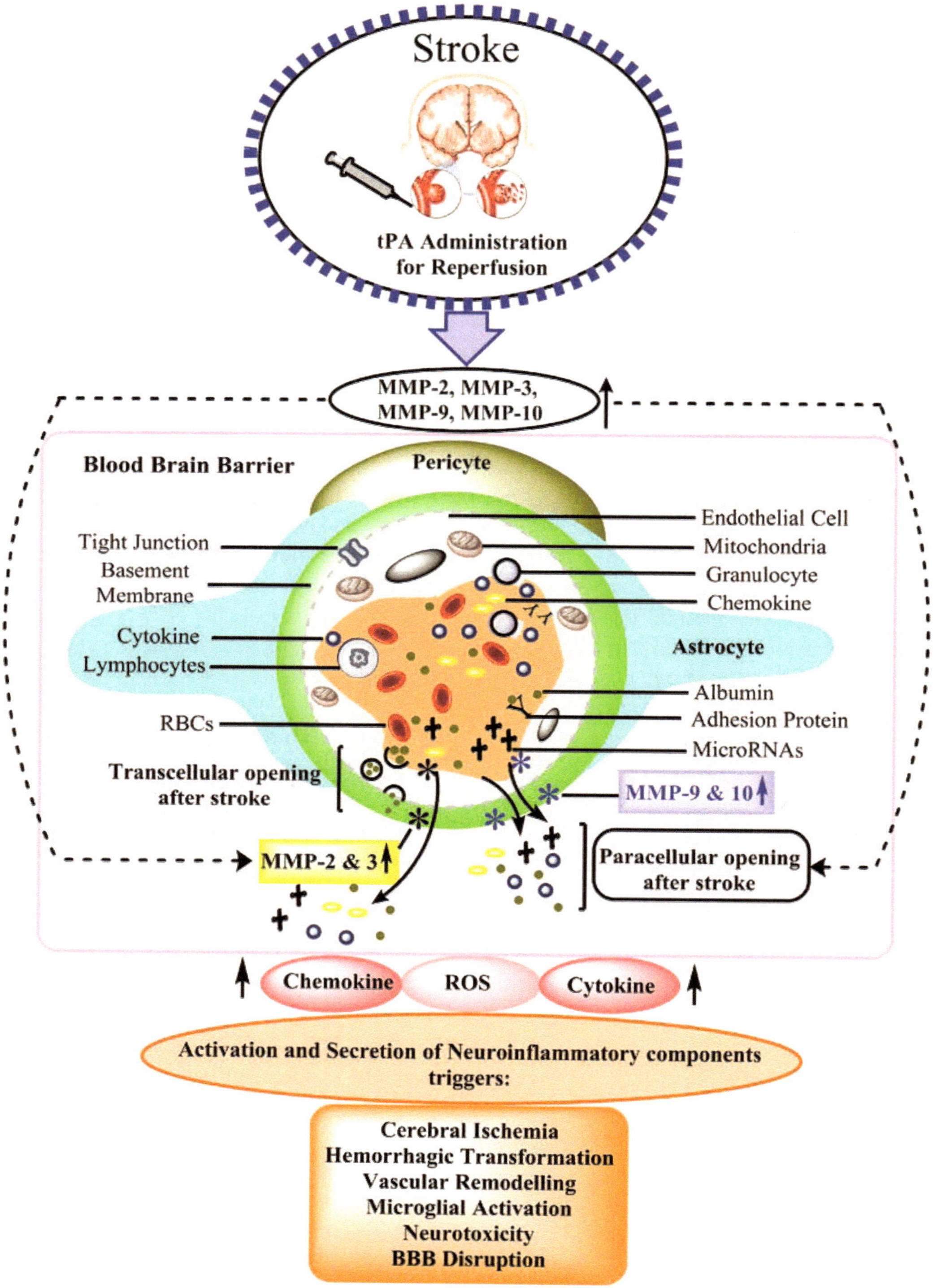

FIGURE 8.2 Pathways involved in MMP-mediated blood-brain barrier damage in stroke.

MMP-9 may enhance the safety of the thrombolytic therapy [75]. In addition to MMP-9, MMP-3 is also a critical MMP prerequisite for intracranial bleeding post tPA administration. It was noticed that MMP-3 levels were significantly elevated in the areas with ischemia and were further elevated in endothelial cells with tPA treatment. However, tPA administration to MMP-3 knockout mice showed reduced hemorrhagic transformation as compared to wild-type animals. The tPA was thus found to deteriorate the intracranial bleeding via inducing MMP-3 in cells. The inhibition of MMP-3 by LRP or NF-kB activation led to a dampening increase in the intracranial bleeding by the tPA [42].

8.7 NEUROPROTECTIVE EFFECTS OF MATRIX METALLOPROTEINASES

The MMPs were also found to have some neuroprotective effects in IS recovery phase [27, 76, 77]. Neuro recovery occurs in 3 phases which are angiogenesis, neurogenesis, and synaptic plasticity. MMPs refine the angiogenesis process. These participate in the remodeling of basement membranes and extracellular matrix [27, 78] which activates microglia in the penumbra creating a protective zone to reduce infarct volume after IS. The early inhibition of MMPs was found to delay the recovery from ischemia [79].

8.8 MATRIX METALLOPROTEINS AS A THERAPEUTIC TARGET IN ISCHEMIC STROKE

The potential therapeutic regimens to date alleviate the complications associated with tPA-mediated neurotoxicity target MMPs. Thus, in light of MMPs' critical role in mediating BBB disruption along with tPA administration, they are the golden target for enhancing the activity of tPA. The drug regimens explored till now, utilize MMP-inhibitors as a combination therapy to aid in the enhancement of the therapeutic window of tPA along with imparting other neuroprotective benefits (Table 8.1). Multiple combination strategies have been tested *in-vitro* and *in-vivo* for their significant contribution to expanding the thrombolytic window for ischemic stroke. Thus, maximizing the therapeutic benefits and alleviating the side effects associated with the thrombolytic treatment. Minocycline, which is widely explored for neuroprotective effects in the cerebral ischemia and neurodegenerative models was found to decrease the infarct size and significantly enhanced the neurological outcome when administered with tPA [80, 81]. The mechanism of action behind the therapeutic activity of minocycline was the inhibition of MMP-9 via an extracellular signaling-related kinase pathway. Minocycline was found to inhibit the MMP-9 at various levels and is sufficient in low doses [82–85]. The combination treatment of tPA along with minocycline was administered at 6 h which also resulted in a reduction in infarct volumes by 50% [82–84, 86]. It was also found to ameliorate brain hemorrhage. Another MP inhibitor investigated in the tPA-related hemorrhagic transformation of ischemic stroke was Cilostazol which is a phosphodiesterase-III inhibitor known for its pleiotropic effects as antiplatelet, vasodilatory, and protective effects of endothelial cells. The drug decreased stroke recurrences, intracranial hemorrhage post-stroke, and extracranial bleeding. Cilostazol is known to protect the entire

TABLE 8.1 Beneficial Effects of MMPs Modulation Along with tPA Administration

Adjuctive Intervention (Dose, Route, Timing)	tPA (Dose, Timing, Route)	Species Which Evaluated	Outcome	References
Minocycline (3 mg/kg, Intravenous, 4 hours PS).	10 mg/kg, IV, 6 hours PS.	Male SHR; embolic.	Reduced hemorrhagic transformation, decreased infarct volume, decreased MMP-9 plasma level.	[82–84]
Cilostazole (10 mg/kg, i.p., administered before tPA).	10 mg/kg, IV, 6 hours PS.	Male ddY rats, intraluminal filament/reperfusion.	Reduced hemorrhagic transformation, infarct volume, MMP-9 expression, and improved locomotion.	[88, 89]
GM-6001 (100 mg/kg; i.p., administered along with tPA).	10 mg/kg IV, administered 6 hours post-stroke.	Male ddY mice with intraluminal filament/reperfusion.	Decreased hemorrhagic transformation; decreased MMP-9, enhanced occluding, enhanced ZO-1.	[91]
Fingolimod (0.5 mg/kg, i.p., administered along with tPA).	10 mg/kg IV, administered 6 hours post-stroke.	C57BL/6 male mice with middle cerebral artery occlusion.	Reduced infarct volume, reduced hemorrhagic transformation.	[93–95]
Fasudil (3 mg/kg, i.p., before tPA).	10 mg/kg, IV, 6 hours PS.	Male SD rats; intraluminal filament/reperfusion.	Reduced hemorrhagic transformation, infarct volume remains unchanged, decreased MMP-9 levels (*in-vivo*), improved locomotor behavior.	[94, 95, 99]
Candesartan (1 mg/kg, IV, 3 hours PS).	10 mg/kg, IV, 6 hours PS).	Male Wistar rats; embolic.	Reduced hemorrhagic transformation; unchanged infarct volume; MMP-9, 2; decreased MMP-3, NF-κB, TNF-α.	[100]
IMM-H004 (6 mg/kg, IV, along with tPA).	10 mg/kg, IV, poststroke.	Male SD rats; embolic and male SD rats with intraluminal filament/reperfusion.	Decreased hemorrhagic transformation, reduced along with Akt, Occludin, and pro-MMP-9.	[102]
Bryostatin	5 mg/kg, IV, 6 hours post-stroke.	Female SD rats; embolic.	Decreased hemorrhagic transformation, decreased MMP-9, reduced infarct volume, MMP-2, PKC-α, unchanged PKCδ, and increased PKCε.	[103]

Abbreviations: PS: Post-stroke; SHR: Spontaneously hypertensive rats; IV: Intravenous; tPA: Tissue plasminogen activator; MMP: Matrix metalloproteinase; i.p.: intraperitoneal; PKC: Protein kinase C; and SD: Sprague-Dawley rats.

BBB as *in-vitro* examinations demonstrated its protective effects on endothelial cells, pericytes, TJ proteins, AJ proteins, and the basement membrane following the collagenase injury [87]. Combination treatment of cilostazol with tPA downregulated the MMP-9 activity which prevented the hemorrhagic transformation, edema, endothelial injury and the opening of BBB by inhibiting claudin-5 expression leading to overall decreased morbidity and mortality incidences at 18 h and 7 h post reperfusion [88, 89] The administration of both the tPA and cilostazol halted injury to brain pericytes and endothelial cells and, thus increasing the therapeutic window for treatment by tPA [90]. GM6001, a MMP inhibitor, showed protection against tPA induced hemorrhagic complications and increase survival rate [91]. Further, co-adminsitration of tPA and GM6001 decreased the elevated brain hemoglobin, and prevented degradation of occludin and ZO-1 [91].

Another MMP inhibitor, p-aminobenzoyl-gly-pro-d-leu-d-ala-hydroxymate (AHA) prevents interstitial collagenases, gelatinases, and stromelysin. The results analyzed at 24 hours demonstrated the lowering of cerebral edema and risk of intracerebral hemorrhage along with improved neurological outcomes when MMP was inhibited post 3 hours of the onset of ischemia in rats. The results validated the involvement of MMP in hemorrhagic transformation and combination treatment aided in widening the therapeutic window for thrombolysis [3, 92].

Fingolimod was found to decrease the infarct size in the animal model of IS and aid in improving the integrity of BBB by targeting MMP9 activity. It was found to impede the neurological deficit and infarct size in thromboembolically induced IS. Moreover, when administered with tPA, fingolimod complements the thrombolytic therapy by improving neurological outcomes and reducing the chances of hemorrhagic transformation [93–95].

Activated Protein C (APC), a vitamin K-dependent serine protease possesses anticoagulant and direct vasculoprotective activity. In rodents, it has been found that APC inhibits tPA-mediated brain hemorrhage post IS via impeding the NF-κB-dependent MMP-9 pathway [96]. A neuroprotective analogie of APC, known as 3K3A-APC possessing lesser anticoagulant activity, was found to decrease the infarct volume when administered in combination with tPA widening its therapeutic window manifold [97]. Another research study found the APC to promote neurogenesis in the post-ischemic brain PAR-1 [98].

Fasudil–a rho kinase inhibitor, subdued the MMP-9 activity in endothelial cells. When it was combined with tPA, it reduced the risk of the development of hemorrhagic transformation. The combination greatly reduced death rates and improved locomotive post-reperfusion in mice stroke models. Further administration of combination in human brain endothelial cells prevented the occurrence of injury by downregulating the MMP-9 activity [94, 95, 99]. Administration of candesartan combined with tPA decreased the risk of brain hemorrhage and enhanced the neurological outcome in comparison to the rats treated with tPA alone after induction of IS via embolism. The combination was found to upregulate MMP-9 activity and downregulate the expressions of MMP-3, TNF-α, and NF-κB and increased endothelial nitric oxide synthase activation leading to amelioration of tPA-induced hemorrhage post embolic stroke [100, 101].

IMM-H004, a coumarin compound, inhibited hemorrhagic transformation and decreased the death rate, cerebral edema, and infarct volume by decreasing the MMP9 and co-localization of MMP-2 with astrocytes and increasing occluding. It also improved neurological functions and preserved the integrity of BBB [102]. Bryostatin, a Protein kinase-C modulator widens the thrombolytic therapeutic window to six hours following ischemic insult in rodents. Rats treated with bryostatin before recombinant tPA demonstrated reduced mortality and hemispherical swelling as compared to rats administered recombinant tPA alone. It downregulated MMP-9 expression, BBB disruption, and hemorrhagic transformation while upregulating the PKCε expression [103–105].

8.9 CONCLUSIONS AND FUTURE PROSPECTS

The chapter gives insights into the intricate role of MMPs in modulating the BBB which is one of the prime events during the journey of ischemic injury. MMPs play a dual role by worsening the pathology or the stage of the disease as well as imparting neuroprotection. Efforts have been made to curtail the deleterious roles of MMPs by the development of MMP-9 and MMP-2 inhibitors and there is hope for the development of MMP agonists or agents that can increase the concentration of neuroprotective MMPs like MMP-10 agonists or those agents that can initiate recovery from the prolonged MMP inhibition. The MMP-10 known for its protective properties can be explored further for its therapeutic applications. Also, many novel molecules like GM6001, Fasudil, minocycline, IMM-H004, etc., that have targeted MMPs as a supplement to the tPA for enhancing the treatment outcome can be examined further. However, the road to the development of a concrete treatment for acute ischemic injury is still long. Since there are only a handful of MMPs whose role has been completely explored the potential of many of the MMPs is yet to be investigated as a prospective target site for the development of a successful therapeutic regimen. A deeper understanding of these areas along with enhancements and developments in the designing of preclinical and clinical studies can pave the way towards the goal.

KEYWORDS

- **blood-brain barrier disruption**
- **ischemic stroke**
- **matrix metalloproteinases**
- **neurodegeneration matrix metalloproteinases**
- **neuroprotective effects**
- **oxidative stress**

REFERENCES

1. Khan, H., Kashyap, A., Kaur, A., & Singh, T. G., (2020). Pharmacological postconditioning: A molecular aspect in ischemic injury. *Journal of Pharmacy and Pharmacology, 72*(11), 1513–1527.

2. Thapa, K., Khan, H., Singh, T. G., & Kaur, A., (2021). Traumatic brain injury: Mechanistic insight on pathophysiology and potential therapeutic targets. *Journal of Molecular Neuroscience, 71*(9), 1725–1742.
3. Lakhan, S., Kirchgessner, A., Tepper, D., & Aidan, L., (2013). Matrix metalloproteinases and blood-brain barrier disruption in acute ischemic stroke [internet]. *Frontiers in Neurology*, 4.
4. Saklani, P., Khan, H., Gupta, S., Kaur, A., & Singh, T. G., (2022). Neuropeptides: Potential neuroprotective agents in ischemic injury. *Life Sciences, 288*, 120186.
5. Garg, N., Singh, T. G., Khan, H., Arora, S., Kaur, A., & Mannan, A., (2021). Mechanistic interventions of selected *Ocimum* species in management of diabetes, obesity, and liver disorders: Transformative developments from preclinical to clinical approaches. *Biointerface Res. Appl. Chem., 12*(1), 1304–1323.
6. Cucullo, L., Hossain, M., Puvenna, V., Marchi, N., & Janigro, D., (2011). The role of shear stress in blood-brain barrier endothelial physiology. *BMC Neurosci., 12*(1), 40.
7. Singh, V., Krishan, P., Singh, N., Kumar, A., & Shri, R., (2017). Amelioration of ischemia-reperfusion induced functional and biochemical deficit in mice by *Ocimum kilimandscharicum* leaf extract. *Biomed Pharmacother., 85*, 556–563.
8. Singh, V., Chauhan, G., & Shri, R., (2021). Anti-depressant like effects of quercetin 4'-O-glucoside from *Allium cepa* via regulation of brain oxidative stress and monoamine levels in mice subjected to unpredictable chronic mild stress. *Nutr. Neurosci., 24*(1), 35–44.
9. Kaur, R., Singh, V., & Shri, R., (2017). Anti-amnesic effects of *Ganoderma* species: A possible cholinergic and antioxidant mechanism. *Biomed. Pharmacother., 92*(1), 1055–1061.
10. Singh, V., Krishan, P., & Shri, R., (2018). Amelioration of ischemia reperfusion-induced cerebral injury in mice by liposomes containing *Allium cepa* fraction administered intranasally. *Artif. Cells, Nanomedicine, Biotechnol., 46*(sup3), S982–992.
11. Jin, R., Yang, G., & Li, G., (2010). Molecular insights and therapeutic targets for blood–brain barrier disruption in ischemic stroke: Critical role of matrix metalloproteinases and tissue-type plasminogen activator. *Neurobiol. Dis., 38*(3), 376–385.
12. Abbott, N. J., Patabendige, A. A. K., Dolman, D. E. M., Yusof, S. R., & Begley, D. J., (2010). Structure and function of the blood–brain barrier. *Neurobiol. Dis., 37*(1), 13–25.
13. Bhatia, S., Rawal, R., Sharma, P., Singh, T., Singh, M., & Singh, V., (2021). Mitochondrial dysfunction in Alzheimer's disease: Opportunities for drug development. *Curr. Neuropharmacol., 19*.
14. Sharma, V. K., Singh, T. G., & Mehta, V., (2021). Stressed mitochondria: A target to intrude Alzheimer's disease. *Mitochondrion, 59*, 48–57.
15. Rosenberg, G. A., Estrada, E. Y., & Dencoff, J. E., (1998). Matrix metalloproteinases and TIMPs are associated with blood-brain barrier opening after reperfusion in rat brain. *Stroke, 29*(10), 2189–2195.
16. Turner, R. J., & Sharp, F. R., (2016). Implications of MMP9 for blood brain barrier disruption and hemorrhagic transformation following ischemic stroke [internet]. *Frontiers in Cellular Neuroscience, 10*, 56.
17. Mun-Bryce, S., & Rosenberg, G. A., (1998). Gelatinase B modulates selective opening of the blood-brain barrier during inflammation. *Am. J. Physiol. Integr. Comp. Physiol., 274*(5), R1203–1211.
18. Rosenberg, G. A., & Yang, Y., (2007). Vasogenic edema due to tight junction disruption by matrix metalloproteinases in cerebral ischemia. *Neurosurg. Focus, 22*(5), 1–9.
19. Romanic, A. M., White, R. F., Arleth, A. J., Ohlstein, E. H., & Barone, F. C., (1998). Matrix metalloproteinase expression increases after cerebral focal ischemia in rats: Inhibition of matrix metalloproteinase-9 reduces infarct size. *Stroke, 29*(5), 1020–1030.
20. Fujimura, M., Gasche, Y., Morita-Fujimura, Y., Massengale, J., Kawase, M., & Chan, P. H., (1999). Early appearance of activated matrix metalloproteinase-9 and blood–brain barrier disruption in mice after focal cerebral ischemia and reperfusion. *Brain Res., 842*(1), 92–100.
21. Rosenberg, G. A., Cunningham, L. A., Wallace, J., Alexander, S., Estrada, E. Y., Grossetete, M., et al., (2001). Immunohistochemistry of matrix metalloproteinases in reperfusion injury to rat brain: Activation of MMP-9 linked to stromelysin-1 and microglia in cell cultures. *Brain Res., 893*(1, 2), 104–112.
22. Klein, T., & Bischoff, R., (2011). Physiology and pathophysiology of matrix metalloproteases. *Amino Acids, 41*(2), 271–290.
23. Rempe, R. G., Hartz, A. M. S., & Bauer, B., (2016). Matrix metalloproteinases in the brain and blood–brain barrier: Versatile breakers and makers. *J. Cereb. Blood Flow. Metab., 36*(9), 1481–1507.

24. Tallant, C., Marrero, A., & Gomis-Rüth, F. X., (2010). Matrix metalloproteinases: Fold and function of their catalytic domains. *Biochim. Biophys. Acta – Mo.l Cell Res., 1803*(1), 20–28.
25. Sharma, A., Khan, H., Singh, T. G., Grewal, A. K., Najda, A., Kawecka-Radomska, M., Kamel, M., et al., (2021). Pharmacological modulation of ubiquitin-proteasome pathways in oncogenic signaling. *International Journal of Molecular Sciences, 22*(21), 11971.
26. Clark, I. M., Swingler, T. E., Sampieri, C. L., & Edwards, D. R., (2008). The regulation of matrix metalloproteinases and their inhibitors. *Int. J. Biochem. Cell Biol., 40*(6, 7), 1362–1378.
27. Rehni, A. K., Singh, T. G., Kalra, R., & Singh, N., (2009). Pharmacological inhibition of inducible nitric oxide synthase attenuates the development of seizures in mice. *Nitric Oxide., 21*(2), 120–125.
28. Rosell, A., & Lo, E. H., (2008). Multiphasic roles for matrix metalloproteinases after stroke. *Curr. Opin. Pharmacol., 8*(1), 82–89.
29. Kaur, A., Singh, T. G., Khan, H., Kumar, M., Singh, N., & Abdel-Daim, M. M., (2022). Neuroprotective effect of piclamilast-induced post-ischemia pharmacological treatment in mice. *Neurochemical Research*, 1–4.
30. Solé, S., Petegnief, V., Gorina, R., Chamorro, Á., & Planas, A. M., (2004). Activation of matrix metalloproteinase-3 and Agrin cleavage in cerebral ischemia/reperfusion. *J Neuropathol. Exp. Neurol., 63*(4), 338–349.
31. Kalra, P., Khan, H., Kaur, A., & Singh, T. G., (2022). Mechanistic insight on autophagy modulated molecular pathways in cerebral ischemic injury: From preclinical to clinical perspective. *Neurochemical Research*, 1–9.
32. Cunningham, L. A., Wetzel, M., & Rosenberg, G. A., (2005). Multiple roles for MMPs and TIMPs in cerebral ischemia. *Glia., 50*(4), 329–339.
33. Kurzepa, J., Kurzepa, J., Golab, P., Czerska, S., & Bielewicz, J., (2014). The significance of matrix metalloproteinase (MMP)-2 and MMP-9 in the ischemic stroke. *Int. J. Neurosci., 124*(10), 707–716.
34. Malemud, C. J., (2017). Chapter 7–matrix metalloproteinases and synovial joint pathology. In: Raouf A. Khalil (eds.), *Matrix Metalloproteinases and Tissue Remodeling in Health and Disease: Target Tissues and Therapy* (pp. 305–325). Academic Press.
35. Clark, A. W., Krekoski, C. A., Bou, S. S., Chapman, K. R., & Edwards, D. R., (1997). Increased gelatinase A (MMP-2) and gelatinase B (MMP-9) activities in human brain after focal ischemia. *Neurosci. Lett., 238*(1), 53–56.
36. Tiwari, P., Khan, H., Singh, T. G., & Grewal, A. K., (2022). Poly (ADP-ribose) polymerase: An overview of mechanistic approaches and therapeutic opportunities in the management of stroke. *Neurochemical Research*, 1–23.
37. Pan, R., Liu, W., & Liu, K. J., (2021). MMP-2/9-cleaved occludin promotes endothelia cell death in ischemic stroke. *Brain Hemorrhages, 2*(2), 63–70.
38. Hernandez-Guillamon, M., Martinez-Saez, E., Delgado, P., Domingues-Montanari, S., Boada, C., Penalba, A., et al., (2012). MMP-2/MMP-9 plasma level and brain expression in cerebral amyloid angiopathy-associated hemorrhagic stroke. *Brain Pathol., 22*(2), 133–141.
39. Corbin, Z. A., Rost, N. S., Lorenzano, S., Kernan, W. N., Parides, M. K., Blumberg, J. B., et al., (2014). White matter hyperintensity volume correlates with matrix metalloproteinase-2 in acute ischemic stroke. *J. Stroke Cerebrovasc. Dis., 23*(6), 1300–1306.
40. Suzuki, Y., Nagai, N., Umemura, K., Collen, D., & Lijnen, H. R., (2007). Stromelysin-1 (MMP-3) is critical for intracranial bleeding after t-PA treatment of stroke in mice. *J. Thromb. Haemost., 5*(8), 1732–1739.
41. Rihal, V., Khan, H., Kaur, A., & Singh, T. G., (2022). Vitamin D as therapeutic modulator in cerebrovascular diseases: A mechanistic perspectives. *Critical Reviews in Food Science and Nutrition, 5*, 1–23.
42. Suzuki, Y., Nagai, N., Yamakawa, K., Kawakami, J., Lijnen, H. R., & Umemura, K., (2009). Tissue-type plasminogen activator (t-PA) induces stromelysin-1 (MMP-3) in endothelial cells through activation of lipoprotein receptor–related protein. *Blood, 114*(15), 3352–3358.
43. Gurney, K. J., Estrada, E. Y., & Rosenberg, G. A., (2006). Blood–brain barrier disruption by stromelysin-1 facilitates neutrophil infiltration in neuroinflammation. *Neurobiol. Dis., 23*(1), 87–96.
44. Gasche, Y., Fujimura, M., Morita-Fujimura, Y., Copin, J. C., Kawase, M., Massengale, J., et al., (1999). Early appearance of activated matrix metalloproteinase-9 after focal cerebral ischemia in mice: A possible role in blood—Brain barrier dysfunction. *J. Cereb. Blood Flow Metab., 19*(9), 1020–1028.

45. Gidday, J. M., Gasche, Y. G., Copin, J. C., Shah, A. R., Perez, R. S., Shapiro, S. D., et al., (2005). Leukocyte-derived matrix metalloproteinase-9 mediates blood-brain barrier breakdown and is proinflammatory after transient focal cerebral ischemia. *Am. J. Physiol. Circ. Physiol., 289*(2), H558–568.
46. Abdelnaseer, M. M., Elfauomy, N. M., Esmail, E. H., Kamal, M. M., & Elsawy, E. H., (2017). Matrix metalloproteinase-9 and recovery of acute ischemic stroke. *J. Stroke Cerebrovasc. Dis., 26*(4), 733–740.
47. Ramos-Fernandez, M., Bellolio, M. F., & Stead, L. G., (2011). Matrix metalloproteinase-9 as a marker for acute ischemic stroke: A systematic review. *J. Stroke Cerebrovasc. Dis., 20*(1), 47–54.
48. Khan, H., Gupta, A., Singh, T. G., & Kaur, A., (2021). Mechanistic insight on the role of leukotriene receptors in ischemic–reperfusion injury. *Pharmacological Reports, 73*(5), 1240–1254.
49. Gill, J. H., Kirwan, I. G., Seargent, J. M., Martin, S. W., Tijani, S., Anikin, V. A., et al., (2004). MMP-10 is overexpressed, proteolytically active, and a potential target for therapeutic intervention in human lung carcinomas. *Neoplasia, 6*(6), 777–785.
50. Duffy, M. J., Maguire, T. M., Hill, A., McDermott, E., & O'Higgins, N., (2000). Metalloproteinases: Role in breast carcinogenesis, invasion, and metastasis. *Breast Cancer Res., 2*(4), 252.
51. Orbe, J., Barrenetxe, J., Rodriguez, J. A., Vivien, D., Orset, C., Parks, W. C., et al., (2011). Matrix metalloproteinase-10 effectively reduces infarct size in experimental stroke by enhancing fibrinolysis via a thrombin-activatable fibrinolysis inhibitor–mediated mechanism. *Circulation, 124*(25), 2909–2919.
52. Roncal, C., Martinez De, L. S., Salicio, A., Chevilley, A., Rodriguez, J. A., Rosell, A., et al., (2017). New thrombolytic strategy providing neuroprotection in experimental ischemic stroke: MMP10 alone or in combination with tissue-type plasminogen activator. *Cardiovasc Res., 113*(10), 1219–1229.
53. Navarro-Oviedo, M., Roncal, C., Salicio, A., Belzunce, M., Rabal, O., Toledo, E., et al., (2019). MMP10 promotes efficient thrombolysis after ischemic stroke in mice with induced diabetes. *Transl. Stroke Res., 10*(4), 389–401.
54. Heim-Riether, A., Taylor, S. J., Liang, S., Gao, D. A., Xiong, Z., Michael A. E., et al., (2009). Improving potency and selectivity of a new class of non-Zn-chelating MMP-13 inhibitors. *Bioorg. Med. Chem. Lett., 19*(18), 5321–5324.
55. Cuadrado, E., Rosell, A., Borrell-Pagès, M., García-Bonilla, L., Hernández-Guillamon, M., Ortega-Aznar, A., et al., (2009). Matrix metalloproteinase-13 is activated and is found in the nucleus of neural cells after cerebral ischemia. *J. Cereb. Blood Flow Metab., 29*(2), 398–410.
56. Ma, F., Martínez-San, S. P., Barceló, V., Morancho, A., Gabriel-Salazar, M., Giralt, D., et al., (2016). Matrix metalloproteinase-13 participates in neuroprotection and neurorepair after cerebral ischemia in mice. *Neurobiol. Dis., 91*, 236–246.
57. Rosell, A., Alvarez-Sabín, J., Arenillas, J. F., Rovira, A., Delgado, P., Fernández-Cadenas, I., et al., (2005). A matrix metalloproteinase protein array reveals a strong relation between MMP-9 and MMP-13 with diffusion-weighted image lesion increase in human stroke. *Stroke, 36*(7), 1415–1420.
58. Ozkul, A., Akyol, A., Yenisey, C., Arpaci, E., Kiylioglu, N., & Tataroglu, C., (2007). Oxidative stress in acute ischemic stroke. *J. Clin. Neurosci., 14*(11), 1062–1066.
59. Shri, R., Singh, V., Kaur, R., & Bhatia, S., (2021). *Natural BACE1 Inhibitors: Promising Drugs for the Management of Alzheimer's Disease* (pp. 137–171).
60. Allen, C. L., & Bayraktutan, U., (2009). Oxidative stress and its role in the pathogenesis of ischemic stroke. *Int. J. Stroke, 4*(6), 461–470.
61. Khan, H., Sharma, K., Kumar, A., Kaur, A., & Singh, T. G., (2022). Therapeutic implications of cyclooxygenase (COX) inhibitors in ischemic injury. *Inflammation Research, 17*, 1–6.
62. Maier, C. M., Hsieh, L., Crandall, T., Narasimhan, P., & Chan, P. H., (2006). Evaluating therapeutic targets for reperfusion-related brain hemorrhage. *Ann. Neurol., 59*(6), 929–938.
63. Harada, K., Suzuki, Y., Yamakawa, K., Kawakami, J., & Umemura, K., (2012). Combination of reactive oxygen species and tissue-type plasminogen activator enhances the induction of gelatinase B in brain endothelial cells. *Int. J. Neurosci., 122*(2), 53–59.
64. Khan, H., Tiwari, P., Kaur, A., & Singh, T. G., (2021). Sirtuin acetylation and deacetylation: A complex paradigm in neurodegenerative disease. *Molecular Neurobiology, 58*(8), 3903–3917.
65. Singh, T., & Goel, R. K., (2015). Neuroprotective effect of *Allium cepa* L. in aluminum chloride induced neurotoxicity. *Neurotoxicology, 49*, 1–7.

66. Siao, C. J., Fernandez, S. R., & Tsirka, S. E., (2003). Cell type-specific roles for tissue plasminogen activator released by neurons or microglia after excitotoxic injury. *J. Neurosci., 23*(8), 3234 LP – 3242.
67. Sallés, F. J., & Strickland, S., (2002). Localization and regulation of the tissue plasminogen activator–plasmin system in the hippocampus. *J. Neurosci., 22*(6), 2125 LP – 2134.
68. Yepes, M., Sandkvist, M., Coleman, T. A., Moore, E., Wu, J. Y., Mitola, D., et al., (2002). Regulation of seizure spreading by neuroserpin and tissue-type plasminogen activator is plasminogen-independent. *J. Clin. Invest., 109*(12), 1571–1578.
69. Suzuki, Y., Mogami, H., Ihara, H., & Urano, T., (2009). Unique secretory dynamics of tissue plasminogen activator and its modulation by plasminogen activator inhibitor-1 in vascular endothelial cells. *Blood, 113*(2), 470–478.
70. Morton, P. A., Owensby, D. A., Sobel, B. E., & Schwartz, A. L., (1989). Catabolism of tissue-type plasminogen activator by the human hepatoma cell line Hep G2: Modulation by plasminogen activator inhibitor type 1. *J. Biol. Chem., 264*(13), 7228–7235.
71. Yepes, M., Sandkvist, M., Moore, E. G., Bugge, T. H., Strickland, D. K., & Lawrence, D. A., (2003). Tissue-type plasminogen activator induces opening of the blood-brain barrier via the LDL receptor–related protein. *J. Clin. Invest., 112*(10), 1533–1540.
72. Yepes, M., Sandkvist, M., Wong, M. K. K., Coleman, T. A., Smith, E., Cohan, S. L., et al., (2000). Neuroserpin reduces cerebral infarct volume and protects neurons from ischemia-induced apoptosis. *Blood, J. Am. Soc. Hematol., 96*(2), 569–576.
73. Zhang, Z., Zhang, L., Yepes, M., Jiang, Q., Li, Q., Arniego, P., et al., (2002). Adjuvant treatment with neuroserpin increases the therapeutic window for tissue-type plasminogen activator administration in a rat model of embolic stroke. *Circulation, 106*(6), 740–745.
74. Rosell, A., Cuadrado, E., Ortega-Aznar, A., Hernández-Guillamon, M., Lo, E. H., & Montaner, J., (2008). MMP-9–positive neutrophil infiltration is associated to blood–brain barrier breakdown and basal lamina type IV collagen degradation during hemorrhagic transformation after human ischemic stroke. *Stroke, 39*(4), 1121–1126.
75. Ning, M., Furie, K. L., Koroshetz, W. J., Lee, H., Barron, M., Lederer, M., et al., (2006). Association between tPA therapy and raised early matrix metalloproteinase-9 in acute stroke. *Neurology, 66*(10), 1550–1555.
76. Zhao, B. Q., Wang, S., Kim, H. Y., Storrie, H., Rosen, B. R., Mooney, D. J., et al., (2006). Role of matrix metalloproteinases in delayed cortical responses after stroke. *Nat. Med., 12*(4), 441–445.
77. Zhao, B. Q., Tejima, E., & Lo, E. H., (2007). Neurovascular proteases in brain injury, hemorrhage, and remodeling after stroke. *Stroke, 38*(2), 748–752.
78. Rundhaug, J. E., (2005). Matrix metalloproteinases and angiogenesis. *J. Cell Mol. Med., 9*(2), 267–285.
79. Zhang, X., Zhao, H. H., Li, D., & Li, H. P., (2019). Neuroprotective effects of matrix metalloproteinases in cerebral ischemic rats by promoting activation and migration of astrocytes and microglia. *Brain Res. Bull., 146*, 136–142.
80. Fagan, S. C., Waller, J. L., Nichols, F. T., Edwards, D. J., Pettigrew, L. C., Clark, W. M., et al., (2010). Minocycline to improve neurologic outcome in stroke (MINOS). *Stroke, 41*(10), 2283–2287.
81. Fagan, S. C., Cronic, L. E., & Hess, D. C., (2011). Minocycline development for acute ischemic stroke. *Transl. Stroke Res., 2*(2), 202–208.
82. Lee, C. Z., Yao, J. S., Huang, Y., Zhai, W., Liu, W., Guglielmo, B. J., et al., (2006). Dose–response effect of tetracyclines on cerebral matrix metalloproteinase-9 after vascular endothelial growth factor hyperstimulation. *J. Cereb. Blood Flow Metab., 26*(9), 1157–1164.
83. Machado, L. S., Kozak, A., Ergul, A., Hess, D. C., Borlongan, C. V., & Fagan, S. C., (2006). Delayed minocycline inhibits ischemia-activated matrix metalloproteinases 2 and 9 after experimental stroke. *BMC Neurosci., 7*(1), 56.
84. Switzer, J. A., Hess, D. C., Ergul, A., Waller, J. L., Machado, L. S., Portik-Dobos, V., et al., (2011). Matrix metalloproteinase-9 in an exploratory trial of intravenous minocycline for acute ischemic stroke. *Stroke, 42*(9), 2633–2635.
85. Singh, T., Thapliyal, S., Bhatia, S., Singh, V., Singh, M., Singh, H., et al., (2022). Reconnoitering the transformative journey of minocycline from an antibiotic to an antiepileptic drug. *Life Sci., 293*, 120346.
86. Murata, Y., Rosell, A., Scannevin, R. H., Rhodes, K. J., Wang, X., & Lo, E. H., (2008). Extension of the thrombolytic time window with minocycline in experimental stroke. *Stroke, 39*(12), 3372–3377.

87. Takagi, T., & Hara, H., (2016). Protective effects of cilostazol against hemorrhagic stroke: Current and future perspectives. *J. Pharmacol. Sci., 131*(3), 155–161.
88. Tomoda, M., Gonda, R., Kasahara, Y., & Hikino, H., (1986). Glycan structures of ganoderans B and C, hypoglycemic glycans of *Ganoderma lucidum* fruit bodies. *Phytochemistry, 25*(12), 2817–2820.
89. Nagai, N., Yoshioka, C., Ito, Y., Funakami, Y., Nishikawa, H., & Kawabata, A., (2015). Intravenous administration of cilostazol nanoparticles ameliorates acute ischemic stroke in a cerebral ischemia/reperfusion-induced injury model. *International Journal of Molecular Sciences, 16.*
90. Ishiguro, M., Mishiro, K., Fujiwara, Y., Chen, H., Izuta, H., Tsuruma, K., et al., (2010). Phosphodiesterase-III inhibitor prevents hemorrhagic transformation induced by focal cerebral ischemia in mice treated with tPA. *PLoS One, 5*(12), e15178.
91. Mishiro, K., Ishiguro, M., Suzuki, Y., Tsuruma, K., Shimazawa, M., & Hara, H., (2012). A broad-spectrum matrix metalloproteinase inhibitor prevents hemorrhagic complications induced by tissue plasminogen activator in mice. *Neuroscience, 205*, 39–48.
92. Copin, J. C., Merlani, P., Sugawara, T., Chan, P. H., & Gasche, Y., (2008). Delayed matrix metalloproteinase inhibition reduces intracerebral hemorrhage after embolic stroke in rats. *Exp. Neurol., 213*(1), 196–201.
93. Campos, F., Qin, T., Castillo, J., Seo, J. H., Arai, K., Lo, E. H., et al., (2013). Fingolimod reduces hemorrhagic transformation associated with delayed tissue plasminogen activator treatment in a mouse thromboembolic model. *Stroke, 44*(2), 505–511.
94. Knecht, T., Borlongan, C., & Dela, P. I., (2018). Combination therapy for ischemic stroke: Novel approaches to lengthen therapeutic window of tissue plasminogen activator. *Brain Circ., 4*(3), 99–108.
95. Peña, I. D., Borlongan, C., Shen, G., & Davis, W., (2017). Strategies to extend thrombolytic time window for ischemic stroke treatment: An unmet clinical need. *J. Stroke, 19*(1), 50–60.
96. Khan, H., Singh, A., Thapa, K., Garg, N., Grewal, A. K., & Singh, T. G., (2021). Therapeutic modulation of the phosphatidylinositol 3-kinases (PI3K) pathway in cerebral ischemic injury. *Brain Research, 1761*, 147399.
97. Wang, Y., Zhang, Z., Chow, N., Davis, T. P., Griffin, J. H., Chopp, M., et al., (2012). An activated protein C analog with reduced anticoagulant activity extends the therapeutic window of tissue plasminogen activator for ischemic stroke in rodents. *Stroke, 43*(9), 2444–2449.
98. Thiyagarajan, M., Fernández, J. A., Lane, S. M., Griffin, J. H., & Zlokovic, B. V., (2008). Activated protein C promotes neovascularization and neurogenesis in postischemic brain via protease-activated receptor 1. *J. Neurosci., 28*(48), 12788 LP–12797.
99. Ishiguro, M., Kawasaki, K., Suzuki, Y., Ishizuka, F., Mishiro, K., Egashira, Y., et al., (2012). A rho kinase (ROCK) inhibitor, fasudil, prevents matrix metalloproteinase-9-related hemorrhagic transformation in mice treated with tissue plasminogen activator. *Neuroscience, 220*, 302–312.
100. Ishrat, T., Pillai, B., Ergul, A., Hafez, S., & Fagan, S. C., (2013). Candesartan reduces the hemorrhage associated with delayed tissue plasminogen activator treatment in rat embolic stroke. *Neurochem. Res., 38*(12), 2668–2677.
101. Puri, V., Nagpal, M., Sharma, A., Thakur, G. S., Singh, M., & Aggarwal, G., (2020). Development of binary dispersions and nanocomposites of irbesartan with enhanced antihypertensive activity. *Bioimpacts, 10*(4), 269–278.
102. Zuo, W., Chen, J., Zhang, S., Tang, J., Liu, H., Zhang, D., et al., (2014). IMM-H004 prevents toxicity induced by delayed treatment of tPA in a rat model of focal cerebral ischemia involving PKA-and PI3K-dependent Akt activation. *Eur. J. Neurosci., 39*(12), 2107–2118.
103. Tan, Z., Lucke-Wold, B. P., Logsdon, A. F., Turner, R. C., Tan, C., Li, X., et al., (2015). Bryostatin extends tPA time window to 6 h following middle cerebral artery occlusion in aged female rats. *Eur. J. Pharmacol., 764*, 404–412.
104. Arora, A., Aggarwal, G., Singh, T. G., Singh, M., Arora, G., & Manju, N., (2019). Inclusion complexes of atorvastatin calcium–sulfobutyl ether β cyclodextrin with enhanced hypolipidemic activity. *J. Appl. Pharm. Sci., 9*(11), 60–68.
105. Gu, Z., Cui, J., Brown, S., Fridman, R., Mobashery, S., Strongin, A. Y., et al., (2005). A Highly specific inhibitor of matrix metalloproteinase-9 rescues laminin from proteolysis and neurons from apoptosis in transient focal cerebral ischemia. *J. Neurosci., 25*(27), 6401 LP–6408.

CHAPTER 9

Acute Ischemic Stroke in COVID-19 Patients

PIYUSH VATSHA,[1] KAJAL BAGRI,[1] SAKSHI RANA,[1] RAHUL DESHMUKH,[1] and THAKUR GURJEET SINGH[2]

[1]*Maharaja Ranjit Singh Punjab Technical University, Bathinda, Punjab, India*

[2]*Chitkara College of Pharmacy, Chitkara University, Punjab, India*

ABSTRACT

The appearance and spread of the extremely infectious new coronavirus illness (COVID-19) have sparked the century's biggest public health issue. Acute ischemic stroke (AIS) has developed as a consequence of the condition, in addition to being a main respiratory disorder. While current research suggests that COVID-19 may induce ischemic stroke, particularly in severe cases, complications for severe illness COVID-19 and ischemic stroke are comparable, highlighting the complicated link between the two disorders. The epidemic has made acute stroke care more difficult. COVID-19 has hampered the rapid evaluation and time-sensitive therapies necessary for optimal results in acute stroke care owing to the necessity for disease transmission prevention measures. The purpose of this study is to examine the possible causes of ischemic stroke in COVID-19 patients, as well as the clinical characteristics of those who have an ischemic stroke. In addition, we review current care recommendations and assess the challenges of AIS treatment in the context of COVID-19.

9.1 INTRODUCTION

The emergence of coronavirus 2 causing severe acute respiratory syndrome (SARS-CoV-2; formerly known as 2019 novel coronavirus or 2019-nCoV) at the end of 2019 resulted in a significant worldwide epidemic and a serious public health concern. According to statistics from the WHO, there have been 8,769,844 documented cases of COVID-19 in 214 countries/regions, with 4,63,745 fatalities as of June 21, 2020. Human-to-human transmission by direct touch or aerosols is how it spreads [1]. The SARS-CoV-2 infection resulted in clusters of severe respiratory disease, comparable to the coronavirus that resulting in severe acute respiratory syndrome and was linked to ICU admission and high mortality [2]. Fever was the

Ischemic Injury: A Molecular Insight. Thakur Gurjeet Singh & Heena Khan (Eds.)

most prevalent symptom among patients with SARS-CoV-2 pneumonia, followed by cough [3]. The most prevalent observation on computed tomography imaging of the chest was bilateral involvement of the lungs with ground-glass opacity. Lymphopenia is a prevalent symptom in COVID-19 patients, and it may be a key factor linked to disease severity and death [4]. Acute respiratory distress syndrome (29%) was the most common consequence, followed by anemia (15%), acute cardiac damage (12%), and secondary infection (10%). Around 26.1% of COVID-19 patients were admitted to the ICU due to problems [5]. During the clinical treatment of COVID-19, we discovered an intriguing instance in which COVID-19 was followed by the start of ischemic stroke, even though pneumonia is a common consequence following stroke [6, 7]. A 79-year-old man was hospitalized for right leg weakness and subsequently diagnosed with COVID-19 in this case report. Age, oxidative stress, endothelial dysfunction, inflammatory state, and other vascular risk factors all might play a vital role in stroke [8–11]. COVID-19-induced hypoxia and inflammation may have a role in the occurrence, development, and diagnosis of ischemic stroke. This common occurrence shows that neurologists should remain vigilant for COVID-19, particularly in patients who present with neurological symptoms as their first symptom, and take appropriate protective measures.

Acute ischemic stroke (AIS) is a potentially fatal consequence of infection with the coronavirus disease 2019 (COVID-19) [12]. While much about the new coronavirus remains unclear, the association between COVID-19 and ischemic stroke has been widely reported in recent research and the media, whether in young COVID-19 patients or older COVID-19 patients [13] or those without major comorbidities or pre-existing cardiovascular risk factors [14]. The exact prevalence of AIS in COVID-19 patients is yet unknown. The stroke incidence was 2.34% in a study of 214 individuals in Wuhan, the pandemic's first epicenter [15]. In more recent research done in New York, however, the incidence was far lower, at 0.9% [16]. Furthermore, the fundamental mechanism of COVID-19 stroke remains unknown. The elevated concentrations of fibrinogen, factor VIII, and d-dimer [17] support the idea that COVID-19 may create a prothrombotic condition [18]. Various case studies and publications also point to a high prevalence of large vascular occlusion (LVO) AIS. Given the rapidly expanding frequency of COVID-19 infections and AIS problems across the world, a better understanding of their relationship is critical. The pathogenesis of AIS linked with COVID-19 is thought to be related to an associated coagulopathy caused by antiphospholipid antibodies [19] or endotheliopathy [19]. This idea has been seen in critically sick individuals as well as in younger people who have a major artery obstruction [13]. However, it's unclear if people who had an asymptomatic or minimally symptomatic COVID-19 infection are at the same risk for AIS as those who had a severe COVID-19 infection.

9.2 CORONAVIRUS AND SARS-COV-2

There are at least six coronaviruses proven to infect humans include the beta coronaviruses SARS, MERS, and the novel SARS-CoV-2 coronavirus [20]. Coronaviruses are single-stranded RNA viruses with an enclosed positive sense strand. Spike surface glycoprotein

(S), an envelope protein (E), membrane protein (M), and nucleocapsid protein (N) are the four major structural proteins (N). The S protein is responsible for initiating attachment of the virus to the host cell receptor and allowing the virus to infect the cell (Figure 9.1). The virus infects cells through the ACE-2 receptor, which is also the principal inducer of neutralizing antibodies in the case of SARS-CoV-2 [21]. It is believed that the virus spreads through respiratory droplets such as sneezing and coughing as well as through aerosol transmission following prolonged exposure to aerosolized virus in confined spaces. There is a 5.2-day incubation period during which infected people might continue to release considerable quantities of the virus [1]. Tissue factor (TF) expression is known to be induced by viral infections in endothelial cells, lung epithelial cells, and monocytes. There are at least six coronaviruses known to infect humans, including three beta coronaviruses: SARS, MERS, and the novel SARS-CoV-2 coronavirus. Coronaviruses are encased, single-stranded, positive-sense RNA viruses that are transmitted via airborne droplets such as sneezing and coughing as well as aerosol transmission following continued exposure to the aerosolized virus in enclosed environments [22].

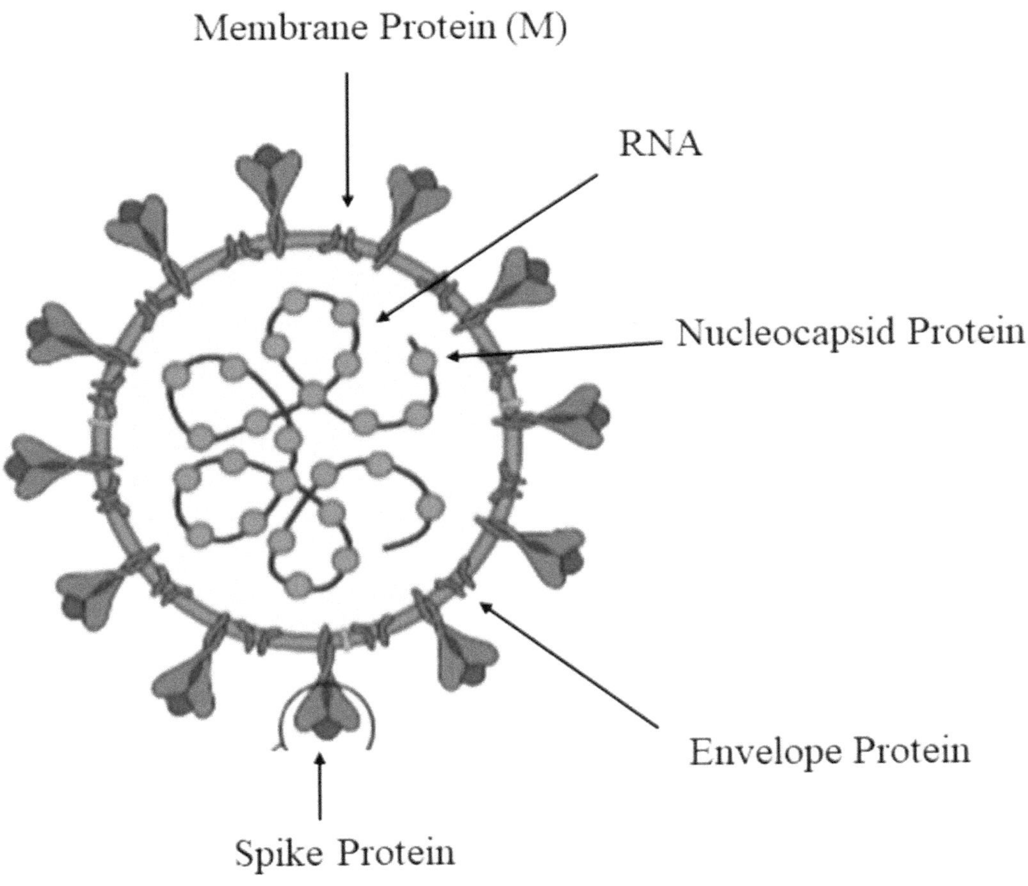

FIGURE 9.1 Structural elements of SARS-CoV-19.

9.3 PATHOGENESIS OF ISCHEMIC STROKE AND COVID-19

The coronavirus that causes severe acute respiratory syndrome (SARS-CoV-2) is an enclosed, single-stranded, positive-sense RNA virus with a spike-protein-containing phospholipid bilayer capsule. The interaction of SARS-spike CoV-2's (S) protein with angiotensin-converting enzyme 2 (ACE2), which is abundant in numerous organs including the lung, kidney, heart, and intestines, allows the virus to enter the body [23]. While the pathogenesis of ischemic stroke in COVID-19 is still being elucidated, existing research points to potential pathways [24–26].

9.3.1 COAGULOPATHY

COVID-19 individuals have coagulopathy with an elevated risk of thromboembolic consequences, according to current research. Severe COVID-19 disease has been linked to "sepsis-induced coagulopathy" (SIC) [27]. SIC is a kind of disseminated intravascular coagulopathy (DIC) that is defined by a high D-dimer, a prolonged prothrombin time, and a low platelet count, but it does not have hypofibrinogenemia [28]. (This is followed by a systemic inflammatory reaction syndrome produced by an immunological response to the virus, which results in endothelial dysfunction and microcirculation thrombosis) [29]. Furthermore, increased synthesis of plasminogen activator inhibitor-1 suppresses fibrinolysis too much, increasing the risk of thrombosis [30]. There is no bleeding, unlike DIC. COVID-19 patients are increasingly reporting arterial and venous thrombosis [31]. Antiphospholipid (APL) antibodies have also been related to a prothrombotic condition for a long time. In a study of 56 patients were diagnosed COVID-19, 14 participants were positive for lupus anticoagulant, whereas 5 exhibited anti-cardiolipin or anti-2-glycoprotein-1 IgM or IgG antibodies [32]. Antibodies to APL are linked to arterial and venous thrombosis, but they're also seen in healthy people, and higher levels have been documented in viral and bacterial infections [33]. It's uncertain if the APL antibodies presence in COVID-19 individuals causes coagulopathy or raises the risk of ischemic stroke (Figure 9.2).

9.3.2 ENDOTHELIAL DYSFUNCTION

The genesis of arterial and venous thrombosis is predisposed by endothelial dysfunction. Pro- and antithrombotic factors are generally in balance in normal vascular endothelium to avoid bleeding and clot formation. The loss of ACE2 receptors, which are abundant in vascular endothelial cells, is one proposed explanation of endothelial dysfunction in COVID-19 patients. The renin-angiotensin-aldosterone system includes a homolog of ACE2 and ACE1. (The ACE2/angiotensin (1–7)/mitochondrial assembly receptor (MasR) axis, that has been recognized as a down regulator of angiotensin II, and the renin/ACE1/angiotensin II/angiotensin II (RAAA) type 1 receptor axis, which has pathological and physiological roles in blood vessel, kidney, brain, and heart diseases [34]. Renin is generated by the renal juxtaglomerular cells and transforms the liver's angiotensinogen

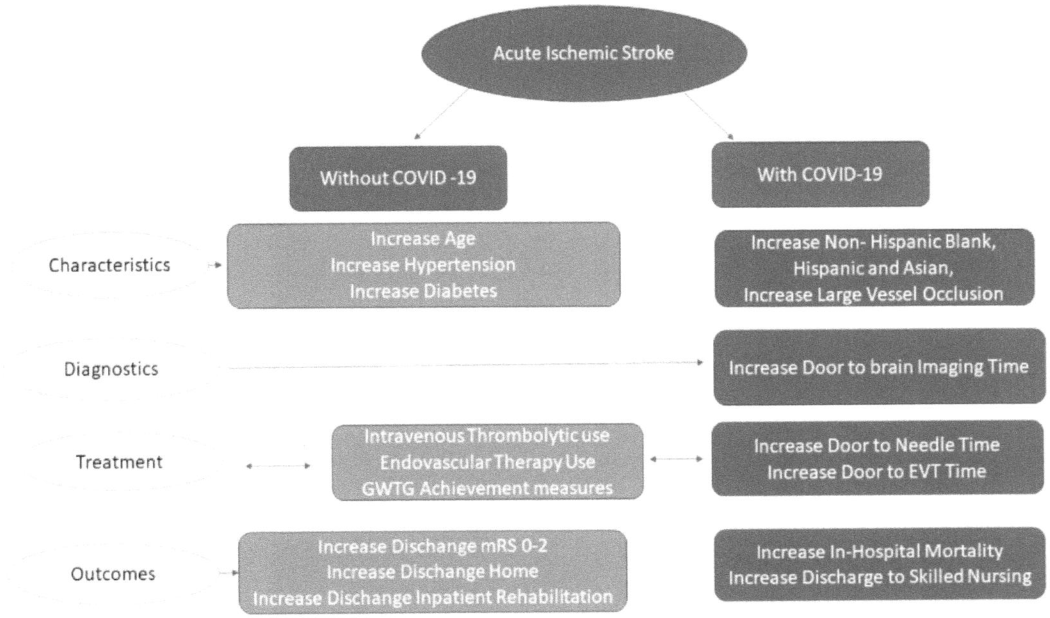

FIGURE 9.2 Pathogenesis of ischemic stroke and COVID-19.

(AGT) to angiotensin I. ACE1 converts angiotensin I to angiotensin II. When it binds to its receptors, Angiotensin II exerts vasoconstrictive, procoagulant, and proinflammatory effects. Angiotensin II receptor inhibitors have been demonstrated to have antihypertensive and neuroprotective properties. ACE2 works by changing angiotensin (AT)II to AT (1–7), an anti-inflammatory and vasodilator, and AT I to AT (1–9), which is then converted to angiotensin (1–7) [35]. The Mas receptor is used by AT (1–7) to modulate its action. SARS-CoV-2 causes receptor endocytosis via binding to ACE2, resulting in depletion of the "protective" endothelium ACE2 and a shift in the balance toward ACE1 and angiotensin II, with heightened proinflammatory disposition and endothelial damage. The immune hyper response known as a cytokine storm is another probable cause of endothelial damage in COVID-19 patients [36, 37]. Unregulated and excessive production of cytokines in COVID-19 is a primary factor in various organ dysfunction. COVID-19 has been shown to have elevated amounts of cytokines such as IL-1, IL-7, IL-8, IL9, IL-10, granulocyte-macrophage colony-stimulating factor (GMCSF), IFN-c, monocyte chemoattractant protein, and tumor necrosis factor (TNF) [38]. Endothelial activation and increased production of endothelial leukocyte adhesion molecules (ELAMs) such as vascular cell adhesion molecule 1 (VCAM-1), E-selectin, and intercellular adhesion molecule 1 (ICAM-1) that interact with leukocyte surface receptors are caused by inflammatory cytokines [39, 40]. The action of cytokines and leukocyte-mediated damage on endothelial cells might compromise their integrity. The exact amount of this mechanism that leads to ischemia in patients with COVID-19 is yet unknown.

9.3.3 CARDIO EMBOLISM

After respiratory failure, cardiac damage has been recognized as one of the primary reasons for mortality in COVID-19 [41, 42]. Increased indicators of cardiac damage, such as creatinine kinase-MB and troponin I&T, have been linked to disease severity in COVID-19 patients. The influence of the cytokine storm and direct virus-mediated damage have been hypothesized as causes of cardiac injury in COVID-19 [43]. The risk of cardiac thromboembolism and stroke is increased by cardiac endothelial dysfunction and arrhythmia. (At this time, the function of this in COVID-19 ischemic stroke is yet speculative.)

9.3.4 DIRECT VIRAL INVASION OF THE CENTRAL NERVOUS SYSTEM

For SARS-CoV-2, two possible CNS invasion pathways have been proposed:

i. Systemic dispersion causes hematogenous dissemination to the cerebrovascular system; and
ii. Transmission from the olfactory epithelium to the olfactory bulb through the cribriform plate.

The slow cerebral microcirculation promotes the interaction between the viral spike proteins and ACE2 are found on cerebral endothelial cells since the virus is prevalent in the general circulation, including the brain circulation [44].

While the specific mechanism of SARS-transcribiform CoV-2's invasion is unknown, a mouse model of SARS-CoV-1 infection, a close "relative" of SARS-CoV-2, displays the virus in parts of the brain adjacent to the olfactory bulb following trans nasal exposure, which is linked to neuronal death [45]. Hypogeusia, Anosmia, and other neuropsychological findings in COVID-19 patients are consistent with the theory that the virus has a direct effect on the brain.

9.3.5 NEUROIMAGING FEATURES

Acute ischemic stroke (AIS) is the most prevalent neuro-radiologic abnormality identified in COVID-19 individuals with neurological symptoms, including the major arteries in 60–65% of cases [13]. Multiple vascular region infarction was common, impacting 27/103 (26%) of patients, but small vessel blockage occurred in just 9/103 (9%) of cases [46]. AIS in the vertebrobasilar area was also prevalent, with 6/17 (35%) of individuals experiencing it. Other publications have verified the common incidence of posterior circulation stroke. Patients with high D-dimer levels, CTA confirmed blockage of the anterior cerebral artery or middle cerebral artery, and episode of suspended thrombi in the aortic valve, common carotid artery strongly imply the participation of hypercoagulable condition. A young man with bilateral cerebellar ischemia lesions attributed to CTA-confirmed occlusion of the left vertebral artery, posterior inferior cerebellar artery (PICA), and bilateral anterior inferior cerebellar artery (AICA) [47]. Three adult and two pediatric patients with AIS were found to have vasculitis or a vasculitis-like phenotype (including pediatric localized cerebral arteriopathy). On magnetic resonance imaging (MRI), two of these instances revealed arterial vessel wall enhancement, which is consistent with inflammation. The vasculitic process was widespread in the adult patient, affecting the ACA bilaterally, the MCA, the vertebral arteries, and the basilar artery. Arteriopathy has been linked to several viral infections, including VZV, HIV, herpesviruses, and influenza A. Multiple ischemic strokes in vascular territories are a common feature of VZV vasculopathy [48]. Although data on intracerebral hemorrhage in COVID-19 patients is scarce, existing data indicate that hemorrhagic episodes occur in 0.5% to 0.9% of individuals. Hemorrhagic stroke has been documented in SARS-CoV-2 infected individuals in the range of 21.7% (5/23) to 25.7% (9/35). Multiple hematomas may develop in both supra- and infra-tentorial sites, and hemorrhage can be substantial with widespread hemisphere involvement [48]. In critically ill patients, parenchymal hemorrhages can occur spontaneously, especially in the context of circulation instability and multi-organ failure (MOF) [48], or as a problem of other pathologic conditions, such as the hemorrhagic transformation of acute stroke, rupture of pseudo-aneurism, or hemorrhagic infarction linked with cerebral venous sinus thrombosis (CVT) [49]. With just 18 individuals recorded, CVT seems to be uncommon in COVID-19 [50] revealed concurrent arterial and venous thrombosis in one elderly patient with ischemic lesions owing to CTA-demonstrated bilateral blockage of the MCA and a right sigmoid sinus filling defect consistent with thrombosis. Surprisingly, several case reports imply a link between COVID-19 and extracranial vertebral artery or carotid artery dissection in people with otherwise normal medical histories or cardiovascular risk factors. In one instance, cervical vertebral artery dissection was linked to reversible cerebral vasoconstriction syndrome (RCVS) affecting the anterior circulation and bilateral frontal subarachnoid hemorrhage as established by digital subtraction angiography (SAH). They presented a case of serious SAH in the posterior fossa and fourth ventricle as a result of a ruptured PICA dissecting aneurysm [51]. It's likely that SARS-CoV-2 causes dissection of the arterial wall or rupture with bleeding by interacting with ACE2 on the surface of endothelial and arterial smooth muscle cells. Furthermore, as previously described, dysregulation of the renin-angiotensin system leads to increased availability of angiotensin II, which causes vasoconstriction and immunological activation, resulting in a loss of self-control, a change in the BBB, vasogenic edema and cerebral hypoperfusion [52, 53]. These processes give a theoretical foundation for understanding the neuro-radiological results mentioned in the literature.

9.4 MECHANISMS OF STROKE IN COVID-19

Many theories have been offered to explain ischemic stroke in the context of COVID-19 infection, including the link between severe COVID-19 illness happening particularly to older people who are more likely to have additional comorbidities that increase their stroke risk. However, data from outside of China reveal that COVID-19-positive individuals with LVO have fewer pre-existing cardiac comorbidities and are younger than COVID-19-negative patients [54]. There is an interaction of the suggested processes to explain COVID-19-associated stroke, rather than a single route, which would account for the variation found in the literature between sites. The development of a cytokine storm and activation of the innate defense system, embolic events perpetuated by pre-existing or onset arrhythmias, hypoxia-induced secondary ischemia to severe respiratory disease, thromboembolic microangiopathy, endothelialitis, and multifactorial coagulation stimulation are among the key proposed mechanisms [55, 56]. The findings which showed that stroke occur in 5.7% of patients with critical illness compared to 0.8% of patients with milder COVID-19 disease support the hypothesis that COVID-19 contributes to risk factors associated with stroke depends differently on the severity of illness, with patients with more serious illness having a higher risk of stroke [57]. Elevated D-dimer levels in COVID-19-infected patients experiencing an AIS, suggesting activation of the coagulation and innate immune systems, seem to be a unifying element in the presently available material. More than 90% of COVID-19-positive patients in New York who arrived with urgent LVO had increased D-dimer values, and 42% showed AIS symptoms rather than normal respiratory symptoms [58]. Furthermore, individuals who had a stroke before developing COVID-19 symptoms had a higher risk of thrombosis, showing that even in the early stages of the illness, there may be an elevated risk of thrombosis. According to a recent study, individuals with severe COVID-19 illness had higher D-dimer levels than those with mild COVID-19 disease [59]. While D-dimer levels seem to be much higher in patients with severe illness, and stroke appears to occur more often in very sick COVID-19-infected individuals, it is probable that they are also raised in patients with less serious COVID-19 infection. Another factor to consider is the effect of a cytokine storm that results in raised levels of IL-6 and C-reactive protein, both of which have been linked to an increased risk of stroke and myocardial infarction in healthy people [60]. COVID-19 may enhance the risk of stroke by a two-fold mechanism involving the expression and viral uptake of the ACE-2 receptor. First, viral-induced vasculitis might be caused by direct infection of the brain endothelium that expresses the ACE-2 receptor [61, 62]. Second, ACE-2 receptor inhibition caused by a viral infection and cell entry endocytosis might reduce ACE-2 function, reducing angiotensin (1–7) production and compromising cerebral vascular autoregulation. Finally, hypoxia caused by severe COVID-19 lung infection may increase the risk of stroke by reducing oxygen supply (Figure 9.3).

9.5 RISK OF STROKE AND COVID-19 PATIENTS

Stroke is an uncommon, but potentially life-threatening, COVID-19 consequence, affecting hospitalized patients, i.e., 1–3 and up to 6% of those in the ICU [63]. Due to the fact that male patients are more prone to experience severe conditions like SARS-CoV-2 symptoms that need ICU hospitalization, it's no surprise that the majority of patients who suffer a stroke during COVID-19 are male (62%) and have a median age of 63 years [46]. Importantly, most of the patients already had significant vascular risk factors (in particular, HT and DM) [1, 64, 65], thus the infection may merely be a trigger rather than an independent cause in this group. Nonetheless, the consequence of COVID-19 as a stroke trigger should not be overlooked, given that SARS-CoV-2 infection was linked to a 7.6-fold upsurge in the risk of cerebrovascular consequences when compared to influenza [66]. Symptoms of COVID-19 infection, such as respiratory symptoms and fever, often showed before the beginning of the stroke, which occurred 10 days later (delayed presentation) [66]. Race/ethnicity is a significant contributing factor for COVID-19 infection, perhaps because cardiovascular risk factors are overrepresented in particular ethnic groups (e.g., African-Americans). In research covering 7,868 individuals hospitalized with COVID-19 [67], black patients had the greatest rate of obesity, hypertension, and diabetes, presumably due to long-standing racial/ethnic health and socioeconomic imbalances (Figure 9.4). The unequal distribution of various ethnic groups between patients with COVID-19-associated stroke is a result of these conditions. In recent research comprising 83 instances, 47% were Black, 28% Hispanic, and 16% White. Young-onset stroke patients [13] were also found, and this group have specific clinical features:

- Prior risk factors and comorbidity appear uncommon or nonexistent;
- Stroke typically occurs prior to the advent of COVID-19 symptoms;
- Occlusion of the large-vessel appears to be common; and
- It is hypothesized that COVID-19 plays a major role in causing stroke in this cohort of otherwise healthy individuals.

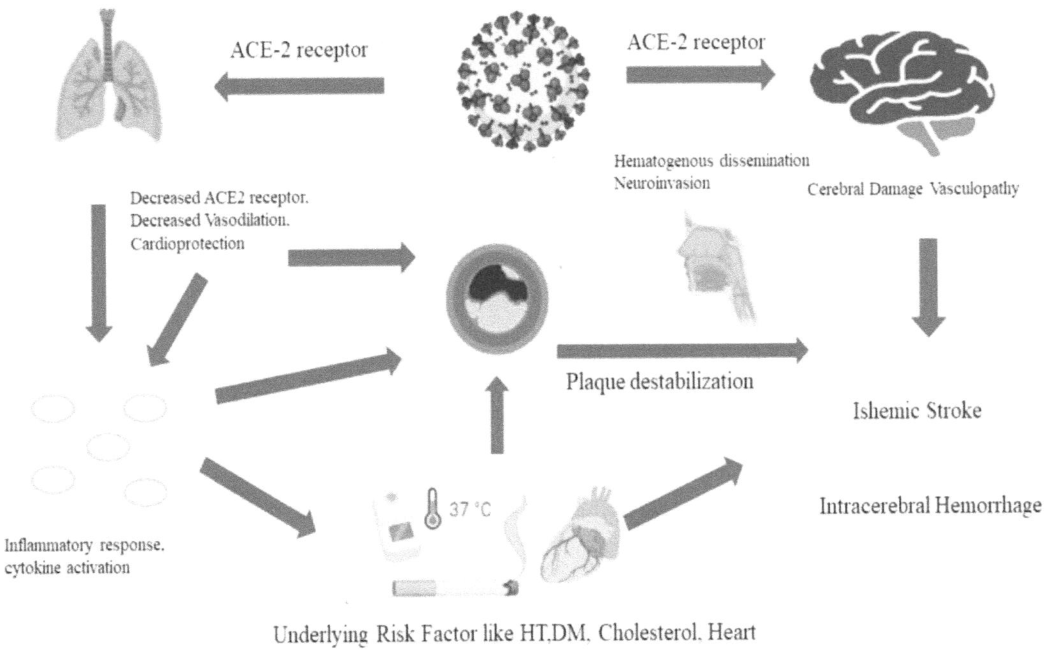

FIGURE 9.3 Management of stroke in the setting of COVID-19.

9.6 CLINICAL PRESENTATION

Patients with COVID-19 appeared especially susceptible to (i) Occlusion of large vessel (including internal carotid artery occlusion, M1 and M2 segments of the MCA, and basilar artery); (ii) involvement of multi-territory; and (iii) involvement of otherwise uncommonly affected vessels [13, 48], including for example the occlusion of the pericallosal artery [48],

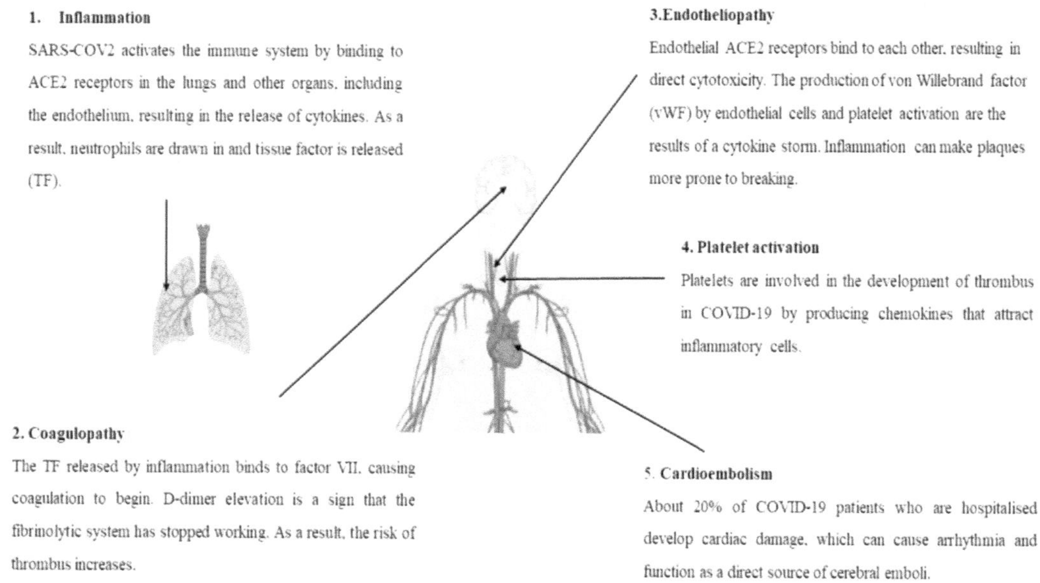

FIGURE 9.4 Risk assessment of stroke in COVID-19 patients.

or the presence of multiple focal stenoses in the V4 segment of the vertebral artery [13]. The subsequent neurological loss was often severe (NIHSS, ranging from 19 to 20), and almost a quarter of the cases exhibited signs of systemic thrombosis, such as venous thrombosis, pulmonary embolism, and spleen embolism [46, 48]. More than 40% of the patients were diagnosed with cryptogenic stroke, which was typically accompanied by an embolic radiologic appearance. A small vessel pattern, on the other hand, was noted occasionally [46]. Coagulopathy (raised fibrinogen, elevated D-dimer), a significant prevalence of hepatic and renal failure, and elevated lactate dehydrogenase (LDH) were also found in the lab [48]. Antiphospholipid antibodies (lupus anticoagulant, anticardiolipin, and anti-2-glycoprotein antibodies) were found in a large number of instances [46]. SARS-CoV-2 coagulopathy differed from that reported in infections with other RNA viruses like Ebola and other hemorrhagic fever viruses in the absence of bleeding signs [68]. The unique features of stroke in COVID-19 individuals need an ad hoc approach to diagnosis and treatment. The prerequisite for a "protected code stroke," as well as the necessity to promptly identify patients with infection (by nasopharyngeal swab) and those with concomitant pneumonia (pulmonary imaging utilizing chest CT may be integrated into the first imaging for stroke) are underlined [69]. Given the relatively high prevalence of COVID-19-related acute renal damage, extra caution should be taken to avoid (and, ultimately, treat) contrast-induced nephropathy in CT angiography (CTA) patients [69]. Patients with suspected major artery blockage who are considering mechanical thrombectomy, including those with SARS-CoV-2 infection, should have a CTA. The identification of clinical indications indicative of stroke in ICU patients (intubated and sedated) might be especially difficult, hence regular sedation interruption for neurologic examination ("wake up test") should be scheduled [69]. According to available resources, the incidence of the wake-up test should be increased, and concentrated training may be required. Unless there are particular contraindications, it is fair to recommend prompt preventive anticoagulation with low-molecular-weight heparin due to the extremely prothrombotic condition. Full-intensity anticoagulation, on the other hand, is not indicated for all COVID-19 patients in the hospital, but only for those with a clinical justification (e.g., pulmonary embolism) [68]. Furthermore, quantifying the stroke burden/size and assessing hemorrhagic conversion are two critical considerations to consider before initiating anticoagulation. Because of the unique characteristics of COVID-19-related stroke, researchers devised distinct classifications of potential and likely instances.

9.7 TREATMENT OF ACUTE ISCHEMIC STROKE (AIS) IN COVID-19

9.7.1 *THROMBOPROPHYLAXIS*

Although there are insufficient data to recommend an ideal medical treatment regimen for patients with severe ischemia or to inhibit cerebral infarction among hospitalized patients with COVID-19, treatment paradigms to prevent ischemic stroke have remained largely unchanged since the advent of COVID. One research indicated that hospitalized COVID-19 patients had AIS despite being on anticoagulant therapy [70]. Table 9.1 shows the current antiplatelets and anticoagulants in use, as well as their mechanisms of action and suggested CAC mitigation mechanisms. In individuals with COVID-19, it's uncertain whether anticoagulants or antiplatelet medications provide better primary or secondary stroke protection. Further studies might examine the anticoagulation effectiveness in preventing stroke in the context of COVID-19, especially considering the worry about activation of the TF-dependent coagulation pathway. A case series survey of patients who had mild COVID-19 experienced macrothrombosis and stroke used apixaban 5 mg twice per day in one patient and enoxaparin 1 mg/kg every 12 hours in another patient, both of whom had non-occlusive thrombus involving the internal carotid artery. At 10 and 7 days, respectively, the thrombus had nearly completely resolved [71]. This is not to argue that anticoagulation should be used for all stroke patients, especially as anticoagulation has been found as an independent risk factor for cerebral hemorrhage in COVID-19 patients [72]. Instead, it emphasizes the need of focusing future research on the anticoagulants usage in treating acute ischemia in COVID-19-positive individuals who are not candidates for presently indicated treatments (such as rtPA or thrombectomy). They provided a recommended clinical treatment route for COVID-19 patients in the hospital, which directs therapy with preventive vs therapeutic enoxaparin and heparin, as well as rtPA, in COVID-19 patients in the hospital [73]. Low molecular weight heparin, in particular, seems to be linked with a better prognosis than other anticoagulants in hospitalized patients with severe illness or increased D-dimer [27]. The function of platelet hyperactivity in platelet activation/propagation of platelet aggregation through the inflammatory and immunological response, however, may support the notion that antiplatelet therapy is a better therapeutic option. Other hypotheses propose that addressing the underlying hyperinflammatory state with immune modifying medications may operate synergistically with antithrombotic or antiviral medicines in individuals with severe illness by treating the thrombosis's primary source. In patients with ARDS, another option is to utilize 'targeted' COVID-19 treatment that targets ACE-2 and the renin-angiotensin system [28].

TABLE 9.1 Antithrombotic Mediators and Possible Mechanisms of Mitigating CAC

	Drug	Mechanism of Action	Mechanism of Inhibiting CAC	References
Antiplatelet	Aspirin	Thromboxane A2	Platelet activation	[28]
	–	–	Neutrophil activation	
	Clopidogrel	P2Y12	NETosis	
	Prasugrel	P2Y12	Cytokine storm	
Anticoagulant	Heparin	Factor IIa	Thrombin activity	[70]
	–	–	Thrombin production	
	LMWH	Factor Xa	Platelet activation	
	Warfarin	Factor II, VII, IX, X, proteins C and S	Neutrophil activation	
	Apixaban	Factor Xa	Thrombin activity	
	Rivaroxaban	Factor Xa	Thrombin production	
	Edoxaban	Factor Xa	Platelet activation	
			Neutrophil activation	
			Virus spike protein activity	

9.7.2 INTRAVENOUS THROMBOLYSIS

In selected patients who can be cured within 3 hours and substantially chosen individuals who can be treated within 3–4.5 hours of onset, existing global standards advise intravenous thrombolysis [69]. This suggestion is competent by the finding that patients with COVID-19 have a higher incidence of rising levels of inflammation-related and hypercoagulability indicators. These indicators have been linked to an increased risk of death as well as post-thrombolytic intracranial hemorrhage in patients with acute ischemia but without COVID-19 virus. Before administering rtPA, the authors advise evaluating coagulopathy and liver function to help determine the risk/benefit ratio.

9.7.3 ANESTHESIA FOR MECHANICAL THROMBECTOMY

Regarding mechanical thrombectomy, it is advised to stick to the most recent clinical recommendations for inclusion criteria and to take the necessary safety precautions, such as using personal protective equipment for everyone inside the angiography suite. Criteria to better distinguish patients as candidates for general anesthesia versus monitored anesthesia care have been suggested in a consensus statement from the Society for Neuroscience in Anesthesiology and Critical Care (SNAC) [74]. Whereas many centers adopted a policy of intubating every patient prior to mechanical thrombectomy, preliminary data suggest that the use of general anesthesia and intubation enhanced the door-to-reperfusion time

(138 vs 100 minutes) and was linked to a greater likelihood of in-hospital mortality and a lower risk of functional capacity at discharge [75]. In contrast to previous research, which revealed no difference in long-term results between individuals undergoing mechanical thrombectomy with general anesthesia and those receiving monitored anesthetic treatment, this study found a significant difference [76]. Intriguingly, COVID-19 status was not statistically linked with in-hospital mortality or functional independence, despite the fact that almost half of patients were not evaluated and a discharge modified Rankin Scale score was unavailable for 23.6% of patients. It is advised that patients requiring general anesthesia be intubated in a negative pressure environment, which may require intubation in the emergency room prior to transport to the angiography lab.

9.7.4 IMMUNOGLOBULIN THERAPY

Due to the lack of a viable vaccination or treatment for COVID-19, convalescent plasma is being tested on critically ill COVID-19 patients [77]. IVIg medication (Intravenous immunoglobulin) has been utilized both for postexposure prophylaxis and for the treatment of virus infections. The probability of an intravenous immunoglobulin-associated thromboembolic event is estimated to be between 0.6% and 4.5% with IVIg therapy. Five patients with severe acute respiratory distress syndrome (SARS) who had IVIg therapy developed large-artery cerebral infarctions, according to a study of 206 SARS patients [78]. Critically ill patients are already at a higher risk of thrombosis; hence, the use of IVIg in these patients may raise the probability of thromboembolic events. Currently, limited data is here to support the use of IVIg treatment in patients with COVID-19 and ischemic stroke.

9.8 CHALLENGES OF MANAGING STROKE IN THE SETTING OF COVID-19

COVID-19 has an impact on stroke care both directly and indirectly. Since the disease's onset, numerous steps have been put in place to battle the epidemic of this infection, including lockdown procedures, the suspension of elective surgeries and regular outpatient clinics, as well as the redeployment of medical resources [79]. The World Stroke Organization (WSO) reports that many nations are dealing with overcrowded hospital systems, which forces hospitals to divert beds for neurology and stroke patients as well as ICU space intended for stroke care to COVID-19 patients [80, 81]. Additionally, many hospitals redeployed stroke doctors and nurses to provide COVID-19 care [82, 83]. According to an observatory study including 280 Chinese facilities, hospital admissions for stroke decreased by 40% in February 2020 compared to February 2019 while the number of instances of thrombolysis and thrombectomy decreased by 25% [79]. Additionally, healthcare professionals who come into contact with patients who are infected run the danger of contracting COVID-19. There have been cases of COVID-19 infection and fatalities among medical professionals reported across the globe. Redeployment could result in quarantine or even hospital admission if infection spreads among stroke care team members who are already in limited supply. This would further deplete the pool of human resources available for stroke care.

9.9 CONCLUSION AND RECOMMENDATION FOR FUTURE RESEARCH

Coagulopathy, endothelial dysfunction, Cardioembolism, and direct viral-mediated neuronal damage have all been recognized as COVID-19 consequences, with endothelial dysfunction, coagulopathy, cardioembolism, and direct viral-mediated neuronal injury as plausible underlying causes. This consequence is more prevalent in older individuals with severe illness, although large-vessel blockage is sometimes seen in younger people. COVID-19 has hindered the rapid evaluation and management necessary for optimal results in hyperacute stroke therapy. While providing proper care for these individuals, infection control measures should be implemented. Substantial research is required to better comprehend how COVID-19 causes stroke and the association between younger age and large-vessel stroke. Furthermore, the role of anticoagulants in the prevention and treatment of stroke in COVID-19 patients should be investigated.

KEYWORDS

- **acute ischemia**
- **angiotensin-converting enzyme-II**
- **cardio embolism**
- **COVID-19**
- **immunoglobulin therapy**
- **intravenous thrombolysis**
- **stroke**

REFERENCES

1. Li, Q., Guan, X., Wu, P., Wang, X., Zhou, L., Tong, Y., Ren, R., et al., (2020). Early transmission dynamics in Wuhan, China, of novel coronavirus–infected pneumonia. *New England Journal of Medicine*.
2. Huang, C., Wang, Y., Li, X., Ren, L., Zhao, J., Hu, Y., Zhang, L., et al., (2020). Clinical features of patients infected with 2019 novel coronavirus in Wuhan, China. *The Lancet*, *395*(10223), 497–506.
3. Li, Y., Li, M., Wang, M., Zhou, Y., Chang, J., Xian, Y., Wang, D., et al., (2020). Acute cerebrovascular disease following COVID-19: A single center, retrospective, observational study. *Stroke and Vascular Neurology*, *5*(3).
4. Chan, J. F. W., Yuan, S., Kok, K. H., To, K. K. W., Chu, H., Yang, J., Xing, F., Liu, J., Yip, C. C. Y., & Poon, R. W. S., (2020). A familial cluster of pneumonia associated with the 2019 novel coronavirus indicating person-to-person transmission: A study of a family cluster. *The Lancet*, *395*(10223), 514–523.

5. Wang, D., Hu, B., Hu, C., Zhu, F., Liu, X., Zhang, J., Wang, B., et al., (2020). Clinical characteristics of 138 hospitalized patients with 2019 novel coronavirus–infected pneumonia in Wuhan, China. *JAMA, 323*(11), 1061–1069.
6. Hoffmann, M., Kleine-Weber, H., Krüger, N., Müller, M., & Drosten, C., (n.d.). Pöhlmann, VOP 2020 *The Novel Coronavirus (2019-nCoV) Uses the SARS-Coronavirus Receptor ACE2 and the Cellular Protease TMPRSS2 for Entry into Target Cells*. BioRxiv.
7. Khan, H., Kashyap, A., Kaur, A., & Singh, T. G., (2020). Pharmacological postconditioning: A molecular aspect in ischemic injury. *Journal of Pharmacy and Pharmacology, 72*(11), 1513–1527.
8. Regenhardt, R. W., Das, A. S., Lo, E. H., & Caplan, L. R., (2018). Advances in understanding the pathophysiology of lacunar stroke: A review. *JAMA Neurology, 75*(10), 1273–1281.
9. Ham, I. I. I., P. B., & Raju, R., (2017). Mitochondrial function in hypoxic ischemic injury and influence of aging. *Progress in Neurobiology, 157*, 92–116.
10. Prabhakar, N. K., Khan, H., Grewal, A. K., & Singh, T. G., (2022). Intervention of neuroinflammation in the traumatic brain injury trajectory: *In vivo* and clinical approaches. *International Immunopharmacology, 108*, 108902.
11. Kaur, A., Singh, T. G., Khan, H., Kumar, M., Singh, N., & Abdel-Daim, M. M., (2022). Neuroprotective effect of piclamilast-induced post-ischemia pharmacological treatment in mice. *Neurochemical Research*, 1–14.
12. Avula, A., Nalleballe, K., Narula, N., Sapozhnikov, S., Dandu, V., Toom, S., Glaser, A., & Elsayegh, D., (2020). COVID-19 presenting as stroke. *Brain, Behavior, and Immunity, 87*, 115–119.
13. Oxley, T. J., Mocco, J., Majidi, S., Kellner, C. P., Shoirah, H., Singh, I. P., De Leacy, R. A., et al., (2020). Large-vessel stroke as a presenting feature of COVID-19 in the young. *New England Journal of Medicine, 382*(20), e60.
14. Viguier, A., Delamarre, L., Duplantier, J., Olivot, J. M., & Bonneville, F., (2020). Acute ischemic stroke complicating common carotid artery thrombosis during a severe COVID-19 infection. *Journal of Neuroradiology, 47*(5), 393–394.
15. Mao, L., Jin, H., Wang, M., Hu, Y., Chen, S., He, Q., Chang, J., et al., (2020). Neurologic manifestations of hospitalized patients with coronavirus disease 2019 in Wuhan, China. *JAMA Neurology, 77*(6), 683–690.
16. Yaghi, S., Ishida, K., Torres, J., Mac, G. B., Raz, E., Humbert, K., Henninger, N., et al., (2020). SARS-CoV-2 and stroke in a New York healthcare system. *Stroke, 51*(7), 2002–2011.
17. Panigada, M., Bottino, N., Tagliabue, P., Grasselli, G., Novembrino, C., Chantarangkul, V., Pesenti, A., et al., (2020). Hypercoagulability of COVID-19 patients in intensive care unit: A report of thromboelastography findings and other parameters of hemostasis. *Journal of Thrombosis and Haemostasis, 18*(7), 1738–1742.
18. Valderrama, E. V., Humbert, K., Lord, A., Frontera, J., & Yaghi, S., (2020). Severe acute respiratory syndrome coronavirus 2 infection and ischemic stroke. *Stroke, 51*(7), e124–e127.
19. Zhang, Y., Xiao, M., Zhang, S., Xia, P., Cao, W., Jiang, W., Chen, H., et al., (2020). Coagulopathy and antiphospholipid antibodies in patients with COVID-19. *New England Journal of Medicine, 382*(17), e38.
20. Arabi, Y. M., Harthi, A., Hussein, J., Bouchama, A., Johani, S., Hajeer, A. H., Saeed, B. T., et al., (2015). Severe neurologic syndrome associated with middle east respiratory syndrome corona virus (MERS-CoV). *Infection, 43*(4), 495–501.
21. Mousavizadeh, L., & Ghasemi, S., (2021). Genotype and phenotype of COVID-19: Their roles in pathogenesis. *Journal of Microbiology, Immunology, and Infection, 54*(2), 159–163.
22. Cascella, M., Rajnik, M., Aleem, A., Dulebohn, S., & Di Napoli, R., (2022). Features, evaluation, and treatment of coronavirus (COVID-19). *StatPearls*.
23. Lu, R., Zhao, X., Li, J., Niu, P., Yang, B., Wu, H., Wang, W., et al., (2020). Genomic characterization and epidemiology of 2019 novel coronavirus: Implications for virus origins and receptor binding. *The Lancet, 395*(10224), 565–574.
24. Khan, H., Garg, N., Singh, T. G., Kaur, A., & Thapa, K., (2022a). Calpain inhibitors as potential therapeutic modulators in neurodegenerative diseases. *Neurochemical Research*, 1–25.
25. Khan, H., Grewal, A. K., & Singh, T. G., (2022b). Pharmacological postconditioning by protocatechuic acid attenuates brain injury in ischemia–reperfusion (I/R) mice model: Implications of nuclear factor erythroid-2-related factor pathway. *Neuroscience, 491*, pp.23–31.

26. Sharma, A., Khan, H., Singh, T. G., Grewal, A. K., Najda, A., Kawecka-Radomska, M., Kamel, M., et al., (2021). Pharmacological modulation of ubiquitin-proteasome pathways in oncogenic signaling. *International Journal of Molecular Sciences*, *22*(21), 11971.
27. Tang, N., Bai, H., Chen, X., Gong, J., Li, D., & Sun, Z., (2020). Anticoagulant treatment is associated with decreased mortality in severe coronavirus disease 2019 patients with coagulopathy. *Journal of Thrombosis and Haemostasis*, *18*(5), 1094–1099.
28. Hess, D. C., Eldahshan, W., & Rutkowski, E., (2020). COVID-19-related stroke. *Translational Stroke Research*, *11*(3), 322–325.
29. Iba, T., Levy, J. H., Warkentin, T. E., Thachil, J., Van, D. P. T., & Levi, M., (2019). Diagnosis and management of sepsis-induced coagulopathy and disseminated intravascular coagulation. *Journal of Thrombosis and Haemostasis: JTH*.
30. Koyama, K., Madoiwa, S., Nunomiya, S., Koinuma, T., Wada, M., Sakata, A., Ohmori, T., Mimuro, J., & Sakata, Y., (2014). Combination of thrombin-antithrombin complex, plasminogen activator inhibitor-1, and protein C activity for early identification of severe coagulopathy in initial phase of sepsis: A prospective observational study. *Critical Care*, *18*(1), 1–11.
31. Klok, F. A., Kruip, M., Van, D. M. N. J. M., Arbous, M. S., Gommers, D., Kant, K. M., Kaptein, F. H. J., et al., (2020). Incidence of thrombotic complications in critically ill ICU patients with COVID-19. *Thrombosis Research*, *191*, 145–147.
32. Harzallah, I., Debliquis, A., & Drénou, B., (2020). Lupus anticoagulant is frequent in patients with COVID-19: Response to reply. *Journal of Thrombosis and Haemostasis*.
33. Asherson, R. A., & Cervera, R., (2003). Antiphospholipid antibodies and infections. *Annals of the Rheumatic Diseases*, *62*(5), 388–393.
34. Xu, J., Fan, J., Wu, F., Huang, Q., Guo, M., Lv, Z., Han, J., et al., (2017). The ACE2/angiotensin-(1–7)/mas receptor axis: Pleiotropic roles in cancer. *Frontiers in Physiology*, *8*, 276.
35. Mecca, A. P., Regenhardt, R. W., O'Connor, T. E., Joseph, J. P., Raizada, M. K., Katovich, M. J., & Sumners, C., (2011). Cerebroprotection by angiotensin-(1–7) in endothelin-1-induced ischemic stroke. *Experimental Physiology*, *96*(10), 1084–1096.
36. Mehta, P., McAuley, D. F., Brown, M., Sanchez, E., Tattersall, R. S., & Manson, J. J., (2020). COVID-19: Consider cytokine storm syndromes and immunosuppression. *The Lancet*, *395*(10229), 1033–1034.
37. Thapa, K., Khan, H., Singh, T. G., & Kaur, A., (2021). Traumatic brain injury: Mechanistic insight on pathophysiology and potential therapeutic targets. *Journal of Molecular Neuroscience*, *71*(9), 1725–1742.aaa.
38. Ifeanyi, O. E., Mercy, O. H., Prayer, N. N., & Chijindu, O. H., (2020). Cytokines, coagulation profile and haematological changes in covid 19 patients as indicators of their health staus: A review. *European Journal of Biomedical*, *7*(7), 724–729.
39. Zhang, C., (2008). The role of inflammatory cytokines in endothelial dysfunction. *Basic Research in Cardiology*, *103*(5), 398–406.
40. Khan, H., Tiwari, P., Kaur, A., & Singh, T. G., (2021a). Sirtuin acetylation and deacetylation: A complex paradigm in neurodegenerative disease. *Molecular Neurobiology*, *58*(8), 903–3917.
41. Ruan, Q., Yang, K., Wang, W., Jiang, L., & Song, J., (2020). Clinical predictors of mortality due to COVID-19 based on an analysis of data of 150 patients from Wuhan, China. *Intensive Care Medicine*, *46*(5), 846–848.
42. Khan, H., Sharma, R., Kaur, A., & Singh, T. G., (2018). The endocannabioids system and their implications in various disorders. *Int. J. Pharm. Sci. Rev. Res.*
43. Akhmerov, A., & Marbán, E., (2020). COVID-19 and the heart. *Circulation Research*, *126*(10), 1443–1455.
44. Baig, A. M., Khaleeq, A., Ali, U., & Syeda, H., (2020). Evidence of the COVID-19 virus targeting the CNS: Tissue distribution, host–virus interaction, and proposed neurotropic mechanisms. *ACS Chemical Neuroscience*, *11*(7), 995–998.
45. Kalra, P., Khan, H., Kaur, A., & Singh, T. G., (2022). Mechanistic insight on autophagy modulated molecular pathways in cerebral ischemic injury: From preclinical to clinical perspective. *Neurochemical Research*, 1–19.
46. Tan, Y. K., Goh, C., Leow, A. S. T., Tambyah, P. A., Ang, A., Yap, E. S., Tu, T. M., et al., (2020). COVID-19 and ischemic stroke: A systematic review and meta-summary of the literature. *Journal of Thrombosis and Thrombolysis*, *50*(3), 587–595.

47. Cavallieri, F., Marti, A., Fasano, A., Dalla, S. A., Ghirarduzzi, A., Moratti, C., Bonacini, L., et al., (2020). Prothrombotic state induced by COVID-19 infection as trigger for stroke in young patients: A dangerous association. *eNeurologicalSci., 20.*
48. Morassi, M., Bagatto, D., Cobelli, M., D'Agostini, S., Gigli, G. L., Bnà, C., & Vogrig, A., (2020). Stroke in patients with SARS-CoV-2 infection: Case series. *Journal of Neurology, 267*(8), 2185–2192.
49. Poillon, G., Obadia, M., Perrin, M., Savatovsky, J., & Lecler, A., (2021). Cerebral venous thrombosis associated with COVID-19 infection: Causality or coincidence? *Journal of Neuroradiology, 48*(2), 121.
50. Malentacchi, M., Gned, D., Angelino, V., Demichelis, S., Perboni, A., Veltri, A., Bertolotto, A., & Capobianco, M., (2020). Concomitant brain arterial and venous thrombosis in a COVID-19 patient. *European Journal of Neurology.*
51. Al Saiegh, F., Ghosh, R., Leibold, A., Avery, M. B., Schmidt, R. F., Theofanis, T., Mouchtouris, N., et al., (2020). Status of SARS-CoV-2 in cerebrospinal fluid of patients with COVID-19 and stroke. *Journal of Neurology, Neurosurgery & Psychiatry, 91*(8), 846–848.
52. Bartynski, W. S., (2008). Posterior reversible encephalopathy syndrome, part 2: Controversies surrounding pathophysiology of vasogenic edema. *American Journal of Neuroradiology, 29*(6), 1043–1049.
53. Gupta, A., Khan, H., Kaur, A., & Singh, T. G., (2021). Novel targets explored in the treatment of alcohol withdrawal syndrome. *CNS & Neurological Disorders-Drug Targets (Formerly Current Drug Targets-CNS & Neurological Disorders), 20*(2), 158–173.
54. Sweid, A., Hammoud, B., Weinberg, J. H., Oneissi, M., Raz, E., Shapiro, M., DePrince, M., et al., (2020). Thrombotic neurovascular disease in COVID-19 patients. *Neurosurgery, 87*(3), E400–E406.
55. Khan, H., Gupta, A., Singh, T. G., & Kaur, A., (2021). Mechanistic insight on the role of leukotriene receptors in ischemic–reperfusion injury. *Pharmacological Reports, 73*(5), 1240–1254.
56. Khan, H., Sharma, K., Kumar, A., Kaur, A., & Singh, T. G., (2022). Therapeutic implications of cyclooxygenase (COX) inhibitors in ischemic injury. *Inflammation Research*, 1–16.
57. MW, M. L., (2020). Neurological manifestations of hospitalized patients with COVID-19 in Wuhan. *China: A Retrospective Case Series Study.*
58. Majidi, S., Fifi, J. T., Ladner, T. R., Lara-Reyna, J., Yaeger, K. A., Yim, B., Dangayach, N., et al., (2020). Emergent large vessel occlusion stroke during New York City's COVID-19 outbreak: Clinical characteristics and paraclinical findings. *Stroke, 51*(9), 2656–2663.
59. Violi, F., Pastori, D., Cangemi, R., Pignatelli, P., & Loffredo, L., (2020). Hypercoagulation and antithrombotic treatment in coronavirus 2019: A new challenge. *Thrombosis and Haemostasis, 120*(06), 949–956.
60. Tzoulaki, I., Murray, G. D., Lee, A. J., Rumley, A., Lowe, G. D. O., & Fowkes, F. G. R., (2007). Relative value of inflammatory, hemostatic, and rheological factors for incident myocardial infarction and stroke: The Edinburgh artery study. *Circulation, 115*(16), 2119–2127.
61. Berger, J. R., (2020). COVID-19 and the nervous system. *Journal of Neurovirology, 26*(2), 143–148.
62. Tiwari, P., Khan, H., Singh, T. G., & Grewal, A. K., (2022). Poly (ADP-ribose) polymerase: An overview of mechanistic approaches and therapeutic opportunities in the management of stroke. *Neurochemical Research*, 1–23.
63. Iadecola, C., Anrather, J., & Kamel, H., (2020). Effects of COVID-19 on the nervous system. *Cell, 183*(1), 16–27.
64. Garg, N., Singh, T. G., Khan, H., Arora, S., Kaur, A., & Mannan, A., (2021). Mechanistic interventions of selected *Ocimum* species in management of diabetes, obesity, and liver disorders: Transformative developments from preclinical to clinical approaches. *Biointerface Res. Appl. Chem., 12*(1), 1304–1323.
65. Firdous, S. M., Ghosh, S., Nath, P., & Khan, H., (2021). Molecular basis of diabetic nephropathy. *Diabetes and Diabetic Complications.* Nova Science Publishers, Hauppauge.
66. Merkler, A. E., Parikh, N. S., Mir, S., Gupta, A., Kamel, H., Lin, E., Lantos, J., et al., (2020). Risk of ischemic stroke in patients with coronavirus disease 2019 (COVID-19) vs patients with influenza. *JAMA Neurology, 77*(11), 1366–1372.
67. Rodriguez, F., Solomon, N., De Lemos, J. A., Das, S. R., Morrow, D. A., Bradley, S. M., Elkind, M. S. V., et al., (2021). Racial and ethnic differences in presentation and outcomes for patients hospitalized with COVID-19: Findings from the American Heart Association's COVID-19 cardiovascular disease registry. *Circulation, 143*(24), 2332–2342.

68. Connors, J. M., & Levy, J. H., (2020). COVID-19 and its implications for thrombosis and anticoagulation. *Blood, 135*(23), 2033–2040.
69. Qureshi, A. I., Abd-Allah, F., Al-Senani, F., Aytac, E., Borhani-Haghighi, A., Ciccone, A., Gomez, C. R., et al., (2020). Management of acute ischemic stroke in patients with COVID-19 infection: Report of an international panel. *International Journal of Stroke, 15*(5), 540–554.
70. Beyrouti, R., Adams, M. E., Benjamin, L., Cohen, H., Farmer, S. F., Goh, Y. Y., Humphries, F., et al., (2020). Characteristics of ischemic stroke associated with COVID-19. *Journal of Neurology, Neurosurgery & Psychiatry, 91*(8), 889–891.
71. Fara, M. G., Stein, L. K., Skliut, M., Morgello, S., Fifi, J. T., & Dhamoon, M. S., (2020). Macrothrombosis and stroke in patients with mild COVID-19 infection. *Journal of Thrombosis and Haemostasis, 18*(8), 2031–2033.
72. Melmed, K. R., Cao, M., Dogra, S., Zhang, R., Yaghi, S., Lewis, A., Jain, R., et al., (2021). Risk factors for intracerebral hemorrhage in patients with COVID-19. *Journal of Thrombosis and Thrombolysis, 51*(4), 953–960.
73. Gavioli, E. M., Sikorska, G., Man, A., Rana, J., & Vider, E., (2020). Current perspectives of anticoagulation in patients with COVID-19. *Journal of Cardiovascular Pharmacology, 76*(2), 146–150.
74. Sharma, D., Rasmussen, M., Han, R., Whalin, M. K., Davis, M., Kofke, W. A., Venkatraghvan, L., et al., (2020). Anesthetic management of endovascular treatment of acute ischemic stroke during COVID-19 pandemic: Consensus statement from society for neuroscience in anesthesiology & critical care (SNACC): Endorsed by society of vascular & interventional neurology (SVI. *Journal of Neurosurgical Anesthesiology, 32.*
75. Al Kasab, S., Almallouhi, E., Alawieh, A., Levitt, M. R., Jabbour, P., Sweid, A., Starke, R. M., et al., (2020). International experience of mechanical thrombectomy during the COVID-19 pandemic: Insights from STAR and ENRG. *Journal of Neurointerventional Surgery, 12*(11), 1039–1044.
76. Löwhagen, H. P., Rentzos, A., Karlsson, J. E., Rosengren, L., Leiram, B., Sundeman, H., Dunker, D., et al., (2017). General anesthesia versus conscious sedation for endovascular treatment of acute ischemic stroke: The an stroke trial (anesthesia during stroke). *Stroke, 48*(6), 1601–1607.
77. Bloch, E. M., Shoham, S., Casadevall, A., Sachais, B. S., Shaz, B., Winters, J. L., Van, B. C., et al., (2020). Deployment of convalescent plasma for the prevention and treatment of COVID-19. *The Journal of Clinical Investigation, 130*(6), 2757–2765.
78. Umapathi, T., Kor, A. C., Venketasubramanian, N., Lim, C. C., Pang, B. C., Yeo, T. T., Lee, C. C., et al., (2004). Large artery ischemic stroke in severe acute respiratory syndrome (SARS). *Journal of Neurology, 251*(10), 1227–1231.
79. Zhao, J., Li, H., Kung, D., Fisher, M., Shen, Y., & Liu, R., (2020). Impact of the COVID-19 epidemic on stroke care and potential solutions. *Stroke, 51*(7), 1996–2001.
80. Muttenthaler, M., King, G. F., Adams, D. J., & Alewood, P. F., (2021). Trends in peptide drug discovery. *Nature Reviews Drug Discovery, 20*(4), 309–325. https://doi.org/10.1038/s41573-020-00135-8.
81. Saklani, P., Khan, H., Gupta, S., Kaur, A., & Singh, T. G., (2022). Neuropeptides: Potential neuroprotective agents in ischemic injury. *Life Sciences, 288*, 120186.
82. Kursumovic, E., Lennane, S., & Cook, T. M., (2020). Deaths in healthcare workers due to COVID-19: The need for robust data and analysis. *Anaesthesia.*
83. Khan, H., Singh, A., Thapa, K., Garg, N., Grewal, A. K., & Singh, T. G., (2021b). Therapeutic modulation of the phosphatidylinositol 3-kinases (PI3K) pathway in cerebral ischemic injury. *Brain Research, 1761*, 147399.
84. Grewal, A. K., Singh, N., & Singh, T. G., (2019). Effects of resveratrol postconditioning on cerebral ischemia in mice: Role of the sirtuin-1 pathway. *Canadian Journal of Physiology and Pharmacology, 97*(11), 1094–1101.
85. Zhao, J., Rudd, A., & Liu, R., (2020). Challenges and potential solutions of stroke care during the coronavirus disease 2019 (COVID-19) outbreak. In: *Stroke. Am. Heart Assoc., 51*(5), 1356–1357.

CHAPTER 10

Neurotransmission Imbalance: A Major Factor in the Worsening of Ischemic Injury

KAJAL BAGRI,[1] VIPASHA MEHTA,[1] AARCHI SINGLA,[1] RAHUL DESHMUKH,[1] and THAKUR GURJEET SINGH[2]

[1]*Maharaja Ranjit Singh Punjab Technical University, Bathinda, Punjab, India*

[2]*Chitkara College of Pharmacy, Chitkara University, Punjab, India*

ABSTRACT

Ischemic Injury is characterized by diminished/absent blood flow in the brain. There are numerous factors and molecular cascades that are linked with pathological hallmarks of Ischemic Injury. There is a neurotransmission imbalance after the injury. Numerous neurotransmissions regulate the essential signaling for normal physiology and the working of the brain. After the injury, there is impairment in the balance of major neurotransmitters, i.e., Glu, which shows expression which leads to neuronal excitotoxicity compromising the normal physiology. GABA is reversely related to Glu, post-injury the reduced level of GABA contributes to neuronal excitability and neuronal death. Monoamines are responsible for the happiness, mood swings and the movement of the individual, post-injury creates a deficit in the level of the monoamine which could generate the state of altered mood, movement impairment, mood swings, etc. The cognitive impairment is observed after the brain insult which could be correlated with the level of Ach. AChE activity is increased in the ischemic patients which decreases the ACh level which could be correlated with the memory impairment. These neurotransmission imbalances could be beneficial to target pathological alterations and may give new insight to find novel treatments to manage associated complications.

10.1 INTRODUCTION

The brain is the most active organ of the human system. To operate the various functions of the body, it is mandatory to circulate the appropriate blood and nutrition to the brain. Reduction in the supply of cerebral blood at the focal area or deprivation of the oxygen delivery and glucose to the neurons results in ischemic injury. Ischemic injury is identified

Ischemic Injury: A Molecular Insight. Thakur Gurjeet Singh & Heena Khan (Eds.)

by the loss of blood supply to the brain. Normally the level of blood flow in the brain is 50 mL/100 g per minute but under ischemic conditions, it falls from 50 ML/100 [1, 2]. Ischemia is the restriction of blood flow in the brain due to injury. It is accounted for more than 90% of the people after TBI. Ischemia affects the cerebral hemispheres, hippocampus, cerebellum, and basal ganglia part of the brain. It causes edema formation, swelling, an increase in intracranial pressure, and leads to ischemia. Dizziness, blurring vision, difficulty in speaking, loss of memory, tremor, and loss of consciousness are the major symptoms of cerebral ischemia. Ischemia has two types based on the site and extent of the injury: focal and global. The blockage of specific arteries in the brain causes focal cerebral ischemia (FCI) or stroke, whereas global cerebral ischemia (GCI) is caused by a drop-off in blood flow overall [1]. On the report 2016 Global Burden of Disease data, which was published in 2019, one in every four people will have a stroke throughout their lifetime. Each year, there are approximately 96 million ischemic strokes and 41 million hemorrhagic strokes (including intracerebral and subarachnoid hemorrhage) worldwide [2]. Elevated blood pressure is the most influential danger sign for stroke, and it relates to both ischemic stroke and intracerebral hemorrhage. Smoking, diabetes, metabolic disorders, and lack of physical activity are major concerns that are necessary for the regulation of the community to change lifestyle and the environment [3]. Stroke is mainly classified into two forms according to pathological mechanism, i.e., ischemic stroke and hemorrhagic stroke. Ischemic stroke constitutes about 85% of all strokes whereas 15% of hemorrhage strokes become the reason for stroke [4]. The barrier in cerebral circulation leads to the closedown of neuronal electrical activity within seconds and the worsening of the energy state and ion homeostasis within minutes [5].

10.2 EPIDEMIOLOGY

Stroke is a life-threatening and devastating illness that affects 15 million people globally each year. It is an aging malady, in which mostly the individuals 65 years old have been affected [6]. After a stroke, older individuals had a greater death rate and a worse quality of life than younger patients. The younger patients had a better quality of life and chances of living as compared to the old patients affected after a stroke. Although men have a greater incidence of stroke over most of their lives, women have a larger stroke prevalence overall due to an increase in stroke risk with age and a longer average lifetime in females. Ischemic strokes, which occur when the blood supply to the brain is cut off, account for 87% of all strokes [7]. Ischemic stroke incidence, pathogenesis, and treatment effectiveness are all influenced by age and gender. The revised overall incidence of stroke in rural areas is 84-262/1,00,000, while it is 334-424/1,00,000 in urban areas. The overall incidence, according to recent population-based studies, is 119-145/1,00,000 [8]. There's also a big difference in rates of morbidity and mortality, with Kolkata having the largest at 42%. Stroke units are exclusively found in private clinics and are mostly located in urban areas. Intravenous (IV) and intra-arterial (IA) thrombolytics are widely used in India. In the continuing Indo USA National stroke registry, the rate of IV thrombolytic is 11%.

Around the turn of the golden year, the age-based annual extent of stroke in Europe ranged from 95 to 290 per 1,00,000, with a month case fatality rates ranging from 13 to 35%. Each year, roughly 1.1 million Europeans suffer a stroke, with ischemic stroke accounting for nearly 80% of cases [9]. However, global stroke rates are declining and the rates among young adults are rising, emphasizing the need for more effective prevention measures. Furthermore, officially the count of strokes is looking forward to climbing significantly in the future as a result of the older population; around 1.5 million Europeans will suffer from stroke every year by 2025 (10, 11).

10.3 TYPES OF STROKES

There are numerous types of stroke and among them, ischemic stroke, hemorrhagic stroke, transient ischemic attack (mini-stroke), brain stem stroke, and cryptogenic stroke are the major types [12, 13] (Table 10.1).

TABLE 10.1 Classification of Stroke

SL. No.	Types	Major Symptom	References
1.	Ischemic stroke	When a blood clot blocks or narrows an artery leading to the brain.	[14]
2.	Hemorrhagic stroke	When blood from an artery suddenly begins bleeding into the brain.	[15]
3.	Transient ischemic	It happens when the blood supply to part of the brain is briefly blocked.	[16]
4.	Attack (mini-stroke)	Occurs when part of the brain experiences a temporary lack of blood flow.	[17]
5.	Brain stem stroke	When there is a blockage obstructing the flow of blood to the brain stem.	[18]
6.	Cryptogenic stroke	When an artery to your brain becomes blocked, usually by a blood clot and this may be of unknown cause.	[19]

10.3.1 HEMORRHAGIC SHOCK

It is also known as intracranial bleeding which involves bleeding within the intracranial vault, functional tissues in the brain, and subarachnoid space [20]. ICH is the most common and well-known feature of TBI and is also a high risk of mortality.

10.3.2 BRAINSTEM HEMORRHAGE STROKE

After blunt, a head injury is a rare event. It affects around 0.75%–3.6% of patients suffering from a head injury while admitted to the hospital. It is caused mainly by direct contact of the object at the end of the head, usually due to acceleration and deacceleration of the head or may be due to backward falls, impact on the head which leads to the displacement of the brainstem structure [21].

10.4 RISK FACTORS

10.4.1 SMOKING

Persons who chewed tobacco or smoked have a higher chance of having a stroke later in life. This is because nicotine raises blood pressure levels in the body. Smoking a cigarette for an extended period produces fatty buildup in the neck's major artery. It also thickens your blood over time, making it more prone for clots to develop, increasing your stroke risk.

10.4.2 ALCOHOL CONSUMPTION

Heavy drinking can have a bad impact on every element of one's life. This is especially true when it comes to a person's health. Alcohol abuse has been linked to a variety of health problems, ranging from minor to life-threatening. Alcohol may increase the risk of Blood pressure, diabetes, liver damage, atrial fibrillation (AF), and obesity.

10.4.3 MOTOR VEHICLE ACCIDENTS (MVA)

The collision of the motor vehicle with an obstacle, and crash accidents on road mainly at the night are the leading cause of severe ischemic injury [22].

10.4.4 VIOLENCE

Domestic violence, child abuse, family issues, fights, etc., are the major cause of inducing ischemia. It is also linked with Shaken baby syndrome in case of child abuse. It is a syndrome associated with vigorously shaking of the head by external manual force which leads to intracranial bleeding without external bleeding. Shaken baby syndrome is mainly seen in children under two years [23].

10.4.5 EXPLOSIVE BLASTS

Explosive blasts are the common cause of Ischemic patients, especially in the case of active-duty military personnel. Whereas, how these blasts affect the brain was not fully understood but many researchers believe that the pressure wave passing through the brain might disrupt the functioning of the brain.

10.4.6 FIREARMS

They are another highly effective cause of Ischemia (12% of all TBIs). They are usually seen in individuals aged 25–34 years. The fatalities by TBIs and gunshots are highly seen in men than in women. They are more prevalent in Africa as compared to white Americans [24].

10.4.7 GENDER

Males are at approximately twice the risk of Ischemia compared to females. But the gender ratio of TBI varies according to age, in children (<5 years of age) the rate of incidence is equal in both genders. After this the rate of incidence increase in males than females at double speed mainly during adolescence [25, 26] (Table 10.2).

TABLE 10.2 Symptoms of Ischemia

SL. No.	Symptoms of Ischemia	References
1.	Loss of consciousness	–
2.	Transient confusion	–
3.	Disorientation	–
4.	Amnesia	–
5.	Seizures	–
6.	Headache	–
7.	Dizziness	–
8.	Irritated behavior	–
9.	Fatigue	–
10.	Poor concentration	–
11.	Anxiety	–
12.	Severe headache	–
13.	Nausea/vomiting	–
14.	Pupil dilation	–
15.	Slurred speech	–
16.	Aphasia	–
17.	Sensory deficits	–
18.	Memory loss	[27, 28]
19.	Emotional dysfunction	–
20.	Loss of muscle control	–
21.	Bleeding from head	–
22.	Locomotor impairment	–
23.	Lack of concentration	–

10.4.8 HYPOXIA

There is deprivation of oxygen demand in the brain tissues. The prolonged duration of hypoxia may be responsible for various complications such as coma, seizures, and ultimately brain damage. The indications include suffocation of oxygen, lack of awareness, blackout or amnesia, and motor ataxia. Long-term hypoxia can cause coma, seizures, and even loss of neurons in the brain.

10.4.9 ANOXIA

Anoxic brain due to injury defines the brain as completely lacking oxygen supply. The severity of destruction and damage depends upon the duration of the absence of oxygen in that part. The brain is one of the most sensitive parts of the body and can be dead within five minutes after the oxygen supply has been cut off. This leads to permanent damage to the brain. This is the major cause of death of patients after TBI and can be rescued by instantly providing an administering oxygen supply [29].

10.5 NEUROTRANSMISSION: THE GAME-CHANGER IN ALL ASPECTS

Neurotransmitters transmit signals from nerve cells to target cells. It is a well-known aspect that neurotransmitters usually receive a response from their effector cells through a mechanism known as synaptic transmission (Figure 10.1). The transmission across the synapse is the first step and relatively new step for the neurons to communicate [30]. The nervous system is responsible for the majority of human body processes such as awareness, movement command, and cognitive behavior. They are governed by brain cell networks that generate and propagate electrochemical signals, mostly through the release and diffusion of neurotransmitters through synaptic connections, which facilitate the interchange of those signals between neural cells [31]. The chemical synapse has given the framework for comprehending fundamental concepts of fast chemical communication across the synapse through an extended series of exquisite experiments. The synaptic vesicles holding Ach, merge with the cell membrane and discharge into the synaptic cleft. It is distributed across the synaptic gap to enchain and activate nAChRs on the cell membrane of skeletal muscle. When nAChRs on postsynaptic are activated, they depolarize and allow contraction, a process known as "excitation–contraction coupling" [30] (Table 10.3).

TABLE 10.3 Role of Neurotransmitters in Normal Brain vs. Ischemic Brain

SL. No.	Neurotransmitter	Normal Functioning	Impaired Neurotransmission Effects During Ischemic	References
1.	Glutamate	Synaptic plasticity, repair of the CNS.	Over activation of glutamate.	[32]
2.	GABA	Decreases the cell's excitability.	Inhibitory action on GABA.	[33]
3.	Acetylcholine	Alters neuronal excitability, influences synaptic transmission.	Decrease in acetylcholine level.	[34]
4.	Monoamines	Maintain mood changes, psychomotor functions, and CVS functions.	Lead to changes in behavioral, and mood order.	[35]

The neural membrane is polarized and has a negative charge while at rest. The binding of neurotransmitters activates the ion fluxes via regulating specific ion channels. Depending on which channels are engaged, either 'hyperpolarizing' or 'depolarizing' negative charges may invade the cell. The stasis of two charges, i.e., negative and positive—is combined inside the neuron's dendrites and cell body. When the edge of depolarization is achieved, specific voltage-gated Na^+ channels respond to open, causing a loss of polarization down the axon, known as an action potential. After the generation of the action potential, it reaches the distal end of the axon where an enormous positive charge activates voltage-sensitive Ca^{2+} channels. The entrance of Ca^{2+} then pushes the vesicles to fuse with the presynaptic cell membrane and release the neurotransmitter to the synaptic cleft. When neurotransmitters, such as ACh or Glu, activate cation (for example, Na+ or Ca^{2+}) channels and thereby depolarize the cell, they are said to be excitatory; whereas neurotransmitters such as GABA activate anion (for example, Cl^-) channels, they are said to be inhibitory (36, 37). Because chemical synapses are the basis for rapid information transfer in the nervous system, neurotransmitters and neurotransmitter receptors are essential components of neuronal identity. They are essential to the reason for the sign, intensity, and dynamics of signal transmission when combined with release and uptake machinery [38].

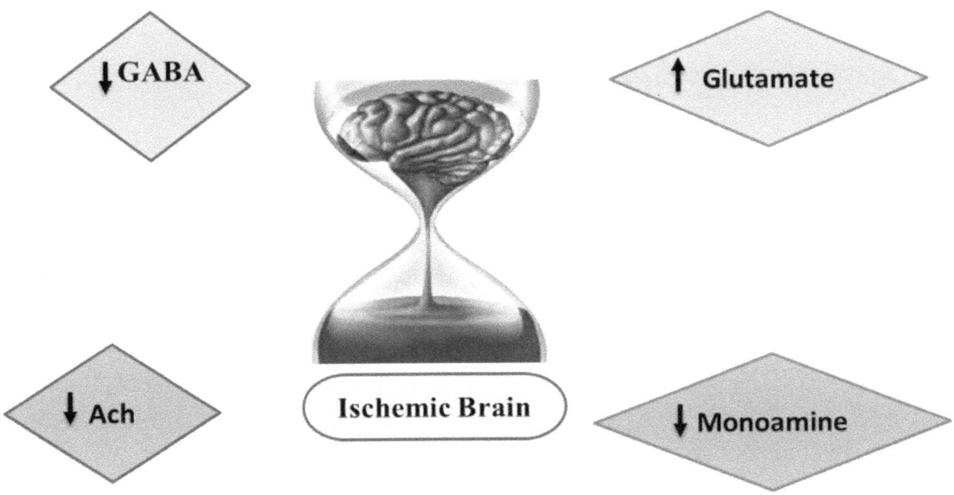

FIGURE 10.1 Various neurotransmitter activities during ischemia.

10.6 GLUTAMATE: HALLMARK OF ISCHEMIC INJURY

Excitotoxicity is a type of neurotoxicity caused by a sustained release of the neurotransmitter Glu. It affects approximately 30–40% of synapses. Glu is a key neurotransmitter in the brain that is required for synaptic transmission, neural communication, synaptic growth, neuronal development, brain progression as well as synaptic plasticity [39]. Glu is the prominent excitatory neurotransmitter present in the brain. This non-essential amino acid serves a variety of activities in different grades, both inside and outside the CNS. Glu signaling is frequently connected with the synaptic transmission in the CNS. The excitotoxicity of Glu can be seen in various events like Alzheimer's disease, traumatic brain injury, ischemic injury, and Huntington's disease [1]. The fundamental neuronal information hub is the synapse, which is defined by a presynaptic and postsynaptic location separated by the synaptic cleft [40]. Under physiological conditions, Glu's primary mechanism is synaptic release however, there are several significant non-vesicular mechanisms (non-exocytotic) in charge of Glu release in pathological circumstances. Excitotoxicity caused by excessive Glu release is vital in the pathophysiology of brain ischemia [41].

At normal conditions, Glu is dictated by active ATP-dependent transporters present in neurons and glial cells. In case, any disturbance in these mechanisms leads to an excessive accumulation of Glu, which results in stimulation of Na^+ and Ca^+ ions disrupting cells. Intracellular Glu is produced by the hydrolysis of glutamine under the catalysis of glutaminase. It can be transported to the extracellular fluid in response to specific physiological stimuli. To a significant part, the conservation of extracellular Glu level is dependent on the influence of excitatory amino acid transporters on the astrocyte membrane, i.e., EAAT1 or EAAT2, which transport extracellular Glu into astrocytes (42, 43). Neurons and astrocytes are involved in Glu regulation, and any dysfunction within the system might result in an imbalance in excitation and inhibition [44].

10.6.1 ROLE OF GLU RECEPTORS

Glu activates a variety of ion-channel receptors out of which ionotropic and GPCR, i.e., metabotropic play a vital role in brain functioning. The alpha-amino-3-hydroxy-5-methylisoxazole-4-propionate (AMPA), NMDA, and kainic acid receptors are the major ionotropic receptors regulated by Glu signaling. Ionotropic receptors are cation-permeable ligand-gated ion channels. Overactivation of these receptors increases intracellular Ca^{2+} load and catabolic enzyme activity, which can set off a chain reaction of events that leads to apoptosis and necrosis [45]. When potassium (K+) levels rise, Glu is produced, which stimulates N^{a+}/Ca^{2+} channels related to NMDARs. Extracellular Glu also stimulates the amino-3-hydroxy-5-methyl-4-isoxazole propionic acid (AMPA) and metabotropic Glu receptors, which is an important step in the inflammatory cascade. Metabotropic and NMDA receptors function via mono-ionic channels and, as a result, increase intracellular Ca^{2+} levels. Several deleterious mechanisms, including voltage and receptor gated Ca^{2+} influx, cause a large increase in free cytosolic calcium levels, resulting in mitochondrial dysfunction. Overactivation of Ca^{2+} levels results in the activation of lipases, kinases, phosphatases, proteases, and free radicals that cause cell damage. Mg^{2+} normally clogs NMDA receptor channel pores, however, when Glu is released from presynaptic sites and AMPA receptors are activated, Mg^{2+} is entirely removed from NMDARs due to partial depolarization of the post-synaptic membrane. This results in an influx of Na^+ and CA^+ into the cell, which plays a significant role in ischemia cell death [46]. Ca^{2+} influx through NMDARs is not only essential for normal physiological activities in neurons but also plays a significant role in the initiation of ischemic cell death. Excess release of Glu causes overactivation of NMDARs and calcium inside neurons during excitotoxicity. Overflow of calcium results in activation of downstream pro-death signaling. It activates the formation of reactive oxygen species (ROS), calpain activation, and mitochondrial damage that results in cell necrosis or apoptosis [47]. In the adult brain, GluN2A and GluN2B are present as major NMDAR. Activation of synaptic and extrasynaptic GluN2B-containing NMDARs after an ischemic event causes excitotoxicity followed by neuronal death [46]. After ischemic event excitotoxicity occurs by the activation of GluN2B in the synaptic and extrasynaptic NMDARs. Extracellular Glu levels in the central nervous system rise dramatically within the conclusion following acute ischemia and subsequently reduce promptly after reperfusion, while not recovering to pre-ischemia levels [42].

10.7 GABA: KEY IN THE PATHOGENESIS OF ISCHEMIC INJURY

GABA, an essential neurotransmitter, was first synthesized in 1883 before being discovered as a metabolic product in plants and microbes. It was discovered in mammalian brains in 1950. GABA is now known to largely suppress neuronal excitabilities of postsynaptic membranes in the adult central nervous system. GABA has the ability to relax the muscle at neuromuscular synapses [48]. The stability between the excitatory and inhibitory neurotransmission is important for the functioning of the neuronal circuit. As a result, controlling GABAergic synaptic strength and transmission is critical for maintaining excitatory/inhibitory synaptic balance [49]. GABA is seen as a neuroprotective by reducing excitotoxicity causing cell death in the phase of severe stroke [50].

GABA is a key neurotransmitter employed by approximately 20%–50% of all synapses in the central nervous system in both slow and fast inhibitory synaptic transmission (Figure 10.2). GABA binds to transmembrane receptors in the plasma membrane of presynaptic and postsynaptic neurons [51]. GABA is synthesized by both prokaryotes and eukaryotes by the decarboxylation of Glu by GAD. GAD can be found in Gram-positive as well as Gram-negative bacteria [52]. GABA is primarily generated in the CNS by the enzymes GAD65 or GAD67, which both need pyroxidine as a cofactor to convert Glu to GABA. GAD65, which is found in nerve terminals, generates GABA for classic equilibrium neurotransmission. GABA generated by GAD 67, which is mostly expressed in neuronal somata, has a non-neurotransmitter, metabolic role that contributes to synaptogenesis as well as oxidation-reduction (redox) regulation. Glutamine is a substrate of GABA as well as a precursor for Glu production via phosphate-activated glutaminase. GABA is packed for release into synaptic vesicles by GABA-specific vesicular transporters (VGATs) in a manner dependent on both the electrochemical and pH gradients [53]. GABA-A are pentameric proteins that belong to the Cysloop ligand-gated ion channel superfamily.

GABAAR's physiologic and pharmacological characteristics vary due to variations in subunit composition [54]. GABAA receptors moderate the rapid GABA responses which are predominantly permeable to chloride channels. It also participates in synaptic plasticity by modifying the transmembrane chloride gradient, hence influencing synaptic strength. GABAA operates as an excitatory neurotransmitter during early brain development due to a high intracellular than extracellular Cl concentration and then switches to an inhibitory activity due to a decrease in intracellular Cl levels after the first postnatal week

[52]. Alternatively, GABABRs are G-protein linked receptors that can block voltage-gated calcium channels and activate G protein-coupled inwardly-rectifying potassium channels (GIRK) potassium channels [54]. GABA can cause neuronal hyperpolarization via anion channels (GABA-A) and presynaptic G-protein coupled receptors (GABA-B). This hyperpolarization combats the depolarization that initiates the biochemical ischemic cascade. Second, there is no shortage of GABA under ischemia conditions, but the affinity of GABA receptors is reduced. GABA activation lowers the respiratory rate, retains glucose, and reduces acidity, facilitating local cerebral blood flow [55]. During cerebral ischemia, GABA levels in the extracellular space rise rapidly and transiently, like with other neurotransmitters. In contrast, the level of both the GABA-A and GABA-B receptors reduced to vary after ischemia in different areas of the brain of rodents subjected to transient MCAo. During reperfusion, early impairment of GABA-mediated neurotransmission may result in sustained neuronal excitability and ultimately neuronal death [5].

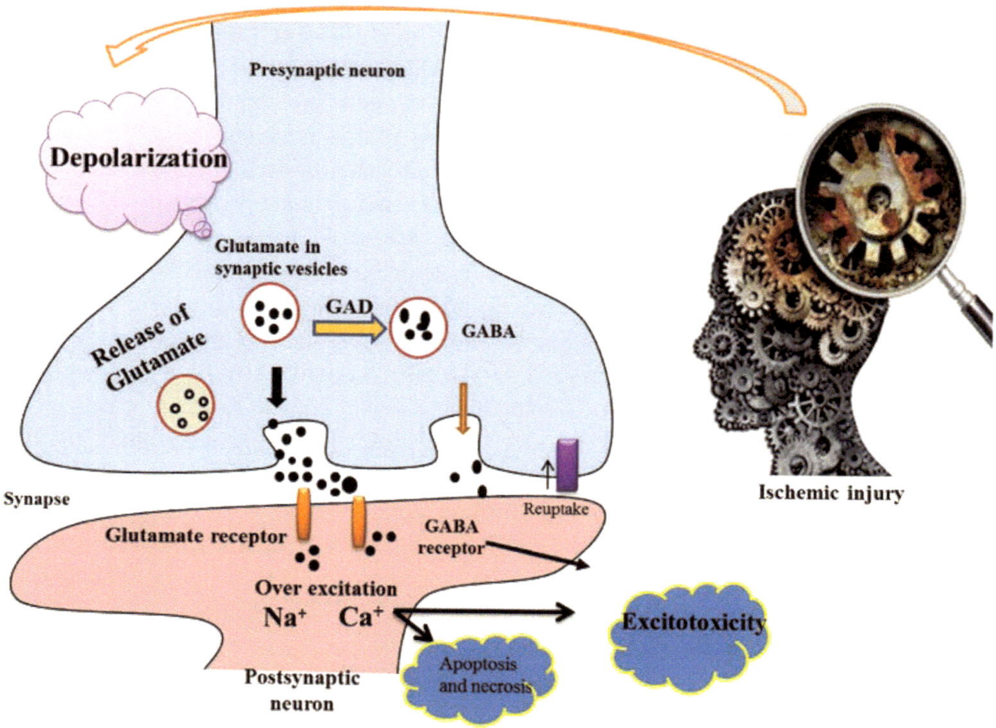

FIGURE 10.2 The imbalance between the excitatory and inhibitory neurotransmitters by the trigger of brain insult leading to ischemic injury.

10.8 ACETYLCHOLINE (ACH)

Although neurons and glia were described as nervous system cells in the late nineteenth century, the shifting of impulses among neurons was thought to be a solely electrifying activity for many years. ACh has been recognized as one of the most important neurotransmitters in the body for more than eight decades, and current studies continue to uncover novel and diversified activities of this signaling molecule in health and disease [56]. ACh and the action of cholinergic neurons have proven to improve learning and reminiscence development, raise alertness and attentiveness, and signal behaviorally apposite sensorial processes [57]. The acetyl group from acetyl coenzyme A interacts with choline, which is mediated by ChAT and synthesizes ACh. Furthermore, the vesicular ACh transporter transports ACh into vesicles, where it is destroyed by acetylcholinesterase (AChE) and butyrylcholinesterase (BChE). nAChR and muscarinic (mAChR) cholinergic receptors, called after their consistently occurring alkaloid agonists, are ACh receptors [56]. Inflammatory responses have been linked to the development and advancement of ischemic stroke. The central nervous system is an important controller of the resistant reaction and regulates swelling and redness. nAChRs are abundant in skeletal muscle and the neurological system.

Lately, the expression of nAChRs in immune cells and glial cells has sparked interest as a possible therapeutic target in inflammation and neurodegenerative illnesses [43]. By'cholinergic anti-inflammatory route' depending on 7nAChR, the vagus nerve can influence the immunologic response and control inflammation. ACh, the main neurotransmitter of the vagus nerve, can restrict the building up of pro-inflammatory cytokines by human macrophages via a process involving the 7nAChR (58, 59). Because Ach is a major arbitrator of communication between the brain and the immune system, the parasympathetic nervous system (primarily the vagus nerve) engages in the systemic tuning up of the immune response after CNS injury, alongside the sympathetic nervous system and the HPA axis [56]. A rise in choline acetyltransferase function or a drop inAChE activity can both elevate Ach levels in the brain [34]. Recent findings focused on the level of ACh in the ischemic model. The Ach can be increased by increasing choline acetyltransferase activity or decreasing AChEactivity. A decrease in the Ach level has been seen in the preclinical ischemic model, which gives a direct view that a reduction in the level of Ach results in the pathophysiology of ischemic damage (60, 34). Researchers have demonstrated that ACh (produced by AChE) mediates a variety of actions on the vascular endothelium, including the production of endothelium-derived relaxing factors. Endothelial dysfunction has been linked to the pathophysiology of hypertension, causing structural and mechanical changes that may provide vital insights into the onset and progression of cardio- and cerebrovascular damage. This study discovered a decrease in AChE plasma activity in acute ischemia individuals [61].

10.9 ROLE OF MONOAMINES

Ischemia is the restriction of blood circulation in the brain due to injury. The risk factors of IR injury could be related to lifestyle, environmental factors, any brain dysfunction, Traumatic brain injury, etc. It is accounted for more than 90% of the people after TBI [62]. Ischemia affects the delegated part of the brain like as the cerebral hemispheres, hippocampus, cerebellum, and basal ganglia affecting the neurotransmission. Monoamines are important for maintaining the various physiological activities of the body including emotions, learning, memory, locomotion, etc. [63].

When neurons turn acutely ischemic as a result of thrombus formation, edema, or vasculitis that might occur after hemorrhage or injury, they, furthermore, are likely to "leak" numerous intracellular elements that they ordinarily reserve in great concentrations. The abnormal depletion of some of these chemicals, including the monoamine neurotransmitters DA, norepinephrine, and 5-HT, may aggravate the etiological alterations generated by the acute ischemia. If these molecules remain within the synaptic clefts, they are likely to overflow the receptors of post-synaptic neurons, causing abiding hyperpolarization or depolarization. By relying on whether the monoamine neurotransmitter was inhibitory or excitatory, it interferes with whatever brain functions the post-synaptic neurons serve (64, 65). The monoamine neurotransmitters have been linked to the etiology of neuronal damage during ischemia. Ischemia-induced dopamine release was discovered to play a significant part in the development of striatal neuronal injury following ischemia in the dorsolateral striatum. Transient ischemia causes a huge release of norepinephrine in the hippocampus, which has a part in CA1 neural damage concluding ischemia while the roles of DA and norepinephrine are well established and widely studied, the role of serotonin in transient global ischemia remains unknown [66].

Monoamine neurotransmitters such as 5-HT, norepinephrine (NE), and DA are the essential neurotransmitters in the pathophysiology of major mental diseases as well as the molecular mechanisms of action of many psychiatric medicines. Presynaptic and somato dendritic auto receptors allow serotonin, norepinephrine, and dopamine to regulate their own release (Figure 10.3). There are also numerous ways in which these monoamine neurotransmitters interact with one another to regulate one another. Alterations in the neurotransmission of monoamines may indicate both the downstream primary problems in signal transmission as well as upstream effects on neuron survival, neuroplasticity, and neuroprotection [67].

DA is the main monoamine being synthesized from tyrosine by the action of various enzymes and the final formed compound, i.e., DA, 5-HT, Nor epinephrine are primarily used to exhibit mood stability and motor coordination. The ischemic injury which may be due to various reasons including lifestyle, environmental factors, any brain dysfunctioning, and accidents or TBI [68] leads to a sudden a burst of monoamines and then a sudden fall in the level of the monoamine due to its decline in the activity of dopamine receptors or increased the reuptake of dopamine and NE by monoamines transporters or due to metabolism of the chemical by Catechol-O-Methyltransferase (COMT), Monoamine oxidase (MAO). This decreased or abolished level of monoamines is purely linked with the altered mood, emotions, learning, memory, locomotion, etc. [69]. Depression is a common consequence of stroke or ischemic Injury which could be hypothetically and preclinically relates to the altered or abolished level of DA, 5-HT, and Nor-Epinephrine [70]. Monoamines nerve terminals are highly sensitive to ischemic damage [71].

Post Stroke Depression is the major challenge faced by the stroke/ischemic injured patients, and the contribution of 5-HT remains on top as this mainly affects the mood and mental state of the patient. After the ischemic there is over activation of the 5-HT reuptake co-transporters increasing the reuptake of the serotonin, the biotransformation of the 5-HT also get a boost by MAO and the binding of post synaptic serotonin receptor gets decline which abolishes the serotonergic neuronal availability.

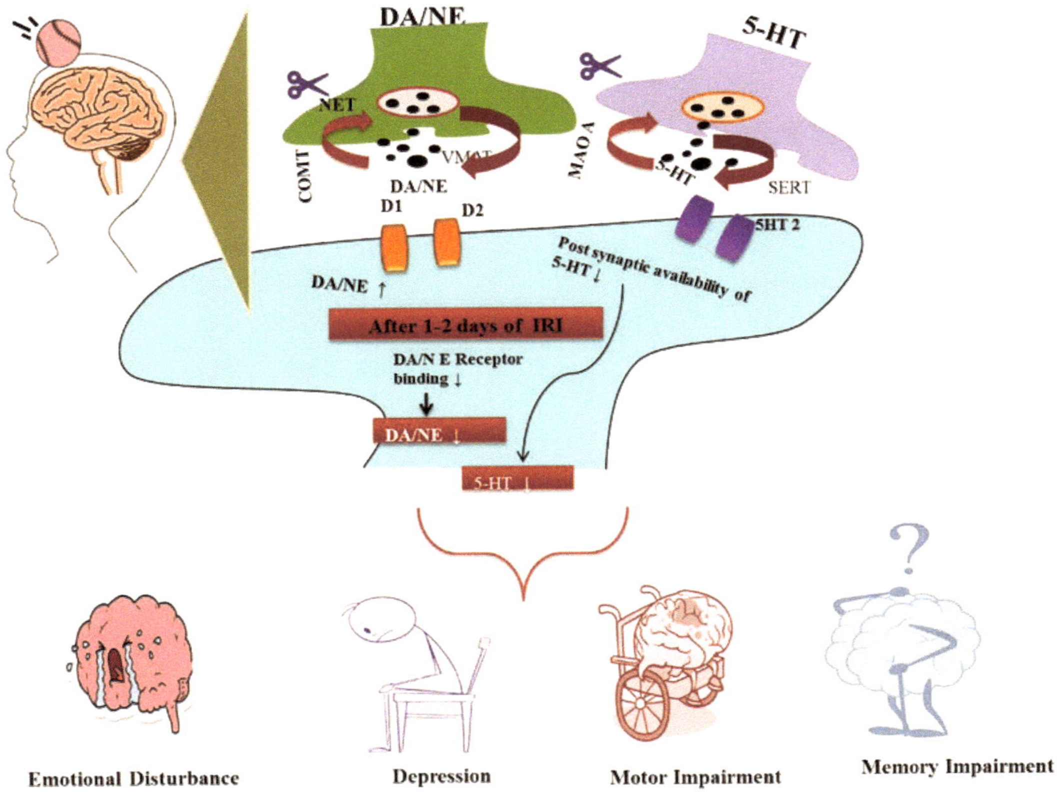

FIGURE 10.3 The imbalance between monoamines leads to various post-stroke symptoms.

10.10 CONCLUSION

Stroke/ischemic injury remains a hot topic as it affects the population at a high rate. The secondary associated symptoms after stroke are irregular body homeostasis, cognitive impairment, depressive mood, mental stability, and an emotional breakdown which are all due to the neurotransmission imbalance that leads to such disturbing and may be lifelong co-morbidities. So this huge expansion in the knowledge to understand the neurotransmission imbalance in the pathology of Ischemic injury would be beneficial in introducing novel therapeutic targets. Ischemic injury is a multifaceted complex disease that can be targeted only when all these factors are kept in the background.

KEYWORDS

- **acetylcholine**
- **cholinergic**
- **cognition**
- **depression**
- **dopamine**
- **gamma-aminobutyric acid**
- **glutamate**
- **ischemia**
- **monoamine**
- **neurotransmission**
- **serotonin**

REFERENCES

1. Belov, K. D., Kriska, J., Tureckova, J., & Anderova, M., (2020). Ischemia-triggered glutamate excitotoxicity from the perspective of glial cells. *Frontiers in Cellular Neuroscience, 14*.
2. Hughes, J. D., Bond, K. M., Mekary, R. A., Dewan, M. C., Rattani, A., Baticulon, R., et al., (2018). Estimating the global incidence of aneurysmal subarachnoid hemorrhage: A systematic review for central nervous system vascular lesions and meta-analysis of ruptured aneurysms. *World Neurosurg., 115*, 430–447.
3. Campbell, B. C. V., & Khatri, P., (2020). Stroke. *Lancet, 396*(10244), 129–142.
4. Khoshnam, S. E., Winlow, W., Farzaneh, M., Farbood, Y., & Moghaddam, H. F., (2017). Pathogenic mechanisms following ischemic stroke. *Neurological Sciences, 38*, 1167–1186.
5. Amantea, D., & Bagetta, G., (2017). Excitatory and inhibitory amino acid neurotransmitters in stroke: From neurotoxicity to ischemic tolerance. *Curr. Opin. Pharmacol., 35*, 111–119.
6. Holroyd, S., (2019). Urinary incontinence after stroke. *Br. J. Community Nurs., 24*(12), 590–594.
7. Leigh, R., Knutsson, L., Zhou, J., & Van, Z. P. C. M., (2018). Imaging the physiological evolution of the ischemic penumbra in acute ischemic stroke. *J. Cereb. Blood Flow Metab., 38*(9), 1500–1516.
8. Kanase, S. B., (2020). Effect of motor relearning program and conventional training on functional mobility in post-stroke patients. *Indian J. Public Health, 11*(05), 497.
9. Bredelyte, Aelita & Skarbalius, Egidijus & Rapoliene, Lolita. (2020). Analysis of the local contexts/Report My Way a European Collaborative And Innovative Partnership To Promote Physical Activity After Stroke Event.
10. Feigin, V. L., Norrving, B., & Mensah, G. A., (2017). Global burden of stroke. *Circ. Res., 120*(3), 439–448.
11. Béjot, Y., Bailly, H., Durier, J., & Giroud, M., (2016). Epidemiology of stroke in Europe and trends for the 21st century. *Presse. Med., 45*(12), e391–398.
12. Alkuraishy, H. M., Al-Gareeb, A. I., & Waheed, H. J., (2018). Lipoprotein-associated phospholipase A2 is linked with poor cardio-metabolic profile in patients with ischemic stroke: A study of effects of statins. *J. Neurosci. Rural Pract., 9*(04), 496–503.
13. Saklani, P., Khan, H., Gupta, S., Kaur, A., & Singh, T. G., (2022). Neuropeptides: Potential neuroprotective agents in ischemic injury. *Life Sciences, 288*, 120186.
14. Wang, Y., (2021). The study of ischemic stroke. *J. Clin. Nurs. Res., 5*(3).
15. Ręba, P., (2020). Stroke-diagnosis and treatment methods. *J. Educ. Heal Sport, 10*(5), 169–174.

16. Chen, Y. R., Hsieh, F. I., Lien, L. M., Hu, C. J., Jeng, J. S., Peng, G. S., et al., (2018). Rheumatoid arthritis significantly increased recurrence risk after ischemic stroke/transient ischemic attack. *J. Neurol., 265*(8), 1810–1818.
17. Amarenco, P., (2020). Transient ischemic attack. *New England Journal of Medicine, 382*(20), 1933–1941.
18. De Mendivil, A. O., Alcalá-Galiano, A., Ochoa, M., Salvador, E., & Millán, J. M., (2013). Brainstem stroke: Anatomy, clinical and radiological findings. In: *Seminars in Ultrasound, CT and MRI* (pp. 131–141). Elsevier.
19. Saver, J. L., (2016). Cryptogenic stroke. *N. Engl. J. Med., 374*(21), 2065–2074.
20. Caceres, J. A., & Goldstein, J. N., (2012). Intracranial hemorrhage. *Emerg. Med. Clin. North Am., 30*(3), 771.
21. Thapa, K., Khan, H., Singh, T. G., & Kaur, A., (2021). Traumatic brain injury: Mechanistic insight on pathophysiology and potential therapeutic targets. *Journal of Molecular Neuroscience, 71*(9), 1725–1742.
22. Javouhey, E., Guerin, A. C., & Chiron, M., (2006). Incidence and risk factors of severe traumatic brain injury resulting from road accidents: A population-based study. *Accid. Anal. Prev., 38*(2), 225–233.
23. Paiva, W. S., Soares, M. S., Amorim, R. L. O., De Andrade, A. F., Matushita, H., & Teixeira, M. J., (2011). Traumatic brain injury and shaken baby syndrome. *Acta Med. Port., 24*(5), 805–808.
24. Prabhakar, N. K., Khan, H., Grewal, A. K., & Singh, T. G., (2022). Intervention of neuroinflammation in the traumatic brain injury trajectory: *In vivo* and clinical approaches. *International Immunopharmacology, 108*, 108902.
25. McKinlay, A., Grace, R. C., Horwood, L. J., Fergusson, D. M., Ridder, E. M., & MacFarlane, M. R., (2008). Prevalence of traumatic brain injury among children, adolescents, and young adults: Prospective evidence from a birth cohort. *Brain Inj., 22*(2), 175–181.
26. Kraus, J. F., Rock, A., & Hemyari, P., (1990). Brain injuries among infants, children, adolescents, and young adults. *Am. J. Dis. Child., 144*(6), 684–691.
27. Khan, H., Kashyap, A., Kaur, A., & Singh, T. G., (2020). Pharmacological postconditioning: A molecular aspect in ischemic injury. *Journal of Pharmacy and Pharmacology, 72*(11), 1513–1527.
28. Wall, H. K., Beagan, B. M., O'Neill, H. J., Foell, K. M., & Boddie-Willis, C. L., (2008). Addressing stroke signs and symptoms through public education: The stroke heroes act FAST campaign. *Prev. Chronic Dis., 5*(2).
29. Godoy, D. A., Lubillo, S., & Rabinstein, A. A., (2018). Pathophysiology and management of intracranial hypertension and tissular brain hypoxia after severe traumatic brain injury: An integrative approach. *Neurosurg. Clin., 29*(2), 195–212.
30. Miller, S. L., & Yeh, H. H., (2017). Neurotransmitters and neurotransmission in the developing and adult nervous system. *Conn's Translational Neuroscience* (pp. 49–84). Elsevier Inc.
31. Niyonambaza, S. D., Kumar, P., Xing, P., Mathault, J., Koninck, P. De., Boisselier, E., et al., (2019). A Review of neurotransmitters sensing methods for neuro-engineering research. *Appl. Sci., 9*(21), 1–31.
32. Nicolo, J. P., O'Brien, T. J., & Kwan, P., (2019). Role of cerebral glutamate in post-stroke epileptogenesis. *Neuro Image Clin., 24*, 102069.
33. Blicher, J. U., Near, J., Næss-Schmidt, E., Stagg, C. J., Johansen-Berg, H., Nielsen, J. F., et al., (2015). GABA levels are decreased after stroke and GABA changes during rehabilitation correlate with motor improvement. *Neurorehabil. Neural Repair, 29*(3), 278–286.
34. Yuan, Y., Shan, X., Men, W., Zhai, H., Qiao, X., Geng, L., et al., (2020). The effect of crocin on memory, hippocampal acetylcholine level, and apoptosis in a rat model of cerebral ischemia. *Biomed Pharmacother., 130*, 110543.
35. Oliva, I., Fernández, M., & Martín, E. D., (2013). Dopamine release regulation by astrocytes during cerebral ischemia. *Neurobiol. Dis. [Internet], 58*, 231–241. Available from: Http://dx.doi.org/10.1016/j.nbd.2013.06.007.
36. Wardhan, R., & Mudgal, P., (2017). Nerve transmission. In: *Textbook of Membrane Biology* (pp. 205–222). Springer.
37. Olesen, S. P., Bentzen, B. H., & Timmermann, D. B., (2016). Structure and function. *Textb. Drug Des. Discov., 213*.
38. Spitzer, N. C., (2017). Neurotransmitter switching in the developing and adult brain. *Annu. Rev. Neurosci., 40*, 1–19.

39. Khan, H., Singh, A., Thapa, K., Garg, N., Grewal, A. K., & Singh, T. G., (2021). Therapeutic modulation of the phosphatidylinositol 3-kinases (PI3K) pathway in cerebral ischemic injury. *Brain Research, 1761*, 147399.
40. Spitzer, S., Volbracht, K., Lundgaard, I., & Káradóttir, R. T., (2016). Glutamate signaling: A multifaceted modulator of oligodendrocyte lineage cells in health and disease. *Neuropharmacology, 110*, 574–585.
41. Krzyzanowska, W., Pomierny, B., Budziszewska, B., Filip, M., & Pera, J., (2016). N-acetylcysteine and ceftriaxone as preconditioning strategies in focal brain ischemia: Influence on glutamate transporters expression. *Neurotox. Res., 29*(4), 539–550.
42. You, J., Feng, L., Xin, M., Ma, D., & Feng, J., (2018). Cerebral ischemic postconditioning plays a neuroprotective role through regulation of central and peripheral glutamate. *Biomed Res Int., 2018.*
43. Akaike, A., & Izumi, Y. (2018). Overview. In: Akaike, A., Shimohama, S., Misu, Y., eds. Nicotinic Acetylcholine Receptor Signaling in Neuroprotection [Internet]. Singapore: Springer; 2018. Chapter 1. Available from: https://www.ncbi.nlm.nih.gov/books/NBK543551/ doi: 10.1007/978-981-10-8488-1_1.
44. Gupta, A., Khan, H., Kaur, A., & Singh, T. G., (2021). Novel targets explored in the treatment of alcohol withdrawal syndrome. *CNS & Neurological Disorders-Drug Targets (Formerly Current Drug Targets-CNS & Neurological Disorders), 20*(2), 158–173.
45. Quillinan, N., Herson, P. S., & Traystman, R. J., (2016). Neuropathophysiology of brain injury. *Anesthesiol Clin., 34*(3), 453–464.
46. Khan, H., Sharma, K., Kumar, A., Kaur, A., & Singh, T. G., (2022). Therapeutic implications of cyclooxygenase (COX) inhibitors in ischemic injury. *Inflammation Research*, 1–6.
47. Sharma, V. K., Singh, T. G., & Mehta, V., (2021). Stressed mitochondria: A target to intrude Alzheimer's disease. *Mitochondrion, 59*, 48–57.
48. Chen, C., Zhou, X., He, J., Xie, Z., Xia, S., & Lu, G., (2019). The roles of GABA in ischemia-reperfusion injury in the central nervous system and peripheral organs. *Oxid. Med. Cell Longev., 2019.*
49. Mele, M., Costa, R. O., & Duarte, C. B., (2019). Alterations in GABAA-receptor trafficking and synaptic dysfunction in brain disorders. *Front Cell Neurosci., 13*, 1–16.
50. Hiu, T., Farzampour, Z., Paz, J. T., Wang, E. H. J., Badgely, C., Olson, A., et al., (2016). Enhanced phasic GABA inhibition during the repair phase of stroke: A novel therapeutic target. *Brain, 139*(2), 468–480.
51. Rashmi, D., Zanan, R., John, S., Khandagale, K., & Nadaf, A., (2018). γ-aminobutyric acid (GABA): Biosynthesis, role, commercial production, and applications. *Studies in Natural Products Chemistry* (1st edn. Vol. 57, pp. 413–452). Elsevier B.V.
52. Behuet, S., Cremer, J. N., Cremer, M., Palomero-Gallagher, N., Zilles, K., & Amunts, K., (2019). Developmental changes of glutamate and GABA receptor densities in Wistar rats. *Front Neuroanat., 13*, 1–14.
53. Sears, S. M. S., & Hewett, S. J., (2021). Influence of glutamate and GABA transport on brain excitatory/inhibitory balance. *Exp. Biol. Med., 246*(9), 1069–1083.
54. Beltrán, G. A. N., López, P. M. I., & Calvo, D. J., (2020). Reactive oxygen species in the regulation of the GABA mediated inhibitory neurotransmission. *Neuroscience, 439*, 137–145.
55. Liu, J., Zhang, J., & Wang, L. N., (2018). Gamma aminobutyric acid (GABA) receptor agonists for acute stroke. *Cochrane Database Syst, Rev., 2018*(10).
56. Winek, K., Soreq, H., & Meisel, A., (2021). Regulators of cholinergic signaling in disorders of the central nervous system. *J. Neurochem., 158*(6), 1425–1438.
57. Granger, A. J., Mulder, N., Saunders, A., & Sabatini, B. L., (2016). Cotransmission of acetylcholine and GABA. *Neuropharmacology, 100*, 40–46.
58. Guo, J. M., Zhang, L., Niu, X. C., Shu, H., Wang, L., Su, D. F., et al., (2017). Involvement of arterial baroreflex and nicotinic acetylcholine receptor α7 subunit pathway in the protection of metformin against stroke in stroke-prone spontaneously hypertensive rats. *Eur. J. Pharmacol., 798*, 1–8.
59. Kalra, P., Khan, H., Kaur, A., & Singh, T. G., (2022). Mechanistic insight on autophagy modulated molecular pathways in cerebral ischemic injury: From preclinical to clinical perspective. *Neurochemical Research, 7*, 1–9.
60. Mehla, J., Reeta, K. H., Gupta, P., & Gupta, Y. K., (2010). Protective effect of curcumin against seizures and cognitive impairment in a pentylenetetrazole-kindled epileptic rat model. *Life Sci., 87*(19–22), 596–603.

61. Mesut, I., Beydemir, Ş., Aslan, Y., Emin, M., & Esra, H., (2017). *Oxidative Stress and mRNA Expression of Acetylcholinesterase in the Leukocytes of Ischemic Patients, 87*, 561–567. ScienceDirect.
62. Grewal, A. K., Singh, N., & Singh, T. G., (2019). Effects of resveratrol postconditioning on cerebral ischemia in mice: Role of the sirtuin-1 pathway. *Canadian Journal of Physiology and Pharmacology, 97*(11), 1094–1101.
63. Liang, F., Xu, Q., Jiang, M., Feng, R., Jiang, S., Yuan, B., et al., (2021). Emotion induced monoamine neuromodulator release affects functional neurological disorders. *Front Cell Dev. Biol., 154*.
64. Rihal, V., Khan, H., Kaur, A., & Singh, T. G., (2022). Vitamin D as therapeutic modulator in cerebrovascular diseases: A mechanistic perspectives. *Critical Reviews in Food Science and Nutrition, 5*, 1–23.
65. Ellis, H., Calne, R., & Watson, C., (2016). *General Surgery*. John Wiley & Sons.
66. Kim, Y., (2018). *The Role of Neuroinflammation and Neurovascular Dysfunction in Major Depressive Disorder*, 179–192.
67. Fišar, Z., (2016). Drugs related to monoamine oxidase activity. *Prog. Neuro-Psychopharmacology Biol. Psychiatry., 69*, 112–124.
68. Mayor, D., & Tymianski, M., (2018). Neurotransmitters in the mediation of cerebral ischemic injury. *Neuropharmacology, 134*(Pt B), 178–188.
69. Khan, H., Gupta, A., Singh, T. G., & Kaur, A., (2021). Mechanistic insight on the role of leukotriene receptors in ischemic–reperfusion injury. *Pharmacological Reports, 73*(5), 1240–1254.
70. Komsiyska, D., (2018). Oxydative stress and post-stroke depression. *Tr. J. Sci., 3*, 249–253.
71. Costiniti, V., Spera, I., Menabo, R., Palmieri, E. M., Menga, A., Scarcia, P., et al., (2018). Monoamine oxidase-dependent histamine catabolism accounts for post-ischemic cardiac redox imbalance and injury. *Biochim. Biophys. Acta (BBA)-Molecular Basis Dis., 1864*(9), 3050–3059.

CHAPTER 11

Role of the Renin-Angiotensin-Aldosterone System (RAAS) in Amelioration of Stroke Pathophysiology

RAJENDER SANDHU,[1] SHILPA KUMARI,[1] SMRITI THAKUR,[1] RAHUL DESHMUKH,[1] and THAKUR GURJEET SINGH[2]

[1]*Department of Pharmaceutical Sciences and Technology, Maharaja Ranjit Singh Punjab Technical University, Bathinda, Punjab, India*

[2]*Chitkara College of Pharmacy, Chitkara University, Punjab, India*

ABSTRACT

A stroke is a neurological deficit caused by an acute focal injury to the central nervous system (CNS). It is one of the major causes of death and dysfunctionality worldwide. Cerebral infarction, intracerebral hemorrhage (ICH), or subarachnoid hemorrhage (SAH) are the vascular events that lead to stroke. According to figures from the World Health Organization (1996), cardiovascular disease is the most common illness worldwide, accounting for at least 15 million deaths per year, or 30% of the total. It has been shown that recurring cardiac episodes can be decreased by using the RAAS system modulation, still, its mechanism in preventing re-occurring cerebrovascular mortality remains unknown. RAAS is a circulating tissue-dependent endocrine system. Through angiotensin II (Ang II) activity, it controls body equilibrium and tissue feedback to an infraction in CNS as well as in the periphery area. The RAAS system has been related to the development and growth of cerebrovascular diseases in several animal studies. The risks of a stroke may be diminished by suppressing the RAS system and it may be useful in cerebrovascular events that aren't connected to blood pressure fluctuations. Despite contradicting evidence, numerous clinical studies show that the type I receptor antagonists of angiotensin II, ACE inhibitors, and ARBs in addition to decreasing blood pressure, give cerebroprotective benefits. This chapter looked at how the RAS systems in the brain and heart are involved in stroke, as well as how the RAAS system is manipulated in stroke treatment. Also, the potential therapeutic effects of ACE inhibitors and ARBs in lowering stroke risks and managing ischemic-reperfusion damage.

Ischemic Injury: A Molecular Insight. Thakur Gurjeet Singh & Heena Khan (Eds.)

11.1 INTRODUCTION

Stroke is a leading cause of disability and death across the globe. It is described as a neurological impairment produced by an immediate focused injury to the central nervous system (CNS) caused by a vascular source, such as cerebral infarction, intracerebral hemorrhage (ICH), or subarachnoid hemorrhage (SAH) [1]. Despite its frequent usage, the term "stroke" isn't defined consistently in clinical practice, clinical research, or public health assessments. After ischemic heart disease, stroke is the second biggest cause of death globally, and it is anticipated to remain so until 2030. In addition, stroke survivors may have challenges, demanding temporary or permanent assistance, resulting in a significant financial and human costs. Evidence suggests that socioeconomic deprivation affects not just the severity and mortality of strokes, but also the incidence of strokes at a younger age [1]. Although stroke is the third largest cause of mortality after heart disease and cancer, it leaves 80% of survivors with lasting disability.

The two forms of strokes are ischemic and hemorrhagic, with ischemic strokes accounting for 85% of all strokes. Cerebrovascular disease, especially stroke, accounts for a significant portion of disability-adjusted life years and deaths globally (47–67%). Obstruction of an artery caused by an embolus or thrombus is known as an ischemic stroke. The blocked artery determines the size of the ischemic area in the brain, which in turn determines the functional and metabolic abnormalities that occur during an ischemic stroke [2]. To reduce the burden of stroke in the community, it is necessary to identify modifiable risk factors and demonstrate the effectiveness of risk reduction initiatives. There are various modifiable (e.g., diet) and nonmodifiable (e.g., concomitant conditions) risk factors for stroke (e.g., age, race). Risk factors for stroke include short-term hazards or triggers (e.g., viral events, sepsis, and stress), intermediate-term risk factors (e.g., hypertension and hyperlipidemia), and long-term risk factors (e.g., hypertension and hyperlipidemia) (e.g., sex and race). Young individuals' stroke risk factors are likely to vary from those of older adults [3]. Ischemia/reperfusion injury (IRI) is a term that describes the functional and anatomical changes that occur once blood flow is restored after a period of ischemia. Blood flow repair may have potentially serious side effects, such as necrosis of irreversibly wounded cells, substantial cell growth, and nonuniform flow restoration to all parts of the tissue, in addition to treating ischemia.

The renin-angiotensin-aldosterone system (RAAS) is a circulating hormonal system that controls tissue homeostasis and injury responses. This system regulates extracellular fluid volume and arterial pressure to maintain vascular tonicity. Water, blood, plasma, lymph, and interstitial fluid are strictly managed by this system for the pulsing organ heart and filtering organ kidney to operate properly. This system's disruption may cause blood pressure to fluctuate, resulting in chronic or acute disorders, as well as abrupt death. The RAS in the brain and the RAS in the heart both have a role in modulating ischemia damage and reperfusion by lowering angiotensin II activity, which is linked to the pathophysiology. ACE inhibitors and ARBs are often used to avoid ischemic stroke and reperfusion damage, and thereby aid in the overactivation of the RAAS system in stroke pathogenesis as well depicted in Figure 11.1. In this chapter, we will attempt to explore the major activities of the RAAS system and how to improve the actions mediated by angiotensin II, which are typically pathogenic, and how medicines might help avoid stroke recurrence.

11.2 RAAS AND ISCHEMIC REPERFUSION INJURY

Ischemia-reperfusion injury is a dangerous disorder that requires doctors to control cell destruction while still preserving organ function. The term "ischemia" refers to tissue hypoperfusion. A range of situations, including sepsis, acute coronary syndrome (ACS), organ transplantation, and limb damage, may result in tissue hypoperfusion. When there is a shortage of oxygen in cellular components, it damages blood vessels and the blood perfused in the ischemic organ or cell, resulting in biomembrane damage and mitochondrial malfunction, as well as a fall in calcium ion concentration in the cytoplasm and mitochondria. Caspases are activated when a mitochondrial malfunction causes a reduction in ATP synthesis. There is a rise in reactive oxygen species and free radicals as a result of reperfusion, which increases autophagy, endoplasmic reticulum stress, and inflammation [4]. All of these processes eventually result in cell death and the process is well depicted in Figure 11.1. To maintain organ function, previous research has focused on reducing the time period that organs are hypo-perfused. Early goal-directed therapy for prompt resuscitation in sepsis is recommended by the Surviving Sepsis Campaign, which involves early antibiotic treatment and resuscitation with enough fluids and vasopressors to minimize peripheral tissue hypoperfusion. In patients with the ACS, the ACCF/AHA/SCAI guidelines advocate early revascularization to reduce myocardial damage. Both organ transplantation and limb amputation have received similar advice. However, clinical results after the restoration of blood flow to ischemic tissue are not optimum. Recent research has indicated that reperfusion has the potential to injure ischemic tissue in the future, a phenomenon known as ischemia-reperfusion injury. The RAAS is a circulating and tissue-based hormonal system that controls tissue homeostasis and damage responses [5]. In both systems, angiotensin II is the principal effector hormone. Ang II controls endothelial function and promotes inflammatory, proliferative, fibrotic, and thrombotic processes in the vasculature, as well as controlling salt and water balance via aldosterone [6]. The Ang II Type 1 receptor (AT1), which is found in a variety of organ systems and on a variety of cells, including vascular smooth muscle, endothelium, and macrophages, is principally responsible for Ang II's well-known effects [7]. The activation of a second Ang II receptor, type II (AT2), which is present mostly in the brain, vasculature, heart, and a few other tissues, may counterbalance the effects of AT1 receptor stimulation and initiate tissue regeneration or apoptosis [8].

Mechanism of cell death initiates when there is a lack of oxygen in cellular components and thus it damages blood vessels and the blood perfused in the ischemic organ or cell and causes biomembrane damage which induces mitochondrial dysfunction as well as a decrease in calcium ion concentration in cytoplasm and mitochondria as well. Mitochondria dysfunction causes a decrease in ATP production which activates caspases. Due to reperfusion, there is an increase in reactive oxygen species and free radicals as well which increase autophagy and endoplasmic reticulum stress and inflammation. All these pathways ultimately lead to cell death.

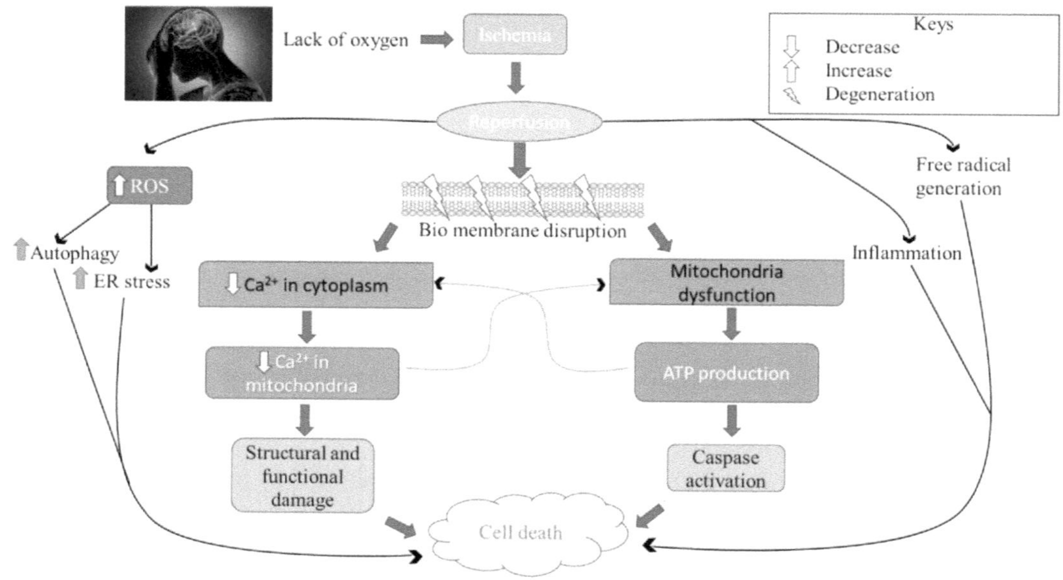

FIGURE 11.1 Stroke-induced pathological changes lead to cell death.

11.3 CEREBROVASCULAR SYSTEM AND STROKE

After a stroke event, the primary brain injury initiates which is oftentimes non-reversible. In ischemic reperfusion injury, immediate blockage of blood flow in vascular sections causes brain cell death, resulting in an ischemic scar surrounding a hypo-perfused zone known as the penumbra. Intracerebral hemorrhage is caused when a blood vessel ruptures, permitting vascular components to enter the brain directly, resulting in a hematoma and structural damage. This initial damage sets off a chain reaction that includes excitotoxicity, oxidative stress, and mitochondrial dysfunction, all of these factors lead to brain injury [9] followed by a brain hemorrhage, the activation of thrombin, and release of clot components (e.g., hemoglobin and iron) are also implicated [10]. In cerebrovascular research, the RAAS system plays a remarkable role. It regulates blood pressure, vasoconstriction, thrombosis, and arterial wall damage [11]. Angiotensin I (ATI) is derived from the cleavage of angiotensinogen (AGT) via the action of renin release from the juxtaglomerular cells in the kidney. The ATI is converted into angiotensin II (ATII) by the action of the angiotensin-converting enzyme (ACE). ATII has a fast onset of action and has a more intrinsic ability. ATII also increases intravascular volume by promoting the synthesis of aldosterone, which is responsible for salt absorption. Plasminogen activating inhibitor levels are stimulated by the action of aldosterone and ATII receptors which lead to coagulation and activate the platelets in the damaged area. In several studies, the RAAS has been associated with brain injury (BI) [12]. The Heart Outcomes Prevention Evaluation (HOPE), Losartan Intervention for Endpoint Reduction in Hypertension Study (LIFE), and acute candesartan cilexetil evaluation in stroke survivors (ACCESS) studies show that by inhibiting RAAS the chances of cardiovascular and brain injury could be reduced, by numerous pathways rather than lowering blood pressure [13] (Figure 11.2).

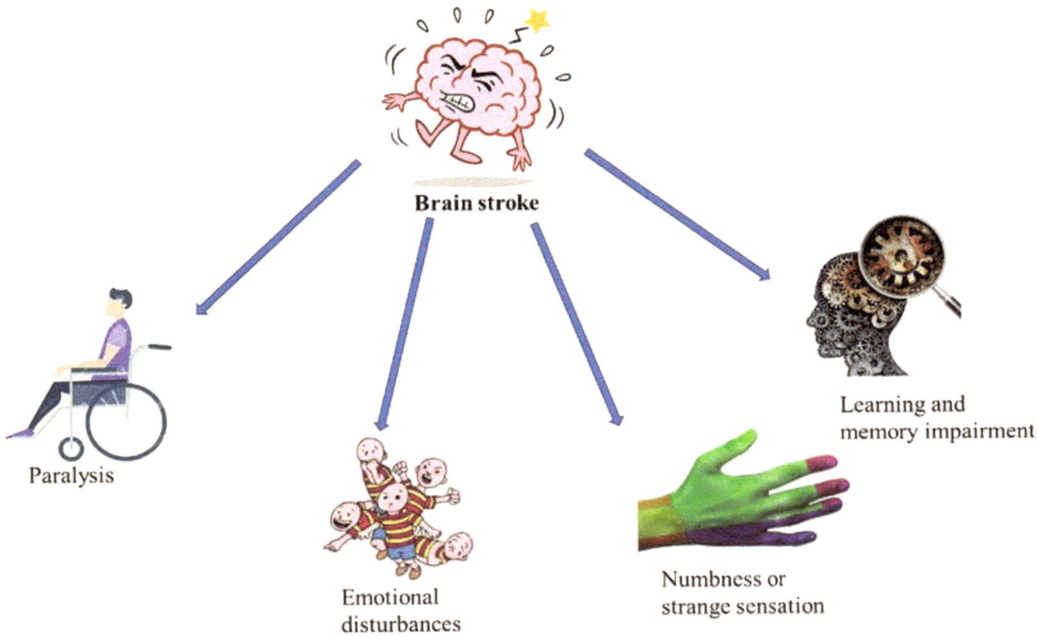

FIGURE 11.2 Various secondary comorbidities after stroke.

After the stroke, the expected events can be of following such as paralysis, emotional disturbances, numbness or strange sensation, and impairment of learning and memory.

11.3.1 CEREBRAL INFLAMMATION FOLLOWING STROKE

A powerful inflammatory response is triggered after a stroke. Even though the pathogenic pathways of initial damage differ, both ischemic and hemorrhagic strokes activate the immune system in the same way. The production of damage-associated molecular patterns (DAMPs) from damaged cells triggers the immunological response [14, 15]. In AIS and intracerebral hemorrhage, various factors influence the pathogenesis such as heat shock proteins, interleukins (IL), and high mobility group box 1 (HMGB1) [16] shows involvement in both AIS and intracerebral hemorrhage [15]. Intracellular signaling pathways are activated by the (DAMPs). The pattern recognition receptors (PRRs) present on the immune cells detect the DAMPs and initiate transduction pathways. Immune responses in the brain are triggered by the DAMPs such as coagulation factors, blood components, and mainly intracerebral hemorrhage [17]. The very first cell population in the CNS is microglia which react to these death signals. Microglia are activated following an injury and undergo physiological changes and secrete cytokines. Widely distributed glial cells in the brain are astrocytes. These astrocytes take part in the immunity reactions and cellular damage by engaging outermost immune cells and via microgliosis releases chemokines and cytokines. Additionally, ischemic neuron survival is increased by the astrocytes during the acute phase [18].

11.3.2 ROLE OF BRAIN RAS

The renin enzyme cleaves AGT into the Ang I peptide. Renin is plentiful in the blood, but only traces of it are present in the brain. According to numerous studies, renin and AGT are both present in distinct brain parts. Renin was shown to be expressed both intracellularly and as pro-renin released by neurons. Pro-renin binds to pro-renin receptors (PRRs) in the brain with a higher affinity than renin, causing AGT cleavage. Activation of renin/pro-renin signaling, as well as stimulation of angiotensin receptors, causes cognitive impairment. According to Van Thiel et al. intracellular renin is thought to participate in the production of angiotensins in the brain as well as the coordination of non-RAS activities [19]. When the blood volume or blood pressure decreases it stimulates the juxtaglomerular cells in the kidney to produce renin which rises the blood volume via the aldosterone system and RAS system but in normal physiology increased blood volume inhibits the release of renin. Renin hydrolyzes circulating AGT to create Angiotensin I (Ang I), which is then converted to physiologically active octapeptide, and Ang II, produced by ACE. Ang II operates by activating Ang II type I and type II receptors (AT1R and AT2R). Because Ang II has a greater affinity for the AT1R, it constricts blood vessels, raises blood pressure, and stimulates aldosterone synthesis in the adrenal glands, boosting water and salt reabsorption in the kidneys and thereby increasing blood volume levels [20]. The brain RAAS appears to be involved in heart dysfunction and ischemia, and experimental investigations have indicated that blocking RAAS has an anti-ischemic effect. In fact, in hereditary hypertensive rats, the overactivation of cerebral AT1 receptors can cause cerebral blood vessels inflammation and increases nitric oxide levels. Blocking the AT1 receptor, on the other hand, may minimize the risk of cerebrovascular ischemia and neuronal damage [21]. The ACE2/A1-7/MAS1 axis is found in the central nervous systems of both animals and humans, and its activation results in baroreflex facilitation, hypotension, the release of bradykinin, vasopressin, and nitric oxide, as well as norepinephrine inhibition. A1-7 therapy may have a prospective influence on cerebral protection through the synthesis of bradykinin and nitric oxide, possibly enhancing cerebrovascular reserve following ischemic damage, based on these processes. Data showed that a prolonged central injection of A1-7 reduces the size of infarct and provided neuroprotection in a mouse stroke model [22]. Thus, implicating role of brain RAS in stroke and its possible mechanisms via which this pathology can be targeted and RAS could be helpful in modulation of stroke events.

11.3.3 ROLE OF AT1 RECEPTORS

Angiotensin receptors are G-protein superfamily receptors such as AT1 and AT2 receptors, although their expression and signaling routes differ. AT1 receptor is found all over the body and is completely expressed in adults [23]. The ATI receptor is responsible for the various major activities of Ang II such as salt and water retention, vasoconstriction, and blood pressure management. All these actions are governed by the G-protein signaling pathways of phospholipase A29, D, and C [24]. Stimulation of Ang II via ATI receptor might cause cessation of cerebral blood flow and local vasoconstriction which leads to ischemic injury [25]. In ischemic injury cells, death and apoptosis are caused by increased load or reactive oxygen species and inflammation mediated by the Ang II via ATI receptor. According to findings neuronal apoptosis and necrosis have been involved After an ischemic stroke. Cells near the periphery of the lesion (penumbra) survive but are mostly damaged, those damaged or half-survived cells can be preserved with the restoration of the local blood supply [26]. The volume and amount of brain damage are reduced when these cells are revived. Therefore, disruption of the activity of Ang II by inhibition of angiotensin type 1 receptors may minimize the size of the brain infarct [6]. Thus, all these data suggest that inhibition of Ang II actions via ATI receptors could be a beneficial target in reducing the ischemic stroke pathologies (Figure 11.3).

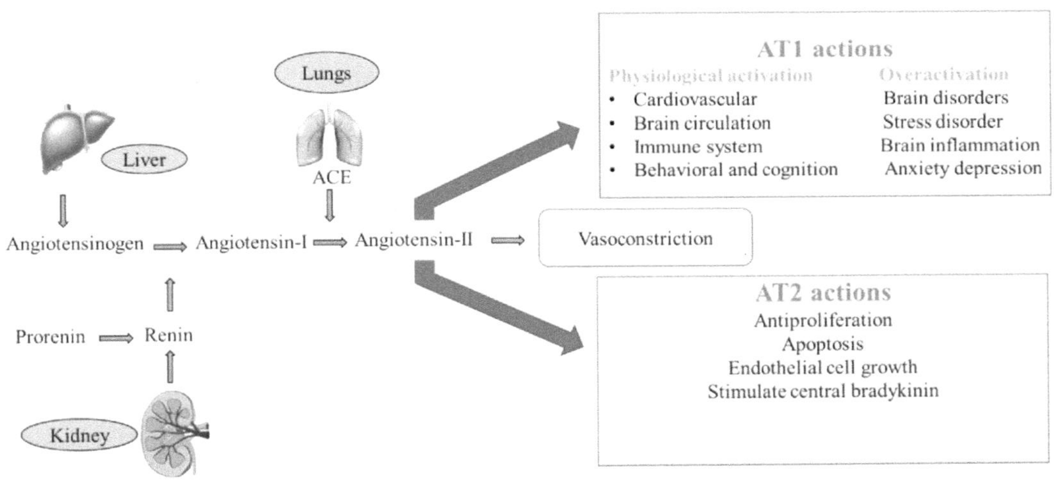

FIGURE 11.3 RAAS induced pharmacological changes via angiotensin I and II, and role of AT1 and AT2 receptors.

The liver releases the AGT enzyme which is converted by renin into angiotensin-1. Ang I is converted by ACE enzyme into Ang II. Angiotensin II causes vasoconstriction and aldosterone activation which regulates salt and water balance in the body it also binds to AT1 and T2 receptors which initiates further pathways like antiproliferation apoptosis via activation of angiotensin type 1 receptor and cerebrovascular and cardiac events via stimulation of angiotensin type 1 receptors.

11.3.4 ROLE OF AT2 RECEPTORS

AT2 receptors are less abundant when compared to AT1 receptors which are widely distributed throughout the body. The quantity of AT2 receptors decreases with an increase in age. Although they may be found in vascular endothelial cells, the heart, central and peripheral neurons, and other organs [23]. AT2 receptors increase when there is brain damage, mitochondrial infraction, and tissue injury [27]. Western blot research revealed a 2.1-fold overactivation of AT2 receptors in the ischemic region of the brain following middle cerebral artery occlusion in rats, according to Li and colleagues. The AT1 receptors, on the other side, exhibited no significant alterations. The AT2 receptor's signaling mechanisms are unknown, although they include phospholipase A2 stimulation, which contributes to lipid messenger modulation, and elevated levels of RAS p21, a protein kinase engaged in synaptic function and neurite growth [28]. Other researchers have discovered that stimulating AT2 receptors in rats protect cognitive function. As a result, the AT2 receptor appears to play a key part in Ang II's stroke and neural protective actions [29]. According to new investigations, the peptide AT2-R agonist CGP42112 was administered intracerebroventricularly before or after a stroke, with a considerable reduction in the infarct area and recovery in neurological deficits [30]. When the AT1 receptors in the brain are inhibited, the AT2 receptor causes vasodilation, which can minimize neuronal cell death and promote neurite regeneration after ischemia [31]. Unlike ACE inhibitors, ARBs have a different pharmacodynamic action on the AT2 receptor. ACE inhibition decreases angiotensin II synthesis while concurrently activating AT1 and AT2 receptors. AT1 receptor inhibition enhances angiotensin II formation and renin release. In the brain, where AT2 receptor antagonists were demonstrated to have a large influence when AT1 receptors are inhibited, these reflective strategies of AT2 receptor stimulation seem to be efficacious [32].

11.4 CARDIOVASCULAR DISORDERS AND IR INJURY

Lack of oxygen supply causes ischemia in the body. Reperfusion is the restoration of blood flow in ischemic tissues, whereas deprivation is tissue deprivation. The myocardial ischemia-reperfusion injury occurs when tissue is damaged during reperfusion following an ischemic Injury (I/R). It causes cardiac muscle injury by increasing the overexpression of oxygen, sodium, and calcium ions, all leading to protease activation and cell death. Myocardial ischemia and reperfusion damage are complicated by the renin-angiotensin system (RAS) in the heart. Vasodilation is caused by angiotensin 1–7, whereas vasoconstriction is caused by angiotensin II [33]. RAAS system is the main pharmacological target in cardiovascular treatment. In cardiovascular unfavorable remodeling and morbidity, RAAS works as a 'non-specific amplifier' of risk factors like hypertension and insulin resistance, as well as a 'local amplifier' by generating deleterious cardiac and vascular remodeling directly. In terms of cardiovascular endocrine regulation, plasma renin concentration is the rate-limiting stage in the production of angiotensin-II (Ang-II) from its circulating substrate, AGT. As a result of renin activity, AGT, which is largely released by the liver, circulates at concentrations that are equal to or lesser than the Michaelis constant (KM) for renin and has an estimated plasma half-life of several hours. Although AGT has no acute regulatory features, its basic activity may play a role in RAAS activity modulation over time. ACE is a non-rate limiting enzyme that creates angiotensin-II from the renin-produced decapeptide angiotensin-I (Ang-I). It is abundantly distributed on practically all endothelial cells and has a soluble unit that circulates in plasma. The G-protein family receptors such as AT1R and AT2R are responsible for the Ang-mediated cardiovascular effects. As a result, the RAAS is a unique endocrine system in that the fundamental point of control is the release of an enzyme rather than the active form of the hormone [34].

11.4.1 ROLE OF HEART RAS

Heart AGT is an α2-globulin glycoprotein produced in the heart muscle. AGT gene expression has been discovered in mice and rats. The ventricles have a lower concentration of AGT than the heart atria [35]. The level of cardiac AGT mRNA has been recorded. The liver content of rats and mice is roughly 5% and 1%, respectively. Glucocorticoid In the rat heart, hormone therapy stimulates the formation of hepatic AGT. There are two kinds of angiotensin: angiotensin I and angiotensin II. Angiotensin I is a hormone that regulates blood pressure. Have no physiologic effect on the renin-angiotensin system in the heart They are (Angiotensin II and angiotensin III precursors) [36]. Renin is an effector hormone that plays a key function in the heart which causes hypertension by elevating blood pressure. Vasoconstriction and salt and fluid retention in the body. Angiotensin II is also responsible for inducing inflammatory responses. NADPH oxidase, phospholipase A2, and JAK/STAT are examples of mediators (STAT is a signaling pathway). Angiotensin II was discovered in 1999 by De Leeuw. Treatment with converting enzymes lowers angiotensin II levels [33]. In myocardial ischemia and reperfusion damage (I/R), the renin-angiotensin system (RAS) plays a critical role. Angiotensin-II induced preconditioning (APC) and ischemia preconditioning (IPC) displayed cardioprotection by influencing mitochondrial respiration and cardiac functions, according to the cardiac RAS system, which includes AGT, Ang I, Ang II, ACE, and ACE 2. Angiotensin (1–7) inhibits the progression of heart failure following myocardial infarction in rats, according to Loot et al. Angiotensin (1–7) affects cardiac functions, blood pressure, cardiac hypertrophy, heart failure, and cell proliferation, according to Trask et al. By producing bradykinin and nitric oxide, angiotensin (1–7) has an anti-proliferative effect on vascular smooth muscles, cardiac muscle cells, and enhances endothelial function [37]. The AT1 receptor is primarily responsible for angiotensin II's harmful cardiovascular effects. Vasoconstriction, aldosterone secretion, oxygen radical generation, vascular and myocardial hypertrophy, fibrosis, and sympathetic activation are all examples of this [38]. Angiotensin II type 2 (AT2) receptor is involved in the proliferation, apoptosis, and differentiation of embryonic tissues, as well as pathophysiologic processes such as wound healing in adults [39]. Furthermore, AT2 receptors have been found in differentiated tissues such as the endothelium, fibroblasts, myocytes, and neurons, where they appear to counteract the activities of the AT1 receptor subtype [40].

11.5 MODULATION OF RAAS

According to growing data based on recent clinical trials, blocking RAAS could help to prevent strokes. The current data on the best antihypertensive medication strategy for initial stroke prevention is still not clear. To obtain solid conclusions, larger, better-planned clinical investigations are required [41]. Furthermore, antihypertensive therapy may be employed in the future to prevent cognitive decline or possibly to improve mildly compromised brain function. Some clinical evidence suggests that calcium channel antagonists, ACE inhibitors, and ARBs have benefits in stroke reduction. National authorities say that all main types of antihypertensive medications are beneficial for primary stroke prevention based on existing evidence [42]. The Losartan Intervention for Endpoint study (LIFE) assigned 9,193 hypertension patients to once-daily atenolol or losartan with 4.8 years of average follow-up [43]. No remarkable differences in blood pressure values between the two groups were observed. When compared to atenolol, losartan reduced the incidence of lethal and non-lethal stroke by 25% in people with primary hypertension with left ventricular hypertrophy [44]. However, it was unclear whether this reduction was due only to losartan's beneficial benefits or also to atenolol's detrimental vascular effects. According to a study of the Heart Outcomes Prevention Evaluation (HOPE), ramipril prevents stroke in most vulnerable patients without affecting their blood pressure. The Valsartan Antihypertensive Long Term Use Evaluation (VALUE) study, in which amlodipine with valsartan treatment regimens are compared in hypertensive patients that are at high cardiovascular risk. The study results found no difference in stroke incidence between the two groups [45]. Furthermore, very minimal differences in systolic and diastolic blood pressure were observed across the therapy groups, and it was also observed that amlodipine achieved blood pressure control more quickly than valsartan. When comparing candesartan-treated elderly patients to controls, the Study on Cognition and Prognosis in the Elderly (SCOPE) discovered a significant decrease in the rate of non-fatal strokes (27.8%). The candesartan group had a somewhat higher blood pressure decrease, although it's unclear if this difference accounted for the reported stroke reduction [46].

11.5.1 ACE INHIBITORS IN THE MANAGEMENT OF STROKE

According to experimental research, a 28-day pre-treatment with the ACE inhibitor captopril successfully reduced the infarct region by 25%. In a pre-treatment study with ARB candesartan for 5–10 days, but not with ACEIs such as ramipril or enalapril, the reduction in infarct size and enhanced neurological function were observed. When ramipril comparison was done to ARB telmisartan, similar findings were obtained, and the combination of the two did not give any significant advantage above telmisartan alone. Another study demonstrated that, despite lower blood pressure, ACEI pre-treatment reduced infarct volume in normotensive Wistar–Kyoto (WKY) rats yet not in spontaneously hypertensive rats (SHRs). It was found from studies that ACEIs are less effective than ARBs in AIS treatment. A drop in Ang II levels results in reduced activity of Ang II and AT2R in the brain [47]. In the Perindopril Protection Against Recurrent Stroke Study (PROGRESS), treatment with perindopril contributed to only a 5% non-significant decline in recurrent stroke for a 5/3 mm Hg BP reduction. However, when compared to a combination of perindopril and the diuretic indapamide for a 12/5 mm Hg BP reduction, the decline in stroke occurrence was more than 43% [48]. Because a 1.0-mm Hg decline in systolic BP (SBP) is thought to account for a 3–4% reduction in strokes, the additional 7-mm Hg decline in SBP with combination treatment could have been responsible for a 28% decrease in strokes rather than the 43%. The antihypertensive and lipid-lowering treatment to prevent heart attack trial (ALLHAT) found similar effects in black individuals. The intent-to-treat analysis found a 25% rise in stroke events rate in individuals of the captopril-based treatment regimen, compared to a 43% reduced prevalence of stroke in those medicated with α-blocker/diuretic regimen, according to the Captopril Prevention Project (CAPP). However, the increased stroke risk in the captopril group can be attributable to worse blood pressure management in this group [50]. The 25% increased stroke rate may not be due to a 3-mm Hg difference in SBP. However, among individuals with lower blood pressure, the drop appeared to be comparable in both treatment groups (Table 11.1).

TABLE 11.1 Various Drugs with Their Class and Mechanism of Action Are Used in Stroke Prevention

Class	Drug	Mechanism	References
Direct renin inhibitors	Aliskiren	Inhibit production of angiotensin I by angiotensinogen.	[51]
ACEI	Captopril	Blocks the conversion of angiotensin I to angiotensin II.	[52]
ARBs	Telmisartan	Blocks the action of angiotensin II type 1 receptor.	[53]
Aldosterone receptor antagonists	Spironolactone	Inhibits the action of aldosterone in distal renal tubules.	[54]
Calcium channel blockers	Amlodipine	Reduces calcium entry in cardiac and muscular cells and causes vasodilation.	[55]
Beta-blockers	Atenolol	Blocks beta I receptors in the heart decreases heart rate and workload.	[56]
Thrombolytic drugs	Streptokinase	It binds to plasminogen for the generation of plasmin and produces fibrin degradation products.	[57]

11.5.2 ROLE OF ARBS

In AIS therapy of ARBs has long been investigated. In pre-and post-therapy of stroke using ARBs has shown neuroprotective, antioxidant, and anti-inflammatory properties according to pre-clinical investigations when compared to ACEIs. Prolonged pre-treatment using ARB protected against cerebral ischemia and boosts stroke recovery in both non-hypertensive and hypertensive rats. Following a stroke, ARB treatment provided consistent neuroprotection and behavioral improvement [58]. According to studies, normotensive rats were given a single dose of 1 mg/kg of candesartan for hypotensive action. After both permanent and temporary middle artery cerebral blockage, treatment with candesartan after stroke resulted in neuroprotective action and functional recovery. Candesartan preventive advantages were only seen at a hypotensive dose in hypertensive rats, whereas a similar dose resulted in an extreme BP drop following a stroke in hypertensive rats [32]. As per investigations, sub-hypotensive ARB doses, while combined with tissue-type plasminogen activator (tPA), enhance stroke survival, indicating clear protection independent of blood pressure impacts. *In vitro* evidence demonstrates that cell damage was larger in cortical neurons derived from WT mice than in AT1aR KO mice after 2 hours of oxygen-glucose deprivation (OGD), as measured by lactate dehydrogenase (LDH) formation. Interestingly, ARB losartan pre-treatment declined the injury of WT neurons in comparison with AT1a KO neurons. The health benefits of pre-treatment of ARB towards glutamate excitotoxicity or OGD/reperfusion were due to AT1R blockade but not due to indirect activation of AT2R. Although it has been shown that long-term ARB medication decreased ischemic stroke risk, immediate posttreatment prevention for stroke patients has yet to be proved [59]. In ACCESS research, candesartan treatment on 342 participants for 12-month decreased mortality rate and vascular diseases while being ineffective on blood pressure [60]. This was followed by the Scandinavian Candesartan Acute Stroke Trial (SCAST), a bigger multicenter experiment on 2029 patients from nine northern European nations. Acute therapy of stroke patients with candesartan was shown to have no effect and may cause damage in the experiment. The related BP decrease was partly to blame for the unfavorable result. This has moved focus in recent years to the RAS protective axis to gain the advantages of ARB in ischemic stroke without reducing blood pressure [47]. Many clinical trial studies suggested that ARB-treated hypertensive patients result in the prevention of stroke [61]. In the study of losartan treatment for endpoint reduction, the left ventricular hypertrophy patients of number 27 treated with a losartan treated regimen had a 25% stroke decline as compared to patients treated with an atenolol treatment regimen which resulted in equal blood pressure reductions [62]. In a sub-study of isolated hypertension in left ventricular hypertrophy elderly patients, there was a reduction in stroke by 40% in patients who underwent with losartan regimen when compared to the patients receiving the atenolol regimen. In a study SCOPE, 29 hypertension patients undergone candesartan-based regimen treatment (ARBs) reduces chances of the non-fatal stroke to 27% and total stroke to 23% while a comparative study was done with candesartan-treated patients had a similar reduction in BP [44]. It should be emphasized that for starting 3 months of the study, the control group was given a placebo before switching to active medication. The ARB-treated group had a – 3.2/1.6 mm hg difference in BP initially which might be considered for a few variations [63]. In another sub-study of isolated systolic hypertension patients, the candesartan-treated group have a 42% reduction in stroke occurrence when compared with a placebo or active drug regimen [64]. In the ACCESS pilot study, there is the prevention of stroke reoccurrence with the treatment of candesartan Cilexetil in hypertensive patients having 12 months stroke history. This resulted in 52% decline in reoccurring stroke and cerebrovascular disease when studied in comparison to antihypertensives other than ARBs and placebo but no significant difference in BP was there in both groups. Thus ARBs treatment plays crucial role in AIS and reduces stroke risks, reoccurrence, and provides much therapeutic efficacy [65].

11.6 CONCLUSION

The renin-angiotensin-aldosterone system (RAAS) is effective in preventing repeated cardiac events, and its involvement in avoiding recurrent cerebrovascular mortality is uncertain. The involvement of the brain and heart RAS in stroke, as well as the manipulation of the RAAS system in stroke care. The brain RAAS plays a role in both cardiovascular and neurological diseases. However, the development of compounds able to act on these components and, at the same time, able to cross the blood-brain barrier is challenging, especially if the metabolic stability and the half-life are taken into consideration. However, the RAAS could represent in the coming years a promised land to obtain new pharmacological available alternatives in several clinical conditions.

KEYWORDS

- **angiotensin receptor blockers**
- **angiotensin-converting enzyme inhibitors**
- **blood pressure**
- **cerebroprotective**
- **cerebrovascular system**
- **ischemic reperfusion injury**
- **renin angiotensin aldosterone system**
- **stroke**

REFERENCES

1. Avan, A., Digaleh, H., Di Napoli, M., Stranges, S., Behrouz, R., Shojaeianbabaei, G., et al., (2019). Socioeconomic status and stroke incidence, prevalence, mortality, and worldwide burden: An ecological analysis from the Global Burden of Disease Study 2017. *BMC Med., 17*(1).
2. Prabhakar, N. K., Khan, H., Grewal, A. K., & Singh, T. G., (2022). Intervention of neuroinflammation in the traumatic brain injury trajectory: *In vivo* and clinical approaches. *International Immunopharmacology, 108*, 108902.
3. Boehme, A. K., Esenwa, C., & Elkind, M. S. V., (2017). Stroke risk factors, genetics, and prevention. *Circ. Res., 120*(3), 472–495.
4. Khan, H., Bangar, A., Grewal, A. K., Bansal, P., & Singh, T. G., (2022). Caspase-mediated regulation of the distinct signaling pathways and mechanisms in neuronal survival. *International Immunopharmacology, 110*, 108951.
5. Mascolo, A., Scavone, C., Rafaniello, C., De Angelis, A., Urbanek, K., Di Mauro, G., et al., (2021). The role of renin-angiotensin-aldosterone system in the heart and lung: Focus on COVID-19. *Front. Pharmacol., 12*.
6. Forrester, S. J., Booz, G. W., Sigmund, C. D., Coffman, T. M., Kawai, T., Rizzo, V., et al., (2018). Angiotensin II signal transduction: An update on mechanisms of physiology and pathophysiology. *Physiol. Rev., 98*(3), 1627–1738.
7. Juillerat-Jeanneret, L., (2020). The other angiotensin II receptor: AT2R as a therapeutic target. *J. Med. Chem., 63*(5), 1978–1995.
8. Singh, K. D., & Karnik, S. S., (2016). Angiotensin receptors: Structure, function, signaling and clinical applications. *J. Cell Signal, 1*(2).
9. Sharma, V. K., Singh, T. G., & Mehta, V., (2021). Stressed mitochondria: A target to intrude Alzheimer's disease. *Mitochondrion, 59*, 48–57.
10. Khan, H., Kashyap, A., Kaur, A., & Singh, T. G., (2020). Pharmacological postconditioning: A molecular aspect in ischemic injury. *Journal of Pharmacy and Pharmacology, 72*(11), 1513–1527.

11. Ghafar, M. T. A., (2020). An overview of the classical and tissue-derived renin-angiotensin-aldosterone system and its genetic polymorphisms in essential hypertension. *Steroids, 163*, 108701.
12. Dart, A. B., Wicklow, B., Scholey, J., Sellers, E. A., Dyck, J., Mahmud, F., et al., (2020). An evaluation of renin-angiotensin system markers in youth with type 2 diabetes and associations with renal outcomes. *Pediatr. Diabetes, 21*(7), 1102–1109.
13. Hong, K. S., (2017). Blood pressure management for stroke prevention and in acute stroke. *J. Stroke, 19*(2), 152.
14. Liesz, A., Dalpke, A., Mracsko, E., Roth, S., Zhou, W., Yang, H., et al., (2015). DAMP signaling is a key pathway inducing immune modulation after brain injury. *J. Neurosci., 35*(2), 583–598.
15. Shichita, T., Ito, M., Morita, R., Komai, K., Noguchi, Y., Ooboshi, H., et al., (2017). MAFB prevents excess inflammation after ischemic stroke by accelerating clearance of damage signals through MSR1. *Nat. Med., 23*(6), 723–732.
16. Khan, H., Garg, N., Singh, T. G., Kaur, A., & Thapa, K., (2022). Calpain inhibitors as potential therapeutic modulators in neurodegenerative diseases. *Neurochemical Research*, 1–25.
17. Tiwari, P., Khan, H., Singh, T. G., & Grewal, A. K., (2022). Poly (ADP-ribose) polymerase: An overview of mechanistic approaches and therapeutic opportunities in the management of stroke. *Neurochemical Research*, 1–23.
18. Khan, H., Singh, A., Thapa, K., Garg, N., Grewal, A. K., & Singh, T. G., (2021). Therapeutic modulation of the phosphatidylinositol 3-kinases (PI3K) pathway in cerebral ischemic injury. *Brain Research, 1761*, 147399.
19. Abiodun, O. A., & Ola, M. S., (2020). Role of brain renin angiotensin system in neurodegeneration: An update. *Saudi J. Biol. Sci., 27*(3), 905–912.
20. Khan, H., Sharma, K., Kumar, A., Kaur, A., & Singh, T. G., (2022). Therapeutic implications of cyclooxygenase (COX) inhibitors in ischemic injury. *Inflammation Research*, 1–6.
21. Kangussu, L. M., Marzano, L. A. S., Souza, C. F., Dantas, C. C., Miranda, A. S., & Simoes, E. S. A. C., (2020). The renin-angiotensin system and the cerebrovascular diseases: Experimental and clinical evidence. *Protein Pept. Lett., 27*(6), 463–475.
22. Mascolo, A., Sessa, M., Scavone, C., De Angelis, A., Vitale, C., Berrino, L., et al., (2017). New and old roles of the peripheral and brain renin–angiotensin–aldosterone system (RAAS): Focus on cardiovascular and neurological diseases. *Int. J. Cardiol., 227*, 734–742.
23. Li, X. C., Zhang, J., & Zhuo, J. L., (2017). The vasoprotective axes of the renin-angiotensin system: Physiological relevance and therapeutic implications in cardiovascular, hypertensive and kidney diseases. *Pharmacol. Res., 125*, 21–38.
24. Su, C., Xue, J., Ye, C., & Chen, A., (2021). Role of the central renin-angiotensin system in hypertension. *Int. J. Mol. Med., 47*(6), 1–16.
25. Urmila, A., Rashmi, P., Nilam, G., & Subhash, B., (2021). Recent advances in the endogenous brain renin-angiotensin system and drugs acting on it. *J. Renin-Angiotensin-Aldosterone Syst., 2021*.
26. Eckenstaler, R., Sandori, J., Gekle, M., & Benndorf, R. A., (2021). Angiotensin II receptor type 1–An update on structure, expression, and pathology. *Biochem. Pharmacol., 192*, 114673.
27. Kalogeris, T., Baines, C. P., Krenz, M., & Korthuis, R. J., (2016). Ischemia/reperfusion. *Compr. Physiol., 7*(1), 113.
28. Sumners, C., Peluso, A. A., Haugaard, A. H., Bertelsen, J. B., & Steckelings, U. M., (2019). Anti-fibrotic mechanisms of angiotensin AT2-receptor stimulation. *Acta Physiol., 227*(1), e13280.
29. Noureddine, F. Y., Altara, R., Fan, F., Yabluchanskiy, A., Booz, G. W., & Zouein, F. A., (2020). Impact of the renin–angiotensin system on the endothelium in vascular dementia: Unresolved issues and future perspectives. *Int. J. Mol. Sci., 21*(12), 4268.
30. Schwengel, K., Namsolleck, P., Lucht, K., Clausen, B. H., Lambertsen, K. L., Valero-Esquitino, V., et al., (2016). Angiotensin AT2-receptor stimulation improves survival and neurological outcome after experimental stroke in mice. *J. Mol. Med., 94*(8), 957–966.
31. Vadhan, J. D., & Speth, R. C., (2021). The role of the brain renin-angiotensin system (RAS) in mild traumatic brain injury (TBI). *Pharmacol. Ther., 218*, 107684.
32. Arendse, L. B., Danser, A. H. J., Poglitsch, M., Touyz, R. M., Burnett, J. C., Llorens-Cortes, C., et al., (2019). Novel therapeutic approaches targeting the renin-angiotensin system and associated peptides in hypertension and heart failure. *Pharmacol. Rev., 71*(4), 539–570.

33. Rihal, V., Khan, H., Kaur, A., & Singh, T. G., (2022). Vitamin D as therapeutic modulator in cerebrovascular diseases: A mechanistic perspectives. *Critical Reviews in Food Science and Nutrition,* 1–23.
34. Nehme, A., & Zibara, K., (2017). Efficiency and specificity of RAAS inhibitors in cardiovascular diseases: How to achieve better end-organ protection? *Hypertens Res., 40*(11), 903–909.
35. Morato, M., (2017). Renin-angiotensin system. *Renin-Angiotensin Syst Present Future.*
36. Hussain, M., & Awan, F. R., (2018). Hypertension regulating angiotensin peptides in the pathobiology of cardiovascular disease. *Clin. Exp. Hypertens., 40*(4), 344–352.
37. Vargas, R. A. V., Millán, J. M. V., & Bonilla, E. F., (2022). Renin–angiotensin system: Basic and clinical aspects—A general perspective. *Endocrinol Diabetes y Nutr. (English ed), 69*(1), 52–62.
38. De Angelis, E., Pecoraro, M., Rusciano, M. R., Ciccarelli, M., & Popolo, A., (2019). Cross-talk between neurohormonal pathways and the immune system in heart failure: A review of the literature. *Int. J. Mol. Sci., 20*(7), 1698.
39. Danigo, A., Rovini, A., Bessaguet, F., Bouchenaki, H., Bernard, A., Sturtz, F., et al., (2021). The angiotensin II type 2 receptor, a target for protection and regeneration of the peripheral nervous system? *Pharmaceuticals, 14*(3), 175.
40. Eguchi, S., Kawai, T., Scalia, R., & Rizzo, V., (2018). Understanding angiotensin II type 1 receptor signaling in vascular pathophysiology. *Hypertension, 71*(5), 804–810.
41. Briasoulis, A., Ruiz, D. E., Mouselimis, D., Tsarouchas, A., Bakogiannis, C., & Alvarez, P., (2020). The role of renin-angiotensin system in patients with left ventricular assist devices. *J. Renin-Angiotensin-Aldosterone Syst., 21*(4), 1470320320966445.
42. Chen, F. A., Chien, C. C., Chen, Y. W., Wu, Y. T., & Lin, C. C., (2016). Angiotensin converting-enzyme inhibitors, angiotensin receptor blockers, and calcium channel blockers are associated with prolonged vascular access patency in uremic patients undergoing hemodialysis. *PLoS One, 11*(11), e0166362.
43. Bavishi, C., Bangalore, S., & Messerli, F. H., (2016). Renin angiotensin aldosterone system inhibitors in hypertension: Is there evidence for benefit independent of blood pressure reduction? *Prog. Cardiovasc. Dis., 59*(3), 253–261.
44. Alloubani, A., Saleh, A., & Abdelhafiz, I., (2018). Hypertension and diabetes mellitus as a predictive risk factors for stroke. *Diabetes. Metab. Syndr. Clin. Res. Rev., 12*(4), 577–584.
45. Mehlum, M. H., Liestøl, K., Kjeldsen, S. E., Wyller, T. B., Julius, S., Rothwell, P. M., et al., (2020). Blood pressure–lowering profiles and clinical effects of angiotensin receptor blockers versus calcium channel blockers. *Hypertension, 75*(6), 1584–1592.
46. Parsons, C., Murad, M. H., Andersen, S., Mookadam, F., & Labonte, H., (2016). The effect of antihypertensive treatment on the incidence of stroke and cognitive decline in the elderly: A meta-analysis. *Future Cardiol., 12*(2), 237–248.
47. Fouda, A. Y., Artham, S., El-Remessy, A. B., & Fagan, S. C., (2016). Renin-angiotensin system as a potential therapeutic target in stroke and retinopathy: Experimental and clinical evidence. *Clin. Sci., 130*(4), 221–238.
48. Chalmers, J., & MacMahon, S., (2021). Progress in blood pressure control for the prevention of secondary stroke. *Cerebrovasc. Dis., 50*(6), 617–621.
49. Corrêa, J. W. Do. N., Prado, C. M., Riul, M. E., Araujo, A. V., Rossi, M. A., & Bendhack, L. M., (2020). Reversion of cardiovascular remodeling in renovascular hypertensive 2K-1C rats by renin–angiotensin system inhibitors. *Clin. Exp. Pharmacol. Physiol., 47*(12), 1965–1977.
50. Cipolla, M. J., Liebeskind, D. S., & Chan, S. L., (2018). The importance of comorbidities in ischemic stroke: Impact of hypertension on the cerebral circulation. *J. Cereb. Blood Flow Metab., 38*(12), 2129–2149.
51. Panahpour, H., Terpolilli, N. A., Schaffert, D., Culmsee, C., & Plesnila, N., (2019). Central application of aliskiren, a renin inhibitor, improves outcome after experimental stroke independent of its blood pressure lowering effect. *Front Neurol.,* 942.
52. Hussaini, S., & Sayed, S., (2021). Superiority of angiotensin II receptor blockers over angiotensin converting enzyme inhibitors in stroke prophylaxis. *Int. Pharm. Sci., 12*, 113. doi 1031531/2231. 5896(2).
53. Ernawati, I., & Hidayati, H. B., (2020). The effects neuroprotection of telmisartan on stroke with hypertension. *Malang Neurol. J., 6*(1), 41–46.

54. Shrestha, A., Che, R. C., Zhang A-H. (2019). Role of aldosterone in renal fibrosis. *Ren. Fibros. Mech. Ther.,* 325–346.
55. Graudins, A., Lee, H. M., & Druda, D., (2016). Calcium channel antagonist and beta-blocker overdose: Antidotes and adjunct therapies. *Br. J. Clin. Pharmacol., 81*(3), 453–461.
56. Khilnani, G., Khilnani, A. K., & Thaddanee, R., (2017). Learning pharmacology by metaphors: A tale of beta-blockers. *J. Pharmacol. Pharmacother., 8*(4), 172.
57. Liu, S., Feng, X., Jin, R., & Li, G., (2018). Tissue plasminogen activator-based nanothrombolysis for ischemic stroke. *Expert. Opin. Drug Deliv., 15*(2), 173–184.
58. Bennion, D. M., Steckelings, U. M., & Sumners, C., (2018). Neuroprotection via AT2 receptor agonists in ischemic stroke. *Clin. Sci., 132*(10), 1055–1067.
59. Jackson, L., Dong, G., Althomali, W., Sayed, M. A., Eldahshan, W., Baban, B., et al., (2020). Delayed administration of angiotensin II type 2 receptor (AT2R) agonist compound 21 prevents the development of post-stroke cognitive impairment in diabetes through the modulation of microglia polarization. *Transl. Stroke Res., 11*(4), 762–775.
60. McManus, M., & Liebeskind, D. S., (2016). Blood pressure in acute ischemic stroke. *J. Clin. Neurol., 12*(2), 137–146.
61. Fares, H., DiNicolantonio, J. J., O'Keefe, J. H., & Lavie, C. J., (2016). Amlodipine in hypertension: A first-line agent with efficacy for improving blood pressure and patient outcomes. *Open Hear., 3*(2), e000473.
62. Gerholt, L., Bang, C., Gerdts, E., Larstorp, A. C., Kjeldsen, S., Julius, S., et al., (2022). Left atrial systolic force in hypertensive patients with left ventricular hypertrophy: A predictor of incident atrial fibrillation. The LIFE Study. *OBM Geriatr., 6*(1), 1.
63. Datta, G. A., Visvanathan, R., Cameron, I., Koblar, S. A., Howell, S., & Wilson, D., (2019). Efficacy of botulinum toxin in modifying spasticity to improve walking and quality of life in post-stroke lower limb spasticity-a randomized double-blind placebo-controlled study. *BMC Neurol., 19*(1), 1–7.
64. Sprigg, N., Flaherty, K., Appleton, J. P., Salman, R. A. S., Bereczki, D., Beridze, M., et al., (2018). Tranexamic acid for hyperacute primary intra cerebral haemorrhage (TICH-2): An international randomized, placebo-controlled, phase 3 superiority trial. *Lancet, 391*(10135), 2107–2115.
65. Rodríguez-Yañez, M., Gómez-Choco, M., López-Cancio, E., Amaro, S., De Leciñana, M. A., Arenillas, J. F., et al., (2021). Stroke prevention in patients with arterial hypertension: Recommendations of the Spanish society of neurology's stroke study group. *Neurol. (English Ed.), 36*(6), 462–471.

CHAPTER 12

Stroke and Diabetes in Vitamin D Deficiency: Prospects of Intertwined Pathobiologies

RAVINDER SINGH, SANJANA MEHTA, VERTIKA AWASTHI, PRATIMA KUMARI, VARINDER SINGH, and THAKUR GURJEET SINGH

Chitkara College of Pharmacy, Chitkara University, Punjab, India

ABSTRACT

Vitamin D deficiency has reached epidemic proportions in India, with new epidemiologic research indicating that the pandemic of vitamin D deficiency is affecting about half of the world's population, across all age categories and ethnicities. A substantial correlation between hypovitaminosis of vitamin D concentrations in the body and the development of cardiovascular disease has been established over the last several decades through its association with risk factors such as diabetes and stroke. A stroke is defined as brain damage caused by a disruption in the brain's blood supply. Vitamin D deficiency, which is caused by relative hyperparathyroidism, activates the RAAS system, disturbs lipid metabolism, and affects astrocytes in the brain, resulting in endothelial dysfunction and stroke. Furthermore, type-1 diabetes is caused by beta-cell destruction in the pancreas due to a variety of factors including glucotoxicity, lipotoxicity, and inflammation, whereas type-2 diabetes is caused by insulin resistance caused by an inflammatory state associated with elevated cytokinase levels, which causes endothelial cell dysfunction through oxidative stress. It is called "Adult–Onset Diabetes" since it usually develops after the age of 40. Diabetes is expected to impact over 109 million people worldwide by 2035. In all age categories, however, men have a higher frequency of diabetes and stroke than women, but females often develop more acute complications and are at higher mortality risk. It could be suggested that vitamin D deficiency is the possible reason behind these complications and therefore, it is quite evident that vitamin D supplementation fairly reduces the chances of getting diabetes.

12.1 INTRODUCTION

Vitamin D (VD) is an Ecosteroid-based chemical substance that is soluble in fats and is primarily responsible for calcium and phosphorus control, among other physiological

Ischemic Injury: A Molecular Insight. Thakur Gurjeet Singh & Heena Khan (Eds.)

tasks [1, 2]. The actively operating conformation of VD is calcitriol [1,25(OH)$_2$D$_3$], which adheres to the VD receptor. More than 30–32 ng/mL of calcidiol or 25 hydroxyvitamin D must be present in the bloodstream, as recommended by experts to maintain a healthy vitamin D level [3]. Despite getting plenty of sun exposure, sub-clinical deficiency of vitamin D is more common than ever in both advanced and emerging economies, with an estimated global prevalence of up to 1 billion people [5, 6] and several investigations ranging from basic sciences to practical applications have found its clear link with chronic conditions and acute disorders, such as cancer, cardiovascular diseases including stroke and diabetes [7, 8]. It's also worth noting that hyperglycemia is associated with a higher chance of developing stroke [9] and that hyperglycemia is now known to be caused by vitamin D insufficiency [10]. Furthermore, pathological links of vitamin D with energy homeostasis, as well as immunological and endocrine system modulation, are complemented by the huge amount of observational data currently available [8] (Table 12.1).

TABLE 12.1 General Vitamin D Diagnosis Classification based on 25-Hydroxyvitamin D3 Levels as Stated by Alshahrani et al. [4]

Concentration of 25-Hydroxyvitamin D3 (nmol/L)	Concentration of 25-Hydroxyvitamin D3 (ng/ml)	Classification	References
Less than 50	Less than 20	Deficiency	[4]
50 to 80	20 to 32	Insufficiency	[4]
135 to 225	54 to 90	Normal for sunny countries.	[4]
More than 250	More than 100	Excess	[4]
More than 325	More than 150	Intoxication	[4]

12.2 CORRELATION BETWEEN VITAMIN D DEFICIENCY AND STROKE

The second leading cause of mortality worldwide is stroke, attributing to more than 10% of all deaths (approx. 5,700,000) each year, with the total amount of cases expected to climb in the coming future decades [11]. Approximately 85% of stroke cases are triggered by clot formation in the cerebral artery with a dysfunctional or defective endothelium [12]. Despite significant advances in basic prevention as well as peracute treatment in recent years, stroke remains a deadly illness [13].

The number of stroke cases is on a rise chiefly because of the increasing elderly population as well as other risk factors including HTN and type 2 diabetes [14]. It has been established that due to their disrupted cardio-metabolic profiles and predisposition to cardio-metabolic insufficiency, the South Asian population, especially Indians, have a considerably increased risk of developing cardiovascular diseases [15]. During the previous decade, stroke's total combined incidence rate fluctuated around 105 to 152 in 1,00,000 people annually, and the gross prevalence of the disease varied from 44.29 to 559 in 1,00,000 people in various sections across the nation. These figures outperformed the statistical data from higher-income countries [16]. Deficiency in vitamin D is regarded to be a unique aspect of risk for mortality due to CVS events including ischemic stroke [17].

Furthermore, data from numerous prospective population-based studies revealed that chances of future stroke rise significantly in the absence or deficit of Vit D [18–23]. A meta-analysis carried out on 25 hydroxyvitamin D and indicative stroke due to ischemia has discovered that a decrease in serum 25(OH)D levels increase the likelihood of symptomatic stroke in a linear fashion [18]. This link is also supported by findings from various cross-sectional studies based on patients, however, the evidence is limited [24–26].

12.2.1 PLAUSIBLE MECHANISMS

Vitamin D affects a variety of hormone systems and pathways, which in turn alter cardiovascular functioning through a variety of processes. Increased production of parathyroid hormone or PTH is caused by a chronic vitamin D deficit, which induces myocardial calcification, and hypertrophy in the left ventricle, along with HTN. The potential of vitamin D to improve cardiac function is mostly due to its ability to lower PTH levels [27].

Rise in inflammatory reactions, endothelial functioning disruption together with activation of the RAS, which has been proven to make a significant contribution to maintaining electrolyte balance, arterial blood pressure (BP), and blood volume, are all linked to the insufficiency of vitamin D. There have been reports that VD receptor null mice having overactive RAS had elevated blood pressure (BP) and cardiac hypertrophy [28]. In normal and hypertensive participants, an inverse relationship was also established involving $1{,}25(OH)_2D_3$ in the bloodstream and RAS mechanism along with blood pressure [29].

VD has also been demonstrated to control the cell cycle of cardiomyocytes [30] via its receptor, limiting the proliferation of the cardiac cell, and safeguarding the integrity and functions of cardiomyocyte cells, in addition to impacting cardiac function through hormonal systems [31].

In both healthy and diseased settings, epidemiological data suggested that vitamin D insufficiency is linked to endothelial dysfunction [32, 33]. In experimental atherosclerosis, calcitriol regulates endothelial nitric oxide synthase or eNOS by direct transcriptional regulation, promoting the normalization of eNOS-mRNA expression along with its enzymatic activity [34]. In mice with something like a mutated, physiologically dormant VDR reduced eNOS expression causes greater stiffness in the arteries, endothelial dysfunction, as well as decreased NO bioavailability [35]. In mice showing selective endothelial obliteration of VDR, the effect observed included decreased expression of eNOS, inhibited endothelium-dependent vasodilation, and increased angiotensin 2 pressor effects [36].

Additionally, by regulating NF-κB and unfolded protein response pathway (UPR pathway), vitamin D can decrease inflammatory reactions as well as enhance the homeostasis of proteins, respectively [37, 38]. VD is also involved in the homeostasis of the matrix by controlling various matrix metalloproteinases (MMPs) in addition to, tissue inhibitors of metallo-proteinases (TIMPs) [39], hence the VD deficiency has the potential to become a cause of serious cerebrovascular illnesses along with malformations like aneurysms in brain and stroke, where destabilization of the matrix is a problem [40–42].

Altogether, a decrease in circulating VD levels may lead to poor functionality of the endothelium, which raises the likelihood of developing CVS diseases and fatalities [18, 43].

Increased non-HDL cholesterol levels in children (i.e., total cholesterol excluding HDL) have also been related to a higher risk of developing atherosclerosis-related death in adulthood [44]. The NHLBI, situated in Maryland, America considers non-HDL cholesterol in children to be an essential measure for CVD risk. Stronger 25-hydroxyvitamin D levels have recently been linked to reduced non-HDL cholesterol levels in pediatrics, and there is a proposed hypothesis that states, that immediate VD therapy given to pediatrics can significantly decrease the chances of developing cardiovascular diseases by lowering the level of non-HDL cholesterol [45].

As a result, severe VD deficit, regardless of age or functional status, is emerging as a substantial unfavorable predictor of survival following Ischemic Stroke [17] (Figure 12.1).

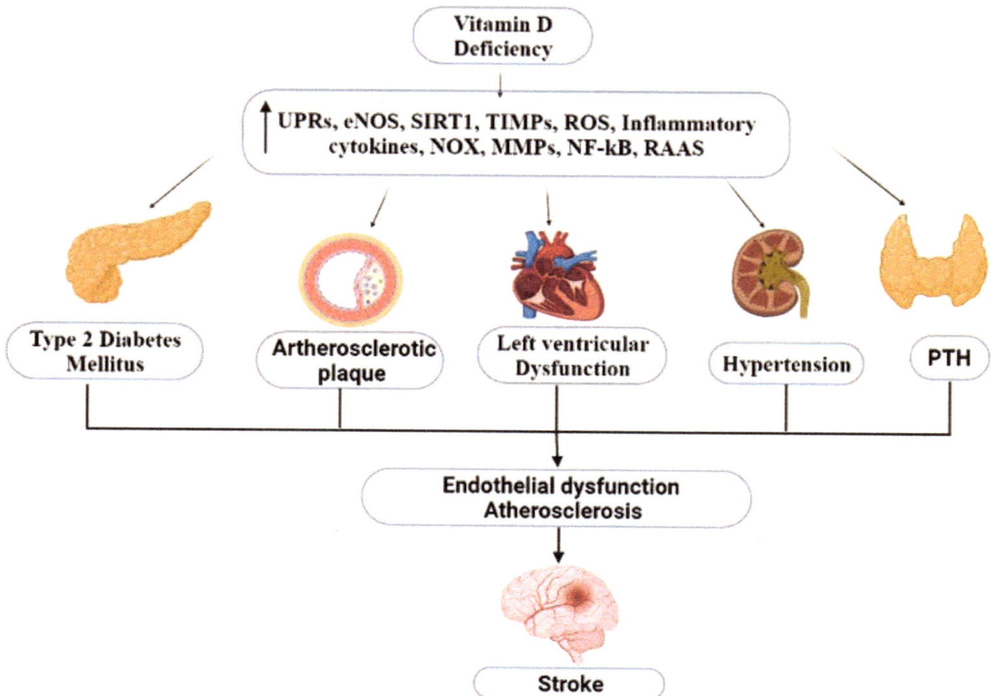

FIGURE 12.1 Vitamin D deficiency can lead to stroke through various pathways.

Abbreviations: UPR pathway: Unfolded protein response pathway; eNOS: Endothelial nitric oxide synthase; SIRT1: Sirtuin 1; TIMPs: Tissue inhibitors of metalloproteinases; ROS: Reactive oxygen species; MMPs: Matrix metallo-proteinases; NF kB: Nuclear factor-kappa B; and RAAS: Renin-angiotensin-aldosterone system.

12.3 HYPERGLYCEMIA AND VITAMIN D DEFICIT: THE NEW INTERWOVEN RELATIONSHIP

T2DM is growing to be a global epidemic, with WHO projecting, that a current worldwide estimate of 170 million diabetic individuals would more than double to 370 million by 2030 [46].

It has been well established that the pathogenesis of T2DM entails a continuous decline in insulin production, often in conjunction with insulin-resistance [47]. Alongside the quintessential involvement of vitamin D in calcium homeostasis as well as bone metabolism [48, 49], multiple pieces of evidence show a correlation linking insufficiency of VD with a group of disorders known as the "metabolic syndrome," which includes obesity, hyperlipidemia, insulin-resistance, HTN as well as an elevated risk about heart disease together with T2DM [50]. The linkage between VD deficiency, insulin resistance in addition to its decreased production, and their significant pathophysiological implications has led to the supposition that VD might have a significant part in the etiology of T2DM [51–60]. Polymorphisms taking place in Vit D Receptor are found to be related to the development of T2DM, supporting this theory [61].

Several studies have recorded seasonal differences in metabolic glycemic management in the pediatric population having type 1 DM [62, 63]. It was seen in a summer camp organized for children with type 1 DM, good glycemic management was demonstrated by a reduction in the needed dose of insulin. Furthermore, just like in patients suffering from type 1 DM [64, 65], levels of fasting blood glucose and glycated hemoglobin (HbA1c) were also seen to be fluctuating seasonally in type 2 diabetes patients [66]. Besides the presence of VDRs, the enzymes responsible for Vitamin D metabolization have also been recently discovered in numerous types of cells, including β islet cells and insulin-responsive cells like adipocyte cells. It is well known that fat cells are a key storehouse of the sunshine vitamin or VD and a major source of cytokines and adipokines that have a crucial role in the development of systemic inflammation [67]. Obesity, particularly visceral one, is considered to be very well-known for increasing the chances of developing T2DM. Vitamin D insufficiency coexisting with obesity has also been postulated as a possible connection between diabetes and obesity. It has been suggested in various clinical research works, VD seems to control a myriad of chain events that enable beta (β) cells in producing insulin, thus controlling the glucose levels in the blood [68].

12.4 MOLECULAR MECHANISMS

12.4.1 VIT D MAINTAINS THE FUNCTIONING OF PANCREATIC BETA CELLS

According to various published evidence, VD appears as a major modulator in regards to insulin release, Ca^{2+} ion levels, and pancreatic beta-cell survival. Its insufficiency has been linked by several studies, to impaired glucose-mediated insulin production in pancreatic beta-cells [69–73]. In the β islet-cells, both VDR plus CYP27B1 are expressed, which denotes that vitamin D activity in the beta-cells of the pancreas appears to be mediated directly by vitamin D binding to VDR [74, 75]. It was also noticed that mice not having a functioning VD receptor have reduced insulin production after receiving a glucose load. It was also linked to lower insulin generation by β-cells, resulting in a reduced amount of insulin stored [76]. Because VDRE (Vitamin D Receptor Element) is a known promoter of the insulin gene in pancreatic beta-cells, calcitriol may directly trigger insulin secretion [77]. VDRE stimulates the transcription of multiple genes associated with cytoskeletal architecture, intra-cellular junctions, and pancreatic beta-cell cellular proliferation, in addition to the insulin gene [78].

In the pancreatic beta-cells, VD also controls the Ca^{2+} flow. Also, a swift, non-genomic process of VD activity has been discovered to be engaged in boosting Ca^{2+} levels in the cytoplasm, which accelerates the mechanism of exocytosis related to insulin production from the beta cells [79, 80]. CaMKII was shown to be activated by increasing Ca^{2+} levels, resulting in insulin secretion. Its primary purpose is to promote protein phosphorylation related to insulin vesicle mobilization and exocytosis. It's also been suggested that higher intracellular Ca^{2+} levels could trigger gene expression of insulin via CREB or the 'cAMP-responsive Element binding' Protein [81]. All this valid data indicates clearly that, vitamin D appears to control depolarization-stimulated insulin production via intracellular Ca^{2+} modulation [82] (Figure 12.2).

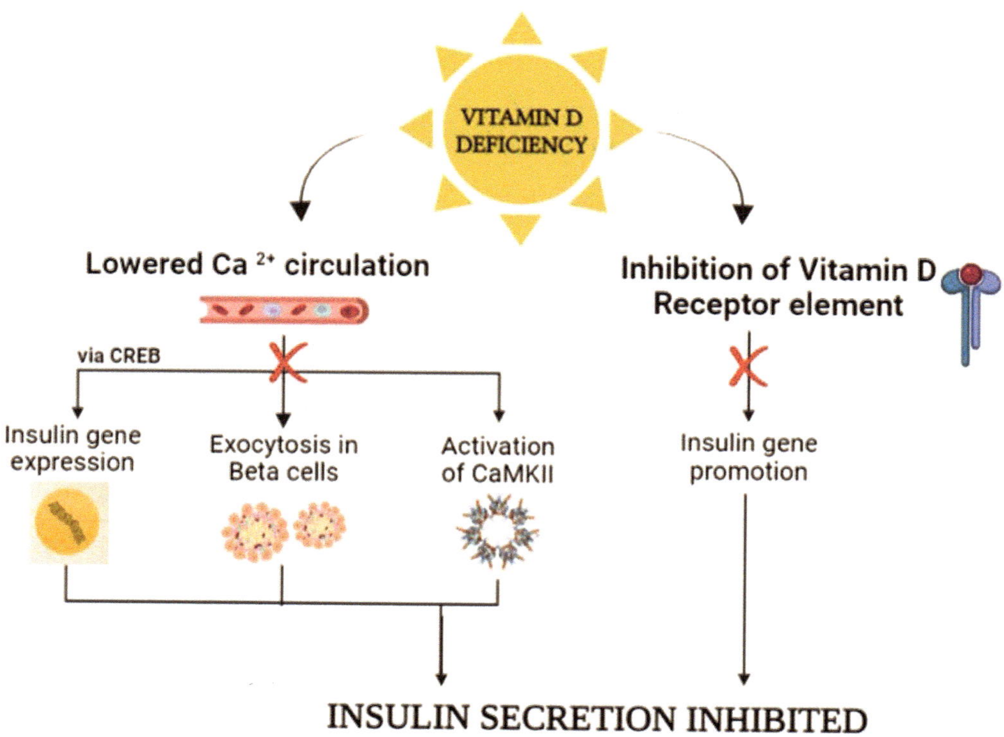

FIGURE 12.2 Vitamin D deficiency can lead to inhibition of insulin secretion via lowering the levels of circulating calcium ions in the blood flow and through inhibition of VDRE or vitamin D receptor element.

12.4.2 VITAMIN D IN RELATION TO INSULIN SIGNALING AND ITS SENSITIVITY

VD engagement is seen in the functioning of tissues responding to insulin such as fats, and liver [80], in addition to pancreatic beta-cells. It is known to enhance insulin sensitivity [83], promotion of insulin receptors activity in the target cells [84] and activates the PPAR delta

[85]. Vit D attaches to its receptor, then joins Retinoid X Receptor in insulin-responsive cells. The complex then joins the VDR element. Thus, the I R gene's transcriptional activation is improved, and the quantity of IRs increases [86]. The appropriate insulin signaling pathway is maintained by increased IR gene expression. As a result, vitamin D's active metabolite appears to be an IR stimulator, which enhances insulin sensitivity [84, 87, 88]. Thus, Vit D insufficiency has a part in the formation of insulin resistance due to IR down-regulation [89].

It was also discovered that supplementing with 1,25(OH)2D3 increases glucose metabolism and uptake in high glucose-treated C_2C_{12} myotubes via upregulating the SIRT1/IRS1/GLUT4 signaling cascade [90]. The transcription factor peroxisome proliferator-activated receptor delta (PPAR-δ) is involved in fatty acid metabolism and mobilization in adipose tissue and skeletal muscle. 1,25(OH)2D has been demonstrated to activate PPAR-δ, which improves insulin sensitivity. FFA-induced insulin resistance in skeletal muscle is reduced by 1,25(OH)2D3-mediated PPAR activation [85]. Vitamin D's ability to reduce insulin resistance in skeletal muscle is also linked with regulating Ca^{2+} levels intracellularly [91]. Along with this, increased levels of PTH, which is linked to insulin resistance, are also linked to vitamin D insufficiency [92, 93]. In insulin-responsive tissues, such as skeletal muscle and adipose tissue, PTH may raise the concentration of free intracellular Ca^{2+} [94, 95]. PTH treatment of osteoblast-like cells resulted in decreased insulin-stimulated glucose transfer, according to the findings of a study [96]. PTH may cause insulin resistance by reducing the amount of GLUT1 and GLUT4 in cell membranes, lowering glucose uptake. As a result, PTH induces insulin resistance by lowering glucose absorption in adipose tissue, the liver, and muscle [97–99].

It's worth noting that vitamin D can influence insulin resistance indirectly through the renin-angiotensin-aldosterone system (RAAS). The inhibitory effects of RAAS on insulin signaling in peripheral tissues and the maintenance of cellular Ca^{2+} levels in skeletal muscle cells are well-known. As a result of GLUT4 recruitment, this modulation may increase glucose transport across the membrane [91, 100, 101]. Furthermore, angiotensin 2 causes the production of reactive oxygen species (ROS) via activating nuclear factor kappa B (NF-κB), which leads to resistance of insulin in skeletal muscles [100]. In VDR-null mice, renin expression and angiotensin 2 production were increased, while 1,25(OH)2D3 treatment reduced renin biosynthesis [102–104]. As a result, vitamin D may increase insulin sensitivity by inhibiting the RAAS pathway [105]. Intriguingly, vitamin D deficiency is linked to increased fat infiltration in skeletal muscle, which occurs regardless of body mass and appears to contribute to insulin resistance [106]. It is also worth mentioning that increased production of reactive oxygen species (ROS) is a key promoter of insulin resistance [107, 108]. Increased levels of FFAs that act on mitochondria to promote ROS generation (like hydrogen peroxide, hydroxyl radical ion, and superoxide ions) may cause oxidative stress to coexist with diabetes [109, 110].

Vitamin D deficiency has also been linked to a decrease in mitochondrial respiration, which is caused by a decrease in the nuclear mRNA molecules and proteins involved with the process [111, 112]. Decreased respiration lowers mitochondrial bioenergetics through altering mitochondrial oxidative phosphorylation, lowering ATP synthesis, and raising the production of reactive oxygen species [113]. An increase in reactive oxygen species (ROS) reduces the insulin signaling pathway functionality by phosphorylating IRS on serine/

threonine, reducing GLUT4 gene transcription, disrupting insulin signaling redistribution throughout the cell, and altering mitochondrial activity [114]. The association between diabetes and vitamin D insufficiency could be explained by vitamin D's involvement in maintaining appropriate mitochondrial activity [115].

The findings of fundamental and clinical investigations combined demonstrated that Vitamin D insufficiency is a crucial risk factor that may hasten the onset of insulin resistance. A better knowledge of vitamin D's molecular role in insulin signaling mechanisms could lead to new therapeutic techniques to prevent the onset of insulin-resistance-related diseases (Figure 12.3).

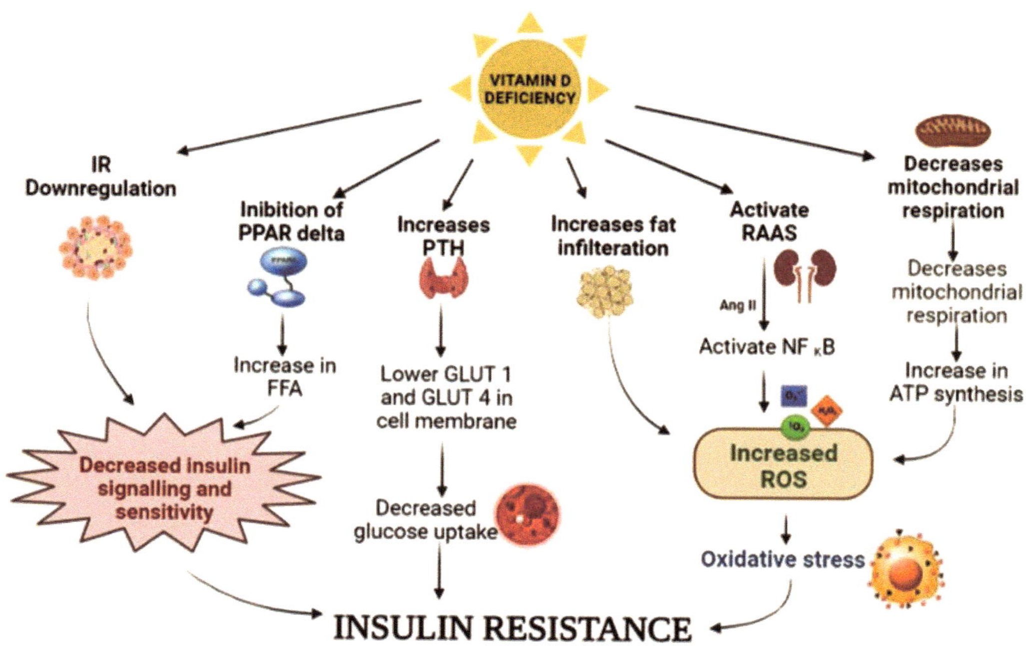

FIGURE 12.3 The deficit or lack of vitamin D can lead to insulin resistance through a multitude of processes including, downregulation of insulin receptor (IR), inhibition of peroxisome proliferator-activated receptor (PPAR), increase in parathyroid hormone (PTH) levels, elevated infiltration of fats, activation of renin-angiotensin aldosterone system (RAAS) and reduced mitochondrial respiration. To induce insulin resistance, they either impair insulin signaling and sensitivity or increase oxidative stress.

12.5 CONCLUSION

It is now well understood that sub-clinical deficiency of vitamin D is more common than ever, with an estimated prevalence of around 1 billion people worldwide. A lot of studies have discovered a clear association between this hypovitaminosis of VD and cardiovascular disorders such as stroke and diabetes. It has long been understood that hyperglycemia is connected to an increased risk of stroke and that hyperglycemia is now recognized to be caused by a lack of vitamin D in the body. This deficiency is regarded to be a unique aspect of risk for mortality due to CVS events including ischemic stroke. The rise in inflammatory reactions increased production of PTH, endothelial functioning disruption together with activation of the RAS, are all linked to the insufficiency of vitamin D, which has the potential to become a cause of serious cerebrovascular illnesses along with malformations like aneurysms in brain and stroke.

Moreover, multiple evidences show a correlation linking insufficiency of VD and the "metabolic syndrome," which includes obesity, hyperlipidemia, insulin resistance, HTN together with T2DM. Polymorphisms in Vit D Receptor are found to be related to the development of T2DM. Further, VD insufficiency coexisting with obesity has also been postulated as a possible connection between diabetes and obesity. It has been suggested in various studies that VD seems to control a myriad of chain events that enable beta (β) cells in producing insulin, thus controlling the glucose levels in the blood. Vitamin D deficiency leads to inhibition of insulin secretion via lowering circulating calcium ions in the blood flow and through inhibition of VDRE. Similarly, this deficit also leads to Insulin Resistance through a multitude of processes including, the downregulation of insulin receptors (IR), inhibition of Peroxisome proliferator-activated receptor (PPAR), an increase in parathyroid hormone (PTH) levels, and many more as explained above. To induce insulin resistance, they either impair insulin signaling and sensitivity or increase oxidative stress.

These results act as substantial proof that a lack of vitamin D is a significant risk factor that can hasten the progression of insulin resistance and, ultimately, stroke. A deeper understanding of vitamin D's biochemical role in hyperglycemia and brain abnormalities such as stroke could lead to novel therapeutic strategies in the future to avoid their occurrence.

KEYWORDS

- **beta cells**
- **diabetes**
- **heart disease**
- **hyperglycemia**
- **inflammation**
- **molecular mechanisms**
- **pancreatic beta cells**
- **stroke**
- **vitamin D**

REFERENCES

1. Al Mheid, I., & Quyyumi, A. A., (2017). Vitamin D and cardiovascular disease: Controversy unresolved. *Journal of the American College of Cardiology, 70*(1), 89–100.

2. Alkhatatbeh, M. J., Abdul-Razzak, K. K., Khasawneh, L. Q., & Saadeh, N. A., (2017). High prevalence of vitamin D deficiency and correlation of serum vitamin D with cardiovascular risk in patients with metabolic syndrome. *Metabolic Syndrome and Related Disorders, 15*(5), 213–219.
3. Thapa, K., Khan, H., Singh, T. G., & Kaur, A., (2021). Traumatic brain injury: Mechanistic insight on pathophysiology and potential therapeutic targets. *Journal of Molecular Neuroscience, 71*(9), 1725–1742.
4. Alshahrani, F., & Aljohani, N., (2013). Vitamin D: Deficiency, sufficiency, and toxicity. *Nutrients, 5*(9), 3605–3616.
5. Nair, R., & Maseeh, A., (2012). Vitamin D: The "sunshine" vitamin. *Journal of Pharmacology and Pharmacotherapeutics, 3*(2), 118–126.
6. Van, S. N., & Lips, P., (2017). Global overview of vitamin D status. *Endocrinology and Metabolism Clinics, 46*(4), 845–870.
7. Holick, M. F., (2005). Vitamin D: Important for prevention of osteoporosis, cardiovascular heart disease, type 1 diabetes, autoimmune diseases, and some cancers. *Southern Medical Journal, 98*(10), 1024–1028.
8. Crowe, F. L., Jolly, K., MacArthur, C., Manaseki-Holland, S., Gittoes, N., Hewison, M., Scragg, R., & Nirantharakumar, K., (2019). Trends in the incidence of testing for vitamin D deficiency in primary care in the UK: A retrospective analysis of the health improvement network (THIN), 2005–2015. *BMJ Open, 9*(6), p.e028355.
9. Kim, A. S., & Johnston, S. C., (2013). Temporal and geographic trends in the global stroke epidemic. *Stroke, 44*(6_suppl_1), S123–S125.
10. Go, A. S., Mozaffarian, D., Roger, V. L., Benjamin, E. J., Berry, J. D., Borden, W. B., Bravata, D. M., et al., (2013). Heart disease and stroke statistics—2013 update: A report from the American Heart Association. *Circulation, 127*(1), e6–e245.
11. Béjot, Y., Bailly, H., Durier, J., & Giroud, M., (2016). Epidemiology of stroke in Europe and trends for the 21st century. *La Presse Médicale, 45*(12), e391–e398.
12. Singh, V., Prabhakaran, S., Chaturvedi, S., Singhal, A., & Pandian, J., (2017). An examination of stroke risk and burden in South Asians. *Journal of Stroke and Cerebrovascular Diseases, 26*(10), 2145–2153.
13. Katan, M., & Luft, A., (2018). Global burden of stroke. In: *Seminars in Neurology* (Vol. 38, No. 2, pp. 208–211). Thieme Medical Publishers.
14. Kamalakannan, S., Gudlavalleti, A. S., Gudlavalleti, V. S. M., Goenka, S., & Kuper, H., (2017). Incidence & prevalence of stroke in India: A systematic review. *The Indian Journal of Medical Research, 146*(2), 175.
15. Wajda, J., Świat, M., Owczarek, A. J., Brzozowska, A., Olszanecka-Glinianowicz, M., & Chudek, J., (2019). Severity of vitamin D deficiency predicts mortality in ischemic stroke patients. *Disease Markers, 2019.*
16. Brøndum-Jacobsen, P., Nordestgaard, B. G., Schnohr, P., & Benn, M., (2013). 25-hydroxyvitamin D and symptomatic ischemic stroke: An original study and meta-analysis. *Annals of Neurology, 73*(1), 38–47.
17. Kojima, G., Bell, C., Abbott, R. D., Launer, L., Chen, R., Motonaga, H., Ross, G. W., Curb, J. D., & Masaki, K., (2012). Low dietary vitamin D predicts 34-year incident stroke: The Honolulu heart program. *Stroke, 43*(8), 2163–2167.
18. Sun, Q., Pan, A., Hu, F. B., Manson, J. E., & Rexrode, K. M., (2012). 25-hydroxyvitamin D levels and the risk of stroke: A prospective study and meta-analysis. *Stroke, 43*(6), 1470–1477.
19. Drechsler, C., Pilz, S., Obermayer-Pietsch, B., Verduijn, M., Tomaschitz, A., Krane, V., Espe, K., et al., (2010). Vitamin D deficiency is associated with sudden cardiac death, combined cardiovascular events, and mortality in haemodialysis patients. *European Heart Journal, 31*(18), 2253–2261.
20. Anderson, J. L., May, H. T., Horne, B. D., Bair, T. L., Hall, N. L., Carlquist, J. F., Lappé, D. L., et al., (2010). Relation of vitamin D deficiency to cardiovascular risk factors, disease status, and incident events in a general healthcare population. *The American Journal of Cardiology, 106*(7), 963–968.
21. Rihal, V., Khan, H., Kaur, A., & Singh, T. G., (2022). Vitamin D as therapeutic modulator in cerebrovascular diseases: A mechanistic perspectives. *Critical Reviews in Food Science and Nutrition*, 1–23.
22. Poole, K. E., Loveridge, N., Barker, P. J., Halsall, D. J., Rose, C., Reeve, J., & Warburton, E. A., (2006). Reduced vitamin D in acute stroke. *Stroke, 37*(1), 243–245.
23. Buell, J. S., Dawson-Hughes, B., Scott, T. M., Weiner, D. E., Dallal, G. E., Qui, W. Q., Bergethon, P., et al., (2010). 25-hydroxyvitamin D dementia, and cerebrovascular pathology in elders receiving home services. *Neurology, 74*(1), 18–26.

24. Kendrick, J., Targher, G., Smits, G., & Chonchol, M., (2009). 25-hydroxyvitamin D deficiency is independently associated with cardiovascular disease in the third national health and nutrition examination survey. *Atherosclerosis*, *205*(1), 255–260.
25. Andersson, P., Rydberg, E., & Willenheimer, R., (2004). Primary hyperparathyroidism and heart disease—A review. *European Heart Journal*, *25*(20), 1776–1787.
26. Xiang, W., Kong, J., Chen, S., Cao, L. P., Qiao, G., Zheng, W., Liu, W., et al., (2005). Cardiac hypertrophy in vitamin D receptor knockout mice: Role of the systemic and cardiac renin-angiotensin systems. *American Journal of Physiology-Endocrinology and Metabolism*, *288*(1), E125–E132.
27. Lind, L., Hänni, A., Lithell, H., Hvarfner, A., Sörensen, O. H., & Ljunghall, S., (1995). Vitamin D is related to blood pressure and other cardiovascular risk factors in middle-aged men. *American Journal of Hypertension*, *8*(9), 894–901.
28. O'Connell, T. D., Berry, J. E., Jarvis, A. K., Somerman, M. J., & Simpson, R. U., (1997). 1, 25-Dihydroxyvitamin D3 regulation of cardiac myocyte proliferation and hypertrophy. *American Journal of Physiology-Heart and Circulatory Physiology*, *272*(4), H1751–H1758.
29. Bae, S., Singh, S. S., Yu, H., Lee, J. Y., Cho, B. R., & Kang, P. M., (2013). Vitamin D signaling pathway plays an important role in the development of heart failure after myocardial infarction. *Journal of Applied Physiology*, *114*(8), 979–987.
30. Al Mheid, I., Patel, R., Murrow, J., Morris, A., Rahman, A., Fike, L., Kavtaradze, N., et al., (2011). Vitamin D status is associated with arterial stiffness and vascular dysfunction in healthy humans. *Journal of the American College of Cardiology*, *58*(2), 186–192.
31. Yiu, Y. F., Chan, Y. H., Yiu, K. H., Siu, C. W., Li, S. W., Wong, L. Y., Lee, S. W., et al., (2011). Vitamin D deficiency is associated with depletion of circulating endothelial progenitor cells and endothelial dysfunction in patients with type 2 diabetes. *The Journal of Clinical Endocrinology & Metabolism*, *96*(5), E830–E835.
32. Talmor, Y., Golan, E., Benchetrit, S., Bernheim, J., Klein, O., Green, J., & Rashid, G., (2008). Calcitriol blunts the deleterious impact of advanced glycation end products on endothelial cells. *American Journal of Physiology-Renal Physiology*, *294*(5), F1059–F1064.
33. Andrukhova, O., Slavic, S., Zeitz, U., Riesen, S. C., Heppelmann, M. S., Ambrisko, T. D., Markovic, M., et al., (2014). Vitamin D is a regulator of endothelial nitric oxide synthase and arterial stiffness in mice. *Molecular Endocrinology*, *28*(1), 53–64.
34. Ni, W., Watts, S. W., Ng, M., Chen, S., Glenn, D. J., & Gardner, D. G., (2014). Elimination of vitamin D receptor in vascular endothelial cells alters vascular function. *Hypertension*, *64*(6), 1290–1298.
35. Khan, H., Kashyap, A., Kaur, A., & Singh, T. G., (2020). Pharmacological postconditioning: A molecular aspect in ischemic injury. *Journal of Pharmacy and Pharmacology*, *72*(11), 1513–1527.
36. Zhu, J., Bing, C., & Wilding, J. P., (2018). Vitamin D receptor ligands attenuate the inflammatory profile of IL-1β-stimulated human white preadipocytes via modulating the NF-κB and unfolded protein response pathways. *Biochemical and Biophysical Research Communications*, *503*(2), 1049–1056.
37. López-López, N., González-Curiel, I., Cruz, T. S., Beatriz, M., Rivas-Santiago, B., Trujillo-Paez, V., Enciso-Moreno, J. A., & Serrano, C. J., (2014). Expression and vitamin D-mediated regulation of matrix metalloproteinases (MMPs) and tissue inhibitors of metalloproteinases (TIMPs) in healthy skin and in diabetic foot ulcers. *Archives of Dermatological Research*, *306*(9), 809–821.
38. Choquet, H., Trapani, E., Goitre, L., Trabalzini, L., Akers, A., Fontanella, M., Hart, B. L., et al., (2016). Cytochrome P450 and matrix metalloproteinase genetic modifiers of disease severity in cerebral cavernous malformation type 1. *Free Radical Biology and Medicine*, *92*, 100–109.
39. Prabhakar, N. K., Khan, H., Grewal, A. K., & Singh, T. G., (2022). Intervention of neuroinflammation in the traumatic brain injury trajectory: *In vivo* and clinical approaches. *International Immunopharmacology*, *108*, 108902.
40. Khan, H., Gupta, A., Singh, T. G., & Kaur, A., (2021). Mechanistic insight on the role of leukotriene receptors in ischemic–reperfusion injury. *Pharmacological Reports*, *73*(5), 1240–1254.
41. Park, K. Y., Chung, P. W., Kim, Y. B., Moon, H. S., Suh, B. C., Won, Y. S., Kim, J. M., et al., (2015). Serum vitamin D status as a predictor of prognosis in patients with acute ischemic stroke. *Cerebrovascular Diseases*, *40*(1, 2), 73–80.

42. Raitakari, O. T., Juonala, M., Kähönen, M., Taittonen, L., Laitinen, T., Mäki-Torkko, N., Järvisalo, M. J., et al., (2003). Cardiovascular risk factors in childhood and carotid artery intima-media thickness in adulthood: The cardiovascular risk in young Finns study. *JAMA, 290*(17), 2277–2283.
43. Birken, C. S., Lebovic, G., Anderson, L. N., McCrindle, B. W., Mamdani, M., Kandasamy, S., Khovratovich, M., et al., (2015). Association between vitamin D and circulating lipids in early childhood. *PLoS One, 10*(7), e0131938.
44. Wild, S., Roglic, G., Green, A., Sicree, R., & King, H., (2004). Global prevalence of diabetes: Estimates for the year 2000 and projections for 2030. *Diabetes Care, 27*(5), 1047–1053.
45. Giaccari, A., Sorice, G., & Muscogiuri, G., (2009). Glucose toxicity: The leading actor in the pathogenesis and clinical history of type 2 diabetes–mechanisms and potentials for treatment. *Nutrition, Metabolism and Cardiovascular Diseases, 19*(5), 365–377.
46. Firdous, S. M., Dhang, P., Koley, A., & Khan, H., (2021). Diabetic retinopathy: Pathogenesis and therapeutic management. *Diabetes and Diabetic Complications.* Nova Science Publishers, Hauppauge.
47. Martini, L. A., & Wood, R. J., (2006). Vitamin D status and the metabolic syndrome. *Nutrition Reviews, 64*(11), 479–486.
48. National Cholesterol Education Program (US). Expert Panel on Detection and Treatment of High Blood Cholesterol in Adults, (2002). *Third Report of the National Cholesterol Education Program (NCEP) Expert Panel on Detection, Evaluation, and Treatment of High Blood Cholesterol in Adults (Adult Treatment Panel III)* (No. 2). The Program.
49. Mezza, T., Muscogiuri, G., Sorice, G. P., Prioletta, A., Salomone, E., Pontecorvi, A., & Giaccari, A., (2012). Vitamin D deficiency: A new risk factor for type 2 diabetes. *Annals of Nutrition and Metabolism, 61*(4), 337–348.
50. Scragg, R., Sowers, M., & Bell, C., (2004). Serum 25-hydroxyvitamin D, diabetes, and ethnicity in the third national health and nutrition examination survey. *Diabetes Care, 27*(12), 2813–2818.
51. Mathieu, C., Gysemans, C., Giulietti, A., & Bouillon, R., (2005). Vitamin D and diabetes. *Diabetologia, 48*(7), 1247–1257.
52. Palomer, X., González-Clemente, J. M., Blanco-Vaca, F., & Mauricio, D., (2008). Role of vitamin D in the pathogenesis of type 2 diabetes mellitus. *Diabetes, Obesity and Metabolism, 10*(3), 185–197.
53. Takiishi, T., Gysemans, C., Bouillon, R., & Mathieu, C., (2010). Endocrinol. *Metab. Clin. N. Am, 39,* 419–446.
54. Garg, N., Singh, T. G., Khan, H., Arora, S., Kaur, A., & Mannan, A., (2021). Mechanistic interventions of selected *Ocimum* species in management of diabetes, obesity, and liver disorders: Transformative developments from preclinical to clinical approaches. *Biointerface Res. Appl. Chem, 12*(1), 1304–1323.
55. Wacker, M., & Holick, M. F., (2013). Vitamin D—Effects on skeletal and extra skeletal health and the need for supplementation. *Nutrients, 5*(1), 111–148.
56. Lips, P., Eekhoff, M., Van, S. N., Oosterwerff, M., De Jongh, R., Krul-Poel, Y., & Simsek, S., (2017). Vitamin D and type 2 diabetes. *The Journal of Steroid Biochemistry and Molecular Biology, 173,* 280–285.
57. Alam, U., Arul-Devah, V., Javed, S., & Malik, R. A., (2016). Vitamin D and diabetic complications: True or false prophet? *Diabetes Therapy, 7*(1), 11–26.
58. Wimalawansa, S. J., (2018). Associations of vitamin D with insulin resistance, obesity, type 2 diabetes, and metabolic syndrome. *The Journal of Steroid Biochemistry and Molecular Biology, 175,* 177–189.
59. Sentinelli, F., Bertoccini, L., Barchetta, I., Capoccia, D., Incani, M., Pani, M. G., Loche, S., et al., (2016). The vitamin D receptor (VDR) geners11568820 variant is associated with type 2 diabetes and impaired insulin secretion in Italian adult subjects, and associates with increased cardio-metabolic risk in children. *Nutrition, Metabolism and Cardiovascular Diseases, 26*(5), 407–413.
60. Nordfeldt, S., & Ludvigsson, J., (2000). Seasonal variation of HbA1c in intensive treatment of children with type 1 diabetes. *Journal of Pediatric Endocrinology and Metabolism, 13*(5), 529–535.
61. Braatvedt, G. D., Mildenhall, L., Patten, C., & Harris, G., (1997). Insulin requirements and metabolic control in children with diabetes mellitus attending a summer camp. *Diabetic Medicine, 14*(3), 258–261.
62. Gikas, A., Sotiropoulos, A., Pastromas, V., Papazafiropoulou, A., Apostolou, O., & Pappas, S., (2009). Seasonal variation in fasting glucose and HbA1c in patients with type 2 diabetes. *Primary Care Diabetes, 3*(2), 111–114.
63. Tseng, C. L., Brimacombe, M., Xie, M., Rajan, M., Wang, H., Kolassa, J., Crystal, S., et al., (2005). Seasonal patterns in monthly hemoglobin A1c values. *American Journal of Epidemiology, 161*(6), 565–574.

64. Karvonen, M., Jäntti, V., Muntoni, S., Stabilini, M., Stabilini, L., Muntoni, S., & Tuomilehto, J., (1998). Comparison of the seasonal pattern in the clinical onset of IDDM in Finland and Sardinia. *Diabetes Care, 21*(7), 1101–1109.
65. Abbas, M. A., (2017). Physiological functions of vitamin D in adipose tissue. *The Journal of Steroid Biochemistry and Molecular Biology, 165*, 369–381.
66. Khan, H., Bangar, A., Grewal, A. K., Bansal, P., & Singh, T. G., (2022). Caspase-mediated regulation of the distinct signaling pathways and mechanisms in neuronal survival. *International Immunopharmacology, 110*, 108951.
67. Norman, A. W., Frankel, J. B., Heldt, A. M., & Grodsky, G. M., (1980). Vitamin D deficiency inhibits pancreatic secretion of insulin. *Science, 209*, 823–825.
68. Cade, C., & Norman, A. W., (1986). Vitamin D3 improves impaired glucose tolerance and insulin secretion in the vitamin D-deficient rat *in vivo*. *Endocrinology, 119*(1), 84–90.
69. Tanaka, Y., Seino, Y., Ishida, M., Yamaoka, K., Yabuuchi, H., Ishida, H., Seino, S., et al., (1984). Effect of vitamin D3 on the pancreatic secretion of insulin and somatostatin. *European Journal of Endocrinology, 105*(4), 528–533.
70. RUBIN, B. S., & BARFIELD, R. J., (1983). Progesterone in the ventromedial hypothalamus facilitates estrous behavior in ovariectomized, estrogen-primed rats. *Endocrinology, 113*(2), 797–804.
71. Kadowaki, S., & Norman, A. W., (1984). Dietary vitamin D is essential for normal insulin secretion from the perfused rat pancreas. *The Journal of Clinical Investigation, 73*(3), 759–766.
72. Boucher, B. J., Mannan, N., Noonan, K., Hales, C. N., & Evans, S. J. W., (1995). Glucose intolerance and impairment of insulin secretion in relation to vitamin D deficiency in east London Asians. *Diabetologia, 38*(10), 1239–1245.
73. Borissova, A. M., Tankova, T., Kirilov, G., Dakovska, L., & Kovacheva, R., (2003). The effect of vitamin D3 on insulin secretion and peripheral insulin sensitivity in type 2 diabetic patients. *International Journal of Clinical Practice, 57*(4), 258–261.
74. Inomata, S., Kadowaki, S., Yamatani, T., Fukase, M., & Fujita, T., (1986). Effect of 1 alpha (OH)-vitamin D3 on insulin secretion in diabetes mellitus. *Bone and Mineral, 1*(3), 187–192.
75. Al-Shoumer, K. A., & Al-Essa, T. M., (2015). Is there a relationship between vitamin D with insulin resistance and diabetes mellitus? *World Journal of Diabetes, 6*(8), 1057–1064.
76. Nyomba, B. L., Auwerx, J., Bormans, V., Peeters, T. L., Pelemans, W., Reynaert, J., Bouillon, R., et al., (1986). Pancreatic secretion in man with subclinical vitamin D deficiency. *Diabetologia, 29*(1), 34–38.
77. Firdous, S. M., Ghosh, S., Nath, P., & Khan, H., (2021). Molecular basis of diabetic nephropathy. *Diabetes and Diabetic Complications*. Nova Science Publishers, Hauppauge.
78. Johnson, J. A., Grande, J. P., Roche, P. C., & Kumar, R., (1994). Immunohistochemical localization of the 1, 25 (OH) 2D3 receptor and calbindin D28k in human and rat pancreas. *American Journal of Physiology-Endocrinology and Metabolism, 267*(3), E356–E360.
79. Bland, R., Markovic, D., Hills, C. E., Hughes, S. V., Chan, S. L., Squires, P. E., & Hewison, M., (2004). Expression of 25-hydroxyvitamin D3-1α-hydroxylase in pancreatic islets. *The Journal of Steroid Biochemistry and Molecular Biology, 89*, 121–125.
80. Zeitz, U., Weber, K., Soegiarto, D. W., Wolf, E., Balling, R., & Erben, R. G., (2003). Impaired insulin secretory capacity in mice lacking a functional vitamin D receptor. *The FASEB Journal, 17*(3), 509–511.
81. Altieri, B., Grant, W. B., Della, C. S., Orio, F., Pontecorvi, A., Colao, A., Sarno, G., & Muscogiuri, G., (2017). Vitamin D and pancreas: The role of sunshine vitamin in the pathogenesis of diabetes mellitus and pancreatic cancer. *Critical Reviews in Food Science and Nutrition, 57*(16), 3472–3488.
82. Wolden-Kirk, H., Overbergh, L., Gysemans, C., Brusgaard, K., Naamane, N., Van, L. L., Schuit, F., et al., (2013). Unraveling the effects of 1, 25 (OH)2D3 on global gene expression in pancreatic islets. *The Journal of Steroid Biochemistry and Molecular Biology, 136*, 68–79.
83. Reusch, J. E. B., Begum, N., Sussman, K. E., & Draznin, B., (1991). Regulation of GLUT-4 phosphorylation by intracellular calcium in adipocytes. *Endocrinology, 129*(6), 3269–3273.
84. Draznin, B., Lewis, D., Houlder, N., Sherman, N., Adamo, M., Garvey, W. T., Leroith, D., & Sussman, K., (1989). Mechanism of insulin resistance induced by sustained levels of cytosolic free calcium in rat adipocytes. *Endocrinology, 125*(5), 2341–2349.

85. Altieri, B., Grant, W. B., Della, C. S., Orio, F., Pontecorvi, A., Colao, A., Sarno, G., & Muscogiuri, G., (2017). Vitamin D and pancreas: The role of sunshine vitamin in the pathogenesis of diabetes mellitus and pancreatic cancer. *Critical Reviews in Food Science and Nutrition, 57*(16), 3472–3488.
86. Sharma, V. K., & Singh, T. G., (2020). CREB: A multifaceted target for Alzheimer's disease. *Current Alzheimer Research, 17*(14), 1280–1293.
87. Sooy, K., Schermerhorn, T., Noda, M., Surana, M., Rhoten, W. B., Meyer, M., Fleischer, N., et al., (1999). Calbindin-D28k controls $[Ca^{2+}]$ I and insulin release: Evidence obtained from calbindin-d28k knockout mice and β cell lines. *Journal of Biological Chemistry, 274*(48), 34343–34349.
88. Simpson, R. U., Thomas, G. A., & Arnold, A. J., (1985). Identification of 1, 25- dihydroxy vitamin D3 receptors and activities in muscle. *Journal of Biological Chemistry, 260*(15), 8882–8891.
89. MAESTRO, B., CAMPIÓN, J., DÁVILA, N., & CALLE, C., (2000). Stimulation by 1, 25-dihydroxyvitamin D3 of insulin receptor expression and insulin responsiveness for glucose transport in U-937 human promonocytic cells. *Endocrine Journal, 47*(4), 383–391.
90. Dunlop, T. W., Väisänen, S., Frank, C., Molnár, F., Sinkkonen, L., & Carlberg, C., (2005). The human peroxisome proliferator-activated receptor δ gene is a primary target of 1α, 25-dihydroxyvitamin D3 and its nuclear receptor. *Journal of Molecular Biology, 349*(2), 248–260.
91. Maestro, B., Molero, S., Bajo, S., Dávila, N., & Calle, C., (2002). Transcriptional activation of the human insulin receptor gene by 1, 25-dihydroxyvitamin D3. *Cell Biochemistry and Function: Cellular Biochemistry and its Modulation by Active Agents or Disease, 20*(3), 227–232.
92. Maestro, B., Dávila, N., Carranza, M. C., & Calle, C., (2003). Identification of a vitamin D response element in the human insulin receptor gene promoter. *The Journal of Steroid Biochemistry and Molecular Biology, 84*(2, 3), 223–230.
93. Berridge, M. J., (2017). Vitamin D deficiency and diabetes. *Biochemical Journal, 474*(8), 1321–1332.
94. Manna, P., Achari, A. E., & Jain, S. K., (2018). 1, 25 (OH) 2-vitamin D3 upregulates glucose uptake mediated by SIRT1/IRS1/GLUT4 signaling cascade in C2C12 myotubes. *Molecular and Cellular Biochemistry, 444*(1), 103–108.
95. Wright, D. C., Hucker, K. A., Holloszy, J. O., & Han, D. H., (2004). Ca^{2+} and AMPK both mediate stimulation of glucose transport by muscle contractions. *Diabetes, 53*(2), 330–335.
96. Reis, J. P., Von, M. D., Kritz-Silverstein, D., Wingard, D. L., & Barrett-Connor, E., (2007). Vitamin D, parathyroid hormone levels, and the prevalence of metabolic syndrome in community-dwelling older adults. *Diabetes Care, 30*(6), 1549–1555.
97. Chiu, K. C., Chuang, L. M., Lee, N. P., Ryu, J. M., McGullam, J. L., Tsai, G. P., & Saad, M. F., (2000). Insulin sensitivity is inversely correlated with plasma intact parathyroid hormone level. *Metabolism-Clinical and Experimental, 49*(11), 1501–1505.
98. Baczynski, R., Massry, S. G., Magott, M., El-Belbessi, S., Kohan, R., & Brautbar, N., (1985). Effect of parathyroid hormone on energy metabolism of skeletal muscle. *Kidney International, 28*(5), 722–727.
99. Ni, Z., Smogorzewski, M., & Massry, S. G., (1994). Effects of parathyroid hormone on cytosolic calcium of rat adipocytes. *Endocrinology, 135*(5), 1837–1844.
100. Thomas, D. M., Rogers, S. D., Sleeman, M. W., Pasquini, G. M., Bringhurst, F. R., Ng, K. W., Zajac, J. D., & Best, J. D., (1995). Modulation of glucose transport by parathyroid hormone and insulin in UMR 106–01, a clonal rat osteogenic sarcoma cell line. *Journal of Molecular Endocrinology, 14*(2), 263–275.
101. Teegarden, D., & Donkin, S. S., (2009). Vitamin D: Emerging new roles in insulin sensitivity. *Nutrition Research Reviews, 22*(1), 82–92.
102. Sung, C. C., Liao, M. T., Lu, K. C., & Wu, C. C., (2012). Role of vitamin D in insulin resistance. *Journal of Biomedicine and Biotechnology, 2012*.
103. Wei, Y., Sowers, J. R., Clark, S. E., Li, W., Ferrario, C. M., & Stump, C. S., (2008). Angiotensin II-induced skeletal muscle insulin resistance mediated by NF-κB activation via NADPH oxidase. *American Journal of Physiology-Endocrinology and Metabolism, 294*(2), E345–E351.
104. Muscogiuri, G., Chavez, A. O., Gastaldelli, A., Perego, L., Tripathy, D., Saad, M. J., Velloso, L., & Folli, F., (2008). The crosstalk between insulin and renin-angiotensin- aldosterone signaling systems and its effect on glucose metabolism and diabetes prevention. *Current Vascular Pharmacology, 6*(4), 301–312.

105. Kong, J., & Li, Y. C., (2003). Effect of ANG II type I receptor antagonist and ACE inhibitor on vitamin D receptor-null mice. *American Journal of Physiology-Regulatory, Integrative and Comparative Physiology, 285*(1), R255–R261.
106. Yuan, W., Pan, W., Kong, J., Zheng, W., Szeto, F. L., Wong, K. E., Cohen, R., et al., (2007). 1, 25-dihydroxyvitamin D3 suppresses renin gene transcription by blocking the activity of the cyclic AMP response element in the renin gene promoter. *Journal of Biological Chemistry, 282*(41), 29821–29830.
107. Li, Y. C., Kong, J., Wei, M., Chen, Z. F., Liu, S. Q., & Cao, L. P., (2002). 1, 25-dihydroxyvitamin D3 is a negative endocrine regulator of the renin-angiotensin system. The *Journal of Clinical Investigation, 110*(2), 229–238.
108. Angellotti, E., & Pittas, A. G., (2017). The role of vitamin D in the prevention of type 2 diabetes: To D or not to D? *Endocrinology, 158*(7), 2013–2021.
109. Gilsanz, V., Kremer, A., Mo, A. O., Wren, T. A., & Kremer, R., (2010). Vitamin D status and its relation to muscle mass and muscle fat in young women. *The Journal of Clinical Endocrinology & Metabolism, 95*(4), 1595–1601.
110. Newsholme, P., Cruzat, V. F., Keane, K. N., Carlessi, R., & De Bittencourt, Jr. P. I. H., (2016). Molecular mechanisms of ROS production and oxidative stress in diabetes. *Biochemical Journal, 473*(24), 4527–4550.
111. Fridlyand, L. E., & Philipson, L. H. (2006). *Reactive Species and Early Manifestation of Insulin Resistance in Type, 2,* 136–145.
112. Lowell, B. B., & Shulman, G. I., (2005). Mitochondrial dysfunction and type 2 diabetes. *Science, 307*(5708), 384–387.
113. Carlsson, C., Borg, L. A., & Welsh, N., (1999). Sodium palmitate induces partial mitochondrial uncoupling and reactive oxygen species in rat pancreatic islets *in vitro*. *Endocrinology, 140*, 3422–3428.
114. Kim, H. K., Andreazza, A. C., Yeung, P. Y., Isaacs-Trepanier, C., & Young, L. T., (2014). Oxidation and nitration in dopaminergic areas of the prefrontal cortex from patients with bipolar disorder and schizophrenia. *Journal of Psychiatry and Neuroscience, 39*(4), 276–285.
115. Scaini, G., Rezin, G. T., Carvalho, A. F., Streck, E. L., Berk, M., & Quevedo, J., (2016). Mitochondrial dysfunction in bipolar disorder: Evidence, pathophysiology, and translational implications. *Neuroscience & Biobehavioral Reviews, 68*, 694–713.
116. Berridge, M. J., (2017). Vitamin D deficiency and diabetes. *Biochemical Journal, 474*(8), 1321–1332.
117. Rains, J. L., & Jain, S. K., (2011). Oxidative stress, insulin signaling, and diabetes. *Free Radical Biology and Medicine, 50*(5), 567–575.
118. Szymczak-Pajor, I., & Śliwińska, A., (2019). Analysis of association between vitamin D deficiency and insulin resistance. *Nutrients, 11*(4), 794.

CHAPTER 13

Remote Ischemic Postconditioning: An Overview

HEENA KHAN,[1] SHIVANI BAGGA,[1] MANISH KUMAR,[1] THAKUR GURJEET SINGH,[1] and RAHUL DESHMUKH[2]

[1]*Chitkara College of Pharmacy, Chitkara University, Punjab, India*

[2]*Department of Pharmaceutical Sciences and Technology, Maharaja Ranjit Singh Punjab Technical University, Bathinda, Punjab, India*

ABSTRACT

RIC is defined as short, tedious, and sublet Ischemia-Reperfusion applicable to an organ or tissue in order towards activating global endogenous patience preventing distant objective organs or tissues from subsequent prolongation Ischemic remote conditioning, implying especially the conservational effect of Ischemic conditioning in situ and in faraway organs also. Early blood supply restoration was shown to provide a significant advantage by rescuing possible tissues from ischemic injury. Remote ischemic conditioning has proved to provide intramyocardial cardioprotection across various arteries. Nevertheless, soon after this discovery, it was determined that remote ischemic conditioning stimulation was enforced on an organ distant from the various organs that could protect the heart, brain, intestine, kidney, liver, lung. In this chapter, we have provided evidence with respect to these.

13.1 INTRODUCTION

Early blood supply restoration was shown to provide a significant advantage by rescuing possible tissues from ischemic injury [1]. Ischemia-reperfusion injury is commonly accustomed in analytic settings such as anticoagulants following a cerebrovascular accident or myocardial infarction caused by CABG, and surgical operation [2]. Nevertheless, the clinical outcomes of current access to IRI control remain undesirable. Furthermore, human duration has increased, so the deteriorating disorders on human beings, thus frequently affect collaborative organs of the aged thereby decreasing the physiological curve.

In recent times, awareness should be concentrating on the imaginative path, known as ischemic conditioning, especially remote ischemic conditioning, experiencing to be a tedious, ephemeral, and sublethal sequence of Ischemia-reperfusion can activate endogenous

Ischemic Injury: A Molecular Insight. Thakur Gurjeet Singh & Heena Khan (Eds.)

shielding forbearance to counteract subsequent ischemic threats [3]. Remote ischemic conditioning might benefit different organs of the body. It resembles promising medicines and non-abscission remedies as hindering the therapeutic age-affiliated vascular illness persuaded neurons related conditions. The elegant approach was first described as remote ischemic preconditioning and paved the way for simple analytic application [4]. The authors demonstrated in their canine experiments that inducing short circumstances of ischemia and reperfusion in the circumflex coronary artery territory was capable of reducing the latterly amount of the myocardial infarct caused by blockage of the left anterior descending coronary artery. Intriguingly, the concept related to myocardial protection was further expanded to non-cardiac units, and it was demonstrated especially myocardial infarct amount might also act significantly which is further decreased for short repetitive ischemia and reperfusion stimuli in tissue or organs distant from the heart, such as the kidney, liver, and intestine [5].

The protective mechanisms of RIC that give protection to distantly addressed organs against Ischemia-reperfusion injury are quite mature and have yet to be detected. Present data show that the available basic mechanisms are associated with a variety of factors, most notably both humoral and neuronal which have integral interactions with one another, and the effective one may vary [6]. Despite the majority of conclusions arising from diagnosing preclinical studies clarifying heart-related conditions, they may cast a detail regarding the investigation of Remote ischemic conditioning mechanisms in individuals [6].

13.2 SIGNAL TRANSDUCTION

The remote stimulus, by the term, operates at the interval from the objective organ. Collection of distant stimuli induces objective target organ protection – (i) electrical incitement or electro-acupuncture; (ii) chemical/pharmacological agents, such as autocoids (adenosine, bradykinin); (iii) mechanical incentive, related to trauma, i.e., surgical procedure alternatively; and (iv) rounds related to ischemic reperfusion about margin alternate organ, which again include chemical and mechanical stimuli [7]. The activation of incidental sensory nerves is common to all of these stimuli. It is unclear to what extent nerve stimulation at the peripheral sites or other operations such as oxygen deficiency/acidosis or vascular stress devote towards remote conditioning in response to aplastic anemia/reperfusion stimulus [8]. This currently refrains from analyzing the various stimulus and estimating the direct remote incentive because no stimulus strength-response relationship has been authorized for any of the stimuli. As a result, all previous comparisons are of little value [9].

Both components are concerned with the transfer of signal out of external incentive location towards the objective unit; working together and thus collaborating at three steps of levels. (i) Incentive level: peripheral afferents remain straightforwardly activated and humoral factors are discharged, activating the peripheral sensory nerves. (ii) Fundamental surface–tangential sensational afferent nerves move towards the spinal cord, where they design into autonomic centers in CNS, activating efferent vagal nerves [10]. Activation of the vagal nerve (a) secretes acetylcholine, which activates muscarinic receptors in the objective unit, actuates muscarinic sense organs in non-objective units, causing humoral factors to be released, that act on the target organs; (b) secretes humoral factors that affect the target organs [11]. Neither the quantity nor precise natures of both mediators in remote ischemic conditioning are known at this time [12]. Since it happens in all organs tested thus, the feedback to remote ischemic conditioning is structural. Remote ischemic conditioning in various organs including, the heart, and abdominal reduces cellular damage, the result of aplastic anemia reperfusion. Familiar intracellular signal transduction routes are involved in various cell types [13]. Cardiomyocyte intracellular signaling is most likely species-specific. Regardless of the specific signaling paths, a powerhouse of cell mitochondria is an essential end effector of prevention [14]. Cerebrovascular accident injury causes cell death via various processes (necrosis, necroptosis, apoptosis, and/or autophagy) independent of cell pattern and unit-distinct liable to the injury, and preserved mitochondrial function defend against apoptosis [15].

13.2.1 REMOTE STIMULUS

Remote stimuli such as automatic, synthetic/pharmacological, mechanical, and aplastic anemia/reperfusion rounds together act as deleterious triggers that operate external sensation nerves. The response to the incentive that exclusively operates peripheral sensory nerves, such as automatic nerve stimulation or synthetic/pharmacological nociceptor incentive like capsaicin, demonstrates a causal involvement of neuronal activation. Peripheral nerve transection or local anesthesia removes the resulting cardio protection [16]. It is currently unknown whether the humoral element is discharged nearby as a straight response to the incentive being discharged from other neuronally activated units [17]. Rodent research found that occlusion of femoral vessel exhausting the limb experience RIC by three rounds of thigh artery occlusion along reperfusion lead to abolishment in size reduction.

13.2.2 AUTOMATIC AXON INCENTIVE

In rabbits, four cycles of five minutes of femoral nerve stimulation and five minutes of rest resulted in the discharging of the humoral element in the direction of fundamental circulation which reduce the infarct amount in heart of a rabbit. In rodents, electro acupuncture at the forelimb reduced myocardial infarct size [18] actuating the discharging of humoral elements into rabbit fundamental flow, which then decreased infarct amount after transfer to an isolated acceptor rabbit heart [19]. Femoral nerve transection thus prevents discharging of cardioprotective humoral elements following transcutaneous automatic nerve stimulation, further not preventing the discharge related to NO and additionally infarct amount reduction before RIC, implying a preservative action of shear stress as an incentive [20].

13.2.3 LOCAL SYNTHETIC/PHARMACOLOGICAL INCENTIVE

Confined alkaloid carefully operates non-myelinated afferent C-fibers in rodents, alkaloid capsaicin applicable towards abdominal skin reduced infarct amount [21]. Local anesthesia with lidocaine or dimethyl sulfoxide (DMSO) barrier of non-myelinated C-fiber activation abolished the protection afforded by capsaicin application towards the abdominal membrane seen among rodents [22]. Fibers that trigger NO-sensitive are C-fiber, and before nerve stimulation, nearby application of a nitric oxide donor blocked discharge of transferable cardioprotective humoral elements. Mechanical deleterious incentive additionally stimulates C-fiber nociceptors [23]. Different organs can be stimulated to reduce infarct amounts by adenosine or bradykinin [24]. In rats, mesenteric duct infusions of adenosine or bradykinin, femoral artery infusions of adenosine in rodents and individuals come across angiography, although the operation of bradykinin to the stomach skin of rodents decreased the closure. Within rodents, 2 protein kinase C isozymes (PKC, PKC) remained absorbed in the local wave response to bradykinin application to the intestinal skin, and their refrainment abolished the cardio-protection by bradykinin [25].

Transverse intestinal skin clefts trauma, and mechanical incentive, and decreased the volume of a heart attack in rodents and dogs [26]. This type of intestinal skin incision undoubtedly triggers sensory nerves [27]. In mice and dogs, blocking bradykinin 2 receptors prevented protection in response to skin incision. Because bradykinin two receptors were systemically closed in these studies, actually moving is not clear even if bradykinin is discharged its action is restrictedly in response to the nociceptive incentives as a choice yet proximate [28].

In rodents, pigs, and humans, cerebrovascular accident [29] of an extremity or parenchymal organ decreases myocardial closure amount actuating the discharge of humoral elements into the fundamental distribution, which results in cardio protection after moving to an isolated acceptor heart [30]. In rabbits, thigh nerve transection prevented closure amount decreased by three rounds of five-minute ischemia and five-minute reperfusion of one hind limb [31]. However, even if the incentive as a choice for signal transfer is absent in diabetic neuritis is unknown [32].

The absence of discharge of interchangeable humoral cardioprotective elements in diabetic neuritis in contrast to diabetics except for neuritis and healthy commands suggested sensory nerve stimulation via RIC with rounds of hypertension cuff expansion/collapsing of the arm. However, even if incentives as a choice for relocation of the signal are absent in diabetic neuritis is unknown [33].

TRPV1 is a non-exclusive, acid-sensitive cation passage that is explicated for more sensory nerves, containing sensory C-fibers [34]. Corporeal or synthetic impulses operate Transient receptor potential vanilloid 1, causing neuropeptides such as CGRP and neuropeptide substance P to be released from peripheral nerve terminals. In rodents, remote ischemic conditioning including three rounds of five minutes of aplastic anemia and five minutes of reperfusion on 1 hind limb was associated with TRPV1-mediated calcitonin gene-related peptide and substance P discharge. The Transient receptor potential vanilloid 1 antagonist capsazepine prevented the discharge of Calcitonin gene-related protein and substance P as well as an observed reduction in infarct amount [35]. Nevertheless, because capsazepine was administered systemically, the act of Transient receptor potential vanilloid 1 activation to another fragment of the organism was unable to rule out. In rodents, IT and I.V administration of Calcitonin gene-related peptide receptor antagonists abrogated the infarct amount decrease caused by intrathecal morphine administration, denoting that

CGRP- and opioid-mediated cardioprotection interacts [36]. Oxygen deficiency of the perivascular parenchymal tissue also activates the substance related to hypoxia-sensitive factors and further gets discharged into the bloodstream [37]. During reperfusion, reactive hyperemia condition elevates vascular shear stress and actuates the discharge of humoral elements out of the endothelium, namely NO [38].

13.3 REMOTE ISCHEMIC CONDITIONING

RIC is defined as short, tedious, and sublet Ischemia-Reperfusion applicable to an organ or tissue in order towards activating global endogenous patience preventing distant (remote) objective organs or tissues [39] from subsequent prolongation Ischemic remote conditioning, implying especially the conservational effect of Ischemic conditioning can be actuated not only *in situ* but likewise in faraway organs also [40]. According to previous research, there are 3 methods of Remote Ischemic Conditioning intervention focusing on the timing of Remote ischemic conditioning in affiliation to Ischemia-reperfusion injury: remote ischemic preconditioning (RIPreC, commenced before Ischemia-reperfusion injury), (RIPerC, commenced adjacent to time of ischemia), (RIPerC, initiated after IRI) (RIPostC, initiated at reperfusion stage) [41].

13.3.1 REMOTE ISCHEMIC PRECONDITIONING

Remote ischemic preconditioning was first observed in an animal cardiac study [41]. Remote ischemic preconditioning is a method that performs a transient, precedent cerebrovascular accident incentive applicable to a faraway site from the objective organ or tissue that defends it from the eventually enduring aplastic anemia or Ischemia-reperfusion injury [42]. In cerebrovascular accident stroke studies, Remote ischemic preconditioning generally carried out hours or immediately before cerebrovascular accident onset, may relate according to the foregoing two protective phases of Remote ischemia preconditioning: 1st phase (early, short phase) and 2nd phase (later, prolonged phase), the outcome being accomplished from a clinical investigation analyzing prevention related to Remote ischemic pre Conditioning against the lining of endothelium abnormality of the brachial artery lead by arm stroke [43]. 1st protective stage was demonstrated to be initiated immediately after preconditioning began and abided four hours [43] notified that remote limb ischemic preconditioning including 3 rounds of 15 15-minute femoral blockage immediately before cerebrovascular accident beginning significantly decreased cortex infarct amount actuated by 30 minutes of (MCAO) in rodents [44]. While others exhibit that limb-remote ischemia preconditioning performed one hour before Middle cerebral artery occlusion in rodents was still protective, along with consideration four-hour stage [45]. 2nd trial was once deemed to start 24 hours after preconditioning, but later it was corrected to be 12 h, and the period of delaying protection proved to carry for at minimum 48 hours [46]. For example, Remote ischemic pre-conditioning started at 24 hours and 48 hours, but not 72 hours, preceding the cerebrovascular accident were both protective in preclinical trials. Nevertheless, a distinct result was obtained that showed no protective effect was acquired when Remote ischemic pre-conditioning was performed 48 hours before the cerebrovascular accident [47]. Despite some disputes exist; the prolongation of protection of Remote ischemic pre-conditioning discloses especially may be a potential area of remedial time windows related to cerebrovascular accidents. Moreover, to reduce infarct amount, Remote ischemic pre-conditioning was established to be fitted to decrease transient ischemic attack repetition, increase individuals healing rate and augment cerebral perfusion in analytical research [48].

A subsequent rabbit study found that motivating the gastric-related muscle while limiting femoral plasma flow could significantly reduce Ischemic reperfusion injury-actuated cardiac injury, which indicates the possibility of actuating preconditioning activation using clinically more appliable methods namely limb tourniquets or other blocking devices [49]. Another conducted two experiments in 2002 and discovered that: three bouts of five-minute aplastic anemia reperfusion actuated by human upper limb Remote Ischemic Pre-Conditioning thus preventing the contralateral forearm out of endothelial Ischemic reperfusion injury, four bouts of five-minute lower limb Remote ischemic preconditioning could reduce the extent of myocardial injury in a swine model [50] of previously sustained infarction. These discoveries aided in the conversion of Remote Ischemic Preconditioning from experimental models to clinical trials [50]. The Remote ischemic Pre-Conditioning group had decreased potential myocardial injury than the control batch, according to researchers. Subsequent studies on Remote Ischemic Pre-Conditioning revealed that it has the potential to improve Ischemic reperfusion injury in a variety of organs.

13.3.2 REMOTE ISCHEMIC PRECONDITIONING

Remote ischemic preconditioning was observed by Schmidt and coworkers in 2007, a hog model of coronary artery blockage [51], called the action that Remote ischemic Conditioning was operated during a cerebrovascular accident before reperfusion of the objective organ. Later, Remote ischemic preconditioning was stated effective in a cerebrovascular accident and even superior to Remote ischemic preconditioning in preventing the brain resistant Ischemic reperfusion injury. Another experiment performed by Hahn Limb-Remote ischemic preconditioning and remote limb ischemic preconditioning in discrepancy groups of rodents' Middle cerebral artery occlusion models actuated by 120 minutes Middle cerebral artery occlusion and subsequent reperfusion, and discovered that both remote ischemic postconditioning and remote ischemic preconditioning decreased brain infarct volume, but the preceding has been greater with decreased in infarct amount [52]. Remote ischemic preconditioning is another subtype of RIC that refers to the triggering of sublethal conditioning incentive during an episode of cerebrovascular accidents. Similar to Remote ischemic postconditioning, Remote ischemic preconditioning was considered to confer two preventive age or time windows: pre-clinical investigations exhibited that the 1st period begins following the application of Remote ischemic preconditioning elongated for three-six hours, and the 2nd period started 24 hours and prolonged up to four-day period [53, 54]. Voxel-wise examination of investigation showed Remote ischemic preconditioning decreased tissue hazard of brain infarction, though switch the study of penumbral retrieval, ultimate infarct amount at one month, infarct growth among baseline, one month, and clinical result following three months the investigators pointed out prehospital limb-Remote ischemic preconditioning can possess instant neuroprotective results, further investigation from a prolonged individual population will be required for finalization [55]. Schmidt first demonstrated remote aplastic anemia preconditioning in experimental pigs using cyclical hindlimb Ischemic injury with a tourniquet during LADCA occlusion, with a satisfactory outcome [56]. An authoritative single-center randomized controlled trial published in 2010 stated that Remote ischemic preconditioning ameliorates myocardial rescue in patients with suspected segment elevated heart attack although used at emergency before health center admission [57].

13.3.3 REMOTE ISCHEMIC POSTCONDITIONING

Remote ischemic postconditioning was first described in a study of coronary heart disease [58]. Remote ischemic postconditioning is a maneuver of performing Remote ischemic conditioning in isolated organs or tissue following the cerebrovascular accident era of the brain to prevent the brain from Ischemic reperfusion damage [59]. A preclinical investigation showed that the isolated limb ischemic postconditioning actuated instantly or three hours beyond brain reperfusion beginning decreased the brain infarction by about 67% or 43% in rodent cerebrovascular accident model actuated by a complete blockage of the left distal middle of the cerebral artery for 30-minute blockage of the bilateral common carotid arteries, although no preventive result was seen among limb-Remote ischemic postconditioning has been accomplished six hours afterward reperfusion [60]. Moreover, utilizing a different stroke model related to the 90-minute discrepancy of experimental middle cerebral artery ischemia and subsequent reperfusion, showed that bilateral hind limb-remote ischemia postconditioning applied six hours after Middle cerebral artery reperfusion exerted better neuroprotection than that performed at three hours [61]. The discrepancy of investigations among the two results is due to the discrepancy models with the dissimilar method of performing Remote ischemic conditioning (Figure 13.1). The same outcome has been viewed in another preclinical model with optic nerve damage: declared that limb-remote ischemic postconditioning performed six hours following optic nerve injury still conferred prevention across promoting ganglion cell survival at seven days post-damage, the prolonged neuroprotective result was received when limb-Remote ischemic postconditioning was done for 10 minutes [62]. The first report on the cardioprotection of cerebrovascular conditioning was published in 2003 and shows that short repetitive cerebrovascular accidents before fully beginning the rise in a canine pattern of coronary blockage-reperfusion may weaken reperfusion-related myocardial injury [63]. Further in 2005, published the first clinical case, indicating that postconditioning used on individuals enduring coronary angioplasty for STEMI may prevent heart protection [64]. Succeeding publications also showed promising results of ischemic postconditioning in individuals undergoing cardiac surgery, evidenced in lower cardiac enzyme levels and ICU stay life span [65].

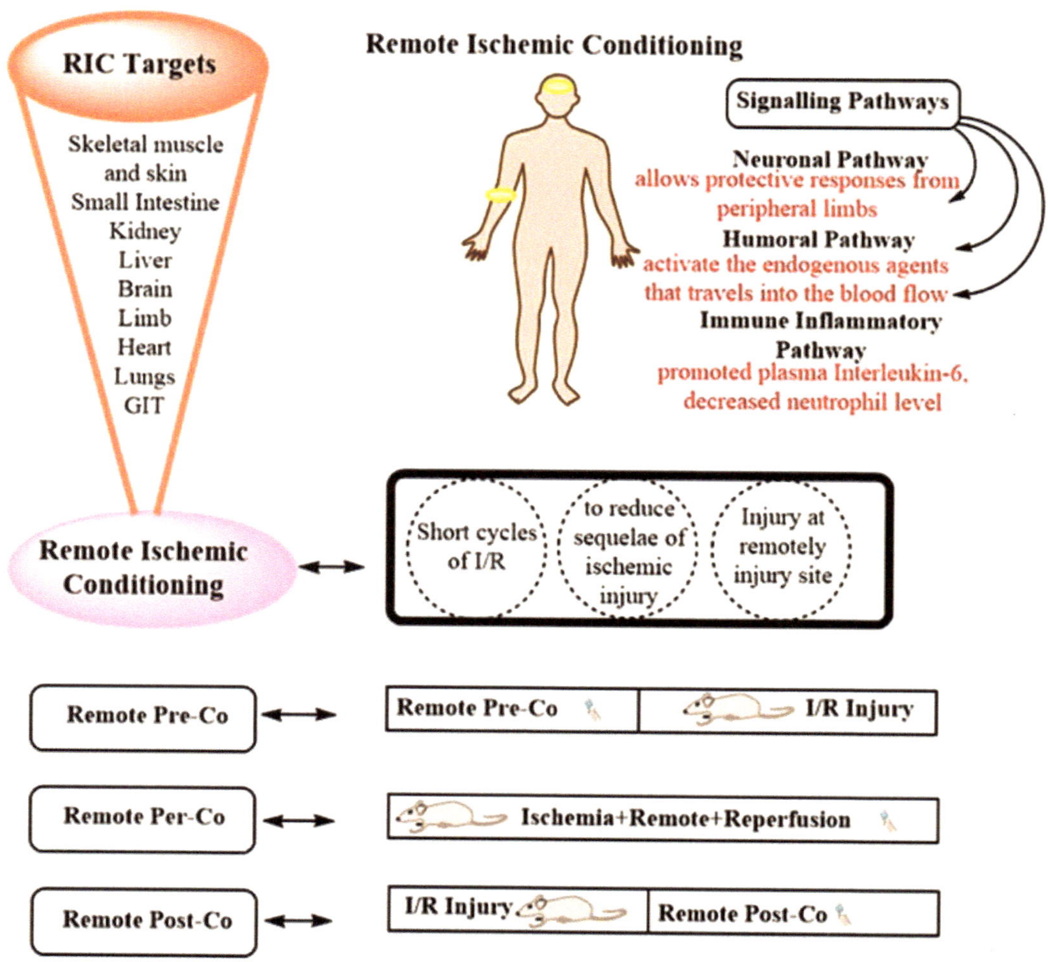

FIGURE 13.1 Illustration of signaling types and target organ involved in remote ischemic conditioning.

13.4 SIGNAL TRANSFER

13.4.1 NEURONAL PATHWAY

The neuronal mechanism, which allows protective responses from peripheral limbs to be transferred to cerebrovascular accident organs, is expected to play a necessary function in Remote ischemic conditioning. This neuronal hypothetic notion of Remote ischemic conditioning is supported by a growing body of evidence [66]. Furthermore, spinal cord reflexes may play important roles in neuronal pathways, as evidenced by the fact that transection of the backbone at the T9–T10 level and selective occlusion of the spinal opioid receptor with naloxone methiodide abrogated the Remote ischemic conditioning process [67]. In terms of spiral outflow, cardiac investigations have proved that the parasympathetic nervous system plays a critical role in the cardioprotection provided by Remote ischemic conditioning whereas there is still a lack of proof considering cerebrovascular diseases [68], tentative investigations regarding rodent cerebrovascular accident model demonstrated especially parasympathetic nerve activation, particularly vagus sense nerve fiber, might have a neuroprotective effect demonstrated the purpose of parasympathetic outflow in Remote ischemic conditioning in clinical trials [69].

13.4.2 HUMORAL PATHWAY

The humoral route hypothesis states that a repetition round of Ischemic conditioning at a distant site might activate the discharge of material that travels into the blood flow and extended to the cerebrovascular accident organs/tissues to bring about a preventive result [70]. Further, the presence of required factors following Remote ischemic conditioning has also been investigated in humans, with the following results: blood dialysate among diabetic patients shows peripheral mononeuritis which underwent Remote ischemic Conditioning can confer a cardioprotective result on rodents organ affecting hearts grieving from Ischemic reperfusion injury [71], although another impact diminished in hearts submerged in Remote ischemic conditioning dialysate out of diabetic patients who also had peripheral neuronitis. Current investigation found especially sub-diaphragmatic vagotomy, and selective sectioning of the behind gastric branch abolished Remote ischemia conditioning cardioprotection, implying that the posterior gastric branch regarding vagus nerve plays a potentially critical role in the awareness of distributing areas educed after Remote ischemic conditioning [72]. Remote ischemic conditioning may promote the discharge of one or more distributing protective factors that are transferrable among several species; an entire neural pathway may be required in Remote ischemic conditioning to actuate protection; and a stiff interaction between both pathways, though details need to be made clear [73]. Before existing clinical application, these probable biomarkers have to be reassessed and confirmed including their effectiveness, possible MOA, the adjustment mode of genomics and proteomics succeeding distinct conditioning procedure [74], time intermission further assessed, compatible related to Remote ischemic conditioning-arbitrator prevention, the whole of which are intriguing but adequate of claims [75].

13.4.3 IMMUNE-INFLAMMATORY PATHWAY

Evolving confirmation plunks as long as the actual trigger of endogenous prevention via Remote ischemic conditioning is moderately connected to the inflection of immuno-inflammatory responses of special attentiveness [76], transient unilateral hindlimb cerebrovascular accident persuaded in rodents could enhance myocardial organic phenomenon in return to free radical anxiety, inflammation, and mitochondrial function at both stages followed by the method [77, 78]. Remote Ischemia Conditioning arbitrated prevention was reflected as decreased neutrophil stimulation together with leukocyte-endothelium interactivity [79]. Moreover, it's going to ensue a few points and additional investigational subjects or protocols deliberately for analyzing these arbitrators. Although Interleukin-6 is a familiar member of the proinflammatory family member, cerebrovascular accident preconditioning is antiquated to reinforce Interleukin-six expression along with blockage of Interleukin-six would elongate by first preconditioning effect. Moreover increased level of Interleukin-6 is to some level to blame for the Remote ischemic conditioning arbitrated cardioprotective effects declared as the primary duration for Interleukin-10 [80]. The modern study declared results regarding Remote ischemic conditioning upon positive inflammatory/anti-inflammatory arbitrator outlines and immune cells, along with the processes lying under Remote ischemic conditioning arbitrated neuroprotection [81]. Remote ischemic conditioning without following cerebrovascular accident promoted plasma interleukin-6 expression with no impact on the concentration of Interleukin-10 and TNF-α [82]. Nevertheless, it had been also confined that the immune system Part spleen plays a vital role in Remote ischemic conditioning arbitrated alterations within the peripheral system, and immunomodulation of the splenic response by Remote ischemic conditioning may have a good immune response that affects the progression of cerebrovascular accident [81].

13.5 RIC TARGETS

The authentic experimental study that described Remote ischemic conditioning demonstrated intramyocardial cardioprotection across various arteries [82]. Nevertheless, soon after this discovery, it was determined that Remote ischemic conditioning stimulation was enforced on an organ distant from the various organs that could protect the heart.

13.5.1 KIDNEY

Kidney injury caused by acute Ischemia-reperfusion injury is a common cause of prevalence and extinction after a variety of medical problems and surgical operations. Ates [83] was the first to show that short liver ischemia significantly reduced biochemical (TNF-associated material) and histopathological markers of kidney-related cerebrovascular accident injury. Analogous renoprotection was notified using limb condition aplastic anemia as the RIC stimulus, which was caused by a brief infrarenal artery occlusion [82]. This is consistent with a recent study that found renoprotection via an adenosine-independent mechanism via remote limb preconditioning [83]. The author demonstrated that bilateral Remote ischemic postconditioning was more effective than unilateral Remote ischemic postconditioning in reducing kidney injury, implying protective threshold can be decided across the volume/mass of primed tissue [83].

13.5.2 SMALL INTESTINE

Acute IRI can have a negative impact on the intestine, stomach, and pancreas. The abdomina are the most commonly affected by acute Ischemia-reperfusion injury in a variety of medical settings, including cardiac arrest, hemorrhagic shock, and major vascular and cardiac surgery [84], with the effects being boost intestinal epithelial barrier permeability and translocation of pathogenic bacteria and endotoxins 112 with consecutive inflammation, sepsis, and multiorgan failure [85]. It was first determined 15-minute blockage of the anterior mesenteric artery could reduce Myocardial infarct amount in a rodents model of Ischemia-reperfusion injury, a finding that was recently confirmed. Unlike Remote ischemic conditioning caused by short renal ischemia, it was discovered that 15 minutes of mesenteric aplastic anemia succeeded by 10 minutes of reperfusion prevented the hog heart along with subsequent Ischemia remote injury under several conditions [86].

13.5.3 LIVER

Limb RIC has been shown in animal models of hepatic IRI to decrease liver damage by developing hepatic plasma flow, lowering serum aminotransferase amount, asserting arterial pressure, and decreasing neutrophil adhesion further apoptosis. Some analytical accomplishments have investigated the liver being an area for applying a Remote ischemic conditioning incentive [87]. Similarly, Remote ischemic conditioning induced by blockage and rushes of common arteria hepatica and vena was noticed to attenuate gastric mucosal lesion actuated by prolonged Ischemia remote injury, a possessive result which looked as if it would be interfered by the cyclooxygenase pathway [88]. Therefore, stimulation of sensory afferent nerves additionally to sensory nerves, also stated that the prevention actuated by Remote ischemic conditioning of the liver was attenuated in vagotomized rats suggesting a very effective role of vagal nerves within the brain-gut axis. Furthermore, the gastroprotection afforded by Remote ischemic conditioning of the liver shows to be akin to those actuated by Remote ischemic conditioning of the myocardium and also the immediate stomach Ischemic preconditioning.

Later mechanistic studies have identified nitric oxide and eNOS as possible arbitrators of Remote ischemic conditioning activating liver protection [89]. It was demonstrated that following hepatic IRI, short hepatic cerebrovascular accident derived in decreased neutrophil infiltration, limited oxidative damage, and less vascular discharge in the pancreas, possibly by abolishing fundamental TNF discharge out of the liver, therefore, preventing sticking molecule P-selectin upregulation [90].

13.5.4 BRAIN

The brain, being a crucial organ susceptible to aplastic anemia injury, is not the best alternative for the RIC incentive application site. An early study found that causing 10 or 30 minutes of brain cerebrovascular accident with reperfusion by increasing intracranial pressure has not yet prevented the heart through afterward acute myocardial Ischemia-reperfusion injury [91]. Stroke caused by cerebral accident or intracranial hemorrhage is the world's 2nd leading cause of death and impairment. It was discovered that Remote ischemia of postconditioning of the limb area decreased brain edema, hemorrhage, and cerebrovascular accident generated by the hypothermic arrest in pigs, these additional facts are associated with significant changes in brain-related function [92]. Neuroprotection appeared towards animals being exposed to either RIPC or RIPost of the limb in a rodent model related to cerebral accident [93].

13.5.5 LIMB

The analytical application of Remote ischemic conditioning would necessitate a less invasive method of delivering the Remote ischemic conditioning incentive. The researchers described a decreased interfering method of actuating hindlimb ischemia as a Remote ischemic conditioning incentive, demonstrating that applying reperfusion to the hindlimb to activate 10 minutes of limb cerebrovascular accident reduced reperfusion arrhythmias in a rat heart [94]. Nevertheless, a great number of exploratory accomplishments have declared the use of the lower limb area as a location for applying a Remote ischemic conditioning incentive due to its ease of access and ability to be applicable with a simple tourniquet. However, limb Remote ischemic conditioning accomplished through tourniquet demand to the area associated with limb or direct blockage of the thigh artery, and antiquated being adequate and reproducible in decreasing organ damage in both animals and humans [95].

13.5.6 HEART

Remote ischemic conditioning twice day-to-day for a week improved coronary microcirculation including chronic heart failure individuals not showing any adverse effects; Remote ischemic conditioning once day-to-day for 4 weeks improved function of muscle and thus lower blood pressure [96–98].

13.5.7 LUNG

Another organ the lung is dominated by sudden aplastic anemia reperfusion injury amongst coronary artery bypass graft surgery, at present suggests being replicated exploratory by coronary artery blockage and reperfusion to actuate pulmonary Ischemia-reperfusion injury. A Study demonstrated Remote ischemic conditioning of the limb conserved lung function in sheep using this experimental model [99]. Furthermore, RIC reduced lung damage after liver IRI or liver transplantation. Recently, intermittent limb cerebrovascular accident has been shown to reduce lung injury caused by the shock of hemorrhagic type and resuscitation in an HO-1-dependent manner. In addition to other lung injuries, Remote ischemic conditioning thus protects the lung from direct Ischemia-reperfusion injury, evidenced by improved lung function and oxygenation capacity associated with it as well as reduced hypertension [100].

13.5.8 GASTROINTESTINAL TRACT

Acute IRI can have a negative impact on the intestine, stomach, and pancreas. The abdomina are the most regularly affected by acute Ischemia-reperfusion injury in a variety of clinical settings, including cardiac arrest, burn trauma, and major vascular and cardiac surgery [101], with the effects being increased intestinal epithelial barrier permeability and translocation of pathogenic bacteria and endotoxins-112 with subsequent inflammation, sepsis, and multiorgan failure [102]. It was demonstrated remote ischemic conditioning actuates gastroprotection in a rat model of gastric Ischemia-reperfusion injury. Furthermore, Dickson [100] discovered that the coronary effluent of cerebrovascular accidents in preconditioned rodent hearts contained humoral factors which improve jejunum sensitivity to cerebrovascular damage via opioid receptor stimulation and KATP channel opening. Some have found that Remote ischemic conditioning reduces intestinal injury caused by hepatic Ischemia-reperfusion injury [101].

13.5.9 FLAP OF SKELETAL MUSCLE AND SKIN

Soft tissue coverage is typically performed in trauma, plastic, and reconstructive surgery along with the utilization of local or free skin flaps, an abscission method that is connected with significant prevalence due to the damage of necrosis of the implanted flap in 1–5% of cases [102]. As long as the flap is exposed during surgery, this is most likely brought about by prolonged postoperative flap cerebrovascular accidents. RIC has been shown in several studies to have cytoprotective effects on muscle flaps. Remote ischemic conditioning shown in pre-clinical studies to prevent tissue death of epigastric adipocutaneous flaps and anatomical muscles. The protection was linked to advanced microcirculation and endothelial role decreased leukocyte sticking and its collection, and conservation of ATP matter [103].

13.6 SPECIFIC MEDIATORS IN REMOTE ISCHEMIC CONDITIONING

13.6.1 ADENOSINE

Adenosine is involved in a variety of biosynthetic processes, including energy transfer, impulse transduction, and neuromodulation [104]. In rabbits, dogs, and pigs, stimulation of adenosine receptors produces a heart attack preconditioning effect. Adenosine stimulates the toxin-sensitive G protein, which in turn activates the ATP-sensitive K+ channel [105]. Surendra and colleagues formerly demonstrated that RIPC-induced cardioprotection is mediated by opioid and -opioid-related sense organs interacting with adenosine A1 receptors [106]. Nevertheless, stimulation of the Adenosine triphosphate-sensitive potassium channel obsoletely shown towards playing a job in canine myocardial preconditioning. Takaoka and colleagues, using an adenosine receptor inhibitor, demonstrated the role of adenosine in RIPC via artery Ischemia-reperfusion in an exceeding rabbit model and notified a discount in Myocardial Infarct amount [107]. Post hoc analysis, individuals who accepted PCI within a few hours of the start of manifestations accepted by the adenosine infusion had decreased prevalence of heart failure [108].

13.6.2 THROMBOCYTE-DEPENDENT SIGNALING

Thrombocyte stimulation caused mainly by pressure, aplastic anemia, or tissue impairment triggers comprehensive results, which are related to defense. Oberkofler investigated the role of actuated platelet-derived serotonin in Remote ischemic postconditioning-mediated I/R protection of the liver and kidney [109]. It was discovered that Remote ischemic pre-conditioning actuated Ischemia-reperfusion injury decreasing the level associated with platelet serotonin release and serotonin-actuated stimulation of the VEGF/IL-10/MMP8 pathway [110]. RIPC reduced platelet numbers by 40%, and although thrombocyte stimulation was inhibited along platelet-activating factor receptor antagonist SM-12502, the impact of Remote ischemic preconditioning was evacuated. Further demonstrates the significance of thrombocyte stimulation in RIPC. The investigation found additionally serotonin is important in thrombocyte-dependent prevention mechanisms [111] When tryptophan hydrolase one rodent lacks platelet serotonin was confined to Remote ischemic postconditioning prior to hepatic Ischemia-reperfusion injury, no reduction in Ischemia-reperfusion injury was observed. Nevertheless, injecting 5HTP, a serotonin precursor, reversed this effect. Prior to I/R, recombinant VEGF-A treatment produced similar RIPC effects, whereas anti-VEGF antibody cure blocked the Remote ischemia pre-conditioning effect, indicating the usefulness of circulating Vascular endothelial growth factor in Remote ischemic postconditioning [112].

13.6.3 ERYTHROPOIETIN

Erythropoietin (EPO) is a glycoprotein accomplished with a molecular weight of 30 kDa that circulates in the blood and is a key controller of RBC production. These 65 amino acid proteins claimed four glycosylated chains that give it biological action and are protected from oxygen radicals [113]. Erythropoietin prevented apoptotic cell death in confined cardiomyocytes by actuating the JAK2-PI3K-AKT signaling pathway and abolishing glycogen synthase kinase-3, a serine-threonine kinase that actuates apoptosis in a variety of cell characters such as vascular smooth muscle cells, cardiomyocytes, and neurons. Oba previously stated particularly renal nerve-interposing erythropoietin release is required for Remote ischemia postconditioning actuating cardioprotection [114]. In a rodent model including four cycles of five-minute anoxia and reoxygenation, serum erythropoietin levels increased in a time-dependent manner, peaking one hour after Remote ischemic postconditioning and returning baseline 24 hours later. A human study of Remote ischemic conditioning with three cycles of 5 minutes of upper arm ischemia and 5 minutes of reperfusion yielded similar results [115]. The production of EPO and

EPO receptors increases in hypoxic tissue (EPOR). Later 2 hours of permanent blockage of the left coronary artery, RIPC activated the HIF-1-Erythropoietin pathway and reduced MI size. Furthermore, administration of the anti-Erythropoietin neutralizing antibody 20 hours before Remote ischemic postconditioning prevented the Remote ischemic conditioning-actuated attenuation of infarct amount. Nevertheless, disagreement about the capacity of renal nerves in Erythropoietin production. Another experiment conducted by Oba's team discovered denervation of kidney-related areas eliminated Remote ischemic postconditioning mediated serum EPO elevation and, as a result, the reduction of Myocardial infarct amount [116].

13.6.4 RIBONUCLEIC ACID

Extracellular Ribonucleic acid serves as an immediate indicator of aggregate cell stress or damage, triggering the discharge of cytokines including TNF- and mediating leukocyte trafficking and, as a result, inflammatory procedure [76]. Ribonucleic acid has been shown to protect eRNA-TNF-crosstalk in a mouse cardiac ischemic reperfusion injury model and confined Langendorf-perfused rodent heart model [117]. Ribonucleic acid 1 is widely conveyed in cells, and the existence of highly special Ribonucleic acid inhibitors limits its cytostatic. Aside from the gastrointestinal tract, where RNase1 is primarily produced, vascular endothelial cells also produce ribonucleic acid, first and foremost during stress [118]. Cabrerea-Fuentese and colleagues previously confirmed that Remote ischemic conditioning prior to cardiac operations reduced pathological eRNA and interleukin expression in individuals that more than 20-fold and seven-fold elevated without Remote ischemic postconditioning [119]. Nevertheless, when compared to control, RIPC increased plasma Ribonucleic acid activity by an extra 5-fld. Since proposed that Remote ischemic postconditioning dependent endothelial RNase1 discharge degrades eRNA and protects it from interacting with interleukins and other proteins, thereby limiting compulsive features of Ischemic Reperfusion. Nevertheless, more research into Ribonucleic acid 1 activity in the RIPC model is needed for a better understanding of the atomic process of protection arbitrated by Ribonucleic acid 1.

13.6.5 APOLIPOPROTEIN A-I

Apolipoprotein A-I primary constituent of plasma elevated-density lipoprotein. Previously, Hilbert confined ApoA-potential I's arbitrator of Remote ischemic conditioning-activated cardioprotection in Wistar rats [120]. Investigations discovered 10 minutes of limb area reperfusion after 10 minutes of limb cerebrovascular accident increased plasma ApoA-I levels by 30% when contrast to the control group, whereas five minutes of limb area reperfusion after 10 minutes of limb cerebrovascular accident, did not affect blood ApoA-I levels when contrast to the non treated rodents. Apolipoprotein was administered I.V bolus (10 mg/kg) 10 minutes before ischemia for 40 minutes, and a decrease in MI size of nearly 15% was reported after 2 hours of reperfusion. Lang et al. found no difference in ApoA-I expression in blood after Remote ischemic postconditioning in a rat plasma proteomic study [121]. Nevertheless, significant differences between the 2 investigations Hibert used kidney artery occlusion to induce the RIPC effect, and Lang et al. performed occlusion of the hind limb. Furthermore, the reperfusion cycle after ischemia differed among 2 investigations. ApoA-additional I's effect mighty attributed capability to salvage TNF-α discharged by myocardial related tissues. Furthermore, I.V. apolipoprotein injection before renal pedicle cerebrovascular accident preceding reperfusion of myocardial tissues reduces Ischemia reperfusion-mediated injury including decreased cytokine production by the damaged tissue as well as reduced endothelial ICAM-1 expression. Nevertheless, the molecular operation underlying the elevated level of Apolipoprotein among plasma remains unknown, rapid feedback to Remote ischemic preconditioning by increasing ApoA-I blood concentration showed a theory that Remote ischemic postconditioning accomplishes neither elevated the level of new ApoA-I synthesis but rather mobilizes the actual ApoA-I [120–121].

13.6.6 BIOMOLECULES

Several investigations suggested that biomolecules along with spreading arbitrator of RIPC. Egln1, an alpha-ketoglutarate-dependent dioxygenase, widely shows and serves as the primary controller of HIF-1. Ischemia inactivates Egln1, which acts as an oxygen sensor. Another experiment conducted by Olenchock and colleagues showed the constraining alpha – Ketoglutarate-dependent dioxygenase in an anatomical unit provided heart protection against Ischemia-reperfusion injury [122]. The group discovered acute deactivation of Egln1 activity causes triggers of Egln, and that the fundamental position of Egln rises as the Egln1 process declines. The liver converts alpha-ketoglutarate to kynurenic acid. Frezza and colleagues discovered that kynurenine, a precursor to kynurenic acid, is overexpressed in cells [123]. Another experiment conducted by Chao and colleagues found preconditioning of the area of the hind limb further forearm increased flowing kynurenic acid in rodents and humans [124]. However, another acid Kynurenic level was suggested to be elevated after myocardial infarction in both animal models and humans indicating a possible act on kynurenic acid metabolic processes in the RIPC setting.

13.6.7 GCG-RELATED PEPTIDE 1

The GCG-related peptide is an animal incretin hormone that is discharged by the abdominal in return for food intake Vagal efferent activity regulates GLP-1 release. GCG-1 activity can be arbitrated by GCG-1 receptors in both manners either dependent or independent manner [125]. Several investigations have found that GLP-1 has a cardioprotective effect in animal models. GLP-1 therapy has also persistent confirmation to be adequate in limiting reperfusion injury in the analytic background. GCG-1R agonists have also shown to be effective in mediating cardioprotection in humans. The act of stimulation of cardioprotective signaling pathways, on the other hand, is debatable. Basaley and colleagues confined that Remote ischemic conditioning confers heart protection activating GLP-1Rs and requires an M3 muscarinic sense organ-dependent mechanism [126]. Nevertheless, even this increase in plasma GCG-1 is physiologically debatable. GLP-1 had short t1/2, the main part of GLP-1 produced in the intestine is rapidly degraded. GLP-1, on the other hand, provides additional effect, as the flow returns to a constant level. Another experiment conducted by Pyke and colleagues previously proposed such ventricular cardiomyocytes that do not communicate GCG-1R [127]. Still, GCG-1 and GCG-1R agonists are shown to act with units and tissues which lack GCG-1R.

13.6.8 STROMAL UNIT ORIGINATED FACTOR-1A

The stromal unit-originated factor-1 (SDF-1) is a chemokine with a molecular weight of 10 kDa that is involved in the recruitment of tumor-related stem cells at the location of myocardial damage [128]. The stromal unit-originated factor-1 saves the heart through further damage beyond involving stem units, directly intracardiac prior to the cerebrovascular accident *in vivo* or the following perfusion before the cerebrovascular accident in isolated hearts related to experiment performed in *ex vivo* [129]. In response to hypoxia, SDF-1 levels rise. AMD3100, a highly specific CXCR4 inhibitor, abolished this protective effect. Furthermore, AMD3100 inhibited the trigger of Remote ischemic conditioning induced cardioprotection in confined perfused hearts, implying a role for SDF-1 in Remote ischemic conditioning [131]. Proteomic examination of human plasma samples, on the other hand, confirmed that the Stomal unit originating Factor-1 was not overexpressed following Remote ischemic postconditioning stimulation [130]. The stromal unit originated from Factor-1 and has a plasma t1/2 of approximately 25 minutes. RIPC, on the other hand, increased SDF-1 by 1 hour before returning to baseline after 3 hours in female Sprague-Dawley rats. In mice, a similar course of action was reported to increase Stromal unit-originated factor-1 at one hour along with three hours [131].

13.6.9 LYMPHOCYTE ACTIVATING FACTOR ALPHA

Lymphocyte activating factor Alpha (IL-1) is an inflammatory arbitrator that is produced by phagocytes, monocytes, endothelial, and flesh cells. In a mouse MI model, Lugrin and colleagues demonstrated that cardiomyocytes discharge lymphocyte-activating factor-1 into the structural flow without increasing heart lymphocyte-activating factor-1 levels [132]. It has been reported that IL-1 blocking and IL-1 receptor antagonists reduce myocardial ischemia-reperfusion injury. According to Gedik and related colleagues, remote ischemic conditioning on the higher arm of a human notably elevated circulating lymphocyte activating factor-1 compared to the controlled group and remained elevated till further heart-related problems [133]. Because both Coronary artery bypass graft surgery and RIPC induce systemic feedback, Gedik and colleagues [133] investigated whether the observed increase in lymphocyte activating factor-1 is due to Ischemia-reperfusion injury or RIPC.

13.6.10 INTERLEUKIN-10

In circulating human neutrophils, RIPC has shown inflection in chromosome expression in proteins affected in incendiary responses such as chemotaxis related to leukocytes, sticking to cells related to endothelial, exocytosis, cytokine incorporation, response produced by the innate immune, and cell death [134]. Among reperfused tissues, late RIPC constrains inflammatory responses. Interleukin-10 anti-incendiary cytokine inhibits chemokine and cytokine production and can activate pro-survival pathways. The receptor subunits of IL-10 are interleukin-10R1 and interleukin-10R2. Interleukin-10R1 is unconditioned across interleukin-10 sensitive cells, as interleukin-10 R2 is demonstrated by all cells. Cardiomyocytes express IL-10R1 receptors [135]. Late Remote ischemic conditioning decreased infarct amount and increased LVDP, which could be everted by dealing the heart with other antibodies associated with IL-10. Remote ischemic conditioning had no heart-protective effects in interleukin-10 knockout rodents or wild-type rodents treated with IL-ten target molecule antibodies that prevented IL-10 from binding to IL-10R1. Further knockout mice involving IL-ten were recombinant with IL-10 P on the other hand, which significantly reduced embolism [136]. Moreover, the size of IL-10 is compatible with the humoral type of mediators typically 30 kDa Elevated plasma Inter leukin-ten quickly acquired in the body because of Inter leukin-one target molecule expression and cardiomyocyte closeness regarding interleukin-10. On the other hand, interleukin-10 can elevate its expression in tissues.

KEYWORDS

- **biomolecules**
- **erythropoietin**
- **gastrointestinal tract**
- **interleukin**
- **lymphocyte activating factor alpha**
- **remote ischemic conditioning**
- **ribonucleic acid**
- **thrombocyte**

REFERENCES

1. Schmidt, M. R., Pryds, K., & Bøtker, H. E., (2014). Novel adjunctive treatments of myocardial infarction. *World J. Cardiol., 6*, 434–443. 10.4330/wjc.v6.i6.434.
2. Yellon, D. M., & Hausenloy, D. J., (2007). Myocardial reperfusion injury. *N. Engl. J. Med., 357*, 1121–1135. 10.1056/NEJMra071667.

3. Hess, D. C., Blauenfeldt, R. A., Andersen, G., Hougaard, K. D., Hoda, M. N., Ding, Y., & Ji, X., (2015). Remote ischemic conditioning: a new paradigm of self-protection in the brain. *Nat. Rev. Neurol., 11*, 698–710. 10.1038/nrneurol.2015.223 [PubMed] [CrossRef] [Google Scholar].
4. Przyklenk, K., Bauer, B., Ovize, M., Kloner, R. A., & Whittaker, P., (1993). Regional ischemic "preconditioning" protects remote virgin myocardium from subsequent sustained coronaryocclusion. *Circulation, 87*, 893–899.
5. Bolte, C. S., Liao, S., Gross, G. J., & Schultz, J. E. J., (2007). Remote preconditioning-endocrine factors in organ protection against ischemic injury. *EMIDDT, 7*, 167–175.
6. Khan, H., Bangar, A., Grewal, A. K., Bansal, P., & Singh, T. G., (2022). Caspase-mediated regulation of the distinct signaling pathways and mechanisms in neuronal survival. *International Immunopharmacology, 110*, 108951.
7. Lim, S. Y., Yellon, D. M., & Hausenloy, D. J., (2010). The neural and humoral pathways in remote limb ischemic preconditioning. *Basic Res. Cardiol., 105*, 651–655. doi: 10.1007/s00395-010-0099-y.
8. Jeanneteau, J., Hibert, P., Martinez, M. C., Tual-Chalot, S., Tamareille, S., Furber, A., Andriantsitohaina, R., & Prunier, F., (2012). Microparticle release in remote ischemic conditioning mechanism. *Am. J. Physiol. Heart Circ. Physiol., 303*, H871–H877. doi: 10.1152 /ajpheart.00102.2012.
9. Basalay, M., Barsukevich, V., Mastitskaya, S., Mrochek, A., Pernow, J., Sjöquist, P. O., Ackland, G. L., et al., (2012). Remote ischemic pre- and delayed postconditioning – similar degree of cardioprotection but distinct mechanisms. *Exp. Physiol., 97*, 908–917. 10.1113/expphysiol.2012.064923 [PMC free article] [PubMed] [CrossRef] [Google Scholar].
10. Lu, Y., Zhang, Y., Shan, H., Pan, Z., Li, X., Li, B., Xu, C., et al., (2009). MicroRNA-1 downregulation by propranolol in a rat model of myocardial infarction: A new mechanism for ischemic cardioprotection. *Cardiovasc. Res., 84*, 434–441. 10.1093/cvr/cvp232 [PubMed] [CrossRef] [Google Scholar].
11. Schoemaker, R. G., & Van, H. C. L., (2000). Bradykinin mediates cardiac preconditioning at a distance. *Am. J. Physiol. Heart Circ. Physiol., 278*, H1571–H1576.
12. Steensrud, T., Li, J., Dai, X., Manlhiot, C., Kharbanda, R. K., Tropak, M., & Redington, A., (2010). Pretreatment with the nitric oxide donor SNAP or nerve transection blocks humoral preconditioning by remote limb ischemia or intra-arterial adenosine. *Am. J. Physiol. Heart Circ. Physiol., 299*, H1598–H1603. doi: 10.1152/ajpheart.00396.2010.
13. Zhou, C., Li, L., Li, H., Gong, J., & Fang, N., (2014). Delayed remote preconditioning induces cardioprotection: Role of heme oxygenase-1. *J. Surg. Res.* doi: 10.1016/j.jss.2014.03.054.
14. Hu, S., Dong, H., Zhang, H., Wang, S., Hou, L., Chen, S., Zhang, J., & Xiong, L., (2012). Noninvasive limb remote ischemic preconditioning contributes neuroprotective effects via activation of adenosine A1 receptor and redox status after transient focal cerebral ischemia in rats. *Brain Res., 1459*, 81–90. 10.1016/j.brainres.2012.04.017 [PubMed] [CrossRef] [Google Scholar].
15. Kalra, P., Khan, H., Kaur, A., & Singh, T. G., (2022). Mechanistic insight on autophagy modulated molecular pathways in cerebral ischemic injury: From preclinical to clinical perspective. *Neurochemical Research*, 1–9.
16. Pignataro, G., Esposito, E., Sirabella, R., Vinciguerra, A., Cuomo, O., Di, Renzo, G., & Annunziato, L., (2013). nNOS and p-ERK involvement in the neuroprotection exerted by remote postconditioning in rats subjected to transient middle cerebral artery occlusion. *Neurobiol. Dis., 54*, 105–114.
17. Dubin, A. E., & Patapoutian, A., (2010). Nociceptors: The sensors of the pain pathway. *J. Clin. Invest., 120*, 3760–3772. doi: 10.1172/JCI42843.
18. Ovize, M., Baxter, G. F., Di Lisa, F., Ferdinandy, P., Garcia-Dorado, D., Hausenloy, D. J., Heusch, G., et al., (2010). Postconditioning and protection from reperfusion injury: Where do we stand? *Cardiovasc. Res., 87*, 406–423. doi: 10.1093 /cvr/cvq129.
19. Silvani, A., Calandra-Buonaura, G., Dampney, R. A., & Cortelli, P., (2016). Brain-heart interactions: Physiology and clinical implications. *Philos. Trans. A Math. Phys. Eng. Sci., 374*. doi: 10.1098 /rsta. 2015.0181.
20. Whittaker, P., & Przyklenk, K., (2014). From ischemic conditioning to 'hyper conditioning': Clinical phenomenon and basic science opportunity. *Dose Response, 12*, 650–663. doi: 10.2203/ doseresponse.14–035.

21. Kaelin, W. G. Jr., & Ratcliffe, P. J., (2008). Oxygen sensing by metazoans: The central role of the HIF hydroxylase pathway. *Mol. Cell, 30*, 393–402. doi: 10.1016/j.molcel.2008.04.009.
22. Konstantinov, I. E., Arab, S., Li, J., Coles, J. G., Boscarino, C., Mori, A., Cukerman, E., et al., (2005). The remote ischemic preconditioning stimulus modifies gene expression in mouse myocardium. *J. Thorac. Cardiovasc. Surg., 130*, 1326–1332. doi: 10.1016/j.jtcvs.2005.03.050.
23. Leung, C. H., Wang, L., Nielsen, J. M., Tropak, M. B., Fu, Y. Y., Kato, H., Callahan, J., et al., (2014). Remote cardioprotection by transfer of coronary effluent from ischemic preconditioned rabbit heart preserves mitochondrial integrity and function via adenosine receptor activation. *Cardiovasc. Drugs Ther., 28*, 7–17. doi: 10.1007/s10557-013-6489-2.
24. Redington, K. L., Disenhouse, T., Li, J., Wei, C., Dai, X., Gladstone, R., Manlhiot, C., & Redington, A. N., (2013). Electroacupuncture reduces myocardial infarct size and improves post-ischemic recovery by invoking release of humoral, dialyzable, cardioprotective factors. *J. Physiol. Sci., 63*, 219–223. doi: 10.1007/s12576-013-0259-6.
25. Della-Morte, D., Guadagni, F., Palmirotta, R., Ferroni, P., Testa, G., Cacciatore, F., Abete, P., et al., (2012). Genetics and genomics of ischemic tolerance: Focus on cardiac and cerebral ischemic preconditioning. *Pharmacogenomics, 13*, 1741–1757. 10.2217/pgs.12.157 [PubMed] [CrossRef] [Google Scholar].
26. Jones, H., Hopkins, N., Bailey, T. G., Green, D. J., Cable, N. T., & Thijssen, D. H., (2014). Seven-day remote ischemic preconditioning improves local and systemic endothelial function and microcirculation in healthy humans. *Am. J. Hypertens., 27*, 918–925. 10.1093/ajh/hpu004 [PubMed] [CrossRef] [Google Scholar].
27. Zarbock, A., Schmidt, C., Van, A. H., Wempe, C., Martens, S., Zahn, P. K., Wolf, B., et al., (2015). Effect of remote ischemic preconditioning on kidney injury among high-risk patients undergoing cardiac surgery: A randomized clinical trial. *JAMA, 313*, 2133–2141. doi: 10.1001/jama.2015.4189.
28. Stazi, A., Scalone, G., Laurito, M., Milo, M., Pelargonio, G., Narducci, M. L., Parrinello, R., et al., (2014). Effect of remote ischemic preconditioning on platelet activation and reactivity induced by ablation for atrial fibrillation. *Circulation, 129*, 11 – 17. doi: 10.1161/CIRCULATIONAHA.113.005336.
29. Tsai, H. J., Huang, S. S., Tsou, M. T., Wang, H. T., & Chiu, J. H., (2015). Role of opioid receptors signaling in remote electrostimulation–induced protection against ischemia/reperfusion injury in rat hearts. *PLoS One 10*, e0138108. doi: 10.1371/journal.pone.0138108.
30. Yellon, D. M., & Davidson, S. M., (2014). Exosomes: Nanoparticles involved in cardioprotection? *Circ. Res. 114*, 325–332. doi: 10.1152/ajpheart.01102.2009.
31. Heyman, S. N., Leibowitz, D., Mor-Yosef, L. I., Liberman, A., Eisenkraft, A., Alcalai, R., Khamaisi, M., & Rosenberger, C., (2016). Adaptive response to hypoxia and remote ischemia pre-conditioning: A new hypoxia-inducible factors era in clinical medicine. *Acta Physiol., 216*, 395–406. doi: 10.1111/apha.12613.
32. Rihal, V., Khan, H., Kaur, A., & Singh, T. G., (2022). Vitamin D as therapeutic modulator in cerebrovascular diseases: A mechanistic perspectives. *Critical Reviews in Food Science and Nutrition*, 1–23.
33. Contractor, H., Haarup, L. R., Cunnington, C., Li, J., Støttrup, N. B., Manlhiot, C., Bøtker, H. E., et al., (2016). Adenosine receptor activation in the "trigger" limb of remote pre-conditioning mediates human endothelial conditioning and release of circulating cardioprotective factor(s). *J. Am. Coll. Cardiol Basic Trans. Science*. doi: 10.1016/j.jacbts.2016.06.002.
34. Nilius, B., Owsianik, G., Voets, T., & Peters, J. A., (2007). Transient receptor potential cation channels in disease. *Physiol. Rev., 87*, 165–217. doi: 10.1152/physrev.00021.2006.
35. Randhawa, P. K., & Jaggi, A. S., (2015). TRPV1 and TRPV4 channels: Potential therapeutic targets for ischemic conditioning-induced cardioprotection. *Eur. J. Pharmacol., 746*, 180–185.
36. Rassaf, T., Totzeck, M., Hendgen-Cotta, U. B., Shiva, S., Heusch, G., & Kelm, M., (2014). Circulating nitrite contributes to cardioprotection by remote ischemic preconditioning. *Circ. Res., 114*, 1601–1610. doi: 10.1161/CIRCRESAHA.114.303822.
37. Uitterdijk, A., Yetgin, T., Te Lintel, H. M., Sneep, S., KrabbendamPeters, I., Van, B. H. M., Fischer, T. M., et al., (2015). Vagal nerve stimulation started just prior to reperfusion limits infarct size and noreflow. *Basic Res. Cardiol., 110*, 508. doi: 10.1007/s00395-015-0508-3.
38. Foley, R. N., (2008). Erythropoietin: Physiology and molecular mechanisms. *Heart Fail Rev., 13*, 405–414. doi: 10.1007/s10741-008-9083-0.

39. Donato, M., Buchholz, B., Rodriguez, M., Perez, V., Inserte, J., GarciaDorado, D., & Gelpi, R. J., (2012). Role of the parasympathetic nervous system in cardioprotection by remote hindlimb ischemic preconditioning. *Exp. Physiol., 98*, 425–434. doi: 10.1113/expphysiol.2012.066217.
40. Hess, D. C., Blauenfeldt, R. A., Andersen, G., Hougaard, K. D., Hoda, M. N., Ding, Y., & Ji, X., (2015). Remote ischemic conditioning-a new paradigm of self-protection in the brain. *Nat. Rev. Neurol., 11*, 698–710. 10.1038/nrneurol.2015.223 [PubMed] [CrossRef] [Google Scholar].
41. Przyklenk, K., Bauer, B., Ovize, M., Kloner, R. A., & Whittaker, P., (1993). Regional ischemic 'preconditioning' protects remote virgin myocardium from subsequent sustained coronary occlusion. *Circulation, 87*, 893–899.
42. Rehni, A. K., & Singh, T. G., (2012). Involvement of CCR-2 chemokine receptor activation in ischemic preconditioning and postconditioning of brain in mice. *Cytokine, 60*(1), 83–89.
43. Loukogeorgakis, S. P., Panagiotidou, A. T., Broadhead, M. W., Donald, A., Deanfield, J. E., & MacAllister, R. J., (2005). Remote ischemic preconditioning provides early and late protection against endothelial ischemia-reperfusion injury in humans: Role of the autonomic nervous system. *J. Am. Coll. Cardiol., 46*, 450–456.
44. Ren, C., Gao, X., Steinberg, G. K., & Zhao, H., (2008). Limb remote preconditioning protects against focal ischemia in rats and contradicts the dogma of therapeutic time windows for preconditioning. *Neuroscience, 151*, 1099–1103.
45. Hu, S., Dong, H., Zhang, H., Wang, S., Hou, L., Chen, S., Zhang, J., & Xiong, L., (2012). Noninvasive limb remote ischemic preconditioning contributes neuroprotective effects via activation of adenosine A1 receptor and redox status after transient focal cerebral ischemia in rats. *Brain Res., 1459*, 81–90.
46. Loukogeorgakis, S. P., Panagiotidou, A. T., Broadhead, M. W., Donald, A., Deanfield, J. E., & MacAllister, R. J., (2005). Remote ischemic preconditioning provides early and late protection against endothelial ischemia-reperfusion injury in humans: Role of the autonomic nervous system. *J. Am. Coll. Cardiol., 46*, 450–456.
47. Meng, R., Asmaro, K., Meng, L., Liu, Y., Ma, C., Xi, C., Li, G., Ren, C., Luo, Y., Ling, F., et al., (2012). Upper limb ischemic preconditioning prevents recurrent stroke in intracranial arterial stenosis. *Neurology, 79*, 1853–1861.
48. Birnbaum, Y., Hale, S. L., & Kloner, R. A., (1997). Ischemic preconditioning at a distance: Reduction of myocardial infarct size by partial reduction of blood supply combined with rapid stimulation of the gastrocnemius muscle in the rabbit. *Circulation, 96*, 1641–1646. 10.1161/01.CIR.96.5.1641 [PubMed] [CrossRef] [Google Scholar].
49. Kharbanda, R. K., Mortensen, U. M., White, P. A., Kristiansen, S. B., Schmidt, M. R., Hoschtitzky, J. A., Vogel, M., et al., (2002). Transient limb ischemia induces remote ischemic preconditioning *in vivo*. *Circulation, 106*, 2881–2883. 10.1161/01.CIR.0000043806.51912.9B [PubMed] [CrossRef] [Google Scholar].
50. Cheung, M. M., Kharbanda, R. K., Konstantinov, I. E., Shimizu, M., Frndova, H., Li, J., Holtby, H. M., et al., (2006). Randomized controlled trial of the effects of remote ischemic preconditioning on children undergoing cardiac surgery: First clinical application in humans. *J. Am. Coll. Cardiol., 47*, 2277–2282. 10.1016/j.jacc.2006.01.066 [PubMed] [CrossRef] [Google Scholar].
51. Schmidt, M. R., Smerup, M., Konstantinov, I. E., Shimizu, M., Li, J., Cheung, M., White, P. A., Kristiansen, S. B., Sorensen, K., Dzavik, V., et al., (2007). Intermittent peripheral tissue ischemia during coronary ischemia reduces myocardial infarction through a KATP-dependent mechanism: First demonstration of remote ischemic preconditioning. *Am. J. Physiol. Heart Circ. Physiol., 292*, H1883–H1890.
52. Hahn, C. D., Manlhiot, C., Schmidt, M. R., Nielsen, T. T., & Redington, A. N., (2011). Remote ischemic preconditioning: A novel therapy for acute stroke? *Stroke, 42*, 2960–2962.
53. Saxena, P., Newman, M. A., Shehatha, J. S., Redington, A. N., & Konstantinov, I. E., (2010). Remote ischemic conditioning: Evolution of the concept, mechanisms, and clinical application. *J. Card. Surg., 25*, 127–134.
54. Hoda, M. N., Siddiqui, S., Herberg, S., Periyasamy-Thandavan, S., Bhatia, K., Hafez, S. S., Johnson, M. H., Hill, W. D., Ergul, A., Fagan, S. C., et al., (2012). Remote ischemic preconditioning is effective alone and in combination with intravenous tissue type plasminogen activator in murine model of embolic stroke. *Stroke, 43*, 2794–2799.

55. Hougaard, K. D., Hjort, N., Zeidler, D., Sørensen, L., Nørgaard, A., Hansen, T. M., Weitzel-Mudersbach, P. V., Simonsen, C. Z., Damgaard, D., Gottrup, H., et al., (2014). Remote ischemic preconditioning as an adjunct therapy to thrombolysis in patients with acute ischemic stroke: A randomized trial. *Stroke, 45*, 159–167.
56. Schmidt, M. R., Smerup, M., Konstantinov, I. E., Shimizu, M., Li, J., Cheung, M., White, P. A., et al., (2007). Intermittent peripheral tissue ischemia during coronary ischemia reduces myocardial infarction through a KATP-dependent mechanism: First demonstration of remote ischemic preconditioning. *Am. J. Physiol. Heart Circ Physiol., 292*, H1883–1890. 10.1152/ajpheart.00617.2006 [PubMed] [CrossRef] [Google Scholar].
57. Bøtker, H. E., Kharbanda, R., Schmidt, M. R., Bøttcher, M., Kaltoft, A. K., Terkelsen, C. J., Munk, K., Andersen, N. H., Hansen, T. M., Trautner, S., Lassen, J. F., Christiansen, E. H., Krusell, L. R., et al., (2010). Remote ischemic conditioning before hospital admission, as a complement to angioplasty, and effect on myocardial salvage in patients with acute myocardial infarction: A randomized trial. *Lancet, 375*, 727–734. 10.1016/S0140-6736(09)62001-8 [PubMed] [CrossRef] [Google Scholar].
58. Kerendi, F., Kin, H., Halkos, M. E., Jiang, R., Zatta, A. J., Zhao, Z. Q., Guyton, R. A., & Vinten-Johansen, J., (2005). Remote postconditioning. Brief renal ischemia and reperfusion applied before coronary artery reperfusion reduces myocardial infarct size via endogenous activation of adenosine receptors. *Basic Res. Cardiol., 100*, 404–412.
59. Hess, D. C., Hoda, M. N., & Bhatia, K., (2013). Remote limb preconditioning and postconditioning. Will it translate into a promising treatment for acute stroke? *Stroke, 44*, 1191–1197.
60. Khan, H., Kashyap, A., Kaur, A., & Singh, T. G., (2020). Pharmacological postconditioning: A molecular aspect in ischemic injury. *Journal of Pharmacy and Pharmacology, 72*(11), 1513–1527.
61. Sun, J., Tong, L., Luan, Q., Deng, J., Li, Y., Li, Z., Dong, H., & Xiong, L., (2012a). Protective effect of delayed remote limb ischemic postconditioning: Role of mitochondrial K(ATP) channels in a rat model of focal cerebral ischemic reperfusion injury. *J. Cereb. Blood Flow Metab., 32*, 851–859.
62. Liu, X., Sha, O., & Cho, E. Y., (2013). Remote ischemic postconditioning promotes the survival of retinal ganglion cells after optic nerve injury. *J. Mol. Neurosci., 51*, 639–646.
63. Zhao, Z. Q., Corvera, J. S., Halkos, M. E., Kerendi, F., Wang, N. P., & Guyton, R. A., (2003). Vinten-Johansen J. Inhibition of myocardial injury by ischemic postconditioning during reperfusion: Comparison with ischemic preconditioning. *Am. J. Physiol. Heart Circ. Physiol., 285*, H579–588. 10.1152/ajpheart.01064.2002 [PubMed] [CrossRef] [Google Scholar].
64. Staat, P., Rioufol, G., Piot, C., Cottin, Y., Cung, T. T., L'Huillier, I., Aupetit, J. F., et al., (2005). Postconditioning the human heart. *Circulation., 112*, 2143–2148. 10.1161/CIRCULATIONAHA.105.558122 [PubMed] [CrossRef] [Google Scholar].
65. Li, B., Chen, R., Huang, R., & Luo, W., (2009). Clinical benefit of cardiac ischemic postconditioning in corrections of tetralogy of Fallot. *Interact. Cardiovasc. Thorac. Surg., 8*, 17–21. 10.1510/icvts.2008.189373 [PubMed] [CrossRef] [Google Scholar].
66. Gill, R., Kuriakose, R., Gertz, Z. M., Salloum, F. N., Xi, L., & Kukreja, R. C., (2015). Remote ischemic preconditioning for myocardial protection: Update on mechanisms and clinical relevance. *Mol. Cell Biochem., 402*, 41–49. 10.1007/s11010-014-2312-z [PubMed] [CrossRef] [Google Scholar].
67. Xiao, L., Lu, R., Hu, C. P., Deng, H. W., & Li, Y. J., (2001). Delayed cardioprotection by intestinal preconditioning is mediated by calcitonin gene-related peptide. *Eur. J. Pharmacol., 427*, 131–135. 10.1016/S0014-2999(01)01231-6 [PubMed] [CrossRef] [Google Scholar].
68. Wong, G. T., Lu, Y., Mei, B., Xia, Z., & Irwin, M. G., (2012). Cardioprotection from remote preconditioning involves spinal opioid receptor activation. *Life Sci., 91*, 860–865. 10.1016/j.lfs.2012.08.037 [PubMed] [CrossRef] [Google Scholar].
69. Mastitskaya, S., Marina, N., Gourine, A., Gilbey, M. P., Spyer, K. M., Teschemacher, A. G., Kasparov, S., et al., (2012). Cardioprotection evoked by remote ischemic preconditioning is critically dependent on the activity of vagal pre-ganglionic neurones. *Cardiovasc Res., 95*, 487–494. 10.1093/cvr/cvs212 [PMC free article] [PubMed] [CrossRef] [Google Scholar].
70. Eckle, T., Köhler, D., Lehmann, R., El Kasmi, K., & Eltzschig, H. K., (2008). Hypoxia-inducible factor-1 is central to cardioprotection: A new paradigm for ischemic preconditioning. *Circulation, 118*, 166–175. 10.1161/CIRCULATIONAHA.107.758516 [PubMed] [CrossRef] [Google Scholar].

71. Albrecht, M., Zitta, K., Bein, B., Wennemuth, G., Broch, O., Renner, J., Schuett, T., et al., (2013). Remote ischemic preconditioning regulates HIF-1α levels, apoptosis, and inflammation in heart tissue of cardiosurgical patients: A pilot experimental study. *Basic Res. Cardiol., 108*, 314. 10.1007/s00395-012-0314-0 [PubMed] [CrossRef] [Google Scholar].
72. Brandenburger, T., Grievink, H., Heinen, N., Barthel, F., Huhn, R., Stachuletz, F., Kohns, M., Pannen, B., & Bauer, I., (2014). Effects of remote ischemic preconditioning and myocardial ischemia on microRNA-1 expression in the rat heart *in vivo. Shock, 42*, 234–238. 10.1097/SHK.0000000000000201 [PubMed] [CrossRef] [Google Scholar].
73. Schmidt, M. R., Smerup, M., Konstantinov, I. E., Shimizu, M., Li, J., Cheung, M., White, P. A., et al., (2007). Intermittent peripheral tissue ischemia during coronary ischemia reduces myocardial infarction through a KATP-dependent mechanism: First demonstration of remote ischemic preconditioning. *Am. J. Physiol. Heart Circ. Physiol., 292*, H1883–1890. 10.1152/ajpheart.00617.2006 [PubMed] [CrossRef] [Google Scholar].
74. Khan, H., Singh, A., Thapa, K., Garg, N., Grewal, A. K., & Singh, T. G., (2021). Therapeutic modulation of the phosphatidylinositol 3-kinases (PI3K) pathway in cerebral ischemic injury. *Brain Research, 1761*, 147399.
75. Silachev, D. N., Isaev, N. K., Pevzner, I. B., Zorova, L. D., Stelmashook, E. V., Novikova, S. V., Plotnikov, E. Y., et al., (2012). The mitochondria-targeted antioxidants and remote kidney preconditioning ameliorate brain damage through kidney-to-brain cross-talk. *PLoS One., 7*, e51553. 10.1371/journal.pone.0051553 [PMC free article] [PubMed] [CrossRef] [Google Scholar].
76. Meng, R., Ding, Y., Asmaro, K., Brogan, D., Meng, L., Sui, M., Shi, J., et al., (2015). Ischemic conditioning is safe and effective for octo- and nonagenarians in stroke prevention and treatment. *Neurotherapeutics, 12*, 667–677. 10.1007/s13311-015-0358-6 [PMC free article] [PubMed] [CrossRef] [Google Scholar].
77. Khan, H., Sharma, K., Kumar, A., Kaur, A., & Singh, T. G., (2022). Therapeutic implications of cyclooxygenase (COX) inhibitors in ischemic injury. *Inflammation Research*, 1–6.
78. Günaydin, B., Cakici, I., Soncul, H., Kalaycioglu, S., Cevik, C., Sancak, B., Kanzik, I., & Karadenizli, Y., (2000). Does remote organ ischemia trigger cardiac preconditioning during coronary artery surgery? *Pharmacol. Res., 41*, 493–496. 10.1006/phrs.1999.0611 [PubMed] [CrossRef] [Google Scholar].
79. Iliodromitis, E. K., Kyrzopoulos, S., Paraskevaidis, I. A., Kolocassides, K. G., Adamopoulos, S., Karavolias, G., & Kremastinos, D. T., (2006). Increased C reactive protein and cardiac enzyme levels after coronary stent implantation. Is there protection by remote ischemic preconditioning? *Heart, 92*, 1821–1826. 10.1136/hrt.2006.089060 [PMC free article] [PubMed] [CrossRef] [Google Scholar].
80. Prasad, A., Gössl, M., Hoyt, J., Lennon, R. J., Polk, L., Simari, R., Holmes, D. R. Jr., Rihal, C. S., & Lerman, A., (2013). Remote ischemic preconditioning immediately before percutaneous coronary intervention does not impact myocardial necrosis, inflammatory response, and circulating endothelial progenitor cell counts: A single-center randomized sham controlled trial. *Catheter Cardiovasc Interv., 81*, 930–936. 10.1002/ccd.24443 [PubMed] [CrossRef] [Google Scholar].
81. Munk, K., Andersen, N. H., Schmidt, M. R., Nielsen, S. S., Terkelsen, C. J., Sloth, E., Bøtker, H. E., et al., (2010). Remote ischemic conditioning in patients with myocardial infarction treated with primary angioplasty: Impact on left ventricular function assessed by comprehensive echocardiography and gated single-photon emission CT. *Circ. Cardiovasc. Imaging, 3*, 656–662. 10.1161/CIRCIMAGING.110.957340 [PubMed] [CrossRef] [Google Scholar].
82. White, S. K., Frohlich, G. M., Sado, D. M., Maestrini, V., Fontana, M., Treibel, T. A., Tehrani, S., et al., (2015). Remote ischemic conditioning reduces myocardial infarct size and edema in patients with ST-segment elevation myocardial infarction. *JACC Cardiovasc. Interv., 8*(1 Pt B), 178–188. 10.1016/j.jcin.2014.05.015 [PubMed] [CrossRef] [Google Scholar].
83. Wever, K. E., Warle, M. C., Wagener, F. A., Van, D. H. J. W., Masereeuw, R., Van, D. V. J. A., & Rongen, G. A., (2011). Remote ischemic preconditioning by brief hind limb ischemia protects against renal ischemia-reperfusion injury: The role of adenosine. *Nephrol. Dial. Transplant., 26*, 3108–3117.
84. Rupani, B., Caputo, F. J., Watkins, A. C., et al., (2007). Relationship between disruption of the unstirred mucus layer and intestinal restitution in loss of gut barrier function after trauma hemorrhagic shock. *Surgery, 141*, 481–489.

85. Swank, G. M., & Deitch, E. A., (1996). Role of the gut in multiple organ failure: Bacterial translocation and permeability changes. *World J. Surg., 20*, 411–417.
86. Tapuria, N., Kumar, Y., Habib, M. M., Abu, A. M., Seifalian, A. M., & andDavidson, B. R., (2008). Remote ischemic preconditioning: A novel protective method from ischemia reperfusion injury – a review. *J. Surg. Res., 150*, 304–330.
87. Abu-Amara, M., Yang, S. Y., Quaglia, A., Rowley, P., Fuller, B., Seifalian, A., & Davidson, B., (2011b). Role of endothelial nitric oxide synthase in remote ischemic preconditioning of the mouse liver. *Liver Transpl., 17*, 610–619.
88. Kanoria, S., Jalan, R., Davies, N. A., Seifalian, A. M., Williams, R., & Davidson, B. R., (2006). Remote ischemic preconditioning of the hind limb reduces experimental liver warm ischemia-reperfusion injury. *Br. J. Surg., 93*, 762–768.
89. Brzozowski, T., Konturek, P. C., Konturek, S. J., Pajdo, R., Kwiecien, S., Pawlik, M., Drozdowicz, D., et al., (2004a). Ischemic preconditioning of remote organs attenuates gastric ischemia-reperfusion injury through involvement of prostaglandins and sensory nerves. *Eur. J. Pharmacol., 499*, 201–213.
90. De Zeeuw, S., Lameris, T. W., Duncker, D. J., Hasan, D., Boomsma, F., Van, D. M. A. H., & Verdouw, P. D., (2001). Cardioprotection in pigs by exogenous norepinephrine but not by cerebral ischemia-induced release of endogenous norepinephrine. *Stroke, 32*, 767–774.
91. Prabhakar, N. K., Khan, H., Grewal, A. K., & Singh, T. G., (2022). Intervention of neuroinflammation in the traumatic brain injury trajectory: *In vivo* and clinical approaches. *International Immunopharmacology, 108*, 108902.
92. Qi, Z. F., Luo, Y. M., Liu, X. R., Wang, R. L., Zhao, H. P., Yan, F., Song, Z. J., et al., (2012). AKT/GSK3β-dependent autophagy contributes to the neuroprotection of limb remote ischemic postconditioning in the transient cerebral ischemic rat model. *CNS Neurosci Ther., 18*, 965–973. 10.1111/cns.12016 [PMC free article] [PubMed] [CrossRef] [Google Scholar].
93. Kim, Y. H., Yoon, D. W., Kim, J. H., Lee, J. H., & Lim, C. H., (2014). Effect of remote ischemic postconditioning on systemic inflammatory response and survival rate in lipopolysaccharide-induced systemic inflammation model. *J. Inflamm. (Lond.), 11*, 16.
94. Li, S., Hu, X., Zhang, M., Zhou, F., Lin, N., Xia, Q., Zhou, Y., Qi, W., Zong, Y., Yang, H., et al., (2015). Remote ischemic postconditioning improves neurological function by AQP4 downregulation in astrocytes. *Behav. Brain Res., 289*, 1–8.
95. Hougaard, K. D., Hjort, N., Zeidler, D., Sørensen, L., Nørgaard, A., Hansen, T. M., Weitzel-Mudersbach, P. V., Simonsen, C. Z., Damgaard, D., Gottrup, H., Svendsen, K., Rasmussen, P. V., Ribe, L. R., et al., (2014). Remote ischemic preconditioning as an adjunct therapy to thrombolysis in patients with acute ischemic stroke: A randomized trial. *Stroke, 45*, 159–167. 10.1161/STROKEAHA.113.001346 [PubMed] [CrossRef] [Google Scholar].
96. Iliodromitis, E. K., Kyrzopoulos, S., Paraskevaidis, I. A., Kolocassides, K. G., Adamopoulos, S., Karavolias, G., & Kremastinos, D. T., (2006). Increased C reactive protein and cardiac enzyme levels after coronary stent implantation. Is there protection by remote ischemic preconditioning? *Heart, 92*, 1821–1826. 10.1136/hrt.2006.089060 [PMC free article] [PubMed] [CrossRef] [Google Scholar].
97. Gaspar, A., Lourenço, A. P., Pereira, M. Á., Azevedo, P., Roncon-Albuquerque, R. Jr., Marques, J., & Leite-Moreira, A. F., (2018). Randomized controlled trial of remote ischemic conditioning in ST-elevation myocardial infarction as adjuvant to primary angioplasty (RIC-STEMI). *Basic Res. Cardiol., 113*, 14. 10.1007/s00395-018-0672-3 [PubMed] [CrossRef] [Google Scholar].
98. Jan, W. C., Chen, C. H., Tsai, P. S., & Huang, C. J., (2011). Limb ischemic preconditioning mitigates lung injury induced by haemorrhagic shock/resuscitation in rats. *Resuscitation, 82*, 760–766.
99. Waldow, T., Alexiou, K., Witt, W., Albrecht, S., Wagner, F., Knaut, M., & Matschke, K., (2005). Protection against acute porcine lung ischemia/reperfusion injury by systemic preconditioning via hind limb ischemia. *Transpl. Int., 18*, 198–205.
100. Dickson, E. W., Tubbs, R. J., Porcaro, W. A., Lee, W. J., Blehar, D. J., Carraway, R. E., Darling, C. E., & Przyklenk, K., (2002). Myocardial preconditioning factors evoke mesenteric ischemic tolerance via opioid receptors and K(ATP) channels. *Am. J. Physiol. Heart Circ. Physiol., 283*, H22–H28.

101. Zuo, B., Wang, F., Song, Z., Xu, M., & Wang, G., (2015). Using remote ischemic conditioning to reduce acute kidney injury in patients undergoing percutaneous coronary intervention: A meta-analysis. *Curr. Med. Res. Opin., 31*, 1677–1685. 10.1185/03007995.2015.1066766 [PubMed] [CrossRef] [Google Scholar].
102. Jones, H., Nyakayiru, J., Bailey, T. G., Green, D. J., Cable, N. T., Sprung, V. S., Hopkins, N. D., & Thijssen, D. H., (2015). Impact of eight weeks of repeated ischemic preconditioning on brachial artery and cutaneous microcirculatory function in healthy males. *Eur. J. Prev. Cardiol., 22*, 1083–1087. 10.1177/2047487314547657 [PubMed] [CrossRef] [Google Scholar].
103. Freixa, X., Bellera, N., Ortiz-Pérez, J. T., Jiménez, M., Paré, C., Bosch, X., De Caralt, T. M., et al., (2012). Ischemic postconditioning revisited: Lack of effects on infarct size following primary percutaneous coronary intervention. *Eur. Heart. J., 33*, 103–112. 10.1093/eurheartj/ehr297 [PubMed] [CrossRef] [Google Scholar].
104. Auchampach, J. A., & Gross, G. J., (1993). Adenosine A1 receptors, KATP channels, and ischemic preconditioning in dogs. *Am. J. Physiol., 264*, H1327–1336. [PubMed] [Google Scholar].
105. Yokota, R., Fujiwara, H., Miyamae, M., Tanaka, M., Yamasaki, K., Itoh, S., Koga, K., et al., (1995). Transient adenosine infusion before ischemia and reperfusion protects against metabolic damage in pig hearts. *Am. J. Physiol., 268*, H1149–1157. [PubMed] [Google Scholar].
106. Surendra, H., Diaz, R. J., Harvey, K., Tropak, M., Callahan, J., Hinek, A., Hossain, T., et al., (2013). Interaction of delta and kappa opioid receptors with adenosine A1 receptors mediates cardioprotection by remote ischemic preconditioning. *J. Mol. Cell Cardiol., 60*, 142–150. doi: 10.1016/j.yjmcc.2013.04.010. [PubMed] [CrossRef] [Google Scholar].
107. Takaoka, A., Nakae, I., Mitsunami, K., Yabe, T., Morikawa, S., Inubushi, T., & Kinoshita, M., (1999). Renal ischemia/reperfusion remotely improves myocardial energy metabolism during myocardial ischemia via adenosine receptors in rabbits: Effects of "remote preconditioning." *J. Am. Coll. Cardiol., 33*, 556–564. [PubMed] [Google Scholar].
108. Ross, A. M., Gibbons, R. J., Stone, G. W., Kloner, R. A., & Alexander, R. W., (2005). AMISTAD-II Investigators A randomized, double-blinded, placebo-controlled multicenter trial of adenosine as an adjunct to reperfusion in the treatment of acute myocardial infarction (AMISTAD-II). *J. Am. Coll. Cardiol., 45*, 1775–1780. doi: 10.1016/j.jacc.2005.02.061. [PubMed] [CrossRef] [Google Scholar].
109. Oberkofler, C. E., Limani, P., Jang, J. H., Rickenbacher, A., Lehmann, K., Raptis, D. A., Ungethuem, U., et al., (2014). Systemic protection through remote ischemic preconditioning is spread by platelet-dependent signaling in mice. *Hepatology, 60*, 1409–1417. doi: 10.1002/hep.27089. [PubMed] [CrossRef] [Google Scholar].
110. Pryds, K., Nielsen, R. R., Jorsal, A., Hansen, M. S., Ringgaard, S., Refsgaard, J., Kim, W. Y., et al., (2017). Effect of long-term remote ischemic conditioning in patients with chronic ischemic heart failure. *Basic Res. Cardiol., 112*, 67. doi: 10.1007/s00395-017-0658-6. [PubMed] [CrossRef] [Google Scholar].
111. Kerendi, F., Kin, H., Halkos, M. E., Jiang, R., Zatta, A. J., Zhao, Z. Q., Guyton, R. A., & Vinten-Johansen, J., (2005). Remote postconditioning. Brief renal ischemia and reperfusion applied before coronary artery reperfusion reduces myocardial infarct size via endogenous activation of adenosine receptors. *Basic Res. Cardiol., 100*, 404–412. doi: 10.1007/s00395-005-0539-2. [PubMed] [CrossRef] [Google Scholar].
112. Schoen, M., Rotter, R., Gierer, P., Gradl, G., Strauss, U., Jonas, L., Mittlmeier, T., & Vollmar, B., (2007). Ischemic preconditioning prevents skeletal muscle tissue injury, but not nerve lesion upon tourniquet-induced ischemia. *J. Trauma., 63*, 788–797. doi: 10.1097/01.ta.0000240440.85673.fc. [PubMed] [CrossRef] [Google Scholar].
113. Parsa, C. J., Kim, J., Riel, R. U., Pascal, L. S., Thompson, R. B., Petrofski, J. A., Matsumoto, A., et al., (2004). Cardioprotective effects of erythropoietin in the reperfused ischemic heart: A potential role for cardiac fibroblasts. *J. Biol. Chem., 279*, 20655–20662. doi: 10.1074/jbc.M314099200. [PubMed] [CrossRef] [Google Scholar].
114. Oba, T., Yasukawa, H., Nagata, T., Kyogoku, S., Minami, T., Nishihara, M., Ohshima, H., Mawatari, K., Nohara, S., Takahashi, J., Sugi, Y., Igata, S., Iwamoto, Y., et al., (2015). Renal nerve-mediated erythropoietin release confers cardioprotection during remote ischemic preconditioning. *Circ. J., 79*, 1557–1567. doi: 10.1253/circj.CJ-14–1171. [PubMed] [CrossRef] [Google Scholar].
115. Nishihara, M., Miura, T., Miki, T., Sakamoto, J., Tanno, M., Kobayashi, H., Ikeda, Y., et al., (2006). Erythropoietin affords additional cardioprotection to preconditioned hearts by enhanced phosphorylation

of glycogen synthase kinase-3 beta. *Am. J. Physiol. Heart Circ. Physiol., 291*, H748–755. doi: 10.1152/ajpheart.00837.2005. [PubMed] [CrossRef] [Google Scholar].

116. Witthuhn, B. A., Quelle, F. W., Silvennoinen, O., Yi, T., Tang, B., Miura, O., & Ihle, J. N., (1993). JAK2 associates with the erythropoietin receptor and is tyrosine phosphorylated and activated following stimulation with erythropoietin. *Cell, 74*, 227–236. [PubMed] [Google Scholar].
117. Landre, J. B., Hewett, P. W., Olivot, J. M., Friedl, P., Ko, Y., Sachinidis, A., & Moenner, M., (2002). Human endothelial cells selectively express large amounts of pancreatic-type ribonuclease (RNase 1) *J. Cell Biochem., 86*, 540–552. doi: 10.1002/jcb.10234. [PubMed] [CrossRef] [Google Scholar].
118. Fischer, S., Cabrera-Fuentes, H. A., Noll, T., & Preissner, K. T., (2014). Impact of extracellular RNA on endothelial barrier function. *Cell Tissue Res., 355*, 635–645. doi: 10.1007/s00441-014-1850-8. [PubMed] [CrossRef] [Google Scholar].
119. Cabrera-Fuentes, H. A., Niemann, B., Grieshaber, P., Wollbrueck, M., Gehron, J., Preissner, K. T., & Boning, A., (2015). RNase1 as a potential mediator of remote ischemic preconditioning for cardioprotection dagger. *Eur. J. Cardiothorac Surg., 48*, 732–737. doi: 10.1093/ejcts/ezu519. discussion 7. [PubMed] [CrossRef] [Google Scholar].
120. Hibert, P., Prunier-Mirebeau, D., Beseme, O., Chwastyniak, M., Tamareille, S., Lamon, D., Furber, A., et al., (2013). Apolipoprotein a-I is a potential mediator of remote ischemic preconditioning. *PLoS One, 8*, e77211. doi: 10.1371/journal.pone.0077211. [PMC free article] [PubMed] [CrossRef] [Google Scholar].
121. Lang, S. C., Elsasser, A., Scheler, C., Vetter, S., Tiefenbacher, C. P., Kubler, W., Katus, H. A., & Vogt, A. M., (2006). Myocardial preconditioning, and remote renal preconditioning--identifying a protective factor using proteomic methods? *Basic Res. Cardiol., 101*, 149–158. doi: 10.1007/s00395-005-0565-0. [PubMed] [CrossRef] [Google Scholar].
122. Olenchock, B. A., Moslehi, J., Baik, A. H., Davidson, S. M., Williams, J., Gibson, W. J., Chakraborty, A. A., Pierce, K. A., Miller, C. M., Hanse, E. A., Kelekar, A., Sullivan, L. B., Wagers, A. J., et al., (2016). EGLN1 inhibition and rerouting of alpha-ketoglutarate suffice for remote ischemic protection. *Cell, 164*, 884–895. doi: 10.1016/j.cell.2016.02.006. [PMC free article] [PubMed] [CrossRef] [Google Scholar].
123. Frezza, C., Zheng, L., Tennant, D. A., Papkovsky, D. B., Hedley, B. A., Kalna, G., Watson, D. G., & Gottlieb, E., (2011). Metabolic profiling of hypoxic cells revealed a catabolic signature required for cell survival. *PLoS One, 6*, e24411. doi: 10.1371/journal.pone.0024411. [PMC free article] [PubMed] [CrossRef] [Google Scholar].
124. Chao, De. La. B. J. M., Bakhta, O., Kalakech, H., Simard, G., Tamareille, S., Catros, V., Callebert, J., et al., (2016). Metabolic signature of remote ischemic preconditioning involving a cocktail of amino acids and biogenic amines. *J. Am. Heart Assoc., 5*, e003891. doi: 10.1161/JAHA.116.003891. [PMC free article] [PubMed] [CrossRef] [Google Scholar].
125. Bose, A. K., Mocanu, M. M., Carr, R. D., & Yellon, D. M., (2005). Glucagon like peptide-1 is protective against myocardial ischemia/reperfusion injury when given either as a preconditioning mimetic or at reperfusion in an isolated rat heart model. *Cardiovasc. Drugs Ther., 19*, 9–11. doi: 10.1007/s10557-005-6892-4. [PubMed] [CrossRef] [Google Scholar].
126. Basalay, M. V., Mastitskaya, S., Mrochek, A., Ackland, G. L., Del, A. A. G., Sanchez, J., Sjoquist, P. O., et al., (2016). Glucagon-like peptide-1 (GLP-1) mediates cardioprotection by remote ischemic conditioning. *Cardiovasc Res., 112*, 669–676. doi: 10.1093/cvr/cvw216. [PMC free article] [PubMed] [CrossRef] [Google Scholar].
127. Pyke, C., Heller, R. S., Kirk, R. K., Orskov, C., Reedtz-Runge, S., Kaastrup, P., Hvelplund, A., et al., (2014). GLP-1 receptor localization in monkey and human tissue: Novel distribution revealed with extensively validated monoclonal antibody. *Endocrinology, 155*, 1280–1290. doi: 10.1210/en.2013-1934. [PubMed] [CrossRef] [Google Scholar].
128. Askari, A. T., Unzek, S., Popovic, Z. B., Goldman, C. K., Forudi, F., Kiedrowski, M., Rovner, A., et al., (2003). Effect of stromal-cell-derived factor 1 on stem-cell homing and tissue regeneration in ischemic cardiomyopathy. *Lancet, 362*, 697–703. doi: 10.1016/S0140-6736(03)14232-8. [PubMed] [CrossRef] [Google Scholar].
129. Hu, X., Dai, S., Wu, W. J., Tan, W., Zhu, X., Mu, J., Guo, Y., Bolli, R., & Rokosh, G., (2007). Stromal cell derived factor-1 alpha confers protection against myocardial ischemia/reperfusion injury: Role of

the cardiac stromal cell derived factor-1 alpha CXCR4 axis. *Circulation, 116*, 654–663. doi: 10.1161/CIRCULATIONAHA.106.672451. [PMC free article] [PubMed] [CrossRef] [Google Scholar].

130. Ceradini, D. J., Kulkarni, A. R., Callaghan, M. J., Tepper, O. M., Bastidas, N., Kleinman, M. E., Capla, J. M., et al., (2004). Progenitor cell trafficking is regulated by hypoxic gradients through HIF-1 induction of SDF-1. *Nat. Med., 10*, 858–864. doi: 10.1038/nm1075. [PubMed] [CrossRef] [Google Scholar].
131. Davidson, S. M., Selvaraj, P., He, D., Boi-Doku, C., Yellon, R. L., Vicencio, J. M., & Yellon, D. M., (2013). Remote ischemic preconditioning involves signaling through the SDF-1alpha/CXCR4 signaling axis. *Basic Res. Cardiol., 108*, 377. doi: 10.1007/s00395-013-0377-6. [PubMed] [CrossRef] [Google Scholar].
132. Lugrin, J., Parapanov, R., Rosenblatt-Velin, N., Rignault-Clerc, S., Feihl, F., Waeber, B., Muller, O., et al., (2015). Cutting edge: IL-1alpha is a crucial danger signal triggering acute myocardial inflammation during myocardial infarction. *J. Immunol., 194*, 499–503. doi: 10.4049/jimmunol.1401948. [PMC free article] [PubMed] [CrossRef] [Google Scholar].
133. Gedik, N., Kottenberg, E., Thielmann, M., Frey, U. H., Jakob, H., Peters, J., Heusch, G., & Kleinbongard, P., (2017). Potential humoral mediators of remote ischemic preconditioning in patients undergoing surgical coronary revascularization. *Sci. Rep., 7*, 12660. doi: 10.1038/s41598-017-12833-2. [PMC free article] [PubMed] [CrossRef] [Google Scholar].
134. Brown, J. M., White, C. W., Terada, L. S., Grosso, M. A., Shanley, P. F., Mulvin, D. W., Banerjee, A., et al., (1990). Interleukin 1 pretreatment decreases ischemia/reperfusion injury. *Proc. Natl. Acad. Sci. U S A., 87*, 5026–5030. [PMC free article] [PubMed] [Google Scholar].
135. Khan, H., Gupta, A., Singh, T. G., & Kaur, A., (2021). Mechanistic insight on the role of leukotriene receptors in ischemic–reperfusion injury. *Pharmacological Reports, 73*(5), 1240–1254.
136. Lonborg, J., Vejlstrup, N., Kelbaek, H., Botker, H. E., Kim, W. Y., Mathiasen, A. B., Jorgensen, E., Helqvist, S., Saunamaki, K., Clemmensen, P., Holmvang, L., Thuesen, L., Krusell, L. R., et al., (2012). Exenatide reduces reperfusion injury in patients with ST-segment elevation myocardial infarction. *Eur. Heart J., 33*, 1491–1499. doi: 10.1093/eurheartj/ehr309. [PubMed] [CrossRef] [Google Scholar].

CHAPTER 14

Sex-Based Differences in Acute Kidney Injury: Lessons from Renal Ischemia-Reperfusion Injury Model in Rodents

TAJPREET KAUR[1] and AMRIT PAL SINGH[2]

[1]*Department of Pharmacology, Khalsa College of Pharmacy, Amritsar, Punjab, India*

[2]*Department of Pharmaceutical Sciences, Guru Nanak Dev University, Amritsar, Punjab, India*

ABSTRACT

The impact of sex in acquiring renal injury is widely acknowledged in pre-clinical as well as clinical studies. In comparison to females, males are documented to be more prone to kidney injury arising from multiple etiologies. In clinics, renal ischemia-reperfusion (I/R) is observed during kidney transplantation, partial nephrectomy, and urological interventions. The I/R-induced acute kidney injury (AKI) model has been widely used as a tool by researchers to understand the reasons for sex bias towards renal dysfunction in rodents. Various approaches tried to include the use of ovariectomized females and supplementation with estradiol in females as well as in male animals. Estrogen-mediated protection against renal I/R-induced AKI is attributed to its anti-oxidant, anti-inflammatory, anti-apoptotic, and vasodilatory activities. The present review summarizes various mechanisms responsible for the estradiol-mediated cytoprotective effect.

14.1 INTRODUCTION

Acute kidney injury (AKI) is characterized by a sharp decline in kidney function thereby causing an accumulation of nitrogenous and other biochemical waste in the body. It is reported to affect up to 25% of patients in intensive care units (ICUs) and carries a significant mortality rate of 15–60% [1, 2]. If untreated, AKI gradually worsens to chronic kidney disease and end-stage renal disease. Shock and sepsis are the common causes of AKI in ICUs. The data suggest that the prevalence of AKI is one in five adults and one in three children admitted to hospitals with acute illness [3]. Incidence of AKI is observed in

Ischemic Injury: A Molecular Insight. Thakur Gurjeet Singh & Heena Khan (Eds.)

developing as well as developed countries. However, the elderly patients are at higher risk in the developed world, whereas in developing countries, it prevails among young ones and children [4]. Around 85% of cases of AKI are reported in developing countries [5]. Hsu and co-workers reported a 10% annual increase in AKI requiring renal replacement therapy [6]. In India, the incidence of AKI in pediatric intensive care units and in-patient wards has been reported as 25–36% and 5–9%, respectively. A mortality rate of 37.6% was reported in a single-centric study in India involving 500 patients, who were admitted to a hospital with AKI or developed AKI in a hospital [7].

14.2 ISCHEMIA REPERFUSION-INDUCED AKI

Ischemia is the cessation of blood supply and consequent decrease in oxygen and nutrients supply to affected tissues or organs. Although reperfusion regains the supply of nutrients and oxygen, it causes injury to the tissues [8–12]. Ischemia-reperfusion (I/R) injury is characterized by marked oxidative stress and inflammation in the affected organs [9, 13, 14]. In clinics, renal I/R injury is witnessed during renal transplantation, partial nephrectomy, urological interventions, cardio-pulmonary bypass, and hydronephrosis. I/R injury is one of the major risk factors causing significant morbidity and mortality in patients with kidney diseases [15, 16]. I/R injury is essentially observed during surgery and is the major culprit for AKI. It accounts for around 30% of cases of kidney allograft dysfunction following renal transplantation [17].

Renal vasoconstriction, glomerular, and tubular damage, and renal inflammation are major mechanisms involved in I/R-induced AKI [18]. During I/R injury, altered mitochondrial function with impaired electron transport chain causes overproduction of reactive oxygen species (ROS) in the cells. During ischemia, reduced adenosine triphosphate (ATP) generation leads to impairment of membrane-bound exchange pumps and results in intracellular calcium levels. The increased calcium facilitates the conversion of xanthine dehydrogenase to xanthine oxidase through the calcium-dependent enzyme. During reperfusion, hypoxanthine, a breakdown product of ATP, is converted into xanthine along with the production of superoxide radicals, hydrogen peroxide (H_2O_2), and hydroxyl radicals. Overproduction of ROS causes oxidative stress. The raised level of ROS (superoxide radicals, H_2O_2, and hydroxyl radicals) causes the accumulation of neutrophils and the release of lytic enzymes thereby aggravating tissue injury and damage [19]. In normal physiology, the superoxide dismutase (SOD) enzyme converts the superoxide anion to H_2O_2, which is further converted to water and oxygen by the catalase enzyme. During I/R injury, the burst release of superoxide radicals, H_2O_2 and reduced activity of SOD and catalase is responsible for oxidative stress and damage to affected tissues [19, 20].

In the vasculature, I/R-induced oxidative stress and endothelial cell damage cause enhanced production of endothelins thereby increasing vascular congestion and tissue injury [21]. Damaged endothelium supports anchorage of inflammatory cells like neutrophils by up-regulating adhesion molecules like selectins, monocyte chemo-attractant protein-1 (MCP-1), and intercellular adhesion molecule-1 (ICAM-1) [9]. Various cytokines, including tumor necrosis factor-α (TNF-α), and interleukins (IL)-1, 18 play a vital role in renal I/R by activating and transmigrating leukocytes in affected kidneys [22]. Moreover, I/R activates complement proteins and bone marrow-derived cells (macrophages and T cells), which aggravate inflammation and result in tissue injury [22, 23]. The reperfusion phase of I/R involves ROS production and inflammation. Within 30 minutes of reperfusion, the neutrophils start accumulating with activated endothelium in renal tissues and exaggerate inflammatory response by promoting diapedesis. The infiltrated cells produce enzymes (proteases and myeloperoxidase), ROS, and cytokines, which enhances vascular permeability, disruption of tubular cell integrity, and consequent renal damage [19, 24, 25]. Furthermore, leukotriene B4 and platelet-activating factors promote inflammation of renal tissues by increasing vascular permeability and up-regulation of the expression of adhesion molecules [24].

I/R-induced tubular damage includes decreased proximal tubular sodium reabsorption, increased tubular obstruction, and cast formation [21, 26]. I/R-induced AKI promotes apoptosis of renal cells by activating cysteine proteases such as calpains and caspases. Calpains-mediated activation of caspase-1 stimulates inflammation through IL-8, whereas caspase-3-induced cleavage of calpastatin causes apoptosis [14, 21, 27, 28]. Raised levels of intracellular calcium diminish the activity of basolateral Na^+ K^+ ATPase, which further decreases GFR by impeding sodium reabsorption in proximal tubules and aggravates renal injury [21]. Various mechanisms of I/R-induced AKI are summarized in Figure 14.1.

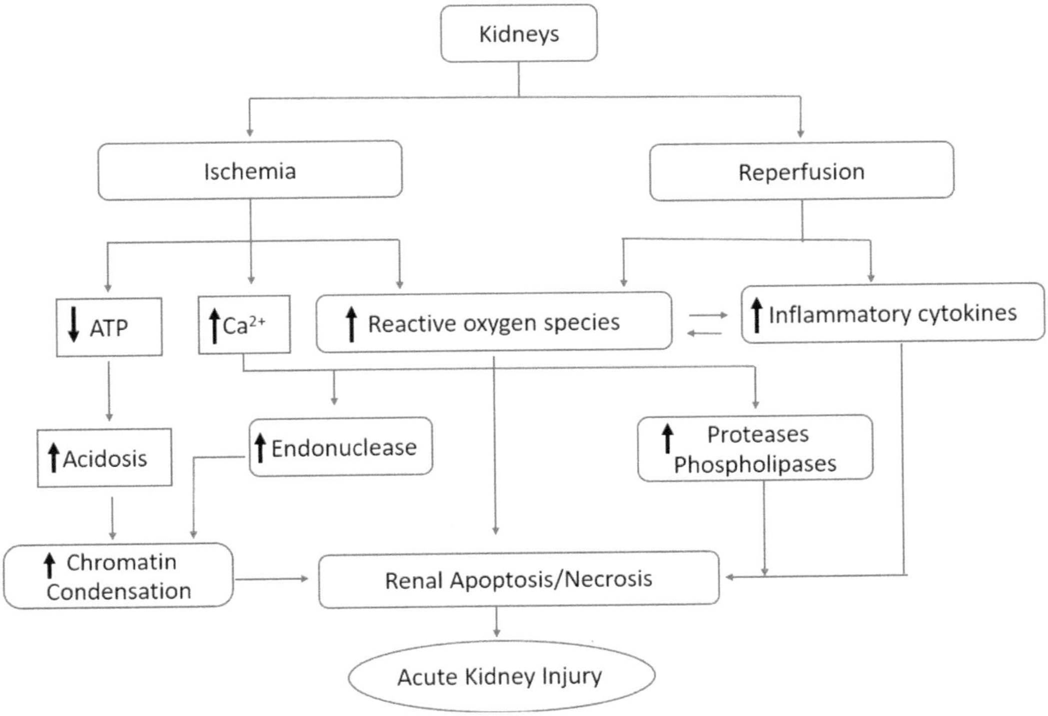

FIGURE 14.1 Various mechanisms of I/R-induced AKI.

14.3 SEX DIFFERENCE AND AKI

The impact of sex difference in acquiring renal complications is well acknowledged in preclinical and clinical studies. Males are noted to be more prone to acquiring renal complications than females. This sex bias is not only restricted to kidneys but also extended to other organs such as the liver, heart, vascular tissues, and brain [29, 30]. Female recipients are reported to have a better outcome with their renal transplants than males [31, 32]. Delayed graft function (DGF) is commonly associated with renal transplantation and is defined as the requirement of dialysis within the first week of transplantation [33]. Analysis of 46,691 patients with kidney transplants suggests that DGF is significantly higher ($P < 0.001$) in males as compared to female recipients [32]. It is reported that females are less likely to undergo dialysis as compared to males [34].

Animal studies have actively taken up the issue of the differential response of males and females toward AKI in models like I/R, diabetes mellitus, hypertension, and nephrotoxin-induced kidney damage [35–39]. In the cited studies, males demonstrated increased severity and higher mortality than females. Muller and his co-workers reported that post-ischemic survival was significantly higher in females (75%) than in males (8%) [35]. Similar sex susceptibility is also demonstrated in I/R injury of the brain and heart [40, 41]. Certain anatomical and physiological differences have been observed among male and female kidneys in humans and rodents. Human data suggest that the decline in GFR is observed in males with age but not in pre-menopausal women [42]. In rats, higher kidney weight is observed in males than females of the same body weight, although, the number of glomeruli is the same in both sexes. It is attributed to higher androgen levels and consequent growth of cortical tubules in male rats [43]. Female rats demonstrated exaggerated vascular resistance and decreased blood flow in kidneys as compared to their male counterparts [44]. The male rats demonstrated kidney dysfunction after 30 minutes of bilateral renal ischemia whereas, ischemia of 60 minutes produced renal dysfunction in female rats [37]. The I/R-induced rise in serum urea, creatinine, and tubular necrosis was markedly higher in male rats as compared to females [45]. Higher levels of cytoprotective heat shock proteins have been demonstrated in the kidneys and heart of female rats in comparison to males [46, 47]. The over-activity of the renal sympathetic nervous system causes kidney damage. Renal denervation or ganglionic blocking agents are reported to curtail I/R-induced AKI by decreasing sympathetic outflow in rodents [48]. Estrogen is noted to suppress sympathetic activity and consequently protect rat kidneys against I/R-induced AKI [49]. Tanaka and co-workers reported that a higher level of monoamine oxidase-A is accountable for decreased sympathetic activity and subsequent renoprotection in female rats [50]. Profound endoplasmic reticulum stress has been noted in kidneys of tunicamycin-treated male mice as compared to females. Interestingly, the administration of testosterone in females resulted in the decline of renal function and induction of endoplasmic stress comparable to males [51]. As compared to females, the male rats witnessed higher binding of angiotensin II with its receptors. Estrogen decreases this binding which may account for reduced renal and cardiovascular complications in females than in males [52].

14.4 MECHANISMS OF ESTROGEN-MEDIATED RENOPROTECTION

Endogenous estrogen in females, as well as exogenously administered estradiol in males, is noted to protect kidneys against a variety of noxious stimuli including I/R injury in rodents. Estrogen attenuated tubulointerstitial fibrosis, glomerulosclerosis, and podocyte injury in diabetic and hypertensive-nephropathy [53–55]. Estrogen produced a marked antioxidant effect in rat kidneys and promoted tubular regeneration in I/R-induced AKI [56, 57]. I/R-induced loss of endothelium causes vasoconstriction. Intact Vascular endothelium-bound nitric oxide synthase (eNOS) produces nitric oxide (NO), which causes vasodilatation and regulates renal blood flow [58, 59]. Endothelial injury causes impaired NO production thereby affecting vascular function and consequent renal function. Moreover, over-activity of inducible nitric oxide synthase (iNOS) causes peroxynitrite generation, which subsequently causes protein nitration and cell damage. The enhanced iNOS activity has been reported to suppress eNOS function [60]. Estrogen increases eNOS activity and consequently results in kidney protection [61]. In comparison to age-matched men, the pre-menopausal women demonstrated increased production of whole-body NO [62]. Protein kinase B or Akt, the key regulator of cell survival, is activated through phosphatidyl inositol 3-kinase (PI3K). Female rodent kidneys have higher basal levels of Akt than males. Estrogen supplementation in males resulted in increased Akt levels [37]. Treatment with estradiol attenuated I/R-induced AKI by activating PI3K/Akt/eNOS pathway in male rats [61]. Following renal I/R injury, expression of the prepro-endothelin gene rose significantly within 2 hours in males, but not in female rats [35]. Pre-treatment with LU 135252, an endothelin receptor antagonist, decreased mortality in males and their survival rate was as good as in females. These findings highlight that estradiol affords renoprotection by regulating levels of endothelin [35]. In cases of kidney transplantation, the inflammation contributes to DGF, graft dysfunction, and rejection [141]. Renal I/R injury increased the renal TNF-α, MCP-1, ICAM-1, and chemokine (C-C motif) ligand-17 (CCL-17) levels resulting in inflammatory cells recruitment at the site of injury and aggravated renal damage in mice [63]. Estrogen supplementation in ovariectomized female mice decreased inflammatory cytokine levels and resulted in renoprotection [63]. Various mechanisms of estrogen-mediated renoprotection are summarized in Figure 14.2.

14.5 ROLE OF PPAR-Γ IN KIDNEY DISEASE

PPAR-γ belongs to a class of ligand-activated transcriptional factors and a family of nuclear receptors. Three isoforms of PPARs, namely PPAR-α, PPAR-β/δ, and PPAR-γ have been identified. PPARs regulate the expression of various genes involved in glucose and lipid metabolism, insulin sensitivity, adipogenesis, cell growth, and differentiation [64]. Endogenous ligands for PPAR-γ include eicosanoids such as 15-deoxy-Δ12,14-prostaglandin J2, oxidized low-density lipoproteins, and oxidized alkyl phospholipids including nitro-linoleic acid and lysophosphatidic acid [65, 66]. The ability of PPAR-γ to regulate adipocyte differentiation, fatty acid storage, and glucose metabolism contributes to its anti-diabetic potential. Synthetic ligands of PPAR-γ such as pioglitazone are used for treating insulin resistance in type 2 diabetes mellitus in clinics.

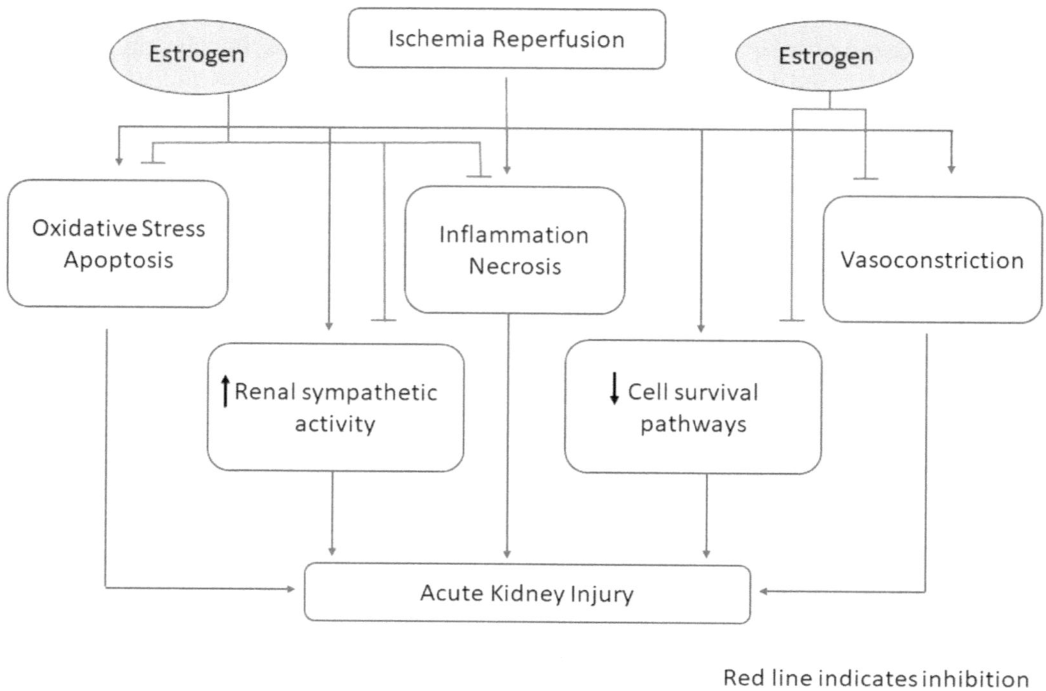

FIGURE 14.2 Various mechanisms of estrogen-mediated renoprotection.

Histological studies revealed the presence of PPAR-γ on human and rodent kidneys. PPAR-γ is expressed on glomeruli, proximal tubules, and medullary collecting duct in the nephron. Moreover, podocytes, glomerular mesangial cells, and epithelial lining of collecting ducts are reported to have PPAR-γ receptors [67]. These findings highlight the involvement of PPAR-γ in renal pathophysiology. Chronic hyperglycemia causes renal glucotoxicity, which involves protein glycation, activation of protein kinase C and accumulation of advanced glycation end products (AGE) causing damage to the kidney and its vasculature. PPAR-γ agonists ameliorate diabetes-induced damage to podocytes, mesangial cells, endothelial cells, and renal tubular epithelial cells [68]. Pioglitazone and rosiglitazone improved diabetes-induced endothelial dysfunction along with a reduction in type IV collagen levels in serum and albuminuria in humans [67, 69]. In type 2 diabetes mellitus patients, PPAR-γ agonists produced comparable diabetic control with standard anti-hyperglycemic agents such as metformin, and glibenclamide, along with better kidney protection [70, 71]. Rosiglitazone abrogated AGE-induced rise in IL-8 and ICAM-1 expression in cultured renal proximal tubular epithelial cells [72]. Nitro-oleic acid, a natural ligand for PPAR-γ, attenuated metabolic syndrome and proteinuria in Zucker fatty rats [73, 74]. In aging rats, PPAR-γ agonist improved GFR and reduced proteinuria [75]. Over-expression of PPAR-γ induces expression of the Klotho gene in renal epithelial cells and mouse kidneys, which improves insulin sensitivity and reduces cellular aging [67, 76]. Various PPAR-γ agonists attenuated aminoglycoside, glycerol, and cyclosporine-induced kidney damage by reducing oxidative stress and inflammation [77–79].

PPAR-γ agonists are protective against I/R injury of various organs such as the lungs, liver, brain, heart, and gastrointestinal tract [80–85]. Administration of 15-deoxy-Δ12,14-prostaglandin J2, an endogenous ligand of PPAR-γ, attenuated I/R-induced AKI in rats [86]. PPAR-γ agonists, rosiglitazone, and ciglitazone, decreased I/R-induced renal damage by attenuating ICAM-1 expression, leukocyte infiltration, and oxidative stress in kidneys [81]. Our laboratory reported that gallic acid, vitamin D, and synthetic molecules like sildenafil attenuated I/R-induced AKI through PPAR-γ agonism [87–89].

14.6 ESTRADIOL AMELIORATES I/R-INDUCED AKI THROUGH PPAR-γ AGONISM

The close interaction between estrogen and PPAR-γ is largely restricted to *in vitro* studies. Selective estrogen receptor modulators such as raloxifene and bazedoxifene are noted to attenuate hypoxia-induced cortical neuronal cells by activating PPAR-γ [90]. Estradiol and genistein, a phytoestrogen, increased PPAR-γ expression in amyloid-beta protein-treated astrocytes and prevented the amyloid-induced increase in inflammatory cytokines in astrocytes [91]. Daidzein activates PPAR-γ and inhibits pro-inflammatory cytokine expression in adipocytes [92, 93]. In mice, ovariectomy decreased the aortic PPAR-γ expression. Moreover, GW9662, the PPAR-γ antagonist, inhibited estrogen-mediated vasorelaxation in mice [142]. Similar results are reported in female rats, where genistein demonstrated vasorelaxation against phenylephrine-induced aortic contraction, and its vasorelaxant effect was abolished by PPAR-γ antagonist [94]. Treatment with estrogen attenuated trauma hemorrhage-induced cytokine production by activating PPAR-γ in male mice [95]. We reported that exogenous administration of estradiol attenuated I/R-induced AKI through PPAR-γ agonism in male rats. Of note, estrogen-mediated anti-oxidant, anti-inflammatory, anti-apoptotic, and the renoprotective effect was significantly abolished by concurrent treatment with PPAR-γ antagonist in rats [96].

14.7 ROLE OF NO IN KIDNEYS

NO is a short-lived gaseous molecule produced by cells exerting a diverse range of biological functions in the human body [97, 98]. In the cells, a family of NOS enzymes catalyzes the conversion of L-arginine into citrulline along with NO production. In normal physiology, NO regulates vascular tone, angiogenesis, neural development, and insulin secretion in the human body. In kidneys, NO regulates renal hemodynamics and sympathetic activity and maintains medullary perfusion, tubular reabsorption of sodium, and natriuresis [99]. NO is synthesized by three isoforms of NOS including neuronal NOS (nNOS), iNOS, and eNOS. Out of three, nNOS and eNOS are calcium-dependent constitutive enzymes producing NO in a pulsative manner, whereas, iNOS is calcium-independent and is activated during inflammation [100]. All three NOS isoforms have been detected in kidneys

[101]. The eNOS is majorly expressed on renal microvasculature, glomerular endothelial cells, proximal tubules, and collecting duct, where it regulates renal vascular tone [58]. The nNOS is expressed in Bowman's capsule, efferent arterioles, macula densa, collecting duct, and controls sodium excretion [58, 97]. The iNOS is weakly expressed in kidneys and is detected in the inner medullary collecting duct.

NO exerts both deleterious as well as cytoprotective effects depending on the involvement of NOS. Various cytokines such as TNF-α, IL-1, and interferon-γ released during inflammation provoke iNOS-mediated burst release of NO, which further combines with free radicals such as superoxide anion to form peroxynitrite thereby causing cell damage [19]. Administration of iNOS inhibitors and peroxynitrite scavengers protected kidneys in the models of AKI [102–104]. On the contrary, eNOS releases short bouts of NO, which is required for physiological functions in the cell. Generally, cellular injury is associated with reduced expression as well as decreased activity of eNOS. The over-activity of iNOS during inflammation decreases L-arginine availability for eNOS, which negatively affects cellular function [143]. During renal I/R, the endothelial injury-induced diminution in eNOS expression and consequent vasoconstriction impairs kidney function in rodents [21]. Treatment with L-arginine, NO donors like molsidomine, and eNOS activation attenuated I/R-induced AKI in rodents [105–107]. Administration of L-arginine increased GFR by enhancing renal perfusion in I/R-induced AKI in rats [108]. It is suggested that the eNOS-mediated release of NO increases blood flow to the tissues which aids their recovery in pathological conditions. Of note, transplantation of functionally competent endothelial cells into circulation afforded significant protection against I/R-induced kidney injury in rats [109]. Molecular studies have demonstrated that phosphorylation of Ser^{1177} in the calmodulin-binding domain of eNOS is a must for its activity. Rho-kinase negatively regulates the phosphorylation at Ser^{1177} and indirectly controls eNOS activity. Administration of U-27632, a Rho kinase inhibitor, demonstrated a profound increase in eNOS activity and attenuation of I/R-induced AKI in rats [110].

14.8 SEX DIFFERENCE AND NO IN RENAL PATHOPHYSIOLOGY

The laboratory and clinical findings suggest that males are more prone to renal damage than females. Since NO maintains the renal vascular tone and maintains kidney function, the differential NOS expression in males and females has emerged as a major contributor to sex bias in acquiring kidney disease [111, 112]. The renal mRNA level of eNOS was 80% higher in female rats than in males. Moreover, NOS inhibitor significantly increased vascular resistance and decreased plasma flow in the kidneys of males than female rats [111]. High salt diet-fed male mice had higher expression of angiotensin receptor-1 and decreased levels of eNOS in renal vasculature as compared to female mice [112]. Male spontaneously hypertensive rats had lower expression of eNOS than female rats [113]. Chronic NOS inhibition resulted in proteinuria in male rats attributed to higher androgen levels, whereas, females were protected, owing to their higher estrogen levels [114]. Estrogen facilitated NOS signaling in kidneys by α7-nicotinic cholinergic receptors and renal nitrite/nitrate levels as well as eNOS expression were significantly higher in estrogen-supplemented rats than in ovariectomized rats [115]. In type 2 diabetes mellitus patients, NO in renal circulation is reported to be higher in females than in males [116]. In our study, estradiol-mediated renoprotection was ameliorated by the NOS inhibitor, L-NAME. Rat kidneys in the I/R group demonstrated decreased eNOS expression, which was corrected with estradiol treatment. Moreover, our results demonstrated that estradiol-mediated activation of eNOS in rat kidneys is PPAR-γ dependent [117].

14.9 NMDA AND KIDNEY

N-methyl-d-aspartate receptor (NMDAR) belongs to a class of ionotropic glutamate receptors using NMDA as its agonist. The binding of the agonist opens the ion channel of NMDAR, which causes calcium influx and subsequent biochemical changes in the cell. The role of NMDAR has been predominantly explored in learning and memory. The NMADR over-activation causes neuronal excitotoxicity and death. NMDAR antagonists are being used in treating neurological disorders like stroke, epilepsy, neuropathic pain, Parkinson's, Huntington, and Alzheimer's disease [118–121].

Besides neuronal tissues, NMDAR is expressed in organs like the pancreas, urogenital tract, heart, parathyroid gland, and bones [122–124]. The presence of NMDAR on kidneys is well documented. NMDAR is present on the glomerulus, podocytes, brush border, and in the outer medulla of the nephron [122, 125]. Diabetes mellitus increased the expression of NMDAR on glomeruli and tubules in mice. NMDAR antagonists, memantine, and dizocilpine, attenuated diabetic nephropathy in mice [126]. Administration of gentamicin increased the expression of NR1 and NR2C subunits of NMDAR in the renal cortex of rats. Treatment with dizocilpine abolished gentamicin-induced renal injury [127]. Both, NMDAR agonists as well as antagonists are reported to protect kidneys against I/R-induced AKI in rodents. NMDAR agonist attenuated I/R-induced renal injury through activation of constitutive NOS followed by vasodilatation [128]. In another study, administration of 2-amino-5-phosphonopentanoic acid, an NMDAR antagonist, protected against I/R-induced AKI [129]. A study from our laboratory indicated that NMDAR activation causes renal damage, whereas, direct channel blockade with magnesium sulfate or antagonism of glycine and polyamine binding site of NMDAR attenuated I/R-induced AKI in rats [122]. Moreover, we demonstrated that glycine aggravates I/R-induced renal damage by activating NMDAR in rat kidneys [130]. Another finding of our laboratory suggested that NMDAR antagonism is one of the key mechanisms in curcumin-mediated protection against I/R-induced AKI in rats [125].

14.10 ESTROGEN, NMDAR, AND KIDNEY INJURY

The interplay between estrogen and NMDAR is primarily illustrated in neurons. Cortical impact-induced traumatic brain injury is associated with increased NMDAR NR2B subunit expression in the hippocampus, which causes oxidative stress and apoptosis in neurons [131]. Estrogen attenuated the enhanced NR2B expression, oxidative stress, apoptosis, and brain injury in mice [131]. Estradiol decreased NR1 expression and abrogated glutamate-induced excitotoxicity in cortical and hippocampal cell cultures [132]. Combined treatment with memantine, the NMDAR antagonist, and estradiol demonstrated synergism against traumatic brain injury-induced neuronal death [133, 134]. Estrogen diminished NMDA-induced seizures in female ovariectomized rats [135]. Furthermore, estrogen ameliorated NMDA-induced retinal neurotoxicity in rats [136]. Estradiol attenuated formalin-induced inflammatory pain in rats. Interestingly, the NMDAR antagonist blocked the estrogen-mediated analgesic effect thereby suggesting crosstalk between NMDAR and estrogen [137]. We published that the estradiol-mediated renoprotective effect is mediated through NMDAR antagonism. The NMDAR agonists, glutamic acid, and spermidine attenuated estrogen-mediated protection against I/R-induced AKI in rats [138]. Another finding of our laboratory suggested that pioglitazone attenuated I/R-induced AKI through NMDAR antagonism [96]. In the light of our previous findings, it may be proposed that estradiol-mediated activation of PPAR-γ negates NMDAR and subsequently produces renoprotection against I/R-induced AKI in rodents. Novel pathways of estrogen-mediated renoprotection against I/R-induced AKI are summarized in Figure 14.3.

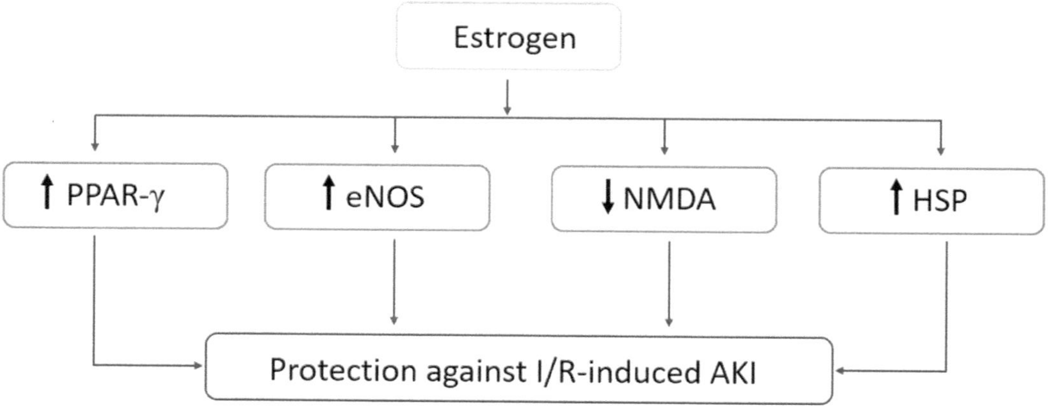

FIGURE 14.3 Novel pathways of estrogen-mediated renoprotection against I/R-induced AKI.

14.11 CONCLUSION

The existing literature highlights the sex-based differences in the models of kidney injury. Females are more resistant than males to AKI, owing to their high estrogen levels. Further mechanistic studies are warranted to investigate the molecular mechanism of estradiol-mediated protection in acute as well as chronic models of kidney injury. It will be interesting to investigate the change in estrogen receptors (ER-α, ER-β) in the models of kidney injury and their possible correction with estradiol in animal experiments. Such an approach shall help us to develop potential therapies against AKI in humans.

KEYWORDS

- **acute kidney injury**
- **estrogen**
- **ischemia-reperfusion**
- **renoprotection**
- **sex difference**
- **tubular damage**

REFERENCES

1. Liaño, F., & Pascual, J., (1996). Epidemiology of acute renal failure: A prospective, multicenter, community-based study. *Kidney Int., 50*, 811–818.
2. Uchino, S., Kellum, J. A., Bellomo, R., Doig, G. S., Morimatsu, H., Morgera, S., Schetz, M., et al., (2005). Beginning and ending supportive therapy for the kidney (BEST kidney) Investigators. acute renal failure in critically ill patients: A multinational, multicenter study. *J. Am. Med. Assoc., 294*, 813–818.
3. Susantitaphong, P., Cruz, D. N., Cerda, J., Abulfaraj, M., Alqahtani, F., Koulouridis, I., Jaber, B. L., & Acute Kidney Injury Advisory Group of the American Society of Nephrology, (2013). World incidence of AKI: A meta-analysis. *Clinical J. Am. Soc. Nephrol., 8*, 1482–1493.
4. Cerdá, J., Bagga, A., Kher, V., & Chakravarthi, R. M., (2008). The contrasting characteristics of acute kidney injury in developed and developing countries. *Nat. Clin. Pract. Nephrol., 4*, 138–153.
5. Ponce, D., & Balbi, A., (2016). Acute kidney injury: Risk factors and management challenges in developing countries. *Int. J. Nephrol. Renovasc. Dis., 9*, 193–200.
6. Hsu, R. K., McCulloch, C. E., Dudley, R. A., Lo, L. J., & Hsu, C. Y., (2013). Temporal changes in incidence of dialysis-requiring AKI. *J. Am. Soc. Nephrol., 24*, 37–42.
7. Eswarappa, M., Gireesh, M. S., Ravi, V., Kumar, D., & Dev, G., (2014). Spectrum of acute kidney injury in critically ill patients: A single center study from South India. *Indian J. Nephrol., 24*, 280–285.
8. Bonventre, J. V., (1993). Mechanism of ischemic acute renal failure. *Kidney Int., 43*, 1160–1178.
9. Laubach, V. E., French, B. A., & Okusa, M. D., (2011). Targeting of adenosine receptors in ischemia-reperfusion injury. *Expert Opin. Ther. Targets, 15*, 103–118.
10. Khan, H., Kashyap, A., Kaur, A., & Singh, T. G., (2020). Pharmacological postconditioning: A molecular aspect in ischemic injury. *J. Pharm. Pharmacol., 72*(11), 1513–1527.
11. Khan, H., Gupta, A., Singh, T. G., & Kaur, A., (2021). Mechanistic insight on the role of leukotriene receptors in ischemic–reperfusion injury. *Pharmacol. Rep., 73*(5), 1240–1254.
12. Khan, H., Garg, N., Singh, T. G., Kaur, A., & Thapa, K., (2022a). Calpain inhibitors as potential therapeutic modulators in neurodegenerative diseases. *Neurochem. Res.*, 1–25.
13. Khan, H., Sharma, K., Kumar, A., Kaur, A., & Singh, T. G., (2022). Therapeutic implications of cyclooxygenase (COX) inhibitors in ischemic injury. *Inflamm. Res.*, 1–16.
14. Khan, H., Bangar, A., Grewal, A. K., Bansal, P., & Singh, T. G., (2022b). Caspase-mediated regulation of the distinct signaling pathways and mechanisms in neuronal survival. *Int. Immunopharmacol., 110*, 108951.
15. Desai, M. M., Gill, I. S., Ramani, A. P., Spaliviero, M., Rybicki, L., & Kaouk, J. H., (2005). The impact of warm ischemia on renal function after laparoscopic partial nephrectomy. *BJU Int., 95*, 377–383.
16. Hernandez, D. J., Roberts, W. B., Miles-Thomas, J., Magheli, A., Saha, S., Schaeffer, E. M., Racusen, L. C., & Allaf, M. E., (2008). Can ischemic preconditioning ameliorate renal ischemia-reperfusion injury in a single-kidney porcine model? *J. Endourol., 22*, 2531–2536.

17. Jassem, W., Fuggle, S. V., Rela, M., Koo, D. D., & Heaton, N. D., (2002). The role of mitochondria in ischemia/reperfusion injury. *Transplantation, 73*, 493–499.
18. Bauerle, J. D., Grenz, A., Kim, J. H., Lee, T. H., & Eltzschig, H. K., (2011). Adenosine generation and signaling during acute kidney injury. *J. Am. Soc. Nephrol., 22*, 14–20.
19. Bonventre, J. V., & Yang, L., (2011). Cellular pathophysiology of ischemic acute kidney injury. *J. Clin. Invest., 121*, 4210–4221.
20. Lee, H. T., & Emala, C. W., (2000). Protective effects of renal ischemic preconditioning and adenosine pretreatment: Role of A1 and A3 receptors. *Am. J. Physiol. Renal Physiol., 278*, F380–F387.
21. Schrier, R. W., Wang, W., Poole, B., & Mitra, A., (2004). Acute renal failure: Definitions, diagnosis, pathogenesis, and therapy. *J. Clin. Invest., 114*, 5–14.
22. Lien, Y. H., Lai, L. W., & Silva, A. L., (2003). Pathogenesis of renal ischemia/reperfusion injury: Lessons from knockout mice. *Life Sci., 74*, 543–552.
23. Prabhakar, N. K., Khan, H., Grewal, A. K., & Singh, T. G., (2022). Intervention of neuroinflammation in the traumatic brain injury trajectory: *In vivo* and clinical approaches. *Int. Immunopharmacol., 108*, 108902.
24. Li, L., & Okusa, M. D., (2006). Blocking the immune response in ischemic acute kidney injury: The role of adenosine 2A agonists. *Nat. Clin. Pract. Nephrol., 2*, 432–444.
25. Saklani, P., Khan, H., Gupta, S., Kaur, A., & Singh, T. G., (2022). Neuropeptides: Potential neuroprotective agents in ischemic injury. *Life Sci., 288*, 120186.
26. Garg, N., Singh, T. G., Khan, H., Arora, S., Kaur, A., & Mannan, A., (2021). Mechanistic interventions of selected *Ocimum* species in management of diabetes, obesity, and liver disorders: Transformative developments from preclinical to clinical approaches. *Biointerface Res. Appl. Chem, 12*(1), 1304–1323.
27. Firdous, S. M., Dhang, P., Koley, A., & Khan, H., (2021). Diabetic retinopathy: Pathogenesis and therapeutic management. *Diabetes and Diabetic Complications*. Nova Science Publishers, Hauppauge.
28. Firdous, S. M., Ghosh, S., Nath, P., & Khan, H., (2021a). Molecular basis of diabetic nephropathy. *Diabetes and Diabetic Complications*. Nova Science Publishers, Hauppauge.
29. Hutchens, M. P., Dunlap, J., Hurn, P. D., & Jarnberg, P. O., (2008). Renal ischemia: Does sex matter? *Anesth. Analg., 107*, 239–249.
30. Rihal, V., Khan, H., Kaur, A., & Singh, T. G., (2022). Vitamin D as therapeutic modulator in cerebrovascular diseases: A mechanistic perspectives. *Crit. Rev. Food Sci. Nutr.*, 1–23.
31. Reed, E., Cohen, D. J., Barr, M. L., Ho, E., Reemtsma, K., Rose, E. A., Hardy, M., & Suciu-Foca, N., (1992). Effect of recipient gender and race on heart and kidney allograft survival. *Transplant. Proc., 24*, 2670–2671.
32. Aufhauser, D. D. Jr., Wang, Z., Murken, D. R., Bhatti, T. R., Wang, Y., Ge, G., Redfield, R. R., et al., (2016). Improved renal ischemia tolerance in females influences kidney transplantation outcomes. *J. Clin. Invest., 126*, 1968–1977.
33. Siedlecki, A., Irish, W., & Brennan, D. C., (2011). Delayed graft function in the kidney transplant. *Am. J. Transplant., 11*, 2279–2296.
34. Hecking, M., Bieber, B. A., Ethier, J., Kautzky-Willer, A., Sunder-Plassmann, G., Säemann, M. D., Ramirez, S. P., et al., (2014). Sex-specific differences in hemodialysis prevalence and practices and the male-to-female mortality rate: The dialysis outcomes and practice patterns study (DOPPS). *PLoS Med., 11*, e1001750.
35. Müller, V., Losonczy, G., Heemann, U., Vannay, A., Fekete, A., Reusz, G., Tulassay, T., & Szabó, A. J., (2002). Sexual dimorphism in renal ischemia-reperfusion injury in rats: Possible role of endothelin. *Kidney Int., 62*, 1364–1371.
36. Tomiyoshi, Y., Sakemi, T., Aoki, S., & Miyazono, M., (2002). Different effects of castration and estrogen administration on glomerular injury in spontaneously hyperglycemic Otsuka Long-Evans Tokushima fatty (OLETF) rats. *Nephron., 92*, 860–867.
37. Park, K. M., Kim, J. I., Ahn, Y., Bonventre, A. J., & Bonventre, J. V., (2004). Testosterone is responsible for enhanced susceptibility of males to ischemic renal injury. *J. Biol. Chem., 279*, 52282–52292.
38. Ji, H., Menini, S., Mok, K., Zheng, W., Pesce, C., Kim, J., Mulroney, S., & Sandberg, K., (2005). Gonadal steroid regulation of renal injury in renal wrap hypertension. *Am. J. Physiol. Renal Physiol., 288*, F513–F520.

39. Nematbakhsh, M., Ebrahimian, S., Tooyserkani, M., Eshraghi-Jazi, F., Talebi, A., & Ashrafi, F., (2013). Gender difference in cisplatin-induced nephrotoxicity in a rat model: Greater intensity of damage in male than female. *Nephrourol. Mon.*, *5*, 818–821.
40. Fukuda, K., Yao, H., Ibayashi, S., Nakahara, T., Uchimura, H., Fujishima, M., & Hall, E. D., (2000). Ovariectomy exacerbates and estrogen replacement attenuates photothrombotic focal ischemic brain injury in rats. *Stroke, 31*, 155–160.
41. Camper-Kirby, D., Welch, S., Walker, A., Shiraishi, I., Setchell, K. D., Schaefer, E., Kajstura, J., et al., (2001). Myocardial Akt activation and gender: Increased nuclear activity in females versus males. *Circ. Res.*, *88*, 1020–1027.
42. Berg, U. B., (2006). Differences in decline in GFR with age between males and females. Reference data on clearances of inulin and PAH in potential kidney donors. *Nephrol. Dial. Transplant.*, *21*, 2577–2582.
43. Sabolić, I., Asif, A. R., Budach, W. E., Wanke, C., Bahn, A., & Burckhardt, G., (2007). Gender differences in kidney function. *Pflugers Arch.*, *455*, 397–429.
44. Munger, K., & Baylis, C., (1988). Sex differences in renal hemodynamics in rats. *Am. J. Physiol.*, *254*, F223–F231.
45. Fekete, A., Vannay, A., Vér, A., Vásárhelyi, B., Müller, V., Ouyang, N., Reusz, G., Tulassay, T., & Szabó, A. J., (2004). Sex differences in the alterations of Na^+K^+ATPase following ischemia-reperfusion injury in the rat kidney. *J. Physiol.*, *555*, 471–480.
46. Voss, M. R., Stallone, J. N., Li, M., Cornelussen, R. N., Knuefermann, P., & Knowlton, A. A., (2003). Gender differences in the expression of heat shock proteins: Effect of estrogen. *Am. J. Physiol. Heart Circ. Physiol.*, *285*, H687–H692.
47. Fekete, A., Vannay, A., Vér, A., Rusai, K., Müller, V., Reusz, G., Tulassay, T., & Szabó, A. J., (2006). Sex differences in heat shock protein 72 expression and localization in rats following renal ischemia-reperfusion injury. *Am. J. Physiol. Renal Physiol.*, *291*, F806–811.
48. Fujii, T., Kurata, H., Takaoka, M., Muraoka, T., Fujisawa, Y., Shokoji, T., Nishiyama, A., et al., (2003). The role of renal sympathetic nervous system in the pathogenesis of ischemic acute renal failure. *Eur. J. Pharmacol.*, *481*, 241–248.
49. Tanaka, M., & Sokabe, M., (2013). Bidirectional modulatory effect of 17β-estradiol on NMDA receptors via ERα and ERβ in the dentate gyrus of juvenile male rats. *Neuropharmacology*, *75*, 262–273.
50. Tanaka, R., Yazawa, M., Morikawa, Y., Tsutsui, H., Ohkita, M., Yukimura, T., & Matsumura, Y., (2017). Sex differences in ischemia/reperfusion-induced acute kidney injury depends on the degradation of noradrenaline by monoamine oxidase. *Clin. Exp. Pharmacol. Physiol.*, *44*, 371–377.
51. Hodeify, R., Megyesi, J., Tarcsafalvi, A., Mustafa, H. I., Hti, L. S. N. S., & Price, P. M., (2013). Gender differences control the susceptibility to ER stress-induced acute kidney injury. *Am. J. Physiol. Renal Physiol.*, *304*, F875–F882.
52. Rogers, J. L., Mitchell, A. R., Maric, C., Sandberg, K., Myers, A., & Mulroney, S. E., (2007). Effect of sex hormones on renal estrogen and angiotensin type 1 receptors in female and male rats. *Am. J. Physiol. Regul. Integr. Comp. Physiol.*, *292*, R794–R799.
53. Kang, D. H., Yu, E. S., Yoon, K. I., & Johnson, R., (2004). The impact of gender on progression of renal disease: Potential role of estrogen-mediated vascular endothelial growth factor regulation and vascular protection. *Am. J. Pathol.*, *164*, 679–688.
54. Maric, C., Sandberg, K., & Hinojosa-Laborde, C., (2004). Glomerulosclerosis and tubulo- interstitial fibrosis are attenuated with 17β-estradiol in the aging Dahl salt sensitive rat. *J. Am. Soc. Nephrol.*, *15*, 1546–1556.
55. Catanuto, P., Doublier, S., Lupia, E., Fornoni, A., Berho, M., Karl, M., Striker, G. E., Xia, X., & Elliot, S., (2009). 17β-estradiol and tamoxifen upregulate estrogen receptor beta expression and control podocyte signaling pathways in a model of type 2 diabetes. *Kidney Int.*, *75*, 1194–1201.
56. Hutchens, M. P., Fujiyoshi, T., Komers, R., Herson, P. S., & Anderson, S., (2012). Estrogen protects renal endothelial barrier function from ischemia-reperfusion *in vitro* and *in vivo*. *Am. J. Physiol. Renal Physiol.*, *303*, F377–F385.
57. Wu, C. C., Chang, C. Y., Chang, S. T., & Chen, S. H., (2016). 17β-estradiol accelerated renal tubule regeneration in male rats after ischemia/reperfusion-induced acute kidney injury. *Shock*, *46*, 158–163.

58. Lee, J., (2008). Nitric oxide in the kidney: Its physiological role and pathophysiological implications. *Electrolyte Blood Press.*, *6*, 27–34.
59. Tiwari, P., Khan, H., Singh, T. G., & Grewal, A. K., (2022). Poly (ADP-ribose) polymerase: An overview of mechanistic approaches and therapeutic opportunities in the management of stroke. *Neurochem. Res.*, 1–23.
60. Schwartz, D., Mendonca, M., Schwartz, I., Xia, Y., Satriano, J., Wilson, C. B., & Blantz, R. C., (1997). Inhibition of constitutive nitric oxide synthase (NOS) by nitric oxide generated by inducible NOS after lipopolysaccharide administration provokes renal dysfunction in rats. *J. Clin. Invest.*, *100*, 439–448.
61. Satake, A., Takaoka, M., Nishikawa, M., Yuba, M., Shibata, Y., Okumura, K., Kitano, K., et al., (2008). Protective effect of 17β-estradiol on ischemic acute renal failure through the PI3K/Akt/eNOS pathway. *Kidney Int.*, *73*, 308–317.
62. Forte, P., Kneale, B. J., Milne, E., Chowienczyk, P. J., Johnston, A., Benjamin, N., & Ritter, J. M., (1998). Evidence for a difference in nitric oxide biosynthesis between healthy women and men. *Hypertension*, *32*, 730–734.
63. Kang, K. P., Lee, J. E., Lee, A. S., Jung, Y. J., Kim, D., Lee, S., Hwang, H. P., et al., (2014). Effect of gender differences on the regulation of renal ischemia-reperfusion-induced inflammation in mice. *Mol. Med. Rep.*, *9*, 2061–2068.
64. Monsalve, F. A., Pyarasani, R. D., Delgado-Lopez, F., & Moore-Carrasco, R., (2013). Peroxisome proliferator-activated receptor targets for the treatment of metabolic diseases. *Mediators Inflamm.*, *2013*, 549627.
65. Lehrke, M., & Lazar, M. A., (2005). The many faces of PPARγ. *Cell*, *123*, 993–999.
66. Khan, H., Singh, T. G., Dahiya, R. S., & Abdel-Daim, M. M., (2022c). α-Lipoic acid, an organosulfur biomolecule a novel therapeutic agent for neurodegenerative disorders: An mechanistic perspective. *Neurochem. Res.*, 1–12.
67. Yang, J., Zhang, D., Li, J., Zhang, X., Fan, F., & Guan, Y., (2009b). Role of PPARγ in renoprotection in type 2 diabetes: Molecular mechanisms and therapeutic potential. *Clin. Sci. (London)*, *116*, 17–26.
68. Qian, Y., Feldman, E., Pennathur, S., Kretzler, M., & Brosius, F. C., (2008). From fibrosis to sclerosis: Mechanisms of glomerulosclerosis in diabetic nephropathy. *Diabetes*, *57*, 1439–1445.
69. Pistrosch, F., Herbrig, K., Kindel, B., Passauer, J., Fischer, S., & Gross, P., (2005). Rosiglitazone improves glomerular hyperfiltration, renal endothelial dysfunction, and microalbuminuria of incipient diabetic nephropathy in patients. *Diabetes*, *54*, 2206–2211.
70. Wu, J., Chen, L., Zhang, D., Huo, M., Zhang, X., Pu, D., & Guan, Y., (2009). Peroxisome proliferator-activated receptors and renal diseases. *Front. Biosci.*, *14*, 995–1009.
71. Mannan, A., Garg, N., Singh, T. G., & Kang, H. K., (2021). Peroxisome proliferator-activated receptor-gamma (PPAR-ɣ): Molecular effects and its importance as a novel therapeutic target for cerebral ischemic injury. *Neurochem. Res.*, *46*(11), 2800–2831.
72. Tang, S. C., Leung, J. C., Chan, L. Y., Tsang, A. W., & Lai, K. N., (2006). Activation of tubular epithelial cells in diabetic nephropathy and the role of the peroxisome proliferator-activated receptor-γ agonist. *J. Am. Soc. Nephrol.*, *17*, 1633–1643.
73. Jia, Z., Sun, Y., Yang, G., Zhang, A., Huang, S., Heiney, K. M., & Zhang, Y., (2014). New insights into the PPARγ agonists for the treatment of diabetic nephropathy. *PPAR Res.*, *2014*, 818530.
74. Garg, C., Kaur, A., Singh, T. G., Sharma, V. K., & Singh, S. K., (2022). Therapeutic implications of sonic hedgehog pathway in metabolic disorders: Novel target for effective treatment. *Pharmacol. Res.*, 106194.
75. Yang, H. C., Deleuze, S., Zuo, Y., Potthoff, S. A., Ma, L. J., & Fogo, A. B., (2009a). The PPAR-γ agonist pioglitazone ameliorates aging-related progressive renal injury. *J. Am. Soc. Nephrol.*, *20*, 2380–2388.
76. Zhang, H., Li, Y., Fan, Y., Wu, J., Zhao, B., Guan, Y., Chien, S., & Wang, N., (2008). Klotho is a target gene of PPAR-gamma. *Kidney Int.*, *74*, 732–773.
77. Yousefipour, Z., Hercule, H., Truong, L., Oyekan, A., & Newaz, M., (2007). Ciglitazone, a peroxisome proliferator-activated receptor gamma inducer, ameliorates renal preglomerular production and activity of angiotensin II and thromboxane A_2 in glycerol-induced acute renal failure. *J. Pharmacol. Exp. Ther.*, *322*, 461–468.
78. Ozbek, E., Ilbey, Y. O., Simsek, A., Cekmen, M., Mete, F., & Somay, A., (2010). Rosiglitazone, peroxisome proliferator receptor-gamma agonist, ameliorates gentamicin-induced nephrotoxicity in rats. *Int. Urol. Nephrol.*, *42*, 579–587.

79. Korolczuk, A., Maciejewski, M., Smolen, A., Dudka, J., Czechowska, G., & Widelska, I., (2014). The role of peroxisome-proliferator-activating receptor gamma agonists: Rosiglitazone and 15-deoxy-delta12,14-prostaglandin J2 in chronic experimental cyclosporine A-induced nephrotoxicity. *J. Physiol. Pharmacol., 65*, 867–876.
80. Cuzzocrea, S., Pisano, B., Dugo, L., Ianaro, A., Patel, N. S., Di Paola, R., Genovese, T., et al., (2003). Rosiglitazone and 15-deoxy-Δ12,14-prostaglandin J2, ligands of the peroxisome proliferator-activated receptor-γ (PPAR-γ), reduce ischemia/reperfusion injury of the gut. *Br. J. Pharmacol., 140*, 366–376.
81. Sivarajah, A., Chatterjee, P. K., Patel, N. S., Todorovic, Z., Hattori, Y., Brown, P. A., Stewart, K. N., et al., (2003). Agonists of peroxisome-proliferator activated receptor-gamma reduce renal ischemia/reperfusion injury. *Am. J. Nephrol., 23*, 267–276.
82. Ito, K., Shimada, J., Kato, D., Toda, S., Takagi, T., Naito, Y., Yoshikawa, T., & Kitamura, N., (2004). Protective effects of preischemic treatment with pioglitazone, a peroxisome proliferator-activated receptor-γ ligand, on lung ischemia-reperfusion injury in rats. *Eur. J. Cardiothorac. Surg., 25*, 530–536.
83. Collino, M., Aragno, M., Mastrocola, R., Gallicchio, M., Rosa, A. C., Dianzani, C., Danni, O., et al., (2006). Modulation of the oxidative stress and inflammatory responses by PPAR-γ agonists in the hippocampus of rats exposed to cerebral ischemia/reperfusion. *Eur. J. Pharmacol., 530*, 70–80.
84. Somi, M. H., Hajipour, B., Asl, N. A., Estakhri, R., Azar, A. N., Zade, M. N., Haghjou, A. G., & Vatankhah, A. M., (2009). Pioglitazone attenuates ischemia/ reperfusion-induced liver injury in rats. *Transplant. Proc., 41*, 4105–4109.
85. Naito, Y., Takagi, T., Katada, K., Tomatsuri, N., Mizushima, K., Handa, O., Kokura, S., Yagi, N., et al., (2011). Gastric peroxisome proliferator activator receptor-γ expression and cytoprotective actions of its ligands against ischemia-reperfusion injury in rats. *J. Clin. Biochem. Nutr., 48*, 170–177.
86. Chatterjee, P. K., Patel, N. S., Cuzzocrea, S., Brown, P. A., Stewart, K. N., Mota-Filipe, H., Britti, D., et al., (2004). The cyclo-pentenone prostaglandin 15-deoxy-Δ12,14-prostaglandin J2 ameliorates ischemic acute renal failure. *Cardiovasc. Res., 61*, 630–643.
87. Kapil, A., Singh, J. P., Kaur, T., Singh, B., & Singh, A. P., (2013). Involvement of peroxisome proliferator-activated receptor gamma in vitamin D-mediated protection against acute kidney injury in rats. *J. Surg. Res., 185*, 774–783.
88. Singh, J. P., Singh, A. P., & Bhatti, R., (2014). Explicit role of peroxisome proliferator- activated receptor gamma in gallic acid-mediated protection against ischemia-reperfusion-induced acute kidney injury in rats. *J. Surg. Res., 187*, 631–639.
89. Mohey, V., Singh, M., Puri, N., Kaur, T., Pathak, D., & Singh, A. P., (2016). Sildenafil obviates ischemia-reperfusion injury–induced acute kidney injury through peroxisome proliferator activated receptor γ agonism in rats. *J. Surg. Res., 201*, 69–75.
90. Rzemieniec, J., Litwa, E., Wnuk, A., Lason, W., & Kajta, M., (2018). Bazedoxifene and raloxifene protect neocortical neurons undergoing hypoxia via targeting ERα and PPARγ. *Mol. Cell. Endocrinol., 461*, 64–78.
91. Valles, S. L., Dolz-Gaiton, P., Gambini, J., Borras, C., Lloret, A., Pallardo, F. V., & Viña, J., (2010). Estradiol or genistein prevent Alzheimer's disease-associated inflammation correlating with an increase PPAR gamma expression in cultured astrocytes. *Brain Res., 1312*, 138–144.
92. Sakamoto, Y., Naka, A., Ohara, N., Kondo, K., & Iida, K., (2014). Daidzein regulates pro-inflammatory adipokines thereby improving obesity-related inflammation through PPARγ. *Mol. Nutr. Food Res., 58*, 718–726.
93. Sakamoto, Y., Kanatsu, J., Toh, M., Naka, A., Kondo, K., & Iida, K., (2016). The dietary isoflavone daidzein reduces expression of pro-inflammatory genes through PPARα/γ and JNK pathways in adipocyte and macrophage co-cultures. *PLoS One, 11*, e0149676.
94. Xiang, Q., Lin, G., Fu, X., Wang, S., & Wang, T., (2010). The role of peroxisome proliferator-activated receptor-γ and estrogen receptors in genistein-induced regulation of vascular tone in female rat aortas. *Pharmacology, 86*, 117–124.
95. Suzuki, T., Kawasaki, T., Choudhry, M. A., & Chaudry, I. H., (2011). Role of PPARγ in the salutary effects of 17β-estradiol on Kupffer cell cytokine production following trauma-hemorrhage. *J. Cell. Physiol., 226*, 205–211.

96. Singh, A. P., Singh, N., & Bedi, P. M., (2016). Pioglitazone ameliorates renal ischemia reperfusion injury through NMDA receptor antagonism in rats. *Mol. Cell. Biochem., 417*, 111–118.
97. Dellamea, B. S., Leitão, C. B., Friedman, R., & Canani, L. H., (2014). Nitric oxide system and diabetic nephropathy. *Diabetol. Metab. Syndr., 6*, 17.
98. Rehni, A. K., Singh, T. G., Kalra, R., & Singh, N., (2009). Pharmacological inhibition of inducible nitric oxide synthase attenuates the development of seizures in mice. *Nitric Oxide, 21*(2), 120–125.
99. Mount, P. F., & Power, D. A., (2006). Nitric oxide in the kidney: Functions and regulation of synthesis. *Acta Physiol. (Oxford), 187*, 433–446.
100. Sharma, S. P., (2004). Nitric oxide and the kidney. *Indian J. Nephrol., 14*, 77–84.
101. Gabbai, F. B., (2001). Effects of nitric oxide synthase blockers on renal function. *Nephrol. Dial. Transplant., 16*, 10–13.
102. Onem, Y., Ipcioglu, O. M., Haholu, A., Sen, H., Aydinoz, S., Suleymanoglu, S., Bilgi, O., & Akyol, I., (2009). Posttreatment with aminoguanidine attenuates renal ischemia/reperfusion injury in rats. *Ren Fail., 31*, 50–53.
103. Abraham, P., & Rabi, S., (2011). Protective effect of aminoguanidine against cyclophosphamide-induced oxidative stress and renal damage in rats. *Redox Rep., 16*, 8–14.
104. Seija, M., Baccino, C., Nin, N., Sánchez-Rodríguez, C., Granados, R., Ferruelo, A., Martínez-Caro, L., et al., (2012). Role of peroxynitrite in sepsis-induced acute kidney injury in an experimental model of sepsis in rats. *Shock, 38*, 403–410.
105. Rodriguez-Peña, A., Garcia-Criado, F. J., Eleno, N., Arevalo, M., & Lopez-Novoa, J. M., (2004). Intrarenal administration of molsidomine, a molecule releasing nitric oxide, reduces renal ischemia-reperfusion injury in rats. *Am. J. Transplant., 4*, 1605–1613.
106. Chander, V., & Chopra, K., (2005). Renal protective effect of molsidomine and L- arginine in ischemia-reperfusion induced injury in rats. *J. Surg. Res., 128*, 132–139.
107. Mahfoudh-Boussaid, A., Hadj, A. T. K., Zaouali, M. A., Roselló-Catafau, J., & Ben, A. H., (2014). Effects of trimetazidine on the Akt/eNOS signaling pathway and oxidative stress in an *in vivo* rat model of renal ischemia- reperfusion. *Ren Fail., 36*, 1436–1442.
108. Schramm, L., Seibold, A., Schneider, R., Zimmermann, J., Netzer, K. O., & Wanner, C., (2008). Ischemic acute renal failure in the rat: Effects of L-arginine and superoxide dismutase on renal function. *J. Nephrol., 21*, 229–235.
109. Brodsky, S. V., Yamamoto, T., Tada, T., Kim, B., Chen, J., Kajiya, F., & Goligorsky, M. S., (2002). Endothelial dysfunction in ischemic acute renal failure: Rescue by transplanted endothelial cells. *Am. J. Physiol. Renal Physiol., 282*, F1140–F1149.
110. Teraishi, K., Kurata, H., Nakajima, A., Takaoka, M., & Matsumura, Y., (2004). Preventive effect of Y-27632, a selective Rho-kinase inhibitor, on ischemia/ reperfusion-induced acute renal failure in rats. *Eur. J. Pharmacol., 505*, 205–211.
111. Reckelhoff, J. F., Hennington, B. S., Moore, A. G., Blanchard, E. J., & Cameron, J., (1998). Gender differences in the renal nitric oxide (NO) system: Dissociation between expression of endothelial NO synthase and renal hemodynamic response to NO synthase inhibition. *Am. J. Hypertens., 11*, 97–104.
112. Schneider, M. P., Wach, P. F., Durley, M. K., Pollock, J. S., & Pollock, D. M., (2010). Sex differences in acute Ang II-mediated hemodynamic responses in mice. *Am. J. Physiol. Regul. Integr. Comp. Physiol., 299*, R899–R906.
113. Sullivan, J. C., Pardieck, J. L., Hyndman, K. A., & Pollock, J. S., (2010). Renal NOS activity, expression, and localization in male and female spontaneously hypertensive rats. *Am. J. Physiol. Regul. Integr. Comp. Physiol., 298*, R61–R69.
114. Verhagen, A. M., Attia, D. M., Koomans, H. A., & Joles, J. A., (2000). Male gender increases sensitivity to proteinuria induced by mild NOS inhibition in rats: Role of sex hormones. *Am. J. Physiol. Renal Physiol., 279*, F664–F670.
115. El-Mas, M. M., El-Gowilly, S. M., Gohar, E. Y., Ghazal, A. R., & Abdel-Rahman, A. A., (2011). Estrogen dependence of the renal vasodilatory effect of nicotine in rats: Role of α7 nicotinic cholinergic receptor/ eNOS signaling. *Life Sci., 88*, 187–193.
116. Schneider, M. P., Ritt, M., Raff, U., Ott, C., & Schmieder, R. E., (2009). Gender is related to alterations of renal endothelial function in type 2 diabetes. *Nephrol. Dial. Transplant, 24*, 3354–3359.

117. Singh, A. P., Singh, N., Pathak, D., & Bedi, P. M. S., (2019). Estradiol attenuates ischemia reperfusion-induced acute kidney injury through PPAR-γ stimulated eNOS activation in rats. *Mol. Cell. Biochem., 453*, 1–9.
118. Qu, X. X., Cai, J., Li, M. J., Chi, Y. N., Liao, F. F., Liu, F. Y., Wan, Y., et al., (2009). Role of the spinal cord NR2B-containing NMDA receptors in the development of neuropathic pain. *Exp. Neurol., 215*, 298–307.
119. Dhawan, J., Benveniste, H., Luo, Z., Nawrocky, M., Smith, S. D., & Biegon, A., (2011). A new look at glutamate and ischemia: NMDA agonist improves long-term functional outcome in a rat model of stroke. *Future Neurol., 6*, 823–834.
120. Gonda, X., (2012). Basic pharmacology of NMDA receptors. *Curr. Pharm. Des., 18*, 1558–1567.
121. Tayeb, H. O., Yang, H. D., Price, B. H., & Tarazi, F. I., (2012). Pharmacotherapies for Alzheimer's disease: Beyond cholinesterase inhibitors. *Pharmacol. Ther., 134*, 8–25.
122. Pundir, M., Arora, S., Kaur, T., Singh, R., & Singh, A. P., (2013). Effect of modulating the allosteric sites of N-methyl-D-aspartate receptors in ischemia reperfusion induced acute kidney injury. *J. Surg. Res., 183*, 668–677.
123. Bozic, M., & Valdivielso, J. M., (2015). The potential of targeting NMDA receptors outside the CNS. *Expert Opin. Ther. Targets, 19*, 399–413.
124. Gupta, A., Khan, H., Kaur, A., & Singh, T. G., (2021). Novel targets explored in the treatment of alcohol withdrawal syndrome. *CNS Neurol. Disord. Drug Targets, 20*(2), 158–173.
125. Kaur, A., Kaur, T., Singh, B., Pathak, D., Buttar, H. S., & Singh, A. P., (2016). Curcumin alleviates ischemia reperfusion-induced acute kidney injury through NMDA receptor antagonism in rats. *Ren. Fail., 38*, 1462–1467.
126. Roshanravan, H., Kim, E. Y., & Dryer, S. E., (2016). NMDA receptors as potential therapeutic targets in diabetic nephropathy: Increased renal NMDA receptor subunit expression in Akita mice and reduced nephropathy following sustained treatment with memantine or MK-801. *Diabetes, 65*, 3139–3150.
127. Leung, J. C., Marphis, T., Craver, R. D., & Silverstein, D. M., (2004). Altered NMDA receptor expression in renal toxicity: Protection with a receptor antagonist. *Kidney Int., 66*, 167–176.
128. Deng, A., Valdivielso, J. M., Munger, K. A., Blantz, R. C., & Thomson, S. C., (2002). Vasodilatory N-methyl-D-aspartate receptors are constitutively expressed in rat kidney. *J. Am. Soc. Nephrol., 13*, 1381–1384.
129. Yang, C. C., Chien, C. T., Wu, M. H., Ma, M. C., & Chen, C. F., (2008). NMDA receptor blocker ameliorates ischemia-reperfusion-induced renal dysfunction in rat kidneys. *Am. J. Physiol. Renal Physiol., 294*, F1433–F1440.
130. Arora, S., Kaur, T., Kaur, A., & Singh, A. P., (2014). Glycine aggravates ischemia reperfusion-induced acute kidney injury through N-methyl-D-aspartate receptor activation in rats. *Mol. Cell. Biochem., 393*, 123–131.
131. Lu, H., Ma, K., Jin, L., Zhu, H., & Cao, R., (2018). 17β-estradiol rescues damages following traumatic brain injury from molecule to behavior in mice. *J. Cell. Physiol., 233*, 1712–1722.
132. Valdés, J. J., & Weeks, O. I., (2009). Estradiol and lithium chloride specifically alter NMDA receptor subunit NR1 mRNA and excitotoxicity in primary cultures. *Brain Res., 1268*, 1–12.
133. Lamprecht, M. R., & Morrison, B., (2015). A combination therapy of 17β-estradiol and memantine is more neuroprotective than monotherapies in an organotypic brain slice culture model of traumatic brain injury. *J. Neurotrauma, 32*, 1361–1368.
134. Day, N. L., Carle, M. S., & Floyd, C. L., (2017). Post-injury administration of a combination of memantine and 17β-estradiol is protective in a rat model of traumatic brain injury. *Neurochem. Int., 111*, 57–68.
135. Kalkbrenner, K. A., & Standley, C. A., (2003). Estrogen modulation of NMDA-induced seizures in ovariectomized and non-ovariectomized rats. *Brain Res., 964*, 244–249.
136. Yamashita, H., Yamada-Nakayama, C., Sugihara, K., Tsuji, S., Sakurai, T., Ban, Y., Tsutsumi, S., Satol Y. (2011). Functional and morphological effects of β- estradiol in eyes with N-methyl-D-aspartate-induced retinal neurotoxicity in rats. *Exp. Eye Res., 93*, 75–81.
137. Khakpay, R., Semnanian, S., Javan, M., & Janahmadi, M., (2010). The effect of intra- locus coeruleus injection of 17β-estradiol on inflammatory pain modulation in male rat. *Behav. Brain Res., 214*, 409–416.
138. Singh, A. P., Singh, N., & Bedi, P. M. S., (2017). Estradiol mitigates ischemia reperfusion-induced acute renal failure through NMDA receptor antagonism in rats. *Mol. Cell. Biochem., 434*, 33–40.

139. Khan, H., Singh, A., Thapa, K., Garg, N., Grewal, A. K., & Singh, T. G., (2021a). Therapeutic modulation of the phosphatidylinositol 3-kinases (PI3K) pathway in cerebral ischemic injury. *Brain Res., 1761*, 147399.
140. Toba, H., Tojo, C., Wang, J., Noda, K., Kobara, M., & Nakata, T., (2012). Telmisartan inhibits vascular dysfunction and inflammation via activation of peroxisome proliferator-activated receptor-γ in subtotal nephrectomized rat. *Eur. J. Pharmacol., 685*, 91–98.
141. Salvadori, M., Rosso, G., & Bertoni, E., (2015). Update on ischemia-reperfusion injury in kidney transplantation: Pathogenesis and treatment. *World Journal of Transplantation, 5*(2), 52.
142. Tiyerili, V., Müller, C. F., Fung, S., Panek, D., Nickenig, G., & Becher, U. M., (2012). Estrogen improves vascular function via peroxisome-proliferator-activated-receptor-γ. *Journal of Molecular and Cellular Cardiology, 53*(2), 268–276.
143. Devarajan, P., (2006). Update on mechanisms of ischemic acute kidney injury. *Journal of the American Society of Nephrology, 17*(6), 1503–1520.

CHAPTER 15

Critical Overview of Conditioning Techniques and Their Current Status in the Treatment of Myocardial Ischemia-Reperfusion Injury

KULDEEP KUMAR, NIRMAL SINGH, and AMTESHWAR S. JAGGI

Department of Pharmaceutical Sciences and Drug Research, Punjabi University, Patiala, Punjab, India

ABSTRACT

Ischemia refers to a condition insufficient blood flow to the tissue with subsequent deprivation oxygen in addition to other nutrients important for optimal cellular metabolism and survival of the tissue. Reperfusion of the ischemic tissue is an immediate intervention to salvage ischemic Injury. Paradoxically, sudden restoration of blood supply after prolonged ischemia causes more severe and lethal injury, known as Ischemia-Reperfusion (IR) injury. Myocardial IR damage is the primary cause of mortality in patients with ischemic heart disease (IHD). Reperfusion damage, which can occur both surgically and non-surgically, is the greatest obstacle for a clinician attempting to treat ischemia. Certain therapies such as anti-platelets, anti-thrombolytic drugs, viability, etc., have been developed to contain IR injury, but with limited success only. Over time, several conditioning techniques such as preconditioning, remote preconditioning, postconditioning, remote postconditioning, and pharmacological conditioning have been developed and used by clinicians to overcome myocardial IR injury. Over the last decade, significant research has been carried out to explore the protective role and understand the molecular mechanism of these conditioning techniques in containing IR injury. Indeed, few of these techniques have been successfully applied at the clinical level also. Further, the studies have documented various triggers, mediators, and end effectors of these conditioning phenomena at the molecular level. The present chapter focuses on myocardial IR injury; various conditioning techniques and their current status in the management of myocardial IR injury. This particular chapter will be of great help to the researchers working in the relevant area.

Ischemic Injury: A Molecular Insight. Thakur Gurjeet Singh & Heena Khan (Eds.)

15.1 INTRODUCTION

Cardiovascular disorders (CVD's) including heart failure, heart attack, hypertension, and coronary artery disease is the major contributors to worldwide mortality and morbidity. A report published by World Health Organization (WHO) in 2019 states that around 17.9 million people die every year due to CVD which represents approximately 32% of global mortality [1]. It is also reported that this total number of deaths can increase up to 23.3 million by end of the year 2030. In addition, it is also documented that four out of five deaths associated with CVD are due to myocardial infarction (MI) and Ischemic Heart disease (IHD) [2, 3].

Ischemia is a state of interrupted perfusion to any tissue followed by deprivation of oxygen and other nutrients important for cellular metabolism and viability of tissue [4]. Several organs such as the brain, heart, intestine, kidney, liver, lungs, etc., can be affected by ischemia [5]. MI is mainly caused by restricted perfusion to the heart mainly due to obstruction in major coronary arteries, primarily due to the deposition of atherosclerotic plaque. Ischemia of longer duration can lead to damage/death of the heart muscle through several pathways such as necrosis, apoptosis, necroptosis, etc. [6, 7]. The size/extent of an infarct is mainly governed by several factors such as the duration of ischemia, the area affected by ischemia, the extent of coronary microvascular dysfunction, and the degree of collateral blood flow [8]. Re-perfusion of the myocardium possibly is an essential intervention to salvage the initial ischemic myocardial injury. However, paradoxically, sudden reperfusion of the myocardium after prolonged ischemia causes a more serious and lethal injury known as myocardial ischemia-reperfusion (IR) injury [9]. Myocardial IR injury is more lethal because reperfusion itself causes the activation of various factors/processes such as oxidative stress, mitochondrial/cytosolic calcium overload, rapid restoration of pH, and, mitochondrial permeability transition pore (MPTP) opening followed by apoptosis and release of cytochrome-C [3].

Many research studies have been carried out to contain myocardial IR injury and its deleterious effects but with limited success. Over the passage of time, several conditioning techniques such as preconditioning, remote ischemic preconditioning (RIPC), postconditioning, and remote ischemic postconditioning (RIPostC) have been developed and applied clinically with variable success. Ischemic conditioning is defined as the endogenous strategy in which one or more short-term cycles of I/R (to the heart or remote organ) are given which is capable of protecting the heart from IR injury and its detrimental effects and eventually improves the clinical outcomes of the patient suffering from IHD [10].

This particular chapter will be focusing on the concept of myocardial IR injury and will provide a critical overview of various types of conditioning techniques along with their different cardioprotective signaling mechanisms. In addition, the present state of the application of various conditioning strategies in the treatment of myocardial IR injury will be discussed. This manuscript will certainly help the researchers/scientists working in the field of IHD to provide a better understanding of various conditioning techniques (along with their cardioprotective signaling mechanisms) and their current status in the management of myocardial IR injury.

15.2 MYOCARDIAL ISCHEMIA-REPERFUSION (IR) INJURY

Restoration of perfusion to the ischemic heart is the only way to reserve the ischemic heart from irreversible damage [11]. Paradoxically, sudden restoration of blood flow of the ischemic heart itself induces cell death of the heart and exaggerates the extent of injury to the heart, a phenomenon known as myocardial IR injury [4, 5]. The concept of IR injury was first introduced in the 1960s and presently occurs in numerous surgical and non-surgical scenarios, such as aneurysm repair, organ transplantation, cardiopulmonary bypass surgery, stroke, myocardial infarction, trauma, shock, hemorrhage, traumatic head injury, and deep hypothermic circulatory arrest [12, 13].

15.2.1 PATHOLOGICAL MECHANISM OF MYOCARDIAL ISCHEMIA AND IR INJURY

Myocardial ischemia occurs in numerous pathogenic conditions, as stated above. Longer period ischemia causes a range of cellular, metabolic, and ultra-structural alterations, including altered membrane potential, elevated serum calcium and sodium overload [14], cellular swelling, a rise in hypoxanthine, and a decline in adenosine triphosphate (ATP), glutathione, and phosphocreatine levels. During the ischemic phase, the heart's aerobic respiration shifts to anaerobic metabolism, which leads to a drop in oxidative phosphorylation and a failure in the resynthesis of energy-rich molecules such as phosphocreatine and ATP. Additionally, it increases hypoxanthine build-up in the heart. The function of ATP-operated ion channels is compromised, leading to an increase in the influx of calcium, sodium, and water into the cell [15, 16]. During cellular homeostasis, xanthine dehydrogenase converts hypoxanthine to xanthine. However, throughout the ischemia phase, xanthine dehydrogenase is changed to xanthine oxidase. Xanthine dehydrogenase utilizes Nicotinamide Adenine Dinucleotide (NAD) as a substrate, whereas xanthine oxidase uses oxygen. Consequently, during the ischemic phase, xanthine oxidase is not able to convert hypoxanthine to xanthine, leading to an accumulation of excessive hypoxanthine in the heart. In contrast, during the reperfusion phase, oxygen re-enters the heart, causing xanthine oxidase [17] to convert excess hypoxanthine (accumulated during the ischemia phase) to xanthine. This process generates enormous quantities of reactive oxygen species (ROS) including superoxide anion (O_2^-), hydrogen peroxide (H_2O_2), and hydroxyl radical (OH^-) [18]. This free radical trigger the intracellular signaling cascade, resulting in membrane damage and IR injury to the myocardium [19]. Moreover, after reperfusion, oxygen will be reintroduced into the cells and begin destroying cellular proteins and Deoxyribose Nucleic Acid (DNA), which are also additional leading causes of myocardial IR injury [20]. Several investigations [21, 22] have demonstrated that oxidative stress, neutrophil and leukocyte activation, and an increased intracellular osmotic load all contribute to the pathophysiology of myocardial IR injury. In complement to these pathogenic pathways, a number of endogenous mediators, including caspase-3, caspase-8 [23], calpains [24], interleukin-6 (IL-6) [25], and tumor necrosis factor-α (TNF-α) [26], play a significant role in the pathophysiology of myocardial IR injury (Figure 15.1).

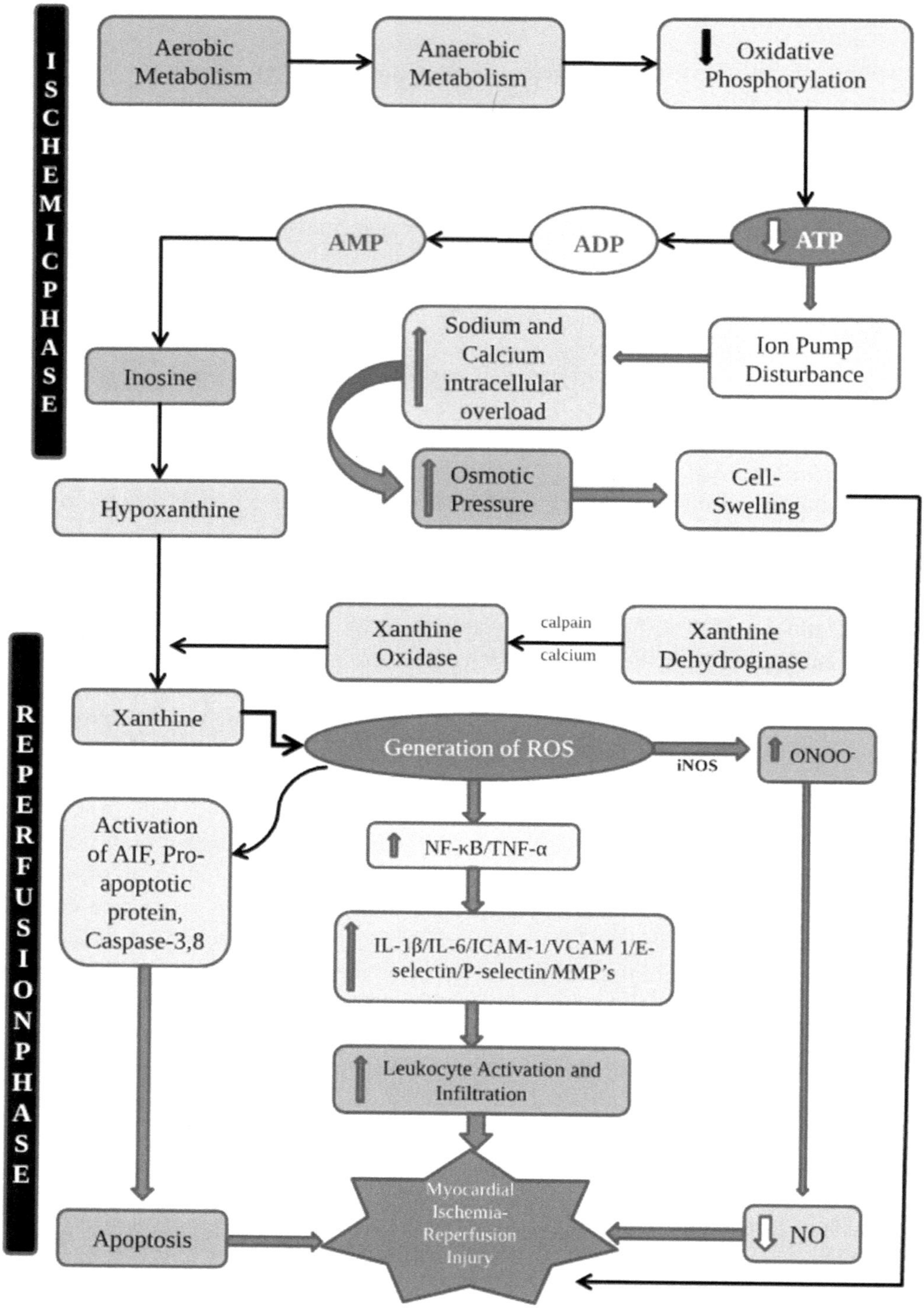

FIGURE 15.1 Pictorial presentation of pathological mechanism of myocardial ischemia and IR injury.

Significant advances have been made in the clinical management of ischemic injury [27, 28] with the development of multiple therapeutic strategies, such as thrombolytic approaches and primary percutaneous coronary intervention (PPCI), including new generations of antiplatelet drugs and antithrombotic agents. To date, however, neither pharmacological medications nor non-pharmacological approaches have proven effective in protecting the heart from IR injury. To mitigate myocardial IR injury, training approaches including brief periods of ischemia accompanied by reperfusion before (preconditioning) or after (postconditioning) restoration of blood flow in the afflicted region have been developed over time [29, 30].

15.3 ISCHEMIC PRECONDITIONING (IPC)

Murry and co-workers introduced the concept of IPC in 1986 to prepare the heart to resist myocardial infarction. In the original study, they subjected the heart to four cycles of 5-minute episodes of I/R to protect the myocardium against prolonged ischemia. The conditioning cycles were given only for 5 min duration because longer duration (such as 10–15 min) for ischemia-reperfusion cycles can lead to arrhythmia as well as increase the death rate. It was predicted that the non-lethal ischemia reduces the rate of ATP loss during prolonged ischemia in the heart. In addition, it was also documented that 5 min intermittent reperfusion was helping in washing out the toxic metabolites (such as lactate) that have been accumulated during the ischemic phase [31–33]. Numerous clinical and preclinical studies since then have given evidence that IPC is capable of reducing the size of infarct of any ischemic organ [34, 35].

15.3.1 PHASES OF CARDIO PROTECTION ACTIVATED BY IPC

It is reported that IPC induces two distinct phases (windows) of cardioprotection.

15.3.1.1 FIRST WINDOW OF CARDIOPROTECTION (FWOC)

This phase of protection is acute and is also known as "classical IPC" or "acute IPC." FWOC begins immediately (within 5 minutes) following the IPC stimulation and lasts up to 3–4 hrs. [31]. It is also reported that the cardioprotective effect of classical IPC occurs due to the activation of several signaling pathways such as Akt, extracellular receptor kinases (ERK ½), protein kinase, tyrosine kinase, etc. [36, 37].

15.3.1.2 SECOND WINDOW OF CARDIOPROTECTION (SWOC)

This is a delayed stage of protection and is also named "delayed IPC." SWOC begins 12–24 hrsafter IPC stimulus and lasts up to 48–72 hrs. [31, 38]. It is also believed that cardioprotective effects of delayed IPC occur due to activation of the several transcription factors such as activation protein (AP-1), hypoxic-inducible factor-1α (HIF-1α), nuclear factor-κβ (NF-κβ), Nrf-2, STAT1/3. These substances enhance the generation of "distal mediators" such as prostaglandin G/H synthase (also referred as COX 2), heat shock proteins (like as Hsp72, and iNOS, which contribute immensely to the cardioprotective actions of delayed IPC [39].

15.3.2 VARIOUS CARDIOPROTECTIVE SIGNALING PATHWAYS (MECHANISMS) ACTIVATED/DEACTIVATED BY IPC

The intracellular signaling pathways/mechanisms responsible for providing cardioprotective effects mediated by IPC are very complex. In IPC, these cardioprotective signaling pathways are categorized into triggers, mediators, and end effectors. Triggers are the major components of signaling pathways that mainly acts before the global ischemic event and further activates downstream signaling mediators. On the other side, mediators and end effectors are the components of signaling pathways that activate during a global ischemic episode or at the time of reperfusion and eventually result in a cardioprotective response. It is also documented that sometimes triggers may act as a mediator or end effectors in providing cardioprotection [40].

As we know that IPC stimulus consists of brief episodes of ischemia and reperfusion that triggers the production of several autocoids/substances such as acetylcholine [41], bradykinin [42], adenosine [43], AG-II [44], CA [45], endothelin [46], and opioids [47] by cardiomyocytes. These autacoids/substances act by binding to their respective membrane-bound G-protein coupled receptors (GPCRs) and further stimulates various signaling pathways that eventually generate cardioprotective signal into the mitochondria. In the mitochondria, ROS formation takes place that further activates several protein kinases (which act as triggers of IPC) such as Akt, ERK ½, Phosphatidylinositol-3-Kinase (PI3K), eNOS, GSK-3β protein kinase-C, and tyrosine kinase. In classical IPC, cardioprotective effects are diminished within 3–4 hrs due to activation of these kinases signaling. On the other side, in delayed IPC, these kinase signaling pathways further activate transcription factors such as AP-1, HIF-1α, NF-κβ, nuclear factor erythroid 2-related factor-2, and STAT 1/3 that eventually lead to the synthesis of distal mediators (end effectors) such as prostaglandin G/H synthase, heat shock proteins (Hsp-72), and iNOS. These aforementioned factors contribute to the cardioprotective effects after 12–24 hrs of IPC [39]. In addition, it is observed that myocardial IR injury, IPC also causes activation of pro-survival signaling pathways such as Survivor Activator Factor Enhancement (SAFE) and Reperfusion Injury Salvage Kinase (RISK) pathways at the onset of reperfusion [48, 49]. Despite this, the final mechanisms of cardioprotection through classical and delayed IPC remain unclear. Although several reports have suggested that restoration of mitochondrial functioning such as decreased calcium overload, decreased ROS production, increased ATP synthesis, increased intracellular pH, increased opening of K_{ATP} channels, and inhibition of mPTP opening might play a chief role in the cardioprotective properties facilitated by IPC (Figure 15.2) [50, 51]. In addition, genes of myocardial tissue have the potential to further unfold the cardioprotective signaling mechanisms activated/deactivated by IPC stimulus [52].

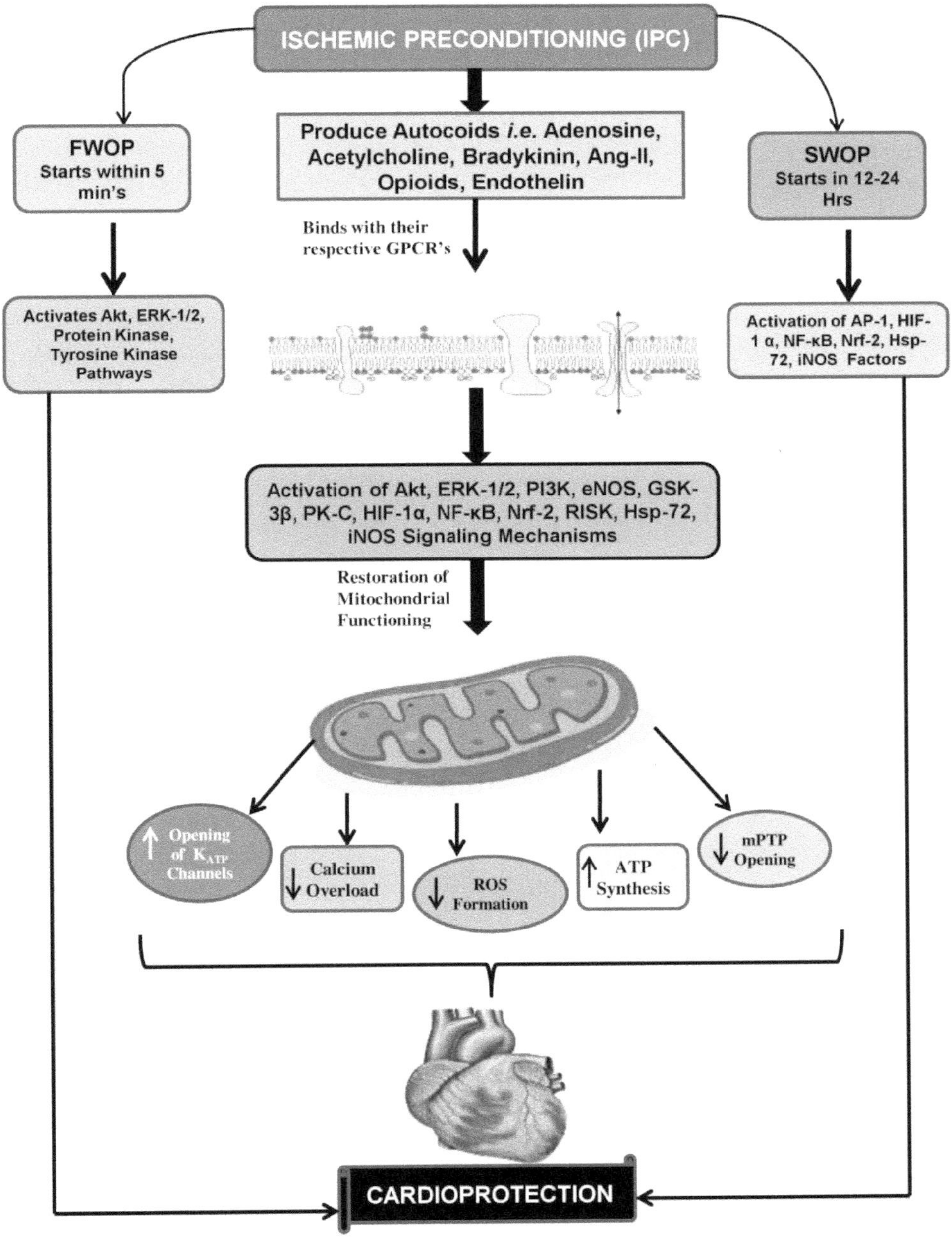

FIGURE 15.2 Modulation of various cardioprotective signaling pathways by IPC.

15.3.3 CLINICAL APPLICATIONS OF IPC IN CARDIOVASCULAR DISORDERS

Several clinical studies have provided evidence of the cardioprotective effects of IPC in patients receiving mitral valve, vascular, and pediatric surgery.

15.3.3.1 EVIDENCE

Yellon and a colleague conducted the first clinical trial to examine the effect of IPC stimulation on patients who undergo coronary artery bypass surgery (CABG) in 1993. They selected 14 patients being subjected to CABG and randomized them into two different groups. Patients of the first group (7 patients) were subjected to the preconditioning stimulus by giving two cycles of 3 min ischemia (through cross-clamping) followed by two cycles of 2 min reperfusion. After performing CABG, all 14 patients were subjected to 10 min global ischemia through cross-clamping. Cardioprotective effects of IPC were assessed in terms of no depletion of ATP in the patients who were subjected to IPC before undergoing CABG as compared to other group patients (who had no IPC stimulus) [53].

Around 30 patients with rheumatic heart disease (RHD) undergoing replacement of the mitral and aortic valves were included in a clinical investigation. Patients underwent preconditioning by receiving two cycles of vena cava and aorta ischemia lasting 2 minutes followed by 3 minutes of reperfusion. Myocardial contractility improved significantly; the production of creatinine kinase-MB (CK-MB) dropped, and the incidence of ventricular fibrillation decreased [54].

In a separate trial, 40 patients with RHD underwent double valve replacement surgery using cardioplegic procedures on cold blood. Around 20 patients were preconditioned with two cycles of 3-minute aortic cross-clamping followed by 2-minute reperfusion prior to cardioplegic arrest. Significant reductions in CK-MB release, ROS generation, and improvement in cardiac function were observed [55].

Besides this, a meta-analysis of 22 clinical trials was conducted that included data from a total of 933 patients. The results were concluded after the application of IPC stimulus and found that IPC was able to reduce ventricular arrhythmias, decrease inotrope requirements, and decrease the time spent in an intensive care unit (ICU) as compared to the control group patients [56].

15.3.4 DISADVANTAGES OF IPC

The key disadvantage of this therapeutic method is that it must be applied before the index ischemia episode, which is impossible to foresee in clinical practice. IPC is an intrusive procedure because it is directly applied to the heart, which may not be possible in all therapeutic circumstances [57].

15.4 REMOTE ISCHEMIC PRECONDITIONING (RIPC)

In 1993, Przyklenk and co-workers made the transformational invention that IPC-induced cardioprotection could be carried from one region of the heart to another. Later, this type of intra-myocardial protection was extended outside of the heart.

RIPC is arguably the most advantageous and safest ischemia conditioning treatment since it does not induce lesions in the heart and non-culprit arteries that may cause additional damage [58]. RIPC may also be a nonsurgical approach in which temporary, repeated episodes of I/R are applied to remote organs other than the heart and renders the heart resistant to prolonged ischemic insult [59, 60]. Several studies were carried out by different researchers to substantiate that the size of MI can be reduced by inducing brief episodes of ischemia and reperfusion to different organs such as the kidney [61] and small intestine [62].

15.4.1 CARDIOPROTECTIVE SIGNALING MECHANISMS UNDERLYING RIPC

The exact mechanism which is responsible for conferring cardioprotection through brief transient episodes of I/R to a remote organ or tissue followed by the subsequent global ischemic insult to the heart remains uncertain. Many researchers have tried to unveil the underlying mechanisms or signaling pathways activated within the heart during RIPC-mediated cardioprotection and that are found to be similar to the case of IPC-mediated cardioprotection. The actual signaling pathways responsible for conveying cardioprotective signals to the heart through remote organs/tissues are remained to be explored. Once the cardioprotective signal reaches the heart, however, similar signaling pathways as underlying in IPC are believed to mediate the protective effect on the heart. Numerous mechanisms/factors such as humoral, neural, and systemic have been suggested to play a key role in mediating cardioprotection through remote organs/tissues. It seems crucial to recognize that different molecular pathways may interact with one another to produce cardioprotection [63].

15.4.1.1 HUMORAL FACTORS IN RIPC-MEDIATED CARDIOPROTECTION

Some experimental studies reported that during RIPC, a brief period of reperfusion may play a crucial role in 'washing out' the substances or transportation of several humoral factors to the heart, released during the preconditioning ischemic phase [64]. This hypothesis was further supported by an experimental study that reported that blood collected from a preconditioned rabbit was able to substantially minimize the size of cardiac infarcts in untreated rabbits by transfusion [65] manifesting the transmission of at least one humoral cardioprotective factor. The same research group conducted another study and demonstrated that coronary effluent collected from an isolated rabbit heart subjected to standard IPC protocol, had the potential to reduce the size of myocardial infarct up to 69% [66] and also improved LV functioning [67] when used to perfuse an untreated secluded rabbit heart. Konstantinov and colleagues conducted an attractive experimental study and provided promising evidence in favor of the humoral mechanism for RIPC. Remote limb preconditioning (RIPre) of a pig that had received a donor's heart was able to reduce the size of myocardial infarct in the denervated donor heart, providing definite evidence that a humoral mediator was essential for RIPC protection [68]. One more similar type of study was conducted by Kristiansen and co-workers and demonstrated that hearts excised from a remote limb preconditioned rat showed a smaller infarction size when exposed to a global ischemia-reperfusion episode of myocardium on a Langendorff'sapparatus [69].

Several studies have been conducted to identify the humoral factors which release during the preconditioning stimulus from remote organs and convey the protective signaling to the heart [65, 70–72]. Serejo and colleagues collected the discharge from preconditioned rat hearts and found temperature-sensitive hydrophobic compounds with a molecular weight of 3.5 kDa that provided cardioprotection by activating protein kinase C (PKC) [73]. However, the actual identification of the humoral factors responsible for mediating cardioprotection remains contentious. In addition to this, many other experimental studies have been conducted to confirm the release of some endogenous substances such as adenosine [74], bradykinin [75], opioids [76], calcitonin gene-related peptide (CGRP) [77], and endocannabinoids [78] from the remote organ during the preconditioning ischemic stimulus (Figure 15.3). Moreover, it was also analyzed that these endogenous substances might be carried to the heart through the bloodstream where they eventually activate some intracellular cardioprotective signaling mechanisms. It is also predicted that these endogenous substances may also engage afferent neural pathways in the distant preconditioned organ to give cardioprotection.

15.4.1.2 NEURAL FACTORS IN RIPC-MEDIATED CARDIOPROTECTION

Studies documented potential evidence for the involvement of neural pathways in bestowing the cardioprotection induced by remote preconditioning of a non-cardiac organ. Gho and colleagues reported that a reduction in the size of myocardial infarct induced by giving a preconditioning stimulus to the anterior mesenteric artery appeared reversible in the presence of a ganglion blocker such as hexamethonium [62]. Endogenous chemicals such as adenosine [74], bradykinin [75], and CGRP [77] are believed to be secreted by the remote preconditioned organ, activated afferent nerve neurons, which subsequently receive and transmit to efferent nerve fibers terminating on the heart to offer cardioprotection (Figure 15.3).

15.4.1.3 SYSTEMIC RESPONSE IN RIPC-MEDIATED CARDIOPROTECTION

Many research studies have analyzed the effect of remote preconditioning of an tissue or organ on the gene transcription profile of the myocardium, and the inflammatory response and have reported that the inflammatory response is suppressed. Moreover, both anti-inflammatory and anti-apoptotic gene transcription profiles appear to be activated through RIPC stimulus. Despite this, the relevance of such a reaction to the cardioprotective effect induced by RIPC is still unknown and requires additional exploration (Figure 15.3) [79, 80].

15.4.1.4 MYOCARDIAL MECHANISM IN CARDIOPROTECTION MEDIATED BY RIPC

It has been documented that after receiving the cardioprotective signals to the heart from remote preconditioned organ/tissue, activation/recruitment of intracellular signaling mechanisms within cardiomyocytes starts which eventually confers cardioprotection just like IPC [50]. This phenomenon of cardioprotection includes the binding of several ligands such as adenosine [74], bradykinin [75], opioids [76], angiotensin [81], and endocannabinoids [78] with their respective GPCRs present on the surface of cardiomyocytes. Once the ligand binds to their receptors then it further activates several intracellular kinases such as PKC-139 and other signaling components including ROS [82], nitric oxide (NO), and the mitochondrial K_{ATP} channel [74]. In addition to this, RIPC also causes activation of some other signaling mechanisms such as the RISK pathway that further inhibits the mPTP as in the case of IPC [63]. In this context, another study was conducted and reported that the activation of the mitogen-activated protein kinases (MAPK's) JNK, p38, and ERK1/2 within the distant organ may also contribute to conferring cardioprotection mediated through RIPC [83] (Figure 15.3). However, it remains contentious whether these mechanisms triggered within the myocardium at the commencement of myocardial reperfusion.

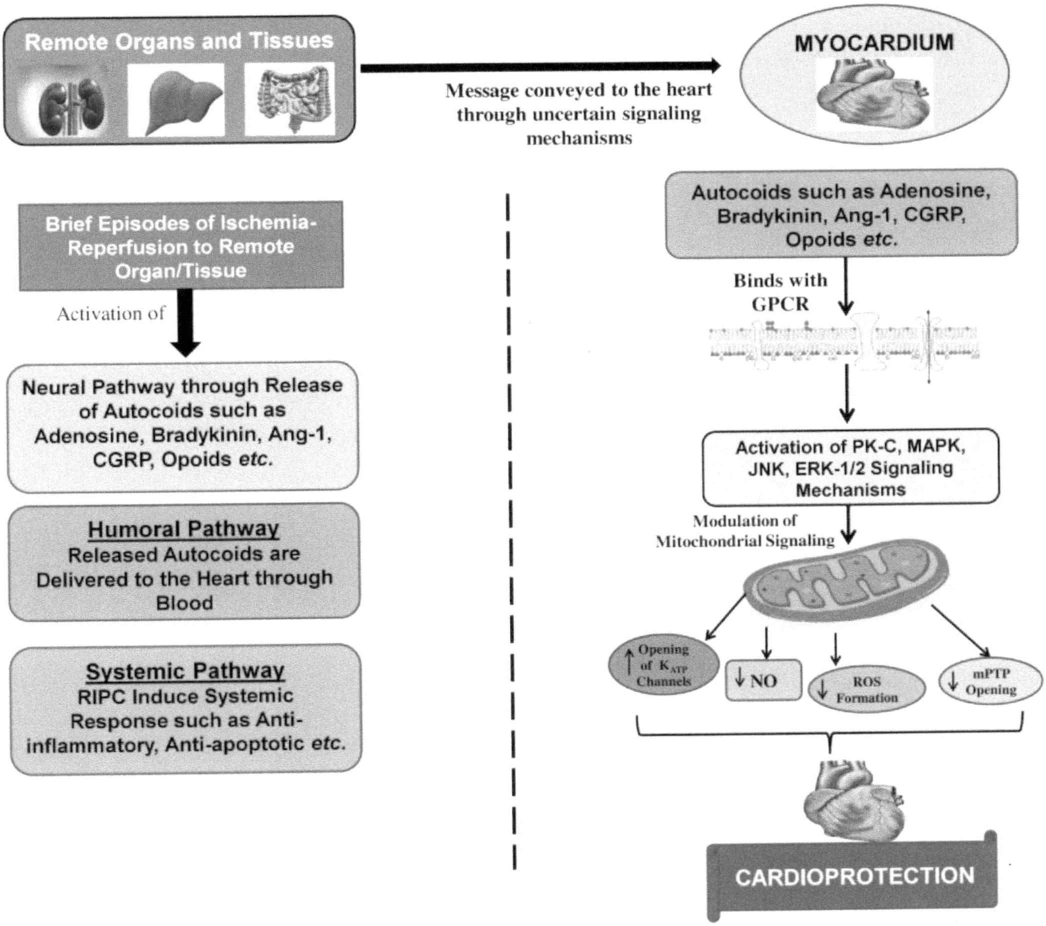

FIGURE 15.3 Various cardioprotective signaling mechanisms underlying RIPC.

15.4.2 CLINICAL APPLICATIONS OF RIPC IN VARIOUS CARDIOVASCULAR DISORDERS

Numerous clinical trials on patients undergoing ventricular septal defect repair, PPCI, coronary angioplasty, and elective abdominal aortic aneurysm (AAA) repair, acute MI, and CABG have been conducted to delineate the role of RIPC [4].

15.4.2.1 EVIDENCE

The first successful clinical study of RIPCwas conducted by Cheung and colleagues on children enduring corrective cardiac surgery for congenital heart disease (CHD). They documented that a standard RIPC stimulus using four 5-minute cycles of lower limb ischemia was able to contain myocardial injury, improve the resistance of the airway, and decrease inotrope score in 17 children subjected to corrective cardiac surgery [84].

Candilio and colleagues studied the positive benefits of RIPC on patients following cardiac surgery in a randomized controlled experiment. The recruitment and randomization into two groups of 180 patients undergoing cardiac bypass surgery or valve replacement. The RIPC stimulus was administered by tying a blood pressure cuff to the patient's upper arm and upper leg and administering two cycles of 5-minute inflation and deflation, followed by reperfusion. Compared to individuals who did not receive RIPC, a substantial decrease in the prevalence of perioperative myocardial damage and postoperative arterial fibrillation was described as the cardioprotective effect of RIPC [85].

Wu and colleagues examined the cardioprotective impact of limb IPC in patients following mitral valve replacement in a study conducted. Patients were treated to three cycles of 5 minutes of inflated and 5 minutes of deflation of the right upper arm pressure cuff. Cardioprotection with RIPC was determined by a substantial decrease in cardiac troponin-1 (C-tPn-1) levels compared to patients in the control group [86].

One more study demonstrated the cardioprotective effect of RIPC (induced by limb ischemia) in elective surgery done for repairing of AAA which was associated with myocardial and renal injury. The beneficial effect of RIPC was manifested in terms of the reduction of serum troponin-1 levels after surgery [87].

15.5 ISCHEMIC POSTCONDITIONING (IPOSTC)

The major drawback associated with IPC as a cardioprotective intervention is that IPC must be applied before the occurrence of the ischemic event, which is very difficult to predict in any clinical setting such as acute MI. In 2003, Zhao and co-researchers made a surprising and exciting invention and reported that the heart could be protected against acute MI by obstructing myocardial blood flow at the time of reperfusion with several brief episodes of myocardial ischemia a phenomenon known as IPostC [40, 88]. The research group also documented that applying three cycles of 30 seconds of LAD occlusion and reperfusion within 1 minute of myocardial reperfusion has the potential to decrease the size of MI up to 44% in canine hearts[88].

In addition to the aforementioned effects, IPostC was also reported to confer a series of protective effects such as reduced levels of myocardial edema, oxidative stress, polymorphonuclear neutrophil accumulation, and maintained endothelial function. These findings further supported the protective effects of IPostC in myocardial IR injury [88]. Surprisingly, the concept of modifying reperfusion as a therapeutic strategy to reduce the size of myocardial infarct had earlier been introduced in the 1980s as 'gentle' or 'gradual' [89] reperfusion. But later on, the phenomenon of IPostC had captured the interest and revitalized efforts to target myocardial reperfusion injury as a therapeutic approach for reducing myocardial infarct size [90, 91].

15.5.1 CARDIOPROTECTIVE SIGNALING PATHWAYS UNDERLYING IPOSTC

It is depicted that IPostC confers the cardioprotection through similar (but not all) signaling mechanisms as activated in the case of IPC at the time of reperfusion. These signaling pathways include activation of cell-surface receptors on the cardiomyocyte (such as adenosine, bradykinin, and opioids) and recruitment of pro-survival signaling pathways including PI3K [92], eNOS [93], ERK1/2 [94], RISK (Akt, protein kinase-B), SAFE, and GSK-3β [95]. Activation of all these signaling mechanisms eventually contributes to the cardioprotection through modulation of various mitochondrial functions such as reduced calcium overload, attenuated oxidative stress, and inhibited MPTP opening. In addition to this, it is also documented that Intermittent reperfusion induced by IPostC also postpones the restoration of intracellular pH, which might contribute to the inhibition of MPTP opening as observed in post-conditioned hearts [96]. Surprisingly, it has been revealed that the signaling pathways underpinning IPostC are species-specific; for example, the RISK pathway mediates IPostC in the hearts of rats but not pigs [97]. Unknown are the reasons behind this disparity. Using human atrial muscle obtained from patients having CABG surgery, the importance of the RISK route for IPostC has been defined [98] (Figure 15.4). IPostC, like IPC, is thought to have two protective processes, namely an early phase, which begins promptly after significant ischemia events, and a delayed phase, which arises 24 hours after ischemic events.

15.5.2 CLINICAL APPLICATIONS OF IPOSTC IN VARIOUS CARDIOVASCULAR DISORDERS

Many clinical studies investigated the effect of postconditioning in patients undergoing cardiac surgery for the treatment of Congestive Heart Diseases (CHD's) [99–101].

15.5.2.1 EVIDENCE

Luo and colleagues revealed the cardioprotective impact of IPostC in patients having valve replacement with rheumatic heart disease. Around 50 adult patients were selected at random and treated to three cycles of 30 seconds of ischemia and 30 seconds of reperfusion.

Reperfusion began 30 seconds after aortic re-clamping and de-clamping following cardioplegic arrest. As a result, their CK-MB and lactate levels were significantly lower than those of the control group, indicating myocardial protection [100].

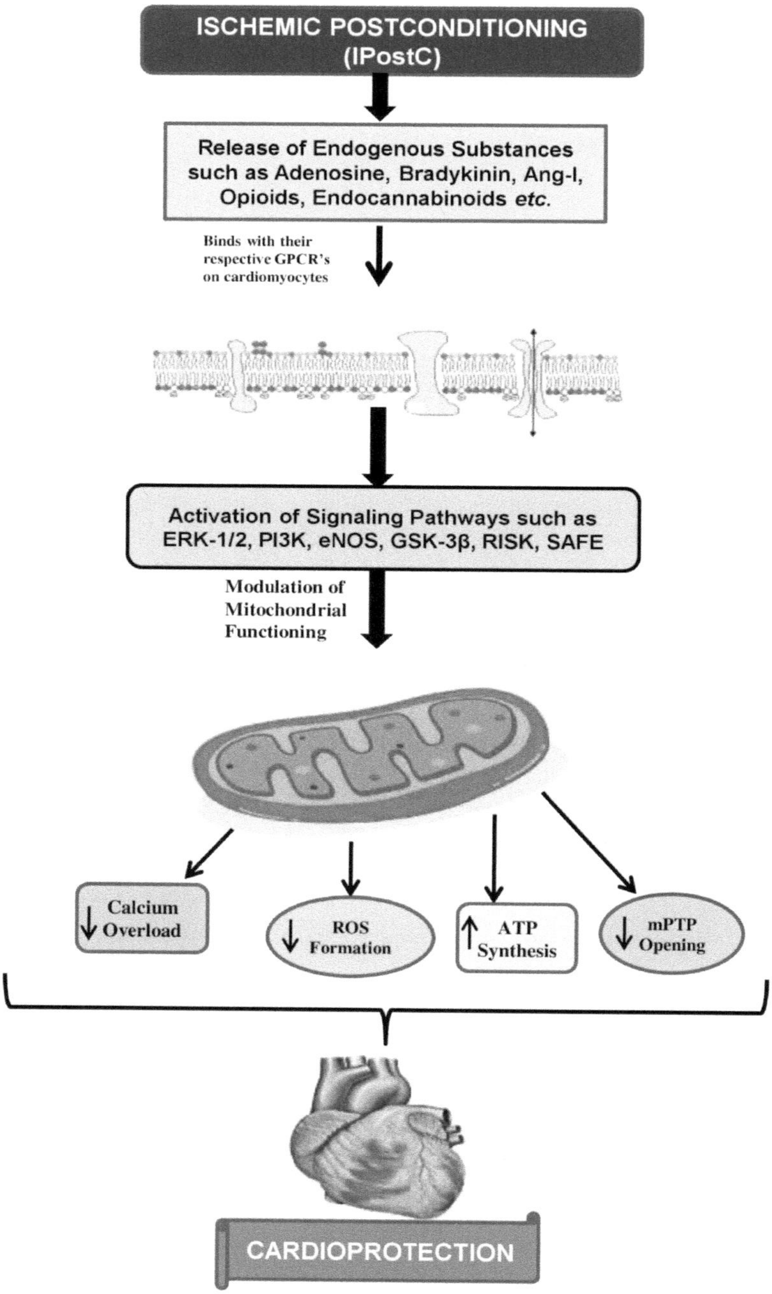

FIGURE 15.4 Modulation of various cardioprotective signaling pathways by IPostC.

In a new investigation, Li and colleagues conducted a randomized experiment to examine the cardioprotective impact of IPostC in patients receiving treatment for tetralogy of Fallot. They treated 99 individuals with tetralogy of Fallot to postconditioning by aortic clamping after reperfusion and discovered a substantial decrease in troponin-1 and lactate concentrations [101].

One more study was carried out by Staat and his group to substantiate the cardioprotective effects of IPostC on the patients undergoing coronary angioplasty for the treatment of acute myocardial infarction (AMI). Around 30 patients were recruited for this study and randomized into the control and postconditioning groups. After angioplasty surgery, within one minute of reperfusion, postconditioning group patients were subjected to four cycles of 1 minute. inflation and 1 min. deflation with an angioplasty balloon. Cardioprotective effects of IPostC were observed in terms of reduction in serum CK-MB levels in post-conditioned patients as compared to control group patients [102].

Xue and his research group also evidenced the cardioprotective effect conferred by IPostC on the patients undergoing primary percutaneous coronary intervention (PPCI) for the correction of ST-segment elevation myocardial infarction (STEMI). They selected 43 patients undergoing PPCI and randomized them into control (20 patients) and postconditioning (23 patients) groups. The postconditioning stimulus was provided by giving four cycles of 1 min. inflation and 1 min. deflation just after performing PPCI. Cardioprotection through IPostC was confirmed by reduced blood levels of CK-MB and C-reactive protein (CRP) in post-conditioned patients as compared to control group patients. Moreover, these beneficial effects were further confirmed by improved results of an electrocardiogram (ECG) and echocardiography in post-conditioned patients [103]. In addition, the cardioprotective effect of IPostC was subsequently confirmed by other studies performed on the patients suffering from MI, tetralogy of Fallot, and other cardiac diseases [104, 105].

15.6 REMOTE ISCHEMIC POSTCONDITIONING (RIPOSTC)

RIPostC is a protective approach in which a conditioning stimulus (short episodes of I/R) is provided to a distant organ at the commencement of reperfusion of the major ischemic organs. Because it is given at the commencement of reperfusion, may be conducted on non-vital organs, can avoid the risk of ischemia postconditioning on critical organs, and is suited for long-term rehabilitation, RIPostC looks to be more useful in clinical settings [4, 106].

15.6.1 CARDIOPROTECTIVE SIGNALING MECHANISMS UNDERLYING RIPOSTC

Basic cardioprotective signaling pathways of RIPostC are the activation of eNOS, PI3K [107], Akt, GSK-3β, T-LAK-cell-originated protein kinase (TOPK) [108, 109] pathways and improvement in endogenous antioxidant enzyme activity [110] and also inhibition of δ-protein kinase-C (δ-PKC) [111] in IR injury model [4] (Figure 15.4).

15.6.2 CLINICAL APPLICATIONS OF RIPOSTC IN CARDIOVASCULAR DISORDERS

15.6.2.1 EVIDENCE

Zhong and colleagues studied the cardioprotective potential of RIPostCin in children undergoing open-heart surgery for the treatment of congenital cardiac abnormalities. Around 69 youngsters receiving open-heart surgery were randomly assigned to one of two groups. At the commencement of aortic declamping, RIPostC was administered during three cycles of 5-minute ischemia and 5-minute reperfusion utilizing a blood pressure cuff on the lower limb. Cardioprotection with RIPostC was measured by a substantial decrease in cardiac troponin-1 and CK-MB levels relative to the control group [112].

Cao and colleagues revealed the cardioprotective impact of RIPostCon on patients receiving PPCI for STEMI with IR damage. Around 80 PPCI patients were selected and randomized into two groups for this study. Patients were treated to RIPostC stimulus by administering four cycles of 5 min ischemia and 5 min reperfusion through the upper arm by inflating and deflating a cuff. Cardioprotection with RIPostC was demonstrated by a significant decrease in blood CK-MB, nitric oxide (NO), creatinine, and stromal cell-derived factor-1a (SDF) levels at 0.5, 8, 24, 48, and 72 hours following PPCI as compared to control group participants [113].

15.7 VARIOUS CHALLENGES FACING CLINICAL RESEARCH

A few decades back, the phenomenon of IPC was first introduced in 1986 but the therapeutic relevance of conditioning techniques is harnessed. Moreover, whether these interventions can provide substantial improvement in clinical outcomes or not has given mixed results and need to be clarified. Numerous cardioprotective therapies have already shown significant results in many preclinical studies but they have not proved their significant efficacy in clinical settings. There are many reasons for this failure to translate the cardioprotective potential of ischemic conditioning techniques and have already been discussed in the cited literature [114–116]. These factors include that the available animal models of acute IRI are inadequate in representing the wide spectrum of co-morbidities and co-existing conditions of patients with IHD such as advanced age, diabetes, hypertension, hyperlipidemia, and pre-existing coronary artery disease [117]. In addition, some clinical studies were not properly designed and investigators did not consider the pre-clinical experimental studies' results [118]. Besides, numerous clinical studies have been carried out to investigate the cardioprotective potential of various therapies without thorough testing in pre-clinical animal models. Therefore, the consortium for pre-clinical assessment of cardioprotective therapies (CAESAR) was formed after discussing this particular issue in two National Heart and Blood Institute workshops. This consortium's primary purpose was to permit the testing of novel cardioprotective medicines employing small-animal and large-animal models of myocardial infarction (MI) in multi-center, randomized, and controlled clinical studies [119, 120]. The major challenge facing clinical cardioprotection research is that clinical outcomes of patients with STEMI after PPCI continue to improve but making it difficult to demonstrate a reduction in the size of MI and benefits in clinical outcomes with a novel cardioprotective therapy. Therefore, there is an emergency to develop novel therapeutic strategies/therapies that can contain IR injury of the myocardium, reduce myocardial infarct size, preserve LV systolic function, and prevent the onset of heart failure. A similar challenge is being faced by clinical research for conferring cardioprotection in patients undergoing CABG surgery. Although, advancements in surgical techniques for the protection of myocardium have significantly reduced the extent of myocardial infarct size along with patient mortality. Despite this, in the case of the aged population and due to the increased prevalence of co-morbidities such as diabetes, obesity, and hypertension, novel cardioprotective therapies/interventions for high-risk patients are needed to be discovered. Therefore, patients who are at high risk of operative complications should perhaps be the focus of future clinical cardioprotection studies.

15.8 CONCLUSION AND FUTURE DIRECTIONS

Ischemic conditioning has been proved to be a potential endogenous cardioprotective strategy for decreasing the size of myocardial infarcts in patients with the incidence of STEMI. Ischemic conditioning has also given significant evidence to mitigate myocardial injury in patients undergoing CABG surgery. Due to some advantages of RIPC such as its simplicity, non-invasive nature, and its flexibility for the timing of application, makes it more feasible to apply in different clinical settings involving acute IRI. Although, the data obtained from various pre-clinical studies have proved that ischemic conditioning; is a highly cardioprotective strategy. However, data obtained from clinical studies have given mixed results. RIPC provided through limb has produced the most significant results in patients with STEMI. On contrary, the clinical studies of pharmacological conditioning have been somewhat disappointing except few pharmacological agents. Therefore, clinical cardioprotection research has been challenging and novel therapies are needed to develop because of the growing occurrence of heart failure in patients with IHD. Such advancement in the development of new effective therapies will certainly open new avenues for reducing myocardial infarct size, and improving cardiac functions, and will also improve clinical outcomes in patients with IHD.

ACKNOWLEDGMENTS

The authors are thankful to the Department of Pharmaceutical Sciences and Drug Research, Punjabi University Patiala for providing the technical support.

CONFLICT OF INTEREST

The authors declare that there is no conflict of interest.

AUTHORS CONTRIBUTION

Kuldeep Kumar carried out the relevant literature research and prepared the primary draft; Dr. Nirmal Singh prepared the final draft of the chapter; Dr. Amteshwar S. Jaggi co-conceptualized and provided valuable inputs in finalizing the chapter.

KEYWORDS

- **cardioprotective signaling**
- **cardiovascular disorders**
- **ischemia**
- **myocardial ischemia-reperfusion injury**
- **postconditioning**
- **preconditioning**
- **remote ischemic postconditioning**
- **remote postconditioning**
- **remote preconditioning**

REFERENCES

1. WHO., (2019). *Cardiovascular Diseases (CVD's).* Fact sheet.
2. Moran, A. E., Forouzanfar, M. H., Roth, G. A., Mensah, G. A., Ezzati, M., Murray, C. J. L., & Naghavi, M., (2014). Temporal trends in ischemic heart disease mortality in 21 world regions, 1980 to 2010: The adenosine, adenosine receptors, and myocardial protection: An updated overview. *Cardiovasc. Res., 52,* 25–39.
3. Khan, H., Sharma, K., Kumar, A., Kaur, A., & Singh, T. G., (2022). Therapeutic implications of cyclooxygenase (COX) inhibitors in ischemic injury. *Inflamm. Res.*, 1–16.
4. Kumar, K., Singh, N., Jaggi, A. S., & Maslov, L., (2020). Clinical applicability of conditioning techniques in ischemia-reperfusion injury: A review of the literature. *Curr. Cardiol. Rev, 16,* 1–13.
5. Randhawa, P. K., Bali, A., & Jaggi, A. S., (2015). RIPC for multiorgan salvage in clinical settings: Evolution of concept, evidences, and mechanisms. *Eur. J. Pharmacol., 746,* 317–332. http://dx.doi.org/10.1016/j.ejphar.2014.08.016 PMID: 25176179.
6. Kalra, P., Khan, H., Kaur, A., & Singh, T. G., (2022). Mechanistic insight on autophagy modulated molecular pathways in cerebral ischemic injury: From preclinical to clinical perspective. *Neurochem. Res.,* 1–19.
7. Khan, H., Bangar, A., Grewal, A. K., Bansal, P., & Singh, T. G., (2022). Caspase-mediated regulation of the distinct signaling pathways and mechanisms in neuronal survival. *Int. Immunopharmacol., 110,* 108951.
8. Tiwari, P., Khan, H., Singh, T. G., & Grewal, A. K., (2022). Poly (ADP-ribose) polymerase: An overview of mechanistic approaches and therapeutic opportunities in the management of stroke. *Neurochem. Res.*, 1–23.
9. Zerva, K., Sanidas, E., Tzanis, G., Chantziara, V., Papadopoulos, D., & Barbetseas, J., (2017). Effect of thrombolytic therapy in plaque inflammation and structural remodeling in acute myocardial infarction patients. *Hellenic. J. Cardiol., 58*(6), 446.

10. Khan, H., Kashyap, A., Kaur, A., & Singh, T. G., (2020). Pharmacological postconditioning: A molecular aspect in ischemic injury. *J. Pharm. Pharmacol., 72*(11), 1513–1527.
11. Papageorgiou, N., Briasoulis, A., & Tousoulis, D., (2018). Ischemia-reperfusion injury: Complex pathophysiology with elusive treatment. *Hellenic. J. Cardiol., 59*(6), 329–330.
12. Reimer, K. A., Lowe, J. E., Rasmussen, M. M., & Jennings, R. B., (1977). The wavefront phenomenon of ischemic cell death. Myocardial infarct size *vs* duration of coronary occlusion in dogs. *Circulation, 56*(5), 786–794. http://dx.doi.org/10.1161/01.CIR.56.5.786 PMID: 912839.
13. Williams, T. M., Waksman, R., De Silva, K., Jacques, A., & Mahmoudi, M., (2015). Ischemic preconditioning-an unfulfilled promise. *Cardiovasc. Revasc. Med., 16*(2), 101–108.
14. Bolli, R., & Marbán, E., (1999). Molecular and cellular mechanisms of myocardial stunning. *Physiol. Rev., 79*(2), 609–634. http://dx.doi.org/10.1152/physrev.1999.79.2.609 PMID:10221990.
15. Maxwell, S. R., & Lip, G. Y., (1997). Reperfusion injury: A review of the pathophysiology, clinical manifestations, and therapeutic options. *Int. J. Cardiol., 58*(2), 95–117. http://dx.doi.org/10.1016/S0167-5273(96)02854-9 PMID: 9049675.
16. Carden, D. L., & Granger, D. N., (2000). Pathophysiology of ischemia-reperfusion injury. *J. Pathol., 190*(3), 255–266. http://dx.doi.org/10.1002/(SICI) 1096-9896 (200002)190:3<255::AID-PATH526>3.0.CO; 2-6 PMID:10685060.
17. Toyokuni, S., (1999). Reactive oxygen species-induced molecular damage and its application in pathology. *Pathol. Int., 49*(2), 91–102. http://dx.doi.org/10.1046/j.1440-1827.1999.00829.x PMID:10355961.
18. Szocs, K., (2004). Endothelial dysfunction and reactive oxygen species production in ischemia/reperfusion and nitrate tolerance. *Gen. Physiol. Biophys., 23*(3), 265–295. PMID: 15638116.
19. Bertuglia, S., Giusti, A., & Del, S. P., (2004). Antioxidant activity of nitro derivative of aspirin against ischemia-reperfusion in hamster cheek pouch microcirculation. *Am. J. Physiol. Gastrointest. Liver. Physiol., 286*(3), G437–443.
20. Mustoe, T., (2004). Understanding chronic wounds: A unifying hypothesis on their pathogenesis and implications for therapy. *Am. J. Surg., 187*(5A), 65S–70S. http://dx.doi.org/10.1016/S0002-9610(03) 00306-4 PMID: 15147994.
21. Jordan, J. E., Zhao, Z. Q., & Vinten-Johansen, J., (1999). The role of neutrophils in myocardial ischemia-reperfusion injury. *Cardiovasc. Res., 43*(4), 860–878.
22. Collard, C. D., & Gelman, S., (2001). Pathophysiology, clinical manifestations, and prevention of ischemia-reperfusion injury. *Anesthesiol., 94*(6), 1133–1138. http://dx.doi.org/10.1097/00000542-200106000-00030 PMID: 11465607.
23. Contreras, J. L., Vilatoba, M., Eckstein, C., Bilbao, G., Anthony, T. J., & Eckhoff, D. E., (2004). Caspase-8 and caspase-3 small interfering RNA decreases ischemia/reperfusion injury to the liver in mice. *Surgery, 136*(2), 390–400. http://dx.doi.org/10.1016/j.surg.2004.05.015 PMID: 15300206.
24. Chen, M., Won, D. J., Krajewski, S., & Gottlieb, R. A., (2002). Calpain and mitochondria in ischemia/reperfusion injury. *J. Biol. Chem., 277*(32), 29181–29186.
25. Aktan, A. O., & Yalcin, A. S., (1998). Ischemia-reperfusion injury, reactive oxygen metabolites, and the surgeon. *Turk. J. Med. Sci., 28*, 1–5.
26. Zhang, M., Xu, Y. J., Saini, H. K., Turan, B., Liu, P. P., & Dhalla, N. S., (2005). Pentoxifylline attenuates cardiac dysfunction and reduces TNF-α level in ischemic-reperfused heart. *Am. J. Physiol. Heart. Circ. Physiol., 289*(2), H832–839.
27. Li, J., Vootukuri, S., & Shang, Y., (2014). RUC-4: A novel αIIbβ3 antagonist for prehospital therapy of myocardial infarction. *Arterioscler. Thromb. Vasc. Biol., 34*(10), 2321–2329. http://dx.doi.org/10.1161/ATVBAHA.114.303724 PMID: 25147334.
28. Bulluck, H., Yellon, D. M., & Hausenloy, D. J., (2016). Reducing myocardial infarct size: Challenges and future opportunities. *Heart, 102*(5), 341–348.
29. Kloner, R. A., (2013). Current state of clinical translation of cardioprotective agents for acute myocardial infarction. *Circ. Res., 113*(4), 451–463. http://dx.doi.org/10.1161/CIRCRESAHA.112.300627 PMID: 23908332.

30. Donato, M., Evelson, P., & Gelpi, R. J., (2017). Protecting the heart from ischemia/ reperfusion injury: An update on remote ischemic preconditioning and postconditioning. *Curr. Opin. Cardiol.*, *32*(6), 784–790. http://dx.doi.org/10.1097/HCO.0000000000000447.
31. Murry, C. E., Richard, V. J., Jennings, R. B., & Reimer, K. A., (1991). Myocardial protection is lost before contractile function recovers from ischemic preconditioning. *Am. J. Physiol.*, *260*, H796–804. PMID: 2000974.
32. Tomai, F., Crea, F., Chiariello, L., & Gioffrè, P. A., (1999). Ischemic preconditioning in humans: Models, mediators, and clinical relevance. *Circulation*, *100*(5), 559–563. http://dx.doi.org/10.1161/01.CIR.100.5.559 PMID: 10430772.
33. Maldonado, Y., Weiner, M. M., & Ramakrishna, H., (2017). Remote ischemic preconditioning in cardiac surgery: Is there proven clinical benefit? *J. Cardiothorac. Vasc. Anesth.*, *31*(5), 1910–1915. http://dx.doi.org/10.1053/j.jvca.2017.03.043 PMID: 28711313.
34. Nakamura, M., Wang, N. P., & Zhao, Z. Q., (2000). Preconditioning decreases Bax expression, PMN accumulation and apoptosis in reperfused rat heart. *Cardiovasc. Res.*, *45*(3), 661–670. http://dx.doi.org/10.1016/S0008-6363(99)00393-4 PMID: 10728387.
35. Rezkalla, S. H., & Kloner, R. A., (2007). Preconditioning in humans. *Heart. Fail. Rev.*, *12*, 201–206.
36. Tong, H., Imahashi, K., Steenbergen, C., & Murphy, E., (2002). Phosphorylation of glycogen synthase kinase-3beta during preconditioning through a phosphatidylinositol-3-kinase--dependent pathway is cardioprotective. *Circ. Res.*, *90*(4), 377–379. http://dx.doi.org/10.1161/01.RES.0000012567.95445.55 PMID:11884365.
37. Jonassen, A. K., Mjøs, O. D., & Sack, M. N., (2004). p70s6 kinase is a functional target of insulin activated Akt cell-survival signaling. *Biochem. Biophys. Res. Commun.*, *315*(1), 160–165.
38. Kuzuya, T., Hoshida, S., Yamashita, N., Fuji, H., Oe, H., Hori, M., Kamada, T., & Tada, M., (1993). Delayed effects of sublethal ischemia on the acquisition of tolerance to ischemia. *Circ. Res.*, *72*, 1293–1299.
39. Khan, H., Singh, A., Thapa, K., Garg, N., Grewal, A. K., & Singh, T. G., (2021). Therapeutic modulation of the phosphatidylinositol 3-kinases (PI3K) pathway in cerebral ischemic injury. *Brain Res.*, *1761*, 147399.
40. Hausenloy, D. J., (2013). Cardioprotection techniques: Preconditioning, postconditioning and remote conditioning (basic science). *Curr. Pharmaceut. Des.*, *19*, 4544–4563.
41. Yao, Z., & Gross, G. J., (1993). Acetylcholine mimics ischemic preconditioning via a glibenclamide-sensitive mechanism in dogs. *Am. J. Physiol.*, *264*, H2221–H2225.
42. Goto, M., Liu, Y., Yang, X. M., Ardell, J. L., Cohen, M. V., & Downey, J. M., (1995). Role of bradykinin in protection of ischemic preconditioning in rabbit hearts. *Circ. Res.*, *77*(3), 611–621.
43. Liu, G. S., Thornton, J., Van, W. D. M., Stanley, A. W., Olsson, R. A., & Downey, J. M., (1991). Protection against infarction afforded by preconditioning is mediated by A1 adenosine receptors in rabbit heart. *Circul.*, *84*(1), 350–356.
44. Liu, Y., Tsuchida, A., Cohen, M. V., & Downey, J. M., (1995). Pretreatment with angiotensin II activates protein kinase C and limits myocardial infarction in isolated rabbit hearts. *J. Mol. Cell. Cardiol.*, *27*(3), 883–892.
45. Tsuchida, A., Liu, Y., Liu, G. S., Cohen, M. V., & Downey, J. M., (1994). Alpha-1 adrenergic agonists precondition rabbit ischemic myocardium independent of adenosine by direct activation of protein kinase C. *Circ. Res.*, *75*(3), 576–585.
46. Wang, P., Gallagher, K. P., Downey, J. M., & Cohen, M. V., (1996). Pretreatment with endothelin-1 mimics ischemic preconditioning against infarction in isolated rabbit heart. *J. Mol. Cell. Cardiol.*, *28*(3), 579–588.
47. Schultz, J. E., Rose, E., Yao, Z., & Gross, G. J., (1995). Evidence for involvement of opioid receptors in ischemic preconditioning in rat hearts. *Am. J. Physiol.*, *268*, H2157–H2161.
48. Hausenloy, D. J., & Yellon, D. M., (2007). Preconditioning and postconditioning: United at reperfusion. *Pharmacol. Ther.*, *116*(2), 173–191.
49. Hausenloy, D. J., Lecour, S., & Yellon, D. M., (2011). Reperfusion injury salvage kinase and survivor activating factor enhancement prosurvival signaling pathways in ischemic postconditioning: Two sides of the same coin. *Antioxid. Redox Signal.*, *14*, 893–907.

50. Yellon, D. M., & Downey, J. M., (2003). Preconditioning the myocardium: From cellular physiology to clinical cardiology. *Physiol. Rev.*, *83*, 1113–1151.
51. Ong, S. B., Dongworth, R. K., Cabrera-Fuentes, H. A., & Hausenloy, D. J., (2015). Role of the MPTP in conditioning the heart — Translatability and mechanism. *Br. J. Pharmacol.*, *172*, 2074–2084.
52. Varga, Z. V., Giricz, Z., Bencsik, P., Madonna, R., Gyongyosi, M., Schulz, R., Mayr, M., et al., (2015). Functional genomics of cardioprotection by ischemic conditioning and the influence of comorbid conditions: Implications in target identification. *Curr. Drug Targets.*, *16*, 904–911.
53. Yellon, D. M., Alkhulaifi, A. M., & Pugsley, W. B., (1993). Preconditioning the human myocardium. *Lancet*, *342*, 276–277.
54. Lu, E. X., Chen, S. X., & Yuan, M. D., (1997). Preconditioning improves myocardial preservation in patients undergoing open heart operations. *Ann. Thorac. Surg.*, *64*(5), 1320–1324. http://dx.doi.org/10.1016/S0003-4975(97)00838-2 PMID: 9386697.
55. Li, G., Chen, S., Lu, E., & Li, Y., (1999). Ischemic preconditioning improves preservation with cold blood cardioplegia in valve replacement patients. *Eur. J. Cardiothorac. Surg.*, *15*(5), 653–657. http://dx.doi.org/10.1016/S1010-7940(99)00070-6 PMID:10386412.
56. Walsh, S. R., Tjun, Y., Kullar, P., Jenkins, D. P., Dutka, D. P., & Gaunt, M. E., (2008). Ischemic preconditioning during cardiac surgery: Systematic review and meta-analysis of perioperative outcomes in randomized clinical trials. *Eur. J. Cardiothorac. Surg.*, *34*, 985–994.
57. Lim, S. Y., & Hausenloy, D. J., (2012). Remote ischemic conditioning: From bench to bedside. *Front. Physiol.*, *3*, 27.
58. Przyklenk, K., Bauer, B., Ovize, M., Kloner, R. A., & Whittaker, P., (1993). Regional ischemic 'preconditioning' protects remote virgin myocardium from subsequent sustained coronary occlusion. *Circulation*, *87*(3), 893–899.
59. Przyklenk, K., & Whittaker, P., (2011). Remote ischemic preconditioning: Current knowledge, unresolved questions, and future priorities. *J. Cardiovasc. Pharmacol. Ther.*, *16*, 255–259. http://dx.doi.org/10.1177/1074248411409040 PMID: 21821525.
60. Randhawa, P. K., & Jaggi, A. S., (2017). Investigating the involvement of TRPV1 ion channels in remote hind limb preconditioning-induced cardioprotection in rats. *Naun. Schmied. Arc. Pharmacol.*, *390*(2), 117–126.
61. McClanahan, T., Nao, B., Wolke, L., Martin, B. J., & Mezt, T. E., (1993). Brief renal occlusion and reperfusion reduces myocardial infarct size in rabbits. *FASEB J.*, *7*, A-18.
62. Gho, B. C., Schoemaker, R. G., Van, D. D. M. A., Duncker, D. J., & Verdouw, P. D., (1996). Myocardial protection by brief ischemia in noncardiac tissue. *Circulation*, *94*, 2193–2200.
63. Hausenloy, D. J., & Yellon, D. M., (2007). Reperfusion injury salvage kinase signaling: Taking a RISK for cardioprotection. *Heart Fail. Rev.*, *12*, 217–234.
64. Hausenloy, D. J., & Yellon, D. M., (2008). Remote ischemic preconditioning: Underlying mechanisms and clinical application. *Cardiovascul. Res.*, *79*, 377–386 doi: 10.1093/cvr/cvn114.
65. Dickson, E. W., Lorbar, M., Porcaro, W. A., Fenton, R. A., Reinhardt, C. P., & Gysembergh, A., (1999). Rabbit heart can be 'preconditioned' via transfer of coronary effluent. *Am. J. Physiol.*, *277*, H2451–H2457.
66. Dickson, E. W., Reinhardt, C. P., Renzi, F. P., Becker, R. C., Porcaro, W. A., & Heard, S. O., (1999). Ischemic preconditioning may be transferable via whole blood transfusion: Preliminary evidence. *J. Thromb. Thrombolysis*, *8*, 123–129.
67. Dickson, E. W., Porcaro, W. A., Fenton, R. A., Heard, S. O., Reindhardt, C. P., & Renzi, F. P., (2000). Preconditioning at a distance' in the isolated rabbit heart. *Acad. Emerg. Med.*, *7*, 311–317.
68. Konstantinov, I. E., Arab, S., Li, J., Coles, J. G., Boscarino, C., & Mori, A., (2005). The remote ischemic preconditioning stimulus modifies gene expression in mouse myocardium. *J. Thorac. Cardiovasc. Surg.*, *130*, 1326–1332.
69. Kristiansen, S. B., Henning, O., Kharbanda, R. K., Nielsen-Kudsk, J. E., Schmidt, M. R., & Redington, A. N., (2005). Remote preconditioning reduces ischemic injury in the explanted heart by a KATP channel-dependent mechanism. *Am. J. Physiol. Heart Circ. Physiol.*, *288*, H1252–H1256.
70. McClanahan, T., Nao, B., Wolke, L., Martin, B. J., & Mezt, T. E., (1993). Brief renal occlusion and reperfusion reduces myocardial infarct size in rabbits. *FASEB J.*, *7*, A18.

71. Xiao, L., Lu, R., Hu, C. P., Deng, H. W., & Li, Y. J., (2001). Delayed cardioprotection by intestinal preconditioning is mediated by calcitonin gene-related peptide. *Eur. J. Pharmacol., 427*, 131–135.
72. Wolfrum, S., Schneider, K., Heidbreder, M., Nienstedt, J., Dominiak, P., & Dendorfer, A., (2002). Remote preconditioning protects the heart by activating myocardial PKCepsilon-isoform. *Cardiovasc. Res., 55*, 583–589.
73. Serejo, F. C., Rodrigues, L. F., Silva, T. K. C., De Carvalho, A. C., & Nascimento, J. H., (2007). Cardioprotective properties of humoral factors released from rat hearts subject to ischemic preconditioning. *J. Cardiovasc. Pharmacol., 49*, 214–220.
74. Pell, T. J., Baxter, G. F., Yellon, D. M., & Drew, G. M., (1998). Renal ischemia preconditions myocardium: Role of adenosine receptors and ATP-sensitive potassium channels. *Am. J. Physiol., 275*, H1542–H1547.
75. Schoemaker, R. G., & Van, H. C. L., (2000). Bradykinin mediates cardiac preconditioning at a distance. *Am. J. Physiol. Heart Circ. Physiol., 278*, H1571–H1576.
76. Patel, H. H., Moore, J., Hsu, A. K., & Gross, G. J., (2002). Cardioprotection at a distance: Mesenteric artery occlusion protects the myocardium via an opioid sensitive mechanism. *J. Mol. Cell. Cardiol., 34*, 1317–1323.
77. Tang, Z. L., Dai, W., Li, Y. J., & Deng, H. W., (1999). Involvement of capsaicin-sensitive sensory nerves in early and delayed cardioprotection induced by a brief ischemia of the small intestine. *Naunyn Schmiedebergs Arch. Pharmacol., 359*, 243–247.
78. Hajrasouliha, A. R., Tavakoli, S., Ghasemi, M., Jabehdar-Maralani, P., Sadeghipour, H., & Ebrahimi, F., (2008). Endogenous cannabinoids contribute to remote ischemic preconditioning via cannabinoid CB(2) receptors in the rat heart. *Eur. J. Pharmacol., 579*, 246–252.
79. Konstantinov, I. E., Arab, S., Kharbanda, R. K., Li, J., Cheung, M. M., & Cherepanov, V., (2004). The remote ischemic preconditioning stimulus modifies inflammatory gene expression in humans. *Physiol. Genomics, 19*, 143–150.
80. Konstantinov, I. E., Li, J., Cheung, M. M., Shimizu, M., Stokoe, J., & Kharbanda, R. K., (2005). Remote ischemic preconditioning of the recipient reduces myocardial ischemia-reperfusion injury of the denervated donor heart via a Katp channel-dependent mechanism. *Transplantation, 79*, 1691–1695.
81. Singh, D., & Chopra, K., (2004). Evidence of the role of angiotensin AT(1) receptors in remote renal preconditioning of myocardium. *Methods Find. Exp. Clin. Pharmacol., 26*, 117–122.
82. Weinbrenner, C., Nelles, M., Herzog, N., Sarvary, L., & Strasser, R. H., (2002). Remote preconditioning by infrarenal occlusion of the aorta protects the heart from infarction: A newly identified non-neuronal but PKC-dependent pathway *Cardiovasc. Res., 55*, 590–601.
83. Heidbreder, M., Naumann, A., Tempel, K., Dominiak, P., & Dendorfer, A., (2008). Remote vs. ischemic preconditioning: The differential role of mitogen-activated protein kinase pathways. *Cardiovasc. Res., 78*, 108–115.
84. Cheung, M. M., Kharbanda, R. K., Konstantinov, I. E., Shimizu, M., Frndova, H., & Li, J., (2006). Randomized controlled trial of the effects of remote ischemic preconditioning on children undergoing cardiac surgery: First clinical application in humans. *J. Am. Coll. Cardiol., 47*, 2277–2282.
85. Candilio, L., Malik, A., & Ariti, C., (2015). Effect of remote ischemic preconditioning on clinical outcomes in patients undergoing cardiac bypass surgery: A randomized controlled clinical trial. *Heart, 101*(3), 185–192.
86. Wu, Q., Gui, P., & Wu, J., (2011). Effect of limb ischemic preconditioning on myocardial injury in patients undergoing mitral valve replacement surgery: A randomized controlled trial. *Circ. J., 75*(8), 1885–1889.
87. Ali, Z. A., Callaghan, C. J., Lim, E., Ali, A. A., Nouraei, S. A., Akthar, A. M., Boyle, J. R., et al., (2007). Remote ischemic preconditioning reduces myocardial and renal injury after elective abdominal aortic aneurysm repair: A randomized controlled trial. *Circulation, 116*, I98–I105.
88. Khan, H., Grewal, A. K., & Singh, T. G., (2022). Pharmacological postconditioning by protocatechuic acid attenuates brain injury in ischemia–reperfusion (I/R) mice model: Implications of nuclear factor erythroid-2-related factor pathway. *Neuroscience, 491*, 23–31.
89. Okamoto, F., Allen, B. S., Buckberg, G. D., Bugyi, H., & Leaf, J., (1986). Reperfusion conditions: Importance of ensuring gentle versus sudden reperfusion during relief of coronary occlusion. *J. Thorac. Cardiovasc. Surg., 92*, 613–620.
90. Tsang, A., Hausenloy, D. J., & Yellon, D. M., (2005). Myocardial postconditioning: Reperfusion injury revisited. *Am. J. Physiol. Heart Circ. Physiol., 289*, H2–H7.

91. Shi, W., & Vinten-Johansen, J., (2012). Endogenous cardioprotection by ischemic postconditioning and remote conditioning. *Cardiovasc. Res.*, *94*, 206–216.
92. Zhu, M., Feng, J., Lucchinetti, E., Fischer, G., Xu, L., Pedrazzini, T., & Zaugg, M., (2006). Ischemic postconditioning protects remodeled myocardium *via* the PI3K-PKB/Akt reperfusion injury salvage kinase pathway. *Cardiovasc. Res.*, *72*(1), 152–162.
93. Guo, J. Y., Yang, T., Sun, X. G., Zhou, N. Y., Li, F. S., Long, D., Lin, T., et al., (2011). Ischemic postconditioning attenuates liver warm ischemia-reperfusion injury through Akt eNOS-NO-HIF pathway. *J. Biomed. Sci.*, *18*, 79. http://dx.doi.org/10.1186/1423-0127-18-79 PMID: 22035453.
94. Darling, C. E., Jiang, R., Maynard, M., Whittaker, P., Vinten-Johansen, J., & Przyklenk, K., (2005). Postconditioning *via* stuttering reperfusion limits myocardial infarct size in rabbit hearts: Role of ERK1/2. *Am. J. Physiol. Heart Circ. Physiol.*, *289*(4), H1618–1626.
95. Tsang, A., Hausenloy, D. J., Mocanu, M. M., & Yellon, D. M., (2004). Postconditioning: A form of "modified reperfusion" protects the myocardium by activating the phosphatidylinositol 3-kinase-Akt pathway. *Circ. Res.*, *95*(3), 230–232.
96. Fujita, M., Asanuma, H., Hirata, A., Wakeno, M., Takahama, H., Sasaki, H., Kim, J., et al., (2007). Prolonged transient acidosis during early reperfusion contributes to the cardioprotective effects of postconditioning. *Am. J. Physiol. Heart Circ. Physiol.*, *292*, H2004–H2008.
97. Skyschally, A., Gent, S., Amanakis, G., Schulte, C., & Heusch, G., (2015). Across-species transfer of protection by remote ischemic preconditioning with species-specific myocardial signal transduction by reperfusion injury salvage kinase and survival activating factor enhancement pathways. *Circ. Res.*, *117*, 279–288.
98. Sivaraman, V., Mudalgiri, N. R., Salvo, C. D., Kolvekar, S., Heyward, M., Yap, J., Hausenloy, D. J., & Yellon, D. M., (2007). Postconditioning protects human atrial muscle through the activation of the RISK pathway. *Basic Res. Cardiol.*, *102*, 453–459.
99. Luo, W., Li, B., Lin, G., & Huang, R., (2007). Postconditioning in cardiac surgery for tetralogy of Fallot. *J. Thorac. Cardiovasc. Surg.*, *133*(5), 1373–1374. http://dx.doi.org/10.1016/j.jtcvs.2007.01.028 PMID: 17467465.
100. Luo, W., Li, B., Chen, R., Huang, R., & Lin, G., (2008). Effect of ischemic postconditioning in adult valve replacement. *Eur. J. Cardiothorac. Surg.*, *33*(2), 203–208. http://dx.doi.org/10.1016/j.ejcts.2007.11.010 PMID: 18078762.
101. Li, B., Chen, R., Huang, R., & Luo, W., (2009). Clinical benefit of cardiac ischemic postconditioning in corrections of tetralogy of Fallot. *Interact. Cardiovasc. Thorac. Surg.*, *8*(1), 17–21.
102. Staat, P., Rioufol, G., Piot, C., Cottin, Y., Tri, C. T., L'Huillier, I., Aupetit, J., et al., (2005). Postconditioning the human heart. *Circulation*, *112*(14), 2143–2148.
103. Xue, F., Yang, X., Zhang, B., Zhao, C., Song, J., Jiang, T., & Jiang, W., (2010). Postconditioning the human heart in percutaneous coronary intervention. *Clin. Cardiol.*, *33*(7), 439–444.
104. Garcia, S., Henry, T. D., Wang, Y. L., Chavez, I. J., Pedersen, W. R., Lesser, J. R., Shroff, G. R., et al., (2011). Long-term follow-up of patients undergoing postconditioning during ST-elevation myocardial infarction. *J. Cardiovasc. Transl. Res.*, *4*(1), 92–98. http://dx.doi.org/10.1007/s12265-010-9252-0 PMID: 21136310.
105. Thuny, F., Lairez, O., Roubille, F., Mewton, N., Rioufol, G., Sportouch, C., Sanchez, I., et al., (2012). Post-conditioning reduces infarct size and edema in patients with ST-segment elevation myocardial infarction. *J. Am. Coll. Cardiol.*, *59*(24), 2175–2181. http://dx.doi.org/10.1016/j.jacc.2012.03.026 PMID: 2267693.
106. Kerendi, F., Kin, H., & Halkos, M. E., (2005). Brief renal ischemia and reperfusion applied before coronary artery reperfusion reduce myocardial infarct size *via* endogenous activation of adenosine receptors. *Basic Res. Cardiol.*, *100*, 404–412.
107. Peng, B., Guo, Q. L., He, Z. J., Ye, Z., Yuan, Y., Wang, N., & Zhou, J., (2012). Remote ischemic postconditioning protects the brain from global cerebral ischemia/reperfusion injury by up-regulating endothelial nitric oxide synthase through the PI3K/Akt pathway. *Brain Res.*, *1445*, 92–102.
108. Qi, Z. F., Luo, Y. M., Liu, X. R., Wang, R., Zhao, H., Yan, F., Song, Z., et al., (2012). AKT/GSK3β-dependent autophagy contributes to the neuroprotection of limb remote ischemic postconditioning in the transient cerebral ischemic rat model. *CNS Neurosci. Ther.*, *18*(12), 965–973. http://dx.doi.org/10.1111/cns.12016 PMID: 23191937.

109. Gao, S., Zhu, Y., Li, H., Xia, Z., Wu, Q., Yao, S., Wang, T., & Yuan, S., (2016). Remote ischemic postconditioning protects against renal ischemia/reperfusion injury by activation of TLAK-cell-originated protein kinase (TOPK)/PTEN/Akt signaling pathway mediated anti-oxidation and anti-inflammation. *Int. Immunopharmacol.*, *38*, 395–401.
110. Danielisová, V., Némethová, M., Gottlieb, M., & Burda, J., (2006). The changes in endogenous antioxidant enzyme activity after postconditioning. *Cell. Mol. Neurobiol.*, *26*(7, 8), 1181–1191.
111. Wang, Q., Zhang, X., Ding, Q., Hu, B., Xie, Y., Li, X., Yang, Q., & Xiong, L., (2011). Limb remote postconditioning alleviates cerebral reperfusion injury through reactive oxygen species-mediated inhibition of delta protein kinase C in rats. *Anesth. Analg.*, *113*(5), 1180–1187.
112. Zhong, H., Gao, Z., Chen, M., Zhao, J., Wang, F., Li, L., Dong, H., et al., (2013). Cardioprotective effect of remote ischemic postconditioning on children undergoing cardiac surgery: A randomized controlled trial. *Paediatr. Anaesth.*, *23*(8), 726–733.
113. Cao, B., Wang, H., Zhang, C., Xia, M., & Yang, X., (2018). Remote ischemic postconditioning (RIPC) of the upper arm results in protection from cardiac ischemia-reperfusion injury following primary percutaneous coronary intervention (PCI) for acute ST-segment elevation myocardial infarction (STEMI). *Med. Sci. Monit.*, *24*, 1017–1026.
114. Bolli, R., Becker, L., Gross, G., Mentzer, Jr. R., Balshaw, D., & Lathrop, D. A., (2004). Myocardial protection at a crossroads: The need for translation into clinical therapy. *Circ. Res.*, *95*, 125–134.
115. Kloner, R. A., & Rezkalla, S. H., (2004). Cardiac protection during acute myocardial infarction: Where do we stand in 2004? *J. Am. Coll. Cardiol.*, *44*, 276–286.
116. Downey, J. M., & Cohen, M. V., (2009). Why do we still not have cardioprotective drugs? *Circ. J.*, *73*, 1171–1177.
117. Ferdinandy, P., Hausenloy, D. J., Heusch, G., Baxter, G. F., & Schulz, R., (2014). Interaction of risk factors, comorbidities, and comedications with ischemia/ reperfusion injury and cardioprotection by preconditioning, postconditioning, and remote conditioning. *Pharmacol. Rev.*, *66*, 1142–1174.
118. Lecour, S., Bøtker, H. E., Condorelli, G., Davidson, S. M., Garcia-Dorado, D., Engel, F. B., Ovize, M., et al., (2014). ESC working group cellular biology of the heart: Position paper: Improving the preclinical assessment of novel cardioprotective therapies. *Cardiovasc. Res.*, *104*, 399–411.
119. Kloner, R. A., & Schwartz, L. L., (2011). State of the science of cardioprotection: Challenges and opportunities—Proceedings of the 2010 NHLBI workshop on cardioprotection. *J. Cardiovasc. Pharmacol. Ther.*, *16*, 223–232.
120. Schwartz, L. L., Kloner, R. A., Arai, A. E., Baines, C. P., Bolli, R., Gottlieb, R. A., Heusch, G., et al., (2011). New horizons in cardioprotection: Recommendations from the 2010 national heart, lung, and blood institute workshop. *Circulation*, *124*, 1172–1179.

CHAPTER 16

Involvement of Platelets in Myocardial Ischemia and Reperfusion Injury

GEETA DESWAL, BENU CHAUDHARY, PREETI ARYA, DEEPAK SINGLA, and AJMER SINGH GREWAL

Guru Gobind Singh College of Pharmacy, Yamunanagar, Haryana, India

ABSTRACT

Ischemia of heart tissue occurs when a coronary artery is blocked, resulting in myocardial infarction. Long-lasting oxygen deficiency can cause tissue necrosis which can lead to heart ailment as well as death. As a result, re-canalizing the negatively affected vessel and restoring coronary blood flow is critical for effective clinical outcomes. Surprisingly, despite the reconstruction of blood circulation, sudden reperfusion causes cellular damage, thereby exacerbating the initial ischemic damage. Myocardial ischemia injury is a complex activity that involves inflammatory cells as well as signals such as inflammatory cytokines, as well as various harmful mechanisms (such as the generation of oxygen species and enhanced calcium ions). Platelets were discovered to be involved in the inflammatory response associated with ischemia/reperfusion injury several years ago, but the underlying mechanisms are still unknown. This chapter focuses on the contribution of platelets to the complex picture of myocardial ischemia injury, platelet exterior receptor up-regulation, immunomodulatory mediators release contribution to the aggravation of myocardial reperfusion injury, and successive cardiac damage. Furthermore, evidence for the cardio-protective roles of various platelet factors during ischemia injury will be addressed.

16.1 INTRODUCTION

Platelets that have been activated have a significant function in cardiovascular disorders, including myocardial infarction (MI). They promote coronary artery blockage by participating in thrombus development and microembolization. Furthermore, an inflammatory factor involving platelets has been identified in numerous cardiovascular illnesses, including MI [1], heart failure [2], and coronary heart disease [3]. Platelets have been proven to have a cardioprotective role in addition to their negative impact on MI. The initial publications

Ischemic Injury: A Molecular Insight. Thakur Gurjeet Singh & Heena Khan (Eds.)

on platelet protection in MI were published in the 1990s, and they mostly emphasized on protective influences of chemicals released by platelets on the stimulation and agglomeration [4–6]. Platelets' cardioprotective characteristics, on the other hand, are still a mystery. For a regular cardiac function to return, ischemic myocardium must be reperfused as immediately as possible. In myocardial ischemia, the major goal of reperfusion therapy seems to be to resume the normal epicardial flow of blood in an infarct-based coronary artery. The restoration of the flow of blood, on the other hand, can hasten the demise of irreversibly injured myocytes, contributing to the development of myocardial infarction and dysfunction [7–10]. Coronary microcirculation Furthermore, even when the epicardial flow is normal, myocardial perfusion could be inadequate, as evidenced by no-reflow myocardial and microcirculation levels [11]. Patients with regular epicardial flow but poor microvascular permeability are more likely to develop arrhythmia, heart failure, and death in many cases [12]. Mechanisms underlying principles of platelet activity in reperfusion may improve therapy options for patients with coronary-related diseases including such myocardial infarction, and improve myocardial recovery. The physiological and pathological involvement of platelets in the injury of cardiac reperfusion by looking at:

- Mechanisms that are involved in the formation of thrombus as well as sites of micro-embolization of the atherosclerotic plaques in the arteries of epicardial coronary.
- Interactions of platelets with leukocytes and endothelium and the implications for microcirculation, inflammation as well as therapy.

16.2 CARDIOPROTECTIVE EFFECT OF PLATELETS

16.2.1 PLATELETS WITH A DIRECT CARDIOPROTECTIVE EFFECT

Platelets have been implicated in the repair of cardiac muscle following a heart attack. αCXCL12 (derived factors of stromal cell-1α (SDF-1α)) is found in platelets and is released when they are activated. The CXCL12 is one of the chemokines that can trigger the migration of cells of progenitor to the injury site, resulting in the healing of tissues and organs. CXCL12 attracts CD34+ cells as well as influences their differentiation and proliferation into foam cells, macrophages, and endothelial cells [13]. Higher expression of platelet CXCL12 has been linked to improved cardiac function and smaller infarct size in MI patients [14]. zSDF-1α expression was shown to be highest in the zone of peri-infarction. The transforming form of growth factor expression beta 1 (TGF-β1) on of platelet surface is linked with the expression of receptors, i.e., SDF-1α whilst CXCR7 and CXCR4 in patients with symptomatic disease of coronary artery (CAD). *In vitro* platelet activation with the recombinant SDF-1α boosted the expression of TGF-β1 considerably [15, 16]. During platelet activation, TGF-β1 is secreted. It plays a role in inflammatory response suppression [17] remodeling and heart repair [18]. In individuals with an acute coronary syndrome (ACS), TGF-β1 expression on the platelet surface was shown to be more than in those with stable CAD reduced TGF-β1 expression of platelet was linked to higher mortality and the occurrence of ST-segment elevated MI (STEMI) [19]. Platelets were found to alleviate cardiac dysfunction caused by I/R in isolated rat heart studies. The protective impact of platelets was visible even when the count of platelets was substantially lower (105 cells/mL) than in normal settings involving 5–10 ×108 cells/mL in the rats. Platelets have a considerable protective capacity for cardiomyocytes against ischemia, according to research [4]. The protective effect of platelets protective upon the heart during IRI is due to chemicals generated by platelets throughout aggregation and activation. Many investigations have revealed that supernatant generated after the aggregation of platelet has cardioprotective characteristics when injected into rat hearts. Platelets' cardioprotective impact is mostly due to the production of adenine nucleotides, serotonin, and TXA2 which cause endothelial NO to be released, which has a cardioprotective impact [6, 20, 21]. As a cardioprotective impact of platelets is there without regarding the contribution of the dense granules of platelet, the chemicals secreted out from alpha granules are partially attributed to them [20]. TGF-1, which is secreted out from alpha granules, has been shown to preserve the rat myocardium and prevent cardiomyocyte apoptosis caused by reoxygenation and hypoxia [21]. SDF1-α, which is produced from the granules of alpha, also has an effect of cardioprotective, delaying the death of cardiomyocytes [20]. CXCR4 and TGFRβ1 receptors for SDF and TGF, respectively are present on cardiomyocytes [22, 23]. During ischemia, SDF1-α and TGF-β1 promote the protein kinase C (PKC) signaling in cardiomyocytes as a mechanism of prosurvival [20] (Figure 16.1).

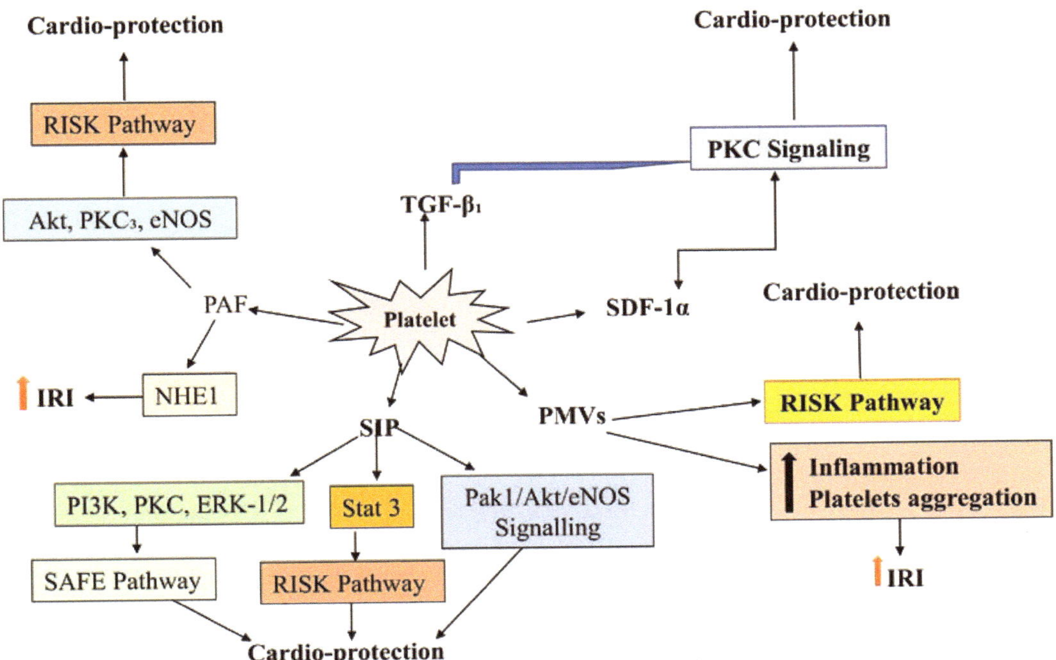

SAFE: survivor activating factor enhancement, Akt: protein kinase B, ERK: extracellular signal-regulated kinase, Pak1: P21-activated kinase; PKC: protein kinase C, PI3 K: phosphatidylinositol 3-kinase, Stat3: signal transducer and activator of transcription 3, eNOS: endothelial nitric oxide synthase, NHE1: Na(+)/H(+) exchanger isoform 1, S1P: sphinosine-1-phosphate, SDF-1α: stromal cell derived factor-1 alpha, TGF-β1: transforming growth factor beta 1.

FIGURE 16.1 During ischemia, protein kinase C signaling in cardiomyocytes acts as a prosurvival strategy.

Platelets have a cardioprotective effect in the case of ischemia-reperfusion damage (IRI). S1P can protect the heart by stimulating the RISK as well as SAFE pathways, which include the kinases PKC, PI3 K, Stat3, and ERK-1/2, and S1P also activates Pak1/NOS3/Akt signaling, which results in cardio-protection. The cardioprotective impact of TGF-β1 and SDF1α is due to the stimulation of the signaling PKC. PMVs and sPAF both affect the cardiac IRI in different ways. This effect is dependent on the PAF level in the context of PAF. RISK: salvage kinases for reperfusion injury. The sphingosine-1-phosphate (S1P) has a cardioprotective impact on the myocardium in MI patients. Sphingosine kinase, found in platelets, transforms membrane sphingosine to S1P. The S1P is a substance that is held in platelets as well as released in large amounts during the activation of platelet, formation of thrombus, and processes of inflammation such as MI [24]. The S1P acts on megakaryocytes to contribute to normal thrombopoiesis, whilst its influence on the platelets is reflected from both pro- and antiaggregatory impacts via G-protein coupled receptors (GPCRs) [25, 26]. The cardioprotective impact of S1P has indeed been confirmed in investigations on individual cardiomyocytes as well as the innervated heart of rats *ex vivo* [27, 28]. After I/R, there were reductions in myocardial infarct size and an increase in viability of cardiomyocytes in hypoxic settings. The favorable benefits of ischemia pretreatments (IPC), as well as postconditioning of undergoing I/R of the heart, were similarly mediated by S1P [29, 30]. The S1P receptors are found in the heart in three forms: S1P3, S1P1, and S1P2. Furthermore, it is thought that S1P limits damages all through I/R by activating the factor of survivor activating enhancer (SAFE) pathways, protein kinase B (Akt), transcription 3 (Stat3) kinases activators, reperfusion injury salvage (RISK) kinases, signal transducer and by S1P3 and S1P2 stimulating with receptors of G-protein coupled receptors [30, 31]. S1P also induces cardioprotection by activating P21-activated kinase (Pak1)/Akt/endothelial nitric oxide synthase (eNOS) signaling [32]. S1P as well as platelets have indeed been demonstrated to be critical components of cangrelor-derived cardio-protection [33], one of the P2Y12 receptor inhibitors.

16.2.2 PLATELETS WITH AN INDIRECT CARDIOPROTECTIVE EFFECT

Leukocytes, endothelial cells, and platelets create PAF, which is a phosphoglyceride [34]. PAF can work on a range of types of cells, such as platelets, endothelial cells, and cardiomyocytes as an autocrine and paracrine and mediator. It works by interacting with particular receptors on the cardiomyocytes surfaces (PAFRs) [35]. PAF can have varying effects on cardiac IRI depending on its levels. PAF at low concentrations (in the range of rapmol/L) has little effect on cardiac contractility, but it can exhibit IPC level of cardio-protection by stimulating RISK signaling kinases such as Akt, eNOS, and PKC [34]. High PAF, i.e., 1–10 nmol/L dosages had a deleterious impact on coronary and cardiac functions, as well as a severe arrhythmogenic effect [35]. After reperfusion, the application of antagonists of the PAF receptor can significantly reduce the size of the infarct, by not only suppressing the receptor of PAR directly by having an antiplatelet impact [35]. PAF appears to possess a clear cardio-depressive impact at high doses. PMVs appear to provide an indirect impact on cardioprotective actions. Table 16.1 summarizes the mechanisms of cardio-protection induced by platelet factors.

TABLE 16.1 Cardioprotection Induced by Platelet Agents

Agent	Receptor of Cardiomyocytes	IRI Model	Cardio-Protection Mechanism	Influence of Proposed Cardio-Protection	References
SDF1α	CXCR7, CXCR4	Mice ventricular cardiomyocytes.	PKC signaling activation.	Direct	[23]
TGF-β1	TGFβR1	Mice ventricular cardiomyocytes.	PKC signaling activation.	Direct	[23]
S1P	S1P3, S1P2, S1P1	Rat ventricular cardiomyocytes, mice hearts.	Receptor (S1P3, S1P2) induced Pak1/Akt/eNOS signaling activation RISK as well as SAFE pathway stimulation.	Direct	[25, 36]
PAF**	PAFRs	Rat hearts	PI3 K and PKC stimulation, similar to IPC.	Indirect	[29]
PMVs	NI	Rat aortic ring model, rat heart.	RISK pathway stimulation.	Indirect	[39]

Abbreviations: RISK: Reperfusion injury salvage kinases; IRI: Ischemia/reperfusion injury; Akt: Protein kinase B; S1P: Sphingosine-1-phosphate; SDF-1α: Stromal cell-derived factor-1 alpha; SAFE: Survivor activating factor enhancement; Pak1: P21-activated kinase; eNOS: Endothelial nitric oxide synthase; PKC: Protein kinase C; PI3 K: Phosphatidylinositol 3-kinase; IPC: Ischemia preconditioning; TGF-β1: transforming growth factor-beta 1; and NI: Not identifiable.

16.3 ANTIPLATELET THERAPY'S CARDIOPROTECTIVE EFFECTS

Cardio-protection was achieved by using platelet component antagonists with cardio-depressant impacts or inhibiting particular receptors. Anti-GPVI antibodies suppressed platelet reactivity to collagen [36] significant amounts of PAF are generated during the IRI, which might have a deleterious impact on heart and coronary function. In the rabbit and sheep model, however, using an antagonist of the PAF receptor before the reperfusion reduces the size of infarct extent and reperfusion injury [37, 38]. Clopidogrel and ticagrelor (P2Y12 receptor), P-selectin inhibitors, Aspirin (COX), GPVI, GPIb, and GPIIb/IIIa are some of the existing platelet-related medicines used to minimize IRI. Except for their antiaggregatory actions and indirectly shielding of the cardiac system against IRI, these medicines have been shown to have a clear cardioprotective impact [39–41]. Aspirin has indeed been demonstrated to limit the MI risk and act as an anti-inflammatory [40]. In patients experiencing atherosclerosis, aspirin reduced dysfunction of endothelial and altered acetylcholine-derived peripheral vasodilation, perhaps via inhibiting COX-dependent vasoconstrictors [41]. Aspirin, on the other hand, inhibited AA-derived platelet aggregation but did not affect the myocardial I/R of rabbit model infarction [42]. P2Y12 inhibitors, such as ticagrelor and clopidogrel have been shown to have a cardioprotective effect against the IRI [43–45]. These medications' protective benefits were formerly thought to be solely due to their antiaggregatory properties. In a model of rabbit myocardial I/R, it was observed that found that clopidogrel and cangrelor safeguarded the hearts from infarction. Its antiaggregatory qualities did not influence this protective effect. The researchers hypothesized that cangrelor's protective effect was due to a similar mechanism as followed by the ischemic postconditioning (IPOC). The protective effect of cangrelor was affected by individual inhibiting signaling elements such as adenosine receptors A2B, signaling of redox, Akt, and mitochondrial ATP sensitive channels of potassium (KATP) [46]. It was found that direct clopidogrel action reduced infarct size in individuals with STEMI in a retrospective study [47]. Because cangrelor did not decrease the size of infarct in rats with thrombocytopenic after half an hour of ischemia and reperfusion for two hours, interactions of platelets seem to be required to start the protective effect of cangrelor [43]. It was found that platelet contact is required for the protection of ticagrelor in the hearts of rats. Furthermore, ticagrelor can cause platelet release of adenosine and S1P, which results *in vivo* cardiac protection. Adenosine and S1P were found in the highest amounts in the ticagrelor-protected heart of rats. Furthermore, in rat hearts, S1P and adenosine both contributed to the activation pathways of RISK [45]. Ticagrelor also prevented endothelial cells from succumbing to hypoxia-induced apoptosis. The adenosine pathway of signaling was linked to ticagrelor's protective action [47].

16.4 CONCLUSION

Although conclusively determining the influence of platelets on the heart in IRI and MI is difficult, there is accumulating evidence that platelets have a beneficial effect on the heart, initiating specific cardioprotective processes. The recently reported effects of ticagrelor and cangrelor which are irrespective of antiaggregatory activities offer strong support for platelet involvement in cardiac IRI protection. The understanding of platelet protective effects in IRI and MI could result in ineffective treatment. Furthermore, the utilization of PMVs as the predictive and prognostic indicators, as well as alterations in the platelets of the proteome, may throw fresh light on MI diagnoses. The exosomes of platelets could be valuable biomarkers for MI in individuals with comorbidities. However, more study is needed to uncover potential novel platelet markers and determine their therapeutic value.

KEYWORDS

- **blood circulation**
- **cardioprotective effect**
- **cytokines**
- **inflammatory cells**
- **myocardial ischemia**
- **oxygen species**
- **platelets**
- **therapeutic value**

REFERENCES

1. Kossmann, H., Rischpler, C., Hanus, F., et al., (2019). Monocyte-platelet aggregates affect local inflammation in patients with acute myocardial infarction. *International Journal of Cardiology, 287*, 7–12.
2. Stumpf, C., Lehner, C., Eskafi, S., et al., (2003). Enhanced levels of CD154 (CD40 ligand) on platelets in patients with chronic heart failure. *European Journal of Heart Failure, 5*(5), 629–637.
3. Danesh, J., Wheeler, J. G., Hirschfield, G. M., et al., (2004). C-reactive protein and other circulating markers of inflammation in the prediction of coronary heart disease. *New England Journal of Medicine, 350*(14), 1387–1397.
4. Yang, B., Mehta, P., & Mehta, J. L., (1998). Platelet-mediated cardioprotective effect against ischemia-reperfusion injury in isolated rat hearts: Role of platelet number and contribution of supernatant of aggregated platelets. *Journal of Cardiovascular Pharmacology and Therapeutics, 3*(1), 23–28.
5. Yang, B. C., Zander, D. S., & Mehta, J. L., (1999). Hypoxia-reoxygenation-induced apoptosis in cultured adult rat myocytes and the protective effect of platelets and transforming growth factor-β1. *Journal of Pharmacology and Experimental Therapeutics, 291*, 733–738.
6. Yang, B. C., & Mehta, J. L., (1994). Platelet-derived adenosine contributes to the cardioprotective effects of platelets against ischemia-reperfusion injury in isolated rat heart. *Journal of Cardiovascular Pharmacology, 24*(5), 779–785.
7. Becker, L. C., & Ambrosio, G., (1987). Myocardial consequences of reperfusion. *Prog. Cardiovasc Dis., 44.*
8. Maxwell, S. R., & Lip, G. Y., (1997). Reperfusion injury: A review of the pathophysiology clinical manifestations and therapeutic options. *Int. J. Cardiol.*, 111.
9. Piper, H. M., Meuter, K., & Scha¨fer, C., (2003). Cellular mechanisms of ischemia– reperfusion injury. *Ann. Thorac. Surg., 8.*
10. Ito, H., (2001). No reflow phenomenon in coronary heart disease. J. *Cardiol., 42.*
11. Gassler, J. P., & Topol, E. J., (1999). Reperfusion revisited: Beyond TIMI 3 flow. *Clin. Cardiol., 9.*
12. Cura, F. A., L'Allier, P. L., Kapadia, S. R., et al., (2001). GUSTO IIb and RAPPORT Investigators. Predictors and prognosis of suboptimal coronary blood flow after primary coronary angioplasty in patients with acute myocardial infarction. *Am. J. Cardiol., 8.*
13. Chatterjee, M. Z., Huang, W., Zhang, et al., (2011). Distinct platelet packaging, release, and surface expression of proangiogenic and antiangiogenic factors on different platelet stimuli. *Blood, 117*(14), 3907–3911.
14. Geisler, T., Fekecs, L., Wurster, T., et al., (2012). Association of platelet-SDF-1 with hemodynamic function and infarct size using cardiac MR in patients with AMI. *European Journal of Radiology, 81*(4), e486–e490.
15. Abbott, J. D., Huang, Y., Liu, D., Hickey, R., Krause, D. S., & Giordano, F. J., (2004). Stromal cell-derived factor-1α plays a critical role in stem cell recruitment to the heart after myocardial infarction but is not sufficient to induce homing in the absence of injury. *Circulation, 110*(21), 3300–3305.
16. Rath, D., Chatterjee, M., Holtkamp, A., et al., (2016). Evidence of an interaction between TGF-β1 and the SDF-1/CXCR4/CXCR7 axis in human platelets. *Rombosis Research, 144*, 79–84.
17. Smith, W. B., Noack, L., Khew-Goodall, Y., Isenmann, S., Vadas, M. A., & Gamble, J. R., (1996). Transforming growth factor beta 1 inhibits the production of IL-8 and the transmigration of neutrophils through activated endothelium. *Journal of Immunology (Baltimore, Md 1950)., 157*, 360–368.
18. Hao, J., Ju, H., Zhao, S., Junaid, A., Scammell-La, F. T., & Dixon, I. M. C., (1999). Elevation of expression of Smads 2, 3, and 4, decorin and TGF-β in the chronic phase of myocardial infarct scar healing. *Journal of Molecular and Cellular Cardiology, 31*(3), 667–678.
19. Rath, D., Chatterjee, M., Müller, I., et al., (2014). Platelet expression of transforming growth factor beta 1 is enhanced and associated with cardiovascular prognosis in patients with acute coronary syndrome. *Atherosclerosis, 237*(2), 754–759.
20. Yang, B. C., Virmani, R., Nichols, W. W., & Mehta, J. L., (1993). Platelets protect against myocardial dysfunction and injury induced by ischemia and reperfusion in isolated rat hearts. *Circulation Research., 72*(6), 1181–1190.
21. Koitabashi, N., Danner, T., Zaiman, A. L., et al., (2011). Pivotal role of cardiomyocyte TGF-β signaling in the murine pathological response to sustained pressure overload. *Journal of Clinical Investigation, 121*(6), 2301–2312.

22. Davidson, S. M., Andreadou, I., Barile, L., et al., (2019). Circulating blood cells and extracellular vesicles in acute cardioprotection. *Cardiovascular Research, 115*(7), 1156–1166.
23. Vito, C. D., Hadi, L. A., Navone, S. E., et al., (2016). Platelet-derived sphingosine-1-phosphate and inflammation: From basic mechanisms to clinical implications. *Platelets, 27*(5), 393–401.
24. Jin, Z. Q., Zhou, H. Z., Zhu, P., et al., (2002). Cardioprotection mediated by sphingosine-1-phosphate and ganglioside GM- 1 in wild-type and PKCε knockout mouse hearts. *American Journal of Physiology-Heart and Circulatory Physiology, 282*(6), H1970–H1977.
25. Jin, Z. Q., Goetzl, E. J., & Karliner, J. S., (2004). Sphingosine kinase activation mediates ischemic preconditioning in murine heart. *Circulation, 110*(14), 1980–1989.
26. Gupta, A., Khan, H., Kaur, A., & Singh, T. G., (2021). Novel targets explored in the treatment of alcohol withdrawal syndrome. *CNS & Neurological Disorders–Drug Targets, 20*(2), 158–173.
27. Vessey, D. A., Li, L., Honbo, N., & Karliner, J. S., (2009). Sphingosine 1-phosphate is an important endogenous cardioprotectant released by ischemic pre- and postconditioning. *American Journal of Physiology-Heart and Circulatory Physiology, 297*(4), H1429–H1435.
28. Means, C. K., Xiao, C. Y., Li, Z., et al., (2007). Sphingosine 1-phosphate S1P2 and S1P3 receptor-mediated Akt activation protects against *in vivo* myocardial ischemia-reperfusion injury. *American Journal of Physiology-Heart and Circulatory Physiology, 292*(6), H2944–H2951.
29. Egom, E. E. A., Mohamed, T. M. A., Mamas, M. A., et al., (2011). Activation of Pak1/Akt/eNOS signaling following sphingosine- 1-phosphate release as part of a mechanism protecting cardiomyocytes against ischemic cell injury. *American Journal of Physiology-Heart and Circulatory Physiology, 301*(4), H1487–H1495.
30. Cohen, M. V., Yang, X. M., White, J., Yellon, D. M., Bell, R. M., & Downey, J. M., (2016). Cangrelor-mediated cardioprotection requires platelets and sphingosine phosphorylation. *Cardiovascular Drugs and Therapy, 30*(2), 229–232.
31. Penna, C., Bassino, E., & Alloatti, G., (2011). Platelet activating factor: The good and the bad in the ischemic/reperfused heart. *Experimental Biology and Medicine, 236*(4), 390–401.
32. Penna, C., Alloatti, G., Cappello, S., et al., (2005). Platelet-activating factor induces cardioprotection in isolated rat heart skin to ischemic preconditioning: Role of phosphoinositide 3-kinase and protein kinase C activation. *American Journal of Physiology-Heart and Circulatory Physiology, 288*(5), H2512–H2520.
33. Montrucchio, G., Alloatti, G., Mariano, F., et al., (1990). Role of platelet-activating factor in the reperfusion injury of rabbit ischemic heart. *The American Journal of Pathology, 137*, 71–83.
34. Khan, H., Gupta, A., Singh, T. G., & Kaur, A., (2021). Mechanistic insight on the role of leukotriene receptors in ischemic–reperfusion injury. *Pharmacology Reports, 73*(5), 1240–1254.
35. Brill, A., Dashevsky, O., Rivo, J., Gozal, Y., & Varon, D., (2005). Platelet-derived microparticles induce angiogenesis and stimulate post-ischemic revascularization. *Cardiovascular Research, 67*(1), 30–38.
36. Penna, C., Aragno, M., Cento, A. S., et al., (2020). Ticagrelor conditioning effects are not additive to cardioprotection induced by direct NLRP3 inflammasome inhibition: Role of RISK, NLRP3, and redox cascades. *Oxidative Medicine and Cellular Longevity*, 970954.
37. Baigent, C., Sudlow, C., Collins, R., & Peto, R., (2002). Collaborative meta-analysis of randomized trials of antiplatelet therapy for prevention of death, myocardial infarction, and stroke in high-risk patients. *The BMJ, 324*, 71–86.
38. Abergel, E., & Nikolsky, E. T., (2010). An investigational oral antiplatelet treatment for reduction of major adverse cardiac events in patients with acute coronary syndrome. *Vascular Health and Risk Management, 6*, 963–977.
39. Khan, H., Bangar, A., Grewal, A. K., Bansal, P., & Singh, T. G., (2022b). Caspase-mediated regulation of the distinct signaling pathways and mechanisms in neuronal survival. *International Immunopharmacology, 110*, 108951.
40. Ridker, P. M., Cushman, M., Stampfer, M. J., Tracy, R. P., & Hennekens, C. H., (1997). Inflammation, aspirin, and the risk of cardiovascular disease in apparently healthy men. *New England Journal of Medicine, 336*(14), 973–979.
41. Husain, S., Andrews, N. P., Mulcahy, D., Panza, J. A., & Quyyumi, A. A., (1998). Aspirin improves endothelial dysfunction in atherosclerosis. *Circulation, 97*(8), 716–720.

42. Yang, X. M., Liu, Y., Cui, L., et al., (2013). Platelet P2Y12 blockers confer direct postconditioning-like protection in reperfused rabbit hearts. *Journal of Cardiovascular Pharmacology and Therapeutics, 18*(3),251–262.
43. Vilahur, G., Guti´errez, M., Casani, L., et al., (2018). P2Y12 antagonists and cardiac repair post-myocardial infarction: Global and regional heart function analysis and molecular assessments in pigs. *Cardiovascular Research, 114*(14), 1860–1870.
44. Montalescot, G., Wiviott, S. D., Braunwald, E., et al., (2009). Prasugrel compared with clopidogrel in patients undergoing percutaneous coronary intervention for ST-elevation myocardial infarction (TRITON-TIMI 38): Double-blind, randomized controlled trial. *The Lancet, 373*(9665), 723–731.
45. Wang, K., Zhou, X., Huang, Y., et al., (2010). Adjunctive treatment with ticagrelor, but not clopidogrel, added to tPA enables sustained coronary artery recanalization with recovery of myocardium perfusion in a canine coronary thrombosis model. *Thrombosis and Haemostasis, 104*, 609–617.
46. Roubille, F., Lairez, O., Mewton, N., et al., (2012). Cardioprotection by clopidogrel in acute ST-elevated myocardial infarction patients: A retrospective analysis. *Basic Research in Cardiology, 107*(4).
47. Feliu, C., Peyret, H., Brassart-Pasco, S., et al., (2020). Ticagrelor prevents endothelial cell apoptosis through the adenosine signaling pathway in the early stages of hypoxia. *Biomolecules, 10*(5), 740.

CHAPTER 17

Recent Advancements in Strategies for the Treatment of Cerebral Ischemia

UDITI HANDA, KAJAL NAGPAL, PEEYUSH KAUSHIK, and AJMER SINGH GREWAL

Guru Gobind Singh College of Pharmacy, Yamunanagar, Haryana, India

ABSTRACT

A stroke is the most important reason for disability-adjusted lifespan growth and a major cause of well-being problems. Stroke, otherwise called cerebral ischemia or a cerebral vascular mishap, is the abrupt passing of synapses brought about by a lacking blood stream, which lessens the accessibility of glucose and oxygen to synapses. Glutamate discharge, glutamate receptor actuation, excitotoxicity, Ca^{2+}-flood in cells, mitochondrial brokenness, initiation of numerous intracellular chemicals, free extreme creation, nitric oxide creation, aggravation, corruption, and apoptosis are a portion of the sub-atomic occasions that can be focused on for stroke. This chapter will deliver a brief outline of the present knowledge about stroke, epidemiology, and pathophysiology of stroke, the targeting of definite pathways and animal as well as human trials, and it will also disclose the emergent targets for newer therapeutic approaches for stroke under trials.

17.1 INTRODUCTION

The WHO recommends that stroke is the leading cause of death and impairment which develops rapidly in low-class and middle-class countries including India. Globally, the clinical indication has increased which resulted in the disturbance of cerebral function [1]. Although it is a nerve-related ailment that causes the obstruction of blood vessels or there is clot formation in the brain. Due to the disturbance of arteries in the brain outcomes in the rapid death of brain cells. Depression and dementia are the main symptoms caused due to stroke. In 2018, the International Classification of Disease II (ICD-II) reported that various diseases of blood vessels induce stroke. This disease is based upon ICD-II coding rationale, clinical evidence of stroke patients, distribution of severity, and disease associated with the burden of stroke [2].

Ischemic Injury: A Molecular Insight. Thakur Gurjeet Singh & Heena Khan (Eds.)

17.2 EPIDEMIOLOGY

The world's second principal reason for mortality and disability is stroke. Stroke affects an estimated 13.7 million people. Strokes caused 5.5 million deaths worldwide each year. About 87% of stroke is ischemic stroke, which is defined as the rapid cessation of blood circulation to a specific area of the brain. The worldwide prevalence of stroke increased between 1990 and 2016. Although the primary hemorrhage (bleeding occurs in the first 24 hours) is more common as compared to the secondary hemorrhage (bleeding occurs after 24 hours) [3, 4]. Their incidences doubled in poorer and middle-income nations between 1990 and 2016 but dropped by 42% in high-income nations over the same period. The Global Burden of Disease Study (GBD) revealed that age-specific stroke, gender-specific stroke, and geographic as well as socioeconomic variation are the main sources to effect the stroke and increased over time [4].

17.2.1 AGE-SPECIFIC STROKE

The occurrence of stroke is depending upon its age because if the age factor is increased, the stroke-related problem is also increased or after 55 years of age, it will be doubled or shows a serious effect. According to a stroke study conducted between 1990 and 2016, for people whose ages in between 20 and 54 years, their effect augmented from 12.9% to 18.6% in entire cases [4–6]. The greatest incidence of stroke has been documented in China, which is around 331–378 people per 1,00,000 life years. Eastern Europe has the 2nd uppermost incidence rate of stroke (181–218 per 1,00,000 life years). Latin America has the lowermost incidence rate (85–100 per 1,00,000 life years) [4].

17.2.2 GENDER-SPECIFIC STROKE

It has been observed that in younger age women its incidence rate is higher whereas in older age men the incidence rate is slightly increased. A high risk of stroke occurs in females due to pregnancy-associated factors like pre-eclampsia, contraceptive use, hormonal changes, migraine, and aura. The risk of stroke rises by 20% in women over the age of 75 who have arterial fibrillation. According to the National Institute of Health Stroke Scale, there have been several types of strokes, including small, large, medium, and severe strokes. The average size of strokes was tested in 10 women and 8.2 men in men. In men's brains, infractions and intracerebral hemorrhage are more common, but a very severe stroke cardioembolic are more prevalent in women. Therefore, the mortality rate is more in the case of women [6–8]. But the common causes of stroke in the case of men are myocardial infraction, alcohol consumption, smoking, and arterial ailments [9].

17.2.3 GEOGRAPHIC VARIATION

According to a Global investigation, the occurrence of stroke and other risk factors are evaluated by population analysis, behavior activity, physical characteristics, medical records, and laboratory reports [10]. On the other hand, emerging nations such as North Eastern China have a variety of diseases, including high blood pressure, which is an important risk factor for stroke, particularly ischemic stroke [11]. However, hypertension is a serious cause not just in China, but also in the United States. It involved a variety of risk factors such as poor physical activity, unhealthy eating habits, and alcohol consumption [12].

17.2.4 SOCIOECONOMIC VARIATION

Stroke and socio-economic change depend on hospital amenities and post-stroke attention in low-income people [13]. In the United States, a case study showed that the better treatment of stroke is responsible for a high class of financial status [14]. In China, a study revealed that the lack of health insurance and the low-income population act as a major reason for secondary stroke attacks [15]. However, thrombolysis is the basic treatment in the case of the Scottish healthcare system that provides economic status for patients [16].

17.3 PATHOPHYSIOLOGY OF STROKE

The study of neurovascular anatomy is essential for understanding the clinical manifestations of stroke. Two internal carotid arteries are located anteriorly in the brain, while two vertebral arteries are located posteriorly. Ischemic stroke is instigated via a deficiency of blood and oxygen delivery to the brain. Hemorrhagic stroke, alternatively, is instigated by bleeding in different areas of the brain. Around 85% of patients account for ischemic stroke and the rest of the patients are responsible for intracerebral bleeding. In the case of ischemic stroke, thrombic and embolic conditions are produced in the brain [17]. Whereas, hemorrhagic stroke is responsible for 10–15% of all stroke cases and its mortality rate is high. In this condition, blood vessels of the brain are ruptured due to internal injury in the brain or damage to brain tissues [18]. It can be subdivided into intracerebral and subarachnoid-type hemorrhages. In intracerebral type hemorrhage, blood vessels rupture leading to an anomalous build-up of blood in the brain. In the case of subarachnoid type hemorrhage, head injury or cerebral aneurysm condition occurs due to accumulation of blood in the subarachnoid space of the brain (Figure 17.1) [19, 20].

17.4 RISK FACTORS

As discussed previously, the risk of stroke upsurges with oldness and doubles above the age of 55 years both in males as well as females. The risk of stroke attack is augmented more once a person is having a prevailing health disorder such as high blood pressure, coronary artery ailment, or hyperlipidemia. Almost 60% of the total stroke cases were reported in persons having a past of "transient ischemic attack." The major risk factors crucial for stroke are classified into modifiable and non-modifiable types (Figure 17.2) [20].

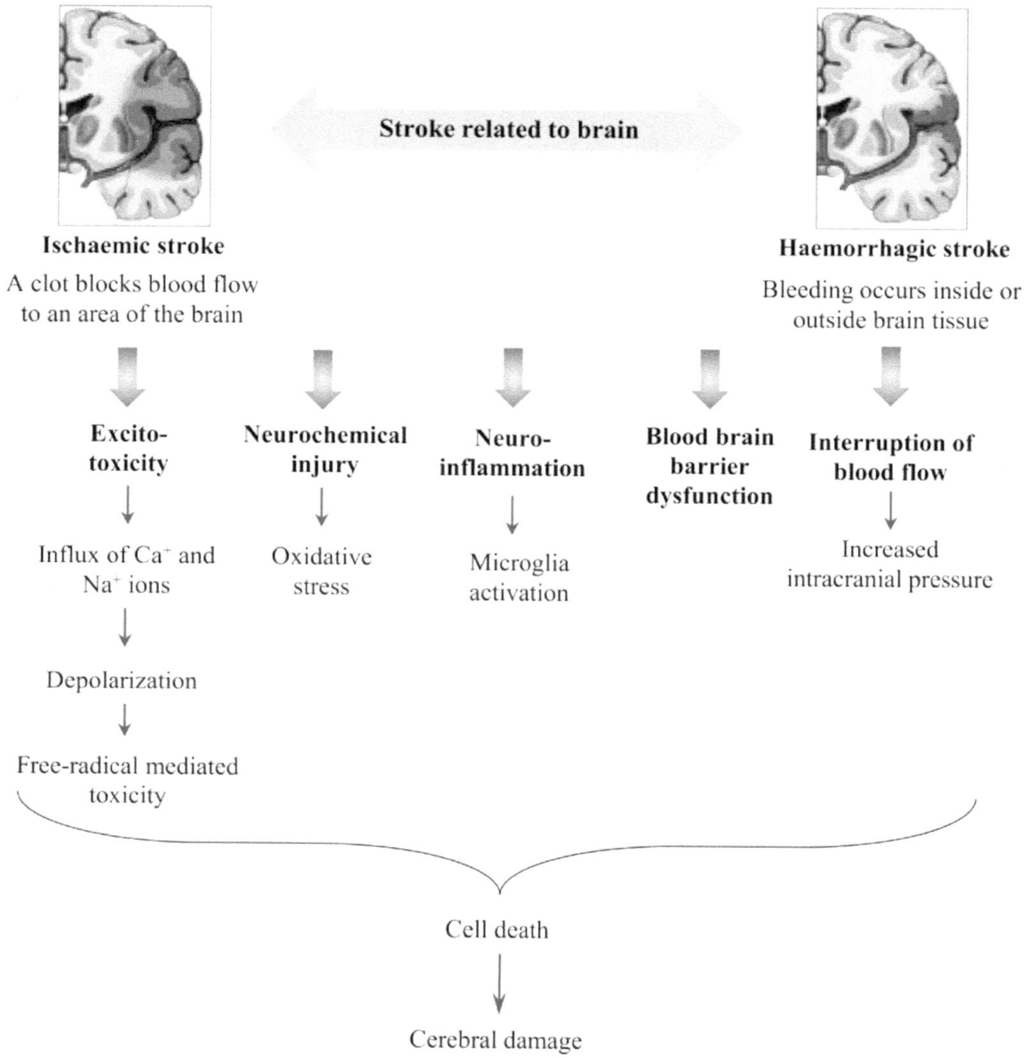

FIGURE 17.1 Pathophysiology of stroke.

17.5 TARGETED DRUGS FOR ISCHEMIC STROKE

Based on different types of therapeutic strategies, the treatment of ischemic strokes has been classified into several types (Figure 17.3).

1. **High-Dose Human Albumin Treatment:** Albumin acts as a promising neuroprotectant that increases microvascular saturation and improves blood flow to highly concentrated brain areas. Human albumin lowers brain edema and total infarct volume. In addition, it improved nerve function in the medical window for 4 hours after the onset of a stroke. Giving human albumin in doses ranging from 0.6–2.5 g/

kg was neuroprotective [21]. N-3 Polyunsaturated fatty acids systemically mobilized by albumin beside the action of haem diluent, polyunsaturated fatty acids disoriented by neural membrane can be reloaded by mobilization process through albumin [22].

Modifiable risk factors	**Non-modifiable risk factors**
❑ Hyperlipoidaemia	❑ Age
❑ Physical inactivity	❑ Sex
❑ Smoking	❑ Race
❑ Hypertension	❑ Ethnicity
❑ Alcoholism	❑ Transient ischemic attack
❑ Drug abuse	❑ Genetics
❑ Atrial fibrillation	
❑ Diet	
❑ Diabetes mellitus	
❑ Genetics	

FIGURE 17.2 Risk factors connected to stroke.

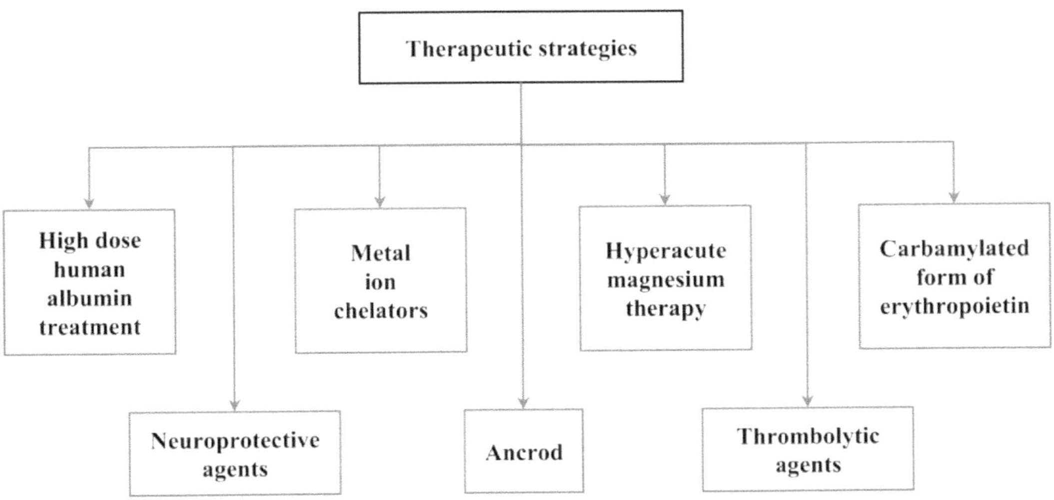

FIGURE 17.3 Classification of therapeutic strategies for ischemic stocks treatment.

2. **Hyperacute Magnesium Therapy:** Magnesium functions as a neuroprotective agent in ischemic stroke with various mechanisms in the brain. It reacts as an antagonist of the calcium endo native channel. Inhibition of stimulant neurotransmitters (glutamate) can be done with it. Magnesium antagonizes the NMDA receptor and has a direct effect on vascular smooth muscles. The investigation is still under process related to the neuroprotective action of hyperacute magnesium therapy [23].
3. **Metal Ion Chelators:** Metal ions are essential components in the regulation of enzymes, co-factors, and cellular transport, including matrix metalloproteinase, calpain, and Cu/Zn superoxide dismutase; the proposed disruption of which is closely related to cell death following acute cerebral ischemia. Disruption of metal ion homeostasis was recommended to be incorporated in numerous neurodegenerative ailments like Parkinson's and Alzheimer's diseases [24].

17.6 NEUROPROTECTIVE AGENTS

Understanding the clinical manifestations of stroke necessitates a thorough understanding of neurovascular anatomy. Ischemic stroke is instigated due to a deficiency of blood and oxygen in the brain. Hemorrhagic stroke, on the other hand, is instigated by bleeding in various regions of the brain [25]. The neuroprotective agents are classified into further sub-categories and these targeted drugs are used to treat ischemic stroke (Table 17.1).

17.7 THROMBOLYTIC AGENTS

Plasminogen activators like streptokinase and urokinase, increase thrombolysis via hydrolyzing the arginine-valine peptide link in plasminogen to generate the active proteolytic enzyme plasmin. Plasmin was quickly suppressed during the fibrinolysis process by a 2-plasmin inhibitor. The mountainous fibrinolytic action of plasmin is done inside the thrombus. Fibrin-plasmin binds inside the thrombus which prevents the inactivation condition by α2-antiplasmin. Thrombolytic agents have both actions, fibrin-plasmin bound and circulation of plasmin non-selectively; such drugs are streptokinase and urokinase [44].

1. **Ancrod:** It has been obtained from the venom of the Malayan ground pit viper (*Calloselasma rhodostoma*). Identical-appearing clear vials of the substance under trial comprising either ancrod (70 WHO IU/mL) or placebo in 10 mmol/L sodium phosphate and 0.9% NaCl solution (pH 6.8), were packaged by an autonomous producer. As a result, after the induction of Ancrod, defibrinogenation showing anticoagulation activity, the viscosity of blood reduces, and indirectly the fibrinolysis recently impacts acute ischemic stroke (AIS) management [45].
2. **Carbamylated Form of Erythropoietin (CEPO):** It may have therapeutic benefits owing to its antioxidant, anti-inflammatory, and neuroprotective properties deprived of augmenting the number of red blood cells. The potential for using CEPO to treat neurodegeneration resulting from ischemia was published by Leist and colleagues [46]. The first demonstration of neuroprotection by CEPO following "ischemia"

TABLE 17.1 Targeted Drug Acting on Ischemic Stocks

Drug's Name	Target Action	Drug Category	Doses	Clinical Phase	References
Phenytoin	Sodium channel blocker.	Antiepileptic	300 mg/day	III	[25, 26]
Nimodipine	Calcium channel blocker.	Antihypertensive	240 mg/day	I or II	[27, 28]
Flunarizine	Calcium channel blocker.	Antihypertensive	50 mg/day	I or II	[29]
Selfotel	Glutamate inhibition (NMDA).	Competitive antagonist	1.5 mg/kg (IV within 6 hrs.)	I or II	[30]
Clomethiazole	Y-Aminobytric acid (GABA) agonist.	Antiepileptic, anticonvulsant, and sedative.	60 mg/kg (IV over 24 hrs.)	III	[31]
Lubeluzole	Nitric oxide inhibitors.	Neuroprotective agents.	10 mg (IV for 5 days)	II	[32]
Trilazad mesylate	Non-glucocorticoid lipid peroxidation inhibitor.	Free radical scavengers (antioxidant).	6 mg/kg (IV for 3 days).	III	[33]
Enlimomb	Murine monoclonal antibody.	Anti-adhesion antibody or anti-inflammatory agent.	10–120 mg range.	III	[34]
Citicoline	GM1-ganglioside	Nootropic agent (nutritional supplements).	2,000 mg (orally)	III	[35]
Nalmefene	Narcotic receptor antagonist.	Opioid antagonist	60 mg (10 mg bolus and 50 mg infusion).	II	[36]
Repinotan	5-HTIA agonist	Serotonin agonist	1.25 mg/day (IV infusion 72 hrs.)	II, III	[37]
Lovastatin	HMG CoA reductase inhibitor.	Statin agent	8 mg/kg/day	II	[38]
Hu23F2G	Humanized monoclonal antibody CD11/CD18.	Anti-leukocytic antibody.	4 mg/kg (saturating dose)	III	[39]
Abciximab	Antiadhersion monoclonal (chimeric) antibody (CD41 7E3) or monoclonal anti-glycoprotein IIb/IIIa receptor antibody.	Antiplatelet fab fragment.	10 μg/min. for 12 hrs. (in combination with aspirin IV bolus dose).	III	[40]
Citicoline	Membrane biosynthesis.	Membrane stabilizer.	50 mg/day for 6 weeks.	II, III	[41]
Fiblast	Neuronal healing	Fibroblast growth factor.	90 and 125 μg/kg	II	[42]
Human albumin infusion	Haemodiluent and systemic mobilization.	Neuroprotective agent.	0.6–2.5 g/kg	II	[25, 27]
Magnesium	Endogenous calcium channel antagonist.	Neuroprotective agent.	Still under investigation.	–	[27]
Arundic acid	S-100β protein synthesis inhibitors.	Astrocyte-modulators.	8 mg/kg/hrs.	II	[43]

used a rodent ischemia model induced by multiple permanent occlusions. The researcher observed the administration of CEPO intravenously till 4 hours after the reduction of cortical cell death by stroke. The same research team, also showed that CEPO could reduce cortical cell death if CEPO treatment was delayed by 6 hours and dosing was repeated at 24 and 48 hours. The treatment of stroke with CEPO will improve the behavior of the patient [47].

17.8 PRE-CLINICAL STUDIES

Heparin sodium is administrated intravenously which is used initially for the treatment of acute stroke, but simultaneously it's enhancing the incidence of intracranial symptomatic brain hemorrhages. According to *in vitro* study, it was observed that heparin which has low molecular weight is refined out under PC12 cells by using Earle's solution to act as a cell damage inhibitor caused by oxygen-glucose deficiency [48]. Heparin also had neuroprotective effects in rats with focal cerebral ischemic injury, significantly reducing infarct volume at doses of 0.5 and 1.0 mg/kg, somewhat via improvement of energy catabolism, inhibition of apoptosis, and reduction of inflammatory response. Administration of enoxaparin reduced the size of infarct with the improvement of neuroscore in both the cases, of permanent and transient focal ischemia in rats which revealed the huge therapeutic window near about 5 h without any vital risk of hemorrhage [49].

17.9 CLINICAL STUDIES

Butylphthalide was studied for the investigation of ischemic stroke in phase 4 clinical trials under large artery, atherosclerosis, and AIS condition. Thus, the drug is allocated a randomized triple-blind and the outcome is still under investigation process. Recombinant human urokinase (rh Pro-UK) was used for the assessment of ischemic stroke in phase 3 clinical trials under AIS conditions. Thus, the medication is to bind in a randomized open-label system. The results are still under processing. Recombinant human tissue kallikrein (DM 199) was used under AIS conditions for the investigation at phase 2 trial allocated a randomized double-blind and still, the results are awaited. Fingolimod (FTY720) is utilized to treat AIS. The medication binding is the non-randomized and open-label at the finished phase 2 level. As a consequence, the outcomes for the treatment of ischemic stroke were successful and well-tolerated. Atorvastatin has also been studied under investigation of phase 4 trial completion which allocated randomized with the open label. Hence, showing better functional response. Edaravone used to evaluate the AIS binds randomly with open-label conditions with the completion of phase 4 trials and gives effective results in acute non-cardioembolic ischemic stroke conditions [50].

17.10 CONCLUSIONS

Stroke is 2nd foremost reason for mortality and impairment all over the world. However, in global health concerns that many therapeutic interventions and post-stroke management are to be more effective. And its graph has to be increased over the last 25 years related to an animal model used in the experiment, therapeutic drugs, clinical trials, and pre-clinical trials, but still the knowledge about the treatment of stroke is remaining. Despite our improved understanding of stroke, epidemiology, and pathophysiology of stroke, its targeted medications, and preclinical and clinical investigations, the translation of research into clinical data has been exceedingly limited in the rapidly growing field of stroke research. Thus, stroke scientists play a tough role in terms of their underlying treatments, generating standard data, doing multi-centered pre-clinical studies, and increasing the constructed value of their data before going on to clinical investigations.

KEYWORDS

- **cerebral ischemia**
- **clinical trial**
- **geographic variation**
- **ischemic stroke**
- **neuroprotective agents**
- **oxidative stress**
- **stroke**

REFERENCES

1. Asplund, K., Tuomilehto, J., Stegmayr, B., Wester, P. O., & Tunstall-Pedoe, H., (1988). Diagnostic criteria and quality control of the registration of stroke events in the MONICA project. *Acta. Med. Scand. Suppl., 728,* 26–39.
2. Shakir, R., (2018). The struggle for stroke reclassification. *Nat. Rev. Neurol., 14*(8), 447–448.
3. Roger, V. L., Go, A. S., Lloyd-Jones, D. M., Adams, R. J., Berry, J. D., Brown, T. M., Carnethon, M. R., et al., (2011). American heart association statistics committee and stroke statistics subcommittee; heart disease and stroke statistics--2011 update: A report from the american heart association. *Circulation, 123*, e18–e209.
4. Collaborators, G. S., (2019). Global, regional, and national burden of stroke, 1990–2016: A systematic analysis for the Global Burden of disease study 2016. *Lancet Neurol., 18*, 439–458.
5. Kelly-Hayes, M., (2010). Influence of age and health behaviors on stroke risk: Lessons from longitudinal studies. *J. Am Geriatr Soc., 58*, S325–S328.
6. Rihal, V., Khan, H., Kaur, A., & Singh, T. G., (2022). Vitamin D as therapeutic modulator in cerebrovascular diseases: A mechanistic perspectives. *Crit. Rev. Food Sci. Nutr.*, 1–23.
7. Appelros, P., Stegmayr, B., & Terént, A., (2009). Sex differences in stroke epidemiology: A systematic review. *Stroke, 40,* 1082–1090.
8. Reeves, M. J., Bushnell, C. D., Howard, G., Gargano, J. W., Duncan, P. W., Lynch, G., Khatiwoda, A., & Lisabeth, L., (2008). Sex differences in stroke: Epidemiology, clinical presentation, medical care, and outcomes. *Lancet Neurol., 7,* 915–926.
9. Saklani, P., Khan, H., Gupta, S., Kaur, A., & Singh, T. G., (2022). Neuropeptides: Potential neuroprotective agents in ischemic injury. *Life Sci., 288*, 120186.

10. Chen, J. C., (2010). Geographic determinants of stroke mortality: Role of ambient air pollution. *Stroke, 41,* 839–841.
11. Zhang, F. L., Guo, Z. N., Wu, Y. H., Liu, H. Y., Luo, Y., Sun, M. S., Xing, Y. Q., & Yang, Y., (2017). Prevalence of stroke and associated risk factors: A population based cross sectional study from northeast China. *BMJ Open, 7,* e015758.
12. Kiefe, C. I., Williams, O. D., Bild, D. E., Lewis, C. E., Hilner, J. E., & Oberman, A., (1997). Regional disparities in the incidence of elevated blood pressure among young adults: The CARDIA study. *Circulation, 96,* 1082–1088.
13. Addo, J., Ayerbe, L., Mohan, K. M., Crichton, S., Sheldenkar, A., Chen, R., Wolfe, C. D., & McKevitt, C., (2012). Socioeconomic status and stroke: An updated review. *Stroke, 43,* 1186–1191.
14. Sandel, M. E., Wang, H., Terdiman, J., Hoffman, J. M., Ciol, M. A., Sidney, S., Quesenberry, C., et al., (2009). Disparities in stroke rehabilitation: Results of a study in an integrated health system in northern California. *PM R, 1,* 29–40.
15. Wang, Y. L., Wu, D., Nguyen-Huynh, M. N., Zhou, Y., Wang, C. X., Zhao, X. Q., Liao, X. L., et al., (2010). The prevention of recurrences of stroke study in China (PRESS-China) investigators; antithrombotic management of ischemic stroke and transient ischemic attack in China: A consecutive cross-sectional survey. *Clin. Exp. Pharmacol Physiol., 37,* 775–781.
16. Kerr, G. D., Higgins, P., Walters, M., Ghosh, S. K., Wright, F., Langhorne, P., & Stott, D. J., (2011). Socioeconomic status and transient ischemic attack/stroke: A prospective observational study. *Cerebrovasc Dis., 31,* 130–137.
17. Khan, H., Kashyap, A., Kaur, A., & Singh, T. G., (2020). Pharmacological postconditioning: A molecular aspect in ischemic injury. *J. Pharm. Pharmacol., 72*(11), 1513–1527.
18. Flaherty, M. L., Woo, D., Haverbusch, M., Sekar, P., Khoury, J., Sauerbeck, L., Moomaw, C. J., et al., (2005). Racial variations in location and risk of intracerebral hemorrhage. *Stroke, 36,* 934–937.
19. Testai, F. D., & Aiyagari, V., (2008). Acute hemorrhagic stroke pathophysiology and medical interventions: Blood pressure control, management of anticoagulant-associated brain hemorrhage and general management principles. *Neurol. Clin., 26,* 963–985.
20. Prabhakar, N. K., Khan, H., Grewal, A. K., & Singh, T. G., (2022). Intervention of neuroinflammation in the traumatic brain injury trajectory: *In vivo* and clinical approaches. *Int. Immunopharmacol., 108,* 108902.
21. Koch, S., Concha, M., Wazzan, T., Romano, J. G., & Forteza, A., (2004). High dose human serum albumin for the treatment of acute ischemic stroke: A safety study. *Neurocrit Care, 1,* 335–341.
22. Huh, P. W., Belayev, L., Zhao, W., Busto, R., Saul, I., & Ginsberg, M. D., (1998). The effect of high-dose albumin therapy on local cerebral perfusion after transient focal cerebral ischemia in rats. *Brain Res., 804,* 105–113.
23. Cadacio, C., Pham, P. T., Bhasin, R., Kamarzarian, A., & Pham, P. C., (2017). Multiple electrolyte and metabolic emergencies in a single patient. *Case Rep Nephrol., 2017,* 4521319.
24. Angel, I., Bar, A., Horovitz, T., Taler, G., Krakovsky, M., Resnitsky, D., Rosenberg, G., et al., (2002). Metal ion chelation in neurodegenerative disorders. *Drug Develop Res., 56,* 300–309.
25. Khan, H., Garg, N., Singh, T. G., Kaur, A., & Thapa, K., (2022). Calpain inhibitors as potential therapeutic modulators in neurodegenerative diseases. *Neurochem. Res.,* 1–25.
26. Khan, H., Sharma, K., Kumar, A., Kaur, A., & Singh, T. G., (2022). Therapeutic implications of cyclooxygenase (COX) inhibitors in ischemic injury. *Inflamm. Res.,* 1–16.
27. Yerram, S., Katyal, N., Premkumar, K., Nattanmai, P., & Newey, C. R., (2018). Seizure prophylaxis in the neuroscience intensive care unit. *J. Intensive Care., 6,* 17.
28. Fagan, S. C., Gengo, F. M., Bates, V., Levine, S. R., & Kinkel, W. R., (1988). Effect of nimodipine on blood pressure in acute ischemic stroke in humans. *Stroke, 19*(3), 401–402.
29. Franke, C. L., Palm, R., Dalby, M., Schoonderwaldt, H. C., Hantson, L., Eriksson, B., Lang-Jenssen, L., & Smakman, J., (1996). Flunarizine in stroke treatment (FIST): A double-blind, placebo-controlled trial in Scandinavia and the Netherlands. *Acta Neurol. Scand., 93*(1), 56–60.
30. Davis, S. M., Albers, G. W., Diener, H. C., Lees, K. R., & Norris, J., (1997). Termination of acute stroke studies involving selfotel treatment. ASSIST steering committed. *Lancet, 349*(9044), 32.

31. Lyden, P., Shuaib, A., Ng, K., Levin, K., Atkinson, R. P., Rajput, A., Wechsler, L., et al., (2002). CLASS-I/H/T Investigators; Clomethiazole acute stroke study in ischemic stroke (CLASS-I): Final results. *Stroke, 33*(1), 122–128.
32. Diener, H. C., Cortens, M., Ford, G., Grotta, J., Hacke, W., Kaste, M., Koudstaal, P. J., & Wessel, T., (2000). Lubeluzole in acute ischemic stroke treatment: A double-blind study with an 8-hour inclusion window comparing a 10-mg daily dose of lubeluzole with placebo. *Stroke, 31*(11), 2543–2551.
33. STIPAS Investigators, (1994). Safety study of tirilazad mesylate in patients with acute ischemic stroke (STIPAS). *Stroke, 25*(2), 418–423.
34. Krams, M., Lees, K. R., Hacke, W., Grieve, A. P., Orgogozo, J. M., Ford, G. A., & ASTIN Study Investigators, (2003). Acute stroke therapy by inhibition of neutrophils (ASTIN): An adaptive dose-response study of UK-279,276 in acute ischemic stroke. *Stroke, 34*(11), 2543–2548.
35. Dávalos, A., Castillo, J., Alvarez-Sabín, J., Secades, J. J., Mercadal, J., López, S., Cobo, E., et al., (2002). Oral citicoline in acute ischemic stroke: An individual patient data pooling analysis of clinical trials. *Stroke, 33*(12), 2850–2857.
36. Clark, W. M., Raps, E. C., Tong, D. C., & Kelly, R. E., (2000). Cervene (Nalmefene) in acute ischemic stroke: Final results of a phase III efficacy study. The cervene stroke study investigators. *Stroke, 31*(6), 1234–1239.
37. Teal, P., Silver, F. L., & Simard, D., (2005). The BRAINS study: Safety, tolerability, and dose-finding of repinotan in acute stroke. *Can. J. Neurol. Sci., 32*(1), 61–67.
38. Elkind, M. S., Sacco, R. L., Macarthur, R. B., Peerschke, E., Neils, G., Andrews, H., Stillman, J., et al., (2009). High-dose lovastatin for acute ischemic stroke: Results of the phase I dose escalation neuroprotection with statin therapy for acute recovery trial (NeuSTART). *Cerebrovasc. Dis., 28*(3), 266–275.
39. Yenari, M. A., Kunis, D., Sun, G. H., Onley, D., Watson, L., Turner, S., Whitaker, S., & Steinberg, G. K., (1998). Hu23F2G, an antibody recognizing the leukocyte CD11/CD18 integrin, reduces injury in a rabbit model of transient focal cerebral ischemia. *Exp Neurol., 153*(2), 223–233.
40. The Abciximab in Ischemic Stroke Investigators, (2000). Abciximab in acute ischemic stroke. A randomized, double-blind, placebo-controlled, dose-escalation study. *Stroke, 31*(3), 601–609.
41. Clark, W. M., Williams, B. J., Selzer, K. A., Zweifler, R. M., Sabounjian, L. A., & Gammans, R. E., (1999). A randomized efficacy trial of citicoline in patients with acute ischemic stroke. *Stroke, 30*(12), 2592–2597.
42. Bogousslavsky, J., Victor, S. J., Salinas, E. O., Pallay, A., Donnan, G. A., Fieschi, C., Kaste, M., et al., (2002). Fiblast (trafermin) in acute stroke: Results of the European-Australian phase II/III safety and efficacy trial. *Cerebrovasc. Dis., 14*(3, 4), 239–251.
43. Pettigrew, L. C., Kasner, S. E., Albers, G. W., Gorman, M., Grotta, J. C., Sherman, D. G., Funakoshi, Y., & Ishibashi, H., (2006). Arundic acid (ONO-2506) stroke study group; safety and tolerability of arundic acid in acute ischemic stroke. *J. Neurol. Sci., 251*(1, 2), 50–56.
44. Khan, H., Singh, A., Thapa, K., Garg, N., Grewal, A. K., & Singh, T. G., (2021). Therapeutic modulation of the phosphatidylinositol 3-kinases (PI3K) pathway in cerebral ischemic injury. *Brain Res., 1761*, 147399.
45. Levy, D. E., Del Zoppo, G. J., Demaerschalk, B. M., Demchuk, A. M., Diener, H. C., Howard, G., Kaste, M., et al., (2009). Ancrod in acute ischemic stroke: Results of 500 subjects beginning treatment within 6 hours of stroke onset in the ancrod stroke program. *Stroke, 40*(12), 3796–3803.
46. Leist, M., Ghezzi, P., Grasso, G., Bianchi, R., Villa, P., Fratelli, M., Savino, C., et al., (2004). Derivatives of erythropoietin that are tissue protective but not erythropoietic. *Science, 305*(5681), 239–242.
47. Lapchak, P. A., (2010). Erythropoietin molecules to treat acute ischemic stroke: A translational dilemma! *Expert Opin. Investig Drugs, 19*(10), 1179–1186.
48. Zhang, Z. G., Sun, X., Zhang, Q. Z., & Yang, H., (2013). Neuroprotective effects of ultra-low-molecular weight heparin on cerebral ischemia/reperfusion injury in rats: Involvement of apoptosis, inflammatory reaction and energy metabolism. *Int. J. Mol. Sci., 14,* 1932–1939.
49. Kalra, P., Khan, H., Kaur, A., & Singh, T. G., (2022). Mechanistic insight on autophagy modulated molecular pathways in cerebral ischemic injury: From preclinical to clinical perspective. *Neurochem. Res.*, 1–19.
50. Dhir, N., Medhi, B., Prakash, A., Goyal, M. K., Modi, M., & Mohindra, S., (2020). Pre-clinical to clinical translational failures and current status of clinical trials in stroke therapy: A brief review. *Curr. Neuropharmacol., 18*(7), 596–612.

CHAPTER 18

Recent Therapeutic Approaches Toward the Management of Ischemic Heart Disease

DEEPIKA PUROHIT, NIRMALA KUMARI YADAV, JYOTI RATHI, SUNIL KUMAR, and SNEHLATA YADAV

Department of Pharmaceutical Sciences, Indira Gandhi University, Meerpur, Rewari, Haryana, India

ABSTRACT

Ischemia is a condition characterized by deprivation of blood supply to an organ of the body causing reversible/irreversible damage to that tissue or sometimes death of the person. Several conditions cause impaired blood flow to various vital organs like the heart, kidney, brain, liver, etc. When this occurs in the heart, the myocardium of the left ventricle may get damaged resulting in losing its contractile ability and progressive dilatation. This ischemic condition is also commonly known as left ventricular remodeling. The prolonged ischemic condition may be difficult to reverse by means of clinical procedures that restore blood flow by use of reperfusion. The emergence of pandemic COVID-19 in the year 2019 throughout the globe has made the condition worse. Previous studies have indicated that the prevalence of ischemic disease is greatly increased in cases of bacterial/viral infections due to the prothrombotic effect of inflammatory response leading to increased mortality. However, numerous traditional and innovative treatments have assisted clinicians in tackling this issue. This chapter looks at the various strategies.

18.1 INTRODUCTION

Human health has been threatened due to disease since time immemorial. Ischemic heart disease is a major global contributor to mortality in youths but its impact is more in Western countries due to diet and sedentary lifestyles [1]. This disease is characterized by the diminished blood supply to the heart muscle due to blockage in coronary arteries by thrombus and atherosclerosis. Arrhythmia (abnormal electrical activity of the heart) [2], angina (chest discomfort of cardiac origin) [3], myocardial infarction (increased cardiac troponin levels with at least one value above the 99th upper reference limit) [4], and heart failure (failure to

Ischemic Injury: A Molecular Insight. Thakur Gurjeet Singh & Heena Khan (Eds.)

pump blood according to body needs) [5] accounts for some of its associated symptoms. The cellular changes in the shape, as well as function of ventricles following chamber dilatation and perivascular and interstitial fibrosis, is known as cardiac remodeling which is the result of chronic heart failure. The remodeling consists of neurohormonal responses, cytokine activation, cardiomyocyte loss owing to necrosis or apoptosis, cardiomyocyte hypertrophy, extracellular matrix disrupton, and collagen buildup followed by scar formation [6]. However, myocardial ischemia remains the primary cause of cardiomyocyte dysfunction and mortality. Myocardial infarction is characterized by apoptotic myocyte death, cardiomyocyte hypertrophy, gradual collagen replacement, and left ventricular enlargement [7]. The term ischemic heart disease thus involves a wide spectrum of gradually worsening pathophysiological conditions that start first from asymptomatic coronary artery disease that gradually progresses to reversible and irreversible myocardial ischemia, necrosis, and impairment of cardiac function [8]. The statistics of 2010 have reported a toll of more than 7.0 million due to ischemic heart disease, making it a foremost cause of death globally [9]. However, in India, more than 10.5 million deaths occur annually due to cardiovascular diseases leading to 20.3% of deaths in men and 16.9% of all deaths in women (Registrar General of India, 2013). Although the age-standardized ischemic heart disease (IHD) mortality has decreased over the past few decades, the global burden of ischemic heart disease has increased by 29% to 129 million from 1990 to 2010 [10]. This review is an effort to combine various aspects of the disease reported in the last five years.

18.2 DISEASE PATHOPHYSIOLOGY

A large amount of information is available regarding the etiology of ischemia based on basic, translational, and clinical research suggesting the intricate biology of ischemic heart disease [11], which goes beyond the conventionally established role of atherosclerosis. Although strategies have evolved to optimize prevention as well as treatment for this disease, the outcomes of IHD are a significant burden on human health globally. This makes it important to overcome the concept that ischemia is always related to atherosclerosis [12]. As per the current research, it has been reported that ischemic heart disease occurs due to various transformed regulating vascular signaling pathways that include inflammation, coronary microvascular dysfunction, atherosclerosis, etc. (Figure 18.1). Numerous IHD patients have low or no epicardial coronary artery disease and as suggested by several clinical trials, the microvascular disease majorly leads to the etiology of ischemia, by regulating the flow of blood and oxygen in the myocardium-microcirculation interaction. The key metabolites involved in this process are H_2O_2 (which causes vasodilation during increased cardiac work), Na_v channels, K_{ATP} channels, and $K_v1.5$ channels, (which represent the final effectors of the mechanisms involved in the coronary blood flow regulation). These metabolites contribute in different ways toward intracellular calcium regulation, affecting the degree of contractile tone, and finally, the coronary blood flow. The ischemic heart disease depends upon impairment of the cross-talk between coronary blood flow and myocardial energy state and thus the diseases associated with coronary microvascular and macrovascular systems may become a part of the multifaceted pathophysiology of IHD [13].

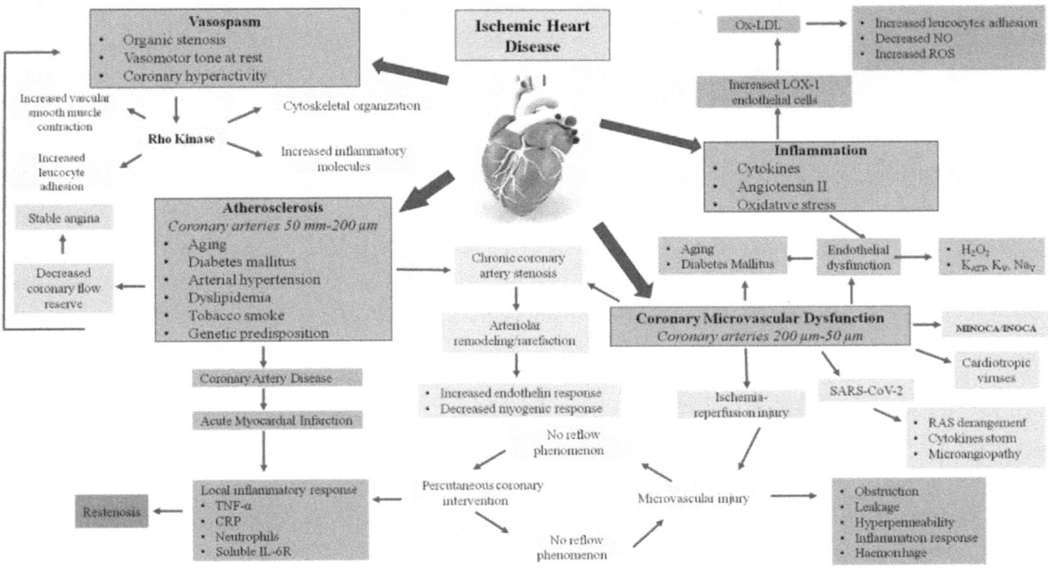

The pathophysiology of ischemic heart disease is multifaceted and complex one and involves several conditions such as coronary microvascular dysfunction, atherosclerosis, vasospasm, and inflammation.

CRP: C-reactive protein
IL-6R: Interleukin-6 receptor
INOCA: Ischemia with non-obstructive coronary arteries
K_{ATP}: ATP-sensitive potassium channel
K_V: Voltage-gated potassium channel
LOX-1: Oxidized low-density lipoprotein receptor-1
MINOCA: Myocardial infarction with non-obstructive coronary arteries
Na_V: Voltage-gated sodium channel
Ox-LDL: Oxidized low-density lipoprotein
RAAS: Renin-angiotensin-aldosterone system
ROS: Reactive oxygen species;
SARS-CoV-2: Severe acute respiratory syndrome coronavirus-2
TNFα: Tumor necrosis factor alpha

FIGURE 18.1 Pathophysiological mechanisms involved in ischemic heart disease.

The conventional pathophysiology of IHD or coronary artery disease (CAD) involves an obstructive plaque inhibiting the blood flow through a coronary artery, thus triggering myocardial ischemia. Normally, the coronary blood flow is maintained by the autoregulation process, the capacity of preserving blood flow at the time of changes in perfusion pressure. The particular mechanisms responsible for autoregulation have not been proven, but the metabolic and myogenic mechanisms have been found to play a significant role. When around 70% of the cross-sectional area of the lumen is obstructed by an atherosclerotic plaque reducing 50% coronary diameter, the proximal resistance significantly increases, and the distal coronary perfusion pressure decrease. In this condition, the autoregulation maintains the basal coronary blood flow but is not capable of maintaining the dilator reserve as it gets greatly compromised leading to non-symptomatic conditions at rest, but deficient flow at high metabolic demands, like while doing physical exercise [14]. Another important factor involved in the pathophysiology of IHD is aging which is an independent cardiovascular risk factor as it impacts arterial hypertension, diabetes, dyslipidemia, etc. Aging greatly affects various changes in both microvasculature and coronary epicardial arteries. It is responsible for the loss of the integrity of the endothelial layer, loss of elasticity of blood vessels, arterial stiffness, and decreased adaptability of vessels to physical forces associated with coronary blood flow [15, 16].

Many years ago, scientists proposed that vasospasm of one of the major coronary arteries can be a cause of IHD leading to total occlusion. Later in the year 1959, Prinzmetal reported a type of angina with an elevation of ST-segment in electrocardiograph in absence of coronary blockade [17]. Prinzmetal angina occurs due to transient vasospasm with an acute decrease in coronary blood supply [18]. In numerous clinical situations involving coronary microcirculation, the relevance of coronary spasm has been shown. Moreover, it is now known that spasms can be initiated by several factors, such as segmental epicardial coronary hyperactivity, organic stenosis, and vasomotor tone at rest. Spasms of atherosclerotic lesions may aggravate IHD, with coronary vasomotor response associated with plaque burden [19]. A downstream vasospastic effector Rho-associated protein kinase is a small universally expressed G protein that is involved in some cellular functions, such as actin cytoskeleton organization, circulating leukocytes adhesion, and smooth muscle contraction [20]. In addition, Rho-associated protein kinase causes increased expression of inflammatory molecules, fibrosis, and thrombosis [21]. Rho-associated protein kinase participates in the pathogenesis of coronary vasospasm, hypertension, angina pectoris, etc. This is why the activity of Rho-associated protein kinase can be an important novel biomarker for assessing the cardiovascular severity of disease and therapeutic responses of medicine used for the treatment [22].

Recently, it has also been reported that inflammation supports the process of atherosclerosis, and a hypothesis has been made that inflammation may act as a stimulating factor for the initiation and evolution of atherosclerosis and ultimately IHD [13]. Inflammatory responses seem to affect the pathogenesis of conventional cardiovascular risk factors, like metabolic syndrome, diabetes mellitus, and arterial hypertension [23]. Some studies have also hypothesized the link of inflammation circulating indicators, like homocysteine and C-reactive protein, with the development and prognosis of IHD [24].

Along with the above-mentioned key biomarkers, i.e., atherosclerosis, vasospasm, and inflammation, involved in the pathogenesis of IHD, a few years back, abnormalities in the coronary microcirculation function and structure have been reported as a significant mechanism involved in the pathophysiology of the ischemic heart disease. Coronary microvascular dysfunction (CMD) causes an increased resistance in blood flow, thus leading to ischemia, as a result of this decreased perfusion pressure [25]. In the year 1985, the term ‘microvascular angina’ was used for such cases, suggesting the altered coronary microcirculation vasoregulation is a major cause of IHD. CMD results in remodeling and rarefaction of arteries, blunting of myogenic responses, and also increased endothelin-induced vasoconstrictor responses due to a lack of endothelin B receptor-mediated vasodilation [26]. In addition, coronary microvascular dysfunction may associate with intramural vessel structural abnormalities, due to hyperplasia and hypertrophy of vascular smooth muscle cells [27], or infection of small blood caused by cardiotropic viruses [28].

Several therapeutic options are available for the treatment of IHD aiming at the reduction of symptoms and prevention of cardiovascular events. The mainly used drugs include β-blockers, short and long-acting nitrates, lipid-lowering drugs, calcium channel antagonists, angiotensin-converting-enzyme inhibitors, and anticoagulants [29]. Some newer treatment approaches are also being used nowadays considering the complex pathophysiology of the disease. Such approaches include targeted drug delivery using novel drug delivery options like nanoparticles. Some repurposed drugs are also being studied to treat this complex disease effectively [30]. The present review focused on discussing the various therapeutic approaches being used for the treatment of IHD along with newer treatment options like nanoparticles. The authors have also compiled the available clinical trials and patent information fulfilling the objective of the review.

18.3 VARIOUS TREATMENTS IN ISCHEMIC HEART DISEASE

1. **Danhong Injection in Ischemic Heart Disease:** Dan Shen and Hong Hua are certified and patented herbs from China. Danhong injection obtained from the extract of these herbs was evaluated for its effect on lipid profile and cardiac function in patients with CHD through a prospective, randomized, double-blind, and placebo-controlled study. Participants were randomly divided into two groups–the treatment group received a Danhong injection and the control group received a placebo treatment. Danhong injection regulated lipid-protein metabolism disorder thereby causing reduced lipid peroxidation products, lower level of blood lipids, and increased activity of antioxidant enzymes [31].
2. **Acarbose in Ischemic Heart Disease:** The study aimed to find out the impact of acarbose, an α-glucosidase inhibitor, in Chinese patients with CHD with impaired glucose tolerance. For this purpose, 6,522 patients from 176 hospital outpatients from China were recruited to carry out a random, placebo-controlled double-blind phase-4 trial. Through a centralized computer system, the patients of both the diseases were assigned (1:1), in blocks by site, to receive 50 mg oral acarbose thrice a day or matched placebo that was added to standardized secondary prevention therapy of cardiovascular diseases. Patients receiving at least one dosage of the studied medication were grouped under the safety population. Around 3,250 patients were assigned to placebo and 3,272 patients to acarbose and the patients were called for a follow-up period of 5·0 years. The primary five-point composite outcomes like cardiovascular death, hospital admission for unstable angina, non-fatal stroke, non-fatal myocardial infarction, and hospital admission for heart failure occurred in 14% of treated patients of 3,272 acarbose group participants and 15% in placebo group participants. The secondary three-point composite outcome, death from any cause, hospital admission for unstable angina, fatal or non-fatal stroke, hospital admission for heart failure, impaired renal function, or cardiovascular death, showed no significant differences between the treatment groups. However, the results demonstrated that there was a decrease in the development of diabetes in acarbose treated group, i.e., 13% of 3,272 in comparison to 16% of 3,250 in the placebo-treated group. Change in dose or discontinuation of the drug was associated with gastrointestinal disorders in 7% of 3,263 patients in the acarbose group *vs* 5% of 3,241 in the placebo group. However, no significant change was observed in the numbers of deaths due to non-cardiovascular deaths and cancer deaths. The

interpretation results of the study demonstrated that acarbose successfully reduced the incidence of diabetes but failed to reduce the risk of adverse cardiovascular events in Chinese patients [32].

3. **Pitavastatin-Nanoparticles in Ischemic Heart Disease:** Innovative cardiac modalities need to be developed for the treatment of acute myocardial infections as the interventional reperfusion therapy in these cases is hampered by ischemic reperfusion injury. Ichimura et al. in a previous study reported that treatment with pitavastatin incorporated in bioabsorbable poly(lactic acid/glycolic acid) nanoparticles, i.e., pitavastatin-nanoparticles, which exerted a cardio-protective effect in a rat ischemic reperfusion injury model. This cardio-protective effect was mediated through the PI3K-Akt pathway activation and inhibition of inflammation. Further to gain preclinical proof-of-concept evidence, the effect of pitavastatin-nanoparticles on myocardial ischemic injury in mindful and anesthetized pig models [33]. For this purpose, left circumflex coronary artery and telemetry transmitters with pneumatic cuff occluder were implanted by surgery in 84 Bama mini-pigs to monitor electrocardiogram, heart rate, and arterial blood pressure continuously. The left circumflex coronary artery was occluded for one hour which was followed by reperfusion under conscious conditions for 24 hours. The infarct size was significantly reduced on intravenous administration for 5 minutes before reperfusion by the nanoparticles incorporated with 8 mg/body of pitavastatin. However, no therapeutic effects were noticed with pitavastatin (8 mg/body) alone. Pitavastatin-nanoparticles also exerted anti-apoptotic effects on cardiomyocytes cultured *in vitro*. Cardiac MRI was performed one month after ischemic reperfusion injury which revealed that pitovastatin-nanoparticles reduced the degree of extent of left ventricle remodeling. However, no observable effects were noticed in blood pressure due to these nanoparticles. The anesthetized pigs were probed for analysis of pharmacokinetics and the effects of pitavastatin-nanoparticles on the no-reflow phenomenon. The results of the study confirmed that the delivery of pitavastatin incorporated in nanoparticles to ischemic reperfusion injured myocardium exerted cardio-protective effects without any observable unfavorable side effects in a conscious preclinical pig model. Thus, nanoparticles incorporated with pitavastatin represented a new therapeutic option for ischemic reperfusion injury in acute myocardial infarction (AMI) [33].

4. **Myeloperoxidase Inhibitors in Ischemic Heart Disease:** It has been revealed through various clinical trials that anti-inflammatory pathways are nowadays of benefit. Myeloperoxidase enzyme is responsible for atherosclerotic plaque formation and destabilization of the fibrous cap and hence increases the menace of atherosclerotic cardiovascular disease especially coronary artery disease. There are numerous prototype compounds like aminobenzoic acid hydrazide, thiouracil derivatives (PF-1355 and PF-0628999), ferulic acid derivative (INV-315), triazolopyrimidines, N-acetyl lysyltyrosylcysteine (KYC), acetaminophen, 2-thioxanthine derivative (AZM198), flavonoids, and other alternative substrates such as nitroxide and thiocyanate radicals that directly or indirectly inhibit the enzymatic activities of myeloperoxidase, hence exerting anti-atherosclerotic effects. Therefore,

targeting myeloperoxidase may be one of the most promising remedial approaches for coronary artery disease and other thrombotic cardiovascular diseases [34].

5. **Combination Therapy with Perindopril and Amlodipine in Ischemic Heart Disease:** The prevention therapy for ischemic heart disease includes aspirin in combination with an angiotensin-converting enzyme (ACE) inhibitor and a statin. To evaluate the impact of the fixed combination of perindopril (an ACE inhibitor) and amlodipine to atorvastatin in patients with coronary artery disease (PAPA-CAD) on the lipid profile of the patient, a 6-month open-label, multicenter, prospective, observational, and non-interventional evaluation was conducted on survey data obtained from patients with hypertension and stable coronary artery disease. The findings from 1,130 patients taking atorvastatin along with a fixed combination of perindopril and amlodipine during each of the four study visits (i.e., at baseline and 1, 3, and 6 months later). The analysis of the subgroup from PAPA-CAD revealed that the clinical benefits of the drugs under study were evident after the 6-month study period. The condition of the patients on persistent statin therapy progressively worsened over the last several years indicating that the statins showed no remarkable reductions in mortality and morbidity. Incorporating the statin into a fixed antihypertensive combination, however, produced practically superior persistence indices. The desired blood pressure of 140/90 mm Hg was achieved in 82.5% of patients with the triple fixed combination of perindopril, amlodipine (a calcium channel blocker), and atorvastatin, compared to 78.8% of individuals who did not take a statin. This fixed triple combination of active substances may improve primary and secondary prevention of secondary CAD due to enhanced patient adherence and persistence as well as the demonstrated synergistic effects of its components. In contrast, the presence of atorvastatin to the fixed combination of perindopril and amlodipine resulted in a considerable improvement in the metabolic parameter. Therefore, it can be concluded from this subgroup analysis that a favorable synergism existed among the three drugs, i.e., perindopril, amlodipine, and atorvastatin [35].
6. **Exosomes in Ischemic Heart Disease:** Bone marrow mesenchymal stem cells used for cardiac regeneration face major drawbacks like low survival rate and difficulty in their retention in the infarcted heart. However, these dilemmas can be overcome by co-delivery exosomes derived from stem cells. Exosomes-derived stem-cells were isolated from mesenchymal stem cells and delivered to an acute myocardial infarcted rat heart through intramyocardial injection with or without an intravenous infusion of atorvastatin-pretreated mesenchymal stem cells on day 1, day 3, or day 7 after infarction. Cardiac function was assessed by echocardiography. However, histological analysis and ELISA tests were performed to assess various parameters like SDF-1, angiogenesis, and expression of an inflammatory factor in the infarcted border region. The anti-apoptosis effect on mesenchymal stem cells due to stem cell-derived exosomes was assessed using flow cytometry and Hoechst 33342 staining assay. The co-delivery of intramyocardial exosomes derived from stem cells and mesenchymal stem cells transplantation into the myocardial heart not only reduced infarct size and improved cardiac function but also increased neovascularization in comparison to controls treated either with exosomes

derived from stem cells or mesenchymal stem cells. The comparison of the three co-transplanting groups showed the highest improvement in heart function when exosomes derived from stem cells were injected intramyocardially half an hour after AMI and followed transplantation with mesenchymal stem cells on day 3 after AMI. The enhanced heart function observed was probably due to an improved microenvironment via injection of exosomes derived from stem cells, which is exemplified by reduced inflammatory responses and better mesenchymal stem cell retention and recruitment. Also, pre-transplantation injection of exosomes derived from stem cells enhanced the survival of mesenchymal stem cells and reduced their apoptosis *in vitro* as well as *in vivo*. The combined delivery of exosomes and stem cells in a sequential manner effectively reduced infarct size and restored heart function after AMI, thereby representing it as an alternative promising remedy for stem cell-based heart repair and therapy [36].

Fibrin-rich thrombus formation is the major cause of myocardial infarction in which coronary arteries are occluded resulting in blockage of blood flow to the heart and triggering fibrin deposition. In the treatment of myocardial infarction, re-establishing blood flow is challenging as ischemia-reperfusion injury can also occur and contribute to cardiac fibrosis. The fibrin-specific poly(N-isopropylacrylamide) nanogels in which the core shell was composed of colloidal hydrogel were utilized in this study to design a dual-delivery system that simultaneously tackles the need to inhibit cardiac fibrosis following ischemia-reperfusion injury and re-established blood flow. Controlling the release of a fibrinolytic protein, tPA, and a small chemical cell contractility inhibitor satisfies these therapeutic requirements (Y-27632). During *in vitro* examination, tissue plasminogen activator and Y-27632-loaded fibrin-specific poly(N-isopropylacrylamide) nanogels rapidly degraded fibrin and reduced cardiac cell stress fiber production and growth of connective tissues factor expression, both of which are elevated in cardiac fibrosis. *In vivo*, fibrin-specific poly(N-isopropylacrylamide) nanogels localized to fibrin in damaged heart tissue and when overloaded with tissue plasminogen activator and Y-27632, showed significant improvement in left ventricular ejection fractions two and four weeks after reperfusion injury. In addition to this, a significant decrease in infarct size, α-smooth muscle actin expression, and expression of connective tissue growth factor four weeks post-ischemic reperfusion were also observed which demonstrated the probability of this targeted therapeutic approach to improve the cardiac function following myocardial infarction [37].

18.4 LIST OF APPROVED DRUGS USED IN ISCHEMIC HEART DISEASE

1. **Beta-Blockers:** The most generally prescribed medicines for angina are beta-adrenergic antagonists, sometimes known as beta-blockers (BBs). BBs reduce heart rate, myocardial contractility, left ventricular wall tension, and blood pressure by inhibiting the β_1 adrenergic receptor. The duration of diastole increases as the heart rate is reduced, thus enhancing coronary perfusion [38].
2. **Nitrate:** Organic nitrates are one of the earliest medications for treating angina. Nitrates primarily cause veno-dilation, which lowers preload and lowers the left

ventricular systolic and diastolic pressures, lowering stress on the left ventricular wall and myocardial oxygen consumption. Furthermore, nitrates produce coronary vasodilation, which causes blood flow to be redistributed to the ischemic myocardium [39].

3. **Calcium Channel Blockers (CAs):** These act by preventing calcium from entering the cell by blocking the L-type calcium receptor. The dihydropyridine CAs, which include nifedipine, act primarily on systemic and coronary vascularization to provide vasodilation and, as a result, a reduction in afterload [40, 41].
4. **Trimetazidine:** It improves cellular tolerance to ischemia by first blocking fatty acid metabolism and then boosting glucose metabolism [42]. Nicorandil is an effective anti-ischemic medication in patients who are unable to take beta-blockers due to contraindications such as bradycardia or aggravated pulmonary illness [43].
5. **Ranolazine:** First designed as an antianginal and anti-ischemic metabolic effector, ranolazine is now used for a variety of indications. Inhibiting fatty acid oxidation shifts myocardial energy generation from fatty acid oxidation to glucose oxidation, which explains ranolazine's positive effects on cardiac metabolism and enhanced exercise tolerance without changing heart rate or blood pressure [44].

Furthermore, nanoparticle (NP) systems containing medications have the potential to improve the cardioprotective potential of pharmaceuticals in patients with myocardial ischemia episodes due to their controlled release properties. Direct intramyocardial injection or intravenous injection with active targets and NPs can provide continuous exposure exactly to the infarcted heart. Nanoparticles (NPs) are nano-sized vesicles that can carry therapeutic medicines in a sustained release system and improve myocardial healing in ischemic heart disease [30]. Nanoparticles are divided into organic, inorganic, and hybrid as presented in Figure 18.2.

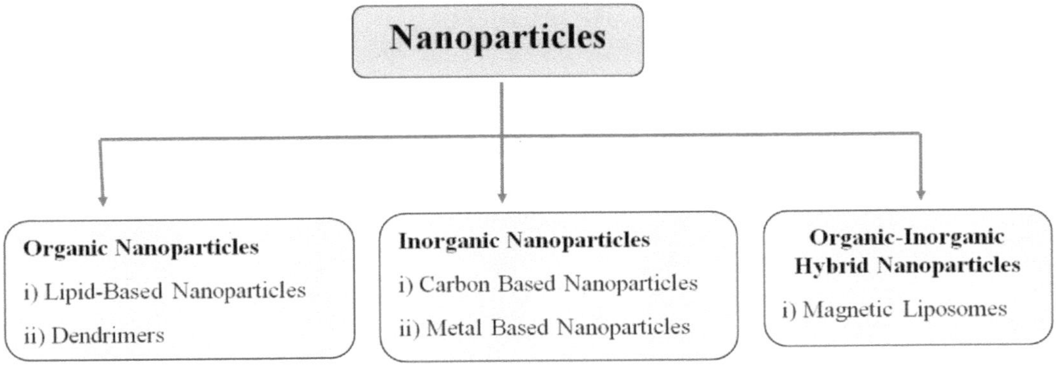

FIGURE 18.2 Types of nanoparticles.

18.4.1 ORGANIC NANOPARTICLES

1. **Lipid-based Nanoparticles:** Solid lipid nanoparticles, nanostructured lipid-drug conjugates, lipid carriers, and nanoemulsions are examples of lipid-based nanoparticle

formulations; all are primarily composed of lipid analogs with surfactants as stabilizers [45], and lipid-based nanoparticles are classified as liposomes (100 nm), micelles (15 nm), or polymeric NPs based on their size [46, 47].

2. **Dendrimers:** These are nanoparticles with diameters ranging from 1 to 15 nanometers. Dendrimers are typically utilized to improve a compound's unique feature [48, 49].

18.4.2 INORGANIC NANOPARTICLES

1. **Carbon-based Nanoparticles:**
   i. **Carbon Nanotubes (CNTs):** These are a type of fullerene consisting of rolled-up graphite sheets that are shaped into tubular shapes [50].
   ii. **Graphene:** The graphene-based family of materials has shown tremendous promise, with a wide range of proposed biosensing applications. "Graphene" refers to two-dimensional (2D) crystal formations made up of single atomic layers of graphite [51].
2. **Metal Nanoparticles:**
   i. **Gold Nanoparticles (GNPs):** These nanoparticles have been studied extensively in several biological applications. Many GNPs are promising candidates in nanomedicine because of their unique physical and chemical features, such as ease of bioconjugation, great stability, and strong biocompatibility [52].
   ii. **Silver Nanoparticles (AgNPs):** These nanoparticles were created as effective antimicrobial agents and have a wide range of uses, including toothpaste, bedding, water purification, and nursing bottles [53]. Humans can absorb roughly 18% of silver after oral contact, according to studies [54].
3. **Organic-Inorganic Hybrid Nanoparticles:**
   i. **Magnetoliposomes (MLs):** These are the first effective hybrid liposome/NP systems for drug delivery, and they are made up of liposomes and magnetic NPs [55]. The advantages and disadvantages of various nanoparticles are summarized in Table 18.1.

Several clinical trials and patents of ischemic heart disease are presented in Tables 18.2 and 18.3.

18.5 CONCLUSION

This review highlights the various clinical and pathological aspects of ischemic heart disease in the last five years. However, despite available drug therapies and newer clinical options, much about the disease remains to be explored as the incidences of IHD have greatly increased due to co-infection with COVID-19. Lifestyle changes in diet and exercise can further help to improve the incidences of this disease. The authors have also tried to compile available information regarding therapeutic options for the treatment of

TABLE 18.1 Advantages and Disadvantages of Different Nanoparticles for the Delivery of Therapeutics for Ischemic Heart Disease

Type	Advantages	Disadvantages	References
Lipid-based NPs	• Higher penetration. • Biodegradable and biocompatible nature. • Increased solubility of the medication. • Better stability during storage.	• Loss of large quantity of drug. • Inadequately controlled drug release. • Toxic effects from burst drug release.	[56]
Dendrimers	• Ease of manufacture. • Possibility of repeated administration.	• Higher batch-to-batch variation and degradation. • When many molecules are attached to each other, a population of conjugates with a wide range of ligand numbers can form.	[57]
Carbon nanotubes	• Extremely compact and lightweight. • Plenty of resources needed. • Temperature resistant.	• Still don't sure how they function. • Insolubility in most solvents. • Difficulty in maintaining good quality and minimizing contaminants.	[58]
Metal nanoparticles	• Extensive plasma absorption. • Imaging of biological systems. • Analyze the chemical composition of a metallic nanoscale substrate.	• Instability. • Impurity. • Concerns about explosions and safety.	[59]

TABLE 18.2 List of Clinical Trials for Ischemic Heart Disease

SL. No.	NCT No.	Conditions	Phase	Interventions	Outcomes	Year	References
1.	NCT04271566	Coronary artery disease	Not applicable	Behavioral: education program	• RAND-short form (SF)-36 questionnaire was used to assess HRQOL. • International physical activity questionnaire (IPAQ)-short form is used to quantify physical activity.	2020	[60]
2.	NCT02771873	Coronary heart disease	Not applicable	Behavioral: integrated cardiovascular disease risk management.	• Multiple cardiovascular risk factors resulted in a composite outcome. • Blood pressure. • Blood lipids. • Blood glucose. • Behavioral factors. • Acute coronary syndrome.	2021	[61]
3.	NCT01507519	Coronary artery disease	Not applicable	Device: sirolimus-eluting coronary stent system (biomime)	• MACE at 30 days clinical F/U. • 8-month F/U angiographic binary restenosis.	2018	[62]
4.	NCT03284229	Coronary artery disease	Not applicable	Device: excimer laser coronary atherectomy.	• Participants who had device success during the procedure, which was defined as the laser catheter successfully crossing the entire length of the stenotic lesion. • Procedural success was defined as target lesions with less than 50% residual stenosis after laser and adjuvant therapy during the procedure. • Safety success–participants free of death, non-Q-wave, and Q-wave myocardial infarction (MI), target lesion revascularization, cardiac tamponade, and life-threatening arrhythmias.	2020	[63]

TABLE 18.2 *(Continued)*

SL. No.	NCT No.	Conditions	Phase	Interventions	Outcomes	Year	References
5.	NCT02406326	Coronary artery disease	–	–	• Thrombosis of the angiographic stent. • Clinically justifiable revascularization of a target lesion.	2018	[64]
6.	NCT02935088	Coronary artery disease	–	Device: fractional flow reserve	• FFR values and resting indices: Number of subjects with 12-month clinical outcomes (MACE) major adverse cardiac events.	2020	[65]
7.	NCT03023020	Coronary artery disease	Not applicable	Drug: Aspirin/P2Y12 inhibitor	• A composite of all-cause death, myocardial infarction, and stroke is characterized as a major adverse cardiac and cerebral event (MACCE). • Cardiovascular causes of death. • Stent thrombosis, either definitive or probable. • Both patients with and without clinically diagnosed over bleeding had high transfusion rates.	2021	[66]
8.	NCT03278301	Coronary artery disease	–	Device: robotic-assisted PCI	• Duration of the entire procedure. • Procedure time for PCI. • Time for fluoroscopy. • Radiation exposure to patients.	2020	[67]
9.	NCT01023789	Coronary artery disease	Not applicable	Device: ABSORB BVS	• The success of a clinical device. • Successful clinical procedure. • Number of people who died from a heart attack.	2018	[68]
10.	NCT02408224	Acute coronary syndrome	Not applicable	–	• Ticagrelore's real-world application pattern. • Ticagrelore's usage pattern.	2018	[69]

TABLE 18.2 *(Continued)*

<table>
<tr><th>SL. No.</th><th>NCT No.</th><th>Conditions</th><th>Phase</th><th>Interventions</th><th>Outcomes</th><th>Year</th><th>References</th></tr>
<tr><td>11.</td><td>NCT02256657</td><td>Acute coronary syndrome</td><td>Not applicable</td><td>Behavioral: quality improvement toolkit</td><td>• Major adverse cardiovascular events in the first 30 days.
• Medication rates in hospital.
• Medication discharge rates.
• Recommendations for a healthy lifestyle following discharge.
• Health-related life quality.</td><td>2018</td><td>[70]</td></tr>
<tr><td>12.</td><td>NCT01492361</td><td>Cardiovascular diseases</td><td>Phase 3</td><td>Drug: AMR101</td><td>• CV death, nonfatal MI (including silent MI), nonfatal stroke, coronary revascularization, and unstable angina caused by myocardial ischemia and requiring emergency hospitalization as determined by invasive/non-invasive testing.
• Non-elective coronary revascularization represents a combination of the emergent and urgent classes.
• Invasive and non-invasive tests reveal unstable angina caused by myocardial ischemia, necessitating emergency hospitalization.</td><td>2019</td><td>[71]</td></tr>
<tr><td>13.</td><td>NCT02415400</td><td>Acute coronary syndromes</td><td>Phase 4</td><td>Drug: Apixaban|Drug: vitamin K antagonist|Drug: Acetylsalicylic acid|Other: Acetylsalicylic acid placebo</td><td>• With aspirin vs. no aspirin, the rate of all-cause death or all-cause rehospitalization.
• Death and ischemic events (stroke, myocardial infarction, stent thrombosis, urgent revascularization) in patients taking aspirin vs. not taking aspirin.</td><td>2020</td><td>[72]</td></tr>
</table>

TABLE 18.3 List of Patents Related to the Treatment of Ischemic Heart Disease

SL. No.	Patent No.	Description of Patent	Assignee	Inventor	Year	References
1.	RU2715209C1	The discovery is related to medicine, specifically medical rehabilitation, balneology, and cardiology, and can be used in rehabilitation centers and departments of multidisciplinary hospitals and sanatoriums to treat patients with cardiovascular diseases following acute coronary syndrome and cardiosurgical interventions.	–	Knyazeva Tatyana Aleksandrovna, Nikiforova Tatyana Ivanovna, Eremushkin Mikhail Anatolevich, Styazhkina Elena Mikhajlovna.	2020	[73]
2.	CN109497992A	The invention relates to a type of coronary heart disease intelligence screening system based on machine learning methods, and it falls within the category of technology for screening coronary artery disease.	Jinan Huiyironggong Technology Co. Ltd.	Zhang Ming, Liu Changchun	2019	[74]
3.	US10357476B1	The invention relates to the treatment of coronary artery disease with a combination of carvedilol, metformin, evolocumab, and one or more statins.	Anis Ahmad	Anis Ahmad	2019	[75]
4.	CN107998305A	The discovery relates to the field of traditional Chinese medicine correlative technology, and more specifically to a Chinese medicine preparation for treating coronary heart disease.	Anhui Nuscan Biological Engineering Co. Ltd.	Wang Yuanbao Han Xing	2018	[76]
5.	RU2725858C1	The invention is related to medicine, specifically pathological anatomy, forensic examination, and cardiovascular surgery, and can be used for macroscopic assessment of coronary artery patency in the near and distant postoperative periods following interventional treatment of coronary heart disease (CHD) with stenting.	–	Todorov Sergej Sergeevich, Sidorov Roman Valentinovich, Todorov Sergej Sergeevich, Berezovskij Dmitrij Pavlovich, Deribas Viktoriya Yurevna, Kazmin Andrej Sergeevich	2020	[77]
6.	CN108774242A	The invention relates to medicinal chemistry, and more specifically, to one type capable of activating lecithin-cholesterol efficiently. The invention also pertains to the preparation methods for the medication acetyltransferase (LCAT).	Mudanjiang Medical University	Kong Juan, Tang Yuli, Hou Liyuan, Yang Chao, Ma Xiaosheng	2018	[78]

IHD along with related clinical trials and patents satisfying the objective of the present review.

KEYWORDS

- **bacteria**
- **clinical therapy**
- **disease pathophysiology**
- **heart disease**
- **inflammation**
- **ischemia**
- **ischemic heart disease**

REFERENCES

1. Virani, S. S., Alonso, A., Benjamin, E. J., Bittencourt, M. S., Callaway, C. W., Carson, A. P., et al., (2020). On behalf of the American Heart Association Council on epidemiology and Prevention Statistics Committee and stroke statistics subcommittee. Heart disease and stroke statistics-2020 update: A report from the American Heart Association. *Circulation, 141*, e139–e596.
2. Humphreys, M., & Warlow, C., (2011). In: Melanie, H., (ed.), *Arrhythmias and Their Management Nursing the Cardiac Patient* (1st edn., pp. 132–155). Blackwell Publishing Ltd. Published by Blackwell Publishing Ltd.
3. Ford, T. J., & Berry, C., (2020). *Heart, 106*, 387–398.
4. Thygesen, K., Alpert, J. S., Jaffe, A. S., Chaitman, B. R., Bax, J. J., Morrow, D. A., & White, H. D., (2018). Fourth universal definition of myocardial infarction. *Circulation, 138*(20), e618–e651.
5. Khan, H., Kashyap, A., Kaur, A., & Singh, T. G., (2020). Pharmacological postconditioning: A molecular aspect in ischemic injury. *Journal of Pharmacy and Pharmacology, 72*(11), 1513–1527.
6. Wang, X. J., & Li, Q. P., (2007). The roles of mesenchymal stem cells (MSCs) therapy in ischemic heart diseases. *Biochemical and Biophysical Research Communications, 359*, 189–193.
7. Templin, C., Kotlarz, D., Faulhaber, J., Schnabel, S., Grote, K., Salguero, G., Luchtefeld, M., et al., (2008). *Ex vivo* expanded hematopoietic progenitor cells improve cardiac function after myocardial infarction: Role of beta-catenin transduction and cell dose. *Journal of Molecular and Cellular Cardiology, 45*, 394–403.
8. Saklani, P., Khan, H., Gupta, S., Kaur, A., & Singh, T. G., (2022). Neuropeptides: Potential neuroprotective agents in ischemic injury. *Life Sciences, 288*, 120186.
9. Moran, A. E., Forouzanfar, M. H., Roth, G. A., Mensah, G. A., Ezzati, M., Murray, C. J., & Naghavi, M., (2014). Temporal trends in ischemic heart disease mortality in 21 world regions, 1980 to 2010: The Global Burden of Disease 2010 Study. *Circulation, 129*, 1483–1492.
10. Moran, A. E., Tzong, K. Y., Forouzanfar, M. H., Roth, G. A., Mensah, G. A., Ezzati, M., Murray, C. J., & Naghavi, M., (2014). Variations in ischemic heart disease burden by age, country, and income: The Global Burden of Diseases, Injuries, and Risk Factors 2010 Study. *Glob. Heart, 9*, 91–99.
11. Thapa, K., Khan, H., Singh, T. G., & Kaur, A., (2021). Traumatic brain injury: Mechanistic insight on pathophysiology and potential therapeutic targets. *Journal of Molecular Neuroscience, 71*(9), 1725–1742.
12. Prabhakar, N. K., Khan, H., Grewal, A. K., & Singh, T. G., (2022). Intervention of neuroinflammation in the traumatic brain injury trajectory: *In vivo* and clinical approaches. *International Immunopharmacology, 108*, 108902.

13. Severino, P., D'Amato, A., Pucci, M., Infusino, F., Adamo, F., Birtolo, L. I., Netti, L., et al., (2020). Ischemic heart disease pathophysiology paradigms overview: From plaque activation to microvascular dysfunction. *International Journal of Molecular Sciences, 21*(21), 8118.
14. Tiwari, P., Khan, H., Singh, T. G., & Grewal, A. K., (2022). Poly (ADP-ribose) polymerase: An overview of mechanistic approaches and therapeutic opportunities in the management of stroke. *Neurochemical Research*, 1–23.
15. Piccirillo, F., Carpenito, M., Verolino, G., Chello, C., Nusca, A., Lusini, M., Spadaccio, C., et al., (2019). Changes of the coronary arteries and cardiac microvasculature with aging: Implications for translational research and clinical practice. *Mechanisms of Ageing Development, 184*, 111161.
16. Nanayakkara, S., Marwick, T. H., & Kaye, D. M., (2018). The aging heart: The systemic and coronary circulation. *Heart, 104*, 370–376.
17. Prinzmetal, M., Kennarner, R., Merliss, B., Wads, T., & Bor, N., (1959). Angina pectoris. I. A variant form of angina pectoris; preliminary report. *The American Journal of Medicine, 27*, 375–388.
18. Rihal, V., Khan, H., Kaur, A., & Singh, T. G., (2022). Vitamin D as therapeutic modulator in cerebrovascular diseases: A mechanistic perspectives. *Critical Reviews in Food Science and Nutrition,* 1–23.
19. Marzilli, M., & Huqi, A., (2012). Coronary vasospasm and coronary atherosclerosis: Do we have to choose? *Journal of the American College of Cardiology, 59*, 663–664.
20. Kalra, P., Khan, H., Kaur, A., & Singh, T. G., (2022). Mechanistic insight on autophagy modulated molecular pathways in cerebral ischemic injury: From preclinical to clinical perspective. *Neurochemical Research*, 1–9.
21. Khan, H., Gupta, A., Singh, T. G., & Kaur, A., (2021). Mechanistic insight on the role of leukotriene receptors in ischemic–reperfusion injury. *Pharmacological Reports, 73*(5), 1240–1254.
22. Noma, K., Oyama, N., & Liao, J. K., (2006). Physiological role of ROCKs in the cardiovascular system. *American Journal of Physiology–Cell Physiology, 290*, C661–C668.
23. Paragh, G., Seres, I., Harangi, M., & Fulop, P., (2014). Dynamic interplay between metabolic syndrome and immunity. *Advances in Experimental Medicine and Biology, 824*, 171–190.
24. Khan, H., Tiwari, P., Kaur, A., & Singh, T. G., (2021). Sirtuin acetylation and deacetylation: A complex paradigm in neurodegenerative disease. *Molecular Neurobiology, 58*(8), 3903–3917.
25. Crea, F., Camici, P. G., & Bairey, M. C. N., (2014). Coronary microvascular dysfunction: An update. *European Heart Journal, 35*, 1101–1111.
26. Duncker, D. J., Koller, A., Merkus, D., & Canty, Jr. J. M., (2015). Regulation of coronary blood flow in health and ischemic heart disease. *Progress in Cardiovascular Diseases, 57*, 409–422.
27. Foà, A., Agostini, V., Rapezzi, C., Olivotto, I., Corti, B., Potena, L., Biagini, E., et al., (2019). Histopathological comparison of intramural coronary artery remodeling and myocardial fibrosis in obstructive versus end-stage hypertrophic cardiomyopathy. *International Journal of Cardiology, 291*, 77–82.
28. Cicalini, S., Petrosillo, N., Russo, M. A., Fedele, F., & Frustaci, A., (2010). High prevalence of intramural coronary infection in patients with drug-resistant cardiac syndrome X: Comparison with chronic stable angina and normal controls. *Heart, 96*, 1926–1931.
29. Kasprzyk, M., Wudarczyk, B., Czyz, R., Szarpak, L., & Jankowska-Polanska, B., (2018). Ischemic heart disease- definition, epidemiology, pathogenesis, risk factors and treatment. *Post. N. Med., XXXI*(6), 358–360.
30. Oduk, Y., Zhu, W., Kannappan, R., Zhao, M., Borovjagin, A. V., Oparil, S., & Zhang, J. J., (2018). VEGF nanoparticles repair the heart after myocardial infarction. *American Journal of Physiology Heart and Circulatory Physiology, 314*, H278–H284.
31. Liang, J., He, X., Zhou, H., & Liang, P., (2021). Effects of Danhong injection on cardiac function and blood lipid in patients with angina pectoris of coronary heart disease: A protocol for randomized, double-blind, placebo controlled clinical trial. *Medicine, 100*(43), e27479.
32. Holman, R. R., Coleman, R. L., Chan, J. C. N., Chiasson, J., Feng, H., Ge, J., Gerstein, H. C., et al., (2017). Effects of acarbose on cardiovascular and diabetes outcomes in patients with coronary heart disease and impaired glucose tolerance (ACE): A randomized, double-blind, placebo-controlled trial. *Lancet Diabetes Endocrinology, 5*(11), 877–886.
33. Ichimura, K., Matoba, T., Nakano, K., Tokutome, M., Honda, K., Koga, J., & Egashira, K., (2016). A translational study of a new therapeutic approach for acute myocardial infarction: Nanoparticle-mediated delivery of pitavastatin into reperfused myocardium reduces ischemia-reperfusion injury in a preclinical porcine model. *PLoS One., 11*(9), e0162425.

34. Chaikijurajai, T., & Wilson, T. W. H., (2020). Myeloperoxidase: A potential therapeutic target for coronary artery disease. *Expert Opinion on Therapeutic Targets, 24*(7), 695–705.
35. Dezsi, C. A., (2018). Treatment with triple combination of atorvastatin, perindopril, and amlodipine in patients with stable coronary artery disease: A subgroup analysis from the PAPA-CAD study. *Journal of International Medical Research, 46*(5), 1902–1909.
36. Huang, P., Wang, L., Li, Q., Xu, J., Xu, J., Xiong, Y., Chen, G., et al., (2019). Combinatorial treatment of acute myocardial infarction using stem cells and their derived exosomes resulted in improved heart performance. *Stem Cell Research & Therapy, 10*, 300.
37. Mihalko, E., Huang, K., Sproul, E., Cheng, K., & Brown, A. C., (2018). Targeted treatment of ischemic and fibrotic complications of myocardial infarction using a dual-delivery microgel therapeutic. *ACS Nanotechnology, 12*(8), 7826–7837.
38. Gayet, J. L., Paganelli, F., & Cohen-Solal, A., (2011). Update on the medical treatment of stable angina. *Archives of Cardiovascular Diseases, 104*, 536–544.
39. Nossaman, V. E., Nossaman, B. D., & Kadowitz, P. J., (2010). Nitrates and nitrites in the treatment of ischemic cardiac disease. *Cardiology in Review, 18*, 190–197.
40. Lin, Y., & Ma, L., (2018). Blood pressure lowering effect of calcium channel blockers on perioperative hypertension: A systematic review and meta-analysis. *Medicine (Baltimore), 97*(48), e13152.
41. Pavasini, R., Camici, P. G., Crea, F., Danchin, N., Fox, K., Manolis, A. J., Marzilli, M., et al., (2019). Anti-anginal drugs: Systematic review and clinical implications. *Int. J. Cardiol., 283*, 55–63.
42. Csaba, A. D., (2016). Trimetazidine in practice: Review of the clinical and experimental evidence. *American Journal of Therapeutics, 23*, e871–e87.
43. Morishita, S., Maeba, H., Takehana, K., & Shiojima, I., (2017). Nicorandil was an effective treatment option for a patient with bland-white-garland syndrome. *Internal Medicine, 56*, 2295–2299.
44. Rouhana, S., Virsolvy, A., & Fares, N., (2022). Ranolazine: An old drug with emerging potential; lessons from pre-clinical and clinical investigations for possible repositioning. *Pharmaceuticals, 15*, 31.
45. Qi, J., Zhuang, J., Lu, Y., Dong, X., Zhao, W., & Wu, W., (2017). *In vivo* fate of lipid-based nanoparticles. *Drug Discovery Today, 22*, 166–172.
46. Paulis, L. E., Geelen, T., Kuhlmann, M. T., Coolen, B. F., Schafers, M., Nicolay, K., & Strijkers, G. J., (2012). Distribution of lipid-based nanoparticles to infarcted myocardium with potential application for MRI-monitored drug delivery. *Journal of Controlled Release, 162*(2), 276–285.
47. Vinhas, R., Mendes, R., Fernandes, A. R., & Baptista, P. V., (2017). Nanoparticles emerging potential for managing leukemia and lymphoma. *Frontiers in Bioengineering and Biotechnology, 5*, 79.
48. Abbasi, E., Aval, S. F., Akbarzadeh, A., et al., (2014). Dendrimers: Synthesis, applications, and properties. *Nanoscale Research Letters, 9*(1), 247.
49. Zahin, N., Anwar, R., Tewari, D., Kabir, M. T., Sajid, A., Mathew, B., Uddin, M. S., Aleya, L., & Abdel-Daim, M. M., (2020). Nanoparticles and its biomedical applications in health and diseases: Special focus on drug delivery. *Environmental Science and Pollution Research International, 27*(16), 19151–19168.
50. Katsuki, S., Matoba, T., Koga, J. I., Nakano, K., & Egashira, K., (2017). Anti-inflammatory nanomedicine for cardiovascular disease. *Frontiers in Cardiovascular Medicine, 4*, 87.
51. Bitounis, D., Ali-Boucetta, H., Hong, B. H., Min, D. H., & Kostarelos, K. (2013). Prospects and challenges of graphene in biomedical applications. *Advanced Materials, 25*, 2258–2268.
52. Zhou, Y., Sun, H., Xu, H., Matysiak, S., Ren, J., & Qu, X., (2018). Mesoporous encapsulated chiral nanogold for use in enantioselective reactions. *Angewandte Chemie International Edition, 57*, 16791–16795.
53. Priyadarsini, S., Mukherjee, S., & Mishra, M., (2018). Nanoparticles used in dentistry: A review. *Journal of Oral Biology and Craniofacial Research, 8*, 58–67.
54. Bostan, H. B., Rezaee, R., Valokala, M. G., Tsarouhas, K., Golokhvast, K., Tsatsakis, A. M., & Karimi, G., (2016). Cardiotoxicity of nano-particles. *Life Sciences, 165*, 91–99.
55. Namdari, M., Cheraghi, M., Negahdari, B., Eatemadi, A., & Daraee, H., (2017). Recent advances in magnetoliposome for heart drug delivery. *Artificial Cells, Nanomedicine and Biotechnology, 45*, 1–7.
56. Ghasemiyeh, P., & Mohammadi-Samani, S., (2018). Solid lipid nanoparticles and nanostructured lipid carriers as novel drug delivery systems: Applications, advantages, and disadvantages. *Research in Pharmaceutical Sciences, 13*(4), 288–303.

57. Pearson, R. M., Sunoqrot, S., Hsu, H. J., Bae, J. W., & Hong, S., (2012). Dendritic nanoparticles: The next generation of nanocarriers? *Therapeutic Delivery, 3*(8), 941–959.
58. Pitroda, J., Jethwa, B., & Dave, S. K., (2016). A critical review on carbon nanotubes. *International Journal of Civil Engineering Research, 2*, 36–42.
59. Kumar, K. H., Venkatesh, N., Bhowmik, H., & Anuttam, K., (2018). Metallic nanoparticle: A review. *Biomedical Journal of Scientific & Technical Research, Biomedical Research Network, 4*(2), 3765–3775.
60. https://ClinicalTrials.gov/show/NCT04271566 (accessed on 9 November 2023).
61. https://ClinicalTrials.gov/show/NCT02771873 (accessed on 9 November 2023).
62. https://ClinicalTrials.gov/show/NCT01507519 (accessed on 9 November 2023).
63. https://ClinicalTrials.gov/show/NCT03284229 (accessed on 9 November 2023).
64. https://ClinicalTrials.gov/show/NCT02406326 (accessed on 9 November 2023).
65. https://ClinicalTrials.gov/show/NCT02935088 (accessed on 9 November 2023).
66. https://ClinicalTrials.gov/show/NCT03023020 (accessed on 9 November 2023).
67. https://ClinicalTrials.gov/show/NCT03278301 (accessed on 9 November 2023).
68. https://ClinicalTrials.gov/show/NCT01023789 (accessed on 9 November 2023).
69. https://ClinicalTrials.gov/show/NCT02408224 (accessed on 9 November 2023).
70. https://ClinicalTrials.gov/show/NCT02256657 (accessed on 9 November 2023).
71. https://ClinicalTrials.gov/show/NCT01492361 (accessed on 9 November 2023).
72. https://ClinicalTrials.gov/show/NCT02415400 (accessed on 9 November 2023).
73. https://patents.google.com/patent/RU2715209C1/en?q=treatment+of+ischemic+heart+diseae&before=priority:20220321&after=priority:20180101 (accessed on 9 November 2023).
74. https://patents.google.com/patent/CN109497992A/en?q=Ischemic+heart+disease&before=priority:20220101&after=priority:20180101 (accessed on 9 November 2023).
75. https://patents.google.com/patent/US10357476B1/en?q=method+for+treating+coronary+artery+disease&before=priority:20220101&after=priority:20180101 (accessed on 9 November 2023).
76. https://patents.google.com/patent/CN107998305A/en?q=kind+of+Chinese+medicine+preparation+for+being+used+to+treat+coronary+heart+disease&before=priority:20220101&after=priority:20180101 (accessed on 9 November 2023).
77. https://patents.google.com/patent/RU2725858C1/en?q=Treatment+of+coronary+artery+disease&before=priority:20220101&after=priority:20180101 (accessed on 9 November 2023).
78. https://patents.google.com/patent/CN108774242A/en?q=Treatment+of+coronary+artery+disease&before=priority:20220101&after=priority:20180101 (accessed on 9 November 2023).

CHAPTER 19

Nanotechnology-Assisted Treatment for Ischemic Heart Disease and Related Recent Patents

MANISH,[1] SUNIL KUMAR,[1] MANU SHARMA,[2] NIRMALA KUMARI YADAV,[1] and DEEKSHA[1]

[1]*Department of Pharmaceutical Sciences, Indira Gandhi University, Meerpur, Rewari, Haryana, India*

[2]*Department of Pharmacy, Banasthali Vidyapith, Rajasthan, India*

ABSTRACT

Ischemic heart disease is now a severe problem for individuals and medical services all over the world. Due to flow-restricting blockage in large-medium-sized coronary arteries, it is usually connected to atherosclerotic plaque. It is among the most common ailments in the United States. Each year, these diseases claim the lives of a large number of people. Despite significant advancements in the treatment of ischemia and infarct heart disease, many patients still experience problems and eventually develop heart failure. The inability to get drugs to the affected location, a lack of successful stem cell implantation, inadequate tissue regeneration, and additional damage induced by reperfusion to ischemic tissues are all factors that influence these results. Recent technical advancements have enabled the development of some nanoparticles that can be used not only to diagnose cardiovascular disorders but also to administer targeted drugs. Because of high selectivity, nanocarriers can transport molecules with poor pharmacokinetics and dynamics, such as peptides, proteins, genes, and even stem cells. The utilization of nanoparticles in the detection and treatment of ischemic heart disease is the goal of this study.

19.1 INTRODUCTION

Cardiovascular diseases (CVDs) are the world's most deadly epidemic, impacting about 37 million people globally. These are also linked to a higher risk of morbidity and death, as well as higher hospitalization rates and a lower quality of life. Among the variety of illnesses that can lead to CVDs, atherosclerosis, and ischemic heart disease are accountable for more than two-thirds of the occurrences that culminate in severe heart failure and mortality [1].

Ischemic Injury: A Molecular Insight. Thakur Gurjeet Singh & Heena Khan (Eds.)

Ischemic heart disease (IHD), also referred to as coronary artery disease (CAD), is a kind of cardiovascular disease that can cause angina (chest discomfort), myocardial infarction (MI), and cardiac death [2]. Ischemia is defined as a lack of oxygen delivery to organ tissues. The obstruction of coronary artery vessels that transport blood and oxygen to heart tissue can be caused by atherosclerosis, a common cause of CAD (Figure 19.1). Lipid deposits that have solidified into plaque are a common cause of the blockage. If the plaque continues to grow, tissues in the surrounding area may become ischemic and die [3].

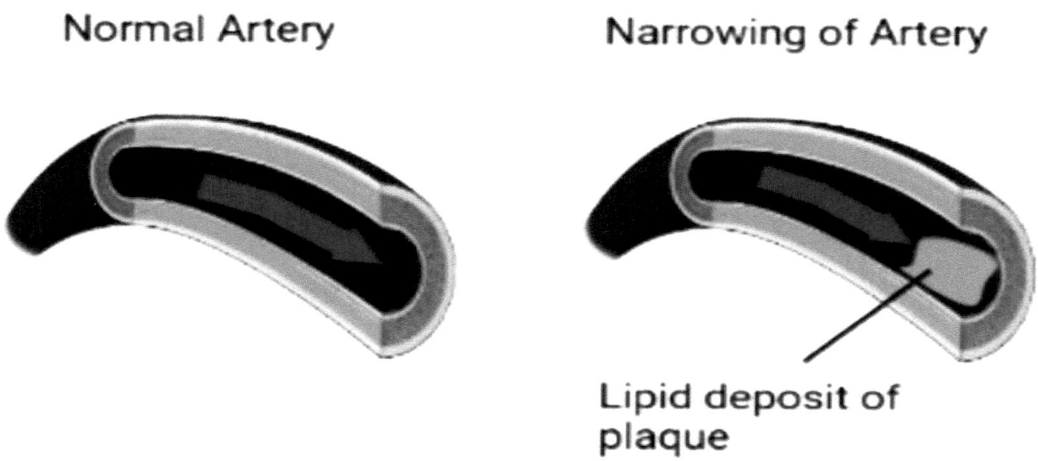

FIGURE 19.1 Lipid deposit in an artery.

Myocardial ischemia impairs heart contractile performance, causing symptoms such as tiredness and angina before potentially progressing to myocardial infarction (MI) and cardiac cell death [4]. Despite current advancements in clinical care for the prevention and treatment of ischemic heart disease (IHD), the consequences of this problem have a substantial impact on public health in terms of mortality and morbidity [5]. The present understanding of the causes of myocardial ischemia is based on basic, translational, and clinical research. Clinical, angiographic, and autoptic findings, on the other hand, indicate the complex pathophysiology of IHD [6–9], which extends beyond atherosclerosis' traditional and basic participation. As a result, it's vital to make the point that IHD is always accompanied by atherosclerosis. Researchers must explore the hypothesized IHD paradigms to better comprehend this complicated disease and provide future perspectives [10–12]. IHD has become a major public health concern in most industrialized and underdeveloped nations, its incidence and mortality rates are decreasing, but absolute numbers are still growing [13–15]. The IHD-related total mortality rate in Europe has decreased, particularly in the United Kingdom, the Netherlands, and Ireland, where it has decreased by more than 60%. Epidemiological studies, on either hand, have shown that the incidence of IHD cases and deaths has increased significantly in China, the Arab States, India, and Latin America [16–19]. Total CVD cases nearly doubled from 271 million (95% UI: 257 to 285 million) in 1990 to 523 million (95% UI: 497 to 550 million) in 2019, and CVD deaths increased steadily from 12.1 million (95% UI: 11.4 to 12.6 million) in 1990 to 18.6 million (95% UI: 17.1 to 19.7 million) in 2019. Years lived with disability quadrupled from 17.7 million (95% UI: 12.9 to 22.5 million) to 34.4 million (95% UI: 24.9 to 43.6 million) throughout that period, according to global trends for disability-adjusted life years (DALYs) and years of life lost. Since 1990, the total number of DALYs attributable to IHD has gradually increased, reaching 182 million (95% UI: 170 to 194 million) DALYs in 2019, 9.14 million (95% UI: 8.40 to 9.74 million) deaths, and 197 million (95% UI: 178 to 220 million) frequent cases of IHD in 2019 [20].

19.2 DISEASE PATHOPHYSIOLOGY

It is a pathophysiological condition caused by differences in the flow of oxygen to the myocardium. The myocardium's nutrition is determined by the blood's oxygen capacity and the amount of coronary flow. A rise in myocardial oxygen demand produced by coronary artery spasm or intravascular blood coagulation at the site of a ruptured atherosclerotic plaque causes ischemia. As a consequence, blood flow to the heart gets restricted. All of those mechanisms can be combined at the same time. The disease is most typically linked to major coronary arteries, where stenosis lowers coronary reserve proportionally to the extent of vasoconstriction. Stenosis can be accompanied by spasms, which amplify the extent of the stenosis. Atherosclerotic plaque that has ruptured is a common site for intravascular clotting, which can contribute to acute coronary events [21]. Because oxygen deficit inhibits glucose and free fatty acid (FFA) oxidation during acute ischemia, enzymatic cytoplasmic glycolysis becomes the primary source of energy. Catecholamines (epinephrine and norepinephrine) secreted by the body speed up the breakdown of fats that reach the heart. Because there is less glucose available, free fatty acid oxidation becomes the only source of energy, causing oxygen demand to surge and the reserve to rapidly deplete, forcing the cell to convert to anaerobic glycolysis. As a consequence, lactates, and hydrogen ions accumulate. Ischemia impairs cardiac contractility and relaxation for a few seconds. The lack of myocardial reperfusion for 45–60 minutes induces cardiac cell necrosis, leading to a heart attack [22]. The pathophysiology of IHD is summarized in the given Figure 19.2.

19.3 RISK PARAMETERS FOR IHD

The risk parameters for IHD have been summarized in Figure 19.3.

Individual variables are the non-modifiable parameters such as male gender, age (risk of disease increases with age), family history, and co-morbidities (kidney disease, thyroid disease, hormone therapy, hypercholesterolemia, type I and II diabetes, and chronic inflammatory disorders) [23]. Whereas on the other hand modifiable factors, are a collection of the most common parameters that can be eradicated or reduced by altering one's lifestyle. Stress, alcohol, smoking, and metabolic syndrome, include high blood sugar, hypertension, inflammation from numerous sources, visceral obesity, lipid abnormalities, and thrombotic diseases [24, 25].

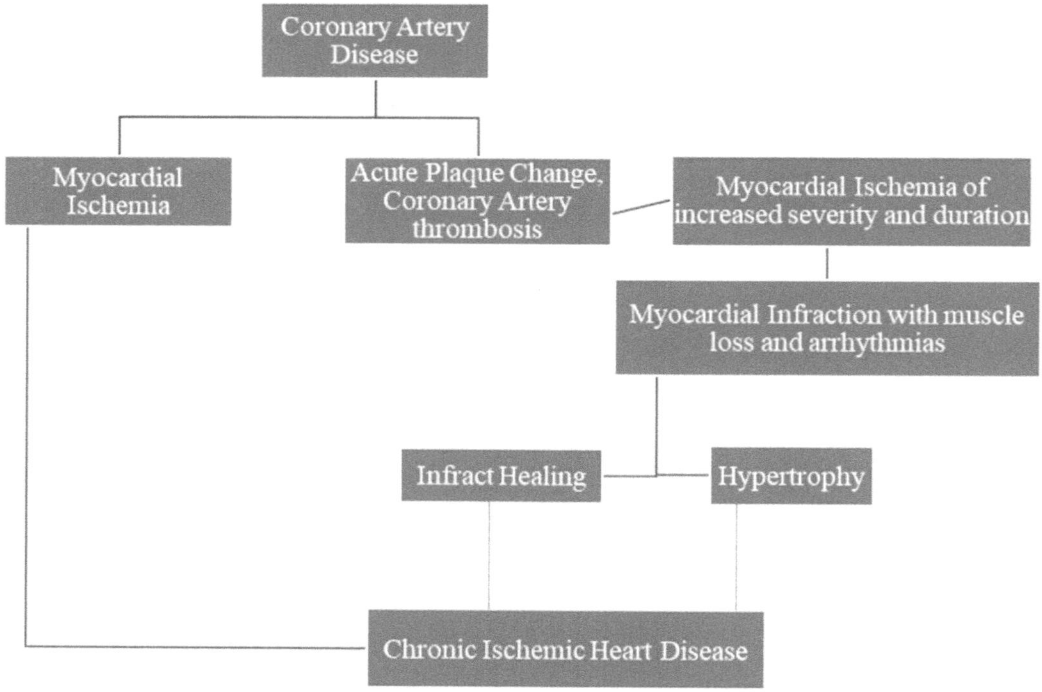

FIGURE 19.2 Pathophysiology of IHD.

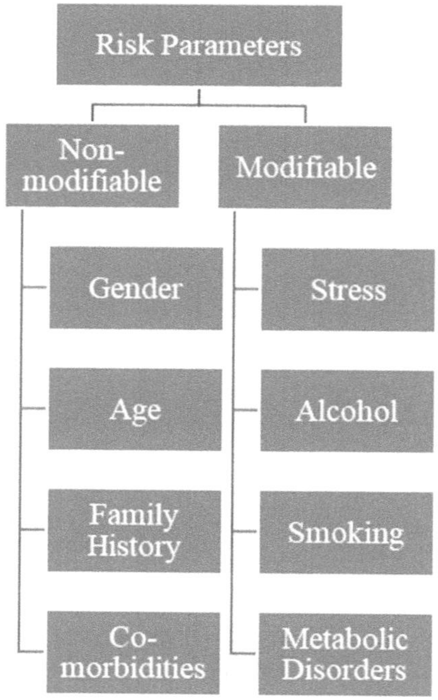

FIGURE 19.3 Illustrates risk parameters for IHD.

Despite advancements in IHD treatment, various drugs might cause extra harm, and many patients continue to develop ventricular hypertrophy and heart failure [26]. This progress might be attributed to the absence of early and effective drug delivery to injured tissue. Moreover, it has already been revealed that, based on whether or not there are any consequences, surgery to repair broken heart tissue can be beneficial or a threatening situation [27]. Nanotechnology is the manipulation of materials and technologies on a 'nano' scale ('nano' in Greek means 'dwarf,' and one nm equals 1 billionth of a meter). From the perspective of numerous cardiovascular illnesses, nanotechnology promises several benefits and technological advancements in the medical industry [28].

Nanotechnology allows for the repurposing of old drugs as well as the creation of new ones. The use of nanotechnology in imaging modalities also results in a greater signal-to-noise ratio [29]. Nanotechnology and its derived materials are becoming increasingly important globally, as seen by the development of numerous national and international organizations to advance nanotechnology and related research [30].

19.4 IMPACT OF NANOTECHNOLOGY ON ISCHEMIC HEART DISEASE

Nanotechnology is the science of atomic and molecular structures with diameters ranging from 0.1 to 100 nanometers. The resulting nanomaterials have specific mechanical, electrical, thermal, magnetic, and imaging properties, allowing for novel and distinctive applications in fields including nanomedicine, nanobiology, and nanobiotechnology. Nanoparticles (NPs) have been utilized in medicine to better strengthen the imaging strategies in the early stages of the disease, to get access to diverse areas, and to deliver therapeutically effective drugs to target cells like cancer cells [31]. Nanomaterials are also commonly used in regenerative drugs to restore damaged tissue [32]. Nanotechnologies have four key applications in cardiovascular diseases (CVDs): diagnostics, molecular imaging, targeted medication delivery, tissue regeneration, and engineering. This is because nano-molecules operate as carriers, allowing for the sustained release of imaging and diagnostic chemicals at the heart damage site [33]. Nanoparticles (NPs) are nanoscale vesicles that can contain therapeutic medications and promote myocardial repair in ischemic heart disease [34]. Nanoparticles represent an ideal drug and chemical carrier because they are engineered to evade the immune system of the host, have adequate biodegradability and biocompatibility, and must target a specific site. Furthermore, these molecules are small enough to cross through the cell membrane and target specific biological components. Liposomes, dendrimers, micelles, polymeric, and inorganic nanocarriers are some of the most investigated nanoparticles in cardiovascular disorders [35]. Carbon nanotubes are examples of inorganic NPs with exceptional characteristics, strength, and electrical capabilities (conducting, semiconducting, or insulating). Metal nanoparticles composed of gold and silver are also included in these inorganic NPs. Organic NPs are usually biocompatible, whereas inorganic NPs are useful for customizing various functions and features [36]. Organic nanoparticles are frequently used NPs in cardiac therapy and are made from carbohydrates, proteins, lipids, and other organic components to a specific size, such as a radius of roughly 100 nanometers [37]. Here we discuss different nanoparticles formulation utilized in ischemic heart disease.

19.5 ORGANIC NANOPARTICLES

1. **Lipid-based NPs:** These containing various types of therapeutics, such as bisphosphonate or prednisolone, have also been shown to suppress neointimal formation in atherosclerotic experimental mice implanted with bare-metal stents. Similar to drug-eluting stents, other polymeric-based nanoparticles, such as polylactic-glycolic acid (PLGA) nanoparticles, were loaded with paclitaxel or sirolimus and showed good outcomes [38]. The results of the systemic nanoparticle paclitaxel (nab-paclitaxel) for in-stent restenosis I (SNAPISTI) experiment were published in 2007, demonstrating that nab-paclitaxel doses of less than 70 mg/m^2 can prevent stent restenosis [39]. When compared to bare-metal stents, drug-eluting stents (DES) have shown to be effective in avoiding stent restenosis. DES, on the other hand, prevents the endothelium layer of the veins from proliferating, increasing the risk of thrombosis and necessitating long-term antithrombotic medication. To address this challenge, experimental studies have offered several ways for creating nanopatterns that facilitate efficient endothelization on the surface of titanium stents [40]. In addition, nano-molecules have been used to create nanoparticle-eluting stents that promote endothelization and acceptable internalization into smooth vascular muscle cells [41]. Lipid-based NP formulations include solid lipid nanoparticles, nanostructured lipid carriers, drug-lipid conjugates, and nanosuspensions; they are essentially made of physiological lipid analogs with surfactants acting as stabilizers [42]. Micelles (15 nm), liposomes (100 nm), and polymeric NPs are the names given to lipid-based nanoparticles based on their size [43]. Micelles are prospective carriers for delivering cardioprotective medicines, which are required during the acute stage of MI, as well as therapies that govern infarct healing during the chronic stage of MI. Alternatively, liposomes are better at delivering pro-angiogenic medicines to the infarct microvasculature [43]. Micelles are made up of lipids and other amphiphilic manmade molecules that self-assemble in an aqueous solution to form a monolayer with a hydrophobic medicinal ingredient inside. The micelle's enclosed space is smaller than that of liposomes. Liposomes are being studied extensively in nanomedicine, and they were the first nanomedicine to receive FDA approval. Liposomes are primarily made up of phospholipids that form bilayers with the aqueous phase within, giving them greater biocompatibility [44]. FDA-approved polymers include polylactic acid (PLA), polyglycolic acid (PGA), and polylactic-co-glycolic acid (PLGA) nanoparticles. PLGA is a copolymer of PLA and PGA that is being investigated for drug delivery methods for difficult-to-treat conditions, such as cardiovascular disease [45]. Lipid NPs are widely regarded as promising therapeutic delivery candidates in the infarcted heart. They have a shape that is comparable to cell membranes and can contain both lipophilic and hydrophilic materials [46]. They have effectively established the ability to deliver low molecular weight medicines, imaging agents, peptides, proteins, and nucleic acids into the target tissue [47].

2. **Inorganic Nanoparticles:** These have a core surrounded by an organic/inorganic shell that is designed to elude the immune system of the host and so boost biocompatibility. They are ideal for theranostic applications since the inorganic component can be employed to improve imaging techniques like CT and MRI. Carbon nanotubes (CNTs) are a type of fullerene made consisting of rolled-up graphite sheets that form tubular structures. To allow selective targeting, nano-carriers include pharmaceuticals in their inner space and display chemically modified external surfaces with biological molecules such as nucleotides and proteins [48]. Carbon nanotubes are categorized as single-walled or multi-walled depending on the total layers [49]. The usage of tubes, which are predominantly employed as drug transporters, can address drug solubility issues, quicker deactivation, and reduced bioavailability. However, one of the CNT's biggest drawbacks is the possibility of dissociation in biological fluids [50]. Carbon nanotubes, on the other hand, are an excellent drug carrier for increased cell penetration and privileged pharmacological activities [51]. They are a good contender for possible therapeutic uses because of their unique optical, electrical, and mechanical features [52]. Furthermore, a few studies have confirmed CNT's promising potential in cardiac tissue engineering, such as supporting cardiomyocyte function and growth and speeding up gap junction creation [53]. Aside from these results, additional research has revealed that a scaffold made of col-hydrogel and CNT could be a promising injectable biomaterial for delivering medications and cells to infarcted myocardial tissues for cardiac tissue regeneration [53, 54].

   Iron oxide NPs is another inorganic NP with good theranostic characteristics (SPION). They can be employed as a contrast agent for MRI of myocardial infarction and myocarditis experimental models due to their superparamagnetic properties, biodegradability, and surface features [55]. In addition, ferumoxytol has been used in clinical phase III trials to characterize infarct pathophysiology by generating hyperenhancement (in T2-weighted images) and signal void (in T2*-mapping images) [56]. SPIONs have also been shown to have cardioprotective properties, with Fe_2O_3 NPs coated with dimercaptosuccinic acid-reducing ROS generation and increasing NO secretion [57].
3. **Metal Nanoparticles:** Nanogold, also known as gold nanoparticles (AuNPs) or colloidal gold, is being studied in some biological applications [58]. Many GNPs are promising candidates in nanomedicine because of their unique physical and chemical features because they are easily synthesized, they present low toxicity and immunogenicity, and have good stability and strong biocompatibility. AuNPs have been conjugated with medications that are already utilized in clinical practice, such as levosimendan and beta-blockers, because of these favorable qualities. These aggregates revealed strong cardio-protective effects in an animal model of doxorubicin-induced HF [59]. Exogenous growth factors were delivered and influenced by Au NPs that accumulated in ischemic heart tissue, resulting in a 1.7-fold increase in blood perfusion at that level [60]. Silver nanoparticles (AgNPs) are antimicrobial compounds that have a wide range of applications, including

toothpaste, bedding, water purification, and nursing bottles [61]. Humans can absorb roughly 18% of silver after oral contact, according to studies [62]. AgNPs exposure causes increased superoxide anion generation and harmful effects in cardiac tissues, according to animal studies. As a result, worries about AgNPs' toxicity have hampered their use in cardiac tissue repair [63].

4. **Nanoparticle-Assisted Stem Cell Therapy:** NPs-based techniques are appealing technology for overcoming the considerable hurdles that stem cell-based heart repair presents. NPs can help stem cells to acquire greater therapeutic potential. Even though different NPs have been created in animal research, biosafety concerns remain a major roadblock to their clinical application. A thorough grasp of the properties of nanoparticles can aid in their biocompatibility design or surface modification. Furthermore, multimodal imaging will be a future trend in stem cell tracking. Different types of NPs could be possibilities for combining numerous labeling properties into a single particle. In addition, improving the resolution and sensitivity of NPs tagging probes in deep tissues, such as the heart, is a challenge. Aside from that, NPs can be a great foundation for integrating multiple apps. The hybrid NPs could be used to deliver therapeutic genes, medicines, and labeling agents into stem cells at the same time, resulting in highly reinforced stem cells that can be used for heart repair and labeling. In addition, NPs with certain properties have been studied for their ability to influence stem cell biology [64].

19.6 LIST OF CLINICAL TRIALS/RECENT PATENTS

Some of the latest clinical trials and patents related to ischemic heart diseases are described in Tables 19.1 and 19.2.

19.7 CONCLUSION

Advances in nanomedicine have enabled the development of new multifunctional nanocarriers that specifically convey therapeutic molecules to ischemia and atherosclerotic areas, not only for diagnosis but also for therapy. The development of new customized nanocomposites for drug-targeted therapy of ischemic heart failure and atherosclerosis has been made possible by recent advances in nanotechnology. Various forms of NPs could be used to combine several labeling properties into a single particle. In addition, improving the resolution and sensitivity of NPs tagging probes in deep tissues, such as the heart, is a challenge. Aside from that, NPs can be a great foundation for combining many applications. This study demonstrated how nanotechnology can be used to improve the treatment of IHD. Given the advances made, it is important to highlight that much of the research presented is still in the early stages of development, and most applications have not completed all rounds of clinical testing.

TABLE 19.1 Ischemic Heart Disease-Related Clinical Trails

SL. No.	NCT Number	Title	Status	Interventions	References
1.	NCT03411369	Creatine, D-ribose, B1 vitamin, and b6 vitamin in ischemic heart disease.	Completed	Dietary supplement: creatine, D-ribose, B1 vitamin, and B6 vitamin other: placebo	[65]
2.	NCT03397095	Sequential treatment of extra-corporeal shock wave combined with autologous bone marrow mesenchymal stem cells on patients with ischemic heart disease: the SCURE study.	Recruiting	Combination Product: CSWT + BMMSCs • Device: CSWT + Sham operation	[66]
3.	NCT02062424	Diet and prevention of ischemic heart disease: a translational approach.	Completed	Other: DIPI	[67]
4.	NCT03959631	Effectiveness and cost-effectiveness of a VCoP to empowerment of patients with ischemic heart disease in PHC: Cluster-RCT.	Recruiting	Behavioral: Virtual communities of practice	[68]
5.	NCT04148820	One versus twice daily administration of multiple cardiovascular agents in patients with ischemic heart disease.	Unknown status	Drug: Aspirin, atorvastatin, perindopril	[69]
6.	NCT02849067	Hemodynamic responses of stable ischemic heart disease patients to a comedy session.	Unknown status	Other: Group treatment Other: Group control	[70]
7.	NCT04974541	Improving wellbeing in ischemic heart disease.	Enrolling by invitation	Other: Heart health yoga Other: Psychosocial education	[71]
8.	NCT03660657	Ozone therapy in refractory ischemic heart disease.	Suspended	Drug: Ozone Drug: Oxygen-pag	[72]
9.	NCT01925352	Safety study of adenovirus hepatocyte growth factor to treat ischemic heart disease.	Unknown status	Drug: Ad-HGF	[73]
10.	NCT05003791	Perceived social support, heart rate variability, and hopelessness in patients with ischemic heart disease.	Recruiting	–	[74]

TABLE 19.2 Recent Patents Related to Ischemic Heart Disease

SL. No.	Patent No.	Description of Patent	Inventor	Current Assignee	Year	References
1.	US20200306343A1	The recent invention directed to the method of administering a therapeutically effective amount of Agrin in an anterograde intracoronary manner, to treat the ischemic heart disease in the subject.	Tzahor, E; Umansky, K.B; Hinkel, R; Kupatt, C;	Yeda Research and Development Co. Ltd.	2020	[75]
2.	US20200400684A1	Described herein are the techniques for early diagnosis of coronary artery disease, unstable angina, stemi/nstemi, and heart failure by using nourin gene-based RNA molecular network biomarkers.	Elgebaly, S.A	Nour Heart Inc.	2020	[76]
3.	US20210298608A1	The inventors have detailed the novel system for screening patients with ischemic heart disease by pattern recognition method.	Anderson, T. S.; Anderson, D. M.; MacCarter, J. D.	Shape Medical Systems Inc.	2021	[77]
4.	US11058707B2	The present disclosure provides a method for treating ischemic heart disease, reducing cardiac fibrosis, and increasing coronary angiogenesis by delivering an effective amount of TRPV4 inhibitor.	Thodeti, C. K.	Northeast Ohio Medical University	2021	[78]
5.	US20200101113A1	This patent explains the concept of utilizing bone marrow-derived neurokinin 1 receptor as a marker for identifying mesenchymal stem cells to treat various cardiovascular diseases such as ischemic cardiomyopathy.	Niece, I. M.; Altman, P.	BioCardia Inc.	2020	[79]
6.	US20220016137A1	This study discloses the methods and compositions of the treatment of various fibrotic and inflammatory diseases such as ischemia through the administration of TR-beta agonists.	Lian, B.; Masamune, H.; Barker, G. E.	Viking Therapeutics Inc.	2022	[80]
7.	WO2019246014A1	The present innovation determines the steps of manufacturing stem cell biomimetic microparticle and also explain their therapeutic applications.	Cheng K.	–	2019	[81]
8.	US20210401834A1	Given herein are the compositions and strategies to treat diseases such as angina, heart failure, and ischemia by utilizing sustained-release formulations of a modified forms of trimetazidine such as CV-8972.	Mooney, G.; Herbig, S.	Imbria Pharmaceuticals Inc.	2021	[82]

TABLE 19.2 *(Continued)*

SL. No.	Patent No.	Description of Patent	Inventor	Current Assignee	Year	References
9.	WO2020247213A1	This study discloses the combination treatment for various diseased conditions such as cancer, diabetes, and a large number of cardiovascular disorders. It includes a substance that changes cellular metabolism from fatty acid oxidation to glucose oxidation and an inhibitor of pyruvate dehydrogenase kinase.	Levin, A. D.	–	2020	[83]
10.	WO2021032955A1	The main aspect of the innovation relates to the treatment of various vascular conditions by using $\alpha v\beta 3$-and/or $\alpha v\beta 5$-integrin targeting agent.	Dilke, K. H.	–	2021	[84]

KEYWORDS

- **cardio-vascular disease**
- **coronary artery disease**
- **disease pathophysiology**
- **ischemic heart disease**
- **myocardial infarction**
- **nanoparticles**
- **risk parameters**

REFERENCES

1. Haba, M. Ş. C., Şerban, D. N., Şerban, L., Tudorancea, I. M., Haba, R. M., Mitu, O., & Tudorancea, I., (2021). Nanomaterial-based drug targeted therapy for cardiovascular diseases: Ischemic heart failure and atherosclerosis. *Crystals, 11*(10), 1172.
2. Bhatia, S. K., (2010). Coronary artery disease. In: *Biomaterials for Clinical Applications* (pp. 23–49). Springer, New York, NY.
3. Ambesh, P., Campia, U., Obiagwu, C., Bansal, R., Shetty, V., Hollander, G., & Shani, J., (2017). Nanomedicine in coronary artery disease. *Indian Heart Journal, 69*(2), 244–251.
4. Ventrelli, L., Ricotti, L., Menciassi, A., Mazzolai, B., & Mattoli, V., (2013). Nanoscaffolds for guided cardiac repair: The new therapeutic challenge of regenerative medicine. *Journal of Nanomaterials, 2013.*
5. Moran, A. E., Forouzanfar, M. H., Roth, G. A., Mensah, G. A., Ezzati, M., Murray, C. J., & Naghavi, M., (2014). Temporal trends in ischemic heart disease mortality in 21 world regions, 1980 to 2010: The Global Burden of Disease 2010 Study. *Circulation, 129*(14), 1483–1492.
6. Tiefenbacher, C. P., & Chilian, W. M., (1998). Heterogeneity of coronary vasomotion. *Basic Research in Cardiology, 93*(6), 446–454.
7. Wolin, M. S., Ahmad, M., & Gupte, S. A., (2005). Oxidant and redox signaling in vascular oxygen sensing mechanisms: Basic concepts, current controversies, and potential importance of cytosolic NADPH. *American Journal of Physiology-Lung Cellular and Molecular Physiology, 289*(2), L159–L173.
8. Tune, J. D., (2006). Withdrawal of vasoconstrictor influences in local metabolic coronary vasodilation. *American Journal of Physiology-Heart and Circulatory Physiology, 291*(5), H2044–H2046.
9. Dzeja, P. P., Bast, P., Pucar, D., Wieringa, B., & Terzic, A., (2007). Defective metabolic signaling in adenylate kinase AK1 gene knock-out hearts compromises post-ischemic coronary reflow. *Journal of Biological Chemistry, 282*(43), 31366–31372.
10. Ottaviani, G., Lavezzi, A. M., Rossi, L., & Matturri, L., (1999). Proliferating cell nuclear antigen (PCNA) and apoptosis in hyperacute and acute myocardial infarction. *European Journal of Histochemistry: EJH, 43*(1), 7–14.
11. Wong, B., Kruse, G., Kutikova, L., Ray, K. K., Mata, P., & Bruckert, E., (2016). Cardiovascular disease risk associated with familial hypercholesterolemia: A systematic review of the literature. *Clinical Therapeutics, 38*(7), 1696–1709.
12. Rihal, V., Khan, H., Kaur, A., & Singh, T. G., (2022). Vitamin D as therapeutic modulator in cerebrovascular diseases: A mechanistic perspectives. *Critical Reviews in Food Science and Nutrition*, 1–23.
13. Hartley, A., Marshall, D. C., Salciccioli, J. D., Sikkel, M. B., Maruthappu, M., & Shalhoub, J., (2016). Trends in mortality from ischemic heart disease and cerebrovascular disease in Europe: 1980 to 2009. *Circulation, 133*(20), 1916–1926.

14. Khan, M. A., Hashim, M. J., Mustafa, H., Baniyas, M. Y., Al Suwaidi, S. K. B. M., AlKatheeri, R., & Lootah, S. N. A. H., (2020). Global epidemiology of ischemic heart disease: Results from the Global Burden of Disease Study. *Cureus, 12*(7).
15. Mensah, G. A., Roth, G. A., & Fuster, V., (2019). The global burden of cardiovascular diseases and risk factors: 2020 and beyond. *Journal of the American College of Cardiology, 74*(20), 2529–2532.
16. Liu, S., Li, Y., Zeng, X., Wang, H., Yin, P., Wang, L., & Zhou, M., (2019). Burden of cardiovascular diseases in China, 1990–2016: Findings from the 2016 Global Burden of Disease Study. *JAMA Cardiology, 4*(4), 342–352.
17. Prabhakaran, D., Jeemon, P., Sharma, M., Roth, G. A., Johnson, C., Harikrishnan, S., & Dandona, L., (2018). The changing patterns of cardiovascular diseases and their risk factors in the states of India: The Global Burden of Disease Study 1990–2016. *The Lancet Global Health, 6*(12), e1339–e1351.
18. Yusuf, S., Rangarajan, S., Teo, K., Islam, S., Li, W., Liu, L., & Dagenais, G., (2014). Cardiovascular risk and events in 17 low-, middle-, and high-income countries. *New England Journal of Medicine, 371*(9), 818–827.
19. Lanas, F., Serón, P., & Lanas, A., (2013). Coronary heart disease and risk factors in Latin America. *Global Heart, 8*(4), 341–348.
20. Roth, G. A., Mensah, G. A., Johnson, C. O., Addolorato, G., Ammirati, E., Baddour, L. M., & GBD-NHLBI-JACC Global Burden of Cardiovascular Diseases Writing Group, (2020). Global burden of cardiovascular diseases and risk factors, 1990–2019: Update from the GBD 2019 study. *Journal of the American College of Cardiology, 76*(25), 2982–3021.
21. Vollmer-Conna, U., Cvejic, E., Smith, I. G., Hadzi-Pavlovic, D., & Parker, G., (2015). Characterizing acute coronary syndrome-associated depression: Let the data speak. *Brain, Behavior, and Immunity, 48*, 19–28.
22. Jankowski, P., Czarnecka, D., Badacz, L., Bogacki, P., Dubiel, J. S., Grodecki, J., & Pająk, A., (2018). Practice setting and secondary prevention of coronary artery disease. *Archives of Medical Science: AMS, 14*(5), 979.
23. Garg, N., Singh, T. G., Khan, H., Arora, S., Kaur, A., & Mannan, A., (2021). Mechanistic interventions of selected *Ocimum* species in management of diabetes, obesity, and liver disorders: Transformative developments from preclinical to clinical approaches. *Biointerface Res. Appl. Chem., 12*(1), 1304–1323.
24. Katz, P., Leiter, L. A., Mellbin, L., & Ryden, L., (2014). The clinical burden of type 2 diabetes in patients with acute coronary syndromes: Prognosis and implications for short-and long-term management. *Diabetes and Vascular Disease Research, 11*(6), 395–409.
25. Wessler, J. D., & Kirtane, A. J., (2013). Patients who require non-cardiac surgery in acute coronary syndrome. *Current Cardiology Reports, 15*(7), 1–9.
26. Evans, C. W., Iyer, K. S., & Hool, L. C., (2016). The potential for nanotechnology to improve delivery of therapy to the acute ischemic heart. *Nanomedicine, 11*(7), 817–832.
27. Begum, M., & Sharma, H., (2016). Scope of nanomedicine against coronary artery disease: A review. *Eur. J. Pharm. Med. Res., 3*, 635–641.
28. Tiwari, P., Khan, H., Singh, T. G., & Grewal, A. K., (2022). Poly (ADP-ribose) polymerase: An overview of mechanistic approaches and therapeutic opportunities in the management of stroke. *Neurochemical Research*, 1–23.
29. Landowski, L. M., Niego, B. E., Sutherland, B. A., Hagemeyer, C. E., & Howells, D. W., (2020). Applications of nanotechnology in the diagnosis and therapy of stroke. In: *Seminars in Thrombosis and Hemostasis* (Vol. 46, No. 5, pp. 592–605). Thieme Medical Publishers.
30. Wimmer, I., Zrzavy, T., & Lassmann, H., (2018). Neuroinflammatory responses in experimental and human stroke lesions. *Journal of Neuroimmunology, 323*, 10–18.
31. Kim, B. Y., Rutka, J. T., & Chan, W. C., (2010). Nanomedicine. *New England Journal of Medicine, 363*(25), 2434–2443.
32. Mahmoudi, M., Hosseinkhani, H., Hosseinkhani, M., Boutry, S., Simchi, A., Journeay, W. S., & Laurent, S., (2011). Magnetic resonance imaging tracking of stem cells *in vivo* using iron oxide nanoparticles as a tool for the advancement of clinical regenerative medicine. *Chemical Reviews, 111*(2), 253–280.
33. Hajipour, M. J., Mehrani, M., Abbasi, S. H., Amin, A., Kassaian, S. E., Garbern, J. C., & Mahmoudi, M., (2019). Nanoscale technologies for prevention and treatment of heart failure: Challenges and opportunities. *Chemical Reviews, 119*(21), 11352–11390.

34. Oduk, Y., Zhu, W., Kannappan, R., Zhao, M., Borovjagin, A. V., Oparil, S., & Zhang, J., (2018). VEGF nanoparticles repair the heart after myocardial infarction. *American Journal of Physiology-Heart and Circulatory Physiology, 314*(2), H278–H284.
35. Sabir, F., Barani, M., Mukhtar, M., Rahdar, A., Cucchiarini, M., Zafar, M. N., & Bungau, S. N., (2021). Nanotreatment of cardiovascular diseases: An overview. *Chemosensors, 9*, 67.
36. Vinhas, R., Mendes, R., Fernandes, A. R., & Baptista, P. V., (2017). Nanoparticles—Emerging potential for managing leukemia and lymphoma. *Frontiers in Bioengineering and Biotechnology, 5*, 79.
37. Pan, K., & Zhong, Q., (2016). Organic nanoparticles in foods: Fabrication, characterization, and utilization. *Annual Review of Food Science and Technology, 7*, 245–266.
38. McDowell, G., Slevin, M., & Krupinski, J., (2011). Nanotechnology for the treatment of coronary in stent restenosis: A clinical perspective. *Vascular Cell, 3*(1), 1–5.
39. Margolis, J., McDonald, J., Heuser, R., Klinke, P., Waksman, R., Virmani, R., & Hilton, D., (2007). Systemic nanoparticle paclitaxel (nab-paclitaxel) for in-stent restenosis I (SNAPIST-I): A first-in-human safety and dose-finding study. *Clinical Cardiology, 30*(4), 165–170.
40. Bassous, N., Cooke, J. P., & Webster, T. J., (2016). Enhancing stent effectiveness with nanofeatures. *Methodist DeBakey Cardiovascular Journal, 12*(3), 163.
41. Nakano, K., Egashira, K., Masuda, S., Funakoshi, K., Zhao, G., Kimura, S., & Sunagawa, K., (2009). Formulation of nanoparticle-eluting stents by a cationic electrodeposition coating technology: Efficient nano-drug delivery via bioabsorbable polymeric nanoparticle-eluting stents in porcine coronary arteries. *JACC: Cardiovascular Interventions, 2*(4), 277–283.
42. Qi, J., Zhuang, J., Lu, Y., Dong, X., Zhao, W., & Wu, W., (2017). *In vivo* fate of lipid-based nanoparticles. *Drug Discovery Today, 22*(1), 166–172.
43. Paulis, L. E., Geelen, T., Kuhlmann, M. T., Coolen, B. F., Schäfers, M., Nicolay, K., & Strijkers, G. J., (2012). Distribution of lipid-based nanoparticles to infarcted myocardium with potential application for MRI-monitored drug delivery. *Journal of Controlled Release, 162*(2), 276–285.
44. Katsuki, S., Matoba, T., Koga, J. I., Nakano, K., & Egashira, K., (2017). Anti-inflammatory nanomedicine for cardiovascular disease. *Frontiers in Cardiovascular Medicine, 4*, 87.
45. Pascual-Gil, S., Simón-Yarza, T., Garbayo, E., Prósper, F., & Blanco-Prieto, M. J., (2017). Cytokine-loaded PLGA and PEG-PLGA microparticles showed similar heart regeneration in a rat myocardial infarction model. *International Journal of Pharmaceutics, 523*(2), 531–533.
46. Saludas, L., Pascual-Gil, S., Roli, F., Garbayo, E., & Blanco-Prieto, M. J., (2018). Heart tissue repair and cardioprotection using drug delivery systems. *Maturitas, 110*, 1–9.
47. Cheraghi, M., Negahdari, B., Daraee, H., & Eatemadi, A., (2017). Heart targeted nanoliposomal/nanoparticles drug delivery: An updated review. *Biomedicine & Pharmacotherapy, 86*, 316–323.
48. Çakmak, H. A., & Demir, M., (2020). MicroRNA and cardiovascular diseases. *Balkan Medical Journal, 37*(2), 60.
49. Sajid, M. I., Jamshaid, U., Jamshaid, T., Zafar, N., Fessi, H., & Elaissari, A., (2016). Carbon nanotubes from synthesis to *in vivo* biomedical applications. *International Journal of Pharmaceutics, 501*(1, 2), 278–299.
50. Raphey, V. R., Henna, T. K., Nivitha, K. P., Mufeedha, P., Sabu, C., & Pramod, K. J. M. S., (2019). Advanced biomedical applications of carbon nanotube. *Materials Science and Engineering: C, 100*, 616–630.
51. Zhang, W., Zhang, Z., & Zhang, Y., (2011). The application of carbon nanotubes in target drug delivery systems for cancer therapies. *Nanoscale Research Letters, 6*(1), 1–22.
52. Gorain, B., Choudhury, H., Pandey, M., Kesharwani, P., Abeer, M. M., Tekade, R. K., & Hussain, Z., (2018). Carbon nanotube scaffolds as emerging nanoplatform for myocardial tissue regeneration: A review of recent developments and therapeutic implications. *Biomedicine & Pharmacotherapy, 104*, 496–508.
53. Ahadian, S., Huyer, L. D., Estili, M., Yee, B., Smith, N., Xu, Z., & Radisic, M., (2017). Moldable elastomeric polyester-carbon nanotube scaffolds for cardiac tissue engineering. *Acta Biomaterialia, 52*, 81–91.
54. Sun, H., Tang, J., Mou, Y., Zhou, J., Qu, L., Duval, K., & Tian, F., (2017). Carbon nanotube-composite hydrogels promote intercalated disc assembly in engineered cardiac tissues through β1-integrin mediated FAK and RhoA pathway. *Acta Biomaterialia, 48*, 88–99.
55. Dadfar, S. M., Roemhild, K., Drude, N. I., Von, S. S., Knüchel, R., Kiessling, F., & Lammers, T., (2019). Iron oxide nanoparticles: Diagnostic, therapeutic and theranostic applications. *Advanced Drug Delivery Reviews, 138*, 302–325.

56. Bietenbeck, M., Florian, A., Sechtem, U., & Yilmaz, A., (2015). The diagnostic value of iron oxide nanoparticles for imaging of myocardial inflammation–quo Vadis? *Journal of Cardiovascular Magnetic Resonance, 17*(1), 1–12.
57. Xiong, F., Wang, H., Feng, Y., Li, Y., Hua, X., Pang, X., & Gu, N., (2015). Cardioprotective activity of iron oxide nanoparticles. *Scientific Reports, 5*(1), 1–8.
58. Zhou, Y., Sun, H., Xu, H., Matysiak, S., Ren, J., & Qu, X., (2018). Mesoporous encapsulated chiral nanogold for use in enantioselective reactions. *Angewandte Chemie, 130*(51), 17033–17037.
59. Zhang, J., Ma, A., & Shang, L., (2018). Conjugating existing clinical drugs with gold nanoparticles for better treatment of heart diseases. *Frontiers in Physiology, 9*, 642.
60. Pala, R., Anju, V. T., Dyavaiah, M., Busi, S., & Nauli, S. M., (2020). Nanoparticle-mediated drug delivery for the treatment of cardiovascular diseases. *International Journal of Nanomedicine, 15*, 3741.
61. Priyadarsini, S., Mukherjee, S., & Mishra, M., (2018). Nanoparticles used in dentistry: A review. *Journal of Oral Biology and Craniofacial Research, 8*(1), 58–67.
62. Bostan, H. B., Rezaee, R., Valokala, M. G., Tsarouhas, K., Golokhvast, K., Tsatsakis, A. M., & Karimi, G., (2016). Cardiotoxicity of nano-particles. *Life Sciences, 165*, 91–99.
63. Lin, C. X., Yang, S. Y., Gu, J. L., Meng, J., Xu, H. Y., & Cao, J. M., (2017). The acute toxic effects of silver nanoparticles on myocardial transmembrane potential, I Na and I K1 channels and heart rhythm in mice. *Nanotoxicology, 11*(6), 827–837.
64. Zhu, K., Li, J., Wang, Y., Lai, H., & Wang, C., (2016). Nanoparticles-assisted stem cell therapy for ischemic heart disease. *Stem Cells International, 2016*.
65. https://clinicaltrials.gov/ct2/show/NCT03411369 (accessed on 9 November 2023).
66. https://clinicaltrials.gov/ct2/show/NCT03397095 (accessed on 9 November 2023).
67. https://clinicaltrials.gov/ct2/show/NCT02062424 (accessed on 9 November 2023).
68. https://clinicaltrials.gov/ct2/show/NCT03959631 (accessed on 9 November 2023).
69. https://clinicaltrials.gov/ct2/show/NCT04148820 (accessed on 9 November 2023).
70. https://clinicaltrials.gov/ct2/show/NCT02849067 (accessed on 9 November 2023).
71. https://clinicaltrials.gov/ct2/show/NCT04974541 (accessed on 9 November 2023).
72. https://clinicaltrials.gov/ct2/show/NCT03660657 (accessed on 9 November 2023).
73. https://clinicaltrials.gov/ct2/show/NCT01925352 (accessed on 9 November 2023).
74. https://clinicaltrials.gov/ct2/show/NCT05003791 (accessed on 9 November 2023).
75. Tzahor, E., Umansky, K. B., Hinkel, R., & Kupatt, C., (2020). *Treatment of an Ischemic Heart Disease.* US z20200306343A1.
76. Elgebaly, S. A., (2020). *Nourin Gene-Based RNA Molecular Network: Novel Early Diagnostic and Prognostic Biomarkers for Coronary Artery Disease, Unstable Angina, Stemi/Nstemi and Heart Failure.* US20200400684A1.
77. Anderson, T. S., Anderson, D. M., & MacCarter, J. D., (2021). *Pattern Recognition System for Identifying Patients with Ischemic Heart Disease.* US20210298608A1.
78. Thodeti, C. K., (2021). *Methods for Treating Ischemic Heart Disease by Targeting TRPV4.* US11058707B2.
79. Niece, I. M., & Altman, P., (2020). *Bone Marrow Derived Neurokinin-1 Receptor Positive (nk1r+) Precursor Cells for Therapeutic Applications.* US20200101113A1.
80. Lian, B., Masamune, H., & Barker, G. E., (2022). *Compositions for the Treatment of Fibrosis and Inflammation.* US20220016137A1.
81. Cheng, K., (2019). *Stem Cell Biomimetic Nanoparticle Therapeutic Agents and Uses Thereof.* WO20192 46014A1.
82. Mooney, G., & Herbig, S., (2021). *Modified Release Formulations of Modified Forms of Trimetazidine.* US20210401834A.
83. Levin, A. D., (2020). *Combination Therapies that Include an Agent that Promotes Glucose Oxidation and an Inhibitor of Pyruvate Dehydrogenase Kinase.* WO2020247213A1.
84. Dilke, K. H., (2021). *Treatment.* WO2021032955A1.

CHAPTER 20

Major Risk Factors Involved in the Pathobiology of Ischemic Brain and Heart Disorders

PRIYANKA SHARMA,[1] SURAY RANJAN,[1] KAJAL BAGRI,[1] RAHUL DESHMUKH,[1] and THAKUR GURJEET SINGH[2]

[1]*Department of Pharmaceutical Sciences and Technology, Maharaja Ranjit Singh Punjab Technical University, Bathinda, Punjab, India*

[2]*Chitkara College of Pharmacy, Chitkara University, Punjab, India*

ABSTRACT

The specific pathophysiology of a stroke affects its risk factors and course of treatment because it is a heterogeneous illness. The two categories of stroke risk factors are: modifiable (those that can be changed) and non-modifiable (those that cannot be changed). Risk factors for ischemic and hemorrhagic stroke include age, sex, and race/ethnicity which cannot be changed, but modifiable risk factors like high blood pressure, smoking, nutrition, and physical inactivity are more frequently documented. Inflammatory conditions, infections, pollutants, and cardiac atrial conditions unrelated to atrial fibrillation (AF) are the most recent causes and risk factors for stroke. Stroke may be the major symptom of rare, inherited illnesses that are caused by single-gene abnormalities. Additionally, according to recent studies, both common and rare genetic variants may affect the risk of stroke from more common causes may also be affected by both extra risk factors and specific stroke mechanisms, such as AF. It may be more possible to modify genetic factors than previously thought, especially those that interact with the environment. Modifiable risk factors have typically been the focus of stroke prevention. In addition to lowering the risk of stroke, lifestyle, and behavioral modifications including quitting smoking and altering one's diet also lower the risk of other cardiovascular diseases. Other prophylactic methods are recognizing and managing diseases like diabetes and hypertension that raise the risk of stroke. Recent studies into the genetics and risk factors of stroke have revealed those at risk and also targeted the approaches for prevention of stroke in high-risk groups.

Ischemic Injury: A Molecular Insight. Thakur Gurjeet Singh & Heena Khan (Eds.)

20.1 INTRODUCTION

A stroke is a possibly fatal brain injury that results in considerable functional loss and a high death rate. Haemorrhagic stroke and ischemic stroke are the two types of strokes. Memory loss, cognitive impairment, paralysis, etc., are common in patients with stroke [1, 2]. A greater understanding of the biology of the risk of ischemic stroke, death, and loss of function is required for the progress of innovative treatments. Ischemic stroke risk and pathophysiology are influenced in a complex and tangled fashion by age and gender. Age is the most powerful non-modifiable danger issue for ischemic stroke, with older stroke patients experiencing greater death, injury, and practical recovery than younger stroke patients. Importantly, the impact of patient sex on ischemic stroke is changed by patient age [3]. Early in life, ischemic stroke is more common and debilitating in men, whereas stroke turns into more common and devastating in women as they get older. The *in vivo* and *in vitro* research outcomes of experiments show that gender and age have a significant impact on clinical ischemic stroke [4]. Due to sex variances in etiology, clinical symptoms, therapy, and the influence of cardiovascular medications, as recently reviewed, many variations in etiology, clinical symptoms, therapy, and impact of cardiovascular medications have been documented.

20.2 GLOBAL EPIDEMIOLOGY OF STROKE

Stroke is a severe sickness that disables. Around 15 million people annually around the world. Stroke cases are more common and frequent in the age group above 65 years, making it an age-related illness [5, 6]. Older individuals had an increased death rate and lower life quality after a stroke than the young ones. Women experience more strokes overall than men, despite men having a higher rate of stroke for the majority of their lifetimes. This is because women's typical lifespans are longer and their stroke risk rises with age. Around 87% of all stroke cases in the US are ischemic strokes [4]. Between 2012 and 2030, it is anticipated that the cases of ischemic stroke will rise by 3.4 million persons as a result of population aging and a decline in stroke-related case fatalities [7] (Figure 20.1).

20.3 TYPES OF RISK FACTORS FOR STROKE

Various types of risk factors are there for the stroke. Detection of lifestyle modifications and establishment of such effectiveness of risk reduction initiatives are necessary to lower the incidence of stroke in people [8]. High blood pressure, smoke, waist-to-hip ratio, diet, and excessive alcohol use were risk factors for hemorrhagic stroke. A large amount of data is present for the long-term risk factors of the stroke but less is known about the short time risk factors and targets of the stroke [7] (Figure 20.2).

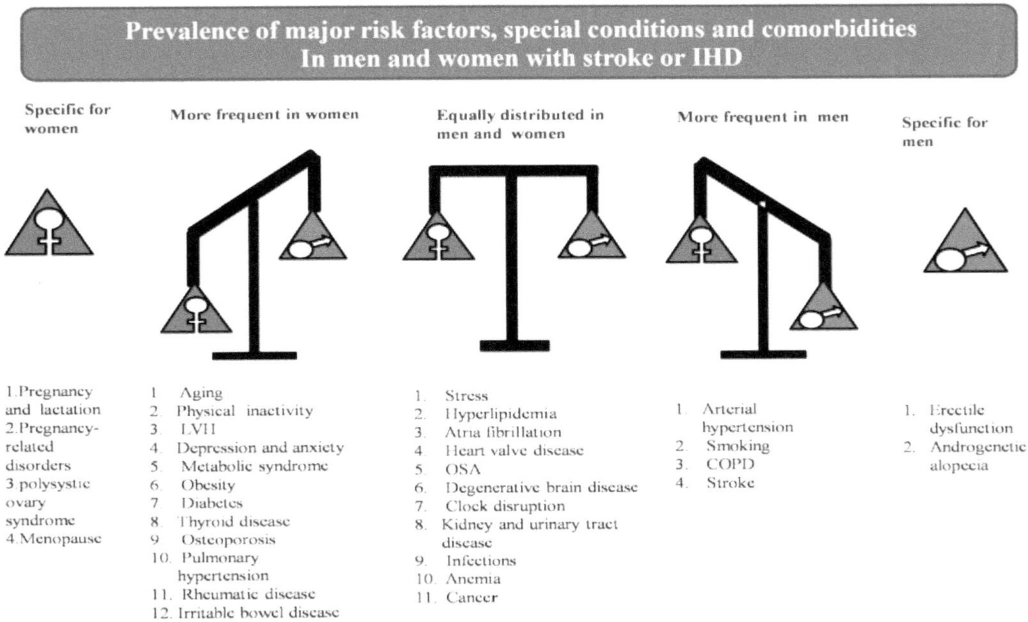

FIGURE 20.1 Comparison of risk factors, special conditions, and co-morbidities in men and women.

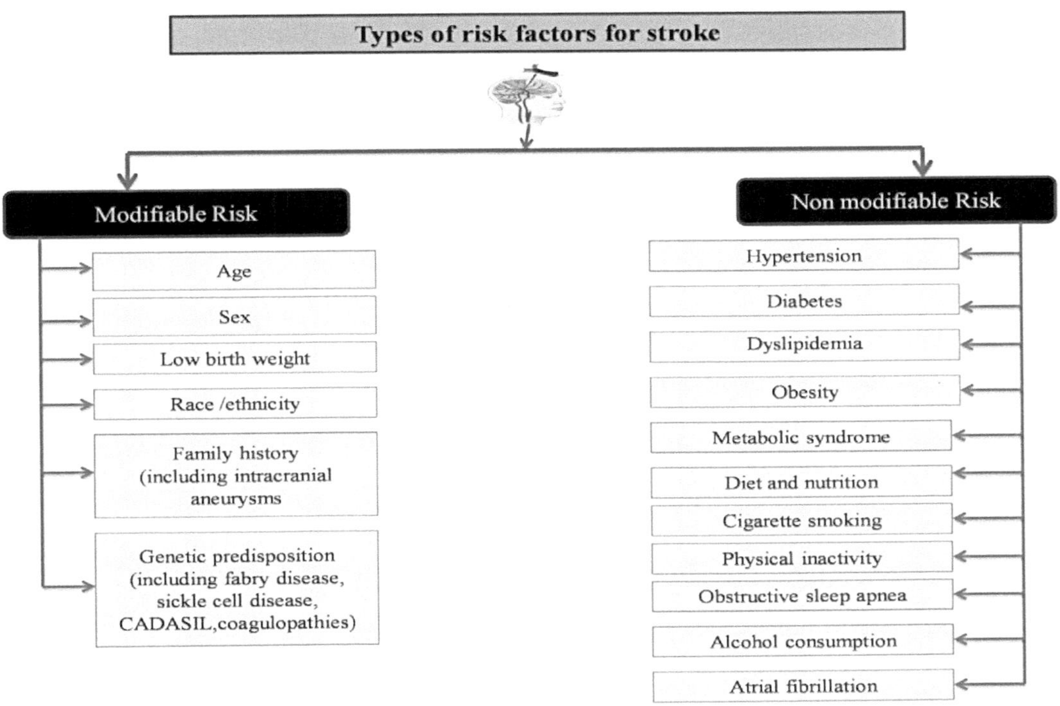

FIGURE 20.2 Type of risk factors for stroke.

20.3.1 MODIFIABLE

Risk factors that can be changed by an individual themself to reduce their chances of acquiring or managing a chronic illness are known as modifiable risk factors. In the following part, different modifiable risk factors like diet, comorbid conditions, obesity, neuro-psychological diseases, smoking, and physical activities are the different risk factors associated with ischemic stroke [9]. All these factors are discussed below one by one.

20.3.1.1 DIET

Along with other risk factors of stroke-like hypertension, diabetes, and dyslipidemia, diet affects the likelihood of having a stroke. The stroke risk has been associated with some dietary aspects. Excess salt consumption in the diet has been linked to hypertension as well as an elevated risk of cardiovascular disease and stroke. Increased intake of fruits and vegetables is linked to a dose-dependent decrease in the stroke risk [10, 11]. It is found in different epidemiological studies that various diets can lower the risk of stroke. Aquatic food contains proteins with lots of hydrogen bond donors and acceptors that attract protons and promote the creation of acids that dissolve insoluble salts and reduce the risk of stroke. Lactic acid-producing bacteria which is found in yogurt also helps to dissolve hard calcium ions. The consumption of this diet limits the expression of risk factors, hence lowering the incidence of stroke. Certain essential amino acids, such as valine, phenylalanine, lysine, and leucine are found in lower amounts in plant-based diets. Thus, this diet intake reduces the expression of risk factors and decreases the risk of stroke. People at stroke risk may benefit from intermittent fasting which restricts the intake of all amino acids [1]. Patients at risk for stroke should decrease their daily consumption of salt to 2–3 grams, decrease their consumption of meat, especially red meat, and avoid egg yolk. Additionally, the patients should be examined for metabolic B12 deficiency and treated for hyper-homocysteinemia, most likely with methyl cobalamin. Such initiatives require further consideration from the general population, and especially from doctors, as they can drastically decline the risk of stroke [12].

20.3.1.2 COMORBID CONDITIONS

Comorbid condition is a term used when two or more two diseases occur together like obesity, infection, inflammation, hypertension, and diabetes [13, 14].

1. **Infection:** Infectious diseases such as viruses, germs, and parasites have been associated with atherosclerosis and stroke. Some illnesses, such as helicobacter pylori, chlamydia pneumonia, and CMV, have a significant correlation with one another, but others, like influenza, need further study. The number of occurrences has not decreased despite this, according to major randomized prospective trials investigating the effectiveness of antibiotic therapy for stroke secondary prevention [15]. It is still completely unknown how gender affects treatment outcomes as well as the link between infections and stroke.

2. **HIV:** Despite recognized risk factors for cardiovascular diseases, such as socio-economic status, HIV infection and certain anti-retroviral treatment consumption are linked to an elevated risk of CVD. HIV infection has been related to two-fold greater stroke risk is higher for females than for males, even though the majority of studies on HIV-positive patients mostly included male subjects [16, 17]. Decreased body mass, a lower drug metabolism, and hormonal regulation could be the causes of the variations in coexisting stroke outcomes and gender-related variations in antiretroviral-related toxicity. Additionally, women with HIV utilize fewer stroke-related treatment methods than men with comparable risk profiles.
3. **Obesity:** Body mass index and obesity are stroke risk factors, although it is still unclear exactly how these factors do so. Obesity is linked to high blood pressure and insulin resistance [18, 19]. In a cohort study of 1.8 million participants which was included in a recent comprehensive meta-analysis, it was found that blood pressure, cholesterol, and glucose levels mediated 76% of the relationship between body mass index, a popular indicator of obesity, and stroke risk. Only blood pressure was responsible for 65% of the risk associated with weight. Differentiating between overall increasing weight, as indicated by the body mass index, and inclined abdominal adiposity, as specified by the waist-to-hip ratio, as the primary contributor to risk, is becoming more and more crucial [20].
4. **Coronary Heart Disease:** In the first two weeks after an acute MI, stroke incidence ranges from 0.7% to 4.7%. The risk of stroke following a heart attack is increased by ventricular dysfunction (lower ejection fraction) and increasing age. The incidence of post-MI stroke is decreased by anticoagulant treatment using aspirin or warfarin, which supports the existence of an embolic process [21]. The most frequent consequence of anterior wall MI is stroke, which affects 2–6% of patients and is brought on by a left ventricular mural thrombus, which is detected in 40% of patients during echocardiographic examinations. In 40% of patients, an echocardiographic finding of a left ventricular mural thrombus may be indicative of inferior wall MI. Mural thrombus or stroke as a result of inferior wall MI is uncommon [22].
5. **Hypertension:** The most significant modifiable factor of risk for stroke is high blood pressure, which is strongly, directly, linearly, and continuously associated with the risk of stroke. Age-related blood pressure rises regardless of the presence of hypertension, raising the lifelong chance of developing it. More than two-thirds of those over 65 have hypertension. The cases of high blood pressure in the United States have sustained at 29%, and hypertension management has increased due to increased knowledge and therapy choices, with 50% controlled in 2008. Patients with high blood pressure are urged to make behavioral and lifestyle changes, such as dietary adjustments and increased physical activity, in addition to taking medication to lower their blood pressure. One of the best ways to reduce the risk of stroke is to treat hypertension, whether with medication or by making lifestyle adjustments, yet this condition is still undertreated in the general population [23].
6. **Diabetes Mellitus:** A stroke accounts for about 20% of mortality in diabetes. Diabetes mellitus is an independent factor of risk for stroke having a 2-fold greater risk of stroke in diabetic individuals. Stroke risk is higher for people with

prediabetes. Patients with diabetes are more likely to be young and to have more additional stroke risk factors [24]. A portion of the rise in the risk of stroke in early age populations may be explained by the rise in diabetes mellitus. It has been demonstrated that combining medical therapy with behavioral adjustment can decline the risk of stroke in diabetics.

7. **Inflammation:** It has been discovered that C-reactive protein is a risk factor for stroke and transient ischemic attack. Taking into account other risk factors, males with CRP levels in the top quartile had a risk of stroke, and TIA was twice as high as men in the lowest quartile; for women, the risk was around three times higher. Because it causes systemic inflammation, gum disease is also associated with a higher risk of stroke [25–27]. Stroke risk may also be correlated with other inflammatory signs of broad or microvascular illness [28].
8. **Dyslipidemia:** Dyslipidemia and the risk of stroke have a complex relationship; increasing total cholesterol is associated with an increased risk of ischemic stroke while elevated high-density lipoprotein cholesterol is associated with a decreased risk of ischemic stroke. There is conflicting evidence about the impact of triglycerides on the risk of stroke. Furthermore, it appears that risk varies depending on the type of stroke, with cholesterol levels having a larger correlation with big artery ischemic stroke than with other ischemic stroke subtypes. Haemorrhagic stroke risk increases when total cholesterol drops, although total cholesterol has an antagonistic relationship with hemorrhagic stroke [29].

Despite the potentially conflicting and opposing results from these studies regarding the relationship between dyslipidemia and risk of acute ischemic stroke (AIS), statin use helps in the prevention of total and ischemic stroke with no distinguishable increased incidence of hemorrhagic stroke in the overall percentage of patients. Furthermore, the relatively significant decrease in the risk of ischemic stroke and other ischemic events caused by statins surpasses any marginally higher risk of hemorrhage in the majority of individuals. However, statins may be linked to an increased risk of intracerebral hemorrhage in some stroke patients, particularly in those with prior bleeding, cerebral amyloid angiopathy, or small vessel disease.

20.3.1.3 NEURO-PSYCHOLOGICAL DISEASES

1. **Degenerative Brain Disease:** Females with IHD are more likely to face dementia since stroke is a danger factor for dementia or memory deficits. Additionally, the prevalence of cognitive impairment and different subtypes of dementia varies between men and women. To minimize the risk of dementia in IHD patients, antiplatelet/antithrombotic medicines have been suggested [30]. Contrarily, antiplatelet medications differed in their protective effects between women and men, emphasizing the significance of sex-linked pathophysiology changes.
2. **Clock Disruption:** Internal molecular clocks control many functions, including hormone production, metabolism, blood pressure, bone marrow function, sleep patterns, heart rate, appetite, and body temperature are all factors. This is reflected

in the diurnal clinical manifestations of disorders like MI, which are more common in the morning. When the normal activity and resting phases are disturbed, cardiovascular measures, healing responses, and remodeling are all affected. The sex and estrogen cycles have been demonstrated to have an impact on changes in the biological clock of plasma corticosterone levels in rats [31]. According to reports, metabolic problems and cardiomyopathy, which were seen in male clock mutant mice, are not present in female clock mutant mice. It is unknown how important this protection is for preventing strokes, although it may be mediated by ovarian hormones through differently controlled metabolic pathways [32, 33].

3. **Depression and Anxiety:** Both male and female stroke patients are prone to depression and anxiety disorders, which have been associated with greater rates of mortality and morbidity as well as increased mortality in patients with coronary artery disease. In research limited to men, depression was a cardiovascular danger equivalent to overweight and high cholesterol levels [34]. In terms of processes, a rat experiment demonstrated a sex-specific variation in (BDNF) and cAMP response element-binding protein in the molecular stress response [35]. Mice with a point mutation in the BDNF protein had a malfunction in the coagulation cascade, and this mutation was strongly linked to MI [36] Intriguingly, the presence of a BDNF polymorphism is linked to either female sex or depressed symptoms, showing a direct connection between altered BDNF function and greater vulnerability to stroke in female carrying this particular variation.

20.3.1.4 SMOKING

The risk of ischemia and hemorrhagic subtypes of stroke is enhanced by the use of illegal drugs, such as cocaine, heroin, amphetamines, and ecstasy. Pack-years and the risk of stroke have a dose-response connection. Cigarette smoking continues to be a significant risk factor for stroke, substantially doubling the risk. Smoking is thought to be a factor in 15% of all stroke deaths annually. Stroke risk is significantly decreased when a smoker stops smoking, and the increased risk almost completely disappears two to four years after quitting. In the REGARDS cohort, second-hand smoke has been found as a solitary stroke risk factor. After controlling for other stroke risk factors, the risk of stroke increases by 30% compared to those of us who have and those who have not been exposed to second-hand smoke [11].

20.3.1.4.1 Alcohol Consumption and Illicit Drug Use

The number of alcoholic drinks has an impact on the link between drinking and stroke risk. Higher risks of cardiovascular disease are linked to heavy alcohol use, whether it be regular daily heavy drinking or binge drinking [37]. On the other hand, the incidence of CHD is negatively correlated with light to moderate alcohol usage. Alcohol consumption and stroke have a complicated association. The amount of alcohol consumed and the chance of having an ischemic stroke are correlated in a U-shaped manner.

The use of illegal substances, such as heroin, cocaine, and amphetamines, is linked to a higher risk of ischemic and hemorrhagic stroke. These medications have been linked to hematologic and metabolic abnormalities, like increased platelet aggregation, blood pressure swings, and vasculopathy, as well as cerebral embolization due to endocarditis or secondary cardiac problems. Strokes can result from side effects of intravenous drug use, such as infective endocarditis [38–41].

20.3.1.5 PHYSICAL ACTIVITY

In addition to improving glycemic control and promoting a healthier life that includes quitting smoking, physical activity can reduce blood pressure, cause reduced loss, slow heart rate, raise High-density Lipid protein and lower Low-Density Lipid triglycerides levels, raise blood sugar levels, and all these things that are risk factors for atherosclerotic disease. Physically healthy men had a 60% reduced ischemic stroke risk than sedentary men after adjusting for possible confounders such as SBP, serum blood cholesterol, hyperglycemia, vital capacity, overweight, left ventricular hypertrophy on the ECG, AE, valvular disease, CHD, and CHE. Conversely, studies have indicated that women can benefit from exercise Similar advantages to those of higher intensive activity levels were provided by moderate physical exercise, such as CHD [11] (Figure 20.3).

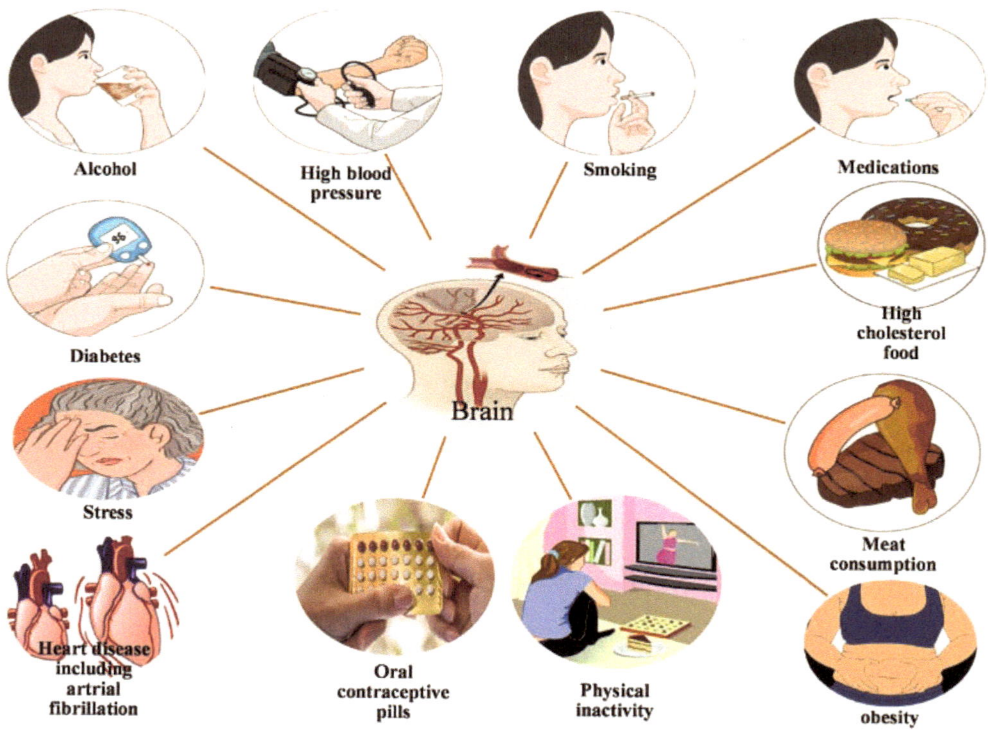

FIGURE 20.3 Common risk factors for IHD/stroke.

20.3.2 NON-MODIFIABLE

These are the aspects that influence an individual's capacity to adjust their chance or likelihood of acquiring or managing chronic illness. Non-modifiable risk factors are also known as risk markers. Non-modifiable risk factors for stroke include sex, age, race-ethnic background, and gene-related factors [7, 42].

20.3.2.1 AGE

Age affects how sex and stroke risk are related. Although the proportionate risk is slightly higher for men as they age, women are just as likely as men to have a stroke when they are young. Younger women have a higher risk of stroke, which is likely brought on by risks related to pregnancy, the postpartum period, and other hormonal factors including the use of hormonal contraceptives more women than men get strokes because women typically live longer than men [43, 44].

20.3.2.2 ETHNIC BACKGROUND

When compared to their contemporaries who are white, black people have a twofold higher prevalence of incident stroke and a higher stroke-related mortality rate. Hispanic/Americans have a higher risk of stroke in some populations. The difference in stroke incidence is especially pronounced among young black individuals, whose incidence of subarachnoid and intracerebral hemorrhage is much greater than that of age-matched whites [23]. The higher incidence of stroke risk factors, such as BP, overweight, and type 2 diabetes, among blacks, as recently shown by the REGARDS trial, may be one explanation for the racial differences. However, the heightened risk seen in these races and ethnicities cannot entirely be attributed to these additional risk factors. Although it has been suggested that the black race influences the link between stroke risk and rurality, this could be due to problems with access to healthcare. Other socioeconomic disease-causing factors, language, and culture are additional variables that may affect racial and differences in the race in stroke risk. The racial differences in stroke incidence are what cause the racial differences in stroke mortality [45].

20.3.2.3 FAMILY HISTORY

The rise in the risk of stroke is related to the family history of stroke (FHS). The difference in the effects of FHS is observed in different studies. An FHS may indicate the development of stroke at a younger age, but the wide variance in age at onset suggests that environmental factors have a role in deciding how far along in the course of stroke a person can be. Different forms of FHS (maternal, paternal, and sibling histories) have been found to differ in incident stroke. Although the difference wasn't always noticeable, sibling history was more strongly connected with incident stroke than parental history. Individuals with maternal vs paternal history had a different chance of getting an incident stroke. In both hemorrhagic stroke and ischemic stroke, young age was more highly correlated with the FHS than old age. In patients of 25–49 age, the presence of family background of an early stroke is a stronger independent risk factor for stroke inheritance [46, 47].

20.3.2.4 SEX

Sex in ischemic stroke etiology depends on patient age since the effect of sex on stroke risk and prognosis varies throughout time. In youth and early adulthood, males had a greater incidence of ischemic stroke and less functional outcomes than females. With the onset of menopause and the disappearance of female sex hormones, ischemic stroke rates in women start to increase in middle age. After middle age, women's stroke rates continue to grow, and according to some studies, women over 85 have a greater stroke rate than older men [48, 49].

Sexually dimorphic risk factors surprisingly change with time, contributing to sex-dependent stroke incidences. Stroke risk factors for women include the use of the oral contraceptive pill (OCP), pregnancy, menopause, and hormone treatment. Despite the low stroke frequency in women of reproductive age, using OCPs significantly increases the risk of stroke, with high-estrogen OCPs carrying the highest risk. Oestrogen promotes coagulation and may raise the chance of coagulation in women who use estrogen-containing OCP. Women with a history of migraines when using Oral contraceptive pills are more prone to stroke [20, 51].

20.3.2.5 GENETIC RISK FACTORS

Stroke risk is influenced by hereditary variables, while it is still difficult to distinguish between risk resulting from genetic mutations and risk resulting from common familial exposures. The endeavor has become more challenging because of the stroke heterogeneity, the abundance of traditional risk factors, and the variation in research and population. However, there are several possible ways that genetic variability may increase the risk of stroke [52] (Figure 20.4).

20.3.2.5.1 Stroke Genetic Variability

Current assessments of heritability based on genomic sequence single nucleotide polymorphism (SNP) data show similar heritability for cardioembolic (32.06%) and big vascular disease (40.03%) but lesser heritability for microvascular disease (16.01%). Stroke risk is 30% higher in families with a history of the disease. Compared to dizygotic twins, monozygotic twins have a fold of 1.65 which increases the chance of stroke. Further influencing stroke heritability include age, sex, and stroke subtype. Adolescent people are much more likely to have a first ancestor who's had a stroke than older patients, and women experiencing stroke seem to be more likely to have a genetic history of stroke. However, concordance rates for MRI measurements of small artery ischemic illness are hereditary susceptibility to small vessel ischemic stroke of 0.61 for monozygotic twins and 0.38 for dizygotic twins [53, 54].

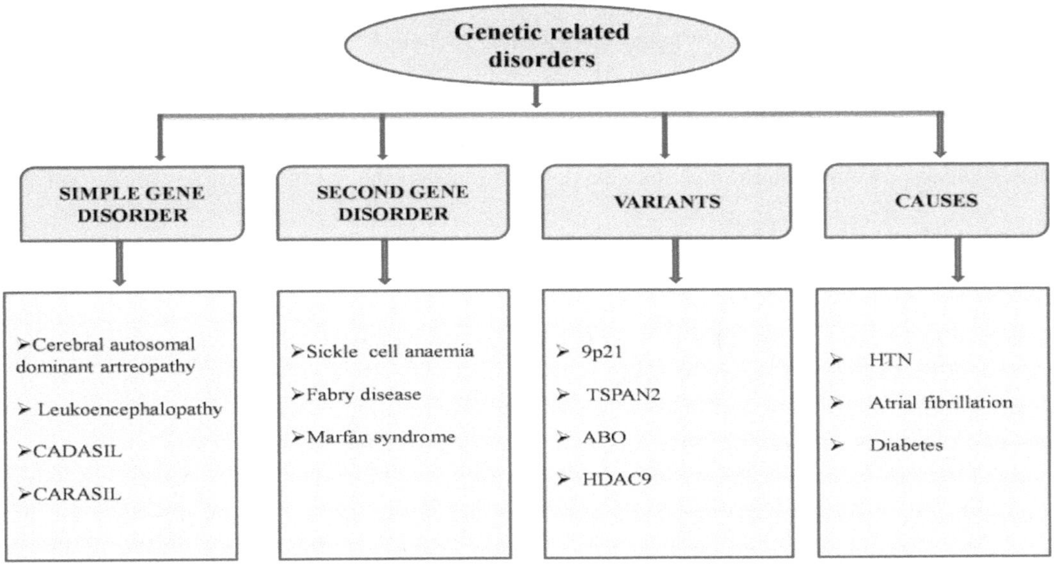

FIGURE 20.4 Genetic disorders affecting stroke.

20.3.2.6.2 Cadasil

CADASIL is a small vessel vasculopathy that affects the brain and skin, though its clinical signs are limited to the nervous system. The Notch3 gene, which is on chromosome 19q12, is linked to CADASIL through mutations. The majority of these mutations are missense ones that change how many extracellular receptor regions express cysteine residues. 19q12 CADASIL only has a clinically significant effect on the nervous system, for unexplained reasons, despite Notch3 being broadly expressed in the body and having a significant impact on development. It has been demonstrated that the faulty receptors gather in the basal lamina of the small arteries, even though they typically do not obstruct phenotypic signaling. Granular osmiophilic material found in a skin biopsy may be pathognomonic for the condition, allowing for the diagnosis of some individuals with very mild imaging signs and unfavorable outcomes of a genetic test for the most prevalent mutations [55].

20.3.2.6.3 Carasil

Cerebral autosomal recessive arteriopathy with subcortical infarcts and leukoencephalopathy (CARASIL), characterized by abnormal HtrA serine peptidase-1 gene, and arterial deformation sickness, resulting from mutations in the SLC2A10 gene, which codes for the glucose carrier GLUT10 [56].

20.3.2.6.4 Sickle Cell Anemia

Cerebrovascular problems from reduced oxygen concentrations and polymerized red blood cells, which cause tiny artery occlusion and sickle-related arterial illness, are well-known in sickle cell anemia (SCA) (moyamoya syndrome). About 6% of children who have a stroke have SCA, but by the time they are 45 years old, 25% of people with SCA will also have a stroke; ischemic strokes most frequently occur in children ages 2 to 5. If left untreated, a stroke can occur again up to 90% of the time. Silent infarcts can affect cognitive function and academic achievement and may be more frequent than symptomatic infarcts [57].

20.3.2.6.5 Fabry Disease

Fabry ailment, an X-linked deficiency in the lysosomal enzyme galactosidase A, is brought on by both misspelling and nonsensical variations in the GLA gene. Globotriaosylceramide builds up in cells all across the body, including the vascular endothelium, when – galactosidase A is no longer functioning. The second most prevalent lysosomal storage disorder is Fabry disease. Young stroke patients may experience cerebrovascular involvement, which can affect both big and small vessels and has a particular impact on the posterior circulation. Some of the molecular mechanisms of stroke include cardiac embolism brought on by cardiac complicity, peripheral vascular venous thrombosis brought by the accumulation of glycolipids in the vascular endothelium and smooth muscle, large vessels develop dolichoectasia and tortuosity, while small vessels develop the occlusive disease [58].

20.3.2.7 ENVIRONMENTAL FACTORS

20.3.2.7.1 Air Pollution

This concern has recently been confirmed in several studies, both in cities with low atmospheric air pollution and ultrafine (PM1.0) particles, as well as potential pathophysiologic pathways [59, 60]. Historically, air pollution has been thought to play a role in an increased chance of CV disease. Stroke risk has been associated with elevated levels of nitrous oxides, ozone, and particulate matter.

20.3.2.7.2 Tobacco Use

In both men and women, smoking has been implicated in the development of ischemic stroke, ICH, and SAH. In the late 1970s, many OC investigations identified smoking as a substantial risk factor for stroke in young women. It was established that the connection between the two was connected to the risk of SAH. Research by the Royal College of Practitioners on OC use in women over 35 who smoked cigarettes and was either current or previous OC users found an elevated incidence of SAH. Tobacco users had a higher prevalence of SAH and angiogenic stroke in the Midwives' National Health Survey, a cohort of more than 12,000 women who were followed prospectively for eight years. The incidence of stroke increased with cigarette smoking in a dose-dependent manner, from four times as high in light smokers to 9.8 times as high in those who consumed 25 or perhaps more cigarettes daily. After adjusting for other risk factors, the risk of SAH in each smoking group was twice that of thromboembolic stroke [61].

20.4 CONCLUSION

The risk, prognosis, and pathophysiology of ischemic stroke are complicated and significantly influenced by sex and age. Young males had a greater rate of stroke than middle-aged men, whereas old women have a greater risk of stroke, mortality, and a worse functional result. Particular emphasis is placed in this chapter on the significance of taking into account sex, age, cardiovascular disease, and numerous environmental factors as biological variables. This can be accomplished in two ways: (i) making sure that clinical research in the future is sufficiently powered to examine the effect of sex and age on treatment outcomes; and (ii) using young and old animals of both sexes in studies that examine the pathogenesis of ischemic stroke and test new therapeutic approaches.

Ischemic heart disease, an epidemic that affects both genders globally, is frequently linked to multimorbidity in adults and the elderly. Sex differences are becoming more widely accepted in scientific and medical organizations, and many ischemic heart disease mechanisms differ between the sexes. Ischemic heart disease has been linked to several risk factors and ischemic heart disease types. Despite this data, there are essentially no studies at all research levels looking into the consequences of comorbidities in ischemic heart disease that are particular to sex. Studies that are too narrowly focused might bias the results of the research and leave out crucial information about human diseases, such as sex differences and how they relate to comorbid disorders. To overcome these obstacles, a virtuous loop firmly combining preclinical, translational, and clinical research would need to account for sex, comorbidities, and their therapies.

KEYWORDS

- **air pollution**
- **alcohol consumption**
- **comorbid conditions**
- **genetic risk factors**
- **neuro-psychological diseases**
- **sickle cell anemia**
- **stroke genetic variability**

REFERENCES

1. Huang, J., Lin, Y., Liao, L., David, C. L., & Liu, Q., (2022). *Molecular Features of Risk Factors for Stroke and Potential Mechanism.* 1–6.
2. Thapa, K., Khan, H., Singh, T. G., & Kaur, A., (2021). Traumatic brain injury: Mechanistic insight on pathophysiology and potential therapeutic targets. *Journal of Molecular Neuroscience, 71*(9), 1725–1742.
3. Kalra, P., Khan, H., Kaur, A., & Singh, T. G., (2022). Mechanistic insight on autophagy modulated molecular pathways in cerebral ischemic injury: From preclinical to clinical perspective. *Neurochemical Research*, 1–19.
4. Aggarwal, N. R., Patel, H. N., Mehta, L. S., Sanghani, R. M., Lundberg, G. P., Lewis, S. J., Mendelson, M. A., et al., (2018). Sex differences in ischemic heart disease: Advances, obstacles, and next steps. *Circulation: Cardiovascular Quality and Outcomes, 11*(2), e004437.
5. Khan, H., Kashyap, A., Kaur, A., & Singh, T. G., (2020). Pharmacological postconditioning: A molecular aspect in ischemic injury. *Journal of Pharmacy and Pharmacology, 72*(11), 1513–1527.
6. Saklani, P., Khan, H., Gupta, S., Kaur, A., & Singh, T. G., (2022). Neuropeptides: Potential neuroprotective agents in ischemic injury. *Life Sciences, 288*, 120186.
7. Boehme, A. K., Esenwa, C., & Elkind, M. S. V., (2017). Stroke risk factors, genetics, and prevention. *Circulation Research, 120*(3), 472–495.
8. Khan, H., Gupta, A., Singh, T. G., & Kaur, A., (2021). Mechanistic insight on the role of leukotriene receptors in ischemic–reperfusion injury. *Pharmacological Reports, 73*(5), 1240–1254.
9. Hill, V. A., & Towfighi, A., (2017). Modifiable risk factors for stroke and strategies for stroke prevention. *Seminars in Neurology, 37*(03), 237–258.
10. Larsson, S. C., Wallin, A., & Wolk, A., (2016). Dietary approaches to stop hypertension diet and incidence of stroke: Results from 2 prospective cohorts. *Stroke, 47*(4), 986–990.
11. Talaei, M., Koh, W., Yuan, J., & Van, D. R. M., (2019). DASH dietary pattern, mediation by mineral intakes, and the risk of coronary artery disease and stroke mortality. *Journal of the American Heart Association, 8*(5), e011054.
12. Spence, J. D., (2019). Nutrition and risk of stroke. *Nutrients, 11*(3), 647.
13. Hu, J. X., Thomas, C. E., & Brunak, S., (2016). Network biology concepts in complex disease comorbidities. *Nature Reviews Genetics, 17*(10), 615–629.
14. Prabhakar, N. K., Khan, H., Grewal, A. K., & Singh, T. G., (2022). Intervention of neuroinflammation in the traumatic brain injury trajectory: *In vivo* and clinical approaches. *International Immunopharmacology, 108*, 108902.
15. Elkind, M. S. V., Boehme, A. K., Smith, C. J., Meisel, A., & Buckwalter, M. S., (2020). Infection as a stroke risk factor and determinant of outcome after stroke. *Stroke, 51*(10), 3156–3168.
16. Bogorodskaya, M., Chow, F. C., & Triant, V. A., (2019). Stroke in HIV. *Canadian Journal of Cardiology, 35*(3), 280–287.
17. Khan, H., Sharma, R., Kaur, A., & Singh, T. G., (2018). The endocannabioids system and their implications in various disorders. *Int. J. Pharm. Sci. Rev. Res.*
18. Garg, N., Singh, T. G., Khan, H., Arora, S., Kaur, A., & Mannan, A., (2021). Mechanistic interventions of selected *Ocimum* species in management of diabetes, obesity, and liver disorders: Transformative developments from preclinical to clinical approaches. *Biointerface Res. Appl. Chem, 12*(1), 1304–1323.
19. Firdous, S. M., Ghosh, S., Nath, P., & Khan, H., (2021). Molecular basis of diabetic nephropathy. *Diabetes and Diabetic Complications.* Nova Science Publishers, Hauppauge.
20. Lee, H. J., Choi, E. K., Lee, S. H., Kim, Y. J., Han, K. D., & Oh, S., (2018). Risk of ischemic stroke in metabolically healthy obesity: A nationwide population-based study. *PLoS One, 13*(3), e0195210.
21. Rihal, V., Khan, H., Kaur, A., & Singh, T. G., (2022). Vitamin D as therapeutic modulator in cerebrovascular diseases: A mechanistic perspectives. *Critical Reviews in Food Science and Nutrition*, 1–23.
22. Benjamin, E. J., Muntner, P., Alonso, A., Bittencourt, M. S., Callaway, C. W., Carson, A. P., Chamberlain, A. M., et al., (2019). Heart disease and stroke statistics—2019 update: A report from the american heart association. *Circulation, 139*(10), e56–e528.

23. Alloubani, A., Saleh, A., & Abdelhafiz, I., (2018). Hypertension and diabetes mellitus as a predictive risk factors for stroke. *Diabetes & Metabolic Syndrome: Clinical Research & Reviews, 12*(4), 577–584.
24. Tini, K., Samatra, I. D. P. G. P., Wiryadana, K. A., & Supadmanaba, I. G. P., (2020). Clinical profile of patients with cerebrovascular disease at stroke unit, Sanglah general hospital, Denpasar, *Bali. Bali Med. J., 9*(1), 129–136.
25. Khan, H., Sharma, K., Kumar, A., Kaur, A., & Singh, T. G., (2022). Therapeutic implications of cyclooxygenase (COX) inhibitors in ischemic injury. *Inflammation Research*, 1–16.
26. Khan, H., Garg, N., Singh, T. G., Kaur, A., & Thapa, K., (2022a). Calpain inhibitors as potential therapeutic modulators in neurodegenerative diseases. *Neurochemical Research*, 1–25.
27. Khan, H., Bangar, A., Grewal, A. K., Bansal, P., & Singh, T. G., (2022c). Caspase-mediated regulation of the distinct signaling pathways and mechanisms in neuronal survival. *International Immunopharmacology, 110*, 108951.
28. Jenny, N. S., Callas, P. W., Judd, S. E., McClure, L. A., Kissela, B., Zakai, N. A., & Cushman, M., (2019). Inflammatory cytokines and ischemic stroke risk: The REGARDS cohort. *Neurology, 92*(20), e2375–e2384.
29. Gainey, J., Blum, B., Bowie, B., Cooley, K., Madeline, L., Ervin, E. L., & Nathaniel, T. I., (2018). Stroke and dyslipidemia: Clinical risk factors in the telestroke versus non-telestroke. *Lipids in Health and Disease, 17*(1), 1–10.
30. Kim, M. Y., Noh, Y., Son, S. J., Shin, S., Paik, H. Y., Lee, S., & Jung, Y. S., (2018). Effect of cilostazol on incident dementia in elderly men and women with ischemic heart disease. *Journal of Alzheimer's Disease, 63*(2), 635–644.
31. Crnko, S., Ernens, I., & Van, L. L. W., (2018). New dimensions in circadian clock function: The role of biological sex. In: *Cardiovascular Research* (Vol. 114, No. 2, pp. 203–204). Oxford University Press.
32. Garg, C., Kaur, A., Singh, T. G., Sharma, V. K., & Singh, S. K., (2022). Therapeutic implications of sonic hedgehog pathway in metabolic disorders: Novel target for effective treatment. *Pharmacological Research*, 106194.
33. Khan, H., Grewal, A. K., & Singh, T. G., (2022b). Pharmacological postconditioning by protocatechuic acid attenuates brain injury in ischemia–reperfusion (I/R) mice model: Implications of nuclear factor erythroid-2-related factor pathway. *Neuroscience, 491*, 23–31.
34. Ladwig, K. H., Baumert, J., Marten-Mittag, B., Lukaschek, K., Johar, H., Fang, X., Ronel, J., et al., (2017). Room for depressed and exhausted mood as a risk predictor for all-cause and cardiovascular mortality beyond the contribution of the classical somatic risk factors in men. *Atherosclerosis, 257*, 224–231.
35. Tiwari, P., Khan, H., Singh, T. G., & Grewal, A. K., (2022). Poly (ADP-ribose) polymerase: An overview of mechanistic approaches and therapeutic opportunities in the management of stroke. *Neurochemical Research*, 1–23.
36. Amadio, P., Colombo, G. I., Tarantino, E., Gianellini, S., Ieraci, A., Brioschi, M., Banfi, C., et al., (2017). BDNFVal66met polymorphism: A potential bridge between depression and thrombosis. *European Heart Journal, 38*(18), 1426–1435.
37. Gupta, A., Khan, H., Kaur, A., & Singh, T. G., (2021). Novel targets explored in the treatment of alcohol withdrawal syndrome. *CNS & Neurological Disorders-Drug Targets (Formerly Current Drug Targets-CNS & Neurological Disorders), 20*(2), 158–173.
38. Falkstedt, D., Wolff, V., Allebeck, P., Hemmingsson, T., & Danielsson, A. K., (2017). Cannabis, tobacco, alcohol use, and the risk of early stroke: A population-based cohort study of 45 000 Swedish men. *Stroke, 48*(2), 265–270.
39. Khan, H., Singh, T. G., Dahiya, R. S., & Abdel-Daim, M. M., (2022d). α-Lipoic acid, an organosulfur biomolecule a novel therapeutic agent for neurodegenerative disorders: An mechanistic perspective. *Neurochemical Research*, 1–12.
40. Saklani, P., Khan, H., Singh, T. G., Gupta, S., & Grewal, A. K., (2022a). Demethyleneberberine, a potential therapeutic agent in neurodegenerative disorders: A proposed mechanistic insight. *Molecular Biology Reports*, 1–13.
41. Kaur, A., Singh, T. G., Khan, H., Kumar, M., Singh, N., & Abdel-Daim, M. M., (2022). Neuroprotective effect of piclamilast-induced post-ischemia pharmacological treatment in mice. *Neurochemical Research*, 1–14.

42. Cui, Q., (2019). Modifiable and non-modifiable risk factors in ischemic stroke: A meta-analysis. *African Health Sciences, 19*(2), 2121–2129.
43. Lasek-Bal, A., Kopyta, I., Warsz-Wianecka, A., Puz, P., Łabuz-Roszak, B., & Zaręba, K., (2018). Risk factor profile in patients with stroke at a young age: Stroke risk factors in young patients. *Neurological Research, 40*(7), 595–601.
44. Roy-O'Reilly, M., & McCullough, L. D., (2018). Age and sex are critical factors in ischemic stroke pathology. *Endocrinology, 159*(8), 3120–3131.
45. George, H., (2016). Where to focus efforts to reduce the black-white disparity in stroke mortality: Incidence versus case fatality? *Stroke, 47*(7), 1893–1898.
46. Pourasgari, M., & Mohamadkhani, A., (2020). Heritability for stroke: Essential for taking family history. *Caspian Journal of Internal Medicine, 11*(3), 237.
47. Tian, T., Jin, G., Yu, C., Lv, J., Guo, Y., Bian, Z., Yang, L., et al., (2017). Family history and stroke risk in China: Evidence from a large cohort study. *Journal of Stroke, 19*(2), 188.
48. Deb-Chatterji, M., Schlemm, E., Flottmann, F., Meyer, L., Alegiani, A., Brekenfeld, C., Fiehler, J., et al., (2021). Sex differences in outcome after thrombectomy for acute ischemic stroke are explained by confounding factors. *Clinical Neuroradiology, 31*(4), 1101–1109.
49. Scrutinio, D., Battista, P., Guida, P., Lanzillo, B., & Tortelli, R., (2020). Sex differences in long-term mortality and functional outcome after rehabilitation in patients with severe stroke. *Frontiers in Neurology, 11*, 84.
50. Sacco, S., Merki-Feld, G. S., Ægidius, K. L., Bitzer, J., Canonico, M., Kurth, T., Lampl, C., et al., (2017). Hormonal contraceptives and risk of ischemic stroke in women with migraine: A consensus statement from the European Headache Federation (EHF) and the European Society of Contraception and Reproductive Health (ESC). *The Journal of Headache and Pain, 18*(1), 1–20.
51. Sheikh, H. U., Pavlovic, J., Loder, E., & Burch, R., (2018). Risk of stroke associated with use of estrogen containing contraceptives in women with migraine: A systematic review. Headache: *The Journal of Head and Face Pain, 58*(1), 5–21.
52. Chauhan, G., & Debette, S., (2016). Genetic risk factors for ischemic and hemorrhagic stroke. *Current Cardiology Reports, 18*(12), 1–11.
53. Aparicio, H. J., & Seshadri, S., (2017). Familial occurrence and heritability of stroke. In: *Stroke Genetics* (pp. 9–20). Springer.
54. Malik, R., & Dichgans, M., (2018). Challenges and opportunities in stroke genetics. *Cardiovascular Research, 114*(9), 1226–1240.
55. Locatelli, M., Padovani, A., & Pezzini, A., (2020). Pathophysiological mechanisms and potential therapeutic targets in cerebral autosomal dominant arteriopathy with subcortical infarcts and leukoencephalopathy (CADASIL). *Frontiers in Pharmacology, 11*, 321.
56. Tan, R. Y. Y., Drazyk, A. M., Urankar, K., Bailey, C., Gräf, S., Markus, H., & Giffin, N. J., (2021). Cerebral autosomal recessive arteriopathy with subcortical infarcts and leukoencephalopathy (CARASIL). *Practical Neurology, 21*(5), 448–451.
57. Hirtz, D., & Kirkham, F. J., (2019). Sickle cell disease and stroke. *Pediatric Neurology, 95*, 34–41.
58. Tapia, D., Floriolli, D., Han, E., Lee, G., Paganini-Hill, A., Wang, S., Zandihaghighi, S., et al., (2021). Prevalence of cerebral small vessel disease in a Fabry disease cohort. *Molecular Genetics and Metabolism Reports, 29*, 100815.
59. Khan, H., Singh, A., Thapa, K., Garg, N., Grewal, A. K., & Singh, T. G., (2021a). Therapeutic modulation of the phosphatidylinositol 3-kinases (PI3K) pathway in cerebral ischemic injury. *Brain Research, 1761*, 147399.
60. Sharma, A., Khan, H., Singh, T. G., Grewal, A. K., Najda, A., Kawecka-Radomska, M., Kamel, M., et al., (2021). Pharmacological modulation of ubiquitin-proteasome pathways in oncogenic signaling. *International Journal of Molecular Sciences, 22*(21), 11971.
61. Etminan, N., Chang, H. S., Hackenberg, K., De Rooij, N. K., Vergouwen, M. D. I., Rinkel, G. J. E., & Algra, A., (2019). Worldwide incidence of aneurysmal subarachnoid hemorrhage according to region, time period, blood pressure, and smoking prevalence in the population: A systematic review and meta-analysis. *JAMA Neurology, 76*(5), 588–597.

62. Grewal, A. K., Singh, N., & Singh, T. G., (2019). Effects of resveratrol postconditioning on cerebral ischemia in mice: Role of the sirtuin-1 pathway. *Canadian Journal of Physiology and Pharmacology, 97*(11), 1094–1101.
63. Sharma, V. K., Singh, T. G., & Mehta, V., (2021). Stressed mitochondria: A target to intrude Alzheimer's disease. *Mitochondrion, 59*, 48–57.

CHAPTER 21

Modulating Mitogen-Activated Protein Kinase (MAPK) Downstream Signaling Pathways in Cerebral Ischemic Hemorrhages

GURPREET SINGH,[1] SHAMSHER SINGH,[2] and SIDHARTH MEHAN[2]

[1]*Department of Pharmaceutical Chemistry, ISF College of Pharmacy, Ghal Kalan, Moga, Punjab, India*

[2]*Department of Pharmacology, ISF College of Pharmacy, Ghal Kalan, Moga, Punjab, India*

ABSTRACT

The MAPK signaling pathway belongs to family of serine/threonine kinases which are stimulated by a wide variety of extracellular stimuli. These enzymes are found in wide variety of mammals, and help in controlling different cellular processes such cell proliferation, differentiation, motility, and survival. They played a key role in the regulation of stroke, ischemic, and hemorrhagic cerebral vascular disease. Due to this they offered a wide prospect of a valuable drug target against these disorders. In stroke there is an involvement of three different pathways which is mediated via the activation of MAPK. The involvement of JNK, p38 and ERK1/2 pathways in the different stages of stroke has led to identification of targeted inhibitors against them. Moreover, research findings have established that inhibition of these signaling pathways could be valuable for the treatment of stroke and can be life-saving. In this chapter, we highlight the significance of MAPK signaling pathways in human diseases, with an emphasis on stroke, ischemic, and hemorrhagic cerebral vascular disease.

21.1 INTRODUCTION

Stroke are medical emergency situations, producing a deadly neurological condition which is characterized by the blockage of blood vessels and are the major cause of death and disabilities in adults, worldwide [1, 2]. The blood clots arise from the brain interrupts the blood supply which leads to clogging arteries and subsequently rupturing of blood vessels

Ischemic Injury: A Molecular Insight. Thakur Gurjeet Singh & Heena Khan (Eds.)

due to increased blood inside of them pressure. The series of event lead to shortening of oxygen and nutrient supply to the brain which results into the sudden death of brain cells and might lead to depression and dementia [3]. Since 1955, stroke was considered as disorder of blood vessels by the International Classification of Diseases-7 (ICD-7) but in 2018, under ICD-11 it was re-revised as serious and life-threatening neurological disorder due to strong advocacy by a group of elite clinicians [3]. The molecular mechanism of stroke is shown below in Figure 21.1.

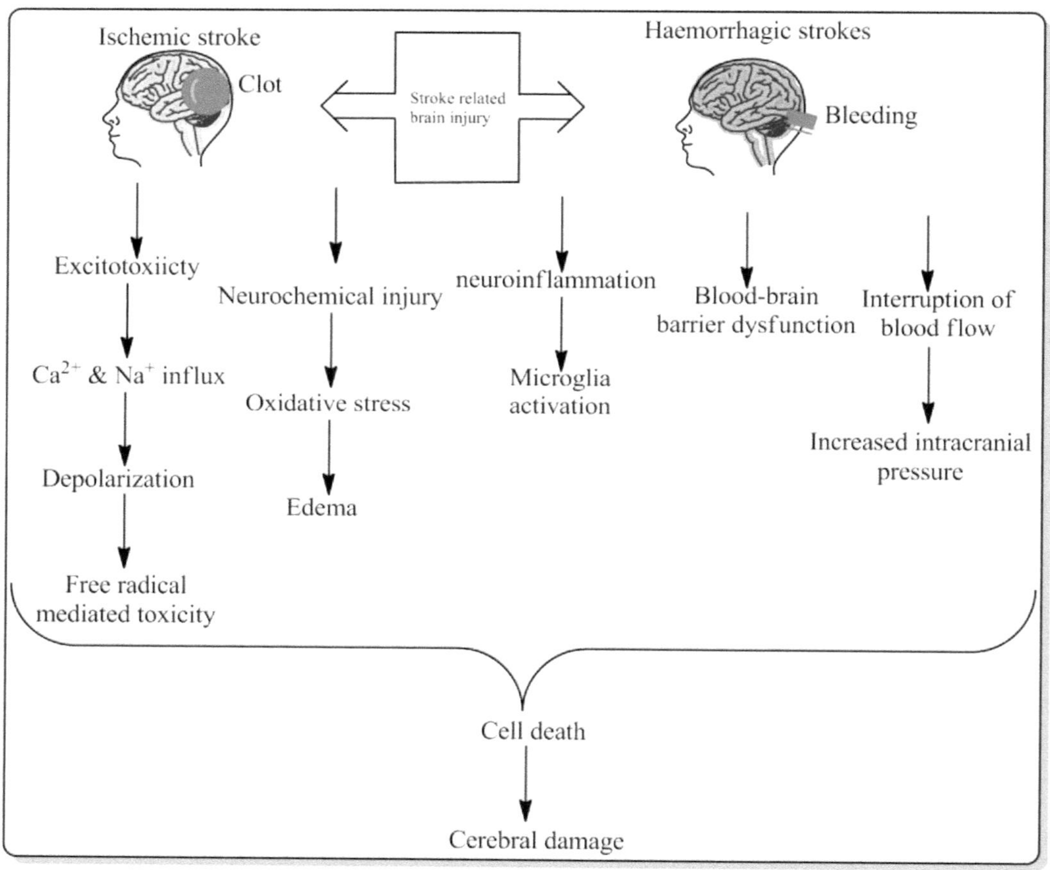

FIGURE 21.1 Mechanism of stroke.

Strokes are categorized into two main categories which are ischemic (IS) and hemorrhagic strokes, nearly, accounting for 85% and 15%, respectively, of cases worldwide [4–7]. The transient ischemic attack (TIA) is a minor stroke which serves as a warning sign for major strokes. The primary reason of the ischemic stroke is focal cerebral ischemia [2, 4] or stenosis [8] whereas hemorrhagic stroke occurs due to cerebral aneurysm rupture [9]. Hemorrhagic stroke resulted from spontaneous intracerebral and subarachnoid hemorrhages which in turn are resulted from the leakage or rupture of a blood artery

[10, 11] while the ischemic stroke happened due to blockage in the supply of blood part of the brain due to the presence of occlusion [12, 13]. Factors like focal hypoperfusion, thrombosis, and embolism can lower the blood flow to the cerebrum and hence reduce the cerebral blood flow (CBF). These in turn lead to lower availability of oxygen and glucose for neurological function and which might turn into ischemic stroke. Moreover, the IS causes two major zones of injury, i.e., the infarct core and the penumbra while the basal ganglia, cerebral lobes, the thalamus, pons, and the brain stem, and the cerebellum are the common site for bleeding in hemorrhagic stroke. Age (More than 55), gender (male > female), high blood pressure, diabetes mellitus, high cholesterol, levels alcoholism, smoking, or other stressors such as Reactive Oxygen Species (ROS) are the primary factors for strokes [1, 14–18]. Blood examination, CT scans, keeping a watch on irregularities in blood pressure and heart beats, presence or absence of fever, MRI scanning of brain, and MR angiography, etc., are common techniques used for detection of the strokes. The pathophysiology of strokes involves variety of the molecular mechanism which are valuable drug targets for drug development. Management of blood pressure and increased intracranial pressure, use of antiplatelet, anticoagulants, and surgeries are the major followed treatment for strokes [19, 20].

21.2 MAPK PATHWAY

The MAPKs or serine/threonine kinases are involved in variety of both transcriptional and non-transcriptional regulation under the influence of extracellular stimuli (Figure 21.3) [21–23]. In turn, MAPK signaling pathway regulates gene expressions and variety of cellular functions such as metabolism, cellular division and apoptosis. The involvement of MAPK signaling in different disorders such as cancer, rheumatoid arthritis, Alzheimer's disease, multiple sclerosis, and strokes has been gaining improved understanding [24–26]. In most of the mammalian cells there are four well-characterized and widely studied subfamilies of the MAPK signaling pathway:

- Extracellular signal-regulated kinase 1/2 (ERK1/2, or p42/44 MAPK);
- p38 MAPK;
- c-Jun amino-terminal kinase (JNK, or stress-activated protein kinase-1 [SAPK1]); and
- ERK5.

The upstream phosphorylation of variety of kinases activated the above stated pathways which initiate the variety of extracellular signaling pathways, among which first three subfamilies are highly studied. Mitogen-activated protein kinases (MAPKs) are associated with the variety of anti-inflammatory or inflammatory cell responses and associated diseases. These serve as key in the development of hemorrhagic and ischemic strokes. Moreover, different pro- and anti-inflammatory cytokines, either stimulated by an external stimuli or mediators, played an important role in the regulation of strokes [27]. MAPKs regulate variety of proinflammatory cytokines which includes TNF-σ and interleukin-1 (IL1), nuclear factor kappa B (NF-κB) and activating protein-1 (AP-1, Figure 21.2), which are somehow involved in pathology of strokes [28–30].

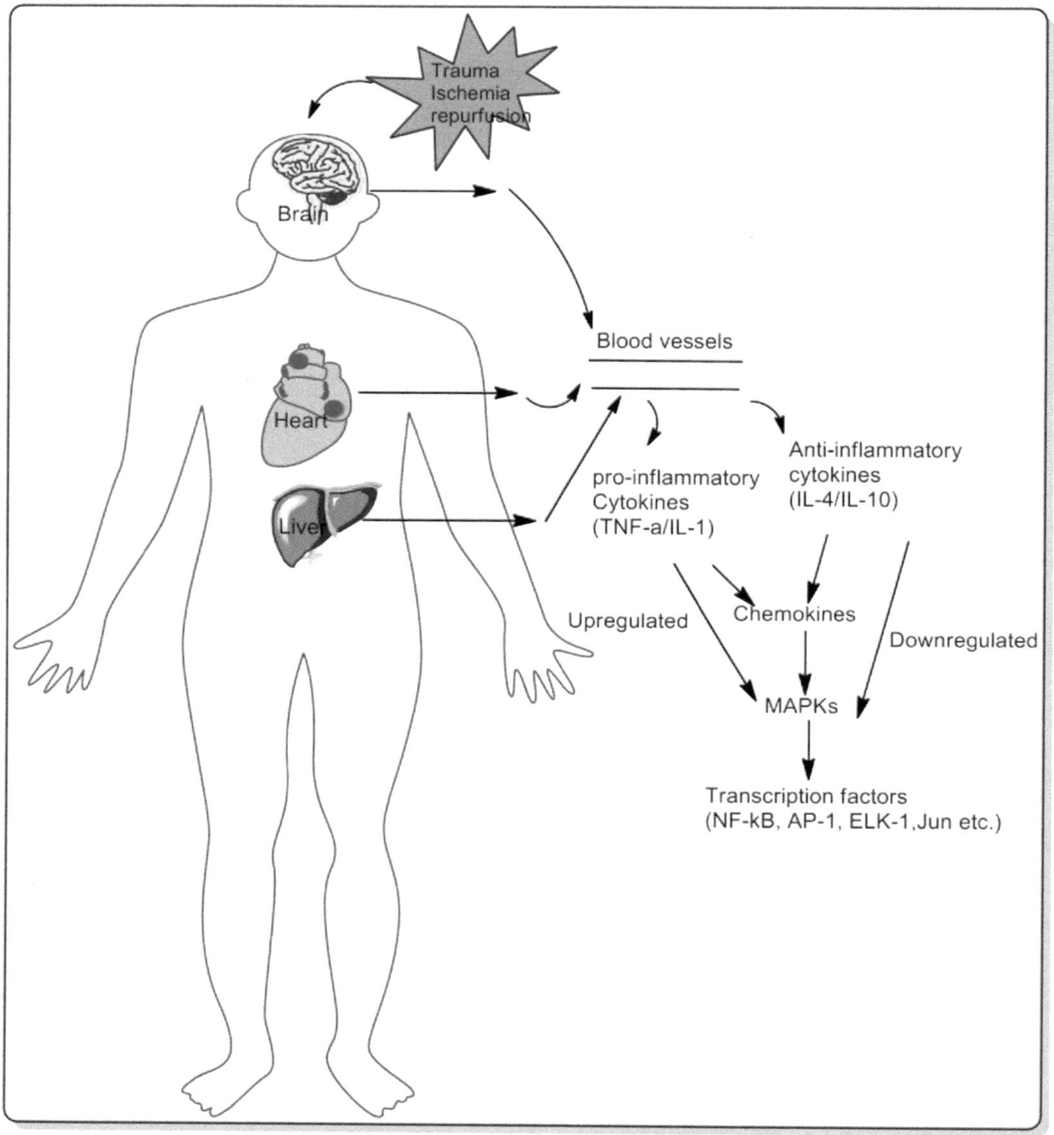

FIGURE 21.2 Molecular response following trauma.

The day-by-day increasing understanding of the involvement of MAPKs in the anti-inflammatory/inflammatory response is of utmost importance. Moreover, the improved understanding about the role of p^{38}-MAPK [31–33] in the development of strokes, has established MAPK as a potential drug target. In general, Homeostasis which is a maintained by a well-regulated system, but sometime due to the higher production or raised levels of cytokines primarily associated with dysfunction of the MAPK system can end up into serious tissue injury or death in some cases [28]. The genome contains nearly 30,000 genes, out of which 1% will constitute protein kinases while 10% will code for molecules involved in signal transduction [34]. The knowledge of the molecular elements responsible for inflammatory response will help in understanding the biological response following trauma, such as hemorrhagic shock and/or ischemia–reperfusion [29].

21.2.1 THE ERK1/2 MAPK PATHWAY

Among different MAPK pathways the first one to characterize and clones were ERK1 and ERK2. Both of the pathways are activated by mitogens and are the key regulators of cell proliferation, differentiation, and survival. Due to their pivotal role in cellular functions, ERK1/2 MAPKs have attracted intense research interest and stood as valuable tool for researchers to develop ERK inhibitors as promising anticancer agents [35]. Growth factors such as brain-derived neurotrophic factor (BDNF) and platelet-derived growth factor (PDGF) and proinflammatory stimuli such as IL-1ß are the basic activators of these pathways [36]. These pathways are also activated by pathogen-associated molecular patterns (PAMPs) and danger-associated molecular patterns (DAMPs). Due to these this pathway finds application cancer and other life-threatening ailments.

The phosphorylation of Thr185 and Tyr187 by MAPKK1 and MAPKK2 activate the ERK1 and ERK2, respectively. After activation ERK1/2 are translocate to the nucleus and control various gene expression by phosphorylation and regulation of various transcription factors. MAPKK1/2 play an important role in the integration of mitogenic signals into the ERK pathway wherein the ERK1/2 are the highly selective substrates for MAPKK [35, 36].

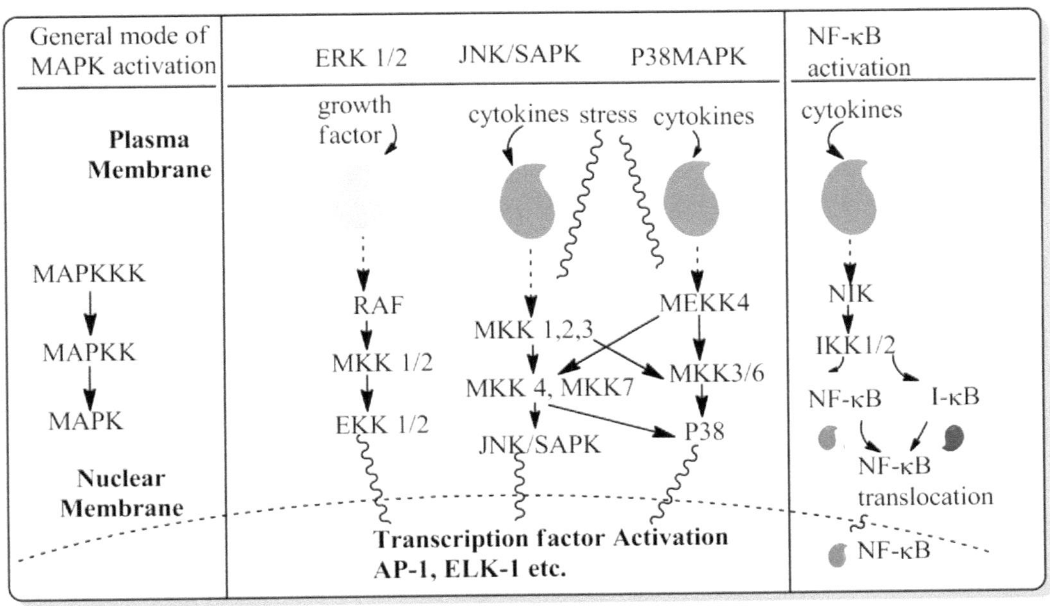

FIGURE 21.3 The activation pathway of MAPK in stress and trauma.

21.2.2 THE P38 MAPK PATHWAY

p38 MAPK is activated by lipopolysaccharide (LPS) [37] and these play central role in inflammation responses and hence serve as valuable drug target [38]. Moreover, it has a pivotal role in the regulation of cell proliferation, apoptosis, environmental stress, and neuropathic pain [39, 40]. These kinases have wide presence in the body and are categorized into four isoforms, i.e., p38α, p38β, p38γ, and p38δ, each being encoded by a separate gene, with high substrate specificity and tissue distribution. p38α and p38β are ubiquitously expressed in the body, while p38γ and p38 δ are expressed in selective tissues only. Stress conditions and inflammatory cytokines, heat shock, hypoxia, IL-1ß, TNF-α, and TGF-ß are responsible for the activation of these isoforms [41–43]. Many of the developed p38 MAPK inhibitors are in advance stages of clinical evaluations as potential treatment for peripheral inflammatory diseases such as rheumatoid arthritis and Crohn's disease. The role of p38 MAPK varies according to the cell type and stimuli. For example, in some cases, p38 MAPK promotes cell death while in some it enhances survival, growth, and differentiation. Furthermore, only the p38α and p38ß isoforms are activated by ultraviolet radiation. The dual phosphorylation of threonine (Thr) and tyrosine (Tyr) is responsible for the activation of p38 MAPK which is then translocated into the nucleus from the cytosol. It may also be transported to the other parts of the cell followed by activation of the downstream kinases or several transcription factors.

21.2.3 THE JNK MAPK PATHWAY

JNK belongs to the family of stress-induced protein kinases that phosphorylate the NH_2 terminus of the transcription factor c-Jun. This pathway is also named as stress-activated protein kinase (SAPK) response. JNK exists in three isoforms, i.e., JNK1, JNK2, and JNK3. Among them, JNK1 and JNK2 are present throughout the body, whereas JNK3 is expressed only in the neuronal tissues such as the testes, and cardiac myocytes. JNK1 and JNK2 play a key role in the regulation of a variety of cancer by controlling cell differentiation and apoptosis. Most of the time JNK isoforms are activated by the stimuli of p38 MAPK due to significant similarities between p38 MAPKs and JNK MAPK. Mitogens are the poor activator of the JNKs while under stress conditions, ischemic reperfusion, injury, and inflammatory cytokines strongly activate the JNKs [44]. Till date, only a few JNK inhibitors have been developed and are evaluated as potential anticancer, anti-inflammatory agents and also for neurodegenerative, and metabolic disorders. The dual phosphorylation of Tyr/Thr within the activation loop (Thr183/Tyr185) also showed to play a significant role in gene expression, cell proliferation, and death.

21.2.4 THE ERK5 MAPK PATHWAY

ERK5 or BMK1, independently cloned by two different groups, is the latest entry to the family of MAPKs. The C-terminus of ERK5 is encoded by MAPK7 while N-terminal catalytic domain share similarities with the ERK1/2 and the protein is twice the size of MAPK7 gene. The ERK5 gene are highly expressed in brain, thymus, and spleen and vital for the development of embryo in the early phase. Furthermore, it is essential for normal development of the vascular system and cell survival. Different growth factors (VEGF, EGF, and NGF) and conditions such as stress, hypoxia, ischemia, and cytokines, i.e., IL-6 and LIF are responsible for the activation of ERK5. After activation it regulate different of cellular functions which includes such as cell survival and proliferation and played a key role in the in pathology of diabetic retinopathy [45]. It is also activated by dual phosphorylation at Thr218/Tyr220 by the upstream kinase MAPKK5. It is present in higher concentration in the in endothelial cells and is vital for the maintenance of endothelial function and blood vessel integrity [46, 47].

21.2.5 THE MAPK SIGNALING PATHWAY AND STROKE

The role of MAPK in the pathology of stroke has been extensively examined by various research groups which have been summarized in Table 21.1. As discussed earlier in the chapter the activation of MAPK is closely associated with the physiology and pathogenesis of stroke and its post-development. There are several incidences that suggest that MAPK contribute significantly towards the regulation of inflammatory responses, cytokines, cell apoptosis, and death in ischemic and hemorrhagic brain injury [48–51]. Hence the inhibition of MAPK signaling pathways could serve as valuable target for drug development in such conditions, especially for the stroke.

21.3 EFFECT OF MAPK IN CEREBRAL ISCHEMIA

The expression, distribution, and activity of MAPK in ischemic stroke has been well documented through the recent past. Moreover, the key findings of various research groups have demonstrated the vital role of MAPK pathway for the regulation of cytokine expression and cell apoptosis after the induction of experimental cerebral ischemia [42, 49, 53, 58, 69]. The researchers have demonstrated that a significant reduction in inflammatory cytokines and cell death in the damage area were achieved by time dependent activation of ERK and JNK and inhibiting MEK [42, 49, 53, 58, 69]. Maddahi and co-workers showed that the Raf-MEK/ERK1/2 signaling transducers played a pivotal role in the regulation of pro-inflammatory mediators in cerebral vessels after induced ischemic stroke [53]. Moreover, they also highlighted that the high levels of matrix metalloproteinase 9 are also promote the cerebral ischemia. In the early stages of cerebral ischemia there is increase expression of p38 gene in the glial cells and neurons [43]. The ischemic stroke and related injuries in the blood vessels can also be aggravated by the release of pro-inflammatory markers post to the activation of p38 MPAK. On the other hand, inflammatory factors strongly activate the p38 MAPK which promote the higher production of ROS which might lead to activation of p38 MAPK and causing cell lipid peroxidation or the induction of apoptosis-related genes and cell death [54, 70]. Nito et al. showed that the blood-brain

TABLE 21.1 Summarization of Studies Modulating MAPK Downstream Signaling Pathway

SL. No.	MAPK Pathway	Model	Effects	Treatment	Outcome	References
Cerebral Ischemia						
1.	ERK1/2	MCAO model	Regulate the expression of IL-1ß3.	ERK inhibitor U0126.	Lowered infarct volume and decrease IL-1ß.	[42]
2.	ERK1/2	MCAO model; organ culture of isolated cerebral arteries.	Regulate the expression IL-1ß, TNF-α, iNOS, IL-6, and cxcl2, and MMP9 and MMP13 genes.	ERK inhibitor U0126	Regulate the inflammatory markers and MM genes in the smooth muscle cells of the cerebral arteries.	[52]
3.	ERK1/2	MCAO model	Modulate the levels of TNF-ß, IL-1ß, IL-6, and iNOS	ERK inhibitor U0126	Reduction in infarct size and improvement of neurological scores.	[53]
4.	ERK1/2	MCAO model; organ culture of isolated cerebral arteries.	Regulate the expression of TNF-α and TNF-α receptors 1 and 2.	ERK inhibitor U0126	Reduce the expression of TNF-a, TNF-R1, and TNF-R2 in the wall of cerebral arteries.	[54]
5.	ERK1/2	MCAO model; primary neuron culture.	Regulate the expression of Arc, Junb, and Fos.	ERK inhibitor PD184161.	Lower excitotoxic neuronal death.	[55]
6.	ERK1/2	Cultured human cerebral arteries.	Modulate vascular contractile receptors.	B-Raf inhibitor SB386023 or SB590885	Decrease vasoconstriction.	[56]
7.	p38	Transient focal cerebral ischemia (tFCI)	Activation and phosphorylation of cPLA2.	p38 inhibitor SB203580	Reduced BBB permeability, edema, and infarct volumes.	[57]
8.	p38	MCAO model.	Regulate the processes mediated iNOS, TNFα, IL-1ß, and COX-2.	p38 inhibitor SB203580	Reduce infarct volume and improve neurological shortages.	[49]
9.	p38	MCAO model; primary astrocyte culture	Regulate the TNF-a, TGF-ß	p38 inhibitor SB239063 and knockout of p38 MAPK	Attenuate astrocyte migration and astrogliosis.	[43]
10.	JNK	NICD1 overexpression in HEK cell cultures under OGD conditions.	Regulate the release of LDH.	JNK inhibitor SP600125	Reduce cell death.	[58]

TABLE 21.1 *(Continued)*

SL. No.	MAPK Pathway	Model	Effects	Treatment	Outcome	References
Intracerebral Hemorrhage						
11.	ERK1/2, JNK, p38	Collagenase-induced ICH model.	Regulation of ICH-induced neuronal injury.	ERK inhibitor PD98059; JNK inhibitor SP600125; p38 inhibitor SB203580	Block striatal tissue injury.	[59]
12.	ERK1/2, p38	Thrombin injection-induced brain injury.	Involved in thrombin-induced striatal neuronal death.	ERK inhibitor. PD98059; p38 inhibitor. SB203580	Reduction in the size of injury.	[50]
13.	p38	Autologous arterial blood-induced ICH model.	Regulate MMP-9	p38 inhibitor SB203580	Improve BBB impairment.	[60]
14.	JNK, p38	Microglial culture.	Regulate LDH release.	p38 inhibitor SB 203580; JNK inhibitor. SP600125	Reduce hemin toxicity in microglia.	[61]
15.	JNK	Neuronal cell culture.	Regulate caspase-3 activation.	JNK inhibitor SP600125	Reduce thrombin-induced neuronal apoptosis.	[62]
Subarachnoid Hemorrhage						
16.	ERK1/2	SAH model	Regulate endothelium B and 5-HT_{1B} receptors.	Raf inhibitor 386023-b.	Prevent ion of cerebral injury.	[63]
17.	ERK1/2	SAH model	Modulate the expression of cerebrovascular smooth muscle cell receptors.	Raf inhibitor 386023-b.	Lower risk of injury to cerebrum.	[64]
18.	ERK1/2	SAH model	Regulate IL-1ß, IL-6, iNOS, MMP-9, and TIMP-1.	Raf inhibitor 386023-b.	Improve neurological function.	[65]
19.	ERK1/2	SAH model	Regulate IL-1ß, IL-6, TNF-α, and MMP-9.	ERK inhibitor U0126	Improve neurological function.	[56]
20.	p38	SAH model	Regulate endothelial cell apoptosis in basilar arteries.	p38 inhibitor	Reduced cerebral vasospasm.	[66]
21.	p38	SAH model	Regulate mitochondrial structure and function; trigger cell death though the mitochondrial pathway.	SB203580	No mitochondrial depolarization and blockage of neuron depletions.	[67]

TABLE 21.1 *(Continued)*

SL. No.	MAPK Pathway	Model	Effects	Treatment	Outcome	References
22.	p38	SAH model	Regulate the TNF-α.	p38 inhibitor	Attenuate cerebral vasospasm.	[68]
23.	JNK	SAH model	Regulate the level of TNF-α.	JNK inhibitor SP600125	Lower apoptotic rates and higher neurological scores.	[66]

barrier disruption could also be promoted by secondary vasogenic edema and superoxide formation via the activation of the p38 MAPK/cPLA2 pathway [57]. Researchers have demonstrated that the inhibition of p38 using specific inhibitors could be beneficial for the treatment of cerebral ischemia [61]. The inhibition of p38 resulted in reduced brain injury via increased cell survival and reduced inflammation and apoptotic cell death. Moreover, the activation of p38 MAPK and ERK1/2 MAPK has also promoted the blood-brain permeability through the phosphorylation of tight junction protein phosphorylation and subsequent redistribution of phosphorylated proteins [71–74]. Furthermore, both ERK1/2 and p38 are implicated in the control of IL-1ß, TNF-, and iNOS in ischemic stroke [49, 52], as well as the breakdown of the blood-brain barrier, and can perform similar biological tasks in cerebral ischemia.

21.4 EFFECT OF MAPK IN INTRACEREBRAL HEMORRHAGE

Another kind of stroke is intracerebral hemorrhage (ICH), which accounts for about 15% of all strokes and has a fatal prognosis. Several experimental studies have revealed possible reasons for the evolution of ICH and its worsening. According to the findings, increased levels of pro-inflammatory cytokines and oxidative stress can contribute to secondary brain damage following ICH [75–77]. The involvement of related pathways has been well documented in the prognosis of ICH. The change in the inflammatory and cytotoxic responses in the ICH are closely associated with the MAPK pathway. The activation of TLR (toll-like receptors) by the heme and iron can lead to serious injuries after an ICH episode [78–80]. The downstream regulation of TLR by p38 MAPK promotes the production TNF-α, IL-1ß, IL-6, and IFN-γ. Moreover, p38 is also involved in thrombin-induced neuronal death and hemin-induced microglial apoptosis. In a nutshell, the p38 pathway can be triggered by a variety of blood proteins and is involved in a number of pathogenic events following ICH [50, 61]. Ac-YVADCMK, a well-known selective caspase-1 inhibitor, was reported to improve BBB tight junction protein degradation by blocking the JNK pathway [64]. JNK has also been demonstrated to have a role in hemin-induced microglial apoptosis and thrombin-induced neuronal apoptosis, respectively, through modulating LDH release and caspase-3 activation [61, 62].

The research findings also highlighted that the thrombin cytotoxicity-induced brain tissue damage is also induced due to the activation of MAPK pathway [50, 62, 81, 82]. The uncontrolled activity of thrombin, following an episode of ICH, resulted into the activation of microglia. By phagocytosing apoptotic cells, active microglia can trigger neuron death and tissue atrophy [83]. Delayed neuronal deaths and induction of apoptotic like features could result from ERK activation in the early phase of thrombin exposure. Moreover, the inhibition of MEK, p38, and JNK has also been shown to drastically reduce thrombin-induced cell death in the cortical region and striatal atrophy. Hence, MAPK pathways could serve as therapeutic targets against thrombin-induced injury in ICH [82].

21.5 EFFECT OF MAPK IN SUBARACHNOID HEMORRHAGE

It is the most fatal type of strokes accounts only for 5–10% cases but conceding 67% of the mortality rate. The high number of deaths have been seen in the patients who survived early SAH. In such patient's early brain injury and subsequent cerebral vasospasm following SAH are responsible for high number of deaths. Researchers around the world have demonstrated association of inflammation and apoptosis in the pathogenesis of SAH upon activation of the MAPK pathway. Also, the involvement of the MAPK pathway in the regulation of onset of early phase of SAH has been documented. Following SAH, the activated ERK1/2 pathway controls cerebral blood flow by modulating the expression of certain cerebrovascular smooth muscle cell receptors and triggering an inflammatory response [63–65].

It has been well documented that following an episode of SAH the activated p38 MAPK in the walls of cortex and arterial after being triggered by SAH. In turn, the activated p38 MAPK led to stimulation of different kind of response in the neuronal and cellular regions which ultimately lead to early brain injury and the development of cerebral vasospasm [66–68]. In a nutshell, it can be concluded that p38 pathway can be activated along the entire process of SAH. However, the p38 plays different roles in the various stages of the disease. p38 can regulate mitochondrial structure and function and induce cell death in the early stages of SAH [67], whereas it can also govern cerebral vasospasm by regulating inflammatory responses and endothelial cell apoptosis in the late stages [66, 68]. Sometimes, after the SAH, there is activation of ERK1/2 and p38 pathways simultaneously in cerebral arteries which collectively regulate the cerebral vasospasm which are mediated by different mechanisms [63, 66]. The use of specific ERK1/2 inhibitor or p38 inhibitor after SAH has been proven beneficial in reducing the after-effects of SAH by inhibiting the activation of the ERK1/2 or p38 pathway and associated markers [54, 55, 66]. The role of JNK activation followed by brain injury after SAH has been documented by various research groups. Zhang and colleagues also showed the similarities between the activation of JNK and TNF-α signaling pathway in the early and late phases of SAH. The elevated TNF-a and caspase-3 levels were found to be lower after the injection of a specific JNK inhibitor [84].

21.6 CONCLUSION

MAPK plays an important role in cerebral and ischemic hemorrhages. In recent years their involvement in progenesis of neurodegenerative diseases such as stroke has been well documented and their role has established them as a valuable drug target. These findings have prompted scientist across the world to develop new small molecules as their specific inhibitors, and many of them have advanced to clinical development. Although some good progresses have been made in the area but still a long way to go further. There is a need to develop more selective and highly potent MAPK pathway inhibitors which can be useful for the treatment of cerebrovascular disorders. Hence the modulation of MAPK. As a result, we believe that modulating MAPK signaling could be used to enhance patient outcomes and could be a target for the development of new stroke medicines.

KEYWORDS

- **cerebral ischemia**
- **hemorrhage**
- **ischemia**
- **ischemic stroke**
- **MAPK pathway**
- **p38 MAPK**
- **stroke**

REFERENCES

1. Spence, J. D., Yi, Q., & Hankey, G. J., (2017). B vitamins in stroke prevention: Time to reconsider. *The Lancet Neurology, 16*(9), 150–760.
2. Thapa, K., Khan, H., Singh, T. G., & Kaur, A., (2021). Traumatic brain injury: Mechanistic insight on pathophysiology and potential therapeutic targets. *Journal of Molecular Neuroscience, 71*(9), 1725–1742.
3. Shakir, R., (2018). The struggle for stroke reclassification. *Nature Reviews Neurology, 14*(8), 147–448.
4. Khan, H., Tiwari, P., Kaur, A., & Singh, T. G., (2021). Sirtuin acetylation and deacetylation: A complex paradigm in neurodegenerative disease. *Molecular Neurobiology, 58*(8), 1903–3917.
5. Khalil, O., et al., (2014). Thyroid dysfunction in acute ischemic stroke in medical intensive care unit in Zagazig University Hospitals. *International Journal of Advanced Research, 2*(7), 185–391.
6. Hinkle, J. L., & Guanci, M. M., (2007). Acute ischemic stroke review. *Journal of Neuroscience Nursing, 39*(5), 285–293.
7. Tiwari, P., Khan, H., Singh, T. G., & Grewal, A. K., (2022). Poly (ADP-ribose) polymerase: An overview of mechanistic approaches and therapeutic opportunities in the management of stroke. *Neurochemical Research*, 1–23.
8. Khan, H., Bangar, A., Grewal, A. K., Bansal, P., & Singh, T. G., (2022). Caspase-mediated regulation of the distinct signaling pathways and mechanisms in neuronal survival. *International Immunopharmacology, 110*, 108951.
9. Altoum, A., Osman, A. L., & Babker, A., (2018). Comparative study of levels of selective oxidative stress markers (malondialdehyde, zinc, and antioxidant vitamins A, E, and C) in ischemic and non-ischemic heart disease patients suffering from type-2 diabetes. *Asian J. Pharm. Clin. Res., 11*(8), 108–510.
10. Panel, et al., (1997). Etiology of stroke. *Stroke, 28*(7), 1501–1506.
11. Krafft, P. R., et al., (2012). Etiology of stroke and choice of models. *International Journal of Stroke, 7*(5), 198–406.
12. Lakhan, S. E., Kirchgessner, A., & Hofer, M., (2009). Inflammatory mechanisms in ischemic stroke: Therapeutic approaches. *Journal of Translational Medicine, 7*(1), 1–11.
13. Aggarwal, A., et al., (2010). Cerebral ischemic stroke: Sequels of cascade. *International Journal of Pharma. and Bio. Sciences, 1*(3), 1–24.
14. Paciaroni, M., et al., (2010). Risk factors for cerebral ischemic events in patients with atrial fibrillation on warfarin for stroke prevention. *Atherosclerosis, 212*(2), 164–566.
15. Garg, N., Singh, T. G., Khan, H., Arora, S., Kaur, A., & Mannan, A., (2021). Mechanistic interventions of selected *Ocimum* species in management of diabetes, obesity, and liver disorders: Transformative developments from preclinical to clinical approaches. *Biointerface Res. Appl. Chem., 12*(1), 1304–1323.
16. Su, X. T., et al., (2020). Mechanisms of acupuncture in the regulation of oxidative stress in treating ischemic stroke. *Oxidative Medicine and Cellular Longevity*, 2020.

17. Firdous, S. M., Ghosh, S., Nath, P., & Khan, H., (2021). Molecular basis of diabetic nephropathy. *Diabetes and Diabetic Complications*. Nova Science Publishers, Hauppauge.
18. Garg, C., Kaur, A., Singh, T. G., Sharma, V. K., & Singh, S. K., (2022). Therapeutic implications of sonic hedgehog pathway in metabolic disorders: Novel target for effective treatment. *Pharmacological Research*, 106194.
19. Smith, S. D., & Eskey, C. J., (2011). Hemorrhagic stroke. *Radiologic Clinics, 49*(1), 17–45.
20. Feske, S. K., (2021). Ischemic stroke. *The American Journal of Medicine.*
21. Cargnello, M., & Roux, P. P., (2011). Activation and function of the MAPKs and their substrates, the MAPK-activated protein kinases. *Microbiology and Molecular Biology Reviews, 75*(1), 10–83.
22. Kyriakis, J. M., & Avruch, J., (2012). Mammalian MAPK signal transduction pathways activated by stress and inflammation: A 10-year update. *Physiological Reviews, 92*(2), 189–737.
23. Munshi, A., & Ramesh, R., (2013). Mitogen-activated protein kinases and their role in radiation response. *Genes & Cancer, 4*(9, 10), 101–408.
24. Culbert, A. A., et al., (2006). MAPK-activated protein kinase 2 deficiency in microglia inhibits pro-inflammatory mediator release and resultant neurotoxicity: Relevance to neuroinflammation in a transgenic mouse model of Alzheimer disease. *Journal of Biological Chemistry, 281*(33), 13658–23667.
25. Krementsov, D. N., et al., (2013). The emerging role of p38 mitogen-activated protein kinase in multiple sclerosis and its models. *Molecular and Cellular Biology, 33*(19), 1728–3734.
26. Gurgis, F. M. S., Ziaziaris, W., & Munoz, L., (2014). Mitogen-activated protein kinase–activated protein kinase 2 in neuroinflammation, heat shock protein 27 phosphorylation, and cell cycle: Role and targeting. *Molecular Pharmacology, 85*(2), 145–356.
27. Holzheimer, R. G., Gross, J., & Schein, M., (1999). Pro-and anti-inflammatory cytokine-response in abdominal aortic aneurysm repair: A clinical model of ischemia-reperfusion. *Shock (Augusta, Ga.), 11*(5), 105–310.
28. Yaffe, M. B., & Fink, M. P., (2000). Cellular signaling in critical care-Putting the pieces together. *Critical Care Medicine, 28*(4), N1, N2.
29. Hu, Y., et al., (1997). Activation of mitogen-activated protein kinases (ERK/JNK) and AP-1 transcription factor in rat carotid arteries after balloon injury. *Arteriosclerosis, Thrombosis, and Vascular Biology, 17*(11), 1808–2816.
30. Lai, E. W., et al., (2004). The role of map kinases in trauma and ischemia–reperfusion. *Journal of Investigative Surgery, 17*(1), 15–53.
31. Mendez, C., et al., (2000). Involvement of p38 mitogen-activated protein kinase in the induction of tolerance to hemorrhagic and endotoxic shock. *Journal of Surgical Research, 91*(2), 165–170.
32. Junger, W. G., et al., (1997). Hypertonic saline activates protein tyrosine kinases and mitogen-activated protein kinase p38 in T-cells. *Journal of Trauma and Acute Care Surgery, 42*(3), 137–445.
33. Sun, Z., & Andersson, R., (2002). NF-κB activation and inhibition: A review. *Shock, 18*(2), 19–106.
34. Oberholzer, A., Oberholzer, C., & Moldawer, L. L., (2000). Cytokine signaling-regulation of the immune response in normal and critically ill states. *Critical Care Medicine, 28*(4), N3–N12.
35. Roberts, P. J., & Der, C. J., (2007). Targeting the Raf-MEK-ERK mitogen-activated protein kinase cascade for the treatment of cancer. *Oncogene, 26*(22), 1291–3310.
36. Ruhul, A. A., et al., (2003). Secretion of matrix metalloproteinase-9 by the proinflammatory cytokine, IL-1β: A role for the dual signaling pathways, Akt and Erk. *Genes to Cells, 8*(6), 115–523.
37. Kotlyarov, A., et al., (1999). MAPKAP kinase 2 is essential for LPS-induced TNF-α biosynthesis. *Nature Cell Biology, 1*(2), 14–97.
38. Borders, A. S., et al., (2008). The p38α mitogen-activated protein kinase as a central nervous system drug discovery target. *BMC Neuroscience, 9*(2), 1–8.
39. Edelmayer, R. M., et al., (2014). Biochemical and pharmacological assessment of MAP-kinase signaling along pain pathways in experimental rodent models: A potential tool for the discovery of novel antinociceptive therapeutics. *Biochemical Pharmacology, 87*(3), 190–398.
40. Koul, H. K., Pal, M., & Koul, S., (2013). Role of p38 MAP kinase signal transduction in solid tumors. *Genes & Cancer, 4*(9, 10), 142–359.
41. Sharma, A., Khan, H., Singh, T. G., Grewal, A. K., Najda, A., Kawecka-Radomska, M., Kamel, M., et al., (2021). Pharmacological modulation of ubiquitin-proteasome pathways in oncogenic signaling. *International Journal of Molecular Sciences, 22*(21), 11971.

42. Wang, Z. Q., et al., (2004). Inhibition of MEK/ERK 1/2 pathway reduces pro-inflammatory cytokine interleukin-1 expression in focal cerebral ischemia. *Brain Research, 996*(1), 15–66.
43. Choudhury, G. R., et al., (2014). Involvement of p38 MAPK in reactive astrogliosis induced by ischemic stroke. *Brain Research, 1551*, 15–58.
44. Khan, H., Garg, N., Singh, T. G., Kaur, A., & Thapa, K., (2022). Calpain inhibitors as potential therapeutic modulators in neurodegenerative diseases. *Neurochemical Research*, 1–25.
45. Wu, Y., & Chakrabarti, S., (2015). ERK5 mediated signaling in diabetic retinopathy. *Medical Hypothesis, Discovery and Innovation in Ophthalmology, 4*(1), 17.
46. Zhao, Q., et al., (2018). Molecular regulation of ERK5 in development of diabetic retinopathy. *Oncotarget, 9*(1), 1229.
47. Wu, H., et al., (2021). The metaflammatory and immunometabolic role of macrophages and microglia in diabetic retinopathy. *Human Cell, 34*(6), 1617–1628.
48. Wang, X., et al., (2002). Mitogen-activated protein kinase-activated protein (MAPKAP) kinase 2 deficiency protects brain from ischemic injury in mice. *Journal of Biological Chemistry, 277*(46), 13968–43972.
49. Piao, C. S., et al., (2003). Administration of the p38 MAPK inhibitor SB203580 affords brain protection with a wide therapeutic window against focal ischemic insult. *Journal of Neuroscience Research, 73*(4), 137–544.
50. Fujimoto, S., et al., (2007). Thrombin induces striatal neurotoxicity depending on mitogen-activated protein kinase pathways *in vivo*. *Neuroscience, 144*(2), 194–701.
51. Wu, B., et al., (2010). Ac-YVAD-CMK decreases blood–brain barrier degradation by inhibiting caspase-1 activation of interleukin-1β in intracerebral hemorrhage mouse model. *Translational Stroke Research, 1*(1), 17–64.
52. Vikman, P., et al., (2007). Cerebral ischemia induces transcription of inflammatory and extracellular-matrix-related genes in rat cerebral arteries. *Experimental Brain Research, 183*(4), 199–510.
53. Maddahi, A., & Edvinsson, L., (2010). Cerebral ischemia induces microvascular pro-inflammatory cytokine expression via the MEK/ERK pathway. *Journal of Neuroinflammation, 7*(1), 1–13.
54. Maddahi, A., et al., (2011). The role of tumor necrosis factor-α and TNF-α receptors in cerebral arteries following cerebral ischemia in rat. *Journal of Neuroinflammation, 8*(1), 1–13.
55. Rihal, V., Khan, H., Kaur, A., & Singh, T. G., (2022). Vitamin D as therapeutic modulator in cerebrovascular diseases: A mechanistic perspectives. *Critical Reviews in Food Science and Nutrition*, 1–23.
56. Ahnstedt, H., et al., (2011). Human cerebrovascular contractile receptors are upregulated via a B-Raf/MEK/ERK-sensitive signaling pathway. *BMC Neuroscience, 12*(1), 1–9.
57. Nito, C., et al., (2008). Role of the p38 mitogen-activated protein kinase/cytosolic phospholipase A2 signaling pathway in blood—Brain barrier disruption after focal cerebral ischemia and reperfusion. *Journal of Cerebral Blood Flow & Metabolism, 28*(10), 1686–1696.
58. Cheng, Y. L., et al., (2014). Evidence that neuronal Notch-1 promotes JNK/c-Jun activation and cell death following ischemic stress. *Brain Research, 1586*, 193–202.
59. Ohnishi, M., et al., (2010). Mitogen-activated protein kinases support survival of activated microglia that mediate thrombin-induced striatal injury in organotypic slice culture. *Journal of Neuroscience Research, 88*(10), 1155–2164.
60. Ma, Q., et al., (2011). PDGFR-α inhibition preserves blood-brain barrier after intracerebral hemorrhage. *Annals of Neurology, 70*(6), 120–931.
61. Rehni, A. K., Singh, T. G., Kalra, R., & Singh, N., (2009). Pharmacological inhibition of inducible nitric oxide synthase attenuates the development of seizures in mice. *Nitric Oxide, 21*(2), 120–125.
62. He, Q., et al., (2015). The protective role of (−)-epigallocatechin-3-gallate in thrombin-induced neuronal cell apoptosis and JNK-MAPK activation. *Neuroreport, 26*(7), 116.
63. Beg, S. A., et al., (2006). ERK1/2 inhibition attenuates cerebral blood flow reduction and abolishes ETB and 5-HT1B receptor upregulation after subarachnoid hemorrhage in rat. *Journal of Cerebral Blood Flow & Metabolism, 26*(6), 146–856.
64. Ansar, S., Maddahi, A., & Edvinsson, L., (2011). Inhibition of cerebrovascular raf activation attenuates cerebral blood flow and prevents upregulation of contractile receptors after subarachnoid hemorrhage. *BMC Neuroscience, 12*(1), 1–17.

65. Maddahi, A., et al., (2011). Blockade of the MEK/ERK pathway with a raf inhibitor prevents activation of pro-inflammatory mediators in cerebral arteries and reduction in cerebral blood flow after subarachnoid hemorrhage in a rat model. *Journal of Cerebral Blood Flow & Metabolism, 31*(1), 144–154.
66. Zhang, X., et al., (2011). Inhibition of the p38 mitogen-activated protein kinase (MAPK) pathway attenuates cerebral vasospasm following experimental subarachnoid hemorrhage in rabbits. *Annals of Clinical & Laboratory Science, 41*(3), 144–250.
67. Huang, L., et al., (2013). Inhibitory effects of p38 inhibitor against mitochondrial dysfunction in the early brain injury after subarachnoid hemorrhage in mice. *Brain Research, 1517*, 133–140.
68. Pan, Y. X., et al., (2013). Intracisternal administration of SB203580, a p38 mitogen-activated protein kinase inhibitor, attenuates cerebral vasospasm via inhibition of tumor-necrosis factor-α. *Journal of Clinical Neuroscience, 20*(5), 126–730.
69. Kovalska, M., et al., (2012). Intracellular signaling MAPK pathway after cerebral ischemia–reperfusion injury. *Neurochemical Research, 37*(7), 1568–1577.
70. Mahtani, K. R., et al., (2001). Mitogen-activated protein kinase p38 controls the expression and posttranslational modification of tristetraprolin, a regulator of tumor necrosis factor alpha mRNA stability. *Molecular and Cellular Biology, 21*(19), 1461–6469.
71. Cardoso, F. L., Brites, D., & Brito, M. A., (2010). Looking at the blood–brain barrier: Molecular anatomy and possible investigation approaches. *Brain Research Reviews, 64*(2), 128–363.
72. Fischer, S., et al., (2005). H_2O_2 induces paracellular permeability of porcine brain-derived microvascular endothelial cells by activation of the p44/42 MAP kinase pathway. *European Journal of Cell Biology, 84*(7), 187–697.
73. Liu, N. Q., et al., (2002). Human immunodeficiency virus type 1 enters brain microvascular endothelia by macropinocytosis dependent on lipid rafts and the mitogen-activated protein kinase signaling pathway. *Journal of Virology, 76*(13), 1689–6700.
74. González-Mariscal, L., Tapia, R., & Chamorro, D., (2008). Crosstalk of tight junction components with signaling pathways. *Biochimica et Biophysica Acta (BBA)-Biomembranes, 1778*(3), 129–756.
75. Aronowski, J., & Zhao, X., (2011). Molecular pathophysiology of cerebral hemorrhage: Secondary brain injury. *Stroke, 42*(6), 1781–1786.
76. Keep, R. F., et al., (2014). Vascular disruption and blood–brain barrier dysfunction in intracerebral hemorrhage. *Fluids and Barriers of the CNS, 11*(1), 1–13.
77. Mracsko, E., & Veltkamp, R., (2014). Neuroinflammation after intracerebral hemorrhage. *Frontiers in Cellular Neuroscience, 8*, 188.
78. Fang, H., et al., (2013). Toll-like receptor 4 signaling in intracerebral hemorrhage-induced inflammation and injury. *Journal of Neuroinflammation, 10*(1), 1–10.
79. Khan, H., Singh, T. G., Dahiya, R. S., & Abdel-Daim, M. M., (2022). α-Lipoic acid, an organosulfur biomolecule a novel therapeutic agent for neurodegenerative disorders: An mechanistic perspective. *Neurochemical Research*, 1–12.
80. Gupta, A., Khan, H., Kaur, A., & Singh, T. G., (2021). Novel targets explored in the treatment of alcohol withdrawal syndrome. *CNS & Neurological Disorders-Drug Targets (Formerly Current Drug Targets-CNS & Neurological Disorders), 20*(2), 158–173.
81. Fujimoto, S., et al., (2006). Thrombin-induced delayed injury involves multiple and distinct signaling pathways in the cerebral cortex and the striatum in organotypic slice cultures. *Neurobiology of Disease, 22*(1), 130–142.
82. Ohnishi, M., et al., (2007). Involvement of thrombin and mitogen-activated protein kinase pathways in hemorrhagic brain injury. *Experimental Neurology, 206*(1), 13–52.
83. Babu, R., et al., (2012). Thrombin and hemin as central factors in the mechanisms of intracerebral hemorrhage–induced secondary brain injury and as potential targets for intervention. *Neurosurgical Focus, 32*(4), E8.
84. Zhang, B. F., et al., (2015). Etanercept alleviates early brain injury following experimental subarachnoid hemorrhage and the possible role of tumor necrosis factor-α and c-Jun N-terminal kinase pathway. *Neurochemical Research, 40*(3), 191–599.

CHAPTER 22

Implication of Mitogen-Activated Protein Kinase (MAPK) in Cerebral Stroke

SHAMLY,[1] SHILPA KUMARI,[1] PARAS ATTRI,[1] RAHUL DESHMUKH,[1] and THAKUR GURJEET SINGH[2]

[1]*Department of Pharmaceutical Sciences and Technology, Maharaja Ranjit Singh Punjab Technical University, Bathinda, Punjab, India*

[2]*Chitkara College of Pharmacy, Chitkara University, Punjab, India*

ABSTRACT

A stroke is a cerebrovascular accident, in which the blood supply to the brain gets obstructed and blood vessels get ruptured. Stroke causes 9% of deaths worldwide. Stroke is the second leading cause of death globally. In 2019, 12.2 million incident cases of stroke and 6.55 million deaths were reported due to stroke. Understanding the intrinsic molecular mechanisms is important for developing effective stroke treatment. MAPK (Mitogen-Activated protein kinase) signaling pathway regulates gene expression and is involved in a variety of cellular processes like growth, inflammation, stress response, differentiation, proliferation, migration, metabolism, and apoptosis. MAPK in response to a variety of different stimuli transmits a signal to the nucleus from the cell membrane. The activation of the MAPK pathway and then to its sub-pathways like JNK, ERK, and p38 through ROS, inflammation, and oxidative stress leads to neuronal cell death. These are the major pathways for autophagy and apoptosis and can play the role of a double-edged sword in stroke. The MAPK pathway is accepted as a therapeutic target in different diseases like cancer and peripheral inflammatory disorders. The evidence is found that signaling is a key regulator of ischemic and hemorrhagic cerebral vascular disease. The upregulation, downregulation, and after-stroke signaling of the MAPK play a crucial role in deciding the associated therapeutic targets in the treatment of stroke through the MAPK signaling pathway. Thus, this signaling has the potential to be treated as a therapeutic target for the ischemic stroke. In this chapter, we have discussed the aggressive and protective sub-signaling pathways of MAPK and the link of different stroke factors with the signaling. Also, the possible therapeutic approaches with MAPK signaling are discussed.

Ischemic Injury: A Molecular Insight. Thakur Gurjeet Singh & Heena Khan (Eds.)

22.1 INTRODUCTION

An ischemic stroke is an event that is related to the blockage of the blood supply in the cerebrovascular system. It is the most common type of stroke and occurs when a blood vessel in the neck or brain gets blocked. The blockage in the blood vessels could be due to embolism (movement of clot from one body part to another), thrombosis (formation of a clot in the blood vessel), or due to stenosis (narrowing of an artery leading to the brain) [1]. In 2016, 13.7 million new cases of stroke occurred worldwide, out of which nearly 87% of cases are of ischemic stroke [2]. In 2019, 12.2 million incidents of stroke and 6.55 million deaths were reported due to stroke. Balance issues, hemiplegia, loss of sensory and vibratory sensation, numbness, decreased reflexes, ptosis (of the eyelid), visual field defects, aphasia, and apraxia are all neurological deficits that occur after a stroke [3]. Age, gender, hypertension, diabetes mellitus, hypercholesterolemia, excessive alcohol use, cigarette smoking, and other stressors are all known to raise the risk of ischemic stroke. Ischemic stroke impairs not only the nervous system's homeostasis but also other tissues and organs as a result of atherosclerotic and inflammatory processes that lead to heart failure [4]. Cerebral ischemia leads to tissue death due to cerebral blood vessel obstruction, and it stimulates mitogen-activated protein kinase (MAPK) and causes the secretion of pro-inflammatory cytokines [5]. The MAPK signaling pathway transmits signals from the cell to the nucleus. This pathway controls a broad spectrum of cell processes, including growth, inflammation, and stress responses. It is accepted as a therapeutic target for cancer and peripheral inflammatory disorders. There is also considerable evidence that MAPK is a key regulator of ischemic and hemorrhagic cerebral vascular disease, suggesting that it could act as newer drug discovery target for treating stroke [6]. In this chapter, we have discussed about the MAPK signaling, protective and aggressive factors of the MAPK signaling in stroke. The events like astrogliosis and glial scar formation which take place after ischemic stroke are discussed in detail. Also, the role of cell apoptosis and JNK MAPK signaling with respect to stroke are discussed. The possible approaches that can obstruct the MAPK signaling in stroke are discussed at the end of the chapter.

22.2 MAPK SIGNALING AND ITS SUBTYPES

The Mitogen-Activated Protein Kinase is the domain of intra-cellular serine/threonine protein kinases and is evolutionary conserved. In mammals, it primarily includes Extracellular signal-regulated kinase 1/2 (ERKs), stress-activated protein kinases/c-Jun N-terminal kinases (SAPKs/JNKs), Nemo-like kinase (NLK), atypical MAPKs such as ERK3/4, ERK7, and p38 MAPKs, ERK5. The MAPK pathway plays an important role in cell differentiation, proliferation, apoptosis, tumor metastasis, and angiogenesis than other intracellular signaling pathways [7]. The activation of MAPKs as it appears to participate in most signal transduction pathways is a prototype of intracellular signaling. In response to various differentiation, stimuli, regulating proliferation, apoptosis, and inflammation, from the cell membrane to the nucleus, the MAPK signaling pathway transmits signals. Several pieces of evidence suggest that MAPK is a key regulator of hemorrhagic and ischemic cerebral vascular disease. It can impair blood-brain barrier (BBB) integrity and intensify neuro-inflammation by releasing pro-inflammatory mediators. Then, after ischemic stroke, it causes neurovascular damage. Several experimental studies have linked apoptosis signal-regulating kinase 1 (ASK1) to cellular mechanisms following cerebral ischemia. ASK1 inhibition, either pharmacologically or genetically, reduces ischemic brain impairment and has additional neuroprotective effects.

22.2.1 APOPTOSIS SIGNAL-REGULATING KINASE 1 (ASK1)

ASK1 belongs to the mitogen-activated protein kinase kinase (MAPKK) family, which initiates mitogen-activated protein kinase (MAP2K: MKK4/MKK7, MKK3/MKK6) and results in the initiation of mitogen-activated protein kinase (MAPK) as a part of the signaling cascade. Threonine phosphorylation (Thr 845 in mice and Thr 838 in humans) is required for ASK1 activation. ASK1 is a homooligomer that attaches to another ASK1 by its C-terminal coiled-coil complex under normal conditions. ASK1's N-terminal coiled-coil domain binds to thioredoxin (Trx) and inhibits ASK1 kinase activity. Oxidized Trx is separated from ASK1, and unbound ASK1 is phosphorylated to activate it under oxidative stress conditions. Oxidative stress and calcium influx can both cause ASK1 Thr phosphorylation. ASK1 is well known to be an early Reactive oxygen species (ROS) responder after cerebral ischemia. Activated ASK1 activates various types of signaling cascades, including p38 and c-Jun N-terminal kinase (JNK), and regulates cellular mechanisms such as cell growth, differentiation, and death. Although ASK1 was previously thought to primarily take part in apoptotic cell death (apoptosis and necrosis), recent research has revealed that it also plays a role in brain edema, reactive gliosis, thrombosis, and inflammatory responses after cerebral ischemia [8].

22.2.2 EXTRACELLULAR SIGNAL REGULATED KINASE ½ (ERK1/2) SIGNALING

Extracellular signal-regulated kinase 1/2 (ERK1/2), is well known and best-studied member of the mitogen-activated protein kinase (MAPK) family or complex. It regulates a wide range of activities, including metabolism, motility, and inflammation, as well as cell death and survival. Cell surface receptors driven by growth factors or cytokines, phosphorylate, and activate it in a three-tiered MEK mechanism. After cerebral ischemia or reperfusion, the amount of phosphorylated ERK1/2 is frequently raised. After development, the ERK1/2 pathway is well known to perform pathological and physiological roles. Significant evidence demonstrates that ERK1/2 regulates inflammatory responses, cytokines, cell apoptosis, and death in ischemia and hemorrhagic brain injury [9]. It has been found in different studies that inhibition of ERK1/2 activity lowers apoptosis and inflammation and also reduces cerebral impairment after a stroke. ERK1/2 signaling is important in neurogenesis after brain ischemia. Significant evidence suggests that the ERK1/2 pathway is involved in the regulation of the alteration in inflammation, cytotoxicity, and cerebral vasospasm which arise immediately after a hemorrhagic stroke. All the above evidence and studies show that the blocking of the ERK1/2 pathway can play an important role as a therapeutic target in stroke (10, 11).

22.3 MAPK SIGNALING IN STROKE: PROTECTIVE OR AGGRESSIVE/A DOUBLE-EDGED SWORD

In early ischemic stroke, activation of the p38 MAPK signaling pathway promotes CHOP 10, Elk1, LEF2C, and the protein kinase MAPKK2/3. These pathways maintain neuronal survival and exhibit anti-apoptotic and anti-inflammatory effects. The p38 MAPK is hyper-activated in the late stage leading to increased expression of desired genes by the activation of proteins and transcription factors like caspases leading to neuronal apoptosis [12]. Thus, this indicates the dual behavior of the MAPK signaling pathway in stroke. At an early stage, it acts as a protective factor whereas in the later stages it acts as an aggressive factor. Thus, lead to cell death through the involvement of the different transcription factors and proteins causing apoptosis. This shows that MAPK act as a double-edged sword in the stroke.

22.3.1 PROTECTIVE MAPK MOLECULAR SIGNALING

At the early stage, MAPK signaling act as an anti-apoptotic and anti-inflammatory pathway which indicates its protective effects in ischemic stroke. Different molecular signaling pathways get activated and involved when MAPK acts as protective signaling in ischemic stroke. The activation of the p38 MAPK signaling pathway in early ischemic stroke promotes Elk1, CHOP10, LEF2C, and protein kinase MAPKK2/3 expression which inhibits autophagic effects and maintains neuron cell survival [13].

22.3.2 AGGRESSIVE MAPK SIGNALING IN STROKE

At the later stage of stroke, various aggressive MAPK signaling gets initiated. ROS, cytokines, and overactivation of p38 MAPK are aggressive factors that act in the later stage of the stroke. Thus, leading to worsening of the conditions. All these aggressive factors are discussed one by one as in subsections.

22.3.2.1 REACTIVE OXYGEN SPECIES (ROS)

ROS are a category of reactive oxygen-containing molecules that freely react with macromolecules, causing irreversible functional changes or even destruction. While ROS plays an important role in human physiological processes, ROS overproduction is a significant aspect of ischemic stroke, and ROS is a key mediator of ischemic damage. ROS can cause both extrinsic and intrinsic apoptosis via a variety of signaling pathways, including stimulation of death receptors, mitochondrial DNA damage, and increased activity of c-Jun N-terminal kinases. During an ischemic stroke, the amount of ROS produced is far greater than what is required for normal physiological functions, resulting in negative effects on the brain. Serum albumin oxidation was increased in stroke patients, which can be attributed to ROS oxidation of amino acid residues [14]. The mitochondrial electron transport chain (ETS) is a main source of ROS. Under normal conditions, cytochrome c oxidase in Complex

IV of the ETS converts O_2 to H_2O. The electron transport chain's ubiquinone-cytochrome b region has been proposed as the main target site for ROS production during ischemia. A recent study reported succinate as a promising mitochondrial metabolite that drives through ROS production during ischemia. During an ischemic stroke, xanthine oxidase (XO) is also a source of ROS. Lipoxygenases (LOXs), cytochrome P450 enzymes, and cyclooxygenases (COXs) are intracellular enzymes that catalyze the generation of ROS. During ischemia, these enzymes are responsible for the metabolism of free arachidonic acid (AA) produced from cell membrane phospholipids [14]. ROS has been likely to be linked to brain injury following an ischemic stroke. Excessive production of reactive oxygen species (ROS) by oxidative stress is a primary mechanism of damage to the brain in stroke and reperfusion-induced stroke. A rapid increase in ROS generation immediately following an acute ischemic stroke (AIS) overwhelms antioxidant defenses, causing additional tissue damage. These ROS can cause cellular macromolecule destruction, resulting in autophagy, apoptosis, and necrosis [14].

22.3.2.2 OVERACTIVATION OF P38MAPK

The p38 MAPK is an important member of the MAPK family that can be phosphorylated against a variety of extracellular stimuli that ultimately activate the MAPK signaling pathway. The p38 MAPK has been linked to cellular differentiation, cell cycle progression, and apoptosis, as well as the regulation of inflammatory responses [15]. There are four distinct isoforms of p38 MAPK (p38α, p38β, p38γ, and p38δ), each with various substrate-specific characteristics and expression sequences, leading to variable p38 MAPK inhibitor sensitivity. Protein kinases, phosphatases, cell-cycle regulators, and transcription factors are all targets of p38 MAPK. The α and β isoforms can autophosphorylate. However, p38 MAPK activation is triggered by MAP3K signaling (such as apoptosis signal-regulating kinase 1 [ASK-1] and MEKK1-4) phosphorylating MAP2K (MKK3 and 6), which then phosphorylates and activates p38 MAPK [16]. The activation of the p38 MAPK pathway is needed for the production of proinflammatory cytokines such as interleukins (IL-1, IL-6) and tumor necrosis factor (TNF). As of now, the amount of research suggests that these cytokines play a role in CNS injury. The p38 MAPK is over-activated in the last stages of ischemic stroke and leads to the activation of transcription factors and proteins such as caspase, which then may promote the expression of target genes leading to neuronal apoptosis. Lipopolysaccharides ultimately stimulate microglia and inhibit ULK1 activity through direct phosphorylation of p38 MAPK, lowering autophagy levels and flux. Thus, it shows the neuroprotective action in ischemic stroke. Sustained p38 activation leads to an increase in ROS generation, which activates ASK1 and p38 [18]. In stress conditions, Daxx (Death Associated protein 6) protein accumulation activates ASK1 and JNK which cause mitochondrial apoptosis. Through AP-1 (activator protein 1) activation, ASK1 stimulates cyclin D1 transcription, which increases ASK1 levels via the Rb-E2F (retinoblastoma-E2F transcription factor) pathway. The increased levels of ASK1 after cerebral ischemia leads to the induction of apoptosis. Thus, ASK1 activation through the p38 MAPK can act as a key target for ischemic stroke (17–19).

22.3.2.3 CYTOKINES IN ISCHEMIC STROKE

In both periphery and CNS, inflammatory mechanisms exhibit an important role in the pathophysiologic processes occurring after the onset of ischemic stroke. Cytokines play an important role in the inflammatory process and contribute to the development of ischemic injury. Cytokines are soluble glycoproteins formed by brain cells in response to injured tissues following ischemia and are responsible for monitoring the innate and adaptive immune response. Pro-inflammatory and anti-inflammatory cytokines can be released into the bloodstream by microglia, astrocytes, endothelial cells, and neurons in the brain [20]. In animal models, a rise in pro-inflammatory cytokine production and a decline in anti-inflammatory cytokine production is associated with larger infarct size. TNF-α, IL-1β, IL-6, and IL-10 are inflammatory cytokines that have been linked to ischemic stroke and have been proposed as targeted therapies and prognostic biomarkers. The action of TNF-α in stroke is through NF-kB signaling. IL-1β is found to be neurotoxic in various models of ischemic stroke. The level of IL-6 depends on the stroke severity and stroke type. Thus, cytokines play a major inflammatory role in the stroke and can be used as a therapeutic target in stroke management (20, 21).

22.4 AFTER ISCHEMIC STROKE SIGNALING THROUGH P38 MAPK SIGNALING

Neurological functional recovery is extremely difficult after a stroke and must be improved. The inhibition of the p38 MAPK pathway blocks inflammatory response. The p38 MAPK pathway is capable of inducing NF-kB cell apoptosis. Typically, the p38 MAPK pathway is important not only in cell apoptosis, but also in a wide range of pathological processes such as inflammation, cell stress, cell cycle, and growth. Furthermore, the p38 MAPK pathway can be activated by a wide range of physical, chemical, and inflammatory factors, as well as stress and gram-positive bacterial cell wall components [22]. The mitogen-activated protein kinase (MAPK) pathways are thought to play a role in neuronal death and survival following a stroke. One of the most common causes of functional recovery difficulty after an ischemic stroke is reactive astrogliosis and glial scar formation, the presence of which restricts neurite regeneration. It was found that at the site of cerebral ischemia, the expression of the p38 MAPK protein is increased in macrophages [22]. TNF-α and IL-1β, which are known to be getting elevated after stroke and are involved in the mechanisms underlying ischemia-induced cell death, strongly activate the p38 MAPK pathway [23].

22.4.1 ASTROCYTE ACTIVATION

Astrocytes are directly engaged in negative responses in mechanical injuries and excito-toxic due to hyperresponsiveness and glial scar formation. However, astrocytes are highly specialized glial cells of the nervous system that manage numerous homeostatic functions for the neurovascular unit's survival and maintenance. Astrocytes possess neuroprotec-tive, angiogenic, immunomodulatory, neurogenic, and antioxidant properties, as well as the ability to modify synaptic function [24]. Astrocytes respond to ischemic stroke injury through a signature process of reactive astrogliosis and subsequent glial scar formation, which is characterized by hypertrophic morphology and astrocyte over-proliferation, as well as increased expression of the intermediate filament protein, glial fibrillary acidic protein (GFAP). The glial scar is primarily composed of reactive astrocytes, though oligodendrocytes and microglia are also present. Depending on the time and situation of the brain injury, astrogliosis can perform both beneficial and adverse functions. Reactive astrocytes are beneficial for tissue repair in the acute stage just after ischemic damage by secreting anti-inflammatory factors, uptake of excess glutamate, and isolation of intact tissue. Concurrently, reactive astrocytes are known to worsen the injury by releasing neurotoxic levels of reactive oxygen species, increasing inflammation through cytokine production, compromising BBB function through vascular endothelial growth factor (VEGF) production, and exacerbating cytotoxic edema [25]. In the case of ischemia, for example, a disruption in brain blood supply results in a lack of oxygen and glucose, and thus a reduction in energy supply. This process results in excessive ion accumulation and deregulation of several signaling pathways. Also lead to overburdening the buffering role of astrocytes and favoring the activation of degradative processes mediated by proteases, lipases, and nucleases, which interrupt neuronal function and cause cell death [24].

22.4.2 GLIAL SCAR FORMATION

The most distinguishing characteristic of the glial scar just after stroke is reactive astrogliosis. A large population of proliferating cells in the peri-infarct tissue includes astrocytes, microglia, and macrophages. Furthermore, a recent study discovered that a subpopulation of perivascular pericytes or related stromal cells is also involved in stroke proliferation. They release extracellular matrix components such as fibronectin and collagen type 1 in the core, resulting in the formation of the fibrotic scar, which is a component of the developing glial scar [26]. A subset of these cells may survive and become a part of the mature glial scar within weeks of the stroke. Proliferation is another important factor in astrocytic accumulation in damaged areas of the CNS, leading to the formation of glial scars. Recent research has shown that reactive astrocytes in the glial scar develop from glial progenitors. In brain injuries, NG2 (polydendrocytes) cells can differentiate into proliferating reactive astrocytes, and in spinal cord injury and stroke, ependymal cell precursors generate scar-forming astrocytes. Glial scars in various pathological conditions such as CNS infection, brain stroke, and traumatic injuries have their patterns for reorganizing nerve tissue and repairing neuronal damage. Furthermore, the glial scar's extracellular matrix influences the prognostic value and the occurrence of neurological complications [27]. Glial scars re-join damaged tissue after a stroke or other CNS injury and are necessary for repair. In recent studies, it is found that glial scars enable axonal regeneration [28]. Recent research indicates that reactive gliosis has both beneficial and detrimental effects on restoring brain functions following ischemic injury. In the acute phase, glial scar formation is critical for sealing the lesion site, remodeling the tissue, and controlling the local immune response both temporally and spatially. The glial barrier seals off the area of damage to prevent further microbial infections and the spread of cellular damage, as well as to maintain extracellular ion and fluid balance, prevent an overwhelming inflammatory and growth factor response, and scavenge free radicals. Furthermore, the glial scar stimulates blood capillary revascularization, which increases the nutritional, trophic, and metabolic support of the nervous tissue. The glial scar, on the other hand, acts as an impediment to axon regeneration and thus prevents CNS function recovery in the chronic phase, because reactive astrocytes secrete several growth-inhibitory molecules that chemically prevent axonal extensions. Furthermore, the basal membrane component creates an additional physical barrier to regenerating axons, preventing them from crossing the lesion [29] (Figure 22.1).

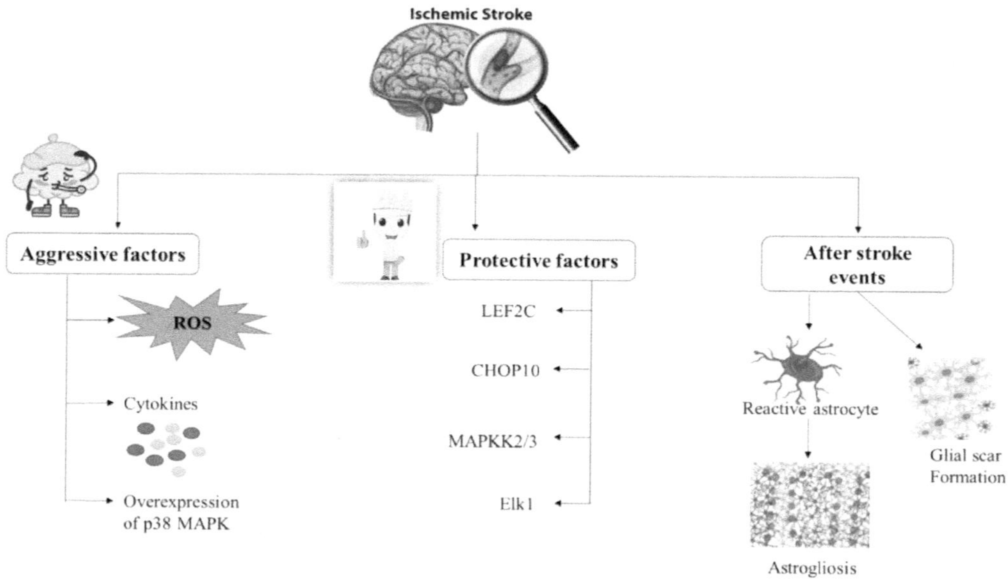

FIGURE 22.1 Events in ischemic stroke.

ROS, Cytokines, and overexpression of p38 MAPK act as aggressive factors in ischemic stroke. ROS and cytokines release is associated with the additional tissue damage in stroke whereas the overactivation of p38 MAPK leads to neuronal apoptosis in Ischemic stroke. LEF2C, CHOP10, Elk1, and MAPKK2/3 are the protective factors in ischemic stroke which get activated at the early stage of the ischemic stroke and thus inhibit autophagic effects and maintain neuronal cell survival. Reactive astrocytes and glial scar formation are the after-stroke events and both are interrelated to each other. Reactive astrocytes are involved in astrogliosis and glial scar formation and both have beneficial as well as detrimental roles in ischemic stroke.

22.5 CELL APOPTOSIS AND MAPK SIGNALING

Apoptosis may be responsible for a significant portion of neuronal death after acute brain ischemia (ABI) [30]. ROS-sensitive apoptosis signal-regulating kinase 1 (ASK1) stimulates apoptosis by activating the MAP kinase JNK. In contrast, ERK, another MAP kinase, maintains cell survival. Ca^{2+} and ROS-mediated signaling pathways activate transcription factors such as NF-kB, AP-1, CREB, STAT, and others, which lead to the activation of proteins associated with cell survival or death [31]. Jannus and Kinase (JNK) and p38 MAPK substrates that promote apoptosis have been identified and validated. JNK and p38 have been shown to regulate several transcription factors, leading to increased expression of pro-apoptotic proteins and reduced expression of anti-apoptotic proteins. The p53 tumor suppressor protein is one of the most well-known transcription factors regulated by the JNK and p38 MAPK cascades in apoptosis. SIRT1 activation is critical for maintaining redox balance and mitochondrial function, as well as suppressing apoptosis via the MAPK signaling pathway. JNKs, like ERKs, are a subfamily of MAPKs that are abundantly expressed in the central nervous system. Furthermore, the JNK pathway has been shown to play a neuroprotective role after an ischemic stroke. Surprisingly, inhibiting the JNK pathway is also involved in reducing autophagy and neuronal apoptosis. As a result, when it comes to autophagy regulation, the JNK pathway may be a downstream pathway of the signaling Information Receiver/Transmitter (SIRT1) [32–35]. SIRT1 suppression has been demonstrated to block IGF-I/IRS-2/Ras/ERK signaling and protect neurons from oxidative stress, implying that SIRT1 and MAPK signaling may play a role in neuronal survival [36].

Caspases are activated when cytochrome c is released from the mitochondria, which is a junction in the cell death pathways. Bcl-2 proteolysis is caused by caspase-9 and caspase-3 activation, which leads to an increase in the release of cytochrome c. Caspase-3 induces proteolysis and continuous activation of MEKK1 (MAPK/ERK). Thus, it enhances cytochrome c release and caspase-9/3 activation via the activation of JNK/p38. Bid proteolysis (tBid) is induced by caspase-3 or caspase-8, which enhances cytochrome c release and caspase activation. Caspase-3 is triggered by caspase-9, which leads to proteolysis and complete activation of caspase-9. A stress signal (Ca^{2+}, ROS, osmotic shock, death receptor, etc.) activates a MAPK cascade, which then feeds back on itself to increase MAPK activity. Additional downstream targets are engaged if a threshold of MAPK activation is exceeded, facilitating cytochrome c release and caspase activation. Other positive response loops downstream of MAPKs will maintain high levels of JNK/p38 and caspase activation, ensuring that the apoptotic process is irreversible [17].

22.6 miRNA TARGETING STROKE VIA MAPK SIGNALING

MicroRNAs (miRNAs) are short non-coding RNAs with a length of 22 nucleotides that regulate gene expression after transcription by attaching to homologous sites in messenger RNAs' 3′-untranslated region. Hypertension, atherosclerosis, atrial fibrillation (AF), diabetes, and dyslipidemia are among stroke risk factors that involve miRNAs. More recent research has focused on the involvement of microRNAs in the pathophysiology of stroke [37].

In multiple studies, stroke has been shown to cause a re-arrangement of the miRNA profile in the cerebral tissue of animals subjected to brain ischemia. In a rat study on middle cerebral artery occlusion, it was observed that 106 transcripts were altered after 24 hours of stroke induction. A new potential stroke neuroprotective target, miR-223-5p, has recently emerged as a putative regulator of the K^+-dependent Na^+/Ca^{2+} exchanger, NCKX2. MAP-kinase, mTOR, Wnt, and GnRH are some of the other key pathways targeted by stroke-regulated miRNAs and are involved in cell regulation, proliferation, and apoptosis. Only let-7a and miR-143, out of the four miRNAs discovered and chosen, would be useful for stroke treatments, according to further focused research. After treatment with a let-7a, there was a reduction in the number of p38 MAPK and p-JNK-immunoreactive cells, as well as decreased levels of TNF-α and IL-6, indicating that knocking out the let-7a gene can prevent cerebral ischemia injury by lowering apoptosis and inflammatory reaction markers. Let-7a has also been demonstrated to target MKP-1 (a mitogen-activated protein kinase phosphatase) which inactivates JNK1/2 and p38, implying that this miRNA has a role in neuroinflammation and apoptosis [38]. It was evidenced that through p38 MAPK/NF-kB pathway inhibition, miRNA22 can reduce ischemic stroke-induced inflammation in rats or *in vitro* models [22]. It was observed that by targeting FGF9/CACNG2 and mediating the activity of the MAPK signaling pathway, miR-339/FGF9/CACNG2 could speed up the onset of Ischemic Stroke, suggesting that miR-339/FGF9/CACNG2 could be used as a new therapeutic target [39].

It was established that certain miRNAs are dysregulated in ischemic stroke, and impact its etiology via altered MAPK pathway regulation. These miRNAs can influence post-stroke processes like inflammation, oxidative stress, and cell death. Through the modification of the MAPK pathway, several miRNAs, including miR-145, miR-137, miR-493, and miR-126, function as regulators of inflammation, apoptosis, angiogenesis, and neurogenesis. Through modulation of various target genes involved in ischemic stroke, management of these miRNAs' expression could reduce permanent neuronal damage [4].

22.7 JNK MAPK PATHWAY

JNK is the most common stress-responsive kinase, and its signaling has been linked to neuroinflammation, disruption of the blood-brain barrier, and oligodendroglia apoptosis in brain injury [40]. JNK has been identified as a promising candidate in studies on new treatment targets for neuroinflammation and neuropathic pain due to its regulatory roles in inducing neuroinflammation *in vivo* and *in vitro*. JNK is a crucial MAPK that upon ischemia brain injury gets activated and plays a remarkably important role in cerebral ischemic injury [41].

JNKs and p38 MAPKs are stress-activated proteins that play critical roles in response to both extracellular and intracellular stresses in balancing cell survival and death [42]. Vast apoptosis findings suggest that these kinases function in a cell context and cell type-specific manner to integrate signals at different transmission points via both transcription-dependent and transcription-independent mechanisms that eventually converge on caspase activation. According to research, the JNK and p38 MAPK pathways are primarily associated with cell stress and apoptosis, whereas the ERK/MAPK signaling pathway, which is the most thoroughly studied MAPK signaling pathway, is associated with cell proliferation and differentiation and plays a critical role in the cell signal transduction network [7]. JNK works via activation of NF-kB signaling and thus increases the levels of the inflammatory cytokine. JNK-In-8 (a JNK inhibitor) was found to suppress the JNK activation and NF-kB signaling and thus lead to the decrease in the levels of phosphorylated JNK and p65 (43–45). This indicated the anti-inflammatory action of the inhibitor. Also, it improves neurological functions leading to the prevention of ischemic stroke. Thus, all the evidence indicates that the inhibition of the JNK MAPK pathway can act as an important therapeutic target for the management of stroke.

ASK1 is an early responder to ROS after cerebral ischemia. ASK1 activates JNK and p38 pathway which alternatively act on the p53 transcription factor. Thus, causing the activation of the apoptotic pathway. ROS and Ca^{2+} act on the mitochondrial functioning and thus causing the release of cytochrome C which in turn activates caspase-3. Caspase-3 is directly responsible for the neuronal damage and also it activates the MAPK signaling pathway leading to the activation of caspase 9. Thus, the regulation of JNK and p38 occurs which leads to apoptosis (Figure 22.2).

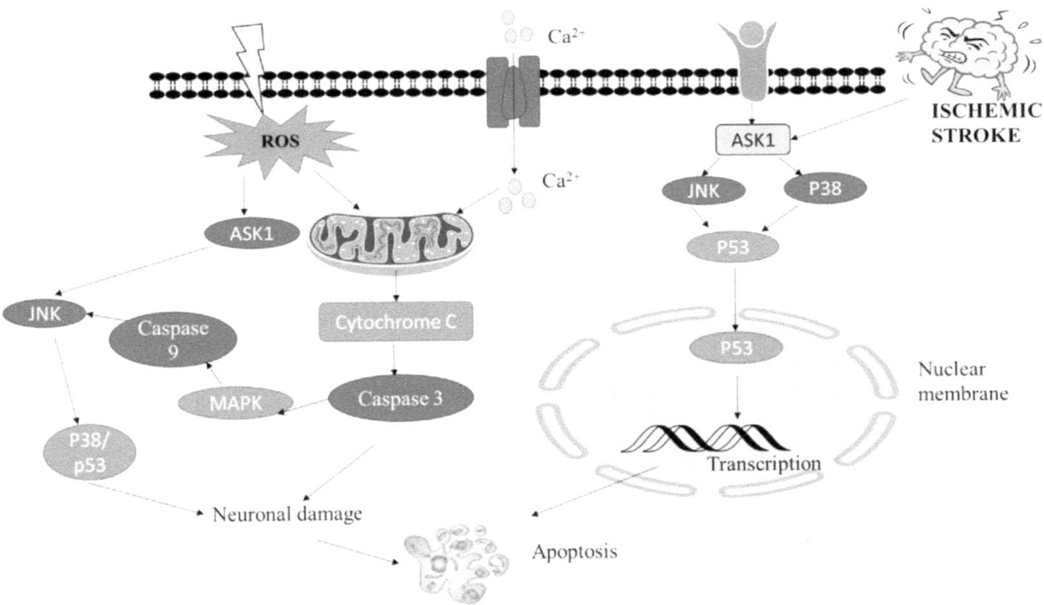

FIGURE 22.2 Role of MAPK/JNK signaling pathway during ischemia.

22.8 APPROACHES TO OBSTRUCT MAPK IN STROKE

MAPK signaling is an important signaling pathway in the pathogenesis of stroke. Thus, it can be used as a favorable therapeutic viewpoint in the management of stroke. By targeting this molecular signaling pathway, newer approaches can be discovered for the treatment of stroke. Inhibitors of the ERK1/2 pathway, such as SB386023-b, PD98059, PD184161, and U0126, block the overexpression of inflammatory mediators and vasospasm-related receptors by decreasing the level of phosphorylated ERK [46]. SB203580, a p38 pathway inhibitor, has been shown to block cPLA2 phosphorylation and activation, as well as decrease the infarct volume and pro-inflammatory mediator expression. PD600125, an anthrapyrazolone derivative of the extremely conserved ATP-binding pocket found in all JNKs, is the most often utilized JNK inhibitor in stroke. It has high efficacy in blocking JNK kinase activity. In hemin and thrombin-induced cell damage, SP600125-mediated JNK inhibition has been shown to prevent neuronal and microglial death. It is found in the research study that Fucoidan which is a complex of sulfated polysaccharide play a protective role in cerebral ischemic-reperfusion injury via the inhibition of the activation of the MAPK pathway [47]. Different approaches for targeting the increased level of ROS, p38MAPK, and cytokines could be used as the therapeutic target for stroke management. Targeting LEF2C, CHOP10, Elk1, and protein kinase MAPKK2/3 expression can be used as a novel therapeutic approach to enhance cell survival chances. From the above studies, it is clear that the inhibition of MAPK signaling can be a possible strategy for the development of the newer therapeutic agents for the treatment of stroke [6].

22.9 CONCLUSION

MAPK signaling now is a subject of intense research for stroke. MAPK can be treated as the therapeutic target for the prevention of stroke pathology. Its dual role in ischemic stroke is still needed to be explored more. More research on the protective factors and signaling involved in early ischemic stroke is required. Following an ischemic stroke, ischemic stroke events such as astrogliosis and glial scar formation can be used to target the ischemic stroke events. MiRNA is a new study topic that can be investigated further as a potential stroke treatment in the future. The inhibitors of different MAPK-related signaling can be used for the treatment of cerebrovascular disorders like stroke. We believe that modulating MAPK signaling could be used to enhance patient outcomes and could be a viable target for the development of new stroke therapeutics.

KEYWORDS

- **apoptosis**
- **autophagy**
- **ERK**
- **inflammation**
- **MAPK**
- **p38**
- **reactive oxygen species**
- **stroke**

REFERENCES

1. Tiwari, P., Khan, H., Singh, T. G., & Grewal, A. K., (2022). Poly (ADP-ribose) polymerase: An overview of mechanistic approaches and therapeutic opportunities in the management of stroke. *Neurochemical Research*, 1–23.
2. Saini, V., Guada, L., & Yavagal, D. R., (2021). Global epidemiology of stroke and access to acute ischemic stroke interventions. *Neurology, 97*(20 Supplement 2), S6–16.
3. Khan, H., Kashyap, A., Kaur, A., & Singh, T. G., (2020). Pharmacological postconditioning: A molecular aspect in ischemic injury. *Journal of Pharmacy and Pharmacology, 72*(11), 1513–1527.
4. Gugliandolo, A., Silvestro, S., Sindona, C., Bramanti, P., & Mazzon, E., (2021). Mirna: Involvement of the MAPK pathway in ischemic stroke. a promising therapeutic target. *Med., 57*(10).
5. Ko, I. G., Jin, J. J., Hwang, L., Kim, S. H., Kim, C. J., Jeon, J. W., et al., (2021). Adenosine A2A receptor agonist polydeoxyribonucleotide ameliorates short-term memory impairment by suppressing cerebral ischemia-induced inflammation via MAPK pathway. *PLoS One [Internet]., 16*(3), 1–14. Available from: http://dx.doi.org/10.1371/journal.pone.0248689.
6. Sun, J., & Nan, G., (2016). The mitogen-activated protein kinase (MAPK) signaling pathway as a discovery target in stroke. *Journal of Molecular Neuroscience, 59*, 90–98.
7. Guo, Y., Pan, W., Liu, S., Shen, Z., Xu, Y., & Hu, L., (2020). ERK/MAPK signaling pathway and tumorigenesis. *Exp. Ther. Med., 19*(3), 1997–2007.
8. Khan, H., Gupta, A., Singh, T. G., & Kaur, A., (2021). Mechanistic insight on the role of leukotriene receptors in ischemic–reperfusion injury. *Pharmacological Reports, 73*(5), 1240–1254.
9. Yang, C. H., Yen, T. L., Hsu, C. Y., Thomas, P. A., Sheu, J. R., & Jayakumar, T., (2017). Multi-targeting andrographolide, a novel NF-κB inhibitor, as a potential therapeutic agent for stroke. *Int. J. Mol. Sci., 18*(8), 1638.
10. Kassouf, T., & Sumara, G., (2020). Impact of conventional and atypical MAPKs on the development of metabolic diseases. *Biomolecules, 10*(9), 1256.
11. Sun, J., & Nan, G., (2017). The extracellular signal-regulated kinase 1/2 pathway in neurological diseases: A potential therapeutic target. *Int. J. Mol. Med., 39*(6), 1338–1346.
12. Kalra, P., Khan, H., Kaur, A., & Singh, T. G., (2022). Mechanistic insight on autophagy modulated molecular pathways in cerebral ischemic injury: From preclinical to clinical perspective. *Neurochemical Research*, 1–9.
13. Wang, X., Fang, Y., Huang, Q., Xu, P., Lenahan, C., Lu, J., et al., (2021). An updated review of autophagy in ischemic stroke: From mechanisms to therapies. *Exp. Neurol. [Internet], 340*, 113684. Available from: https://doi.org/10.1016/j.expneurol.2021.113684.
14. Li, W., & Yang, S., (2016). Targeting oxidative stress for the treatment of ischemic stroke: Upstream and downstream therapeutic strategies. *Brain Circ., 2*(4), 153.
15. Lu, H., Wang, B., Cui, N., & Zhang, Y., (2018). Artesunate suppresses oxidative and inflammatory processes by activating Nrf2 and ROS-dependent p38 MAPK and protects against cerebral ischemia-reperfusion injury. *Mol. Med. Rep., 17*(5), 6639–6646.
16. Gräb, J., & Rybniker, J., (2019). The expanding role of p38 mitogen-activated protein kinase in programmed host cell death. *Microbiol. Insights, 12*, 1178636119864594.
17. Khan, H., Bangar, A., Grewal, A. K., Bansal, P., & Singh, T. G., (2022). Caspase-mediated regulation of the distinct signaling pathways and mechanisms in neuronal survival. *International Immunopharmacology, 110*, 108951.
18. Khan, H., Sharma, K., Kumar, A., Kaur, A., & Singh, T. G., (2022). Therapeutic implications of cyclooxygenase (COX) inhibitors in ischemic injury. *Inflammation Research*, 1–6.
19. Redza-Dutordoir, M., & Averill-Bates, D. A., (2016). Activation of apoptosis signaling pathways by reactive oxygen species. *Biochim. Biophys. Acta (BBA)-Molecular Cell Res., 1863*(12), 2977–2992.
20. Doll, D. N., Barr, T. L., & Simpkins, J. W., (2014). Cytokines: Their role in stroke and potential use as biomarkers and therapeutic targets. *Aging Dis., 5*(5), 294–306.
21. Prabhakar, N. K., Khan, H., Grewal, A. K., & Singh, T. G., (2022). Intervention of neuroinflammation in the traumatic brain injury trajectory: *In vivo* and clinical approaches. *International Immunopharmacology, 108*, 108902.

22. Dong, H., Cui, B., & Hao, X., (2019). MicroRNA-22 alleviates inflammation in ischemic stroke via p38 MAPK pathways. *Mol. Med. Rep., 20*(1), 735–744.
23. Irving, E. A., & Bamford, M., (2002). Role of mitogen-and stress-activated kinases in ischemic injury. *J. Cereb. Blood Flow Metab., 22*(6), 631–647.
24. Khan, H., Singh, A., Thapa, K., Garg, N., Grewal, A. K., & Singh, T. G., (2021). Therapeutic modulation of the phosphatidylinositol 3-kinases (PI3K) pathway in cerebral ischemic injury. *Brain Research, 1761*, 147399.
25. Zhu, Y. M., Lin, L., Wei, C., Guo, Y., Qin, Y., Li, Z. S., et al., (2021). The key regulator of necroptosis, RIP1 kinase, contributes to the formation of astrogliosis and glial scar in ischemic stroke. *Transl. Stroke Res., 12*(6), 991–1017.
26. Sims, N. R., & Yew, W. P., (2017). Reactive astrogliosis in stroke: Contributions of astrocytes to recovery of neurological function. *Neurochem. Int., 107*, 88–103.
27. Wang, H., Song, G., Chuang, H., Chiu, C., Abdelmaksoud, A., Ye, Y., et al., (2018). Portrait of glial scar in neurological diseases. *Int. J. Immunopathol. Pharmacol., 31*, 1–6.
28. Zbesko, J. C., Nguyen, T. V. V., Yang, T., Frye, J. B., Hussain, O., Hayes, M., et al., (2018). Glial scars are permeable to the neurotoxic environment of chronic stroke infarcts. *Neurobiol. Dis. [Internet], 112*(2017), 63–78. Available from: https://doi.org/10.1016/j.nbd.2018.01.007.
29. Huang, L., Wu, Z. B., ZhuGe, Q., Zheng, W. M., Shao, B., Wang, B., et al., (2014). Glial scar formation occurs in the human brain after ischemic stroke. *Int. J. Med. Sci., 11*(4), 344–348.
30. Garg, C., Kaur, A., Singh, T. G., Sharma, V. K., & Singh, S. K., (2022). Therapeutic implications of sonic hedgehog pathway in metabolic disorders: Novel target for effective treatment. *Pharmacological Research*, 106194.
31. Saklani, P., Khan, H., Gupta, S., Kaur, A., & Singh, T. G., (2022). Neuropeptides: Potential neuroprotective agents in ischemic injury. *Life Sciences, 288*, 120186.
32. Xing, Y., Yang, S. D., Wang, M. M., Dong, F., Feng, Y. S., & Zhang, F., (2018). Electroacupuncture alleviated neuronal apoptosis following ischemic stroke in rats via midkine and ERK/JNK/p38 signaling pathway. *J. Mol. Neurosci., 66*(1), 26–36.
33. Khan, H., Garg, N., Singh, T. G., Kaur, A., & Thapa, K., (2022). Calpain inhibitors as potential therapeutic modulators in neurodegenerative diseases. *Neurochemical Research*, 1–25.
34. Becatti, M., Barygina, V., Mannucci, A., Emmi, G., Prisco, D., Lotti, T., et al., (2018). Sirt1 protects against oxidative stress-induced apoptosis in fibroblasts from psoriatic patients: A new insight into the pathogenetic mechanisms of psoriasis. *Int. J. Mol. Sci., 19*(6), 1572.
35. Xue, T. F., Ding, X., Ji, J., Yan, H., Huang, J. Y., Guo, X. D., et al., (2017). PD149163 induces hypothermia to protect against brain injury in acute cerebral ischemic rats. *J. Pharmacol. Sci., 135*(3), 105–113.
36. Khan, H., Tiwari, P., Kaur, A., & Singh, T. G., (2021). Sirtuin acetylation and deacetylation: A complex paradigm in neurodegenerative disease. *Molecular Neurobiology, 58*(8), 3903–3917.
37. Khan, H., Grewal, A. K., & Singh, T. G., (2022). Pharmacological postconditioning by protocatechuic acid attenuates brain injury in ischemia–reperfusion (I/R) mice model: Implications of nuclear factor erythroid-2-related factor pathway. *Neuroscience, 491*, 23–31.
38. Pignataro, G., (2021). Emerging role of microRNAs in stroke protection elicited by remote postconditioning. *Front. Neurol.*, 1833.
39. Gao, X. Z., Ma, R. H., & Zhang, Z. X., (2020). miR-339 promotes hypoxia-induced neuronal apoptosis and impairs cell viability by targeting FGF9/CACNG2 and mediating MAPK pathway in ischemic stroke. *Front. Neurol., 11*, 436.
40. Thapa, K., Khan, H., Singh, T. G., & Kaur, A., (2021). Traumatic brain injury: Mechanistic insight on pathophysiology and potential therapeutic targets. *Journal of Molecular Neuroscience, 71*(9), 1725–1742.
41. Gaire, B. P., Sapkota, A., Song, M. R., & Choi, J. W., (2019). Lysophosphatidic acid receptor 1 (LPA1) plays critical roles in microglial activation and brain damage after transient focal cerebral ischemia. *J. Neuroinflammation, 16*(1), 1–16.
42. Zhou, X., Naguro, I., Ichijo, H., & Watanabe, K., (2016). Mitogen-activated protein kinases as key players in osmotic stress signaling. *Biochim. Biophys. Acta (BBA)-General Subj., 1860*(9), 2037–2052.

43. Wu, Q., Wu, W., Jacevic, V., Franca, T. C. C., Wang, X., & Kuca, K., (2020). Selective inhibitors for JNK signaling: A potential targeted therapy in cancer. *J. Enzyme Inhib. Med. Chem., 35*(1), 574–583.
44. Chen, X., Li, X., Zhang, W., He, J., Xu, B., Lei, B., et al., (2018). Activation of AMPK inhibits inflammatory response during hypoxia and reoxygenation through modulating JNK-mediated NF-κB pathway. *Metabolism, 83*, 256–270.
45. Du, J., Wang, G., Luo, H., Liu, N., & Xie, J., (2021). JNK-IN-8 treatment alleviates lipopolysaccharide-induced acute lung injury via suppression of inflammation and oxidative stress regulated by JNK/NF-κB signaling. *Mol. Med. Rep., 23*(2), 1.
46. Zhu, Z., Ge, M., Li, C., Yu, L., Gu, Y., Hu, Y., et al., (2020). Effects of p38 MAPK signaling pathway on cognitive function and recovery of neuronal function after hypoxic-ischemic brain injury in newborn rats. *J. Clin. Neurosci., 78*, 365–370.
47. Che, N., Ma, Y., & Xin, Y., (2017). Protective role of fucoidan in cerebral ischemia-reperfusion injury through inhibition of MAPK signaling pathway. *Biomol. Ther. (Seoul), 25*(3), 272.

CHAPTER 23

Regulation of JNK Signaling Pathway in Ischemia/Reperfusion Injury

VEERTA SHARMA,[1] HEENA KHAN,[1] AMARJOT KAUR GREWAL,[1] THAKUR GURJEET SINGH,[1] and RAHUL DESHMUKH[2]

[1]*Chitkara College of Pharmacy, Chitkara University, Punjab, India*

[2]*Department of Pharmaceutical Sciences and Technology, Maharaja Ranjit Singh Punjab Technical University, Bathinda, Punjab, India*

ABSTRACT

Ischemic stroke is one of the significant causes of morbidity and mortality in the central nervous system (CNS), affecting millions of people across the globe. c-Jun N-terminal kinase (JNK) plays a vital role among the multiple intracellular signaling pathways that control brain development and disease. JNKs have a role in brain development, regeneration, plasticity, neuroinflammation, and neuronal death. JNK signaling is an attractive target for pharmaceutical intervention due to its unique character. It has been linked to different types of ischemia such as cerebral, myocardial, renal, hepatic, and intestinal. The purpose of this chapter is to give an overview of the research that has been done to relate JNKs to different types of ischemic conditions.

23.1 INTRODUCTION

Extensive attempts have been made to treat the disease, but ischemic stroke remains the leading cause of long-term impairment and continues to be a major type of stroke. In spite of the fact that genetic or transcriptome biomarkers can aid in the diagnosis and prognosis of an ischemic stroke, current stroke therapy [1] is severely limited. Therefore, a thorough comprehension of the regulation of inflammatory processes in response to brain damage can be a prerequisite for the development of an effective treatment for ischemic stroke. Signaling pathways that facilitate signal transmission from the cell surface to the nucleus are activated in neurons in response to extracellular alterations. Signals are sent, amplified, and integrated across these channels, resulting in a physiological response. Such signals mediate

Ischemic Injury: A Molecular Insight. Thakur Gurjeet Singh & Heena Khan (Eds.)

the growth of the brain, healing, neuronal death, and neuroinflammation in the central nervous system (CNS) [2]. The c-Jun N-terminal kinase (JNK) pathway is an example of an evolutionarily conserved family of serine/threonine kinase signaling proteins that play a crucial role in development and disease. JNK has a diverse role in biological functions such as proliferation, differentiation, and death [3, 4]. JNK is upregulated in response to various signals like damage to DNA, oxidative stress, and aging, which control essential pro-apoptotic and anti-apoptotic proteins. Its baseline activity is greater in the brain than in other tissues, indicating that JNK plays a significant part in the regular physiological functions of the nervous system [3, 5]. Due to its unique properties, JNK signaling has been related to several forms of ischemia, including cerebral, myocardial, renal, hepatic, and intestinal ischemia. This review aims to give an overview of the research that has been done to relate JNKs to different types of ischemic conditions.

23.2 JNK SIGNALING PATHWAY

JNK are proteins that belong to the mitogen-activated protein kinase (MAPK) family [6, 7]. The JNK-MAPK pathway is primarily triggered by stress and is involved in the development, apoptosis, cell proliferation, inflammation, and immunological responses [8–10]. Nonetheless, the vital role of the controlled interaction of multiple other proteins that are part of the constructed signaling modules that interact with JNK has emerged as a prominent motif in regulating JNK activity and functioning. The scaffold proteins that act as platforms for assembling the modules and the upstream JNK-2K and JNK-3K, are among them [8].

Furthermore, the existence of numerous JNK forms adds to the complexity of JNK signaling The molecular factors driving JNK signaling molecules assembly and interactions with other proteins in a cell type and signal-specific way and the precise functions of each JNK isozyme in the cell are currently unknown. JNK activation causes apoptosis and at the cellular level promotes cellular survival [11]. The choice of cell fate decision appears to depend on the cell type and the stimulating signal. JNK genes include JNK-1, JNK-2, and JNK-3 which are highly similar, with 10 splice variants, and multiple JNK substrates have been found [12]. JNK-1 and JNK-2 are found throughout the body and JNK-3 are found mainly in the heart and brain [12, 13]. JNK was initially discovered as an enzyme that phosphorylates c-Jun. Therefore, much of the early research was devoted to understanding the kinase-substrate relationship using c-Jun as the model substrate and common functions for JNKs in various developmental processes and disease states.

23.3 PATHOPHYSIOLOGICAL ROLE OF THE JNK PATHWAY IN STROKE

Stroke is one of the leading causes of mortality and disability worldwide. Current models ignore the extensive therapeutical time frame necessary in clinics, so viable medicines have yet to be developed. Ischemic CNS damage can result in neuronal injury and death through various methods. Excitotoxicity, higher intracellular calcium levels, inflammatory processes mediated by cytokines, and loss of survival factors have all been identified to mediate cell death following ischemia [14]. The most common classifications for death processes are necrotic events, including the loss of cellular energy. Apoptotic mechanisms, however, appear to be implicated in ischemic-induced neurodegeneration. Apoptotic bodies, condensed chromatin, shrunken cells, and DNA laddering are morphological markers of neuronal death following ischemia [15]. Degenerating neurons with apoptotic appearance are frequently observed near the penumbra of the ischemic zone and degrade after the initial ischemic event [16]. Anti-apoptotic protein (Bcl-2) overexpression protects mice from permanent and transitory localized ischemia. When mice are administered a medication that inhibits a pro-apoptotic transcription factor (p53), their infarct size is reduced compared to when they are administered a vehicle [17]. In addition, in stroke models, caspases-1 and -3 are activated, and peptide caspase inhibitors reduce the size of the infarct. In animal models of ischemic stroke, suppression of JNK signaling results in neuronal death. Mice lacking JNK-3 or with mutations in the phosphorylation site of c-Jun, on the other hand, are resistant to the hippocampus neurotoxic events caused by the glutamate receptor agonist kainic acid. Mice lacking JNK-1 and JNK-2 are susceptible to kainic acid-induced seizures and neuronal death [18]. These findings imply that distinct JNK isoforms govern diverse responses to neuronal insults, with the JNK-3 gene engaged in glutamate excitotoxicity, a key component of ischemic death, and a brain area susceptible to global ischemia. These findings suggest that JNK signaling is involved in a neuronal type that is susceptible to stroke and is mediated by excitotoxicity. In addition to the signaling processes outlined above, the expression, activation, and localization of MAPK in the CNS of experimental animals suffering from ischemic stroke were investigated.

23.4 TARGETING THE JNK PATHWAY AS A THERAPEUTIC PROTECTIVE STRATEGY FOR DIFFERENT ISCHEMIC INJURY

Extracellular signals are typically relayed to the nucleus via phosphorylation of nuclear substrates by JNK's upstream kinases through the relevant MAPK modules [6]. JNK has been implicated in binding with proteins in the plasma membrane, influencing cellular mobility and regulation of gene expression [12].

23.4.1 JNK SIGNALING PATHWAY AND CEREBRAL ISCHEMIC-REPERFUSION INJURY

There is evidence that MAPK signaling, namely JNK signaling, is activated in the CNS of ischemic stroke-affected experimental animals. The timing of JNK activation, its co-localization in apoptotic neurons, and the resistance of JNK-3-deficient mice to hippocampal excitotoxicity all indicate that this pathway is implicated in neuronal death in this kind of acute neurodegeneration. In the CA1 and CA3 areas of the hippocampus, JNK, p38, and Erk MAPK are active after 5 minutes of carotid artery closure in a gerbil model of acute global ischemia [19]. The rapid and sustained activation of such kinases in response to ischemia in these animals suggested a connection to delayed neuronal death. The numerous JNK inhibitors are listed in Table 23.1, and based on the preceding information, blockage of the JNK pathway appears to be a viable target for avoiding stroke (Table 23.1 and Figure 23.2). Therefore, the preceding evidence led us to infer that JNK pathway inhibition is a promising strategy in stroke (Figure 23.1).

TABLE 23.1 Effect of JNK Pathway Inhibition in Cerebral I/R Injury (Figure 23.2)

SL. No.	Drug	Result	Pathway	References
1.	D-JNKI-1	Protect against cell death.	JNK pathway	[20]
2.	JNK-IN-8	Suppress neuroinflammation.	JNK pathway	[21]
3.	AS601245	Prevention of loss of hippocampus CA1 neurons.	JNK pathway	[22]
4.	SB203580	Decrease mortality (ischemic).	JNK pathway	[23]
5.	SP600125	Protects neuronal injury.	JNK pathway	[24]
6.	PD98059	Decrease mortality (ischemic).	JNK pathway	[23]

23.4.2 JNK SIGNALING PATHWAY AND MYOCARDIAL ISCHEMIC-REPERFUSION INJURY

The JNK-dependent pathway plays a crucial role in the pathophysiology of cardiac hypertrophy and heart ischemia/reperfusion (I/R) damage [25]. JNK is activated in the heart during I/R, and it may be engaged in both protective and detrimental actions [25, 26]. JNK activation is temporary, although it varies depending on the intensity and timing of oxidative damage during I/R [27, 28]. In genetic models, JNK-1 and JNK-2 knockout mice and MAPK-kinase-7 (MKK-7) overexpression animals exhibit protective and detrimental effects on cardiomyocytes during I/R-induced apoptosis in cardiomyocytes *in vivo* [29]. Even though JNK activation is minor during myocardial ischemia caused by coronary artery bypass surgery in people, JNK activity in cardiac tissue increases upon reperfusion [30]. By putting cardiomyocytes in an ischemic buffer and an oxygen-free environment, inevitable consequences of cardiac I/R may be replicated *in vitro* (95% N_2 and 5% CO_2). In neonatal H9c2 cardiomyocytes, in this model ischemia and subsequent reoxygenation causes an increase in the phosphorylation of JNK and its activity [31]. Reactive oxygen species (ROS), mitochondrial dysfunction, and death of cardiomyocytes are all caused by JNK-dependent signaling [32]. Sepsis has been found to cause mechanisms that are linked to ischemia preconditioning. Isolated cardiomyocytes are protected against hypoxia-induced cell death by bacterial lipopolysaccharide via JNK-related signaling pathways [33]. JNK activation contributes considerably to myocardial I/R damage [34]. JNK is involved in regulating the development of mesenchymal stem cells (Figure 23.2). The use of JNK inhibitors to prevent ischemia-mediated inhibition of mesenchymal stem cells' proliferative activity may have therapeutic benefit in mending ischemic lesions, given the relevance of mesenchymal stem cells in the recovery of the heart from post-ischemic injury [35]. JNK regulates cardiomyocyte apoptosis by activating both caspase-dependent and caspase-independent mitochondrial pathways [36, 37]. The regulation of Bad phosphorylation is one of the pathophysiological mechanisms by which JNK activation causes the death of cardiomyocytes in I/R injury [38]. The various JNK inhibitors are listed in Table 23.2 and from the above data, it was prompted to believe that inhibition of the JNK pathway is a promising target for preventing myocardial injury (Table 23.2).

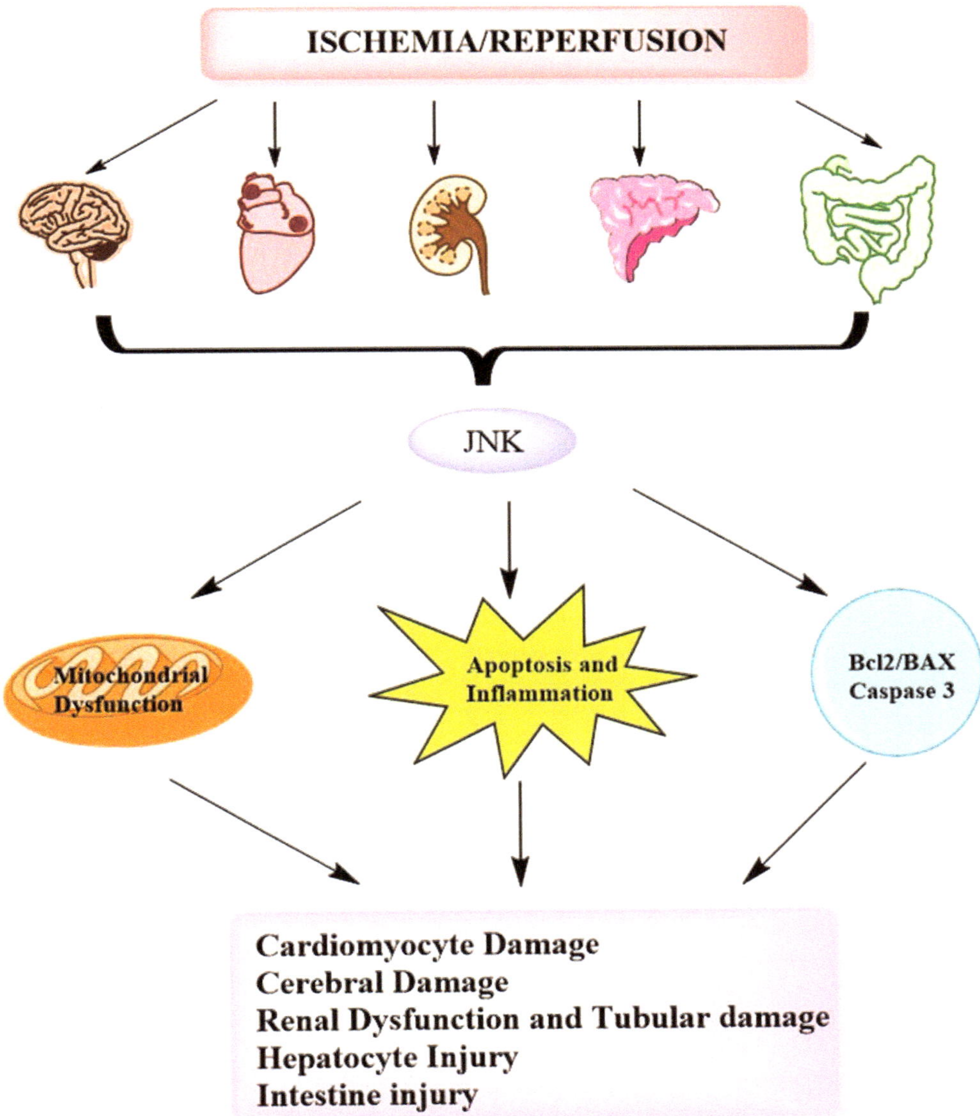

FIGURE 23.1 Schematic diagram showing the involvement of the JNK pathway in various I/R injuries.

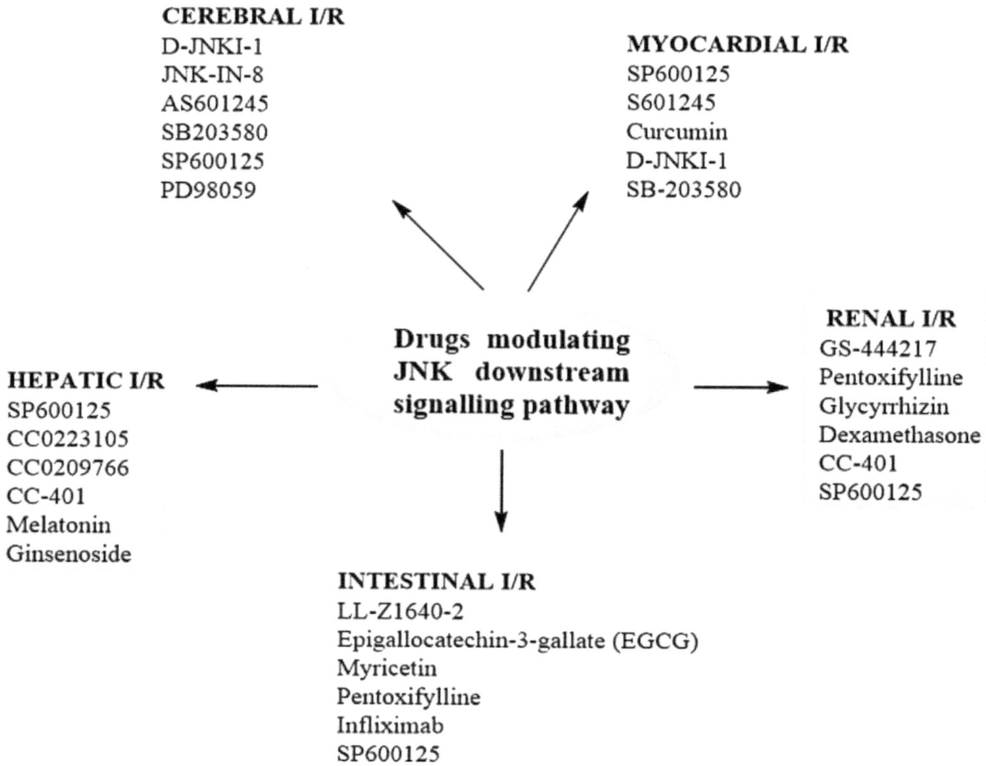

FIGURE 23.2 Illustration of various drugs modulating JNK downstream signaling pathway in different ischemia-reperfusion injuries like cerebral I/R, myocardial I/R, renal I/R, intestinal I/R, and hepatic I/R.

TABLE 23.2 Effect of JNK Pathway Inhibition in Myocardial I/R Injury (Figure 23.2)

SL. No.	Drug	Result	Pathway	References
1.	SP600125	Reduces cardiomyocyte apoptosis.	JNK pathway	[36, 37]
2.	S601245	Reduces cardiomyocyte apoptosis.	JNK pathway	[39]
3.	Curcumin	Inhibition of inflammation.	JNK pathway	[40]
4.	D-JNKI-1	Reduces infarct size.	JNK pathway	[40]
5.	SB-203580	Reduces infarct size.	JNK pathway	[40]

23.4.3 JNK SIGNALING PATHWAY AND RENAL ISCHEMIC-REPERFUSION INJURY

Acute kidney injury (AKI) is now well-known to be associated with chronic renal disease development [41], underlining the necessity of reducing the severity of AKI. AKI can be caused by several things, including sepsis, medication toxicity, and ischemia. AKI is a known side effect of various procedures, including heart bypass surgery, kidney transplantation, and chemotherapy medicines (cisplatin). Within minutes of the ischemic kidney being reperfused, tubular epithelial cells show substantial activation of the JNK pathway. This has been discovered in human kidney transplant biopsies after perfusion [42]. When comparing kidneys from deceased donors to kidneys from surviving donors, the degree of JNK activation was greater in deceased donor allografts, with the degree of JNK activation being proportional to ischemia time. Reduced tubular necrosis and apoptosis, tubular epithelial cell production of pro-inflammatory chemicals (TNF- and CCL2), and interstitial accumulation and activation of macrophages and T-cells are all proposed explanations [42–44] for the protection exhibited by systemic JNK inhibition. AKI induced by the nephrotoxin cisplatin is associated with considerable tubular JNK activation, particularly in older animals who are more susceptible to AKI [45]. Multiple *in vitro* studies [46–49] have established that JNK signaling contributes directly to tubular epithelial cell death in response to painful stimuli such as ATP depletion, ROS, and nephrotoxic substances. Therefore, JNK activation in tubular epithelial cells is a significant element in determining the fate of AKI. The various JNK inhibitors are listed in Table 23.3. Therefore, it is speculated that inhibition of the JNK pathway is a promising target in renal I/R injury (Table 23.3 and Figure 23.2).

TABLE 23.3 Effect of JNK Pathway Inhibition in Renal I/R Injury (Figure 23.2)

SL. No.	Drug	Result	Pathway	References
1.	GS-444217	Prevents renal impairment.	JNK pathway	[50]
2.	Pentoxifylline	Anti-inflammatory and anti-apoptotic effects.	JNK pathway	[51]
3.	Glycyrrhizin	Inhibits inflammatory responses and apoptosis.	JNK pathway	[52]
4.	Dexamethasone	Downregulates the levels of cytokines and inflammation.	JNK pathway	[53]
5.	CC-401	Protects against tubular damage.	JNK pathway	[42, 43]
6.	SP600125	Suppressing extrinsic pathways of apoptosis.	JNK pathway	[42, 43]

23.4.4 JNK SIGNALING PATHWAY AND HEPATIC ISCHEMIC-REPERFUSION INJURY

JNK-1, JNK-2, and JNK-3 are the 3 JNK isoforms found in humans [54]. JNK-1 and JNK-2 are found in the liver. I/R damage in the heart and kidneys was shown to have activated JNK. JNK activation in I/R damage has been reported in other tissues, including the liver. TNF-α and IL-1 cytokines have been demonstrated to activate JNK in experiments. Hepatic I/R damage is mediated by these cytokines. JNK also has a role in inflammatory processes by causing adhesion molecules and inflammatory chemokines to be produced [55]. Hepatocyte apoptosis is also linked to JNK activity. Caspases 3 and cytochrome c JNK activation and increased generation of ROS are also linked to high oxidative stress conditions, such as those seen during I/R [56]. Oxygen-free radicals' inhibitors that modulate JNK activation may offer a novel strategy to counteract the effects of I/R on the liver [57]. JNK inhibition has been established in several investigations to reduce I/R-related damage. After orthotopic liver transplantation, JNK inhibition reduces hepatic necrosis and apoptosis. In a rat model of warm hepatic I/R, JNK inhibition reduces liver damage. JNK inhibitors reduce c-Jun phosphorylation, inflammatory gene expression, and cell proliferation *in vitro*. The various JNK inhibitors are listed in Table 23.4. Therefore, it is speculated that inhibition of the JNK pathway is a promising target in hepatic I/R injury (Table 23.4).

TABLE 23.4 Effect of JNK Pathway Inhibition in Hepatic I/R Injury

SL. No.	Drug	Result	Pathway	References
1.	SP600125	Increase myeloperoxidase, malondialdehyde, and chemokine expression.	JNK pathway	[55]
2.	CC0223105	Decrease necrosis and apoptosis of hepatocytes and sinusoidal endothelial cells.	JNK pathway	[58]
3.	CC0209766	Decrease necrosis and apoptosis of hepatocytes and sinusoidal endothelial cells.	JNK pathway	[58]
4.	CC-401	Decrease necrosis and apoptosis of hepatocytes and sinusoidal endothelial cells.	JNK pathway	[58]
5.	Melatonin	Decrease transaminase levels, leukocyte infiltration, liver damage, and iNOS expression.	JNK pathway	[59]
6.	Ginsenoside	Inhibits inflammation, oxidative stress, and apoptosis.	JNK pathway	[60]

23.4.5 JNK SIGNALING PATHWAY AND INTESTINE ISCHEMIC-REPERFUSION INJURY

The microscopic observations of the colon after ischemia and reperfusion indicated mucosal damage, including denuded villi, loss of villous tissue, and a high degree of monocyte and lymphocyte infiltration into the injured mucosal layer [61]. However, both JNK and p38 were activated during ischemia, after reperfusion the activation of both kinases increased dramatically. Pretreatment with a JNK inhibitor decreased JNK activation, hence substantially lowering histological damage. Multiple studies have demonstrated that JNK activation contributes to I/R damage. Antisense oligonucleotides against JNK inhibited reoxygenation-induced cardiomyocyte death. JNK inhibition was required for superior suppression. As the signaling pathway of the MAPK superfamily is highly interconnected, the inhibition of JNK is critical for achieving additional I/R damage amelioration. The presence of numerous apoptotic cells in the injured epithelium at the tip of the villi, a key site of tissue injury, suggests that apoptosis may considerably contribute to I/R injury. Several research have focused on two potential mechanisms for the apoptosis of the JNK protein. First JNK activation may directly induce apoptosis via the mitochondrial pathway and death signal pathway [61]. Second, JNK activation may be indirectly implicated with apoptosis via pro-inflammatory cytokines and diverse downstream mediators, such as reactive oxygen species. It has been observed that the activation of JNK is essential for the generation of pro-inflammatory cytokines (TNF- and IL-1), which are extracellular stressors for apoptosis [62, 63]. These two distinct pathways are essential to apoptosis and are dependent on JNK activation. JNK is constitutively localized in both the nucleus and cytoplasm and can be activated in both compartments [6]. After I/R, there is a strong activation of JNK in the nucleus of detached villi. JNK also stimulates Elk-1 downstream, suggesting that JNK has a role in differentiation and proliferation in addition to apoptosis [64]. As a result, JNK should be inhibited to prevent tissue injury, especially in the early phase after reperfusion. The various JNK inhibitors are listed in Table 23.5. Therefore, it is speculated that inhibition of the JNK pathway is a promising target in intestinal I/R injury (Table 23.5).

TABLE 23.5 Effect of JNK Pathway Inhibition in Intestine I/R Injury

SL. No.	Drug	Result	Pathway	References
1.	LL-Z1640-2	Decrease apoptosis	JNK pathway	[61]
2.	Epigallocatechin-3-gallate (EGCG)	Decrease malondialdehyde, myeloperoxidase levels, and apoptosis.	JNK pathway	[65]
3.	Myricetin	Reduces pro-inflammatory cytokines and MPO activity.	JNK pathway	[66]
4.	Pentoxifylline	Inhibits cell apoptosis.	JNK pathway	[67]
5.	Infliximab	Inhibits cell apoptosis.	JNK pathway	[67]
6.	SP600125	Inhibits cell apoptosis.	JNK pathway	[68]

23.5 CONCLUSION

Many new studies continue to implicate the JNK pathway in a variety of physiological and pathological processes that are likely at play in CNS disease. In addition, the organization and regulation of the JNK signaling cascade at all levels continue to increase in complexity. Due to the complicated cross-talk within this signaling cascade, as well as its cell type and response-specific regulation, it is difficult to foresee any undesirable consequences that may result from pathway suppression. Developing compounds with adequate specificity for the chosen kinase targets is essential. Compounds that interrupt the JNK pathway and its downstream signaling pathways are advancing in clinical trials, indicating that they have demonstrated sufficient safety and risk-benefit margins. In the future years, clinical evidence will demonstrate the usefulness of targeting this route for therapeutic benefit.

23.6 FUTURE PERSPECTIVES

A literature review on JNK signaling over the past two decades reveals clearly that JNK activity is indeed induced in many disease states. Genetic deletion or chemical inhibition of JNK can ameliorate several disease conditions in animal models. Given these basic pretexts, much work is still required to precisely and efficiently target the JNK pathway in clinical conditions. Therefore, the primary focus should be to determine the causal role of activated JNK in disease manifestation, thereby classifying diseases and conditions dependent on activated JNK signaling. Upon determining diseases that are dominantly JNK dependent, identifying specific substrates/JNK modules for specific disease states, and cell types would be of paramount importance in overcoming the non-specificity of JNK targeting. Once these two primary objectives are achieved, the clinical success of JNK inhibition will become a reality.

KEYWORDS

- **acute kidney injury**
- **apoptosis**
- **c-Jun N-terminal kinases**
- **inflammation**
- **ischemia-reperfusion injury**
- **oxidative stress**
- **reactive oxygen species**

REFERENCES

1. Tiwari, P., Khan, H., Singh, T. G., & Grewal, A. K., (2022). Poly (ADP-ribose) polymerase: An overview of mechanistic approaches and therapeutic opportunities in the management of stroke. *Neurochemical Research*, 1–23.
2. Khan, H., Gupta, A., Singh, T. G., & Kaur, A., (2021). Mechanistic insight on the role of leukotriene receptors in ischemic–reperfusion injury. *Pharmacological Reports*, *73*(5), 1240–1254.
3. Davis, R. J., (2000). Signal transduction by the JNK group of MAP kinases. *Cell, 103*(2), 239–252.
4. Sharma, V., Kaur, A., & Singh, T. G., (2020). Counteracting role of nuclear factor erythroid 2-related factor 2 pathway in Alzheimer's disease. *Biomedicine & Pharmacotherapy*, *129*, 110373.
5. Khan, H., Sharma, R., Kaur, A., & Singh, T. G., (2018). The endocannabioids system and their implications in various disorders. *Int. J. Pharm. Sci. Rev. Res.*
6. Chang, L., & Karin, M., (2001). Mammalian MAP kinase signaling cascades. *Nature, 410*, 37–40.
7. Khan, H., Bangar, A., Grewal, A. K., Bansal, P., & Singh, T. G., (2022). Caspase-mediated regulation of the distinct signaling pathways and mechanisms in neuronal survival. *International Immunopharmacology*, *110*, 108951.
8. Weston, C. R., & Davis, R. J., (2007). The JNK signal transduction pathway. *Current Opinion in Cell Biology*, *19*(2), 142–149.
9. Garg, C., Kaur, A., Singh, T. G., Sharma, V. K., & Singh, S. K., (2022). Therapeutic implications of sonic hedgehog pathway in metabolic disorders: Novel target for effective treatment. *Pharmacological Research*, 106194.
10. Barr, R. K., & Bogoyevitch, M. A., (2001). The c-Jun N-terminal protein kinase family of mitogen-activated protein kinases (JNK MAPKs). *The International Journal of Biochemistry & Cell Biology*, *33*(11), 1047–1063.
11. Kalra, P., Khan, H., Kaur, A., & Singh, T. G., (2022). Mechanistic insight on autophagy modulated molecular pathways in cerebral ischemic injury: From preclinical to clinical perspective. *Neurochemical Research*, 1–19.
12. Bogoyevitch, M. A., & Kobe, B., (2006). Uses for JNK: The many and varied substrates of the c-Jun N-terminal kinases. *Microbiology and Molecular Biology Reviews*, *70*(4), 1061–1095.

13. Khan, H., Garg, N., Singh, T. G., Kaur, A., & Thapa, K., (2022). Calpain inhibitors as potential therapeutic modulators in neurodegenerative diseases. *Neurochemical Research*, 1–25.
14. Mattson, M. P., (2000). Apoptosis in neurodegenerative disorders. *Nature Reviews Molecular Cell Biology*, *1*(2), 120–130.
15. Khan, H., Grewal, A. K., & Singh, T. G., (2022a). Pharmacological postconditioning by protocatechuic acid attenuates brain injury in ischemia-reperfusion (I/R) mice model: Implications of nuclear factor erythroid-2-related factor pathway. *Neuroscience*.
16. Vinters, H. V., & Kleinschmidt-DeMasters, B. K., (2018). General pathology of the central nervous system. In: *Greenfield's Neuropathology-Two Volume Set* (pp. 25–82). CRC Press.
17. Culmsee, C., Bondada, S., & Mattson, M. P., (2001). Hippocampal neurons of mice deficient in DNA-dependent protein kinase exhibit increased vulnerability to DNA damage, oxidative stress, and excitotoxicity. *Molecular Brain Research*, *87*(2), 257–262.
18. Schauwecker, P. E., (2000). Seizure-induced neuronal death is associated with induction of c-Jun N-terminal kinase and is dependent on genetic background. *Brain Research*, *884*(1, 2), 116–128.
19. Thapa, K., Khan, H., Singh, T. G., & Kaur, A., (2021). Traumatic brain injury: Mechanistic insight on pathophysiology and potential therapeutic targets. *Journal of Molecular Neuroscience*, *71*(9), 1725–1742.
20. Hirt, L., Badaut, J., Thevenet, J., Granziera, C., Regli, L., Maurer, F., et al., (2004). D-JNKI1, a cell penetrating c-Jun-N-terminal kinase inhibitor, protects against cell death in severe cerebral ischemia. *Stroke, 35*, 1738–1743.
21. Zheng, J., Dai, Q., Han, K., Hong, W., Jia, D., Mo, Y., Lv, Y., et al., (2020). JNK-IN-8, ac-Jun N-terminal kinase inhibitor, improves functional recovery through suppressing neuroinflammation in ischemic stroke. *Journal of Cellular Physiology*, *235*(3), 2792–2799.
22. Carboni, S., Boschert, U., Gaillard, P., Gotteland, J. P., Gillon, J. Y., & Vitte, P. A., (2008). AS601245, ac-Jun NH2-terminal kinase (JNK) inhibitor, reduces axon/dendrite damage and cognitive deficits after global cerebral ischemia in gerbils. *British Journal of Pharmacology*, *153*(1), 157–163.
23. Sugino, T., Nozaki, K., Takagi, Y., Hattori, I., Hashimoto, N., Moriguchi, T., & Nishida, E., (2000). Activation of mitogen-activated protein kinases after transient forebrain ischemia in gerbil hippocampus. *Journal of Neuroscience*, *20*(12), 4506–4514.
24. Guan, Q. H., Pei, D. S., Liu, X. M., Wang, X. T., Xu, T. L., & Zhang, G. Y., (2006). Neuroprotection against ischemic brain injury by SP600125 via suppressing the extrinsic and intrinsic pathways of apoptosis. *Brain Research*, *1092*(1), 36–46.
25. Javadov, S., Jang, S., & Agostini, B., (2014). Crosstalk between mitogen-activated protein kinases and mitochondria in cardiac diseases: Therapeutic perspectives. *Pharmacology & Therapeutics*, *144*(2), 202–225.
26. Wei, J., Wang, W., Chopra, I., Li, H. F., Dougherty, C. J., Adi, J., Adi, N., et al., (2011). c-Jun N-terminal kinase (JNK-1) confers protection against brief but not extended ischemia during acute myocardial infarction. *Journal of Biological Chemistry*, *286*(16), 13995–14006.
27. Fryer, R. M., Hsu, A. K., & Gross, G. J., (2001). ERK and p38 MAP kinase activation are components of opioid-induced delayed cardioprotection. *Basic Research in Cardiology*, *96*(2), 136–142.
28. Armstrong, S. C., (2004). Protein kinase activation and myocardial ischemia/reperfusion injury. *Cardiovascular Research*, *61*(3), 427–436.
29. Kaiser, R. A., Liang, Q., Huang, Y., Lackey, T., Bueno, O. F., & Molkentin, J. D., (2005). Targeted Deletion of JNK MAPKs Protects Ischemic Myocardium from Infarction by Reducing Myocyte Apoptosis In Vivo. In *Proceedings-Western Pharmacology Society* (Vol. 48, p. 205). [Western Pharmacology Society]; 1998.
30. Talmor, D., Applebaum, A., Rudich, A., Shapira, Y., & Tirosh, A., (2000). Activation of mitogen-activated protein kinases in human heart during cardiopulmonary bypass. *Circulation Research*, *86*(9), 1004–1007.
31. Sun, L., Fan, H., Yang, L., Shi, L., & Liu, Y., (2015). Tyrosol prevents ischemia/reperfusion-induced cardiac injury in H9c2 cells: Involvement of ROS, Hsp70, JNK and ERK, and apoptosis. *Molecules*, *20*(3), 3758–3775.
32. Chambers, J. W., Pachori, A., Howard, S., Iqbal, S., & LoGrasso, P. V., (2013). Inhibition of JNK mitochondrial localization and signaling is protective against ischemia/reperfusion injury in rats. *Journal of Biological Chemistry*, *288*(6), 4000–4011.

33. Walshe, C. M., Laffey, J. G., Kevin, L., & O'Toole, D., (2015). Sepsis protects the myocardium and other organs from subsequent ischemic/reperfusion injury via a MAPK-dependent mechanism. *Intensive Care Medicine Experimental, 3*(1), 1–11.
34. Vassalli, G., Milano, G., & Moccetti, T., (2012). Role of mitogen-activated protein kinases in myocardial ischemia-reperfusion injury during heart transplantation. *Journal of Transplantation, 2012.*
35. Wu, W., Muchir, A., Shan, J., Bonne, G., & Worman, H. J., (2011). Mitogen-activated protein kinase inhibitors improve heart function and prevent fibrosis in cardiomyopathy caused by mutation in Lamin A/C gene. *Circulation, 123*(1), 53–61.
36. Song, Z. F., Ji, X. P., Li, X. X., Wang, S. J., Wang, S. H., & Yun-Zhang, (2008). Inhibition of the activity of poly (ADP-ribose) polymerase reduces heart ischemia/reperfusion injury via suppressing JNK-mediated AIF translocation. *Journal of Cellular and Molecular Medicine, 12*(4), 1220–1228.
37. Zhang, J., Li, X. X., Bian, H. J., Liu, X. B., Ji, X. P., & Zhang, Y., (2009). Inhibition of the activity of Rho-kinase reduces cardiomyocyte apoptosis in heart ischemia/reperfusion via suppressing JNK-mediated AIF translocation. *Clinica Chimica Acta, 401*(1, 2), 76–80.
38. Qi, D., Hu, X., Wu, X., Merk, M., Leng, L., Bucala, R., & Young, L. H., (2009). Cardiac macrophage migration inhibitory factor inhibits JNK pathway activation and injury during ischemia/reperfusion. *The Journal of Clinical Investigation, 119*(12), 3807–3816.
39. Ferrandi, C., Ballerio, R., Gaillard, P., Giachetti, C., Carboni, S., Vitte, P. A., Gotteland, J. P., & Cirillo, R., (2004). Inhibition of c-Jun N-terminal kinase decreases cardiomyocyte apoptosis and infarct size after myocardial ischemia and reperfusion in anaesthetized rats. *British Journal of Pharmacology, 142*(6), 953–960.
40. Milano, G., Morel, S., Bonny, C., Samaja, M., Von, S. L. K., Nicod, P., & Vassalli, G., (2007). A peptide inhibitor of c-Jun NH2-terminal kinase reduces myocardial ischemia-reperfusion injury and infarct size *in vivo. American Journal of Physiology-Heart and Circulatory Physiology, 292*(4), H1828–H1835.
41. Zuk, A., & Bonventre, J. V., (2016). Acute kidney injury. *Annual Review of Medicine, 67*, 293–307.
42. Kanellis, J., Ma, F. Y., Kandane-Rathnayake, R., Dowling, J. P., Polkinghorne, K. R., Bennett, B. L., Friedman, G. C., & Nikolic-Paterson, D. J., (2010). JNK signaling in human and experimental renal ischemia/reperfusion injury. *Nephrology Dialysis Transplantation, 25*(9), 2898–2908.
43. Wang, Y., Ji, H. X., Xing, S. H., Pei, D. S., & Guan, Q. H., (2007). SP600125, a selective JNK inhibitor, protects ischemic renal injury via suppressing the extrinsic pathways of apoptosis. *Life Sciences, 80*(22), 2067–2075.
44. Khan, H., Tiwari, P., Kaur, A., & Singh, T. G., (2021). Sirtuin acetylation and deacetylation: A complex paradigm in neurodegenerative disease. *Molecular Neurobiology, 58*(8), 3903–3917.
45. Guan, Y., Wang, S. R., Huang, X. Z., Xie, Q. H., Xu, Y. Y., Shang, D., & Hao, C. M., (2017). Nicotinamide mononucleotide, an NAD+ precursor, rescues age-associated susceptibility to AKI in a sirtuin 1–dependent manner. *Journal of the American Society of Nephrology, 28*(8), 2337–2352.
46. Ma, Q., & Devarajan, P., (2008). Induction of proapoptotic Daxx following ischemic acute kidney injury. *Kidney International, 74*(3), 310–318.
47. Kaur, A., Singh, T. G., Khan, H., Kumar, M., Singh, N., & Abdel-Daim, M. M., (2022). Neuroprotective effect of piclamilast-induced post-ischemia pharmacological treatment in mice. *Neurochemical Research*, 1–14.
48. Arany, I., Megyesi, J. K., Kaneto, H., Price, P. M., & Safirstein, R. L., (2004). Cisplatin-induced cell death is EGFR/src/ERK signaling dependent in mouse proximal tubule cells. *American Journal of Physiology-Renal Physiology, 287*(3), F543–F549.
49. De Graauw, M., Tijdens, I., Cramer, R., Corless, S., Timms, J. F., & Van De, W. B., (2005). Heat shock protein 27 is the major differentially phosphorylated protein involved in renal epithelial cellular stress response and controls focal adhesion organization and apoptosis. *Journal of Biological Chemistry, 280*(33), 29885–29898.
50. Amos, L. A., Ma, F. Y., Tesch, G. H., Liles, J. T., Breckenridge, D. G., Nikolic-Paterson, D. J., & Han, Y., (2018). ASK 1 inhibitor treatment suppresses p38/JNK signaling with reduced kidney inflammation and fibrosis in rat crescentic glomerulonephritis. *Journal of Cellular and Molecular Medicine, 22*(9), 4522–4533.
51. El-Sadek, H. M., Al-Shorbagy, M. Y., Awny, M. M., Abdallah, D. M., & El-Abhar, H. S., (2021). Pentoxifylline treatment alleviates kidney ischemia/reperfusion injury: Novel involvement of galectin-3 and ASK-1/JNK & ERK1/2/NF-κB/HMGB-1 trajectories. *Journal of Pharmacological Sciences, 146*(3), 136–148.

52. Ye, S., Zhu, Y., Ming, Y., She, X., Liu, H., & Ye, Q., (2014). Glycyrrhizin protects mice against renal ischemia-reperfusion injury through inhibition of apoptosis and inflammation by downregulating p38 mitogen-activated protein kinase signaling. *Experimental and Therapeutic Medicine*, *7*(5), 1247–1252.
53. Zhang, J., Xia, J., Zhang, Y., Xiao, F., Wang, J., Gao, H., Liu, Y., et al., (2016). HMGB1-TLR4 signaling participates in renal ischemia reperfusion injury and could be attenuated by dexamethasone-mediated inhibition of the ERK/NF-κB pathway. *American Journal of Translational Research*, *8*(10), 4054.
54. Tsung, A., Hoffman, R. A., Izuishi, K., Critchlow, N. D., Nakao, A., Chan, M. H., Lotze, M. T., et al., (2005). Hepatic ischemia/reperfusion injury involves functional TLR4 signaling in nonparenchymal cells. *The Journal of Immunology*, *175*(11), 7661–7668.
55. Lee, K. H., Kim, S. E., & Lee, Y. S., (2006). SP600125, a selective JNK inhibitor, aggravates hepatic ischemia-reperfusion injury. *Experimental & Molecular Medicine*, *38*(4), 408–416.
56. Wang, J., & Friedman, E., (2000). Downregulation of p53 by sustained JNK activation during apoptosis. *Molecular Carcinogenesis* (Vol. 29, No. 3, pp. 179–188). Published in Cooperation with the University of Texas MD Anderson Cancer Center.
57. Crenesse, D., Gugenheim, J., Hornoy, J., Tornieri, K., Laurens, M., Cambien, B., Lenegrate, G., et al., (2000). Protein kinase activation by warm and cold hypoxia-reoxygenation in primary-cultured rat hepatocytes–JNK1/SAPK1 involvement in apoptosis. *Hepatology*, *32*(5), 1029–1036.
58. Uehara, T., Peng, X. X., Bennett, B., Satoh, Y., Friedman, G., Currin, R., Brenner, D. A., & Lemasters, J., (2004). c-Jun N-terminal kinase mediates hepatic injury after rat liver transplantation. *Transplantation*, *78*(3), 324–332.
59. Liang, R., Nickkholgh, A., Hoffmann, K., Kern, M., Schneider, H., Sobirey, M., Zorn, M., et al., (2009). Melatonin protects from hepatic reperfusion injury through inhibition of IKK and JNK pathways and modification of cell proliferation. *Journal of Pineal Research*, *46*(1), 8–14.
60. Zhang, X. J., He, C., Li, P., Su, H., & Wan, J. B., (2015). Ginsenoside Rg1, a potential JNK inhibitor, protects against ischemia/reperfusion-induced liver damage. *Journal of Functional Foods*, *15*, 580–592.
61. Murayama, T., Tanabe, M., Matsuda, S., Shimazu, M., Kamei, S., Wakabayashi, G., Kawachi, S., et al., (2006). JNK (c-Jun NH2 terminal kinase) and p38 during ischemia reperfusion injury in the small intestine. *Transplantation*, *81*(9), 1325–1330.
62. Hreniuk, D., Garay, M., Gaarde, W., Monia, B. P., McKay, R. A., & Cioffi, C. L., (2001). Inhibition of c-Jun N-terminal kinase 1, but not c-Jun N-terminal kinase 2, suppresses apoptosis induced by ischemia/reoxygenation in rat cardiac myocytes. *Molecular Pharmacology*, *59*(4), 867–874.
63. Khan, H., Sharma, K., Kumar, A., Kaur, A., & Singh, T. G., (2022). Therapeutic implications of cyclooxygenase (COX) inhibitors in ischemic injury. *Inflammation Research*, 1–16.
64. Krilleke, D., Ucur, E., Pulte, D., Schulze-Osthoff, K., Debatin, K. M., & Herr, I., (2003). Inhibition of JNK signaling diminishes early but not late cellular stress-induced apoptosis. *International Journal of Cancer*, *107*(4), 520–527.
65. Giakoustidis, A. E., Giakoustidis, D. E., Koliakou, K., Kaldrymidou, E., Iliadis, S., Antoniadis, N., Kontos, N., et al., (2008). Inhibition of intestinal ischemia/repurfusion induced apoptosis and necrosis via down-regulation of the NF-kB, c-Jun, and caspace-3 expression by epigallocatechin-3-gallate administration. *Free Radical Research*, *42*(2), 180–188.
66. Sun, Y., Lian, M., Lin, Y., Xu, B., Li, Y., Wen, J., Chen, D., et al., (2018). Role of p-MKK7 in myricetin-induced protection against intestinal ischemia/reperfusion injury. *Pharmacological Research*, *129*, 432–442.
67. Yang, Q., Zheng, F. P., Zhan, Y. S., Tao, J., Tan, S. W., Liu, H. L., & Wu, B., (2013). Tumor necrosis factor-α mediates JNK activation response to intestinal ischemia-reperfusion injury. *World Journal of Gastroenterology: WJG*, *19*(30), 4925.
68. Ramirez-Alcantara, V., LoGuidice, A., & Boelsterli, U. A., (2009). Protection from diclofenac-induced small intestinal injury by the JNK inhibitor SP600125 in a mouse model of NSAID-associated enteropathy. *American Journal of Physiology-Gastrointestinal and Liver Physiology*, *297*(5), G990–G998.

CHAPTER 24

Role of Glycogen Synthase Kinase-3β (Gsk-3β) in Intracranial Hemorrhage

KHADGA RAJ,[1] SUDHANSHU MALLAN,[1] AYANSH KAUSHIK,[1] and SHAMSHER SINGH[2]

[1]*Scholar, Department of Pharmacology, ISF College of Pharmacy, Moga, Punjab, India*

[2]*Professor, Neuroscience Division, Department of Pharmacology, ISF College of Pharmacy, Moga, Punjab, India*

ABSTRACT

Stroke is the world's most common cause of death and disability. It is a sudden loss of brain function caused by an interruption in the brain's blood flow due to ischemia or bleeding. Glycogen synthase kinase-3 (GSK-3) is a versatile protein involved in a wide range of cellular and neurophysiological functions. It is a common serine-threonine kinase comprised of α and β isoforms. GSK3 β has the most predicted substrates of any kinase. After various cell death stimuli, In the cell death/survival pathway, the Akt/GSK3 pathway plays a significant role. Therefore, the activity of GSK-3β is increased, resulting in neuronal death and ischemic stroke. Moreover, GSK-3β has a role in neuronal death, it may be possible to avoid ischemic stroke by modifying this protein kinase. In this chapter, we looked at the potential function of GSK-3β in stroke progression.

24.1 INTRODUCTION

Stroke is a major cause of disease and death globally, behind cancer and heart disease [1]. It is defined as an acute, localized, or generalized dysfunction of the brain caused by arteries that lasts more than a day [2]. The most recent study indicated a decline in stroke-related mortality, and the absolute number of persons afflicted each year continues to climb, with the greatest burden in low-income countries [3]. It has been recommended that addressing modifiable risk factors might prevent up to 80% of recurrent strokes. Therefore, a mass stroke prevention plan may be useful in conjunction with other techniques for lowering the global burden of stroke and related disability. Stroke is categorized as ischemic stroke (87%) and hemorrhagic (13%) based on its etiology [4]. The obstruction of a cerebral artery causes Ischemic stroke due to atherosclerotic or thrombotic, lacunar stroke, and embolic.

Ischemic Injury: A Molecular Insight. Thakur Gurjeet Singh & Heena Khan (Eds.)

The most common causes of hemorrhagic stroke include spontaneous blood vessel ruptures or aneurysms and trauma. Ischemic strokes cause 85% of all strokes, whereas hemorrhagic strokes cause 15% of all cases [5]. The Trial of Org10172 (TOAST) is a commonly used categorization for acute ischemic stroke (AIS), and it is categorized into five subtypes – (i) Atherosclerosis of the major arteries; (ii) thrombosis (blood clots in the arteries); (iii) blockage of a tiny vessel; (iv) stroke of undetermined etiology; and (v) stroke of other determined etiology. Ischemic strokes cause a loss of blood flow, oxygen, nutrients, and metabolic waste elimination. The alterations that occur hinder proper neural function, and occlusion of the vessel causes neuronal death/necrosis [6]. The brain tissue is susceptible to these alterations, and the therapeutic window for preventing reversible ischemia from becoming irreversible infarction [7].

During the last two decades, several stroke-specific structured intervention trials and diagnostic tools were addressed, resulting in improved clinical outcomes with decreased associated mortality and morbidity. The most frequent kind of hemorrhagic stroke is intracerebral hemorrhage (ICH), which accounts for around 20% of all strokes [8]. Patients with ICH have a higher rate of early mortality and long-term disability. Unfortunately, many commonly used ICH treatments, such as acute blood pressure control, coagulopathy reversal, and surgical hematoma evacuation, have had a shortage of randomized controlled trials (RCTs). A subarachnoid hemorrhage occurs when a vessel is located within the subarachnoid space (SAH). An intracerebral or intraparenchymal hemorrhage or hematoma occurs when an artery ruptures within the brain's parenchyma [9]. It is possible for a SAH to have an ICH and vice versa. Either way, blood may accumulate in the ventricles, resulting in an intraventricular hemorrhage (IVH). The causes of cerebral hemorrhage are listed in Table 24.1.

TABLE 24.1 Causes of Intracranial Hemorrhage

Intracerebral Hematoma	Subarachnoid Hemorrhage	References
Arterio-venous malformation	Arterio-venous malformation	[6]
Hypertensive hemorrhage	Aneurysm	[7]
Amyloid angiopathy	Peri-mesencephalic bleed	[7]
Hemorrhagic tumor	Drug abuse	[9]
Aneurysmal hemorrhage	Pituitary apoplexy	[9]
Infarct with hemorrhagic transformation	Pituitary apoplexy	[9]
Dural sinus fistula	Anticoagulant drugs	[9]
Venous infarct with hemorrhagic transformation	Anticoagulant drugs	[9]

24.2 INTRODUCTION TO GLYCOGEN SYNTHASE KINASE-3 (GSK-3B)

GSK-3β is essential to coordinate catabolic and anabolic pathways in various cellular processes [10]. Typically, GSK-3β-targeted phosphorylation decreases the substrate's activity, attenuating the signaling pathway. GSK-3β activation improves cellular homeostasis, whereas GSK-3β inhibition stimulates biological responses [11]. GSK-3β activity is necessary for normal neurodevelopment, and it has been associated with mental illnesses, cognitive impairment, and neurodegeneration [12]. GSK3A and GSK3 are two closely related genes in the GSK3 gene family. GSK3A produces a 51-kDa protein, while GSK3 produces a 47-kDa protein [13]. Although the two GSK isoforms share 84% of their amino acid sequences, while the GSK-3β share 98% of their catalytic domains, They are distinguished by their different N- and C-terminal domains [14]. The two GSK-3 isoforms have a bi-lobular structure, with a large globular domain at the C-terminus containing the catalytic domain and a tiny globular domain at the N-terminus containing the ATP binding site. In certain cells, the GSK-3 isoform controls NF-B and the cAMP response element in the response of bovine endothelial cells to *Staphylococcus aureus* infection [15]. According to the current paradigm, GSK-3β hyperactivity plays a causal role in disease pathogenesis, and GSK-3β inhibition is a possible treatment pathway. According to the research, GSK-3β inhibitors have a variety of chemotypes and modes of action [16].

There are three kinds of inhibitory mechanisms: substrate–competitive inhibitors, non-ATP-competitive inhibitors, and ATP-competitive inhibitors. Therefore, specificity is a critical concern in this market, and advanced structure-based computational studies are being conducted to increase GSK-3 inhibitor specificity and, perhaps, assure targeting of particular GSK-3 isozymes [17].

24.3 STRUCTURE AND SUBUNITS OF GSK-3B

GSK-3α and GSK-3β, as their names suggest, are members of the MAPK, CDK, GSK-3, and CLK families (CMGC). GSK3 is encoded on mouse chromosome 7 and human chromosome 19, whereas GSK-3 is encoded on mouse chromosome 16 and human chromosome 3 [18]. GSK-3 is a protein that is expressed in all mammalian organs since the gene and its product are both present. All eukaryotes, including yeast, plants, and humans, have GSK-3 homologs [19]. Only one GSK-3 isoform, which is the most closely related to GSK-3, is expressed by invertebrates and some birds. GSK-3 has a small N-terminal domain (residues 25–134) and a bigger C-terminal domain (residues 135–380) like other serine/threonine kinases. GSK-3 has seven antiparallel-strands and a helical region at its N-terminus [20].

The DFG and APE motifs compose the activation domain. GSK-3 is most structurally related to mitogen-activated protein kinases (MAPKs). MAPKs are activated by phosphorylating two sites inside their T-loop. There are tyrosine and threonine among these residues [18]. GSK-3, like MAPKs, is phosphorylated (constitutively) at tyrosine 216 and 279 in its active state. However, GSK-3 does not have a phosphorylated threonine in its T-loop, unlike MAPKs. GSK-3 tyrosine phosphorylation, like DYRK, is most likely a cotranslational autophosphorylation mechanism and is not regulated in the same manner as MAPK tyrosine phosphorylation [21]. GSK-3 has been demonstrated to preferentially bind prephosphorylated substrates that are four C-terminal residues away from its phosphorylation site [22]. Casein Kinase-II binds to the GSK-3 priming site on glycogen synthase specifically, which was lucky in retrospect. As a consequence, pure glycogen synthase for *in vitro* GSK-3 phosphorylation was effectively synthesized. It's also worth noting that glucose 6-phosphate's allosteric control of glycogen synthase is likely more crucial than phosphorylation.

24.4 REGULATION OF GSK3-B ACTIVITY

GSK3 is active in resting cells and is reduced quickly and transiently in response to various environmental signals. GSK3 activity is controlled by phosphorylation at specific sites. Phosphorylation of GSK3 at various sites regulates its action, with tyrosine 216 (Tyr216) phosphorylation being the most important, whereas serine 9 (Ser9) phosphorylation suppresses GSK3β function. Ser9 phosphorylation is the most prevalent and important regulating mechanism [23]. Numerous kinases, including p70 S6 kinase, protein kinase B (commonly known as Akt), p90Rsk (also known as MAP-KAP kinase-1), extracellular signal-regulated kinases (ERKs), A, and PKA). In contrast to serine phosphorylation, which suppresses Ser9, tyrosine phosphorylation of GSK3 by specific Protein kinase C (PKC) and cyclic AMP-dependent protein kinase (protein kinase GSK3) isoforms improve the activity of the enzyme. It has been shown that activating Intracellular calcium levels and a calcium-dependent tyrosine kinase, proline-rich tyrosine kinase 2 (PYK2), or Fyn, a member of the Src tyrosine family, are both affected by pGSK3 (Tyr216) [24]. Mitogen-activated protein kinase kinase (MEK1/2) also regulates pGSK3 (Tyr216). At rest, GSK3 is active; however, during cellular processes, it becomes inactive. Consequently, its substrates are dephosphorylated; the majority of these substrates are functionally inhibited by GSK3 [25].

GSK3β appears to act as a general repressor, shutting off or making targets unavailable when they are not in use. GSK3β binds to around 40 proteins involved in various physiological functions, like translation, cytoskeletal regulation, proliferation, transcription, glycogen metabolism, cell differentiation, and death [26]. Autophosphorylation has been shown to alter GSK3β tyrosine phosphorylation. GSK3β can be activated without causing significant changes in the phosphorylation of Tyr216 and Ser9, which is shown in Figure 24.1.

The subcellular location of GSK3β may potentially affect its regulation, while some of GSK3's substrates, such as Tau, are found in the cytoplasm, several transcription factors are located in the nucleus [27]. To regulate these proteins, GSK3 must be present in both compartments. Diehl et al. show that the nuclear GSK3 level rises during the S phase of the cell cycle. Proapoptotic stimuli induce GSK3 translocation to the nucleus. The identification of a nuclear localization sequence (NLS) in GSK3 provided insight into the processes underlying the protein's trafficking between the cytosolic and nuclear compartments [28].

24.5 ROLE OF GSK3-B IN CEREBRAL ISCHEMIC STROKE

Cerebral ischemia stroke occurs due to blood clotting or plaque formation in the cerebral artery, resulting in blood supply disrupted in many brain areas. The substantially reduced cerebral blood flow caused by this blockage results in reduced oxygen and glucose delivery, subsequent energy deprivation, and the buildup of harmful compounds [29]. Ischemia causes nerve cells to lose their ability to store energy, leading to energy generation failure, metabolic stress, and ionic transporter malfunction, including both ATP-dependent and independent manners. Vulnerability after ischemia shock is time-dependent since neurons die within a minute of acute exposure. Several signaling pathways are activated to modify the neuronal morphology depending on the length of the ischemia insult, whether localized or global, transitory or permanent. During the early phase of an injury, protective mechanisms are active and attempt to neutralize the effect. The changes in gene expression (374 genes) of multiple pathways linked with ischemia damage were investigated, and it was discovered that the tumor necrotic factor receptor (TNFR) signaling pathway demonstrated a consistent modification across all periods [30]. In addition to the TNFR signaling pathway genes, changes in protein tyrosine kinase 2 (PyK2), protein kinase C (PKC), E2F transcription factor 1 (E2F1), and vascular endothelial growth factor-A (VEGF-A) were seen over time. In comparison to 3 hours, there were significant morphological alterations and a rise in infarct volume after 12 and 24 hours [31]. Numerous studies have shown the importance of ROS in the early stages of ischemia and their participation in cellular signaling networks. GSK-3β appears to have a function in cerebral ischemia insult, according to the findings. GSK-3β overactivation (phosphorylation of Tyr216) has previously been linked to cerebral ischemia. Furthermore, lithium-induced GSK-β3 inhibition has been demonstrated to be a protective mechanistic mechanism in cerebral ischemia damage. According to certain studies, decreasing GSK-3β activity enhances neuronal cell survival. Lithium activity in cerebral ischemia shows that the PI3K/AKT/GSK-β3 signaling pathway is activated in cerebral ischemia [31].

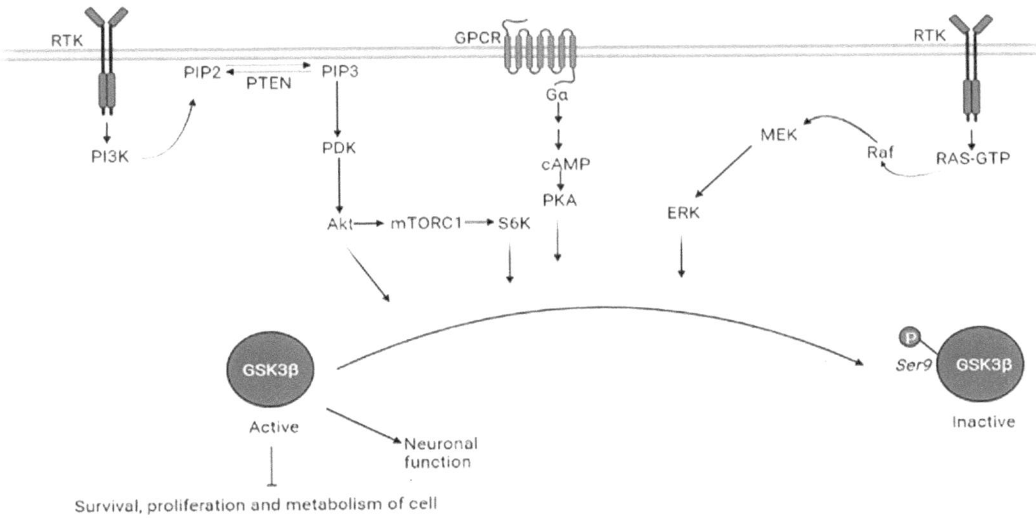

FIGURE 24.1 The figure represents the different proteins that inhibit the GSK3β by phosphorylation of serine 9 positions.

Abbreviations: RTK: Receptor tyrosine kinases; PI3K: Phosphoinositide 3-kinase; PIP2: Phosphatidylinositol 4,5-bisphosphate; PIP3: Phosphatidylinositol (3,4,5)-trisphosphate; PTEN: Phosphatase and tensin homolog; PDK: Pyruvate dehydrogenase kinase; mTORC1: Mammalian target of rapamycin; S6K: S6 kinase; GPCR: G-protein coupled receptor; cAMP: Cyclic adenosine monophosphate; PKA: Protein kinase A; RAS-GTP: Ras-guanosine triphosphate; Raf: Rapidly accelerated fibrosarcoma; MEK: Mitogen-activated protein kinase; ERK: Extracellular signal-regulated kinase; and GSK3β: Glycogen synthase kinase 3 β.

Furthermore, Chir025 showed neuroprotective characteristics by inhibiting GSK-3, leading to decreased infarct size, caspase 3, 9, and Bax activities in the cerebral ischemia model, indicating a function for GSK-3ß in the mitochondrial pathway in disease. Through the ERK and PPAR receptor signaling pathways, GSK-3 inhibition by AR-A 014418, NP00111 has been demonstrated to protect the hippocampus from oxygen-glucose deprivation. This research established a link between GSK-3, PPAR, and ERK in cerebral ischemia [31].

Multiple pathways have direct crosstalk in cerebral ischemia, including calcium signaling kinase pathway (JNK-1,2, MAPK/p38), TNF-alpha, inflammatory pathway (ICAM-1, iNOS, and NF-κB,) and abundant mitochondrial apoptotic markers (cytochrome c, caspase-9, SOD, and Bcl-2). GSK-3 inhibition resulted in neuroprotection by downregulating these numerous pathways. GSK-3 activation can trigger apoptosis via a caspase-independent mechanism involving the p53 protein, which can influence gene transcription. GSK-3β binds to the nuclear p53 c-terminus and phosphorylates Ser376, Ser315, and Ser33 residues. Many post-translational modifications occur when GSK-3β interacts with p53, including phosphorylation, glycosylation, ubiquitination, sumoylation, acetylation, and methylation. GSK-3β regulates p53 by phosphorylating MDM2, an E3 ubiquity ligase specific to p53 [31]. In stroke, GSK-3β is associated with neuronal cell death, and therapy with GSK-3β inhibitors in the acute stage has been shown to enhance neurobehavioral function and decrease infarction volume (Figure 24.2). GSK-3β inhibitors ameliorate post-ischemic stroke sequelae in the chronic state by promoting neurovascular remodeling after stroke [31]. GSK-3β activity suppression is neuroprotective against ischemic stroke Phosphorylation of GSK-3a at Ser21 and/or GSK-3b at Ser9 inhibits GSK-3 β kinase activity. GSK-3 can be inhibited in several ways, including direct binding to the enzyme's ATP-dependent magnesium-sensitive catalytic domain, and GSK-3 isoforms are indirectly inhibited by increased serine phosphorylation. GSK3 β inhibition increases hippocampal-dependent learning as well as neurogenesis. Overactivation of GSK3 β activity causes a range of inflammatory markers and a disruption in the neurogenesis process. These data support the idea that GSK3β inhibition is beneficial and might be used to treat ischemic stroke [32].

24.6 ROLE OF GSK3 B IN BRAIN HEMORRHAGE

GSK3 is a protein kinase that controls cell membrane signaling, neuronal polarity, cell survival/apoptosis, gene transcription/protein translation, and other physiological functions. Ser9 and Tyr216 of GSk3 might be phosphorylated [33]. GSK3 is activated when Ser9 is phosphorylated, and downstream proteins, including eIF2B and glycogen synthase, are phosphorylated. Phosphorylation of the aforementioned products is critical in cellular damage that results in neuron death [34]. After 90 minutes of tFCI, activation of GSK3 β can be initiated, promoting apoptotic cell death. However, after 5 minutes of tGCI, a little insult may result in GSK-3 β inactivation and greater cell survival.

Similarly, after traumatic brain damage, the amount of immunoreactive phospho-GSK-3β at Ser9 peaked at 24 hours and remained stable for one week, but GSK-3 β immunoreactivity did not. GSK-3β inhibition with Chir025, lithium, and insulin-like growth factor-1 reduced neuronal death and infarction significantly [34]. The phosphorylation of β-catenin by GSK-3β leads to its stability and translocation to the nucleus, influencing new studies on gene expression and anti-apoptosis. Phosphorylated GSK-3β and total GSK-3β levels declined in the penumbra and core of the ischemic penumbra 5 to 24 hours after a stroke, and β-catenin phosphorylation was also seen [35]. Although overall β-catenin levels drop dramatically from 5 to 48 hours, hypothermia can develop. β-catenin is stabilized by inhibiting its breakdown and shrinks the infarction's size. GSK3β activation may increase ischemia damage, whereas inhibition may minimize the extent of the infarction. According to studies, the phosphorylation of GSK3 (Ser9) was increased in the injured cortex, which was critical for neuronal survival after traumatic brain injury (TBI) [36]. GSK-3β regulates apoptosis and inflammation, which are recognized factors of stroke-induced cell death. GSK-3β deficiency, rather than GSK-3a deficiency, decreased neuronal death occurs spontaneously in long-term culture models [31]. Lithium-induced non-selective GSK-3β inhibition has shown to be neuroprotective, and GSK3β inhibitors are now commercially available for the therapy of cognitive impairment, clinical studies are underway. GSK-3 has been demonstrated to interact with mitogen-activated protein kinase (MAPK) family members, resulting in enhanced stress signaling.

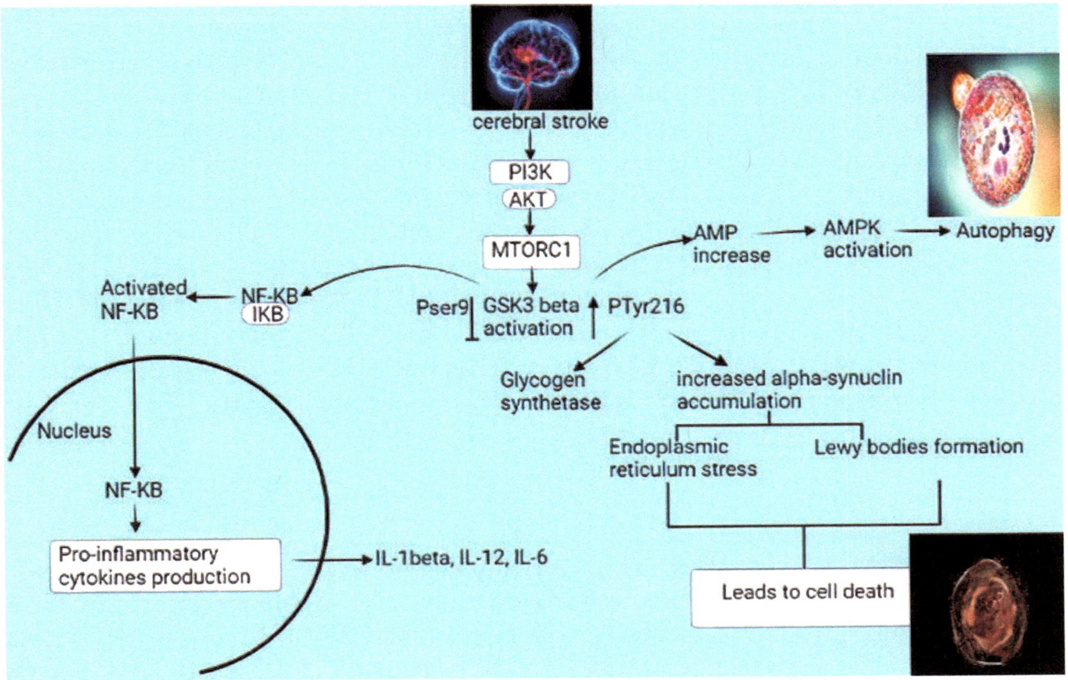

FIGURE 24.2 Role of Gsk-3 β in stroke induces neuronal death.

TAK1 (transforming growth factor-b-activated kinase 1) is a MAPK (mitogen-activated protein kinase) that belongs to the MAPK family of proteins. The cytokines such as interleukin-1, TGF-β, and TNF-α, activate the TAK1 protein (IL-1). TGF-β, TNF-a, and other cytokines, such as interleukin-1 (IL-1), activate TAK1 [37]. TAK1 also functions as an upstream kinase for 5′-adenosine monophosphate-activated protein kinase (AMPK), an essential energy-sensing enzyme-linked to stroke. TAK1 inhibition has been proven to be neuroprotective after localized ischemia in investigations. Inhibiting GSK-3 has increased the number of newborn neural cells [38]. Wnt inhibits GSK-3 and enhances-catenin nuclear translocation, activating several genes involved in cell survival, mitosis, and differentiation. The Wnt/GSK-3β/-catenin pathway also affects the BBB, which causes interface in the pathophysiology of ICH.

The canonical Wnt proteins are necessary to form tight junctions during organ development, and disturbance of the Wnt pathway causes the BBB to collapse. Recently, it was shown that inhibiting GSK-3 protects against microvascular hyperpermeability in hemorrhagic shock. In the aforementioned trials, GSK-3 inhibition decreased hematoma volume during the subacute phase of ICH by restoring-catenin levels [39].

24.7 ROLE OF GSK3B IN SUBARACHNOID HEMORRHAGE

Subarachnoid hemorrhage (SAH) is a clinical condition that is triggered by the rupture of defective cerebral blood arteries, resulting in sudden bleeding into the subarachnoid space. GSK-3β regulates cell survival and death in several clinical circumstances. It is an Akt downstream protein. According to some research studies, phosphorylation at tyrosine-216 (irritant) and serine-9 may regulate GSK3β activity. GSK3β is phosphorylated on serine-9 by activated Akt (serine-473), inhibiting its function and reducing apoptosis. According to the research, the Akt signaling pathway is critical for neuron survival in the biological process. Following SAH, EBI has been associated with Akt and GSK3 phosphorylation. The Akt pathway limits the activity of the pro-apoptotic kinase GSK3 by modifying Bcl2/Bax levels, which, depending on it prevents neuronal death whether the apoptosis-related protein caspase pathway or the mitochondrial route is employed. Perk phosphorylation of GSK-3 (Ser9) is a downstream target of perk, and phosphorylation at ser9 may diminish GSK-3's pro-apoptotic impact. GSK-3 phosphorylates TIP60 on Serd86, increasing p53-mediated apoptosis by activating the PUMA's anti-apoptotic BCL-2 family member (Table 24.2).

24.8 CONCLUSION

Glycogen synthase kinase 3 (GSK-3) is a serine-threonine kinase that controls cell proliferation and tissue protection. Understanding the molecular signaling systems that drive neuronal dysfunctions linked to stroke etiology, such as GSK3β, is crucial for creating innovative stroke therapeutic strategies. GSK-3 dysregulation has been associated with various CNS disorders, including neuroinflammation and neurodegenerative diseases. As a result, the GSK-3β signaling pathway might be linked to stroke. However, the findings are conflicting, indicating that further research is required to understand better the molecular mechanisms that link GSK-3β activation to stroke.

TABLE 24.2 Summarization of the Studies Conducted on the Role of GSK-3β in Stroke

SL. No.	Stroke Type	Therapeutic Intervention/ Dose	Study Type	Mechanism of Action	References
1.	Subarachnoid hemorrhage	MSCs/1.5×10^6	Pre-clinical	Improves sensorimotor function, decreases lesion size, reduce microglia/macrophage associated neuroinflammation, restoration of TH expression.	[40]
2.	Subarachnoid hemorrhage	Salvianolic acid B/10,20, 40 mg/kg.	Pre-clinical	Inhibit overproduction of ROS and enhance SOD and GSH levels, amelioration of brain edema, neural cell, apoptosis, increase Nrf2 nuclear translocation, and upregulation of HO-1 and NQO-1.	[41]
3.	Ischemic cerebral stroke	OX26-PEG-Se NPs	Pre-clinical and *in-vitro*.	Regulate (TSC1/TSC2, p-mTOR, mTORC1, Mst1, ULK1, Bax, Caspase-3 and Bcl-2, rictor/mTORC2).	[42]
4.	Subarachnoid hemorrhage	JNJ16259685/10 mg/kg	Clinical and pre-clinical.	BBB damage and edema were reduced, mGluR1 was inhibited, VASP, p-VASP, and occludin levels were elevated, and AQP levels were decreased.	[43]
5.	Intracerebral hemorrhage	Siponimod (BAF-312)/0.3, 3 mg/kg	Pre-clinical	Attenuation of perihemorrhagic edema and sensorimotor deficits.	[44]
6.	Intracerebral hemorrhage	Bexarotene/5 mg/kg and 0.5 or 1 µmol/L.	Pre-clinical and *in-vitro*.	Increase receptors responsible for erythrophagocytosis by macrophages, the elevation of hematoma clearance, and improve neurobehavioral deficits.	[45]
7.	Cerebral hemorrhage	Resveratrol/10 mg/kg/day	Pre-clinical	Hippocampal cell death and inflammation are inhibited. Downregulation of TGF-β-ERK signaling pathway.	[46]
8.	Subarachnoid hemorrhage	LP 17/1 mg/kg and 10 µM	Pre-clinical and *in-vitro*.	Aβmeliorated microglial pyroptosis and reduced level of GSDMD-N and IL-1β alleviated BBB disruption and neuronal injury.	[47]

KEYWORDS

- **brain hemorrhage**
- **cerebral ischemic stroke**
- **glycogen synthase kinase-3**
- **intracranial hemorrhage**
- **ischemic stroke**
- **stroke**

REFERENCES

1. Towfighi, A., & Saver, J. L., (2011). Stroke declines from third to fourth leading cause of death in the United States: Historical perspective and challenges ahead. *Stroke, 42*(8), 2351–2355.
2. Khan, H., Kashyap, A., Kaur, A., & Singh, T. G., (2020). Pharmacological postconditioning: A molecular aspect in ischemic injury. *Journal of Pharmacy and Pharmacology, 72*(11), 1513–1527.
3. Feigin, V. L., et al., (2014). Global and regional burden of stroke during 1990–2010: Findings from the Global Burden of Disease Study 2010. *The Lancet, 383*(9913), 245–255.
4. Feigin, V. L., et al., (2016). Global burden of stroke and risk factors in 188 countries, during 1990–2013: A systematic analysis for the Global Burden of Disease Study 2013. *The Lancet Neurology, 15*(9), 913–924.
5. Fewel, M. E., Thompson, B. G., & Hoff, J. T., (2003). Spontaneous intracerebral hemorrhage: A review. *Neurosurgical Focus, 15*(4), 1–16.
6. Saklani, P., Khan, H., Gupta, S., Kaur, A., & Singh, T. G., (2022). Neuropeptides: Potential neuroprotective agents in ischemic injury. *Life Sciences, 288*, 120186.
7. Khan, H., Grewal, A. K., & Singh, T. G., (2022). Pharmacological postconditioning by protocatechuic acid attenuates brain injury in ischemia–reperfusion (I/R) mice model: Implications of nuclear factor erythroid-2-related factor pathway. *Neuroscience, 491*, 23–31.
8. Aguilar, M. I., & Brott, T. G., (2011). Update in intracerebral hemorrhage. *The Neurohospitalist, 1*(3), 148–159.
9. Masuda, J., et al., (1992). Histopathological analysis of the mechanisms of intracranial hemorrhage complicating infective endocarditis. *Stroke, 23*(6), 843–850.
10. Rippin, I., & Eldar-Finkelman, H., (2021). Mechanisms and therapeutic implications of GSK-3 in treating neurodegeneration. *Cells, 10*(2), 262.
11. Lal, H., et al., (2015). The GSK-3 family as therapeutic target for myocardial diseases. *Circulation Research, 116*(1), 138–149.
12. Kalra, P., Khan, H., Kaur, A., & Singh, T. G., (2022). Mechanistic insight on autophagy modulated molecular pathways in cerebral ischemic injury: From preclinical to clinical perspective. *Neurochemical Research*, 1–19.
13. Welsh, G. I., Wilson, C., & Proud, C. G., (1996). GSK3: A SHAGGY frog story. Trends in *Cell Biology, 6*(7), 274–279.
14. Forde, J. A., & Dale, T. C. (2007). Glycogen synthase kinase 3: A key regulator of cellular fate. *Cellular and Molecular Life Sciences, 64*(15), 1930–1944.
15. Daly, A., (2007). *Mechanisms of Smad Regulation that Determine the Specificity of TGF-Responses During Tumorigenesis*. University of London, University College London (United Kingdom).
16. Cuny, G., (2009). Kinase inhibitors as potential therapeutics for acute and chronic neurodegenerative conditions. *Current Pharmaceutical Design, 15*(34), 3919–3939.

17. Prabhakar, N. K., Khan, H., Grewal, A. K., & Singh, T. G., (2022). Intervention of neuroinflammation in the traumatic brain injury trajectory: *In vivo* and clinical approaches. *International Immunopharmacology, 108*, 108902.
18. Patel, P., & Woodgett, J. R., (2017). Glycogen synthase kinase 3: A kinase for all pathways? *Current Topics in Developmental Biology, 123*, 277–302.
19. Clouse, S. D., (2002). Brassinosteroid signal transduction: Clarifying the pathway from ligand perception to gene expression. *Molecular Cell, 10*(5), 973–982.
20. Kramer, T., Schmidt, B., & Lo Monte, F., (2012). Small-molecule inhibitors of GSK-3: Structural insights and their application to Alzheimer's disease models. *International Journal of Alzheimer's Disease, 2012.*
21. Kockeritz, L., et al., (2006). Glycogen synthase kinase-3-an overview of an over-achieving protein kinase. *Current Drug Targets, 7*(11), 1377–1388.
22. Palomo, V., et al., (2011). Exploring the binding sites of glycogen synthase kinase 3. Identification and characterization of allosteric modulation cavities. *Journal of Medicinal Chemistry, 54*(24), 8461–8470.
23. Doble, B. W., & Woodgett, J. R., (2003). GSK-3: Tricks of the trade for a multi-tasking kinase. *Journal of Cell Science, 116*(7), 1175–1186.
24. Sharma, V. K., & Singh, T. G., (2020). CREB: A multifaceted target for Alzheimer's disease. *Current Alzheimer Research, 17*(14), 1280–1293.
25. Plattner, F., (2006). *Biochemical Analysis of Genetically-Modified Mice with Learning and Memory Phenotypes.* University of London, University College London (United Kingdom).
26. Luo, J., (2009). Glycogen synthase kinase 3β (GSK3β) in tumorigenesis and cancer chemotherapy. *Cancer Letters, 273*(2), 194–200.
27. Kaur, A., Singh, T. G., Khan, H., Kumar, M., Singh, N., & Abdel-Daim, M. M., (2022). Neuroprotective effect of piclamilast-induced post-ischemia pharmacological treatment in mice. *Neurochemical Research*, 1–14.
28. En Natuurwetenschappen, W., Sterken, E., & Baarsma, H. A. (2012). *Glycogen Synthase Kinase-3 (GSK-3) and β-Catenin: Potential Novel Therapeutic Targets for COPD*. ISBNs: 9789036754682.
29. Khan, H., Singh, A., Thapa, K., Garg, N., Grewal, A. K., & Singh, T. G., (2021). Therapeutic modulation of the phosphatidylinositol 3-kinases (PI3K) pathway in cerebral ischemic injury. *Brain Research, 1761*, 147399.
30. Lahiani, A., et al., (2018). Neuroprotective effects of bioactive compounds and MAPK pathway modulation in "ischemia"—Stressed PC12 pheochromocytoma cells. *Brain Sciences, 8*(2), 32.
31. Khan, H., Gupta, A., Singh, T. G., & Kaur, A., (2021). Mechanistic insight on the role of leukotriene receptors in ischemic–reperfusion injury. *Pharmacological Reports, 73*(5), 1240–1254.
32. Garg, C., Kaur, A., Singh, T. G., Sharma, V. K., & Singh, S. K., (2022). Therapeutic implications of sonic hedgehog pathway in metabolic disorders: Novel target for effective treatment. *Pharmacological Research*, 106194.
33. Wada, A., (2009). Lithium and neuropsychiatric therapeutics: Neuroplasticity via glycogen synthase kinase-3β, β-catenin, and neurotrophin cascades. *Journal of Pharmacological Sciences*, 0905070244–0905070244.
34. Zhao, S., et al., (2012). Activation of Akt/GSK-3beta/beta-catenin signaling pathway is involved in survival of neurons after traumatic brain injury in rats. *Neurological Research, 34*(4), 400–407.
35. Zhang, H., et al., (2008). Hypothermia blocks β-catenin degradation after focal ischemia in rats. *Brain Research, 1198*, 182–187.
36. Khan, H., Sharma, K., Kumar, A., Kaur, A., & Singh, T. G., (2022). Therapeutic implications of cyclooxygenase (COX) inhibitors in ischemic injury. *Inflammation Research*, 1–16.
37. Pera, T., et al., (2012). Role for TAK1 in cigarette smoke-induced proinflammatory signaling and IL-8 release by human airway smooth muscle cells. *American Journal of Physiology-Lung Cellular and Molecular Physiology, 303*(3), L272–L278.
38. Ramamurthy, S., & Ronnett, G., (2012). AMP-activated protein kinase (AMPK) and energy-sensing in the brain. *Experimental Neurobiology, 21*(2), 52.
39. Menet, R., Lecordier, S., & ElAli, A., (2020). Wnt pathway: An emerging player in vascular and traumatic mediated brain injuries. *Frontiers in Physiology*, 1149.

40. Nijboer, C. H., et al., (2018). Intranasal stem cell treatment as a novel therapy for subarachnoid hemorrhage. *Stem Cells and Development, 27*(5), 313–325.
41. Zhang, X., et al., (2018). Cerebroprotection by salvianolic acid B after experimental subarachnoid hemorrhage occurs via Nrf2-and SIRT1-dependent pathways. *Free Radical Biology and Medicine, 124*, 504–516.
42. Amani, H., et al., (2019). Selenium nanoparticles for targeted stroke therapy through modulation of inflammatory and metabolic signaling. *Scientific Reports, 9*(1), 1–15.
43. Zhang, C., et al., (2020). Selective mGluR1 negative allosteric modulator reduces blood–brain barrier permeability and cerebral edema after experimental subarachnoid hemorrhage. *Translational Stroke Research, 11*(4), 799–811.
44. Bobinger, T., et al., (2019). Siponimod (BAF-312) attenuates perihemorrhagic edema and improves survival in experimental intracerebral hemorrhage. *Stroke, 50*(11), 3246–3254.
45. Chang, C. F., et al., (2020). Bexarotene enhances macrophage erythrophagocytosis and hematoma clearance in experimental intracerebral hemorrhage. *Stroke, 51*(2), 612–618.
46. Zhao, R., et al., (2019). Resveratrol ameliorates brain injury via the TGF-β-mediated ERK signaling pathway in a rat model of cerebral hemorrhage. *Experimental and Therapeutic Medicine, 18*(5), 3397–3404.
47. Xu, P., et al., (2021). TREM-1 exacerbates neuroinflammatory injury via NLRP3 inflammasome-mediated pyroptosis in experimental subarachnoid hemorrhage. *Translational Stroke Research, 12*(4), 643–659.

CHAPTER 25

Role of GSK-3 Inhibition in Modulating the Pathology of Stroke

RUPALI SHARMA,[1] SHILPA KUMARI,[1] SATYAN KAPOOR,[1] RAHUL DESHMUKH,[1] and THAKUR GURJEET SINGH[2]

[1]*Department of Pharmaceutical Sciences and Technology, Maharaja Ranjit Singh Punjab Technical University, Bathinda, Punjab, India*

[2]*Chitkara College of Pharmacy, Chitkara University, Punjab, India*

ABSTRACT

Stroke is a cerebroventricular event and serves as a major reason for disability and mortality all over the world. Stroke is characterized by apoptosis, inflammation, cell death, necroptosis, etc. Some important factors increase the incidence of stroke-related pathologies, among which one of the important candidates is Glycogen synthase kinase-3 (GSK-3). It participates in several critical intracellular signaling. GSK-3 activation has a significant part in promoting cell death and inflammation both of which are known contributors to stroke-induced cell death. In other disease models as well, inhibiting GSK-3 has been shown to improve memory. It could be an important target for stroke as a lot of signaling converges to it. Wnt/β-catenin signaling activation followed by focal cerebral ischemia encourages neurogenesis and enhances neurologic activity. That is why, inhibition of GSK-3 might raise the catenin level, further activating the Wnt/β-catenin signaling pathways. Akt inhibition leads to GSK hyperactivity, which contributes to the GSK p53 complex leading to apoptosis. There are various signaling pathways in stroke-related pathologies that interfere with GSK activation and contribute to post-stroke hallmarks. Therefore, inhibiting GSK could be a useful target for stroke as it will alleviate stroke-induced apoptosis. GSK-3β inactivation reduces early ischemia-reperfusion injury and protects the blood-brain barrier. In this chapter, we will acknowledge GSK-3β inhibition through various signaling pathways as a novel therapeutic strategy for post-stroke cognitive deficits.

Ischemic Injury: A Molecular Insight. Thakur Gurjeet Singh & Heena Khan (Eds.)

25.1 INTRODUCTION

Ischemic stroke is an unrecoverable type of brain injury incurred by transitory or consistent hypoxia and ischemia in the brain tissues [1]. Sudden loss of blood flow to brain areas contributes to the leading cause of death, termed ischemic stroke [2]. Ischemia/reperfusion injury (IRI) refers to morphological and anatomical changes that occur during the rejuvenation of blood flow followed by a period of ischemia. In relation to reversing ischemia, blood flow repair can lead to potentially very harmful side effects such as necrosis of irreversibly injured tissue, significant cell swelling, and non-homogeneous flow restoration to all parts of the tissue [3]. As per the ICMR report "India: Health of the Nation's States," stroke has been considered the fourth major cause of mortality and the fifth highest incidence of Disability Adjusted Life Years (DALY). Ischemic stroke is caused by a thrombotic or embolic phenomenon that leads to the reduction of the blood supply into the brain. A thrombotic event comes into play when the flow of blood to the brain is hindered inside a blood vessel because of impairment in the vessel, which is often caused by chronic inflammation, atherosclerotic ailment, arterial evisceration, and fibromuscular dysplasia. When damage from elsewhere in the body obstructs blood circulation through the damaged vessel, an embolism occurs. The pathology of stroke influences its prognostic value and consequences. The movement of a thrombus from the other part of the body, like the heart, to the brain is referred to as embolism, and an extreme obstruction of an artery is referred to as stenosis. Occlusion of cerebral vessels due to thrombosis or embolism leads to decreased blood flow to a specific area causing irreversible damage. Such an area of irreversible damage is known as a core area for stroke followed by an area where brain functions are lost because of low blood supply. Sudden numbness or weakness of body parts including arms, face, and legs along with confusion, trouble speaking, walking, loss of coordination, and severe headache are the main symptoms of a stroke [4]. This pathophysiological method is commonly associated with the onset of critical links like cell death, autophagy, oxidative stress, and mitochondrial impairment. Types of stroke are enlisted as hemorrhagic and ischemic stroke, where the former is further divided into intracerebral hemorrhage (ICH) which is mainly due to hypertension, and subarachadonic hemorrhage (SAH) [5]. One in every six people might have a stroke in their entire lifespan, with more than 13.7 million having a stroke annually and 5.8 million dying as a result [1]. Over 80 million people worldwide have withstood a stroke. About 70% of adverse strokes are ischemic strokes, with the remainder being intracerebral hemorrhage or subarachnoid hemorrhage [6]. Ischemic brain, in the broadest manner, can be classified as multifocal and focal and is resulted due to sudden narrowing or significant reduction in the diameter of an artery that supplies a region of the brain. Brain ischemia is commonly characterized by the loss of blood supply of the brain, culminating in a sudden fall in blood pressure in more proximal causes of hemodynamic dysfunction resulting in a sudden drop in blood pressure [7]. There are various signaling pathways along with the disrupted blood flow mechanism that exaggerates the pathological hallmarks of the stroke. GSK-3β has been found to potentiate the deleterious effects that contribute to stroke-induced pathologies.

25.2 MECHANISTIC APPROACHES CONTRIBUTING TO STROKE

25.2.1 CELL DEATH MECHANISM

Cell death is an important homeostatic and well-regulated event that helps in maintaining tissue, cell, or organ function in the human body. Cell death is a protective mechanism and works in association with cellular genesis. The disturbed equilibrium between generation and death of the cells leads to detrimental conditions such as stroke. Due to hypo-perfused neuronal tissues, oxygen and glucose depleted states occur that initiate various cell death mechanisms in the brain eventually contributing to stroke. The inadequate supply of oxygen and glucose to the neuronal tissues commences mitochondrial dysfunction that further culminates in apoptosis [7]. Stroke also potentiates BBB disruption which increases the NMDA and AMPA activity due to excessive release of excitatory neurotransmitters like glutamate and aspartate, responsible for excitotoxicity. Excitotoxicity leads to conditions like oxidative stress and nitrosative stress that culminate in cellular damage in brain tissue resulting in a stroke. Disrupted blood supply to neurons results in initiating pathophysiological cascades like excitotoxicity, mitochondrial dysfunction, protein misfolding, apoptosis, necrosis, and autophagy leading to cell loss. Post-stroke conditions initiate the various deleterious pathways enumerated as in subsections [8].

25.2.1.1 APOPTOSIS

Despite numerous cell death mechanisms, post-ischemic apoptosis of penumbra has been one of the most prevailed cell death mechanisms and it helps in understanding post-ischemic neuron dysfunction [9]. Apoptosis plays an important part and is come into action through two pathways. The intrinsic apoptotic process occurs through mitochondria, while the extrinsic process is receptor-mediated [10]. Ca^{2+} access into the cell is increased throughout ischemic stroke due to lactic acidosis and excitotoxicity of glutamate via activation of acid-sensing ion channels (ASIC) and AMPA, and NMDA receptors. Increased Ca^{2+} levels result in activated calpains, and enhance the proteolytic processing of BH3-interacting domain death agonist (BID) into its truncated form (tBID) [11]. As a cell receives a death signal, the pro-apoptotic protein tBID transcribes to the mitochondrial membrane and interacts with other pro-apoptotic members of the Bcl-2 family-like Bad, Bak, BclXS, and Bax [12]. This family of pro-apoptotic members might alter the permeability of the mitochondrial membrane, and enhance cyt c release, which ultimately contributes to the creation of apoptosomes through Apoptotic protease activating factor-1 binding (Apaf-1). The dormant procaspase 9 is triggered immediately after the formation of the apoptosomes. This activates caspase-9 ultimately sets off a cascade of caspases comprises of, caspase-6, caspase-3 and caspase-7. This caspases cascade inevitably culminates into nuclear DNA fragmentation, driving toward cell death. Extrinsic apoptosis initiates due to external stimuli like TNF-α, TRAIL, and FasL and binds to death receptors. TNF is released by neuronal cells all through the ischemic injury. When these ligands interact with their specific cell surface receptors, these further employ cytoplasmic adapter protein molecules such as Fas-associated death domain protein (FADD) and tumor necrosis factor receptor type-1-associated death domain (TRADD). The FADD's death effector domain (DED) binds with procaspase 8 prodomain to produce a complex (aprocaspase-8 and Fas ligand-Fas receptor-FADD) known as the death induced signaling complex (DISC). The above complex catalyzes autocatalytic activation and procaspase-8 proteolytic degradation, resulting in active initiation of caspase 8. This activated caspase 8 causes further cleavage of effector caspase-3, which leads to apoptotic cell death through a subsequent cascade [13].

25.2.1.2 NECROSIS

Necrosis refers to the passive, degradative processes that occur as a consequence of disease in cells and tissues. In the brain, necrosis may offer a "double impact" by permanent neuronal death combined with the onset or enhancement of potentially harmful neuroinflammation. Ischemic stroke conditions lead to excessive ATP consumption and mitochondrial dysfunction contributing to the depletion of ATP [13]. Depleted ATP conditions are responsible for Na^{2+} and Ca^{2+} pump failures. In the brain, ATP deprivation causes the release of glutamate from astrocytes and neurons, which stimulates NMDA receptors along with neuronal depolarization which contributes to flooding the cell with calcium and sodium, such conditions lead to cell swelling responsible for rupturing of cell membrane and activation of proteases and phospholipases that eventually leads to necrosis [14].

25.2.1.3 NECROPTOSIS

Necroptosis is known as programed necrosis. It exists at the junction of necrosis and apoptosis. The downstream signaling cascade involved in necroptosis is intrinsically independent of the caspases, however, caspase-8 plays a crucial role in flipping the death mechanisms of cellular molecules from apoptosis to necroptosis at an early stage [15]. Tissue necrosis factor-(TNF-α) helps mediate necroptosis predominantly, which is regulated by the formation of necrosome [16]. Receptor-interacting protein kinase (RIPK1 and RIPK3) and mixed-lineage kinase domain-like protein (MLKL) are two downstream signaling proteins for necroptosis. TNF-α binds to its receptor TNFR1 following cerebral ischemia together with TNF-related apoptosis-inducing ligand (TRAIL) receptors and Fas leads to activation of the necroptotic pathway. When TNF-α interacts with TNFR1, it activates pro-survival complex I, which include TNF receptor-associated death domain (TRADD), ubiquitin 3 ligases and RIPK1. Within complex I, RIPK1 gets polyubiquitinated. When it proceeds to deubiquitinate, it produces complex (IIa, IIb). Caspase 8 gets activated by Complex-IIa, paving the way for apoptosis. It has been discovered that in case of oxygen-glucose deprivation and cerebral ischemia, it inhibits caspase 8 and activates complex IIb. After the activation of complex IIb, receptor-interacting protein kinase-1 interacts with receptor-interacting protein kinase-3, and both are phosphorylated [13]. This linked structure is known as ripoptosome. RIPK3 is oligomerized inside the ripoptosome to further phosphorylate and activate mixed lineage kinase domain protein. Activated mixed lineage kinase domain protein multimerizes inside the cytoplasm, resulting in the development of a necrosome. MLKL and RIPK3 move between the nucleus and the cytoplasm dynamically. The accumulation of necrosomes is responsible for the increase in cellular organelle permeability and plasma membrane as it shows interaction with cardiolipin and phosphatidylinositol which ultimately cause cell death. Moreover, the production of damage-associated membrane proteins (DAMPs) and inflammatory cytokines leads to the activation of higher-order inflammatory signals [14].

25.2.1.4 PYROPTOSIS

Pyroptosis is a pro-inflammatory kind of programmed cell death. Inflammasome activation has this as one of its distinctive traits. Pyroptosis is defined by DNA damage and membrane rupture. Pyroptosis is distinct from apoptosis in that it is triggered by inflammatory caspases like caspase 1 or caspase 4, caspase 5, and caspase 11, followed by the cytosolic release of pro-inflammatory components. Pyroptosis-like cell death in ischemic tissue followed by cerebral ischemia has been observed in recent years, which triggers an inflammatory response and exacerbates brain cellular damage by raising NLRP1 or NLRP3 inflammasome or caspase-1 levels [17]. In contrast to apoptosis and autophagy, pyroptosis is categorized by cell edema, plasma membrane denaturation, DNA fragmentation, and chromatin condensation. Cell lysis leads to the production of inflammatory factors such as interleukins-1, which can trigger inflammation. Pyroptosis is a condition that occurs in the ischemic penumbra. In ischemic stroke, it is a strong inducer of the pro-inflammatory pathway. Necrosis occurs in the ischemic core region after cerebral ischemia for a brief period, and dead cells generate dangerous signals such as peroxidase-associated proteins purine, and heat shock protein. These danger signals trigger the innate immune response by binding to intracellular receptors or receptors on the cell membrane in neighboring cells, resulting in inflammatory reactions. DAMPs refers to endogenous danger signaling molecule. This results in pyroptotic cell death and the release of inflammatory mediators such as IL-1. Although these classical DAMPs perform implicit roles during sterile inflammation, pyroptosis's contribution to stroke is not well-established yet [16]. Independent of IL-1 signaling, the caspase 1/11-dependent cell death pathway may contribute significantly to post-stroke damage and potentiates stroke-induced cell death in the human brain.

25.2.2 INFLAMMATION

Pathological alterations are becoming increasingly clear throughout a stroke, from endothelial activation soon after symptoms manifest through the post-injury wound-healing phase days to months later. Inflammation causes pro-inflammatory cytokines like interleukin-1 (IL-1) to be released during the early stages of a stroke, which then propagate the neuroinflammatory response by activating microglial cells and attracting inflammatory cells. Neutrophils are the first cells to enter the brain after a stroke, followed a few days later by macrophages and lymphocytes [12]. In addition, rather than the necrotic core of ischemic brain tissue, inflammatory cell activity is seen in the peri-infarct zone. The extent of this "inflammatory penumbra" correlates with the severity of the brain damage and, consequently, with the amount of impairment. During both ischemic and hemorrhagic stroke, a nonspecific systemic inflammatory response ensues, either as part of the brain damage process or in reaction to complications such as deep venous thrombosis. Higher levels of inflammatory markers including interleukin-6 (IL-6) and C-reactive protein (CRP) have been linked to a poorer prognosis following both ischemic and hemorrhagic strokes in several studies. Stroke causes an intense inflammatory response, which may be caused by the activation of Toll-like receptors (TLRs) [14]. TLR's clinical significance and therapeutic potential in stroke, on the other hand, are not well established. Post-ischemic inflammation has been identified in animal studies as a multifaceted reaction that includes activation of local glia cells, recruitment of blood-derived leukocytes, and cascades of humoral reactions. Hematogenous cells such as polymorph nuclear neutrophils (PMN) and macrophages rapidly reach the ischemia region in the traditional temporary middle cerebral artery occlusion (tMCAO) paradigm. Among these pathways, phagocyte accumulation has been proposed as a surrogate marker of neuroinflammation [18]. Multiple cell types in the brain may release inflammatory mediators such as cytokines and chemokines after a stroke. In the acute ischemic stroke (AIS) mouse model, an array of cytokines including IL-1, IL-6, transforming growth factor, and tumor necrosis factor (TNF-α) is detected in the contralateral hemisphere during the immediate phase following ischemia. Preclinical research revealed that 2 weeks after a stroke, IL-1, interferon-α, TNF-α, and IL-6 concentrations in total brain tissue are higher, and this increase is related to lymphocyte infiltration [19] as discussed in Figure 25.1.

25.3 STROKE AND GSK

Glycogen synthase kinase 3 (GSK-3) is a component of adenomatous polyposis coli/axin/GSK-3 complex that is involved in the ubiquitin-proteasome degradation of catenin, the primary component of the Wnt/-catenin pathway [20]. GSK3 regulates a large number of functionally varied direct downstream targets, the majority of which are blocked or degraded following GSK3 phosphorylation. Glycogen synthase kinase (GSK)-3, which is present is abundantly in the central nervous system, regulates a variety of biological functions including gene expression, differentiation, and cell proliferation. According to reports On day 7 after cerebral ischemia, phosphorylation of GSK-3 at Ser9 and that of Akt were significantly increased [21]. GSK-3b inhibition has been proved to have neuroprotective effects that may help to prevent cell death after ICH. TWS119 inhibition of GSK-3 lowered neurocognitive deficit rating, brain edema, infarct volume, and BBB disruption in this study. This defense was associated with an initial increase in Wnt/-catenin signaling as well as an increase in ZO-1 expression. TWS119 blocking GSK-3 lowers early ischemia-reperfusion damage and preserves the BBB, according to these findings [22].

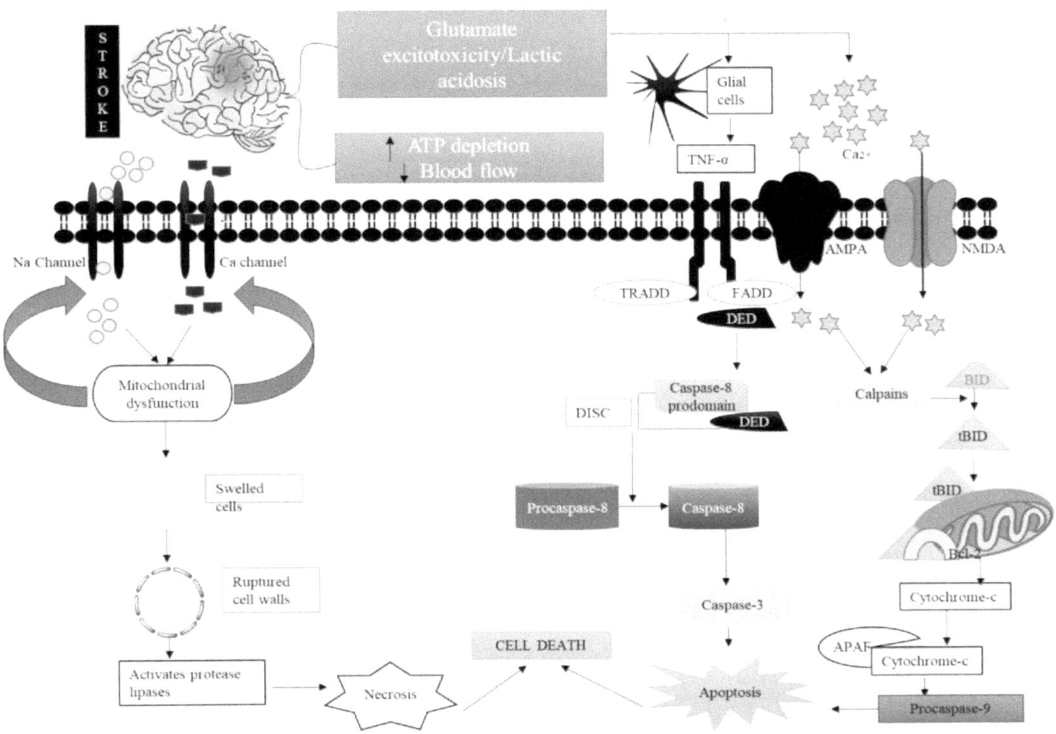

FIGURE 25.1 Following a stroke attack there is an increased level of IL-1, interferon-α, TNF-α, and IL-6 in total brain tissue, and this increase is related to lymphocyte infiltration.

25.4 SIGNALING INVOLVED GSK AND STROKE

25.4.1 WNT/β-CATENIN SIGNALING

Wnt signaling has been derived to perform a cascade of functions in organisms, including the regulation of key operations like proliferation, differentiation, inflammation, neurogenesis, and revitalization [23]. Wnt/β-catenin signaling has a fundamental role in the vascularization of the central nervous system, blood-brain barrier generation, and maturation throughout postnatal and embryonic development. The Wnt/β-catenin signaling is influential in determining adult neurogenesis. Its function in neurovascular maintenance following ischemic injury has not been thoroughly investigated. Research study indicates that Wnt/β-catenin is downregulated following stroke. Ameliorated levels of nuclear catenin and failure of the TJ protein claudin-1 were found within the Endothelial Cells of hemorrhagic stroke patients' brains [24]. Wnt ligands interact with frizzled (FZD) receptor sites, which, stimulates its coreceptor termed as low-density lipoprotein receptor-related protein (LRP), resulting in the destabilization of the adenomatous polyposis coli (APC) complex, axin along with glycogen synthase kinase-3 (GSK-3). Such causes beta-catenin to assemble and transcribe inside the nucleus, where it activates transcription factors responsible for expressing genes. GSK-3, as one of the essential β-catenin inhibitors, is a potent candidate for targeting and modulating the Wnt/β-catenin system. GSK-3β is widely present and accepted in the brain and also is embroiled in several cellular mechanisms, along with the plasticity, migration, survival, differentiation, proliferation, and polarity of neuroblasts or neurons GSK-3 inhibition has been proved to have neuroprotective properties, that could reduce cell death after ICH inactivation of GSK-3 may also start regulating differentiation of neuronal cells, thereby promoting regeneration of neuronal cells. According to research, Catenin, the vital compound in the Wnt/β-catenin pathway, is phosphorylated and degraded by the protein glycogen synthase kinase-3 (GSK-3) according to a research study research. GSK-3 activation prevents catenin cytoplasmic accumulation [25]. GSK-3β, as a critical β-catenin inhibitor, proves to be a potential candidate for Wnt/β-catenin system modulation as discussed in Figure 25.2.

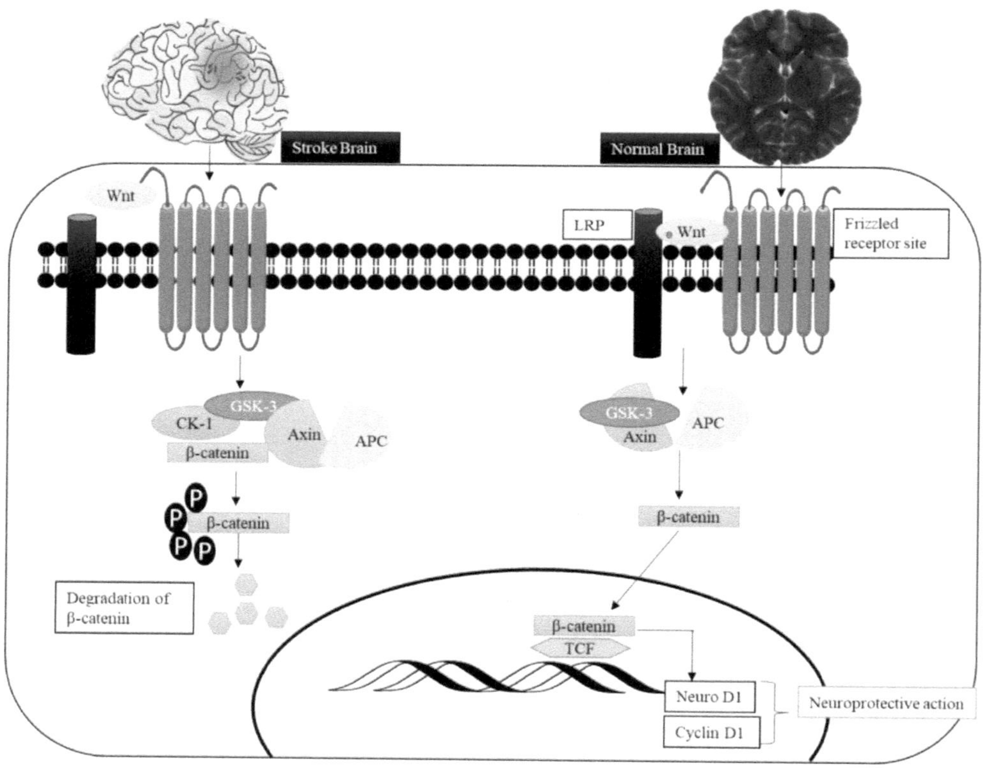

FIGURE 25.2 GSK-3β serves as a critical β-catenin inhibitor having the potential to modulate the Wnt/β-catenin system.

25.4.2 PI3K/AKT/GSK SIGNALING

The PI3K/Akt signaling pathway in the body is a typical pro-proliferation and anti-apoptosis mechanism. PI3K/Akt activation is essential for maintaining appropriate neuron function and apoptosis levels. In several cell types, the PI3K/Akt pathway promotes cell proliferation, differentiation, migration, and death [26]. IGF-1 stimulation of cells results in GSK-3 downregulation via a PI3K-dependent framework. PI3 activates Akt, and Akt is responsible for phosphorylation of GSK at Ser9 residue leading to its inactivation [21]. In less than 12 hours of ischemia stroke or cerebral ischemic-reperfusion, the study discovered significant downregulation of AKT, which is responsible and results in over-activation of glycogen synthase kinase-3β (GSK-3β) [27]. Following I/R injury, the extent of PI3K, Akt phosphorylation, and mTOR phosphorylation get significantly reduced in the mentioned study [28]. Reduced levels of Akt lead to hyperactivation of GSK-3β. Hyperactivity of GSK-3β is mainly responsible for apoptotic pathways in stroke conditions therefore it can be concluded that inhibiting GSK-3β beta would be beneficial in stroke-related injuries. GSK-3 β leads to inhibition of MDM2, leading to hyperactivity of p53. Activating p53 leads to the release of cytochrome-c thereby promoting apoptosis in the post-stroke environment. Thus it can be concluded that PI3/Akt/GSK signaling holds a great significance in stroke and the inactivation of GSK-3β leads to the phosphorylation of the former at ser9 residue through Akt might be an effective target in stroke-related conditions.

25.4.3 NF-κB SIGNALING

The NF-κB pathway has been related to strokes. Penumbral NF-κB gets upregulated when studied in stroke-related patients after 48 hours of the occurrence of stroke. The research noted that mice with p50 subunit knockout of NF-κB have fewer and diminished infarct dimensions preceded by middle cerebral artery occlusion (MCAO), this shows that NF-κB signaling has a great potential in tissue viability and restoration from ischemic injury in both humans and rodent models of ischemic stroke. Previous research found that spontaneously hypertensive rats, a common breed of rats used to mimic hypertension in ischemic stroke studies, had elevated amounts of matrix metalloproteinase activity due to increased NF-κB activation [28]. This demonstrates that NF-κB activation can also diminish the BBB. The NF-κB pathway is a crucial epicenter for inflammatory pathways in the body, and its role in ischemic stroke in both human and animal models of stroke ends up making it a tempting prospect for anti-inflammatory drug development to cure ischemia-reperfusion damage. The preclinical finding demonstrates that repressing NF-κB after ischemia prevents brain tissue in the ischemic penumbra and decreases inflammatory signals within the brain, both of which are promising outcomes for the advancement of stroke therapeutics. During ischemia, TNF-α, IL-1, IL-6, and other proinflammatory cytokines produced by both peripheral immune cells and microglia may activate the classical NF-κB pathway. The NF-κB complex is made up of two subunits: p50 and p65. In its inactive form, NF-κB is linked to the inhibitor of IκB and is found in the cytoplasm of the cell [29]. When cytokines like TNF-α, IL-1, and IL-6 bind to their receptors on the cell's plasma membrane, a signal transduction cascade is initiated, which results in NF-κB activation. The NF-κB essential modulator (NEMO) trimerizes and attaches to the inhibitor of the B kinase (IKK) complex as the first step at the NF-κB level. IKK phosphorylation leads to upregulation of inhibitor of kappa B (IκB). The upregulation of kappa B inhibitor phosphorylation enhances its release from NF-Kb and also promotes its ubiquitination for proteasomal degradation. When IκB degradation takes place, the NF-κB p65 subunit gets phosphorylated, leading to the unmasking of the nuclear localization sequence on the p50 subunit, This allows NF-κB to enter the nucleus and facilitate transcription of proinflammatory and proapoptotic genes. The interaction of various molecules with NF-κB, as well as the subunits that comprise NF-κB itself, provides a variety of therapeutic targets for therapies aimed at inhibiting NF-κB translocation into the nucleus and ameliorating inflammatory actions in disease [29].

25.4.4 MAPK

GSK-3β is known to have interaction with the MAPK family and promotes signaling after stress [30] The MAPK pathways are also linked to ischemic stroke because they are all activated when an ischemic stroke occurs [31]. Their precise roles in ischemic stroke are unknown, but studies have shown that JNK and P_{38} pathways contribute to ischemic injury because inhibiting these pathways reduces the amount of damaged tissue. These pathways aid in the generation of pro-inflammatory cytokines such as TNF-α and IL. Several investigations have been carried out to look into the important function of MAPK in the neurologic curriculum of stroke. MAPK activation has been discovered to influence pathological functions following progression, with a vast body of data indicating that MAPK also contributes to controlling systemic inflammation, cell apoptosis, cytokines, and death in ischemia and hemorrhagic brain damage. Cerebral ischemia leads to generation of oxygen-free radicals that leads to lipid peroxidation eventually contributing to apoptotic activity of the neuronal cells. Post-stroke generated oxygen free radicals and cerebral ischemia leads to activation of P_{38} with increased expression in glial cells. The activation results in secretion of inflammatory cytokines that in addition to increased adhesion activation molecule in endothelial cells also damages the blood-brain barrier. The disrupted blood-brain barrier aggravates the stroke condition more promptly. Cerebral ischemia also initiates the activation of JNK which also contributes to post-stroke pro-apoptotic activity [32]. GSK-3β activation further leads to potentiating deleterious pathways including activation of JNK and P_{38} eventually leading to pro-apoptotic pathways followed by cell death in stroke conditions. Therefore, it can be suggested that GSK-3β activation is responsible for stimulating detrimental pathways culminating in stroke and its related pathological hallmarks [33] as discussed in Figure 25.3.

25.4.5 TAK1 AND GSK SIGNALING

Numerous renowned TAK1-signaling stimuli, such as IL-1, TLR4 TLR2, and TNF-α, are said to be stimulated and exhibit a negative role in cerebral ischemia. TAK1 is also an important upstream regulator with five distinct pathways involved in ischemic brain injury. The transforming growth factor-activated kinase 1 (TAK1; MAP3K7) mitogen-activated protein kinase (MAP3K) is a 67-kDa protein that works as a well-known upstream inducer of numerous kinases, including IKK, JNK, and p38. TAK1 upregulation in response to ischemia is known to boost IL-1, TNF-α, and IL-6 expression. An overabundance of TAK1 causes inflammatory and apoptotic signaling in the brain, resulting in gradual tissue destruction [34]. TAK1 has been identified as a critical component of signaling pathways in stroke that activate the transcription factor NF-κB [35]. According to the research, GSK has been found to potentiate the stabilization of TAK-1 and its interaction with its binding receptor therefore it can be suggested that inhibiting GSK-3β can eventually contribute to stroke deficit effects [30].

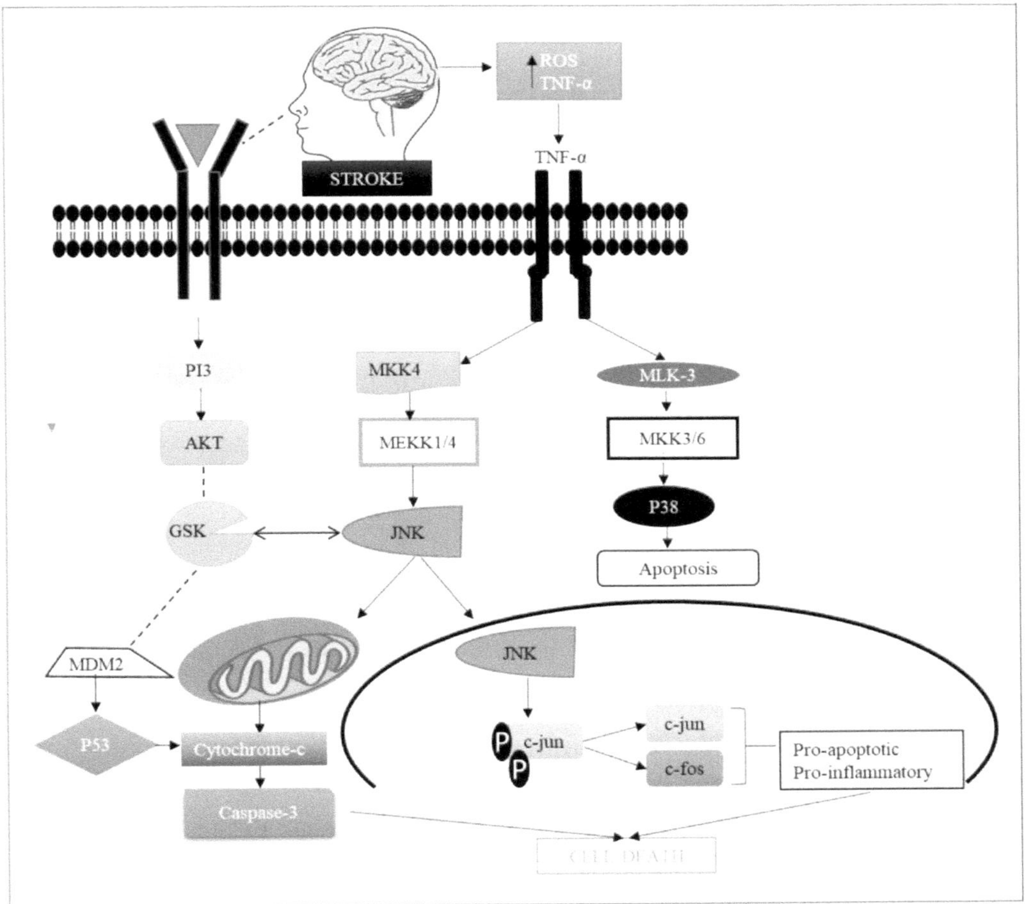

FIGURE 25.3 GSK-3β activation for stimulating detrimental pathways culminating in stroke and its related pathological hallmarks.

25.4.6 DRP-1 AND GSK

Mitochondria has an important participation in various cellular activities, including ATP production, Ca^{2+} homeostasis, ROS production, and apoptosis. In cerebral ischemia, dysfunctional mitochondria have a well-identified role [36]. The principal controller involved in the fission of mitochondria is dynamin-related protein-1 (Drp-1), which is a GTPases dynamin family member that caused the fission of mitochondria when it was translocated within the cytosol to the outer mitochondrial membrane [37]. Mitochondria lead to the generation of extensive ROS during stroke, culminating in dysfunctional mitochondrial along with mitochondrial fission, the excessive ROS generation is responsible for pathological hallmark changes in the functioning of the brain. Drp-1 binds to the mitochondrial membrane and this bounded Drp-1 accelerates oligomer formation of Bax in the outer membrane of mitochondria, thereby helping in enhancing its permeability [38]. When mitochondria are subjected to a challenging situation like ischemic stroke, the characteristics and permeability of mitochondrial inner and outer membranes change, culminates in the transmembrane channel opening, also leads to organelle's membrane potential collapse, and disrupts the respiratory chain along with oxidative phosphorylation, which collectively contributes to depleted cell energy and eventually participating in cell death. Post-stroke Activation of GSK-3β leads to phosphorylation of Drp-1. Drp-1 phosphorylation is responsible for mitochondrial fission followed by mitochondrial fragmentation. Fragmented mitochondria are subjected to cytochrome-c release. Cytochrome-c release regulates the caspases cascade eventually contributing to apoptosis in stroke conditions. GSK-3 inhibition reduced neuronal injury by controlling mPTP opening [39]. Dynamin-related protein 1 (DRP1), which is a cytosolic GTPase, facilitates mitochondrial fission, particular adaptor proteins are considered to be required to assist this protein. Adaptor proteins include mitochondrial elongation factor 2 (MIEF2/MID49), mitochondrial fission factor (MFF), fission protein 1 (FIS1) mitochondrial elongation factor-1 (MIEF1/MID51). The above-mentioned adaptor proteins help recruit DRP-1 to the outer membrane of mitochondria, where it self-assembles into a hexagonal structure that surrounds the mitochondria. Long mitochondria are eventually split into smaller and fragmented mitochondria thereby promoting pro-apoptotic factors [40].

25.4.7 NRF2 AND GSK

Nrf2 is a well-known transcriptor factor that relates to the basic-region leucine family under subfamily cap 'n' collar [41]. This modular protein has seven domains. Every domain of Nrf2 is characterized by distinct functions. Nrf2 is a key regulator for cellular and physiological factors of protection for endogenous and exogenous insult. Its protective action is exhibited in association with the stress-inducible factors and basal stimulation of some protective genes [42]. The Nrf2 is present in the inactive form in association with Keap1. The active Nrf2 is translocated to the nucleus and, in the form of target genes contributes to a diverse set of protective actions [43]. Several ischemia models incorporating Nrf2-deficient mice show that Nrf2 has a significant benefit on reperfusion injury, stroke, neurocognitive deficits, and brain edema. Nrf2 reduces the extent of proteins such as pro-inflammatory cytokines, MMPs, iNOS, cell adhesion molecules apoptotic pathway NF-KB signaling, and COX2 [44]. It is a widely accepted fact that Nrf2 has a significant part in the pathogenic mechanism of ischemia and stroke by increasing the expression of anti-inflammatory mechanisms and hindering inflammatory processes [45]. Physiologically

Nrf2 is constrained in the cytoplasm by its alliance to the Keap1-Cul3-Rbx1 complex, which enhances proteasomal deterioration of Nrf2. Keap-1 releases Nrf2 in the cytoplasm when conditions like oxidative stress due to ROS generation occur [46]. As the concentration of Nrf2 raises in the cytoplasm it contributes to catalyzing the translocation of Nrf2 to the nucleus and in the nucleus, heterodimerization occurs along with maf proteins that are small in size and bind to ARE in a cascade of antioxidative genes that protect from oxidation and other disorders related to the same. Phosphoinositide 3-kinase (PI3K)/(Akt) signaling and extracellular regulated kinases (ERK) mediate its translocation and activated PI3K potentiate its translocation into the nucleus, whereas GSK-3 beta directly inhibits NRF2-ARE binding in the nucleus thereby inhibiting its neuroprotective actions by preventing its transcription in stroke-like condition PI3K, ERK, and AKT inhibition also reduce Nrf2 activation by preventing Nrf2 release and by potentiating GSK-3 activity.

25.5 THERAPIES TO OVERCOME A STROKE

1. **Intravenous Thrombolysis:** Intravenous tPA transfusion, as preferred by the AHA/ASA, is the treatment procedure used most frequently for people diagnosed in less than the first 3 hours of symptom onset. For patients diagnosed, the intervention window can be prolonged to 4.5 hours. The use of intravenous alteplase is based on the landmark National Institute of Neurological Disorders and Stroke (NINDS) [47].
2. **Stem Cells:** Dental pulp stem cells (DPSCs) have indeed been studied for their potential use in the management of neurodegenerative diseases. They have the potential to foster neuroregeneration upon injuries like ischemic stroke. They are ideal for replacing neuronal cells that have been damaged in neurodegenerative diseases like ischemic stroke. Because they arise in the neural crest. With the success of efficient approaches for expanding and differentiating DPSCs, stem cells have emerged as a promising class of potential stroke treatments [48].
3. **Endovascular Thrombectomy:** It has already become one of the advanced therapy for patients who might have an AIS due to massive obstruction in the anterior cerebral circulation [49]. If treatment is initiated within 6 hours of stroke occurrence, Endovascular thrombectomy has been said to be an effective therapy in the management of conditions like an AIS in patients with obstruction into the internal carotid artery [50]. Properly and in time accessibility and availability to Endovascular thrombectomy remains essential to improving health outcomes for people with AIS caused by large vessel occlusion [51]. The reports suggested that an endovascular thrombectomy is an essential tool in the management of stroke.
4. **Antithrombotic Therapy:** Antiplatelet agents are recommended by the AHA/ASA guidelines to minimize the risk of a repetitive event of stroke [52]. The US Food and Drug Administration has approved three alternatives:

    i. Aspirin;
    ii. Clopidogrel; and
    iii. Aspirin/dipyridamole.

Aspirin, in doses ranging from 50 to 325 mg daily, is highly suggested as a primary treatment to preclude reoccurring ischemic stroke. The benefit is the same regardless of dose, but the adverse effect profiles differ considerably. The most serious side effect is gastrointestinal bleeding, which worsens with increased daily dosing. Patients taking 325 mg or less per day have an average yearly risk of gastrointestinal bleeding of 0.4%. There is no strong indication that a higher dose of aspirin reduces the risk of another stroke in patients who have a recurrent stroke while on aspirin therapy. Aspirin is the least expensive therapy option [53].

5. **Drugs:** Various drugs that have been tested against stroke under various pathways that lead to deleterious pathological hallmarks in stroke-like conditions. The following mentioned are the drugs with their action and targets in the pathophysiology of stroke and how they exert their protective action and help in ameliorating diseased conditions in stroke pathologies (Table 25.1).

TABLE 25.1 Drugs in the Management of Stroke

SL. No.	Drugs	Target	Action	References
1.	Ligustilide	Inhibition of Drp-1	ROS ATP production	[54]
2.	TWS119	Target GSK-3 inhibition.	increased protein expression of β-catenin.	[55]
3.	Apelin	Target GSK-3/Nrf2 inhibition.	Decreased apoptosis. Oxidative stress. Brain edema.	[56]
4.	Omaveloxolone	Inhibition of Nrf2	Increased plasma concentration.	[45]
5.	Maraviroc	Antagonizes CCR5	Enhances motor activity.	[57]

25.6 CONCLUSION

Post-stroke cell death mechanisms like apoptosis, necrosis, pyroptosis, etc., emerged as an important and deleterious contributor to stroke-induced cell death. Various factors potentiate these cell death mechanisms culminating in harmful and detrimental effects followed by cerebral ischemia. GSK-3β is one of the important candidates in stimulating different pathways that converge to cell death eventually. GSK-3β upon its activation by its phosphorylation at ser9 residues followed by stroke leads to the breakdown of β-catenin inhibiting its nucleus translocation and ameliorating its neuroprotective properties. Post-stroke conditions also lead to the inactivation of the PI3K/Akt pathway which in return hyperactivates GSK and culminates in apoptotic cell death. GSK has been also found to inhibit Nrf2 directly in the nucleus and indirectly by inhibiting the activity of the PI3K/AKT pathway which potentiates the Nrf2 translocation into the nucleus, thereby supporting its pro-apoptotic and pro-inflammatory properties. From this chapter, it can be concluded that different signaling Inhibiting GSK-3β could prove to be an interesting and potential candidate for ameliorating the stroke-induced pathologies although detailed research studies and evidence are yet to be studied.

KEYWORDS

- **cell death mechanism**
- **GSK**
- **inflammation**
- **ischemic injury**
- **ischemic reperfusion**
- **necrosis**
- **pyroptosis**
- **stroke**

REFERENCES

1. Donkor, E. S., (2018). Stroke in the 21st century: A snapshot of the burden, epidemiology, and quality of life. *Stroke Res. Treat [Internet], 2018*, 3238165. Available from: https://pubmed.ncbi.nlm.nih.gov/30598741 (accessed on 9 November 2023).
2. Alquisiras-Burgos, I., & Aguilera, P., (2022). Involvement of glucose transporter overexpression in the protection or damage after ischemic stroke. *Neural Regen. Res. [Internet], 17*(4), 783–784. Available from: https://pubmed.ncbi.nlm.nih.gov/34472470 (accessed on 9 November 2023).
3. Soares, R. O. S., Losada, D. M., Jordani, M. C., Évora, P., & Castro-e-Silva, O., (2019). Ischemia/ reperfusion injury revisited: An overview of the latest pharmacological strategies. *International Journal of Molecular Sciences, 20*.
4. Randolph, S. A., (2016). Ischemic stroke. *Workplace Health Saf. [Internet], 64*(9), 444. Available from: https://doi.org/10.1177/2165079916665400.
5. Unnithan, A. K. A., & Mehta, P., (2022). *Hemorrhagic Stroke*. In Treasure Island (FL).
6. Phipps, M. S., & Cronin, C. A., (2020). Management of acute ischemic stroke. *BMJ [Internet], 368*, l6983. Available from: http://www.bmj.com/content/368/bmj.l6983.abstract (accessed on 9 November 2023).
7. Saklani, P., Khan, H., Gupta, S., Kaur, A., & Singh, T. G., (2022). Neuropeptides: Potential neuroprotective agents in ischemic injury. *Life Sciences, 288*, 120186.
8. Belov, K. D., Kriska, J., Tureckova, J., & Anderova, M., (2020). Ischemia-triggered glutamate excitotoxicity from the perspective of glial cells. *Front. Cell Neurosci. [Internet], 14*. Available from: https://www.frontiersin.org/article/10.3389/fncel.2020.00051 (accessed on 9 November 2023).
9. Uzdensky, A. B., (2020). Regulation of apoptosis in the ischemic penumbra in the first-day post-stroke. *Neural Regen. Res., 15*(2), 253–254.
10. Kalra, P., Khan, H., Kaur, A., & Singh, T. G., (2022). Mechanistic insight on autophagy modulated molecular pathways in cerebral ischemic injury: From preclinical to clinical perspective. *Neurochemical Research*, 1–9.
11. Khan, H., Garg, N., Singh, T. G., Kaur, A., & Thapa, K., (2022). Calpain inhibitors as potential therapeutic modulators in neurodegenerative diseases. *Neurochemical Research*, 1–25.
12. Khan, H., Gupta, A., Singh, T. G., & Kaur, A., (2021). Mechanistic insight on the role of leukotriene receptors in ischemic–reperfusion injury. *Pharmacological Reports, 73*(5), 1240–1254.
13. Khan, H., Sharma, K., Kumar, A., Kaur, A., & Singh, T. G., (2022). Therapeutic implications of cyclooxygenase (COX) inhibitors in ischemic injury. *Inflammation Research*, 1–6.
14. Prabhakar, N. K., Khan, H., Grewal, A. K., & Singh, T. G., (2022). Intervention of neuroinflammation in the traumatic brain injury trajectory: *In vivo* and clinical approaches. *International Immunopharmacology, 108*, 108902.

15. Bertheloot, D., Latz, E., & Franklin, B. S., (2021). Necroptosis, pyroptosis and apoptosis: An intricate game of cell death. *Cell Mol. Immunol. [Internet], 18*(5), 1106–1121. Available from: https://doi.org/10.1038/s41423-020-00630-3.
16. Khan, H., Bangar, A., Grewal, A. K., Bansal, P., & Singh, T. G., (2022). Caspase-mediated regulation of the distinct signaling pathways and mechanisms in neuronal survival. *International Immunopharmacology, 110*, 108951.
17. Gou, X., Xu, D., Li, F., Hou, K., Fang, W., & Li, Y., (2021). Pyroptosis in stroke-new insights into disease mechanisms and therapeutic strategies. *J. Physiol. Biochem., 77*(4), 511–529.
18. Thapa, K., Khan, H., Singh, T. G., & Kaur, A., (2021). Traumatic brain injury: Mechanistic insight on pathophysiology and potential therapeutic targets. *Journal of Molecular Neuroscience, 71*(9), 1725–1742.
19. Shi, K., Tian, D. C., Li, Z. G., Ducruet, A. F., Lawton, M. T., & Shi, F. D., (2019). Global brain inflammation in stroke. *Lancet Neurol, 18*(11), 1058–1066.
20. Caspi, M., Perry, G., Skalka, N., Meisel, S., Firsow, A., Amit, M., et al., (2014). Aldolase positively regulates of the canonical Wnt signaling pathway. *Mol. Cancer [Internet], 13*(1), 164. Available from: https://doi.org/10.1186/1476-4598-13-164.
21. Khan, H., Singh, T. G., Dahiya, R. S., & Abdel-Daim, M. M., (2022). α-Lipoic acid, an organosulfur biomolecule a novel therapeutic agent for neurodegenerative disorders: An mechanistic perspective. *Neurochemical Research*, 1–2.
22. Wang, W., Li, M., Wang, Y., Wang, Z., Zhang, W., Guan, F., et al., (2017). GSK-3β as a target for protection against transient cerebral ischemia. *Int. J. Med. Sci. [Internet], 14*(4), 333–339. Available from: https://pubmed.ncbi.nlm.nih.gov/28553165 (accessed on 9 November 2023).
23. Liu, J., Xiao, Q., Xiao, J., Niu, C., Li, Y., Zhang, X., et al., (2022). Wnt/β-catenin signaling: Function, biological mechanisms, and therapeutic opportunities. *Signal Transduct. Target Ther., [Internet], 7*(1), 3. Available from: https://doi.org/10.1038/s41392-021-00762-6.
24. Song, S., Huang, H., Guan, X., Fiesler, V., Bhuiyan, M. I. H., Liu, R., et al., (2021). Activation of endothelial Wnt/β-catenin signaling by protective astrocytes repairs BBB damage in ischemic stroke. *Prog. Neurobiol. [Internet], 199*, 101963. Available from: https://www.sciencedirect.com/science/article/pii/S0301008220302185 (accessed on 9 November 2023).
25. Zhao, Y., Wei, Z. Z., Zhang, J. Y., Zhang, Y., Won, S., Sun, J., et al., (2017). GSK-3β inhibition induced neuroprotection, regeneration, and functional recovery after intracerebral hemorrhagic stroke. *Cell Transplant [Internet], 26*(3), 395–407. Available from: https://doi.org/10.3727/096368916X694364.
26. Khan, H., Singh, A., Thapa, K., Garg, N., Grewal, A. K., & Singh, T. G., (2021). Therapeutic modulation of the phosphatidylinositol 3-kinases (PI3K) pathway in cerebral ischemic injury. *Brain Research, 1761*, 147399.
27. Peng, S., Gu, J., Dai, C., Iqbal, K., Liu, F., & Gong, C. X., (2022). AKT/GSK-3β signaling is altered through downregulation of mTOR during cerebral ischemia/reperfusion injury. Mol. *Biol. Rep.* [Internet]. Available from: https://doi.org/10.1007/s11033-022-07247-x.
28. Khan, H., Kashyap, A., Kaur, A., & Singh, T. G., (2020). Pharmacological postconditioning: A molecular aspect in ischemic injury. *Journal of Pharmacy and Pharmacology, 72*(11), 1513–27.
29. Howell, J. A., & Bidwell, G. L. 3rd., (2020). Targeting the NF-κB pathway for therapy of ischemic stroke. *Ther. Deliv., 11*(2), 113–123.
30. Venna, V. R., Benashski, S. E., Chauhan, A., & McCullough, L. D., (2015). Inhibition of glycogen synthase kinase-3β enhances cognitive recovery after stroke: The role of TAK1. *Learn Mem. [Internet], 22*(7), 336–343. Available from: https://pubmed.ncbi.nlm.nih.gov/26077686 (accessed on 9 November 2023).
31. Gugliandolo, A., Silvestro, S., Sindona, C., Bramanti, P., & Mazzon, E., (2021). MiRNA: Involvement of the MAPK pathway in ischemic stroke. A promising therapeutic target. *Medicina (Kaunas), 57*(10).
32. Sun, J., & Nan, G., (2016). The mitogen-activated protein kinase (MAPK) signaling pathway as a discovery target in stroke. *J. Mol. Neurosci. [Internet], 59*(1), 90–98. Available from: https://doi.org/10.1007/s12031-016-0717-8.
33. Ibrahim, S. H., Akazawa, Y., Cazanave, S. C., Bronk, S. F., Elmi, N. A., Werneburg, N. W., et al., (2011). Glycogen synthase kinase-3 (GSK-3) inhibition attenuates hepatocyte lipoapoptosis. *J. Hepatol, 54*(4), 765–772.

34. Zeyen, T., Noristani, R., Habib, S., Heinisch, O., Slowik, A., Huber, M., et al., (2020). Microglial-specific depletion of TAK1 is neuroprotective in the acute phase after ischemic stroke. *J. Mol. Med. [Internet], 98*(6), 833–847. Available from: https://doi.org/10.1007/s00109-020-01916-9.
35. Kaur, A., Singh, T. G., Khan, H., Kumar, M., Singh, N., & Abdel-Daim, M. M., (2022). Neuroprotective effect of piclamilast-induced post-ischemia pharmacological treatment in mice. *Neurochemical Research*, 1–4.
36. Bhatti, J. S., Bhatti, G. K., & Reddy, P. H., (2017). Mitochondrial dysfunction and oxidative stress in metabolic disorders—A step towards mitochondria based therapeutic strategies. *Biochim. Biophys. Acta–Mol. Basis Dis. [Internet], 1863*(5), 1066–1077. Available from: https://www.sciencedirect.com/science/article/pii/S0925443916302927 (accessed on 9 November 2023).
37. Zuo, W., Yang, P. F., Chen, J., Zhang, Z., & Chen, N. H., (2016). Drp-1, a potential therapeutic target for brain ischemic stroke. *Br. J. Pharmacol. [Internet], 173*(10), 1665–1677. Available from: https://doi.org/10.1111/bph.13468.
38. Gao, F., Reynolds, M. B., Passalacqua, K. D., Sexton, J. Z., Abuaita, B. H., & O'Riordan, M. X. D., (2021). The mitochondrial fission regulator DRP1 controls post-transcriptional regulation of TNF-α. *Front. Cell Infect. Microbiol. [Internet], 10*. Available from: https://www.frontiersin.org/article/10.3389/fcimb.2020.593805 (accessed on 9 November 2023).
39. Tanaka, T., Saotome, M., Katoh, H., Satoh, T., Hasan, P., Ohtani, H., et al., (2018). Glycogen synthase kinase-3β opens mitochondrial permeability transition pore through mitochondrial hexokinase II dissociation. *J. Physiol. Sci. [Internet], 68*(6), 865–871. Available from: https://doi.org/10.1007/s12576-018-0611-y.
40. Xie, J. H., Li, Y. Y., & Jin, J., (2020). The essential functions of mitochondrial dynamics in immune cells. *Cell Mol. Immunol. [Internet], 17*(7), 712–721. Available from: https://doi.org/10.1038/s41423-020-0480-1.
41. Khan, H., Grewal, A. K., & Singh, T. G., (2022). Pharmacological postconditioning by protocatechuic acid attenuates brain injury in ischemia–reperfusion (I/R) mice model: Implications of nuclear factor erythroid-2-related factor pathway. *Neuroscience, 491*, 23–31.
42. Tonelli, C., Chio, I. I. C., & Tuveson, D. A., (2018). Transcriptional regulation by Nrf2. *Antioxid Redox Signal, 29*(17), 1727–1745.
43. Liu, L., Locascio, L. M., & Doré, S., (2019). Critical role of Nrf2 in experimental ischemic stroke. *Front Pharmacol., [Internet], 10*. Available from: https://www.frontiersin.org/article/10.3389/fphar.2019.00153 (accessed on 9 November 2023).
44. Khan, H., Sharma, K., Kumar, A., Kaur, A., & Singh, T. G., (2022). Therapeutic implications of cyclooxygenase (COX) inhibitors in ischemic injury. *Inflammation Research*, 1–6.
45. Mazur, A., Fangman, M., Ashouri, R., Arcenas, A., & Doré, S., (2021). Nrf2 as a therapeutic target in ischemic stroke. *Expert Opin. Ther. Targets [Internet], 25*(3), 163–166. Available from: https://doi.org/10.1080/14728222.2021.1890716.
46. Zhang, R., Xu, M., Wang, Y., Xie, F., Zhang, G., & Qin, X., (2017). Nrf2-a promising therapeutic target for defensing against oxidative stress in stroke. *Mol. Neurobiol., 54*(8), 6006–6017.
47. Chen, Y. W., Sung, S. F., Chen, C. H., Tang, S. C., Tsai, L. K., Lin, H. J., et al., (2019). Intravenous thrombolysis administration 3–4.5 h after acute ischemic stroke: A Retrospective, multicenter study. *Front. Neurol. [Internet], 10*. Available from: https://www.frontiersin.org/article/10.3389/fneur.2019.01038 (accessed on 9 November 2023).
48. Khan, H., Tiwari, P., Kaur, A., & Singh, T. G., (2021). Sirtuin acetylation and deacetylation: A complex paradigm in neurodegenerative disease. *Molecular Neurobiology, 58*(8), 3903–3917.
49. Grewal, A. K., Singh, T. G., & Singh, N., (2020). Potential herbal drugs for ischemic stroke: A review. *Plant Arch, 20*(1), 3772–3783.
50. Albers, G. W., Marks, M. P., Kemp, S., Christensen, S., Tsai, J. P., Ortega-Gutierrez, S., et al., (2018). Thrombectomy for stroke at 6 to 16 hours with selection by perfusion imaging. *N. Engl. J. Med. [Internet], 378*(8), 708–718. Available from: https://doi.org/10.1056/NEJMoa1713973.
51. Sarraj, A., Savitz, S., Pujara, D., Kamal, H., Carroll, K., Shaker, F., et al., (2020). Endovascular thrombectomy for acute ischemic strokes. *Stroke [Internet], 51*(4), 1207–1217. Available from: https://doi.org/10.1161/STROKEAHA.120.028850.

52. Kleindorfer, D. O., Towfighi, A., Chaturvedi, S., Cockroft, K. M., Gutierrez, J., Lombardi-Hill, D., et al., (2021). 2021 guideline for the prevention of stroke in patients with stroke and transient ischemic attack: A guideline from the American Heart Association/American Stroke Association. *Stroke [Internet], 52*(7), e364–467. Available from: https://doi.org/10.1161/STR.0000000000000375.
53. Oza, R., Rundell, K., & Garcellano, M., (2017). Recurrent ischemic stroke: Strategies for prevention. *Am. Fam. Physician, 96*(7), 436–440.
54. Wu, Q., Liu, J., Mao, Z., Tian, L., Wang, N., Wang, G., et al., (2022). Ligustilide attenuates ischemic stroke injury by promoting Drp1-mediated mitochondrial fission via activation of AMPK. *Phytomedicine [Internet], 95*, 153884. Available from: https://www.sciencedirect.com/science/article/pii/S0944711321004244 (accessed on 9 November 2023).
55. Wang, W., Li, M., Wang, Y., Li, Q., Deng, G., Wan, J., et al., (2016). GSK-3β inhibitor TWS119 attenuates rtPA-induced hemorrhagic transformation and activates the Wnt/β-catenin signaling pathway after acute ischemic stroke in rats. *Mol. Neurobiol. [Internet], 53*(10), 7028–7036. Available from: https://doi.org/10.1007/s12035-015-9607-2.
56. Duan, J., Cui, J., Yang, Z., Guo, C., Cao, J., Xi, M., et al., (2019). Neuroprotective effect of Apelin 13 on ischemic stroke by activating AMPK/GSK-3β/Nrf2 signaling. *J. Neuroinflammation [Internet], 16*(1), 24. Available from: https://doi.org/10.1186/s12974-019-1406-7.
57. Richards, L. G., & Cramer, S. C., (2021). Advances in stroke. *Stroke [Internet], 52*(1), 348–350. Available from: https://doi.org/10.1161/STROKEAHA.120.033231.

CHAPTER 26

Regulation of Nrf-2 Signaling Pathway in Ischemia/Reperfusion Injury

VEERTA SHARMA,[1] HEENA KHAN,[1] AMARJOT KAUR GREWAL,[1]
THAKUR GURJEET SINGH,[1] and RAHUL DESHMUKH[2]

[1]*Chitkara College of Pharmacy, Chitkara University, Punjab, India*

[2]*Department of Pharmaceutical Sciences and Technology, Maharaja Ranjit Singh Punjab Technical University, Bathinda, Punjab, India*

ABSTRACT

Stroke is a prominent cause of mortality and long-term disability around the world. Effective therapeutic interventions, on the other hand, are still few. Nuclear factor-erythroid factor 2-related factor 2 (Nrf2) is a master regulator that coordinates the basal and stress-inducible activation of numerous cytoprotective genes in cellular and organismal defense against endogenous and exogenous stressors. The Nrf2 network not only regulates various intermediate metabolic pathways but also strictly controls redox homeostasis. As a result, targeting Nrf2 has become a promising therapeutic method for the prevention and treatment of central nervous system (CNS) disorders, such as stroke. The purpose of this review is to give an overview of the research that has been done to relate the Nrf2 pathway in different types of ischemia such as cerebral, myocardial, renal, hepatic, and intestinal.

26.1 INTRODUCTION

Stroke is a prominent cause of mortality and long-term disability around the world [1]. Effective therapeutic interventions, on the other hand, are still few. Using an intravenous recombinant tissue plasminogen activator (tPA) to swiftly restore blood flow after an acute ischemic stroke (AIS) is an effective treatment for preserving the ischemic brain [2, 3], in combination with thrombectomy. Due to its narrow therapeutic window of four hours and limited indications, more than 95% of patients do not get benefit from tPA [4, 5]. Furthermore, no effective therapy has been shown to aid long-term healing. As a result, novel treatment techniques aimed at crucial biological components of the ischemia cascade are urgently needed. By coordinating the basal and stress-inducible activation of many cytoprotective

Ischemic Injury: A Molecular Insight. Thakur Gurjeet Singh & Heena Khan (Eds.)

genes, the transcriptional factor nuclear factor-erythroid factor 2-related factor 2 (Nrf2), a key player in the control of cellular and organismal defense mechanisms against endogenous and external stressors [6–8]. The structure, function, molecular mechanism, control of activity of Nrf2, downstream signaling pathways, and implications as a disease therapy target have all been thoroughly researched [9–11]. It is now recognized as a master regulator of redox homeostasis, ensuring that its activity is enhanced in response to redox perturbation, toxicity, inflammation, and disease conditions by controlling a wide range of target genes that share a common DNA sequence in the promoter region known as antioxidant response element (ARE) [10, 12]. Nrf2 regulates a variety of main and secondary metabolic processes in addition to modulating antioxidant responses [6]. As a result, targeting Nrf2 has become a promising treatment for stroke prevention [7, 13]. Because of its unique properties, Nrf2 has been related to several forms of ischemia, including cerebral, myocardial, renal, hepatic, and intestinal ischemia. This review aims to give an overview of the research that has been done to relate Nrf2 to different types of ischemic conditions.

26.2 STRUCTURE OF NRF2

Nrf2 is a basic leucine zipper (bZIP) protein are class of transcription factor family, which has a cap-n-collar (CNC) structure. From Neh1 to Neh7, Nrf2 is made up of seven functional domains also known as the Nrf2-ECH homology (Neh) domains [6, 7]. The CNC-bZIP region of the Neh1 domain is well conserved and heterodimers with small Maf (musculo-aponeurotic fibrosarcoma) proteins to bind the ARE sequence in DNA. This region (CNC-bZIP) is present in a variety of species and is critical for Nrf2 function. The low-affinity (DLG) and high-affinity (ETGE) motifs in the N-terminal Neh2 domain are used by Nrf2 to attach to its principal negative regulator Kelch-like ECH associating protein 1 (Keap-1). In collaboration with the Neh4 and Neh5 domains, the C-terminal Neh3 region functions as a transactivation domain that binds to chromo-ATPase/helicase DNA-binding protein (CHD-6) and triggers transcription of Nrf2 target genes. The Neh4 and Neh5 domains of Nrf2 are involved in transactivation activity that binds to cysteine/histidine-rich domain 3 (CH3) domains of CBP (CREB (cAMP Responsive Element Binding protein)-binding protein). The Neh6 region which is independent of Keap-1 mediates the repression of Nrf2 stability with another negative regulator the dimeric β-transducin repeat-containing protein (β-TrCP), a substrate adaptor for the Skp1 (S-phase kinase-associated protein 1)-Cul1 (Cullin 1)-Rbx1 core E3 complex, through DSGIS and DSAPGS motifs. The Neh7 region is implicated in the negative regulation of Nrf2 by physically interacting with the retinoid X receptor (RXR) [11].

26.3 NRF2-ARE PATHWAY

Organisms have a defensive system to preserve homeostasis in the face of continual intrinsic and extrinsic insults that cause nucleic acids, proteins, and membrane lipids to be damaged. By regulating a wide spectrum of cytoprotective genes, Nrf2 is a master regulator of the inducible cell defense mechanism [6, 14]. Under normal homeostatic conditions, Nrf2 in the cytoplasm is mostly linked to the Keap-1 ubiquitin E3 ligase complex and is destroyed by the proteasome. Cul3 acts as a scaffolding protein for both Keap-1 and Rbx1. As a result, Nrf2 activity and abundance are kept at a minimum. Under stress, the Keap-1 mediated regulation of Nrf2 protein is broken. The stabilized and accumulated Nrf2 translocates into the nucleus and binds to the ARE in the promoter region of Nrf2 target genes as a heterodimer with one of the tiny Maf proteins, activating a wide range of cytoprotective genes [9, 10]. It is becoming increasingly clear that Keap-1 plays a central role in the regulatory mechanism of Nrf2. Keap-1 acts as a sensor of oxidative and electrophilic stresses for Nrf2 with a subcellular localization in the perinuclear cytoplasm [10]. It is a major repressor of Nrf2 through the activity of the Keap-1-Cul3 complex and specifically targets Nrf2, thereby marking Nrf2 protein for rapid degradation by the ubiquitin-proteasome system [15]. Under normal circumstances, Nrf2 is ubiquitinated by the Keap-1-Cul3 complex and then degraded by proteasomes, resulting in a low level of activity. When the E3 ubiquitin ligase Keap1-Cul3 complex is exposed to cellular stressors (electrophiles and reactive oxygen species (ROS)), its activity is halted, preventing Nrf2 ubiquitination and allowing Nrf2 protein stabilization and transcriptional activation. Normally, it suppresses Nrf2 nuclear translocation in response to cellular insults, but in response to cellular injuries, it stimulates Nrf2 nuclear translocation to activate cytoprotective target gene expression. Keap-1 has a longer half-life (12 hrs.) than Nrf2 (20 min), which contributes to its role in the Nrf2 regulatory network (Figure 26.1).

26.4 NRF2 FUNCTIONS AND ITS TARGET GENES

The complex Nrf2-Maf attaches to the target gene's particular promoter. Using gene expression profiling and chromatin immunoprecipitation (ChIP) techniques, more than 200 Nrf2 target genes have been discovered in the last decade [9, 16]. The Nrf2 target ARE-driven genes have been identified as encoding proteins that regulate a variety of biological functions like detoxification, antioxidation, anti-inflammation, reduced nicotinamide adenine dinucleotide phosphate (NADPH) regeneration, and intermediate metabolisms. By inducing these genes, Nrf2 performs several defensive activities that resist diverse stressors. As a result, Nrf2 is critical for the cell's redox equilibrium to be maintained [11].

26.4.1 XENOBIOTICS DETOXIFICATION

Xenobiotics include hazardous substances and medications which are detoxified by various drug-metabolizing enzymes and transporters. There are 3 stages to this detoxification process [17, 18]. The majority of phase-1 enzymes are cytochrome P450 oxidases, which change xenobiotics through reduction, oxidation, and hydrolysis. Glutathione S-transferases are phase-2 conjugating enzymes that catalyze the conjugation of reactive electrophile species with glutathione (GSH), reducing xenobiotics toxicity. Phase-3 transporters remove GSH conjugates and protect them from harmful chemicals. Many drug-metabolizing enzymes, such as GST subunits and NQO-1 are regulated by Nrf2.

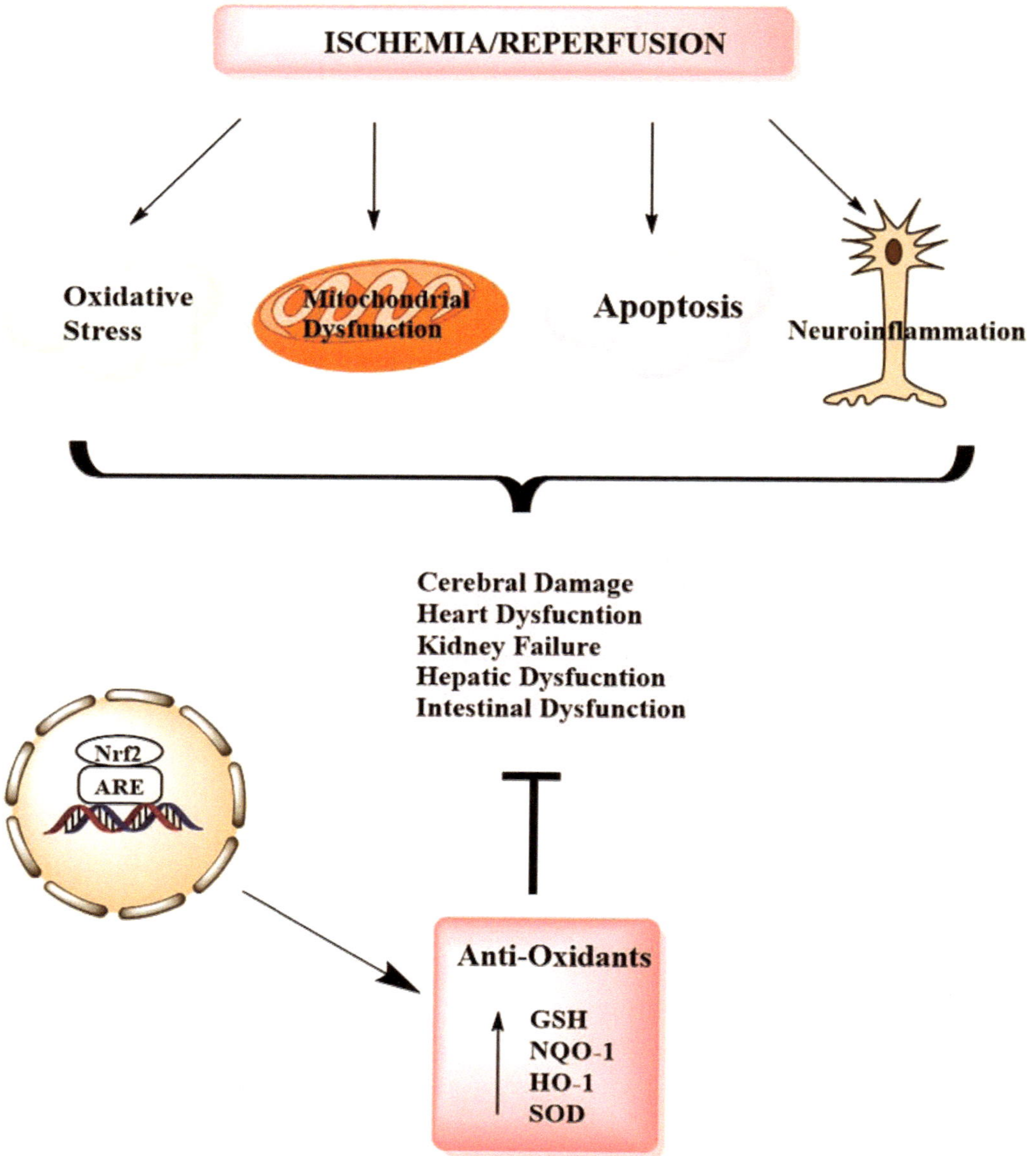

FIGURE 26.1 Pathophysiological role of Nrf-2 in ischemia-reperfusion injury.

26.4.2 ANTIOXIDANT

One of the Nrf2 key functions is to protect the cell from oxidative stress, ensuring redox equilibrium between reactive oxidants and anti-oxidant mechanisms. Anti-oxidant defense and oxidant signaling are modulated by various Nrf2-targeted genes. GSH-based GCL (glutamate-cysteine ligase) is composed of catalytic GCLC and modulatory GCLM (glutamate-cysteine ligase) subunit which controls the entry of cystine into cells. Glutathione peroxidase (GPX-2) produces oxidized glutathione (GSSG) during the reduction of peroxides while glutathione reductase (GSR-1) lowers GSSG to maintain lower intracellular GSH levels. Thioredoxin (TXN-1) and SRXN-1 (sulfiredoxin) are thioredoxin (TXN)-based proteins that decrease oxidized protein thiols. Nrf2 coordinates the regulation of essential anti-oxidant system components, providing precise control of the antioxidant defense at different levels and sufficient response to oxidants in time and space [12, 19].

26.4.3 ANTI-INFLAMMATION

Nrf2 not only protects against xenobiotics and oxidative stress but also plays a function in anti-inflammation [20]. In various mouse models, the lack of Nrf2 causes inflammation to worsen. Nrf2 reduces inflammation by inhibiting the Nf-κB (nuclear factor kappa B) pathway and lowering pro-inflammatory cytokine levels [21]. It was discovered that Nrf2 prevented LPS-induced inflammation by preventing pro-inflammatory cytokine transcriptional upregulation. Nrf2 binds to these genes' promoters, and the Nrf2-mediated suppression is independent of the Nrf2 binding motif or the amount of ROS [22] (Figure 26.2).

26.5 ROLE OF NRF2 IN DIFFERENT TYPES OF ISCHEMIC INJURY

26.5.1 ROLE OF NRF2 IN CEREBRAL ISCHEMIA

One of the main causes of death and long-term disability around the globe is ischemic stroke, although there are few effective treatment therapies. By coordinating the basal and stress-inducible activation of many cytoprotective genes, Nrf2 acts as a main controller in cellular and organismal defense against internal and external stressors. Due to its affinity for Keap-1, Nrf2 is primarily localized in the cytoplasm and maintained at a low basal level. However, during cerebral ischemia, Nrf2 is released from Keap-1, translocated into the nucleus, and bound to the ARE sequence, upregulating the expression of Nrf2 target genes, which code for cell-protective proteins such as anti-oxidative enzymes [11]. During distinct phases of cerebral ischemia, studies have shown strong *in vivo* evidence of dynamic changes in Nrf2 expression, as well as its target genes and cellular distribution of Nrf2. Several studies with brain ischemia models investigated this pathway (Nrf2-ARE) in response to ischemic insults where Nrf2 is beneficial in treating cerebral I/R injury (Table 26.1).

26.5.2 ROLE OF NRF2 IN RENAL I/R INJURY

The interruption and subsequent reestablishment of the renal circulation cause renal I/R damage, in which oxidative stress and inflammation cause tubular shrinkage, microvasculature endothelial monolayer breakdown, and apoptosis [29]. I/R is responsible for half of all acute renal damage hospitalizations. Nrf2-regulated cell defense genes are increased in the kidneys of wildtype but not Nrf2-knockout mice following experimental renal I/R [30]. As a result, Nrf2 plays a critical role in shielding the kidney from the oxidative stress that underpins I/R damage etiology. Increased oxidative stress, mitochondrial malfunction, and inflammation are all symptoms. In this way, Nrf2-based medicines protect against the kidney damage caused by I/R. Several studies with renal ischemia models investigated the Nrf2-ARE pathway in response to ischemic insults where Nrf2 is beneficial in treating renal I/R injury (Table 26.2).

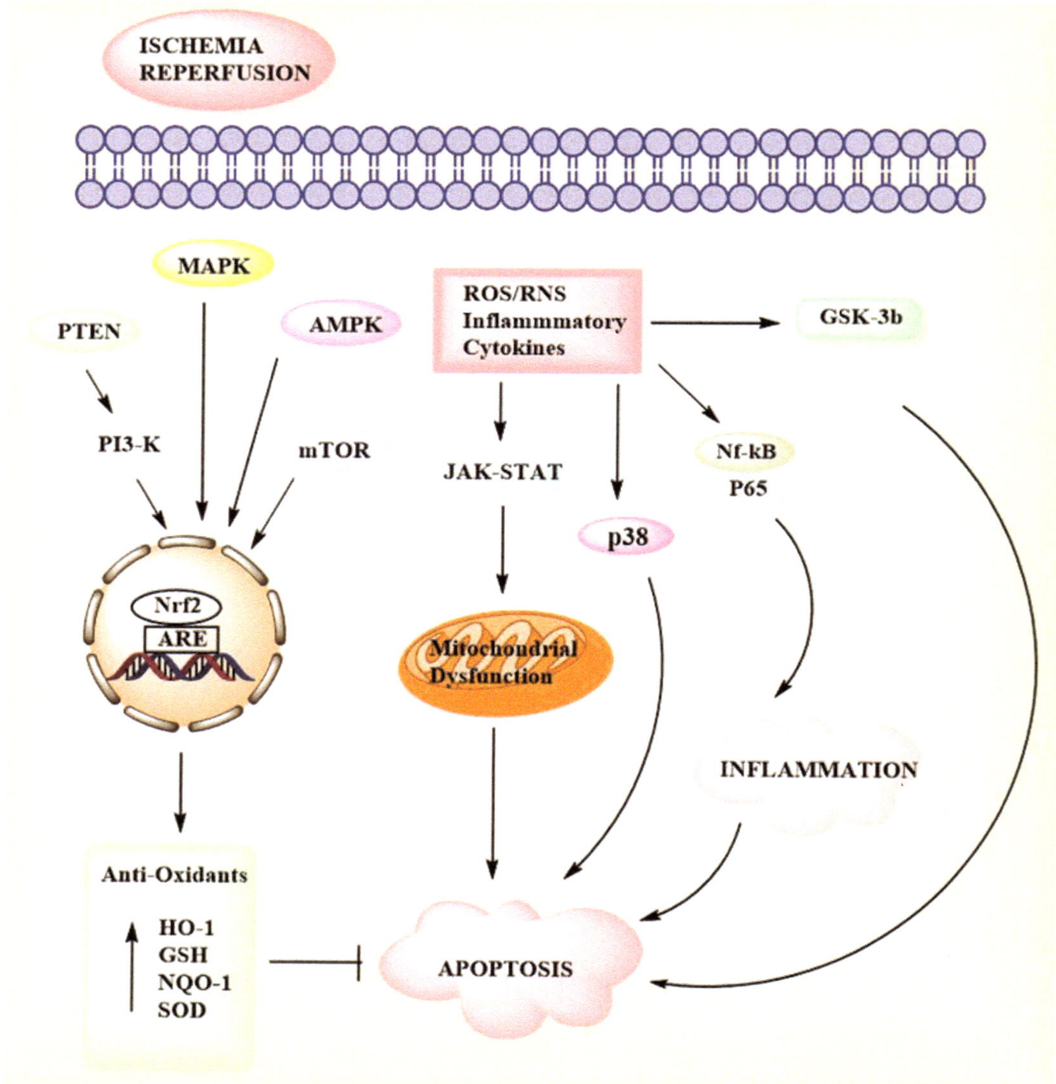

FIGURE 26.2 Molecular pathways involved in upregulation of Nrf2-ARE pathways in the survival of neurons after I/R injury. The various other pathways also get activated after I/R injury which leads to mitochondrial dysfunction, inflammation, and apoptosis.

TABLE 26.1 Beneficial Effects of Nrf2 Inducers in Cerebral I/R Injury

SL. No.	Compound	Beneficial Effects	References
1.	Korean red ginseng	Reduce infarct area via upregulation of HO-1, NQO-1 and SOD-2 protein.	[23]
2.	Trichostatin A	Reduce infarct area and neurological deficits via HO-1 upregulation and Nrf2 nuclear translocation.	[24]
3.	Paeonol	Reduce infarct size, neurological deficits, and brain edema via HO-1 upregulation and nuclear translocation of Nrf2.	[25]
4.	Nobiletin	Reduce infarct size, neurological deficits, and brain edema via HO-1 upregulation and nuclear translocation of Nrf2.	[26]
5.	Docosahexaenoic acid	Reduce infarct size, neurological deficits, and brain edema via HO-1 and Nrf2 upregulation.	[27]
6.	Tetra-methylpyrazine	Reduce infarct size, neurological deficits, and brain edema via HO-1 upregulation and Nrf2 levels.	[28]

TABLE 26.2 Beneficial Effects of Nrf2 Inducers in Renal I/R Injury

SL. No.	Compound	Beneficial Effects	References
1.	Bardoxolone	Decrease inflammation as well as oxidative stress. Enhanced levels of GSH in the kidney, upregulation of NQO-1 and HO-1 expression.	[31]
2.	Resveratrol	Improves renal function. Reduces oxidative stress, cell death, and inflammation.	[32]
3.	Curcumin	Reduce urea and cystatin C plasma levels and renal histological damage, cell death, and ROS.	[33]
4.	RTA-408	Improves renal function and histological damage. Enhances GSH levels and GCL expression.	[34]
5.	Tertbutylhydro-quinone	Decrease oxidative stress and kidney dysfunction and enhanced activation of HO-1.	[35]
6.	Sulforaphane	Decreases histological damage, ROS production, inflammation, and cell death. Enhances HO-1 and NQO-1 expression.	[36]

26.5.3 ROLE OF NRF2 IN INTESTINAL I/R INJURY

The small intestine is the organ most vulnerable to I/R. Ischemia injury to the small intestine is unavoidable in cases of hemorrhagic shock, trauma, and other conditions, and intestinal I/R injury will occur once the ischemic injury is relieved [37]. The pathophysiological process of intestinal I/R damage is complicated, involving oxygen-free radical injury, leukocyte adhesion, the release of inflammatory mediators, energy metabolism disruption, and various other factors. Anti-energy deficiency medications, macrophage activity regulators, oxygen-free radical inhibitors, and other drugs are being employed in the clinical therapy of intestinal I/R injury, but none of these treatments are optimal, and the death rate of intestinal I/R has not decreased. Inflammation plays a critical part in the course of intestinal I/R damage [38]. Recent research has discovered that the transcription factor Nrf2 regulates the HO-1 mediators. The Nrf2 activity is dramatically boosted when cells are exposed to oxidative stress and electrophilic chemical insults and protect the intestine. Several studies with intestinal ischemia models investigated the Nrf2-ARE pathway in response to ischemic insults where Nrf2 is beneficial in treating intestinal I/R injury (Table 26.3).

TABLE 26.3 Beneficial Effects of Nrf2 Inducers in Intestinal I/R Injury

SL. No.	Compound	Beneficial Effects	References
1.	Salvianolic acid	Ameliorate oxidation, inhibit the release of pro-inflammatory cytokines via HO-1 and Nrf2 upregulation.	[39]
2.	Irisin	Enhance the levels of SOD and GSH. Reduce oxidative stress via upregulation of Nrf2.	[40]
3.	Sesamin	Anti-inflammatory and anti-oxidative effects via HO-1 and NQO-1 activation.	[41]
4.	Glutamine	Enhance the levels of SOD and GSH. Reduce oxidative stress via upregulation of Nrf2.	[42]
5.	Lipoxin A4	Inhibit oxidative stress and upregulated Nrf2 nuclear levels.	[43]
6.	Higenamine	Inhibit oxidative stress and upregulated Nrf2 nuclear levels.	[44]

26.5.4 ROLE OF NRF2 IN HEPATIC I/R INJURY

Liver I/R damage can occur in a variety of clinical situations, including liver transplantation and various forms of shock. Due to the formation of ROS and inflammatory cytokines, I/R injury is the leading cause of liver damage and failure [45]. Several studies with hepatic ischemia models investigated the Nrf2-ARE pathway in response to ischemic insults where Nrf2 is beneficial in treating hepatic I/R injury (Table 26.4).

TABLE 26.4 Beneficial Effects of Nrf2 Inducers in Hepatic I/R Injury

SL. No.	Compound	Beneficial Effects	References
1.	Curcumin	Improved histopathological results and upregulation of HO-1 and Nrf2 levels.	[45]
2.	Dimethyl fumarate	Improved histopathological results and upregulation of HO-1 and Nrf2 levels.	[45]
3.	Dexmedetomidine	Decrease apoptosis as well as proinflammatory cytokines and upregulation of HO-1 and Nrf2 expression.	[46]
4.	Cilostazol	Upregulated the expression of HO-1 and Nrf2.	[47]
5.	Tertbutylhydroquinone	Ameliorated tissue damage and upregulated HO-1 and Nrf2 levels.	[48]
6.	Salidroside	Reduce ROS production, prevent apoptosis and upregulation of Nrf2.	[49]

26.5.5 ROLE OF NRF2 IN CARDIAC I/R INJURY

Heart disease has a significant morbidity and mortality rate, and it is still one of the leading causes of death worldwide [50]. The key to preventing myocardial infarction is to restore blood circulation to the myocardium as quickly as feasible [51]. Reperfusion, on the other hand, produces a variety of problems, including the excessive formation of ROS, inflammation, apoptosis, Ca^{2+} overload, and mitochondrial dysfunction [52, 53]. The overproduction of ROS has received a lot of attention, and it's been suggested that lowering ROS levels might significantly reduce myocardial damage [54]. However, there are still just a few viable treatments for I/R injuries. The Nrf2 protein is sensitive to the redox state of the cell and plays an important function in the cell's defense against oxidative stress. The activation of Nrf2 has been linked to a reduction in myocardial ischemia [55]. Several studies with cardiac ischemia models investigated the Nrf2-ARE pathway in response to ischemic insults where Nrf2 is beneficial in treating cardiac I/R injury (Table 26.5).

TABLE 26.5 Beneficial Effects of Nrf2 Inducers in Myocardial I/R Injury

SL. No.	Compound	Beneficial Effects	References
1.	Crocin	Decrease apoptosis via activation of HO-1 levels.	[56]
2.	Luteolin	Ameliorated myocardial viability and upregulated the expression of HO-1, SOD, and GSH.	[57]
3.	*Lycium barbarum*	Repaired cardiomyocyte damage, reduce the levels of pro-inflammatory cytokines, and promoted Nrf2 translocation.	[58]
4.	Ginsenoside	Improved cardiac function, decreased infarct size, and enhanced expression of HO-1 and Nrf2.	[59]
5.	Triptolide	Suppressed pro-inflammatory cytokines and upregulated the levels of SOD, GSH, HO-1, and translocation of Nrf2.	[60]
6.	α-Lipoic acid	Reduces infarct size and prevents cardiac cells via activation of the Nrf2 pathway.	[61]

26.6 CONCLUSION

The Nrf2 network appears to play a critical role in cellular adaptation by modulating a variety of cytoprotective proteins, counteracting numerous internal and external stressors, and giving a prospective optimum beneficial target against a variety of illnesses. It's important for maintaining redox homeostasis and activating and modulating anti-inflammatory, metabolism of drug, antioxidant, detoxification, and radical scavenging capabilities. Indeed, Nrf2 signaling is important for the survival of cells against oxidative stress and the maintenance of appropriate redox balance in cells by increasing anti-oxidative defense that neutralizes ROS and prevents pro-inflammatory gene transcription. There is a growing interest in research into Nrf2's I/R protective effect on maintaining the functional integrity of the blood-brain barrier, avoiding harmful CNS diseases and the development of neuro-inflammatory, as well as the identification of novel approaches targeting Nrf2 to prevent injury.

KEYWORDS

- **antioxidant response element**
- **glutathione**
- **glutamate cysteine ligase**
- **ischemia-reperfusion**
- **neuroinflammation**
- **nuclear factor erythroid 2-related factor 2**
- **oxidative stress**

REFERENCES

1. Benjamin, E. J., Virani, S. S., Callaway, C. W., Chamberlain, A. M., Chang, A. R., Cheng, S., Chiuve, S. E., et al., (2018). Heart disease and stroke statistics—2018 update: A report from the American Heart Association. *Circulation, 137*(12), e67–e492.
2. Khan, H., Kashyap, A., Kaur, A., & Singh, T. G., (2020). Pharmacological postconditioning: A molecular aspect in ischemic injury. *Journal of Pharmacy and Pharmacology, 72*(11), 1513–1527.
3. Khan, H., Gupta, A., Singh, T. G., & Kaur, A., (2021). Mechanistic insight on the role of leukotriene receptors in ischemic–reperfusion injury. *Pharmacological Reports, 73*(5), 1240–1254.
4. Fonarow, G. C., Smith, E. E., Saver, J. L., Reeves, M. J., Bhatt, D. L., Grau-Sepulveda, M. V., Olson, D. M., et al., (2011). Timeliness of tissue-type plasminogen activator therapy in acute ischemic stroke: Patient characteristics, hospital factors, and outcomes associated with door-to-needle times within 60 minutes. *Circulation, 123*(7), 750–758.
5. Emberson, J., Lees, K. R., Lyden, P., Blackwell, L., Albers, G., Bluhmki, E., Brott, T., et al., (2014). Effect of treatment delay, age, and stroke severity on the effects of intravenous thrombolysis with alteplase for acute ischemic stroke: A meta-analysis of individual patient data from randomized trials. *The Lancet, 384*(9958), 1929–1935.
6. Khan, H., Grewal, A. K., & Singh, T. G., (2022). Pharmacological postconditioning by protocatechuic acid attenuates brain injury in ischemia–reperfusion (I/R) mice model: Implications of nuclear factor erythroid-2-related factor pathway. *Neuroscience, 491*, 23–31.
7. Tonelli, C., Chio, I. I. C., & Tuveson, D. A., (2018). Transcriptional regulation by Nrf2. *Antioxidants & Redox Signaling, 29*(17), 1727–1745.
8. Cuadrado, A., Manda, G., Hassan, A., Alcaraz, M. J., Barbas, C., Daiber, A., Ghezzi, P., et al., (2018). Transcription factor NRF2 as a therapeutic target for chronic diseases: A systems medicine approach. *Pharmacological Reviews, 70*(2), 348–383.
9. Kumar, H., Kim, I. S., More, S. V., Kim, B. W., & Choi, D. K., (2014). Natural product-derived pharmacological modulators of Nrf2/ARE pathway for chronic diseases. *Natural Product Reports, 31*(1), 109–139.
10. Yamamoto, M., Kensler, T. W., & Motohashi, H., (2018). The KEAP1-NRF2 system: A thiol-based sensor-effector apparatus for maintaining redox homeostasis. *Physiological Reviews, 98*(3), 1169–1203.
11. Sharma, V., Kaur, A., & Singh, T. G., (2020). Counteracting role of nuclear factor erythroid 2-related factor 2 pathway in Alzheimer's disease. *Biomedicine & Pharmacotherapy, 129*, 110373.
12. Khan, H., Singh, A., Thapa, K., Garg, N., Grewal, A. K., & Singh, T. G., (2021). Therapeutic modulation of the phosphatidylinositol 3-kinases (PI3K) pathway in cerebral ischemic injury. *Brain Research, 1761*, 147399.

13. Wang, Y. C., Lin, S., & Yang, Q. W., (2011). Toll-like receptors in cerebral ischemic inflammatory injury. *Journal of Neuroinflammation*, *8*(1), 1–11.
14. Suzuki, T., & Yamamoto, M., (2015). Molecular basis of the Keap1–Nrf2 system. *Free Radical Biology and Medicine*, *88*, 93–100.
15. Kobayashi, A., Kang, M. I., Okawa, H., Ohtsuji, M., Zenke, Y., Chiba, T., Igarashi, K., & Yamamoto, M., (2004). Oxidative stress sensor Keap1 functions as an adaptor for Cul3-based E3 ligase to regulate proteasomal degradation of Nrf2. *Molecular and Cellular Biology*, *24*(16), 7130–7139.
16. Tiwari, P., Khan, H., Singh, T. G., & Grewal, A. K., (2022). Poly (ADP-ribose) polymerase: An overview of mechanistic approaches and therapeutic opportunities in the management of stroke. *Neurochemical Research*, 1–23.
17. Garg, C., Kaur, A., Singh, T. G., Sharma, V. K., & Singh, S. K., (2022). Therapeutic implications of sonic hedgehog pathway in metabolic disorders: Novel target for effective treatment. *Pharmacological Research*, 106194.
18. Rihal, V., Khan, H., Kaur, A., & Singh, T. G., (2022). Vitamin D as therapeutic modulator in cerebrovascular diseases: A mechanistic perspectives. *Critical Reviews in Food Science and Nutrition,* 1–23.
19. Sharma, V. K., Singh, T. G., & Mehta, V., (2021). Stressed mitochondria: A target to intrude Alzheimer's disease. *Mitochondrion, 59*, 48–57.
20. Kobayashi, E. H., Suzuki, T., Funayama, R., Nagashima, T., Hayashi, M., Sekine, H., Tanaka, N., et al., (2016). Nrf2 suppresses macrophage inflammatory response by blocking proinflammatory cytokine transcription. *Nature Communications*, *7*(1), 1–14.
21. Wardyn, J. D., Ponsford, A. H., & Sanderson, C. M., (2015). Dissecting molecular cross-talk between Nrf2 and NF-κB response pathways. *Biochemical Society Transactions*, *43*(4), 621–626.
22. Ahmed, S. M. U., Luo, L., Namani, A., Wang, X. J., & Tang, X., (2017). Nrf2 signaling pathway: Pivotal roles in inflammation. *Biochimica et Biophysica Acta (BBA)-Molecular Basis of Disease*, *1863*(2), 585–597.
23. Liu, L., Vollmer, M. K., Fernandez, V. M., Dweik, Y., Kim, H., & Doré, S., (2018). Korean red ginseng pretreatment protects against long-term sensorimotor deficits after ischemic stroke likely through Nrf2. *Frontiers in Cellular Neuroscience*, *12*, 74.
24. Wang, B., Cao, W., Biswal, S., & Doré, S., (2011). Carbon monoxide–activated Nrf2 pathway leads to protection against permanent focal cerebral ischemia. *Stroke*, *42*(9), 2605–2610.
25. Zhao, Y., Fu, B., Zhang, X., Zhao, T., Chen, L., Zhang, J., & Wang, X., (2014). Paeonol pretreatment attenuates cerebral ischemic injury via upregulating expression of pAkt, Nrf2, HO-1 and ameliorating BBB permeability in mice. *Brain Research Bulletin*, *109*, 61–67.
26. Zhang, L., Zhang, X., Zhang, C., Bai, X., Zhang, J., Zhao, X., Chen, L., et al., (2016). Nobiletin promotes antioxidant and anti-inflammatory responses and elicits protection against ischemic stroke *in vivo*. *Brain Research*, *1636*, 130–141.
27. Chang, C. Y., Kuan, Y. H., Li, J. R., Chen, W. Y., Ou, Y. C., Pan, H. C., Liao, S. L., et al., (2013). Docosahexaenoic acid reduces cellular inflammatory response following permanent focal cerebral ischemia in rats. *The Journal of Nutritional Biochemistry*, *24*(12), 2127–2137.
28. Kao, T. K., Chang, C. Y., Ou, Y. C., Chen, W. Y., Kuan, Y. H., Pan, H. C., Liao, S. L., et al., (2013). Tetramethylpyrazine reduces cellular inflammatory response following permanent focal cerebral ischemia in rats. *Experimental Neurology*, *247*, 188–201.
29. Kalra, P., Khan, H., Kaur, A., & Singh, T. G., (2022). Mechanistic insight on autophagy modulated molecular pathways in cerebral ischemic injury: From preclinical to clinical perspective. *Neurochemical Research*, 1–19.
30. Leonard, M. O., Kieran, N. E., Howell, K., Burne, M. J., Varadarajan, R., Dhakshinamoorthy, S., Porter, A. G., et al., (2006). Reoxygenation-specific activation of the antioxidant transcription factor Nrf2 mediates cytoprotective gene expression in ischemia-reperfusion injury. *The FASEB Journal*, *20*(14), 2624–2626.
31. Ucar, B. I., Ucar, G., Saha, S., Buttari, B., Profumo, E., & Saso, L., (2021). Pharmacological protection against ischemia-reperfusion injury by regulating the Nrf2-Keap1-ARE signaling pathway. *Antioxidants*, *10*(6), 823.

32. Farkhondeh, T., Folgado, S. L., Pourbagher-Shahri, A. M., Ashrafizadeh, M., & Samarghandian, S., (2020). The therapeutic effect of resveratrol: Focusing on the Nrf2 signaling pathway. *Biomedicine & Pharmacotherapy, 127*, 110234.
33. Asgharpour, M., Tolouian, A. C., Bhaskar, L. V., Tolouian, R., & Massoudi, N., (2020). Herbal antioxidants and renal ischemic-reperfusion injury; an updated review. *Journal of Nephropharmacology, 10*(1).
34. Guerrero-Hue, M., Rayego-Mateos, S., Vázquez-Carballo, C., Palomino-Antolín, A., García-Caballero, C., Opazo-Rios, L., Morgado-Pascual, J. L., et al., (2020). Protective role of Nrf2 in renal disease. *Antioxidants, 10*(1), 39.
35. Zhou, Q., Wang, X., Shao, X., Wang, H., Liu, X., Ke, X., Xiong, C., et al., (2019). Tert-butylhydroquinone treatment alleviates contrast-induced nephropathy in rats by activating the Nrf2/Sirt3/SOD2 signaling pathway. *Oxidative Medicine and Cellular Longevity, 2019*.
36. Yoon, H. Y., Kang, N. I., Lee, H. K., Jang, K. Y., Park, J. W., & Park, B. H., (2008). Sulforaphane protects kidneys against ischemia-reperfusion injury through induction of the Nrf2-dependent phase 2 enzyme. *Biochemical Pharmacology, 75*(11), 2214–2223.
37. Prabhakar, N. K., Khan, H., Grewal, A. K., & Singh, T. G., (2022). Intervention of neuroinflammation in the traumatic brain injury trajectory: *In vivo* and clinical approaches. *International Immunopharmacology, 108*, 108902.
38. Khan, H., Sharma, K., Kumar, A., Kaur, A., & Singh, T. G., (2022). Therapeutic implications of cyclooxygenase (COX) inhibitors in ischemic injury. *Inflammation Research*, 1–16.
39. Zu, G., Zhou, T., Che, N., & Zhang, X., (2018). Salvianolic acid A protects against oxidative stress and apoptosis induced by intestinal ischemia-reperfusion injury through activation of Nrf2/HO-1 pathways. *Cellular Physiology and Biochemistry, 49*(6), 2320–2332.
40. Du, J., Fan, X., Yang, B., Chen, Y., Liu, K. X., & Zhou, J., (2019). Irisin pretreatment ameliorates intestinal ischemia/reperfusion injury in mice through activation of the Nrf2 pathway. *International Immunopharmacology, 73*, 225–235.
41. Wang, Y., Wen, J., Almoiliqy, M., Wang, Y., Liu, Z., Yang, X., Lu, X., et al., (2021). Sesamin protects against and ameliorates rat intestinal ischemia/reperfusion injury with involvement of activating Nrf2/HO-1/NQO1 signaling pathway. *Oxidative Medicine and Cellular Longevity, 2021*.
42. Wang, A. L., Niu, Q., Shi, N., Wang, J., Jia, X. F., Lian, H. F., Liu, Z., & Liu, C. X., (2015). Glutamine ameliorates intestinal ischemia-reperfusion injury in rats by activating the Nrf2/Are signaling pathway. *International Journal of Clinical and Experimental Pathology, 8*(7), 7896.
43. Han, X., Yao, W., Liu, Z., Li, H., Zhang, Z. J., Hei, Z., & Xia, Z., (2016). Lipoxin A4 preconditioning attenuates intestinal ischemia reperfusion injury through Keap1/Nrf2 pathway in a lipoxin A4 receptor independent manner. *Oxidative Medicine and Cellular Longevity, 2016*.
44. Liu, C., Zhu, C., Wang, G., Xu, R., & Zhu, Y., (2015). Higenamine regulates Nrf2-HO-1-Hmgb1 axis and attenuates intestinal ischemia–reperfusion injury in mice. *Inflammation Research, 64*(6), 395–403.
45. Ibrahim, S. G., El-Emam, S. Z., Mohamed, E. A., & Abd, E. M. F., (2020). Dimethyl fumarate and curcumin attenuate hepatic ischemia/reperfusion injury via Nrf2/HO-1 activation and anti-inflammatory properties. *International Immunopharmacology, 80*, 106131.
46. Zhao, Y., Kong, G. Y., Pei, W. M., Zhou, B., Zhang, Q. Q., & Pan, B. B., (2019). Dexmedetomidine alleviates hepatic injury via the inhibition of oxidative stress and activation of the Nrf2/HO-1 signaling pathway. *European Cytokine Network, 30*(3), 88–97.
47. Joe, Y., Zheng, M., Kim, H. J., Uddin, M. J., Kim, S. K., Chen, Y., Park, J., et al., (2015). Cilostazol attenuates murine hepatic ischemia and reperfusion injury via heme oxygenase-dependent activation of mitochondrial biogenesis. *American Journal of Physiology-Gastrointestinal and Liver Physiology*.
48. Zeng, X. P., Li, X. J., Zhang, Q. Y., Liu, Q. W., Li, L., Xiong, Y., He, C. X., et al., (2017). Tert-butylhydroquinone protects liver against ischemia/reperfusion injury in rats through Nrf2-activating anti-oxidative activity. In: *Transplantation Proceedings* (Vol. 49, No. 2, pp. 366–372). Elsevier.
49. Cai, L., Li, Y., Zhang, Q., Sun, H., Yan, X., Hua, T., Zhu, Q., et al., (2017). Salidroside protects rat liver against ischemia/reperfusion injury by regulating the GSK-3β/Nrf2-dependent antioxidant response and mitochondrial permeability transition. *European Journal of Pharmacology, 806*, 32–42.

50. Lu, Y., Hajifathalian, K., Ezzati, M., Woodward, M., Rimm, E. B., Danaei, G., & D'Este, C., (2014). *Metabolic Mediators of the Effects of Body-Mass Index, Overweight, and Obesity on Coronary Heart Disease and Stroke: A Pooled Analysis of 97 Prospective Cohorts with 1.8 Million Participants. 383*(9921), 970–983.
51. Heusch, G., Kleinbongard, P., & Skyschally, A., (2013). Myocardial infarction and coronary microvascular obstruction: An intimate, but complicated relationship. *Basic Research in Cardiology, 108*(6), 1–3.
52. Thapa, K., Khan, H., Singh, T. G., & Kaur, A., (2021). Traumatic brain injury: Mechanistic insight on pathophysiology and potential therapeutic targets. *Journal of Molecular Neuroscience, 71*(9), 1725–1742.
53. Wang, J. X., Zhang, X. J., Li, Q., Wang, K., Wang, Y., Jiao, J. Q., Feng, C., et al., (2015). MicroRNA-103/107 regulate programmed necrosis and myocardial ischemia/reperfusion injury through targeting FADD. *Circulation Research, 117*(4), 352–363.
54. Yang, Y., Duan, W., Jin, Z., Yi, W., Yan, J., Zhang, S., Wang, N., et al., (2013). JAK 2/STAT 3 activation by melatonin attenuates the mitochondrial oxidative damage induced by myocardial ischemia/reperfusion injury. *Journal of Pineal Research, 55*(3), 275–286.
55. Li, W., Wu, M., Tang, L., Pan, Y., Liu, Z., Zeng, C., Wang, J., et al., (2015). Novel curcumin analogue 14p protects against myocardial ischemia reperfusion injury through Nrf2-activating anti-oxidative activity. *Toxicology and Applied Pharmacology, 282*(2), 175–183.
56. Wang, X., Yuan, B., Cheng, B., Liu, Y., Zhang, B., Wang, X., Lin, X., et al., (2019). Crocin alleviates myocardial ischemia/reperfusion-induced endoplasmic reticulum stress via regulation of miR-34a/Sirt1/Nrf2 pathway. *Shock, 51*(1), 123–130.
57. Xiao, C., Xia, M. L., Wang, J., Zhou, X. R., Lou, Y. Y., Tang, L. H., Zhang, F. J., et al., (2019). Luteolin attenuates cardiac ischemia/reperfusion injury in diabetic rats by modulating Nrf2 antioxidative function. *Oxidative Medicine and Cellular Longevity, 2019.*
58. Pan, H., Niu, L., Wu, Y., Chen, L., Zhou, X., & Zhao, Y., (2021). *Lycium barbarum* polysaccharide protects rats and cardiomyocytes against ischemia/reperfusion injury via Nrf2 activation through autophagy inhibition. *Molecular Medicine Reports, 24*(5), 1–14.
59. Zeng, X., Li, J., & Li, Z., (2015). Ginsenoside Rd mitigates myocardial ischemia-reperfusion injury via Nrf2/HO-1 signaling pathway. *International Journal of Clinical and Experimental Medicine, 8*(8), 14497.
60. Yu, H., Shi, L., Zhao, S., Sun, Y., Gao, Y., Sun, Y., & Qi, G., (2016). Triptolide attenuates myocardial ischemia/reperfusion injuries in rats by inducing the activation of Nrf2/HO-1 defense pathway. *Cardiovascular Toxicology, 16*(4), 325–335.
61. Deng, C., Sun, Z., Tong, G., Yi, W., Ma, L., Zhao, B., Cheng, L., et al., (2013). α-Lipoic acid reduces infarct size and preserves cardiac function in rat myocardial ischemia/reperfusion injury through activation of PI3K/Akt/Nrf2 pathway. *PloS One, 8*(3), e58371.

CHAPTER 27

Regulation of PI3K-Akt Signaling Pathway in Ischemia/Reperfusion Injury

KOMAL THAPA,[1] HEENA KHAN,[2] AMARJOT KAUR GREWAL,[2]
THAKUR GURJEET SINGH,[2] and RAHUL DESHMUKH[3]

[1]*Chitkara School of Pharmacy, Chitkara University, Baddi, Himachal Pradesh, India*

[2]*Chitkara College of Pharmacy, Chitkara University, Rajpura, Punjab, India*

[3]*Department of Pharmaceutical Sciences and Technology, Maharaja Ranjit Singh Punjab Technical University, Bathinda, Punjab, India*

ABSTRACT

Stroke, peripheral vascular disease, and myocardial infarction remain among the most common causes of disabling disease and mortality. The early event of ischemia insult is mostly governed by the extent and frequency of an obstruction in the blood supply that leads to tissue injury and/or death, followed by additional damage caused by reperfusion. Induction of cell proliferation, suppression of apoptosis, modulation of tissue inflammation, tumor development, and invasion are all important roles of this PI3K pathway. Several I/R studies have shown that the phosphatidylinositol-3 kinase (PI3K)/protein kinase B (Akt) signaling pathway, which governs a broad variety of cellular functions and provides resistance in response to ischemic injury via activating survival-associated proteins and inactivating apoptosis-associated proteins.

27.1 INTRODUCTION

Stroke, myocardial infarction, and neonatal asphyxia are all common causes of hypoxia-ischemia (H-I). Ischemia can also cause enormous cellular dysfunction and cell death. The drop in cellular oxygen levels during H-I, on the other hand, triggers a slew of compensatory reactions, including neovascularization [2], metabolic control, and the creation of a variety of neurotrophic mediators, which together help to protect neurons against ischemia death. These activities are also an element of an innate adaptive response aimed at defending and assisting tissues in their recovery from ischemia injury [3, 4]. The most significant

Ischemic Injury: A Molecular Insight. Thakur Gurjeet Singh & Heena Khan (Eds.)

mechanism is the quick restoration of blood flow in blocked coronary arteries following H-I. Early repair of an obstructed coronary artery, however, may result in ischemia/reperfusion (I/R) harm [5, 6]. The phosphatidylinositol-3 kinase (PI3K)/Akt signaling pathway is thought to be involved in H-I. In this chapter, we looked at how the PI3K/Akt signaling pathway may play a role in cellular events to H-I resistance.

27.2 PI3K-AKT PATHWAY

PI3K is a multifunctional enzyme that regulates and signals the intracellular cascade [7, 8]. Class I, II, and III are the eight functional isomers of the PI3K family, which are distinguished by their sequence regulation, structural similarity, and substrate selectivity [9, 10]. The PI3K signaling network is involved in several vital physiological functions, including angiogenesis, metabolism, inflammation, cell motility, and cell death [11, 12]. The role of PI3K signaling in the brain, in particular, has a significant impact on aging, spatial memory, cell synaptic plasticity, and the growth of adult neural progenitors [13, 14]. It is well known that PI3K/Akt activation reduces cell death and so confers neuroprotection in cerebral ischemia [15, 16]. The PI3K family has various potential activities throughout the cell. Each protein kinase class plays a role in the regulation of different phosphoinositide pools. PI3Ks play a major role in cell signaling (class I. II) and protein trafficking across membranes (class II, III). Many of the physiological roles of PI3K can be clarified by determining the physiological activities of PtdIns (3,4,5) P3. Cell cycle progression, cell proliferation, cell survival, and apoptosis are all controlled by PtdIns (3,4,5) P3-dependent effector enzyme activation [17]. Metabolic problems are a substantial factor in the occurrence of chronic disease [18]. Insulin resistance is connected to some cardiovascular risk factors, including hypertension, which increases the risk of stroke and mortality from cardiovascular disease. Any alteration in the PI3K signaling pathway is linked to a variety of metabolic illnesses and related repercussions, such as cardiovascular issues, inflammation, and so on. Phosphoinositides are involved in many physiological activities, including cell growth (PI). As a result, enzymes that catalyze both their generation and degradation have become more relevant as biomedical research subjects, as any modification to them has been linked to the spread of pathophysiological illnesses [17]. Phosphatidylinositol (3,4,5)-trisphosphate [PtdIns (3,4,5)P3], phosphatidylinositol (3,4)-bisphosphate [PtdIns (3,4)P2], and phosphatidylinositol 3-phosphate [PtdIns (3) P] are all produced by class I. Class, I PI3Ks are key targets for medications to treat a variety of human disorders and have provided pivotal lessons about the interaction within and between PI3K and other signaling pathways in the search for class I-related therapies. There hasn't been much research done on Class II PI3Ks, thus it's impossible to say what they do. "Vacuolar protein sorting 34 (Vps34)," the solitary PI3K in yeast, has been characterized as a class III PI3K enzyme [18]. Vps34 interacts with Vps15, an adapter protein that controls Vps34's intracellular membrane location and function [19]. The importance of class III phosphatidylinositol 3-kinase (PI3K) in autophagy and human illnesses has inspired basic and applied research into the Vps34 complex's function and control.

27.2.1 PATHOPHYSIOLOGICAL ROLE OF PI3K PATHWAY IN ISCHEMIC INJURY

Certain drugs that act on the PI3K/Akt signaling pathway have been demonstrated to be effective in treating ischemia-induced damage. After treatment with silibinin, a flavonolignan molecule with anti-inflammatory, anti-oxidative, and anti-apoptotic properties, p-Akt 473 and p-Akt 308 upregulation were dramatically enhanced in cerebral ischemia rats [20]. To protect cells from apoptosis, phosphorylated Akt increases the phosphorylation of downstream molecules such as Forkhead box O3 (FoxO3a) transcription factor, mammalian target of rapamycin (mTOR) Bcl-2 apoptosis-related family members, and glycogen synthase kinase-3. In cerebral ischemia rats administered with silibinin, Bcl-2, a neuronal apoptosis inhibitor, is dramatically increased, whereas Bax, a neuronal apoptosis promoter, is significantly downregulated [21]. In developing rat brains, PI3K/Akt/FoxO3a pathway has shown its involvement in neuronal apoptosis according to Li et al. [22]. Activated Akt phosphorylates FoxO3a, resulting in FoxO3a cytoplasmic localization and apoptosis inhibition [23]. Furthermore, bromelain and sodium tanshinone IIA sulfonate protected rat hearts against I/R injury via activating the PI3K/Akt/FoxO3a pathway [24]. mTOR, a member of the phosphoinositide kinase-related kinase family, functions as a Ser/Thr protein kinase in the cytoplasm [25]. The PI3K/Akt signaling pathway has been linked to the regulation of mTOR activity in a prior study [26]. In the PI3K/Akt signaling pathway, fibroblast growth factor-2 is a signaling molecule. By decreasing infarct size and enhancing left ventricular function, fibroblast growth factor-2 inhibits ROS-induced apoptosis and protects the heart against I/R injury via activating PI3K/Akt pathway [27] (Figure 27.1).

27.3 PI3K-AKT SIGNALING PATHWAYS AND DIFFERENT TYPES OF I/R INJURY

27.3.1 PI3K-AKT SIGNALING AND MYOCARDIAL I/R INJURY

PI3K/Akt prevents pathological cardiac hypertrophy while promoting healthy heart development [28]. Akt also phosphorylates Bcl2-associated death promoters, inhibits the activity of caspase-3 [29], upregulates anti-apoptotic B-cell lymphoma protein (Bcl-2) [30], and blocks Bcl-2-like protein 4 (BAX) conformational alterations and mitochondrial membrane incorporation. Importantly, ischemia-reperfusion (I/R) injury causes a reduction in the Bcl-2/BAX ratio, which is linked to myocardial apoptosis and dysfunction [31]. Increased Bcl-2/BAX, on the other hand, implies the initiation of antiapoptotic (or anti-apoptotic) activities, which leads to an anti-infarct action [32]. Activation of the PI3K-Akt pathway has been shown in numerous preclinical investigations to protect against myocardial I/R injury (Table 27.1).

27.3.2 PI3K-AKT SIGNALING AND RENAL I/R INJURY

A major cause of AKI is renal ischemia/reperfusion damage (I/R). It sets off a chain of events in the kidney that leads to renal injury and cell death [42]. The tubular cell damage and endothelial dysfunction damage have all been linked to ATP depletion, intracellular build-up of Ca^{2+}, proinflammatory cytokines, and reactive oxygen species (ROS) as well as activation of apoptotic pathways [43]. Excessive production of reactive oxygen species (ROS) damages and kills renal tubular epithelial cells. The enhanced development of lipid peroxidation and other hazardous products following renal damage supports the role of ROS in the pathogenesis of I/R injury [44]. The PI3K/Akt pathway is important for cell growth and survival, and it has recently been linked to kidney protection against I/R injury by modulating oxidative stress and the inflammatory response [45]. Drugs have been shown to have nephroprotective effects in renal I/R damage models in several investigations (Table 27.2).

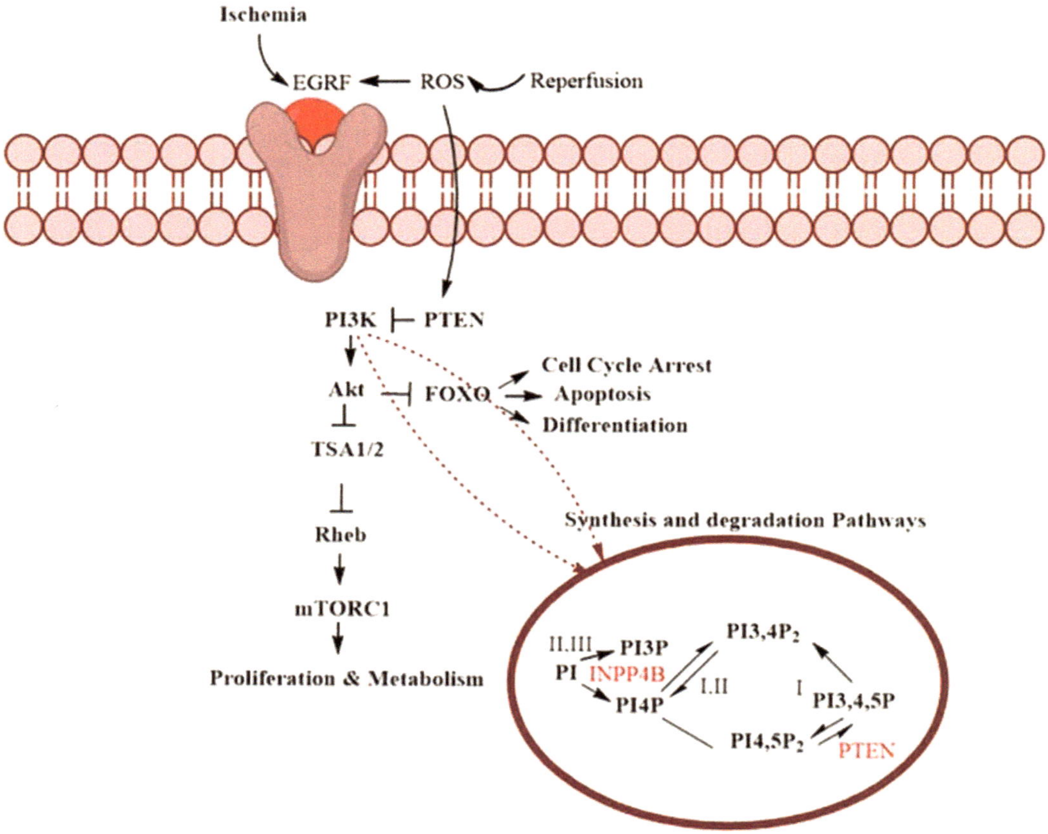

FIGURE 27.1 Properties, synthesis, and degradation pathways of PI3K/Akt pathway.

Abbreviations: INPP4B: Inositol polyphosphate-4-phosphatase, type II; Rheb: Ras homolog enriched in brain; and EGRF: Epidermal growth factor receptor.

27.3.3 PI3K-AKT SIGNALING AND CEREBRAL I/R INJURY

Mitochondrial dysfunction, inflammation, oxidative stress, acidosis, and neuronal apoptosis are all part of the pathogenesis of ischemic stroke [51]. In ischemic stroke, PI3K signaling

TABLE 27.1 Effect of PI3K-Akt Pathway Activation in Myocardial I/R Injury

SL. No.	Drug	Result	Pathway	References
1.	Elabela combined with LY294002	Elabela treatment reduced I/R-induced fibrosis, apoptosis, and mitochondrial dysfunction in cardiac myocytes of rats.	PI3K-Akt pathway	[33]
2.	6-Gingerol (6-G)	6-G significantly inhibited myocardial apoptosis and caspase-3 activation induced by I/R.	PI3K-Akt pathway	[34]
3.	Urolithin A (UA)	UA significantly alleviated myocardial IRI via enhancing antioxidant capacity in cardiomyocytes.	PI3K-Akt pathway	[35]
4.	Troxerutin	Troxerutin preconditioning considerably reduced myocardial infarct size and improved cardiac function in the I/R injury rat model.	PI3K-Akt pathway	[36]
5.	Schisandrin (Sch) B	Sch B decreased the apoptotic index and the serum markers of myocardial infarction.	PI3K-Akt pathway	[37]
6.	Quercetin (Que)	Que postconditioning induced cardioprotection via modulating the expression of Bcl-2 and Bax proteins.	PI3K-Akt pathway	[38]
7.	Epigallocatechin-3-gallate (EGCG)	ECGC offered a cardioprotective effect against myocardial IRI via downregulating the expression of TNF-α, IL-6, and IL-8 in the myocardial tissue.	PI3K-Akt pathway	[39]
8.	Luteolin	Luteolin inhibited apoptosis and improved cardiomyocyte contractile function in simulated ischemia/reperfusion (SI/R).	PI3K-Akt pathway	[40]
9.	Adrenomedullin (AM)	AM significantly attenuated myocardial IRI via exerting antiapoptotic effects.	PI3K-Akt pathway	[41]

plays an important role in the maintenance and control of prosurvival signaling and cell death [52]. Inhibition of apoptosis has a vital role in sparing neurons against damage and death during brain ischemia/reperfusion, indicating therapeutic clinical relevance treatment of brain injury. PI3K signaling inhibition has been found to play a critical function in stroke [53]. As a result, targeted manipulation of the PI3K pathway using activators and inhibitors in pre-clinical and clinical contexts may yield encouraging outcomes (Table 27.3).

TABLE 27.2 Effect of PI3K-Akt Pathway Activation in Renal I/R Injury

SL. No.	Drug	Result	Pathway	References
1.	Nobiletin (NOB)	NOB pre-treatment alleviated renal I/R injury in mice by inhibiting reactive oxygen species production and ERS-induced apoptosis.	PI3K-Akt pathway	[46]
2.	Erythropoietin	Erythropoietin pre-treatment attenuated renal I/R injury in rats by suppressing inflammation.	PI3K-Akt pathway	[47]
3.	Polydatin	PD exhibited prominent nephroprotective effects against renal I/R injury in mice via exerting anti-oxidative and anti-inflammatory effects.	PI3K-Akt pathway	[48]
4.	Complement inhibitor CRIg/FH	CRIg/FH inhibited complement activation at multiple stages in renal IRI-induced rats.	PI3K-Akt pathway	[49]
5.	Oxycodone	Oxycodone provided renoprotective effects against renal IRI injury in rats.	PI3K-Akt pathway	[50]

TABLE 27.3 Effect of PI3K-Akt Pathway Activation in Cerebral I/R Injury

SL. No.	Drug	Result	Pathway	References
1.	Vinpocetine (Vinp)	Vinp reduced cerebral I/R induced oxidative stress, inflammation, and apoptosis induced in brain tissues via Cx43 phosphorylation.	PI3K-Akt pathway	[54]
2.	Aloperine (ALO)	ALO improved cerebral I/R injury and lessened oxidative stress, inflammation, and cell apoptosis.	PI3K-Akt pathway	[55]
3.	Alisol A 24-acetate (24 A)	A24 alleviated GCI/R injury by inhibiting neuroinflammation and apoptosis.	PI3K-Akt pathway	[56]
4.	The TLR9 ligand, CpG-ODN	CpG-ODN significantly reduced cerebral I/R injury via reducing I/R-induced increased levels of Bax and caspase-3.	PI3K-Akt pathway	[57]
5.	Tamibarotene (Am80)	Am80 pre-treatment protected rat's hippocampus against I/R via regulating apoptosis-related protein expressions.	PI3K-Akt pathway	[58]
6.	4-Methylcyclopentadecanone (4-MCPC)	4-MCPC protected against cerebral I/R injury in rats through the inhibition of apoptosis.	PI3K-Akt pathway	[59]
7.	Lupeol	Lupeol showed significant neuroprotective effects against brain I/R injury and neuronal apoptosis in rats.	PI3K-Akt pathway	[60]

27.3.4 PI3K-AKT SIGNALING AND HEPATIC I/R INJURY

Ischemia-reperfusion injury to the liver occurs often following liver surgery, liver transplantation, and different circulatory shock events, and it is one of the leading causes of postoperative hepatic dysfunction, poor disease outcome, and poor patient prognosis [61]. The PI3K/Akt signaling pathway has recently been discovered to be linked to liver ischemia-reperfusion damage. Through downstream associated targeted pathways and proteins, the activated PI3K/Akt signaling system can exert anti-inflammatory, anti-apoptosis, oxidative stress, and autophagy regulating effects reducing liver ischemia-reperfusion injury [62, 63]. As a result, the PI3K/Akt signaling pathway is predicted to become a useful target for the therapeutic prevention and treatment of liver ischemia-reperfusion injury (Table 27.4 and Figure 27.2).

TABLE 27.4 Effect of PI3K-Akt Pathway Activation in Hepatic I/R Injury

SL. No.	Drug	Result	Pathway	References
1.	Carvacrol	Carvacrol alleviated I/R-induced liver injury in rats via exerting anti-oxidative and anti-apoptotic effects.	PI3K-Akt pathway	[64]
2.	β-Arrestin-2	β-Arrestin-2 protected against hepatic IRI via inhibiting apoptosis and autophagy.	PI3K-Akt pathway	[65]
3.	Thymic stromal lymphopoietin (TSLP).	TSLP protected against liver I/R injury in mice via inhibiting autophagy in hepatocytes.	PI3K-Akt pathway	[66]
4.	Shikonin	Shikonin attenuated hepatic I/R injury by inhibiting apoptosis and autophagy.	PI3K-Akt pathway	[67]
5.	Galangin	GA protected against liver I/R injury in rats by inhibiting BRL-mediated cell apoptosis.	PI3K-Akt pathway	[68]
6.	Hesperidin (HDN)	HDN protected against liver I/R injury via ameliorating inflammation, apoptosis, and oxidative stress in liver.	PI3K-Akt pathway	[69]

27.4 CONCLUSION

The PI3K/Akt signaling pathway is essential for the growth and survival of the cell. The PIR/Akt pathway stimulates the phosphorylation of a variety of downstream targets, comprising transcription factors, endothelial nitric oxide (eNOS) synthase, and pro-apoptotic protein which ultimately influences cell survival. By suppressing inflammation, oxidative stress, and apoptosis, the PI3K/Akt pathway protects tissues against I/R injury. The PI3K-Akt pathway appears to play a protective function in ischemia injuries such as myocardial IRI, cerebral IRI, renal IRI, and hepatic IRI, according to new research. It's crucial to create compounds with enough selectivity for the chosen kinase targets. Clinical trials for drugs that may stimulate the PI3K-Akt pathway are progressing, indicating that they have exhibited sufficient safety and risk-benefit margins. Clinical studies will demonstrate the efficacy of addressing this route for therapeutic benefit in the future years.

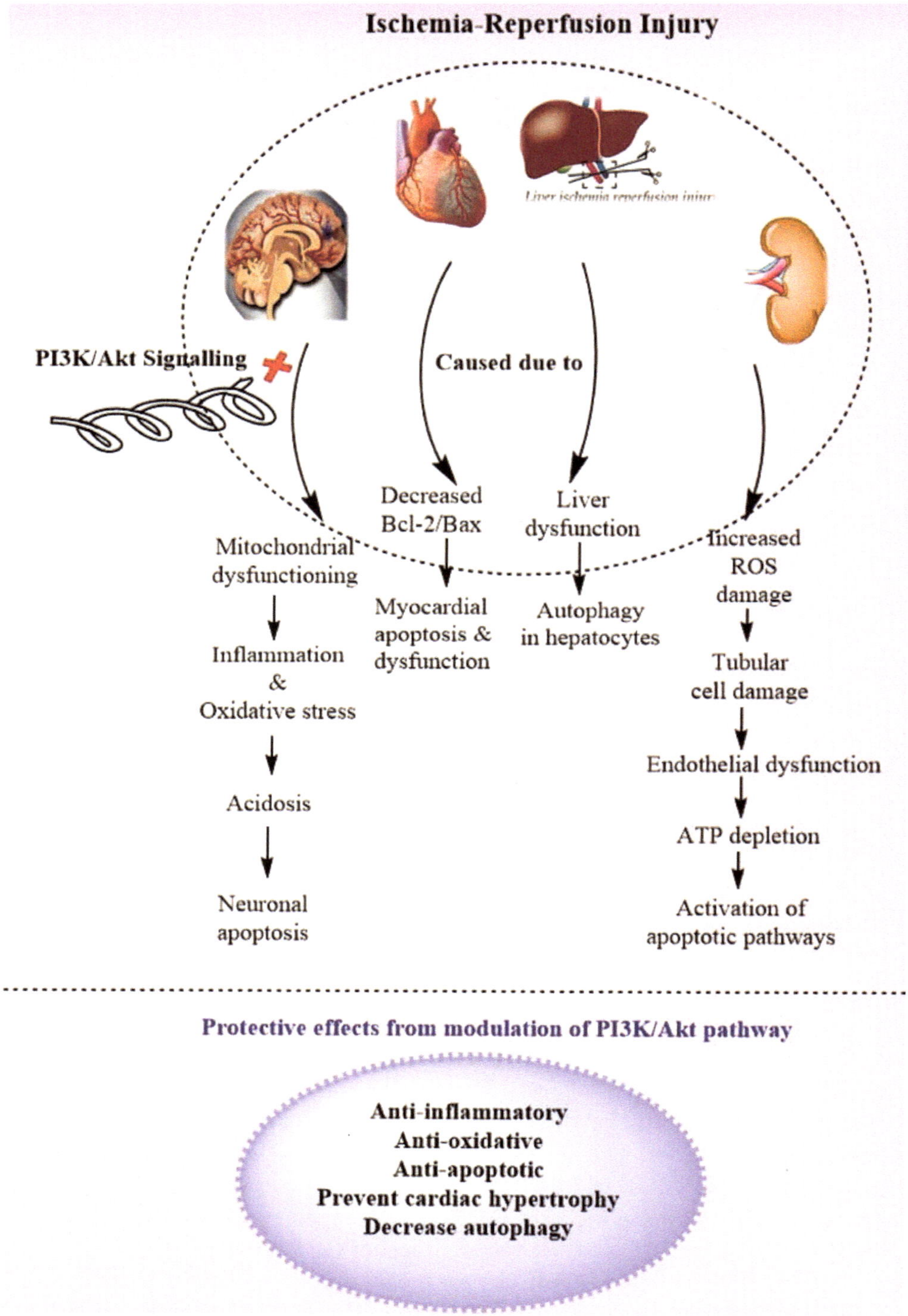

FIGURE 27.2 Pathophysiological illustration of cerebral, myocardial, renal, hepatic I/R injury and how PI3K/Akt pathway attenuate the I/R injury.

27.5 FUTURE PERSPECTIVE

In ischemia damage to numerous organs, the PI3K/Akt pathway and its targets have received a lot of consideration. Novel medicines targeting this pathway are indeed being investigated. In the near future, clinical research on the PI3K pathway might show positive outcomes. As a result, clinical testing of these target-centered medicines is required. Researchers must, however, continue to work on identifying new effective targets and medicines for modulating the PI3K pathway in myocardial IRI, renal IRI, hepatic IRI, and cerebral IRI.

KEYWORDS

- **cerebral IRI**
- **hepatic IRI**
- **ischemic injury**
- **myocardial IRI**
- **PI3K-Akt signaling**
- **renal IRI**
- **signaling pathways**

REFERENCES

1. Zhang, Z., Yao, L., Yang, J., Wang, Z., & Du, G., (2018). PI3K/Akt and HIF-1 signaling pathway in hypoxia-ischemia. *Molecular Medicine Reports*, *18*(4), 3547–3554.
2. Adluri, R. S., Thirunavukkarasu, M., Zhan, L., Akita, Y., Samuel, S. M., Otani, H., Ho, Y. S., et al., (2011). Thioredoxin 1 enhances neovascularization and reduces ventricular remodeling during chronic myocardial infarction: A study using thioredoxin 1 transgenic mice. *Journal of Molecular and Cellular Cardiology*, *50*(1), 239–247.
3. Khan, H., Kashyap, A., Kaur, A., & Singh, T. G., (2020). Pharmacological postconditioning: A molecular aspect in ischemic injury. *Journal of Pharmacy and Pharmacology*, *72*(11), 1513–1527.
4. Tiwari, P., Khan, H., Singh, T. G., & Grewal, A. K., (2022). Poly (ADP-ribose) polymerase: An overview of mechanistic approaches and therapeutic opportunities in the management of stroke. *Neurochemical Research*, 1–23.
5. Gupta, A., Khan, H., Kaur, A., & Singh, T. G., (2021). Novel targets explored in the treatment of alcohol withdrawal syndrome. *CNS & Neurological Disorders-Drug Targets (Formerly Current Drug Targets-CNS & Neurological Disorders)*, *20*(2), 158–173.
6. Mihanfar, A., Nejabati, H. R., Fattahi, A., Latifi, Z., Pezeshkian, M., Afrasiabi, A., Safaie, N., et al., (2019). The role of sphingosine 1 phosphate in coronary artery disease and ischemia reperfusion injury. *Journal of Cellular Physiology*, *234*(3), 2083–2094.
7. Wong, K. K., Engelman, J. A., & Cantley, L. C., (2010). Targeting the PI3K signaling pathway in cancer. *Current Opinion in Genetics & Development*, *20*(1), 87–90.
8. Heras-Sandoval, D., Avila-Muñoz, E., & Arias, C., (2011). The phosphatidylinositol 3-kinase/mTor pathway as a therapeutic target for brain aging and neurodegeneration. *Pharmaceuticals*, *4*(8), 1070–1087.

9. Engelman, J. A., Luo, J., & Cantley, L. C., (2006). The evolution of phosphatidylinositol 3-kinases as regulators of growth and metabolism. *Nature Reviews Genetics*, *7*(8), 606–619.
10. Vanhaesebroeck, B., Guillermet-Guibert, J., Graupera, M., & Bilanges, B., (2010). The emerging mechanisms of isoform-specific PI3K signaling. *Nature Reviews Molecular Cell Biology*, *11*(5), 329–341.
11. Martini, M., De Santis, M. C., Braccini, L., Gulluni, F., & Hirsch, E., (2014). PI3K/AKT signaling pathway and cancer: An updated review. *Annals of Medicine*, *46*(6), 372–383.
12. Sharma, A., Khan, H., Singh, T. G., Grewal, A. K., Najda, A., Kawecka-Radomska, M., Kamel, M., et al., (2021). Pharmacological modulation of ubiquitin-proteasome pathways in oncogenic signaling. *International Journal of Molecular Sciences, 22*(21), 11971.
13. Horwood, J. M., Dufour, F., Laroche, S., & Davis, S., (2006). Signaling mechanisms mediated by the phosphoinositide 3-kinase/Akt cascade in synaptic plasticity and memory in the rat. *European Journal of Neuroscience*, *23*(12), 3375–3384.
14. Khan, H., Garg, N., Singh, T. G., Kaur, A., & Thapa, K., (2022). Calpain inhibitors as potential therapeutic modulators in neurodegenerative diseases. *Neurochemical Research*, 1–25.
15. Abe, E., Fujiki, M., Nagai, Y., Shiqi, K., Kubo, T., Ishii, K., Abe, T., & Kobayashi, H., (2010). The phosphatidylinositol-3 kinase/Akt pathway mediates geranylgeranylacetone-induced neuroprotection against cerebral infarction in rats. *Brain Research*, *1330*, 151–157.
16. Qin-Wei, Z., & Yong-Guang, L. I., (2016). Berberine attenuates myocardial ischemia reperfusion injury by suppressing the activation of PI3K/AKT signaling. *Experimental and Therapeutic Medicine*, *11*(3), 978–984.
17. Jean, S., & Kiger, A. A., (2014). Classes of phosphoinositide 3-kinases at a glance. *Journal of Cell Science*, *127*(5), 923–928.
18. Garg, N., Singh, T. G., Khan, H., Arora, S., Kaur, A., & Mannan, A., (2021). Mechanistic interventions of selected *Ocimum* species in management of diabetes, obesity, and liver disorders: Transformative developments from preclinical to clinical approaches. *Biointerface Res. Appl. Chem.*, *12*(1), 1304–1323.
19. Backer, J. M., (2008). The regulation and function of class III PI3Ks: Novel roles for Vps34. *Biochemical Journal*, *410*(1), 1–17.
20. Xu, X., Chua, C. C., Gao, J., Chua, K. W., Wang, H., Hamdy, R. C., & Chua, B. H., (2008). Neuroprotective effect of humanin on cerebral ischemia/reperfusion injury is mediated by a PI3K/Akt pathway. *Brain Research*, *1227*, 12–18.
21. Khan, H., Grewal, A. K., & Singh, T. G., (2022). Pharmacological postconditioning by protocatechuic acid attenuates brain injury in ischemia–reperfusion (I/R) mice model: Implications of nuclear factor erythroid-2-related factor pathway. *Neuroscience, 491*, 23–31.
22. Li, D., Qu, Y., Mao, M., Zhang, X., Li, J., Ferriero, D., & Mu, D., (2009). Involvement of the PTEN–AKT–FOXO3a pathway in neuronal apoptosis in developing rat brain after hypoxia–ischemia. *Journal of Cerebral Blood Flow & Metabolism*, *29*(12), 1903–1913.
23. Hu, M. C. T., Lee, D. F., Xia, W., Golfman, L. S., Ou-Yang, F., Yang, J. Y., Zou, Y., et al., (2004). IκB kinase promotes tumorigenesis through inhibition of forkhead FOXO3a. *Cell*, *117*(2), 225–237.
24. Zhang, M. Q., Zheng, Y. L., Chen, H., Tu, J. F., Shen, Y., Guo, J. P., Yang, X. H., et al., (2013). Sodium tanshinone IIA sulfonate protects rat myocardium against ischemia-reperfusion injury via activation of PI3K/Akt/FOXO3A/Bim pathway. *Acta Pharmacologica Sinica*, *34*(11), 1386–1396.
25. Hresko, R. C., & Mueckler, M., (2005). mTOR· RICTOR is the Ser473 kinase for Akt/protein kinase B in 3T3-L1 adipocytes. *Journal of Biological Chemistry*, *280*(49), 40406–40416.
26. Han, J. Q., Yu, K. Y., & He, M., (2012). Effects of puerarin on the neurocyte apoptosis and p-Akt (Ser473) expressions in rats with cerebral ischemia/reperfusion injury. *Zhongguo Zhong xi yi jie he za zhi Zhongguo Zhongxiyi Jiehe Zazhi= Chinese Journal of Integrated Traditional and Western Medicine*, *32*(8), 1069–1072.
27. Wang, Z., Zhang, H., Xu, X., Shi, H., Yu, X., Wang, X., Yan, Y., et al., (2012). bFGF inhibits ER stress induced by ischemic oxidative injury via activation of the PI3K/Akt and ERK1/2 pathways. *Toxicology Letters*, *212*(2), 137–146.
28. Walsh, K., (2006). Akt signaling and growth of the heart. *Circulation*, *113*(17), 2032–2034.

29. Hussain, A., Gharanei, A. M., Nagra, A. S., & Maddock, H. L., (2014). Caspase inhibition via A3 adenosine receptors: A new cardioprotective mechanism against myocardial infarction. *Cardiovascular Drugs and Therapy*, *28*(1), 19–32.
30. Pugazhenthi, S., Nesterova, A., Sable, C., Heidenreich, K. A., Boxer, L. M., Heasley, L. E., & Reusch, J. E. B., (2000). Akt/protein kinase B up-regulates Bcl-2 expression through cAMP-response element-binding protein. *Journal of Biological Chemistry*, *275*(15), 10761–10766.
31. Rajtik, T., Carnicka, S., Szobi, A., Mesarosova, L., Matus, M., Svec, P., Ravingerová, T., & Adameová, A., (2012). Pleiotropic effects of simvastatin are associated with mitigation of apoptotic component of cell death upon lethal myocardial reperfusion-induced injury. *Physiological Research*, *61*, S33.
32. Rihal, V., Khan, H., Kaur, A., & Singh, T. G., (2022). Vitamin D as therapeutic modulator in cerebrovascular diseases: A mechanistic perspectives. *Critical Reviews in Food Science and Nutrition*, 1–23.
33. Yu, P., Ma, S., Dai, X., & Cao, F., (2020). Elabela alleviates myocardial ischemia reperfusion-induced apoptosis, fibrosis, and mitochondrial dysfunction through PI3K/AKT signaling. *American Journal of Translational Research*, *12*(8), 4467.
34. Lv, X., Xu, T., Wu, Q., Zhou, Y., Huang, G., Xu, Y., & Zhong, G., (2018). 6-gingerol activates PI3K/Akt and inhibits apoptosis to attenuate myocardial ischemia/reperfusion injury. *Evidence-Based Complementary and Alternative Medicine*, *2018*.
35. Tang, L., Mo, Y., Li, Y., Zhong, Y., He, S., Zhang, Y., Tang, Y., et al., (2017). Urolithin A alleviates myocardial ischemia/reperfusion injury via PI3K/Akt pathway. *Biochemical and Biophysical Research Communications*, *486*(3), 774–780.
36. Shu, L., Zhang, W., Huang, C., Huang, G., & Su, G., (2017). Troxerutin protects against myocardial ischemia/reperfusion injury via Pi3k/Akt pathway in rats. *Cellular Physiology and Biochemistry*, *44*(5), 1939–1948.
37. Zhao, X., Xiang, Y., Cai, C., Zhou, A., Zhu, N., & Zeng, C., (2018). Schisandrin B protects against myocardial ischemia/reperfusion injury via the PI3K/Akt pathway in rats. *Molecular Medicine Reports*, *17*(1), 556–561.
38. Wang, Y., Zhang, Z. Z., Wu, Y., Ke, J. J., He, X. H., & Wang, Y. L., (2013). Quercetin postconditioning attenuates myocardial ischemia/reperfusion injury in rats through the PI3K/Akt pathway. *Brazilian Journal of Medical and Biological Research*, *46*, 861–867.
39. Hu, J., Gu, J., Xu, F., Zhang, H. W., Fan, K. J., Qin, C. Y., Xiao, Z., & Zhang, E. Y., (2016). Epigallocatechin-3-gallate attenuates myocardial reperfusion injury in rats through activation of PI3K/Akt signaling pathway. *Sichuan da xue xue bao. Yi xue ban= Journal of Sichuan University. Medical Science Edition*, *47*(3), 305–309.
40. Fang, F., Li, D., Pan, H., Chen, D., Qi, L., Zhang, R., & Sun, H., (2011). Luteolin inhibits apoptosis and improves cardiomyocyte contractile function through the PI3K/Akt pathway in simulated ischemia/reperfusion. *Pharmacology*, *88*(3, 4), 149–158.
41. Okumura, H., Nagaya, N., Itoh, T., Okano, I., Hino, J., Mori, K., Tsukamoto, Y., et al., (2004). Adrenomedullin infusion attenuates myocardial ischemia/reperfusion injury through the phosphatidylinositol 3-kinase/Akt-dependent pathway. *Circulation*, *109*(2), 242–248.
42. Garg, C., Kaur, A., Singh, T. G., Sharma, V. K., & Singh, S. K., (2022). Therapeutic implications of sonic hedgehog pathway in metabolic disorders: Novel target for effective treatment. *Pharmacological Research*, 106194.
43. Prabhakar, N. K., Khan, H., Grewal, A. K., & Singh, T. G., (2022). Intervention of neuroinflammation in the traumatic brain injury trajectory: *In vivo* and clinical approaches. *International Immunopharmacology*, *108*, 108902.
44. Singh, G., Khan, H., Singh, T. G., Gupta, S., & Kaur, A., (2021). Preclinical and clinical perspectives of *Curculigo orchioides* (Kali musali) as therapeutic modulator. *Plant Cell Biotechnology and Molecular Biology*, 354–372.
45. Xie, L., Zheng, X., Qin, J., Chen, Z., Jin, Y., & Ding, W., (2006). Role of PI3-kinase/Akt signaling pathway in renal function and cell proliferation after renal ischemia/reperfusion injury in mice. *Nephrology*, *11*(3), 207–212.
46. Liu, B., Deng, Q., Zhang, L., & Zhu, W., (2020). Nobiletin alleviates ischemia/reperfusion injury in the kidney by activating PI3K/AKT pathway. *Molecular Medicine Reports*, *22*(6), 4655–4662.

47. Zhang, J., Zou, Y. R., Zhong, X., Deng, H. D. F., Pu, L., Peng, K., & Wang, L., (2015). Erythropoietin pretreatment ameliorates renal ischemia-reperfusion injury by activating PI3K/Akt signaling. *Nephrology, 20*(4), 266–272.
48. Liu, H. B., Meng, Q. H., Huang, C., Wang, J. B., & Liu, X. W., (2015). Nephroprotective effects of polydatin against ischemia/reperfusion injury: A role for the PI3K/Akt signal pathway. *Oxidative Medicine and Cellular Longevity, 2015*.
49. Hu, C., Li, L., Ding, P., Li, L., Ge, X., Zheng, L., Wang, X., et al., (2018). Complement inhibitor CRIg/FH ameliorates renal ischemia reperfusion injury via activation of PI3K/AKT signaling. *The Journal of Immunology, 201*(12), 3717–3730.
50. Wang, Y., Yan, P., Qu, M., & Liu, Y., (2021). Effects of oxycodone on renal ischemia-reperfusion injury in rats through PI3K/Akt signaling pathway. *Panminerva Medica*.
51. Wu, L., Xiong, X., Wu, X., Ye, Y., Jian, Z., Zhi, Z., & Gu, L., (2020). Targeting oxidative stress and inflammation to prevent ischemia-reperfusion injury. *Frontiers in Molecular Neuroscience, 13*, 28.
52. Sharma, V. K., & Singh, T. G., (2020). CREB: A multifaceted target for Alzheimer's disease. *Current Alzheimer Research, 17*(14), 1280–1293.
53. Khan, H., Singh, A., Thapa, K., Garg, N., Grewal, A. K., & Singh, T. G., (2021). Therapeutic modulation of the phosphatidylinositol 3-kinases (PI3K) pathway in cerebral ischemic injury. *Brain Research, 1761*, 147399.
54. Zhao, M., Hou, S., Feng, L., Shen, P., Nan, D., Zhang, Y., Wang, F., et al., (2020). Vinpocetine protects against cerebral ischemia-reperfusion injury by targeting astrocytic connexin43 via the PI3K/AKT signaling pathway. *Frontiers in Neuroscience*, 223.
55. Li, Z., Cao, X., Xiao, L., & Zhou, R., (2021). Aloperine protects against cerebral ischemia/reperfusion injury via activating the PI3K/AKT signaling pathway in rats. *Experimental and Therapeutic Medicine, 22*(4), 1–8.
56. Lu, T., Li, H., Zhou, Y., Wei, W., Ding, L., Zhan, Z., Liu, W., et al., (2022). Neuroprotective effects of alisol A 24-acetate on cerebral ischemia–reperfusion injury are mediated by regulating the PI3K/AKT pathway. *Journal of Neuroinflammation, 19*(1), 1–15.
57. Lu, C., Ha, T., Wang, X., Liu, L., Zhang, X., Kimbrough, E. O., Sha, Z., et al., (2014). The TLR 9 ligand, CpG-ODN, induces protection against cerebral ischemia/reperfusion injury via activation of PI3K/Akt signaling. *Journal of the American Heart Association, 3*(2), p.e000629.
58. Tian, X., An, R., Luo, Y., Li, M., Xu, L., & Dong, Z., (2019). Tamibarotene improves hippocampus injury induced by focal cerebral ischemia-reperfusion via modulating PI3K/Akt pathway in rats. *Journal of Stroke and Cerebrovascular Diseases, 28*(7), 1832–1840.
59. Ma, Y., Li, Y., Zhang, C., Zhou, X., & Wu, Y., (2014). Neuroprotective effect of 4-methylcyclopentadecanone on focal cerebral ischemia/reperfusion injury in rats. *Journal of Pharmacological Sciences*, 14102FP.
60. Wang, Z., Han, Y., Tian, S., Bao, J., Wang, Y., & Jiao, J., (2020). Lupeol alleviates cerebral ischemia–reperfusion injury in correlation with modulation of PI3K/Akt pathway. *Neuropsychiatric Disease and Treatment, 16*, 1381.
61. Wang, M., Zhang, J., & Gong, N., (2021). Role of the PI3K/Akt signaling pathway in liver ischemia reperfusion injury: A narrative review. *Annals of Palliative Medicine*.
62. Xuan, F., & Jian, J., (2016). Epigallocatechin gallate exerts protective effects against myocardial ischemia/reperfusion injury through the PI3K/Akt pathway-mediated inhibition of apoptosis and the restoration of the autophagic flux. *International Journal of Molecular Medicine, 38*(1), 328–336.
63. Khan, H., Sharma, K., Kumar, A., Kaur, A., & Singh, T. G., (2022). Therapeutic implications of cyclooxygenase (COX) inhibitors in ischemic injury. *Inflammation Research*, 1–16.
64. Suo, L., Kang, K., Wang, X., Cao, Y., Zhao, H., Sun, X., Tong, L., & Zhang, F., (2014). Carvacrol alleviates ischemia reperfusion injury by regulating the PI3K-Akt pathway in rats. *PloS One, 9*(8), e104043.
65. Chen, X., Zhang, J., Xia, L., Wang, L., Li, H., Liu, H., Zhou, J., et al., (2021). β-Arrestin-2 attenuates hepatic ischemia-reperfusion injury by activating PI3K/Akt signaling. *Aging (Albany NY), 13*(2), 2251.
66. Li, S., Yi, Z., Deng, M., Scott, M. J., Yang, C., Li, W., Lei, Z., et al., (2019). TSLP protects against liver I/R injury via activation of the PI3K/Akt pathway. *JCI Insight, 4*(22).

67. Liu, T., Zhang, Q., Mo, W., Yu, Q., Xu, S., Li, J., Li, S., et al., (2017). The protective effects of shikonin on hepatic ischemia/reperfusion injury are mediated by the activation of the PI3K/Akt pathway. *Scientific Reports*, *7*(1), 1–13.
68. Li, Y., Tong, L., Zhang, J., Zhang, Y., & Zhang, F., (2018). Galangin alleviates liver ischemia-reperfusion injury in a rat model by mediating the PI3K/AKT pathway. *Cellular Physiology and Biochemistry*, *51*(3), 1354–1363.
69. Li, S., Qin, Q., Luo, D., Pan, W., Wei, Y., Xu, Y., Zhu, J., & Shang, L., (2020). Hesperidin ameliorates liver ischemia/reperfusion injury via activation of the Akt pathway. *Molecular Medicine Reports*, *22*(6), 4519–4530.

CHAPTER 28

COVID-19 and Ischemic Stroke: Advancements and Challenges

RUPINDER KAUR SODHI,[1] RENU ARYA,[1] VINAY KUMAR,[1] JITENDER MADAN,[2] and AROCKIA BABU[1]

[1]*Pharmacology Division, Chandigarh College of Pharmacy, Landran, Mohali, Punjab, India*

[2]*National Institute of Pharmaceutical Education and Research, Hyderabad, Telangana, India*

ABSTRACT

Ischemic stroke has been illustrated to be an intermittent, life-endangering, complexity of COVID-19, affecting approximately 1–3% of admitted patients, and up to 6% of those in the intensive care units. Neuro-radiological abnormalities demonstrate the involvement of large vessels, multiple vascular territory infarction, small vessels, and vertebrobasilar territory occlusion. Studies speculate that cytokine storm-activated hypercoagulable condition, endotheliopathy, cardiac embolism, arterial dissection, and vasculitis-like mechanism may lead to its occurrence. Although the pathogenesis of stroke following a COVID-19 infection is multifarious and incompletely understood, however current evidence indicates crucial involvement of collective effects of virus-associated factors, medical condition of the host, and virus-host interactions. Amongst the hospital-admitted patients with COVID-19 infection, both prophylactic, as well as therapeutic anticoagulants, are capable of improvising hospital mortality rates and prognosis. In this chapter, we summarize the current knowledge on epidemiology, clinical manifestations, and pathophysiology of ischemic stroke in patients with COVID-19. We also aim to discuss the recent advancements and challenges for its diagnosis and rational management.

28.1 STROKE AND COVID-19

Stroke has been indicated in most of the severities of COVID-19 infection, but with elevated propensity in moderate and severe cases. Increasing studies involving SARS-CoV-2 across the globe have illustrated cerebrovascular events in the affected patients. The majority of laboratory tests indicated elevated D-dimer and fibrinogen in most of the cases depicting a

Ischemic Injury: A Molecular Insight. Thakur Gurjeet Singh & Heena Khan (Eds.)

hypercoagulation condition and higher inpatient mortality [1]. The incidences of ischemic stroke in COVID-19 patients varied from as low as 0.1% to 6.9% among hospital-admitted patients, however, the patients having severe infections were five times more afflicted by stroke episodes [2]. Incidences of cerebrovascular events in the context of SARS-CoV-2 infection had a median NIHSS score of 15, mostly with large vessel occlusion (LVO) (79.6%), probably either due to cryptogenic (44.7%) or cardio embolic strokes (21.9%) correspondingly [3]. The radiographic and clinical results of patients infected with SARS-CoV-2 have been disappointing despite judicious treatment with intravenous thrombolytic therapy and post-thrombectomy reperfusion. Stroke severity and mortality in COVID-19 patients may be attributed to LVO, delayed presentation with elevated NIHSS score, and multi-systemic perturbations of COVID-19 [4, 5]. It has been suggested that the typical protocol for the management of such patients is intravenous thrombolysis within 4.5 h of inception of symptoms however administration of recombinant tissue plasminogen activator (tPA) was coupled with enhanced risk of bleeding complications such as severe hemorrhage [6]. Stroke has been illustrated to be a potentially fatal SARS-CoV-2 consequence, affecting 1–3% of hospitalized COVID-19 patients and up to 6% of those within the ICU [7]. It has been demonstrated that men are prone to have severe COVID-19 symptoms requiring hospitalization and most of the patients who suffered an ischemic stroke during COVID-19 were men (62%) with an average age of around 63 years [8]. The majority of the affected individuals suffered from pre-existing adverse health conditions such as hypertension and hyperglycemia [9, 10] suggesting that the virus was a crucial trigger in these patients enhancing the risk of neurovascular problems to about seven times in comparison to influenza [8]. Symptoms of SARS-CoV-19 such as respiratory problems and fever typically occurred 10 days before ischemic stroke [11]. Reports of younger patients with ischemic stroke who lacked established risk factors have also been observed in some cases [12].

28.2 SARS-COV-2: SIGNALING SYSTEM AND IMMUNE RESPONSE

Severe acute respiratory syndrome coronavirus-2 (SARS-CoV-2) is a retrovirus belonging to the family Coronaviridae [1]. The four major structural proteins (N) of corona viruses include surface glycoprotein (S), an envelope protein (E), membrane protein (M), and nucleocapsid protein (N) respectively (Figure 28.1). SARS-CoV-2 has a 79% resemblance with SARS-CoV and 50% resemblance with the Middle East respiratory syndrome corona virus (MERS CoV) respectively [13]. These viruses vary in the length of the spike protein, where the length of spike protein (S protein) is much longer in SARS CoV-2 [13, 14]. The S protein aids in the binding of the virus to the host cell receptor and is hence crucial for virus transmission [13]. Angiotensin-converting enzyme (ACE2) is suggested to be the key receptor for viral transmission [14]. The spike glycoprotein on the SARS-CoV-2 virus that is primed by transmembrane protease serine 2 (TMPRSS2), permits the binding with the ACE2 receptor of the host cell. As the virus attaches to the ACE2 receptor in the pulmonary epithelium, the S protein proteolytic domain causes the fusion of the membrane to enable the virus to infect the host cell [15]. By infecting the ACE2-endothelium, the

SARS-CoV-2 migrates to other organs of the body like the brain [16, 17]. Viral RNA is released into the host cell, and a cascade of virus multiplication ensues. SARS-CoV-2 bypasses the immune system responses by disrupting interferon transcription and allowing the virus to replicate rapidly during the early stages of the infection [15]. Accelerated viral proliferation results in significant pyroptosis, a severely inflammatory lytic form of cellular demise, as well as the production of raises the amount of pro-inflammatory mediators, secreted into the systemic circulation [15, 18, 19]. Increased proinflammatory markers can activate the majority of inflammatory cells in the respiratory tissue, causing cytokine storms and excessive release of inflammatory markers and free radicals which can further damage the respiratory tract, enhance vascular permeability, induce sepsis, septic shock, and organ damage through endothelial dysfunction [15, 18]. Platelet aggregation, intravascular thrombi, and accompanying bleeding in various organs, including the pulmonary and nervous system, indicate endothelial damage [20].

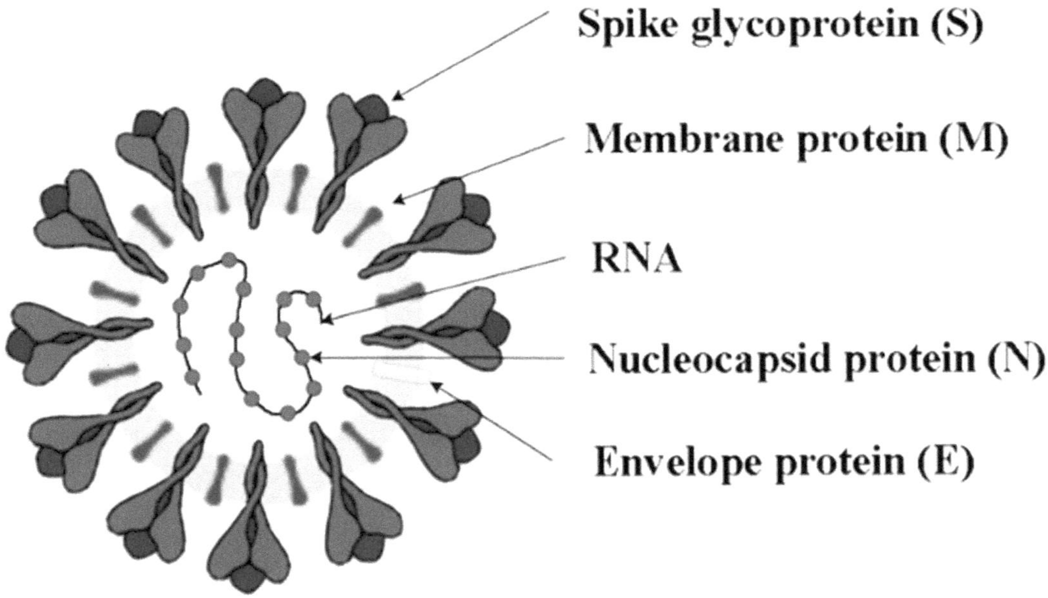

FIGURE 28.1 Structural presentation of severe acute respiratory syndrome coronavirus-2 (SARS-CoV-2).

28.3 ETIOLOGY OF COVID-19 INFLUENCED STROKE

Generally, geriatric patients of COVID-19 with comorbidities suffer from cerebrovascular stroke. In these patients, endothelial dysfunction is caused by thrombotic inflammation and down-regulation of ACE-2 although extrinsic factors significantly influence the intensity and ultimate outcomes of the disease. Most of the COVID-19 patients with ischemic stroke are males, generally falling in the age range of 50 to 80 years [21, 22]. It has also been suggested that higher mortality is seen in aged Individuals suffering from COVID-19 [23]. COVID-19 patients in a hospital setting with cerebrovascular disorder also hold a higher propensity of having cardiovascular risk factors, diabetes, and a higher death incidence [24, 25]. Hence age, gender, and comorbidities all play a role in COVID-19 complications and prognosis and it's plausible to believe that COVID-19 pathophysiology is similarly influenced by these factors. The poor prognosis of aged male patients with SARS-CoV-2 may be attributed to differences in ACE-2 expression on the endothelium [17]. The age-related susceptibility for COVID-19 may also occur as an outcome of steady attenuation in immune responses analogous to influenza and other viral diseases [17]. Aging is also coupled with structural and functional alterations of the vasculature, which eventually causes endothelial dysfunction and smooth muscle cell dysfunction [16]. Race/ethnicity is a significant predictor of severe COVID-19 infection, which is probably credited to overexpression of cardiovascular risk factors in certain ethnicities [26]. The ACE-2 expression in males is almost thrice higher than in females and the protection of sex hormones and X chromosomes may account for the low susceptibility of females to viral diseases [27]. Furthermore, it has been demonstrated that expression of ACE2 considerably increased in the ischemic brain and blood vessels exposed to diabetes, and smoking makes it vulnerable to SARS-CoV-2 infection [28]. In addition to the conventional pulmonary immunological inflammation, ischemic hepatic disorders and aberrant metabolism of iron should also be examined in COVID-19 patients [29, 30]. COVID-19 infection is thought to cause ischemia and enhanced release of inflammatory markers, leading to severe cerebrovascular stroke [31]. This further increases the expression of hypoxia-inducible factor (HIF), which can lead to severe inflammatory response, particularly infiltration of inflammatory cells and release of cytokines [32, 33]. This promotes additional ischemia of tissue rise in blood viscosity and hypercoagulability worsens [34].

28.4 MECHANISMS OF SARS-COV-2 ASSOCIATED STROKE

Several mechanisms have been suggested correlating the occurrence of COVID-19 and stroke. It has been speculated that ACE2 down-regulation, endothelial dysfunction, thromboembolism, hypercoagulation, and occurrence of cardiovascular events lead to the COVID-associated stroke.

28.4.1 RENIN-ANGIOTENSIN SYSTEM (RAS) IMBALANCE

The renin-angiotensin system (RAS) is a pivotal therapeutic option for the management of ischemic stroke since the overactivation of angiotensin-converting enzyme (ACE)/ angiotensin II (Ang II)/angiotensin type 1 (AT1) receptor in RAS is implicated in brain damage after cerebral ischemia [35]. Renin from the kidney converts angiotensinogen to angiotensin I (Ang I). Ang I is further gets transformed into angiotensin II (Ang II) with the help of the Angiotensin converting enzyme (ACE). Ang II can bind to AT1 and AT2 receptors causing vasoconstriction, secretion of aldosterone, inflammation, coagulation, and hypertension. Ang II can worsen cardiac failure and acute respiratory distress syndrome (ARDS) [35, 37]. ACE2, a homolog of ACE is a dipeptidyl carboxydipeptidase and part of the RAS pathway [36, 37]. ACE2 is distributed in the nervous, cardiac, and vascular systems and protects against the damaging actions of Ang II [38], furthermore, ACE2 exhibits antithrombotic actions by inhibiting platelet aggregation and nitric oxide production [35]. ACE2 also catalyzes the conversion of Ang I and to angiotensin 1–7 (ang 1–7) [36]. It has been indicated that Angiotensin (1–7) produces vasodilation and has anti-inflammatory potential by attaching to the Mas receptor to reverse the effects of Ang II [39]. ACE2/Ang (1–7)/Mas axis activation leads to beneficial cardioprotective and neuroprotective actions that counterbalance the deleterious effects of the ACE1/Ang II/AT1 receptor axis [36, 37, 41] (Figure 28.2).

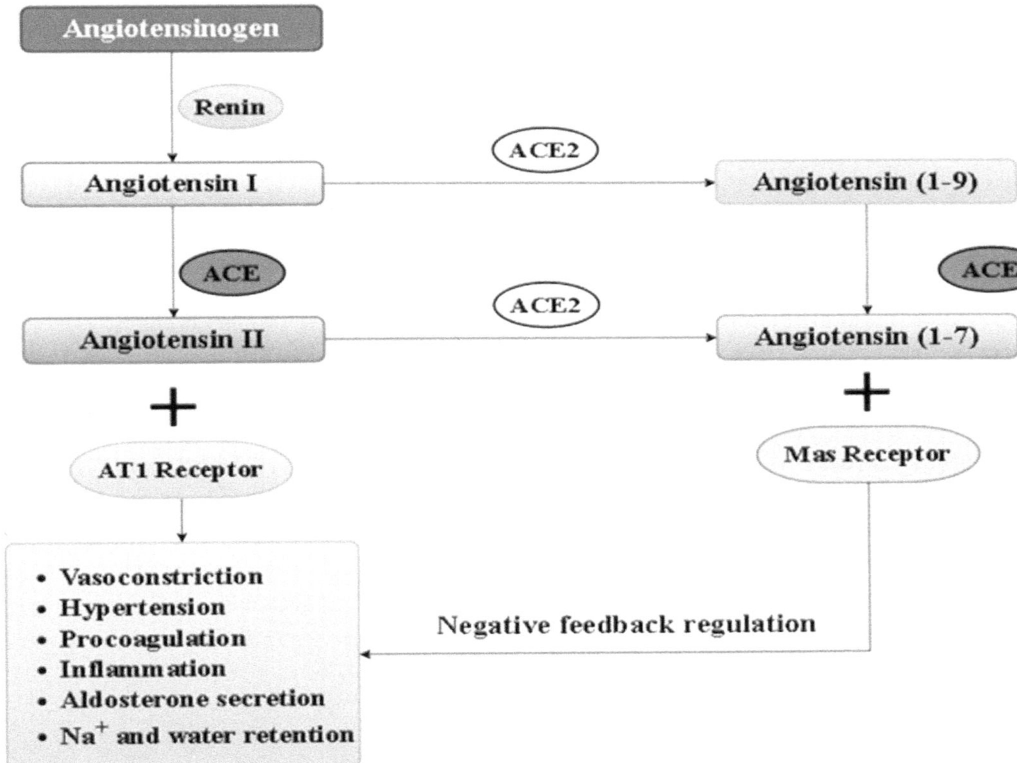

FIGURE 28.2 The RAS pathway.

Hence ACE2 has a critical role in sustaining homeostasis of the cardiovascular and cerebrovascular systems. The SARS-CoV-2 virus is distinguished by its close interaction with ACE2 receptors, which have a major impact on the RAS pathway [27]. It has been suggested that SARS-CoV-2 invades the cell membrane and causes down-regulation of the ACE2 receptor along with a diminution of the catalytic effect of the ACE2 receptor on the outer surface of the membrane [42]. The internalization of ACE2, therefore, occurs followed by its ubiquitination and degradation by Ang II [36, 43, 44]. This SARS-CoV-2-related RAS axis disturbance enacts stroke etiology by promoting inflammatory action, vasoconstriction, and organ injury [39, 45, 46]. The SARS-CoV-2 and ACE2 complex causes down-regulation of ACE2 which elevates the Ang II level in the body. It has been shown that down-regulation of ACE2 and over-activation of tissue factors on endothelial cells together lead to endothelial dysfunction [47]. Cardiovascular hypertrophy, interstitial fibrosis, endothelial dysfunction, increased inflammation, obesity-related high blood pressure, oxidative stress, and enhanced coagulation are all caused by Ang II [42]. In SARS-CoV-2, Ang II also affects acquired immunity by stimulating macrophages and other immune system cells, causing a rise in the production of pro-inflammatory cytokines viz. IL-6, TNF-α, and others [42]. Ang II is capable of enhancing capillary permeability, activating platelets, promote tissue factor transcription and expression on endothelial cells [48]. It has been reported that Ang II can instigate endothelial cells to produce complement system components, indicating the importance of the endothelium in the pathophysiology of thrombosis in COVID-19 patients [49].

28.4.2 ENDOTHELIAL DYSFUNCTION

Endothelial dysfunction is a major part of the thrombogenic condition caused by SARS-CoV-2. This occurs as an outcome of cytotoxic effects on endothelial cells and inflammation of the endothelium, both of which have been observed in COVID-19 patients [50]. Endothelial dysfunction causes endothelial cells to express activated tissue factor (TF), increased thrombin generation, fibrinolytic inhibition, and complement system activation [51]. A combination of thrombotic inflammation, cytokine storm, and down-regulation of ACE2 leads to endothelial dysfunction [39]. Inflammation triggers a coagulation cascade by involving various pathways. The endothelium is at the center of inflammatory action, the coagulative process mediators of inflammation and coagulation factors work in sync with each other to produce the undesired effects. After SARS-CoV-2 manifestation, a severe immune reaction in terms of cytokine storm is observed in the host body turning the endothelium dysfunctional and activating several coagulation pathways, particularly tissue factor pathways promoting thrombosis in microcirculation. Thrombosis in the cerebral circulation can cause a cerebrovascular stroke [39, 52].

28.4.3 THROMBOINFLAMMATION

Thrombotic inflammation is the impairment of the regular antithrombotic and anti-inflammatory activities of the endothelium that causes coagulation disorders, complement system activation, platelet stimulation, and deployment of leukocytes in micro-vessels [53]. The endothelium, a vital regulatory system for thromboinflammation, covers the whole interior cavity of the systemic circulation, from the capillary bed to the cardiac chambers. Microcirculation abnormalities are likely to begin when certain factors induce endothelial dysfunction [53]. Data suggests that COVID-19 infection is coagulative as well as pro-inflammatory since distinct pathways are characterized by high amounts of pro-coagulant markers such as D-dimer and inflammatory markers such as Interleukin-6

(IL-6) are closely connected and significantly increased in such patients [47]. A study demonstrated that COVID-19 patients had high IL-6 levels, demonstrating that IL-6 and fibrinogen have a direct relationship [54]. Furthermore, a retrospective analysis of COVID-19 cases elaborated that coagulation parameters are extensively and closely associated with hepatic function and inflammatory markers, showing that hepatic damage and pro-inflammatory cytokine storm can induce coagulation dysfunction in these patients [55]. Plasma concentrations of soluble fumes-like tyrosine kinase 1 (sFlt-1), which has been shown to increase endothelial dysfunction, were found to be linked to the severity of the pulmonary symptoms, endothelial dysfunction biomarker expression, and the occurrence of organ damage in COVID-19 [56]. Additionally, endothelial disease, as indicated by the elevated level of circulating endothelial cells (CEC) in COVID-19 infection is regarded as an endothelial dysfunction marker [57]. Thus, it is reasonable to consider that thrombotic inflammation contributes to endothelial dysfunction (Figure 28.3).

28.4.4 CYTOKINE STORM IN COVID-19

When endothelial cells are infected by SARS-CoV-2, it results in a significant inflammation that leads to activation of endothelial cells and endothelial dysfunction culminating in a pre-thrombotic state [47]. Mast cells and other immune cells can be activated by SARS-CoV-2 and excessive activation of mast cells causes a rise in inflammatory cytokines and chemokines secretion, aggravating the inflammatory response and increasing the disease severity [58] (Figure 28.3). In COVID-19 cases, IL-6 is a majorly enhanced inflammatory cytokine, released by cells such as monocytes. It causes detrimental endothelial cell activation, triggers massive pro-inflammatory cytokines production, encourages fibroblast and neutrophilic migration to the respiratory epithelium, as well as promotes macrophage activation syndrome (MAS) [47]. Such factors encourage collagen and fibrin deposition, producing injury to the lower pulmonary tissue [47]. The generation of pro-coagulant mediators and injury to the capillary endothelium perturbs the balance of its antithrombotic characteristics resulting in systemic embolism [59].

28.4.5 TISSUE FACTOR COAGULATION PATHWAY

It has been indicated that with the help of exposed tissue factor (TF) and other factors, endothelial dysfunction can aggressively stimulate the coagulation pathway [59]. The tissue factor initiates the coagulation cascade and acts as a cofactor for coagulation factor VIIa, permitting factor Xa to be synthesized [33, 60]. Tissue factors can also act from the nascent complex of TF/FVIIa/FXa, causing alterations in several cells (such as SARS-CoV-2 vegetative cells) via protease-activated receptors (PARs) [60]. Up-regulation of tissue factors has been linked with the activation of FVII (TF-Factor Vila complex) by the extrinsic coagulation cascade, and hence correlated with thrombin production and deposition of fibrin in many organs, including the respiratory system [61]. Vigorous fibrinolysis may cause higher blood plasminogen and plasmin levels [61]. It may be hence summarized that inflammatory mediators activate the tissue factor cascade, which further increases thrombosis, with tissue factor serving as the key link between inflammation and coagulation (Figure 28.3).

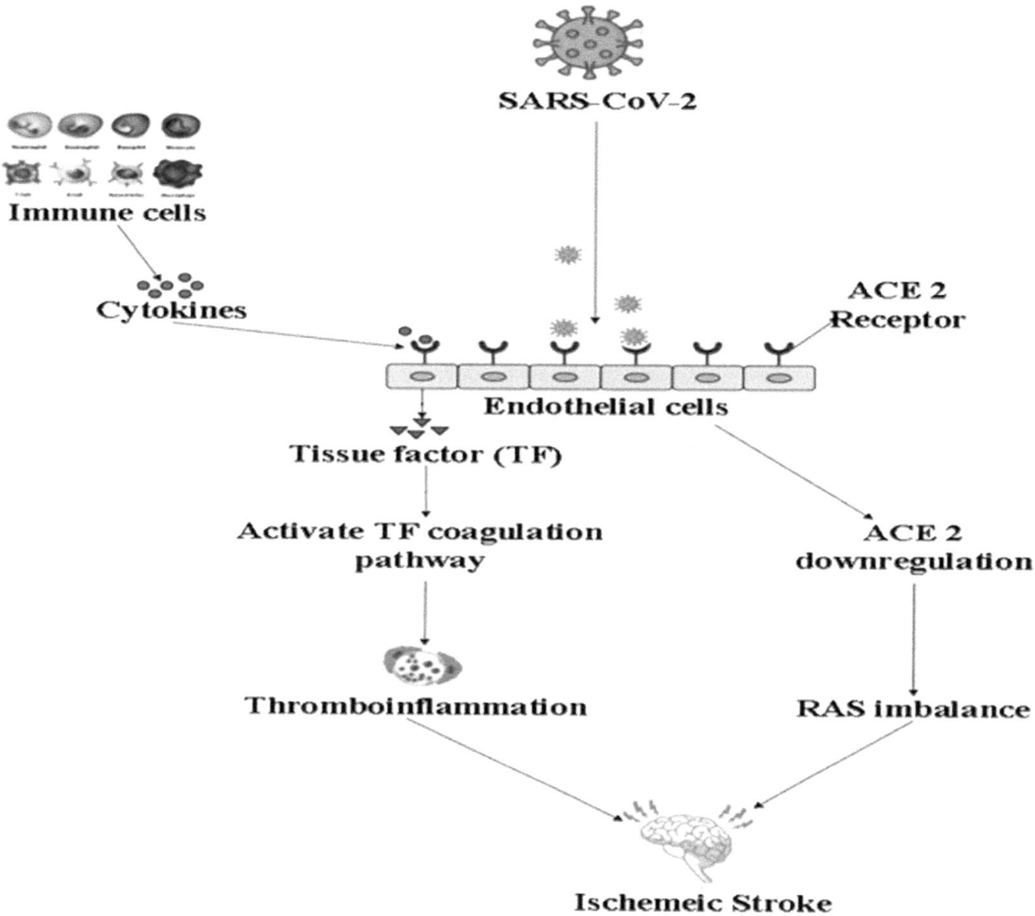

FIGURE 28.3 SARS-CoV-2 and hypercoagulation.

28.4.6 COMPLEMENT PATHWAY ACTIVATION

It has been reported that in the microvasculature of acute COVID-19 patients, substantial accumulation of complement components such as C5b-9, C4d, and mannose-binding lectin (MBL), mannose-associated serine protease-2 have been revealed [62]. In the innate immune system of the host, the complement pathway plays a crucial role to protect against pathogens. After infection, inflammatory response, fibrinolysis, thrombosis, and innate immune response are activated by C5a [63]. Mast cells and neutrophils activated by C5a release pro-inflammatory cytokines. C5a also causes macrophages and endothelial cells activation, resulting in systemic leakage and secretion of enzymes and oxidizing agents [63]. In addition, the von Willebrand factor (vWF) level was also found to be accentuated in patients with COVID-19 [64]. Von Willebrand factor (vWF) is considered an indicator of endothelial cell damage. Endothelial activation, decreased integrity of the vascular system, accumulation of leukocytes, relocation of trans-endothelial, infiltration of tissue, and targeted organ injury are all linked by platelet and vWF accumulation in microvessels [65].

28.4.7 IMMUNE-RESPONSE MEDIATED COAGULOPATHY AND THROMBOSIS

Transmembrane protease serine 2 (TMPRSS2) can process the spike (S) glycoprotein of SARS-CoV-2, to permit its binding with ACE2 and facilitate its passage into the epithelial pulmonary cells [66]. In COVID-19 patients replicated viruses released by dying epithelial cells are detected by Toll-like receptors further causing up-regulation of macrophage, and pro-interleukin (IL)-1 expression [67]. Fragments of the viral and pulmonary cells are recognized as a pathogen and this further promotes inflammasome activation by damage-associated molecular patterns (DAMPs) [67]. A local pulse of proinflammatory cytokines and chemokines by activated inflammasome and macrophages aggravates a persistent rise in interleukin-6 (IL-6) levels. IL-6 regulates the levels of inflammatory mediators by binding to the IL-6r eceptor, sustaining the inflammatory process and the cytokine storm [68]. It has been indicated that hospitalized COVID-19 patients with neurological complications had raised amounts of IL-6, D-dimer, and C-reactive protein [69]. The severely ill COVID-19 patients admitted to intensive care units have elevated levels of inflammatory markers indicating an excessive inflammation process [70]. Circulating viruses also maintain the inflammatory action upon binding to other cells of the body that express ACE2 like endothelial cells [71]. This effect was also seen in COVID-19 patients who were admitted to the ICU, with significantly increased von Willebrand factor, P-selectin, and soluble thrombomodulin levels [72]. Locally activated platelets have been found to promote the release of neutrophil extracellular traps (NETs) coated with tissue factors during this hyper-inflammatory stage, which triggers the extrinsic coagulation cascade, leading to the production of thrombin [73]. Furthermore, histological examination of COVID-19 autopsy of patients elaborated microvascular thrombi including NETs-platelet aggregation in the pulmonary, renal, and cardiac systems whereas a unique pattern of neutrophil-platelet activation predicts severity of disease and systemic hypercoagulability [74]. Histopathological examination also indicates the presence of terminal complement component deposits in the microvascular network, including the C5b-9, C4d, and mannose-binding lectin-associated serine protease 2, indicating a prolonged stimulation of the complement pathway [62]. Furthermore, inflammation of the endothelium has also been related to the accumulation of macrophages, overexpression of the C5a receptor (C5aR1/CD88) across the arteries and in thrombi in COVID-19 cases, implying that complement stimulation is a significant pathophysiological pathway of consistent and diffuse inflammatory action and coagulation among COVID-19 patients [75]. The presence of anti-phospholipid antibodies or lupus anticoagulants in COVID-19 patients has been shown to induce neutrophil extracellular trapsosis further elevating the inflammatory process and thromboembolism [76]. However, whether such antibodies have a role in thrombosis in COVID-19 cases needs to be explored [77].

28.4.8 DAMAGE TO THE NEUROVASCULAR SYSTEM BY SARS-COV-2 VIRUS

It has been demonstrated that SARS-CoV-2 has the potential to harm the neurovascular system. As observed with neurotropic respiratory viruses, systemic inflammation, endothelium inflammation, and microvascular thrombosis may influence BBB permeability and the central nervous system directly [78]. Recent research indicated that olfactory dysfunction is an outcome of direct SARS-CoV-2 infection of the olfactory epithelium and vascular pericytes, whereas olfactory sensory and bulb neurons stay unaffected [79]. Another neuropathological study exhibited serious hypoxic alterations that may be caused by secondary cerebral ischemia and virus transmitted into the brain via the dysregulated blood-brain barrier [80]. SARS-CoV-2 also exists in ischemic areas, implying that central nervous system engagement in COVID-19 could be due to microvascular thrombosis and damage to the neuro-vascular system, culminating in blood-brain barrier disturbances and virus translocation [81]. Apart from a probable direct impact of SARS-CoV-2 in the CNS, which may be co-related to confined cerebral ischemia and inflammation, a secondary impact in individuals with ischemic stroke may not be barred. Excessive release of DAMPs occurs as a result of brain tissue death in stroke, leading to localized and widespread inflammation, which promotes BBB disruption [82, 83]. It has been observed that excessive synthesis of pro-inflammatory acute-response proteins and adhesion molecules in conjunction with circulating activated leukocytes, may lead to a hyper-inflammatory stage in the ischemic brain of these patients. This could be one possibility for the higher ischemic stroke chronicity found in COVID-19 patients [84].

28.4.9 ROLE OF CARDIAC EMBOLISM AND CARDIOPATHY IN COVID-19-RELATED STROKE

It has been suggested that troponin-I levels were reported to be elevated in up to 17% of COVID-19 patients escalating the risk of arrhythmias and mortality [85]. Cardiac injury in COVID-19 can be caused by myocardial infarction (MI), directly by COVID-19 infection, or indirectly by stress and inflammation. It has already been reported that left ventricular impairment, particularly following a MI episode, is the frequent cause of left ventricular thrombi in patients, raising the risk of stroke considerably [86]. Inflammation and increased coagulation in COVID-19, as well as stasis and thrombi development in the weakened left ventricle, are all potential causes of cardiac embolism. Aside from coronary events that can cause myocardial impairment, systemic inflammation or direct SARS-CoV-2 infection can also promote severe cardiac impairment. Numerous investigations have found that COVID-19 patients with cardiac problems like nonspecific myocarditis [87, 88] also suffer from ischemic stroke [89]. SARS-CoV-2-related ACE2 down-regulation in myocardial tissue may provide another possible mechanism of myocardial impairment in COVID-19 patients [90]. COVID-19 cases with increased levels of troponin had a higher incidence of ventricular arrhythmias [91], whereas data suggests that increased inflammation can cause QT prolongation and torsades-de-pointes [92]. Hospital stress-induced atrial fibrillation (AF) can majorly enhance the threat of cerebrovascular stroke or mortality in COVID-19 patients [93].

28.5 CLINICAL PRESENTATION

COVID-19 patients are specifically susceptible to large vessel occlusions like obstruction of the internal carotid artery, middle cerebral artery, and the basilar artery. There is multi-territory participation of otherwise unusually affected vessels [51] comprising the occlusion of the pericallosal artery [9] and the occurrence of multiple focal stenoses in the V4 segment of the vertebral artery [94]. The neurological insufficiency was found to be severe and most of the cases also had systemic thrombosis, venous thrombosis, pulmonary, and spleen embolism [9, 95]. In general, the patients were diagnosed with cryptogenic stroke frequently with an embolic radiologic appearance [8]. Often observed laboratory findings include coagulopathy with elevated fibrinogen and D-dimer, high incidence of hepatic and renal dysfunction, and increased lactate dehydrogenase level [22]. The presence of antiphospholipid antibodies such as lupus anticoagulant, anticardiolipin, and anti-β2-glycoprotein antibodies were also detected in a large number of cases [96]. Bleeding manifestations were seldom observed, distinguishing COVID-19 coagulopathy from that observed in other RNA viruses laden infections such as Ebola and other hemorrhagic fever viruses [96].

28.6 DIAGNOSIS AND MANAGEMENT OF COVID-ASSOCIATED STROKE

The typical features of stroke in patients with COVID-19 require a stringent approach toward diagnosis and management. There should be an urgency to detect the patients with infection and those with concomitant pneumonia using chest computed tomography [97]. Since there are comparatively high occurrences of acute kidney injury (AKI) related to COVID-19, CT angiography should be done in patients with suspected large vessel occlusion. For severely ill ICU patients who are intubated and sedated the detection of clinical signs indicative of stroke can be difficult and frequent sedation interruption should be considered for neurologic evaluation ("wake up test") [9, 97]. Keeping in mind the prothrombotic state, it is recommended to initiate immediate prophylactic anticoagulation with low-molecular-weight heparin, if there are no specific contraindications. On the other hand administration of complete anticoagulation is not suggested in all hospitalized patients with COVID-19 and is particularly essential in patients with pulmonary embolism [96]. In addition, the estimation of stroke burden and proper assessment of hemorrhagic conversion are two important factors to consider before starting anticoagulation.

28.6.1 MANAGEMENT OF ISCHEMIC STROKE IN COVID-19

28.6.1.1 THROMBOPROPHYLAXIS

Treatment options to halt cerebrovascular stroke have remained unchanged since COVID-19 hence modest information is available to certify an optimal treatment regimen for treatment and prevention of stroke in these patients [51]. It is still uncertain if anticoagulants or antiplatelet medications offer superior cerebrovascular stroke protection. Studies have indicated that COVID-19 hospitalized patients suffered from cerebrovascular stroke despite receiving therapeutic anticoagulation [22]. In cases of mild COVID-19, patients with macrothrombosis and cerebrovascular stroke, apixaban in one case and enoxaparin in the other produced complete thrombus resolution on the 10th and 7th day, respectively [98]. Focus has been on the use of anticoagulant agents in managing cerebrovascular stroke in COVID 19 patients who are incompatible with the use of treatments such as thrombectomy or recombinant tissue plasminogen activator (rtPA). Gavioli et al. have suggested treatment with prophylactic versus therapeutic heparin and enoxaparin and also with recombinant tissue plasminogen activator in hospitalized COVID-19 patients [99]. Low molecular weight heparin indicated a better prognosis in comparison to other anticoagulant agents in severe disease or increased D-dimer levels [40, 100]. Treatment of underlying hyperinflammatory conditions with immune modifying agents along with anti-thrombotic and antiviral therapies should also be considered [39, 101, 102]. Table 28.1 represents a summary of present anti-platelets and anti-coagulant agents used, with a proposed mechanism to alleviate COVID-19-related coagulopathy.

28.6.1.2 INTRAVENOUS THROMBOLYSIS

Individuals who have a large-vessel cerebrovascular stroke and meet the intravenous thrombolysis requirements must undergo the surgery performed as per the protocol [95]. Thrombolysis is indicated by present international regulations for certain patients that can be managed within 3 to 4.5 hours after the appearance of symptoms [97]. In COVID-19 patients increased levels of inflammatory and hypercoagulability markers are linked to an increased risk of mortality and post-thrombolytic cerebral bleeding [97]. Treatment with antiplatelet drugs is the primary therapeutic option for ischemic stroke cases that are not suitable for intravenous thrombolysis or endovascular therapy [103, 104]. Before treatment with recombinant tissue plasminogen activator (rtPA), hepatic activity and coagulopathy are assessed to help determine the risk-to-benefit ratio [97].

28.6.1.3 OTHER NECESSARY THERAPEUTIC AND PREVENTIVE MEASURES

The clinical approach for ischemic stroke therapy now includes COVID-19 screening. Patients who are not suspected of having SARS-CoV-2 are treated as normal stroke cases. Patients with suspected SARS-CoV-2 infection who are febrile, coughing, or have flu-like complications should be tested to establish the virus condition without interrupting neuroimaging, other diagnostic procedures, or therapeutic approaches used in stroke

TABLE 28.1 Neutrophil Extracellular Traps (NET), Low Molecular Weight Heparin (LMWH)

Action	Drug	Mechanism of Alleviating COVID-19 Related Coagulopathy
Anti-platelet	• Aspirin • Ticagrelor • Prasugrel • Clopidogrel	• Activation of platelets • Activation neutrophil and NETosis • Cytokine storm
Anti-coagulant	• Heparin • Warfarin • LMWH • Dabigatran • Rivaroxaban • Apixaban • Edoxaban	• Thrombin activity • Production of thrombin • Platelet and neutrophil activation • Thrombin activity • Thrombin production • Platelet and neutrophil activation • Virus spike protein activity

guidelines [103]. When treating patients, physicians must use the necessary protective gear and follow a standard precautionary policy [105, 106]. Because the probability of thromboembolism is significant in severely affected COVID-19 patients, it has been suggested that all hospitalized COVID-19 cases with no contraindications should be administered with a prophylactic dose of low-molecular-weight heparin (LMWH) [107]. Several studies have demonstrated that treatment with low-molecular-weight heparin is correlated with an improved prognosis and lower death rates in COVID-19 cases [108]. To prevent inflammation in SARS-CoV-2, medications must be used at the earliest. Excessive inflammation in COVID-19 infection is the threat to the development of severity in COVID-19 patients rather than the virus itself. Therefore, non-steroidal anti-inflammatory medicines (NSAIDs) should be used to treat early COVID-19 symptoms [109]. Celecoxib or nimesulide is safe and well-tolerated in patients and aspirin may be administered as an alternative treatment to these drugs when toxicity or contraindications are indicated [109]. Though corticosteroids may be utilized to tackle cytokine storms, their use in bronchial virus infections has been linked to an increased fatality rate, a higher risk of subsequent infections, and extended stay in the intensive care unit [110]. Therefore, corticosteroids should not be used for routine therapy of COVID-19 cases [111]. Inhibition of IL-6 is another possible therapeutic option because it plays a crucial role in a cytokine storm. Tocilizumab is a monoclonal antibody targeting the IL-6 receptor and can help COVID-19 patients with their clinical symptoms [110]. Therapy aiming at minimizing platelets and endothelial dysfunction should be discussed as well. Since prostacyclin enhances cAMP concentrations in platelets and endothelial cells, therefore oral medicines that enhance cAMP levels may also be effective COVID-19 therapy options. Phosphodiesterase-3 inhibitors like dipyridamole, which is preventive for stroke and peripheral vascular disease, may also be explored [112]. Nitric oxide and prostacyclin are the two major natural medications that can be advantageous to use [110]. Prostacyclin possesses anti-inflammatory properties and it is an inhibitor of platelet activation, which can prevent thrombosis. Nitric oxide promotes vasodilation, inhibits platelet activation, and reduces white blood cell adherence and inflammatory processes, all of which help to prevent vascular disorders [112]. Anticoagulant agents can be utilized to treat severe large-vessel occlusion in ischemic stroke and severe ischemic stroke even though their efficiency is still unclear [103]. But the administration of anticoagulant drugs for treating cerebrovascular stroke can increase the risk of cerebral hemorrhage [113]. Human recombinant soluble ACE2 (hrsACE2) can be used to provide exogenous ACE2 and thereby prevent SARS-CoV-2 infections [114]. As ACE2 is expressed by both cerebral endothelium and neuronal cells, virus-mediated ACE2 suppression can cause endothelial dysfunction and predispose to or exacerbate ischemic stroke [39]. In comorbid individuals benefits have been observed by treatment with agents such as heparin, low-molecular-weight heparin, dabigatran, or rivaroxaban. However, their effectiveness in reducing impairment and deaths in ischemic stroke has yet to be proven [113]. A preclinical study has focused on statins and found that in addition to their efficacy in reducing cholesterol, they also act as antioxidative, anti-inflammatory, vasodilatory, antithrombotic, and angiogenic agents [115, 149]. Statins can prohibit viral infection from infiltrating cells and reduce the inflammation and activation of the coagulation pathway in COVID-19 patients [116]. These results indicate that statins may facilitate to improvise stroke outcomes, particularly in large vessel occlusion stroke where inflammatory response and thromboembolism are the underlying causes [8, 117]. Sivelestat, a histone inhibitor activated protein C and neutrophil elastase inhibitor and diminished hypercoagulability in ischemic stroke [118]. Tissue plasminogen activator (tPA) used to treat ischemic stroke, is efficacious in patients with COVID 19 as well [119]. Recombinant DNase I can directly degrade neutrophil extracellular traps and has been demonstrated to increase tissue plasminogen activator efficiency *in vitro* [73], it can also minimize hypercoagulability in ischemic stroke [23]. Additional research is necessary to determine the explicit linkage of COVID-19 with ischemic stroke and to establish the efficacy and safety of such treatments. The majority of available evidence indicates that patients with COVID-19 and ischemic stroke have vascular and metabolic conditions that enhance the incidence of ischemic stroke [120]. In-depth studies on the efficacy and overall safety of these drugs for the treatment of ischemic stroke in COVID-19 need to be done [105].

28.7 FUTURE PERSPECTIVES AND CONCLUDING REMARKS

Despite the fact that mounting data has shown numerous pathophysiological pathways of thromboembolism and coagulopathy in these patients, conclusive studies to determine the mechanisms of thrombosis and stroke, as well as potential therapies, are critical [84, 121]. Recent investigations on COVID-19 patients with cerebrovascular stroke do not give long-term outcomes and are defined by retroactive approaches, which vary by country. International prospective multicentric research may provide more accurate data on stroke incidences and functional impairment in individuals with COVID-19 [122, 123]. Healthcare professionals should be alert to stroke symptoms to facilitate timely treatment and avoid further functional loss and mortality [84]. Even though telemedicine has importance in stroke diagnosis and intervention, its unavoidable utilization during the COVID-19 outbreak may result in advancements in stroke assessment at the outpatient and inpatient level [124]. Although telemedicine can never replace the clinical assessment and physical interaction between the patient and the physician, it can bring significant future possibilities to redesign stroke care globally, particularly in the COVID-19 outbreak. Preventive treatment with anticoagulants in COVID-19 patients is might not effective as it increases the risk of bleeding [125]. Comostat mesylate, SARS-CoV-2 specific drug is a potent transmembrane protease serine 2 (TMPRSS2) protease inhibitor that interferes with SARS-CoV-2 invasion into the cell and TRV027, a novel AT-1 selective receptor agonist are being evaluated in human trials. Several experimental stroke models looked at how innate immunity and immunomodulatory drugs affected stroke severity and thrombus development and thrombolysis. Blocking the IL-1 receptor using the canakinumab-surrogate antibody results in superior results in experimental ischemic stroke [126]. Whereas an IL-1 receptor agonist did not affect the functional outcome in cerebrovascular stroke patients compared to a placebo [127]. The hyper-inflammation condition may be partly responsible for the significant re-occlusion incidence of large vessel occlusion strokes and the lower effectiveness of intravenous tissue-type plasminogen activators found in COVID-19 patients. Chloroquine an anti-malarial drug can also be used to prevent thrombosis by suppressing neutrophil extracellular traps and preventing the aggregation of platelet [128]. In numerous diseases, IL 1 and HMGB1 have been shown to trigger the formation of neutrophil extracellular traps. Inhibiting the production of neutrophil extracellular traps by targeting IL 1 or HMGB1 can improve multiple organ functioning [129]. The human recombinant soluble angiotensin-converting enzyme 2 (hrsACE2), is actively being investigated, it binds to viruses competitively, preventing them from invading the cells. The hrsACE2 has been authorized for the management of ARDS [130]. Experiments have indicated that neutrophil extracellular traps enriched thrombi originating from atherosclerotic plaques may be resilient to thrombolytic treatment with DNAase therapy and can provide better outcomes [131]. Multiple immunomodulatory drugs in patients with severe hyperinflammatory conditions, employing either anakinra or canakinumab are helpful [132, 143]. In stroke victims, hyperbaric oxygen treatment has been found to enhance brain functioning [133]. SARS-CoV-2-related respiratory distress has been treated with extracorporeal membrane oxygenation [134]. Nevertheless, considerable investigation is necessary to find out whether oxygen treatment or extracorporeal membrane oxygenation can help patients with ischemic stroke and COVID-19.

KEYWORDS

- **COVID-19**
- **cytokine storm**
- **endothelial dysfunction**
- **intravenous thrombolysis**
- **renin-angiotensin system**
- **signaling system**
- **thromboprophylaxis**
- **tissue factor coagulation pathway**

REFERENCES

1. Su, S., Wong, G., Shi, W., Liu, J., Lai, A. C. K., Zhou, J., Liu, W., Bi, Y., & Gao, G. F., (2016). Epidemiology, genetic recombination, and pathogenesis of coronaviruses. *Trends Microbiol., 24*(6), 490–502.
2. Annie, F., Bates, M. C., Nanjundappa, A., Bhatt, D. L., & Alkhouli, M., (2020). Prevalence and outcomes of acute ischemic stroke among patients ≤50 years of age with laboratory confirmed COVID-19 infection. *Am. J. Cardiol., 130*, 169–170.
3. Nannoni, S., De Groot, R., Bell, S., & Markus, H. S., (2021). Stroke in COVID-19: A systematic review and meta-analysis. *Int. J. Stroke., 16*(2), 137–149.
4. Escalard, S., Maïer, B., Redjem, H., Delvoye, F., Hébert, S., Smajda, S., Ciccio, G., et al., (2020). Treatment of acute ischemic stroke due to large vessel occlusion with COVID-19: Experience from Paris. *Stroke, 51*(8), 2540–2543.
5. Sweid, A., Hammoud, B., Bekelis, K., Missios, S., Tjoumakaris, S. I., Gooch, M. R., Herial, N. A., et al., (2020). Cerebral ischemic and hemorrhagic complications of coronavirus disease 2019. *Int. J. Stroke, 15*(7), 733–742.
6. Sangalli, D., Polonia, V., Colombo, D., Mantero, V., Filizzolo, M., Scaccabarozzi, C., & Salmaggi, A., (2020). A single-center experience of intravenous thrombolysis for stroke in COVID-19 patients. *Neurol. Sci., 41*(9), 2325–2329.
7. Vogrig, A., Gigli, G. L., Bnà, C., & Morassi, M., (2021). Stroke in patients with COVID-19: Clinical and neuroimaging characteristics. *Neurosci. Lett., 743*, 135564.
8. Tan, W. Y. T., Young, B. E., Lye, D. C., Chew, D. E. K., & Dalan, R., (2020). Statin use is associated with lower disease severity in COVID-19 infection. *Sci. Rep., 10*(1), 17458.
9. Morassi, M., Bagatto, D., Cobelli, M., D'Agostini, S., Gigli, G. L., Bnà, C., & Vogrig, A., (2020). Stroke in patients with SARS-CoV-2 infection: Case series. *J. Neurol., 267*(8), 2185–2192.
10. Tan, Y. K., Goh, C., Leow, A. S. T., Tambyah, P. A., Ang, A., Yap, E. S., Tu, T. M., et al., (2020). COVID-19 and ischemic stroke: A systematic review and meta-summary of the literature. *J. Thromb. Thrombolysis, 50*(3), 587–595.
11. Ellul, M. A., Benjamin, L., Singh, B., Lant, S., Michael, B. D., Easton, A., Kneen, R., et al., (2020). Neurological associations of COVID-19. *Lancet Neurol., 19*(9), 767–783.
12. Oxley, T. J., Mocco, J., Majidi, S., Kellner, C. P., Shoirah, H., Singh, I. P., De Leacy, R. A., et al., (2020). Large-vessel stroke as a presenting feature of COVID-19 in the young. *N. Engl. J. Med., 382*(20), e60.
13. Lu, R., Zhao, X., Li, J., Niu, P., Yang, B., Wu, H., et al., (2020). *Genomic Characterization and Epidemiology of 2019 Novel Coronavirus: Implications for Virus Origins and Receptor Binding, 395*, 565–574.
14. Rothan, H. A., & Byrareddy, S. N., (2020). The epidemiology and pathogenesis of coronavirus disease (COVID-19) outbreak. *J. Autoimmun., 109*, 102433.
15. Tay, M. Z., Poh, C. M., Rénia, L., MacAry, P. A., & Ng, L. F. P., (2020). The trinity of COVID-19: Immunity, inflammation, and intervention. *Nat. Rev. Immunol., 20*, 363–374.
16. Amraei, R., & Rahimi, N., (2020). COVID-19, renin-angiotensin system, and endothelial dysfunction. *Cells, 9*(7), 1652.
17. Bourgonje, A. R., Abdulle, A. E., Timens, W., Hillebrands, J. L., Navis, G. J., Gordijn, S. J., Bolling, M. C., et al., (2020). Angiotensin-converting enzyme 2 (ACE2), SARS-CoV-2 and the pathophysiology of coronavirus disease 2019 (COVID-19). *J. Pathol., 251*(3), 228–248.
18. Channappanavar, R., & Perlman, S., (2017). Pathogenic human coronavirus infections: Causes and consequences of cytokine storm and immunopathology. *Semin. Immunopathol., 39*, 529–539.
19. Khan, H., Gupta, A., Singh, T. G., & Kaur, A., (2021). Mechanistic insight on the role of leukotriene receptors in ischemic–reperfusion injury. *Pharmacological Reports, 73*(5), 1240–1254.
20. Xu, Z., Shi, L., Wang, Y., Zhang, J., Huang, L., Zhang, C., et al., (2020). Pathological findings of COVID-19 associated with acute respiratory distress syndrome. *Lancet Respir. Med., 8*, 420–422.
21. Barrios-López, J. M., Rego-García, I., Muñoz, M. C., Romero-Fábrega, J. C., Rivero, R. M., Ruiz, G. J. A., Escamilla-Sevilla, F., et al., (2020). Ischemic stroke and SARS-CoV-2 infection: A causal or incidental association? *Neurologia (English edn.), 35*(5), 295–302.

22. Beyrouti, R., Adams, M. E., Benjamin, L., Cohen, H., Farmer, S. F., Goh, Y. Y., Humphries, F., et al., (2020). Characteristics of ischemic stroke associated with COVID-19. *J. Neurol. Neurosurg Psychiatry, 91*(8), 889–891.
23. Zhou, F., Yu, T., Du, R., Fan, G., Liu, Y., Liu, Z., Xiang, J., et al., (2020). Clinical course and risk factors for mortality of adult in patients with COVID-19 in Wuhan, China: A retrospective cohort study. *Lancet, 395*(10229), 1054–1062.
24. Chen, N., Zhou, M., Dong, X., Qu, J., Gong, F., Han, Y., Qiu, Y., et al., (2020). Epidemiological and clinical characteristics of 99 cases of 2019 novel coronavirus pneumonia in Wuhan, China: A descriptive study. *Lancet, 395*(10223), 507–513.
25. Yan, Y., Yang, Y., Wang, F., Ren, H., Zhang, S., Shi, X., Yu, X., & Dong, K., (2020). Clinical characteristics and outcomes of patients with severe COVID-19 with diabetes. *BMJ Open Diabetes Res. Care, 8*(1), e001343.
26. Rodriguez, F., Solomon, N., De Lemos, J. A., Das, S. R., Morrow, D. A., Bradley, S. M., Elkind, M. S. V., et al., (2021). Racial and ethnic differences in presentation and outcomes for patients hospitalized with COVID-19: Findings from the American Heart Association's COVID-19 cardiovascular disease registry. *Circulation, 143*(24), 2332–2342.
27. Cheng, H., Wang, Y., & Wang, G. Q., (2020). Organ-protective effect of angiotensin-converting enzyme 2 and its effect on the prognosis of COVID-19. *J. Med. Virol., 92*(7), 726–730.
28. Choi, J. Y., et al., (2020). Altered COVID-19 receptor ACE2 expression in a higher risk group for cerebrovascular disease and ischemic stroke. *Biochem. Biophys. Res. Commun., 528*, 413–419.
29. Cavezzi, A., Troiani, E., & Corrao, S., (2020). COVID-19: Hemoglobin, iron, and hypoxia beyond inflammation. A narrative review. *Clin. Pract., 10*(2), 1271.
30. Garg, N., Singh, T. G., Khan, H., Arora, S., Kaur, A., & Mannan, A., (2021). Mechanistic interventions of selected *Ocimum* species in management of diabetes, obesity, and liver disorders: Transformative developments from preclinical to clinical approaches. *Biointerface Res. Appl. Chem., 12*(1), 1304–1323.
31. Zhai, P., Ding, Y., & Li, Y., (2020). The impact of COVID-19 on ischemic stroke. *Diagn. Pathol., 15*(1), 78.
32. Helms, J., Tacquard, C., Severac, F., Leonard-Lorant, I., Ohana, M., Delabranche, X., Merdji, H., et al., (2020). High risk of thrombosis in patients with severe SARS-CoV-2 infection: A multicenter prospective cohort study. *Intensive Care Med., 46*(6), 1089–1098. CRICS TRIGGERSEP Group (clinical research in intensive care and sepsis trial group for global evaluation and research in sepsis).
33. Henry, B. M., Vikse, J., Benoit, S., Favaloro, E. J., & Lippi, G., (2020). Hyperinflammation and derangement of renin-angiotensin-aldosterone system in COVID-19: A novel hypothesis for clinically suspected hypercoagulopathy and microvascular immunothrombosis. *Clin. Chim. Acta, 507*, 167–173.
34. Khan, H., Sharma, K., Kumar, A., Kaur, A., & Singh, T. G., (2022). Therapeutic implications of cyclooxygenase (COX) inhibitors in ischemic injury. *Inflammation Research*, 1–6.
35. Regenhardt, R. W., Desland, F., Mecca, A. P., Pioquinto, D. J., Afzal, A., Mocco, J., & Sumners, C., (2013). Anti-inflammatory effects of angiotensin-(1-7) in ischemic stroke. *Neuropharmacology, 71*, 154–163.
36. Arendse, L. B., Danser, A. H. J., Poglitsch, M., Touyz, R. M., Burnett, J. C. Jr., Llorens-Cortes, C., Ehlers, M. R., & Sturrock, E. D., (2019). Novel therapeutic approaches targeting the renin-angiotensin system and associated peptides in hypertension and heart failure. *Pharmacol. Rev., 71*(4), 539–570.
37. Xu, P., Sriramula, S., & Lazartigues, E., (2011). ACE2/ANG-(1-7)/Mas pathway in the brain: The axis of good. *Am. J. Physiol. RegulIntegr. Comp. Physiol., 300*(4), R804–817.
38. Gan, R., Rosoman, N. P., Henshaw, D. J. E., Noble, E. P., Georgius, P., & Sommerfeld, N., (2020). COVID-19 as a viral functional ACE2 deficiency disorder with ACE2 related multi-organ disease. *Med Hypotheses, 144*, 110024.
39. Hoffmann, M., Hofmann-Winkler, H., Smith, J. C., Krüger, N., Arora, P., Sørensen, L. K., Søgaard, O. S., Hasselstrøm, J. B., Winkler, M., Hempel, T., & Raich, L., (2021). Camostat mesylate inhibits SARS-CoV-2 activation by TMPRSS2-related proteases and its metabolite GBPA exerts antiviral activity. *EBioMedicine*, 65.
40. Tang, N., Bai, H., Chen, X., Gong, J., Li, D., & Sun, Z., (2020). Anticoagulant treatment is associated with decreased mortality in severe coronavirus disease 2019 patients with coagulopathy. *J. Thromb. Haemost., 18*(5), 1094–1099.

41. Saklani, P., Khan, H., Gupta, S., Kaur, A., & Singh, T. G., (2022a). Neuropeptides: Potential neuroprotective agents in ischemic injury. *Life Sciences, 288*, 120186.
42. Verdecchia, P., Cavallini, C., Spanevello, A., & Angeli, F., (2020). The pivotal link between ACE2 deficiency and SARS-CoV-2 infection. *Eur. J. Intern. Med., 76*, 14–20.
43. Labò, N., Ohnuki, H., & Tosato, G., (2020). Vasculopathy and coagulopathy associated with SARS-CoV-2 infection. *Cells, 9*(7), 1583.
44. Sharma, A., Khan, H., Singh, T. G., Grewal, A. K., Najda, A., Kawecka-Radomska, M., Kamel, M., Altyar, A. E., & Abdel-Daim, M. M., (2021). Pharmacological modulation of ubiquitin-proteasome pathways in oncogenic signaling. *International Journal of Molecular Sciences, 22*(21), 11971.
45. Tiwari, P., Khan, H., Singh, T. G., & Grewal, A. K., (2022). Poly (ADP-ribose) polymerase: An overview of mechanistic approaches and therapeutic opportunities in the management of stroke. *Neurochemical Research*, 1–23.
46. Huang, C., Wang, Y., Li, X., et al., (2020). Clinical features of patients infected with 2019 novel coronavirus in Wuhan, China. *Lancet, 395*(10223), 497–506.
47. Tang, X., & Zheng, F., (2022). A review of ischemic stroke in COVID-19: Currently known pathophysiological mechanisms. *Neurol. Sci., 43*(1), 67–79.
48. Sardu, C., Gambardella, J., Morelli, M. B., Wang, X., Marfella, R., & Santulli, G., (2020). Hypertension, thrombosis, kidney failure, and diabetes: Is COVID-19 an endothelial disease? A comprehensive evaluation of clinical and basic evidence. *J. Clin. Med., 9*(5), 1417.
49. Bautista-Vargas, M., Bonilla-Abadía, F., & Cañas, C. A., (2020). Potential role for tissue factor in the pathogenesis of hypercoagulability associated with in COVID-19. *J. Thromb. Thrombolysis, 50*(3), 479–483.
50. Varga, Z., Flammer, A. J., Steiger, P., Haberecker, M., Andermatt, R., Zinkernagel, A. S., Mehra, M. R., et al., (2020). Endothelial cell infection and endotheliitis in COVID-19. *Lancet, 395*(10234), 1417–1418.
51. Zakeri, A., Jadhav, A. P., Sullenger, B. A., & Nimjee, S. M., (2021). Ischemic stroke in COVID-19-positive patients: An overview of SARS-CoV-2 and thrombotic mechanisms for the neurointerventionalist. *J. Neurointerv. Surg., 13*(3), 202–206.
52. Prabhakar, N. K., Khan, H., Grewal, A. K., & Singh, T. G., (2022). Intervention of neuroinflammation in the traumatic brain injury trajectory: *In vivo* and clinical approaches. *International Immunopharmacology, 108*, 108902.
53. Jackson, S. P., Darbousset, R., & Schoenwaelder, S. M., (2019). Thromboinflammation: Challenges of therapeutically targeting coagulation and other host defense mechanisms. *Blood, 133*(9), 906–918.
54. Ranucci, M., Ballotta, A., Di Dedda, U., Baryshnikova, E., Dei, P. M., Resta, M., Falco, M., Albano, G., & Menicanti, L., (2020). The procoagulant pattern of patients with COVID-19 acute respiratory distress syndrome. *J. Thromb. Haemost., 18*(7), 1747–1751.
55. Zhang, Y., He, L., Chen, H., Lu, S., Xiong, Y., Liu, J., Zheng, Y., Wang, S., & Liu, L., (2020). Manifestations of blood coagulation and its relation to clinical outcomes in severe COVID-19 patients: Retrospective analysis. *Int. J. Lab. Hematol., 42*(6), 766–772.
56. Dupont, V., Kanagaratnam, L., Goury, A., Poitevin, G., Bard, M., Julien, G., Bonnivard, M., et al., (2021). Excess soluble fms-like tyrosine kinase 1 correlates with endothelial dysfunction and organ failure in critically ill coronavirus disease 2019 patients. *Clin Infect Dis., 72*(10), 1834–1837.
57. Khider, L., Gendron, N., Goudot, G., Chocron, R., Hauw-Berlemont, C., Cheng, C., Rivet, N., et al., (2020). Curative anticoagulation prevents endothelial lesion in COVID-19 patients. *J. Thromb. Haemost., 18*(9), 2391–2399.
58. Kempuraj, D., Selvakumar, G. P., Ahmed, M. E., Raikwar, S. P., Thangavel, R., Khan, A., Zaheer, S. A., et al., (2020). COVID-19, mast cells, cytokine storm, psychological stress, and neuroinflammation. *Neuroscientist, 26*(5, 6), 402–414.
59. Whittaker, A., Anson, M., & Harky, A., (2020). Neurological manifestations of COVID-19: A systematic review and current update. *Acta Neurol. Scand., 142*(1), 14–22.
60. Jung, F., Krüger-Genge, A., Franke, R. P., Hufert, F., & Küpper, J. H., (2020). COVID-19 and the endothelium. *Clin. Hemorheol. Microcirc., 75*(1), 7–11.
61. Pryzdial, E. L. G., Sutherland, M. R., Lin, B. H., & Horwitz, M., (2020). Antiviral anticoagulation. *Res. Pract. Thromb. Haemost., 4*(5), 774–788.

62. Kasinathan, G., & Sathar, J., (2020). Haematological manifestations, mechanisms of thrombosis and anti-coagulation in COVID-19 disease: A review. *Ann. Med. Surg. (Lond), 56*, 173–177.
63. Magro, C., Mulvey, J. J., Berlin, D., Nuovo, G., Salvatore, S., Harp, J., Baxter-Stoltzfus, A., & Laurence, J., (2020). Complement associated microvascular injury and thrombosis in the pathogenesis of severe COVID-19 infection: A report of five cases. *Transl. Res., 220*, 1–13.
64. Chauhan, A. J., Wiffen, L. J., & Brown, T. P., (2020). COVID-19: A collision of complement, coagulation, and inflammatory pathways. *J. Thromb. Haemost., 18*(9), 2110–2117.
65. Panigada, M., Bottino, N., Tagliabue, P., Grasselli, G., Novembrino, C., Chantarangkul, V., Pesenti, A., Peyvandi, F., & Tripodi, A., (2020). Hypercoagulability of COVID-19 patients in intensive care unit: A report of thromboelastography findings and other parameters of hemostasis. *J. Thromb. Haemost., 18*(7), 1738–1742.
66. Becker, R. C., (2020). COVID-19 update: COVID-19-associated coagulopathy. *J. Thromb. Thrombolysis, 50*(1), 54–67.
67. Hoffmann, M., Kleine-Weber, H., Schroeder, S., Krüger, N., Herrler, T., Erichsen, S., Schiergens, T. S., et al., (2020). SARS-CoV-2 cell entry depends on ACE2 and TMPRSS2 and is blocked by a clinically proven protease inhibitor. *Cell, 181*(2), 271–280.e8.
68. Bauernfeind, F. G., Horvath, G., Stutz, A., Alnemri, E. S., MacDonald, K., Speert, D., Fernandes-Alnemri, T., et al., (2009). Cutting edge: NF-kappaB activating pattern recognition and cytokine receptors license NLRP3 inflammasome activation by regulating NLRP3 expression. *J. Immunol., 183*(2), 787–791.
69. Panigrahy, D., Gilligan, M. M., Huang, S., Gartung, A., Cortés-Puch, I., Sime, P. J., Phipps, R. P., et al., (2020). Inflammation resolution: A dual-pronged approach to averting cytokine storms in COVID-19. *Cancer Metastasis Rev., 39*(2), 337–340.
70. Frontera, J. A., Sabadia, S., Lalchan, R., et al., (2021). A prospective study of neurologic disorders in hospitalized patients with COVID-19 in New York City. *Neurology, 96*(4), e575–e586.
71. Huang, S., Wang, J., Liu, F., Liu, J., Cao, G., Yang, C., Liu, W., Tu, C., Zhu, M., & Xiong, B., (2020). COVID-19 patients with hypertension have more severe disease: A multicenter retrospective observational study. *Hypertens. Res., 43*(8), 824–831.
72. Hamming, I., Timens, W., Bulthuis, M. L., Lely, A. T., Navis, G., & Van, G. H., (2004). Tissue distribution of ACE2 protein, the functional receptor for SARS coronavirus. A first step in understanding SARS pathogenesis. *J. Pathol., 203*(2), 631–637.
73. Goshua, G., Pine, A. B., Meizlish, M. L., Chang, C. H., Zhang, H., Bahel, P., Baluha, A., et al., (2020). Endotheliopathy in COVID-19-associated coagulopathy: Evidence from a single-center, cross-sectional study. *Lancet Haematol., 7*(8), e575–e582.
74. Skendros, P., Mitsios, A., Chrysanthopoulou, A., Mastellos, D. C., Metallidis, S., Rafailidis, P., Ntinopoulou, M., et al., (2020). Complement and tissue factor-enriched neutrophil extracellular traps are key drivers in COVID-19 immunothrombosis. *J. Clin. Invest., 130*(11), 6151–6157.
75. Nicolai, L., Leunig, A., Brambs, S., Kaiser, R., Weinberger, T., Weigand, M., Muenchhoff, M., et al., (2020). Immunothrombotic dysregulation in COVID-19 pneumonia is associated with respiratory failure and coagulopathy. *Circulation, 142*(12), 1176–1189.
76. Carvelli, J., Demaria, O., Vély, F., Batista, L., Chouaki, B. N., Fares, J., Carpentier, S., et al., (2020). Association of COVID-19 inflammation with activation of the C5a-C5aR1 axis. *Nature, 588*(7836), 146–150.
77. Zuo, Y., Estes, S. K., Ali, R. A., Gandhi, A. A., Yalavarthi, S., Shi, H., Sule, G., et al., (2020). Prothrombotic autoantibodies in serum from patients hospitalized with COVID-19. Sci. *Transl. Med., 12*(570), eabd3876.
78. Bowles, L., Platton, S., Yartey, N., Dave, M., Lee, K., Hart, D. P., MacDonald, V., et al., (2020). Lupus anticoagulant and abnormal coagulation tests in patients with COVID-19. *N. Engl. J. Med., 383*(3), 288–290.
79. Bohmwald, K., Gálvez, N. M. S., Ríos, M., & Kalergis, A. M., (2018). Neurologic alterations due to respiratory virus infections. *Front. Cell Neurosci., 12*, 386.
80. Brann, D. H., Tsukahara, T., Weinreb, C., Lipovsek, M., Van, D. B. K., Gong, B., Chance, R., et al., (2020). Non-neuronal expression of SARS-CoV-2 entry genes in the olfactory system suggests mechanisms underlying COVID-19-associated anosmia. *Sci. Adv., 6*(31), eabc5801.

81. Solomon, I. H., Normandin, E., Bhattacharyya, S., Mukerji, S. S., Keller, K., Ali, A. S., Adams, G., et al., (2020). Neuropathological features of COVID-19. *N. Engl. J. Med., 383*(10), 989–992.
82. Song, E., Zhang, C., Israelow, B., Lu-Culligan, A., Prado, A. V., Skriabine, S., Lu, P., et al., (2021). Neuroinvasion of SARS-CoV-2 in human and mouse brain. *J. Exp. Med., 218*(3), e20202135.
83. Kaur, A., Singh, T. G., Khan, H., Kumar, M., Singh, N., & Abdel-Daim, M. M., (2022). Neuroprotective effect of piclamilast-induced post-ischemia pharmacological treatment in mice. *Neurochemical Research*, 1–4.
84. Shi, K., Tian, D. C., Li, Z. G., Ducruet, A. F., Lawton, M. T., & Shi, F. D., (2019). Global brain inflammation in stroke. *Lancet Neurol., 18*(11), 1058–1066.
85. Sagris, D., Papanikolaou, A., Kvernland, A., Korompoki, E., Frontera, J. A., Troxel, A. B., Gavriatopoulou, M., et al., (2021). COVID-19 and ischemic stroke. *Eur. J. Neurol., 28*(11), 3826–3836.
86. Wang, D., Hu, B., Hu, C., Zhu, F., Liu, X., Zhang, J., Wang, B., et al., (2020). Clinical characteristics of 138 hospitalized patients with 2019 novel coronavirus-infected pneumonia in Wuhan, China. *JAMA, 323*(11), 1061–1069.
87. Merkler, A. E., Diaz, I., Wu, X., Murthy, S. B., Gialdini, G., Navi, B. B., Yaghi, S., et al., (2018). Duration of heightened ischemic stroke risk after acute myocardial infarction. *J. Am. Heart Assoc., 7*(22), e010782.
88. Tavazzi, G., Pellegrini, C., Maurelli, M., Belliato, M., Sciutti, F., Bottazzi, A., Sepe, P. A., et al., (2020). Myocardial localization of coronavirus in COVID-19 cardiogenic shock. *Eur. J. Heart Fail., 22*(5), 911–915.
89. Zeng, J. H., Liu, Y. X., Yuan, J., Wang, F. X., Wu, W. B., Li, J. X., Wang, L. F., et al., (2020). First case of COVID-19 complicated with fulminant myocarditis: A case report and insights. *Infection, 48*(5), 773–777.
90. Kariyanna, P. T., Chandrakumar, H. P., Jayarangaiah, A., Khan, A., Vulkanov, V., Ashamalla, M., Salifu, M. O., & McFarlane, S. I., (2020). Apical takotsubo cardiomyopathy in a COVID-19 patient presenting with stroke: A case report and pathophysiologic insights. *Am. J. Med. Case Rep., 8*(10), 350–357. Epub 2020 Jul 3.
91. Oudit, G. Y., Kassiri, Z., Jiang, C., Liu, P. P., Poutanen, S. M., Penninger, J. M., & Butany, J., (2009). SARS-coronavirus modulation of myocardial ACE2 expression and inflammation in patients with SARS. *Eur. J. Clin. Invest., 39*(7), 618–625.
92. Guo, T., Fan, Y., Chen, M., Wu, X., Zhang, L., He, T., Wang, H., Wan, J., Wang, X., & Lu, Z., (2020). Cardiovascular implications of fatal outcomes of patients with coronavirus disease 2019 (COVID-19). *JAMA Cardiol., 5*(7), 811–818.
93. Lazzerini, P. E., Boutjdir, M., & Capecchi, P. L., (2020). COVID-19, arrhythmic risk, and inflammation: Mind the gap! *Circulation, 142*(1), 7–9.
94. Walkey, A. J., Hammill, B. G., Curtis, L. H., & Benjamin, E. J., (2014). Long-term outcomes following development of new-onset atrial fibrillation during sepsis. *Chest, 146*(5), 1187–1195.
95. Hernández-Fernández, F., Sandoval, V. H., Barbella-Aponte, R. A., Collado-Jiménez, R., Ayo-Martín, Ó., Barrena, C., Molina-Nuevo, J. D., et al., (2020). Cerebrovascular disease in patients with COVID-19: Neuroimaging, histological and clinical description. *Brain, 143*(10), 3089–3103.
96. Khosravani, H., Rajendram, P., Notario, L., Chapman, M. G., & Menon, B. K., (2020). Protected code stroke: Hyperacute stroke management during the coronavirus disease 2019 (COVID-19) pandemic. *Stroke, 51*(6), 1891–1895.
97. Connors, J. M., & Levy, J. H., (2020). COVID-19 and its implications for thrombosis and anticoagulation. *Blood, 135*(23), 2033–2040.
98. Qureshi, A. I., Abd-Allah, F., Al-Senani, F., Aytac, E., Borhani-Haghighi, A., Ciccone, A., Gomez, C. R., et al., (2020). Management of acute ischemic stroke in patients with COVID-19 infection: Report of an international panel. *Int. J. Stroke, 15*(5), 540–554. doi: 10.1177/1747493020923234. Epub.
99. Fara, M. G., Stein, L. K., Skliut, M., Morgello, S., Fifi, J. T., & Dhamoon, M. S., (2020). Macrothrombosis and stroke in patients with mild COVID-19 infection. *J. Thromb. Haemost., 18*(8), 2031–2033.
100. Gavioli, E. M., Sikorska, G., Man, A., Rana, J., & Vider, E., (2020). Current perspectives of anticoagulation in patients with COVID-19. *J. Cardiovasc. Pharmacol., 76*(2), 146–150.
101. Xiong, T. Y., Redwood, S., Prendergast, B., & Chen, M., (2020). Coronaviruses and the cardiovascular system: Acute and long-term implications. *Eur. Heart J., 41*(19), 1798–1800.

102. Grewal, A. K., Singh, N., & Singh, T. G., (2019). Effects of resveratrol postconditioning on cerebral ischemia in mice: Role of the sirtuin-1 pathway. *Canadian Journal of Physiology and Pharmacology, 97*(11), 1094–1101.
103. Mannan, A., Garg, N., Singh, T. G., & Kang, H. K., (2021). Peroxisome proliferator-activated receptor-gamma (PPAR-γ): Molecular effects and its importance as a novel therapeutic target for cerebral ischemic injury. *Neurochemical Research, 46*(11), 2800–2831.
104. Powers, W. J., Rabinstein, A. A., Ackerson, T., Adeoye, O. M., Bambakidis, N. C., Becker, K., Biller, J., et al., (2019). *Guidelines for the Early Management of Patients with Acute Ischemic Stroke: 2019 Update to the 2018 Guidelines for the Early Management of Acute Ischemic Stroke: A Guideline for Healthcare Professionals COVID-19 and Ischemic Stroke: A Systematic Review and Meta-summary of the Literature, 50*(12), e344–e418.
105. Grewal, A. K., Singh, T. G., & Singh, N., (2020). Potential herbal drugs for ischemic stroke: A review. *Plant Arch., 20*(1), 3772–3783.
106. Rahmawati, P. L., Tini, K., Susilawathi, N. M., Wijayanti, I. A. S., & Samatra, D. P., (2021). Pathomechanism and management of stroke in COVID-19: Review of immunopathogenesis, coagulopathy, endothelial dysfunction, and downregulation of ACE2. *J. Clin. Neurol., 17*(2), 155–163.
107. Baracchini, C., Pieroni, A., Viaro, F., Cianci, V., Cattelan, A. M., Tiberio, I., Munari, M., & Causin, F., (2020). Acute stroke management pathway during Coronavirus-19 pandemic. *Neurol. Sci., 41*(5), 1003–1005.
108. Dogra, S., Jain, R., Cao, M., Bilaloglu, S., Zagzag, D., Hochman, S., Lewis, A., et al., (2020). Hemorrhagic stroke and anticoagulation in COVID-19. *J. Stroke Cerebrovasc. Dis., 29*(8), 104984.
109. Thachil, J., Tang, N., Gando, S., Falanga, A., Cattaneo, M., Levi, M., Clark, C., & Iba, T., (2020). ISTH interim guidance on recognition and management of coagulopathy in COVID-19. *J. Thromb. Haemost., 18*(5), 1023–1026.
110. Suter, F., Consolaro, E., Pedroni, S., Moroni, C., Pastò, E., Paganini, M. V., Pravettoni, G., et al., (2021). A simple, home-therapy algorithm to prevent hospitalization for COVID-19 patients: A retrospective observational matched-cohort study. *eClinical Medicine, 37*, 100941.
111. Bhaskar, S., Sinha, A., Banach, M., Mittoo, S., Weissert, R., Kass, J. S., Rajagopal, S., Pai, A. R., & Kutty, S., (2020). Cytokine storm in COVID-19-immunopathological mechanisms, clinical considerations, and therapeutic approaches: The REPROGRAM consortium position paper. *Front. Immunol., 11*, 1648.
112. Sahu, K. K., Borogovac, A., & Cerny, J., (2021). COVID-19 related immune hemolysis and thrombocytopenia. *J. Med. Virol., 93*(2), 1164–1170.
113. Gu, S. X., Tyagi, T., Jain, K., Gu, V. W., Lee, S. H., Hwa, J. M., Kwan, J. M., et al., (2021). Thrombocytopathy and endotheliopathy: Crucial contributors to COVID-19 thromboinflammation. *Nat. Rev. Cardiol., 18*(3), 194–209.
114. Sandercock, P. A., Counsell, C., & Kane, E. J., (2015). Anticoagulants for acute ischemic stroke. *Cochrane Database Syst. Rev., 2015*(3), CD000024.
115. Monteil, V., Kwon, H., Prado, P., Hagelkrüys, A., Wimmer, R. A., Stahl, M., Leopoldi, A., et al., (2020). Inhibition of SARS-CoV-2 infections in engineered human tissues using clinical-grade soluble human ACE2. *Cell, 181*(4), 905–913.e7.
116. Berger, C., Xia, F., Maurer, M. H., & Schwab, S., (2008). Neuroprotection by pravastatin in acute ischemic stroke in rats. *Brain Res. Rev., 58*(1), 48–56.
117. Rodrigues-Diez, R. R., Tejera-Muñoz, A., Marquez-Exposito, L., Rayego-Mateos, S., Santos, S. L., Marchant, V., Tejedor, S. L., et al., (2020). Statins: Could an old friend help in the fight against COVID-19? *Br. J. Pharmacol., 177*(21), 4873–4886.
118. Rossi, R., Talarico, M., Coppi, F., & Boriani, G., (2020). Protective role of statins in COVID 19 patients: Importance of pharmacokinetic characteristics rather than intensity of action. *Intern. Emerg. Med., 15*(8), 1573–1576.
119. Zhou, P., Li, T., Jin, J., Liu, Y., Li, B., Sun, Q., Tian, J., et al., (2020). Interactions between neutrophil extracellular traps and activated platelets enhance procoagulant activity in acute stroke patients with ICA occlusion. *EBio Medicine, 53*, 102671.
120. Wang, J., Hajizadeh, N., Moore, E. E., McIntyre, R. C., Moore, P. K., Veress, L. A., Yaffe, M. B., et al., (2020). Tissue plasminogen activator (tPA) treatment for COVID-19 associated acute respiratory distress syndrome (ARDS): A case series. *J. Thromb. Haemost., 18*(7), 1752–1755.

121. Avula, A., Nalleballe, K., Narula, N., Sapozhnikov, S., Dandu, V., Toom, S., Glaser, A., & Elsayegh, D., (2020). COVID-19 presenting as stroke. *Brain Behav. Immun., 87*, 115–119.
122. Khan, H., Grewal, A. K., & Singh, T. G., (2022b). Pharmacological postconditioning by protocatechuic acid attenuates brain injury in ischemia–reperfusion (I/R) mice model: Implications of nuclear factor erythroid-2-related factor pathway. *Neuroscience, 491*, 23–31.
123. Frontera, J., Mainali, S., Fink, E. L., Robertson, C. L., Schober, M., Ziai, W., Menon, D., et al., (2020). Global consortium study of neurological dysfunction in COVID-19 (GCS-NeuroCOVID): Study design and rationale. *Neurocrit Care, 33*(1), 25–34.
124. Gunst, J. D., Staerke, N. B., Pahus, M. H., Kristensen, L. H., Bodilsen, J., Lohse, N., Dalgaard, L. S., et al., (2021). Efficacy of the TMPRSS2 inhibitor camostatmesilate in patients hospitalized with COVID-19-a double-blind randomized controlled trial. *eClinical Medicine, 35*, 100849.
125. Shoirah, H., Wechsler, L. R., Jovin, T. G., & Jadhav, A. P., (2019). Acute stroke trial enrollment through a telemedicine network: A 12-year experience. *J. Stroke Cerebrovasc. Dis., 28*(7), 1926–1929. doi: 10.1016/j.jstrokecerebrovasdis.2019.03.046. Epub 2019 Apr 20. PMID: 31010762.
126. Tremblay, D., Van, G. M., Alsen, M., Thibaud, S., Kessler, A., Venugopal, S., Makki, I., et al., (2020). Impact of anticoagulation prior to COVID-19 infection: A propensity score-matched cohort study. *Blood, 136*(1), 144–147.
127. Liberale, L., Diaz-Cañestro, C., Bonetti, N. R., Paneni, F., Akhmedov, A., Beer, J. H., Montecucco, F., et al., (2018). Post-ischemic administration of the murine canakinumab-surrogate antibody improves outcome in experimental stroke. *Eur. Heart J., 39*(38), 3511–3517.
128. Smith, C. J., Hulme, S., Vail, A., Heal, C., Parry-Jones, A. R., Scarth, S., Hopkins, K., et al., (2018). SCIL-STROKE (subcutaneous interleukin-1 receptor antagonist in ischemic stroke): A randomized controlled phase 2 trial. *Stroke, 49*(5), 1210–1216.
129. Quiros, R. E., Biasiotto, G., Magro, P., & Zanella, I., (2020). The possible mechanisms of action of 4-aminoquinolines (chloroquine/hydroxychloroquine) against Sars-Cov-2 infection (COVID-19): A role for iron homeostasis? *Pharmacol. Res., 158*, 104904.
130. Mistry, P., Carmona-Rivera, C., Ombrello, A. K., Hoffmann, P., Seto, N. L., Jones, A., Stone, D. L., et al., (2018). Dysregulated neutrophil responses and neutrophil extracellular trap formation and degradation in PAPA syndrome. *Ann. Rheum. Dis., 77*(12), 1825–1833.
131. Alhenc-Gelas, F., & Drueke, T. B., (2020). Blockade of SARS-CoV-2 infection by recombinant soluble ACE2. *Kidney Int., 97*(6), 1091–1093.
132. Peña-Martínez, C., Durán-Laforet, V., García-Culebras, A., Ostos, F., Hernández-Jiménez, M., Bravo-Ferrer, I., Pérez-Ruiz, A., et al., (2019). Pharmacological modulation of neutrophil extracellular traps reverses thrombotic stroke tPA (tissue-type plasminogen activator) resistance. *Stroke, 50*(11), 3228–3237.
133. Urwyler, P., Charitos, P., Moser, S., Heijnen, I. A. F. M., Trendelenburg, M., Thoma, R., Sumer, J., et al., (2021). Recombinant human C1 esterase inhibitor (*Conestat alfa*) in the prevention of severe SARS-CoV-2 infection in hospitalized patients with COVID-19: A structured summary of a study protocol for a randomized, parallel-group, open-label, multi-center pilot trial (protect-COVID-19). *Trials, 22*(1), 1.
134. Hadanny, A., Rittblat, M., Bitterman, M., May-Raz, I., Suzin, G., Boussi-Gross, R., Zemel, Y., et al., (2020). Hyperbaric oxygen therapy improves neurocognitive functions of post-stroke patients – a retrospective analysis. *Restor. Neurol. Neurosci., 38*(1), 93–107.
135. Beyls, C., Huette, P., Abou-Arab, O., Berna, P., & Mahjoub, Y., (2020). Extracorporeal membrane oxygenation for COVID-19-associated severe acute respiratory distress syndrome and risk of thrombosis. *Br. J. Anaesth.*, e260–e262.
136. Al-Sarraj, S., Troakes, C., Hanley, B., Osborn, M., Richardson, M. P., Hotopf, M., Bullmore, E., & Everall, I. P., (2021). Invited review: The spectrum of neuropathology in COVID-19. *Neuropathol. and Appl. Neurobiol., 47*, 3–16.
137. Bryce, C., Grimes, Z., Pujadas, E., Ahuja, S., Beasley, M. B., Albrecht, R., Hernandez, T., et al., (2021). Pathophysiology of SARS-CoV-2: The mount Sinai COVID-19 autopsy experience. *Mod. Pathol., 34*(8), 1456–1467.
138. De Wit, E., Van, D. N., Falzarano, D., & Munster, V. J., (2016). SARS and MERS: Recent insights into emerging coronaviruses. *Nat. Rev. Microbiol., 14*, 523–534.

139. Ford, J. S., Holmes, J. F., & Jones, R. F., (2020). Cardioembolic stroke in a patient with coronavirus DISEASE of 2019 (COVID-19) myocarditis: A case report. *Clin. Pract. Cases Emerg. Med., 4*(3), 332–335.
140. Fridman, S., Bres, B. M., Jimenez-Ruiz, A., Costantini, P., Shah, P., Just, C., Vela-Duarte, D., et al., (2020). Stroke risk, phenotypes, and death in COVID-19: Systematic review and newly reported cases. *Neurology, 95*(24), e3373–e3385.
141. Gerotziafas, G. T., Catalano, M., Colgan, M. P., Pecsvarady, Z., Wautrecht, J. C., Fazeli, B., Olinic, D. M., et al., (2020). Guidance for the management of patients with vascular disease or cardiovascular risk factors and COVID-19: Position paper from VAS-European independent foundation in angiology/vascular medicine. *Thromb. Haemost., 120*(12), 1597–1628.
142. John, S., Kesav, P., Mifsud, V. A., Piechowski-Jozwiak, B., Dibu, J., Bayrlee, A., Elkambergy, H., et al., (2020). Characteristics of large-vessel occlusion associated with COVID-19 and ischemic stroke. *AJNR Am. J. Neuroradiol., 41*(12), 2263–2268.
143. Kawashima, S., Yamashita, T., Miwa, Y., Ozaki, M., Namiki, M., Hirase, T., Inoue, N., et al., (2003). HMG-CoA reductase inhibitor has protective effects against stroke events in stroke-prone spontaneously hypertensive rats. *Stroke, 34*(1), 157–163.
144. Kyriakopoulou, K., Dekkers, B., & van der Goot, A.J., (2019). Plant-based meat analogues. In *Sustainable Meat Production and Processing* (pp. 103–126). Academic Press.
145. Li, Y., Li, M., Wang, M., Zhou, Y., Chang, J., Xian, Y., Wang, D., Mao, L., Jin, H., & Hu, B., (2020). Acute cerebrovascular disease following COVID-19: A single center, retrospective, observational study. *Stroke Vasc. Neurol., 5*(3), 279–284.
146. Melmed, K. R., Cao, M., Dogra, S., Zhang, R., Yaghi, S., Lewis, A., Jain, R., et al., (2021). Risk factors for intracerebral hemorrhage in patients with COVID-19. *J. Thromb. Thrombolysis, 51*(4), 953–960.
147. David Owen & Katrina Pollock (2021). *RAS and Coagulopathy in COVID19.* Clinical Trials. Gov.
148. Paniz-Mondolfi, A., Bryce, C., Grimes, Z., Gordon, R. E., Reidy, J., Lednicky, J., et al., (2020). Central nervous system involvement by severe acute respiratory syndrome coronavirus-2 (SARS-CoV-2). *J. Med. Virol., 92*, 699–702.
149. Vivas, D., Roldán, V., Esteve-Pastor, M. A., Roldán, I., Tello-Montoliu, A., Ruiz-Nodar, J. M., Cosín-Sales, J., et al., (2020). Revisores expertos. Recomendaciones sobre el tratamiento antitrombótico durante la pandemia COVID-19. Posicionamiento del grupo de trabajo de trombosis cardiovascular de la Sociedad Española de cardiología [Recommendations on antithrombotic treatment during the COVID-19 pandemic. Position statement of the working group on cardiovascular thrombosis of the Spanish society of cardiology]. *Rev. Esp. Cardiol., 73*(9), 749–757.
150. Kawashima, T., (2005). The autonomic nervous system of the human heart with special reference to its origin, course, and peripheral distribution. *Anatomy and Embryology, 209,* 425–438.

Index

D

I

N

O